GRANITIC SYSTEMS

ILMARI HAAPALA VOLUME

GRANITIC SYSTEMS

ILMARI HAAPALA VOLUME

Edited by

O.T. RÄMÖ

University of Helsinki, Helsinki, Finland

2005

ELSEVIER

Amsterdam - Boston - Heidelberg - London - New York - Oxford
Paris - San Diego - San Francisco - Singapore - Sydney - Tokyo

ELSEVIER B.V.   ELSEVIER Inc.   ELSEVIER Ltd   ELSEVIER Ltd
Radarweg 29   525 B Street, Suite 1900   The Boulevard, Langford Lane   84 Theobalds Road
P.O. Box 211, 1000 AE Amsterdam   San Diego, CA 92101-4495   Kidlington, Oxford OX5 1GB   London WC1X 8RR
The Netherlands   USA   UK   UK

First edition 2005
This book is reprinted from the Special Issue of Lithos (volume 80, nos. 1—4), which contains papers based on the invited presentations given at the symposium "Granitic Systems – State of the Art and Future Avenues" on the occasion of Ilmari Haapala's retirement, held at the Department of Geology, University of Helsinki, Finland, 12--14 January 2003.

ISBN 9780444518828

♾ The paper used in this publication meets the requirements of ANSI/NISO Z39.48-1992 (Permanence of Paper).

Working together to grow
libraries in developing countries

www.elsevier.com | www.bookaid.org | www.sabre.org

ELSEVIER   BOOK AID International   Sabre Foundation

Printed and bound in the United Kingdom

Transferred to Digital Print 2011

Available online at www.sciencedirect.com

Lithos 80 (2005) vii–viii

LITHOS

www.elsevier.com/locate/lithos

Contents

Special Issue: Granitic Systems
        Ilmari Haapala Volume

Available online at www.sciencedirect.com

SCIENCE @ DIRECT°

Lithos 80 (2005) ix

LITHOS

www.elsevier.com/locate/lithos

ELSEVIER

Professor Ilmari Haapala at field work in the Newberry Mountains, Clark County, Nevada, USA. Photo: Tapani Rämö (October 1995).

doi:10.1016/j.lithos.2005.00015-0

Available online at www.sciencedirect.com

Lithos 80 (2005) xi–xix

www.elsevier.com/locate/lithos

Preface

Granitic systems—a special issue in honor of Ilmari Haapala

O. Tapani Rämö*

Department of Geology, P.O. Box 64, FI-00014 University of Helsinki, Finland

Available online 7 December 2004

Abstract

This paper is an introduction to the special issue of *Lithos* honoring Ilmari Haapala, the 1982–2002 Professor of Geology and Mineralogy of the Department of Geology, University of Helsinki. The articles in the issue are based on invited presentations given at a January 12–14, 2003 symposium "Granitic Systems – State of the Art and Future Avenues" that was held at the Department of Geology, University of Helsinki on the occasion of Ilmari Haapala's December 31, 2002 retirement. The scientific achievements of Professor Haapala are briefly reviewed and the main results of the papers that comprise the volume are discussed.
© 2004 Elsevier B.V. All rights reserved.

Keywords: Granite; Granitic pegmatite; Petrology; University of Helsinki; Finland

1. Introduction

This special issue stems from an international symposium that was held at the Department of Geology, University of Helsinki, Finland on January 12–14, 2003 to mark the retirement of Professor Ilmari Haapala. The symposium was attended by a versatile group of earth scientists with granitic systems as a common denominator (Fig. 1) and was aimed at giving a comprehensive account of the current status of granite-oriented research. The topics of the presentations ranged from mineralogy, petrology, and geochemistry to tectonics and crustal

evolution. Thirty-three invited papers by 61 authors were submitted to the symposium (cf. Rämö et al., 2003), and 20 were developed into the full papers that constitute the present volume. In this introductory article, Ilmari Haapala's career in granite petrology is briefed, and the results of the 20 papers comprising the volume are summarized.

2. Ilmari Haapala: rapakivi granites, tin mineralization, granitic pegmatites

Ilmari Haapala retired after a long and productive career that, in terms of science, focused on granites, granitic pegmatites, and tin mineralization associated with granitic systems. Ilmari had his PhD at the Department of Geology, University of Helsinki in

---

* Tel.: +358 40 526 0636; fax: +358 9 191 50826.
E-mail address: tapani.ramo@helsinki.fi.

0024-4937/$ - see front matter © 2004 Elsevier B.V. All rights reserved.
doi:10.1016/j.lithos.2004.09.015

1967. His thesis (Haapala, 1966) dealt with the mineralogy, petrology, and internal structure of rare mineral-bearing Paleoproterozoic granitic pegmatites in the Peräseinäjoki–Alavus area in western Finland and was followed by a series of mineralogical studies related to various types of mineral occurrences in Finland. In his early professional years (1966–1981), Ilmari was affiliated with the Exploration Department of the Geological Survey of Finland. He served as an Exploration Geologist, Senior Exploration Geologist, and acting Head of the Department and, among other duties, led a project that assessed the metallogenic potential of the locus classicus rapakivi granites of southern Finland. This led to the recognition of the late-stage topaz-bearing alkali feldspar granites of rapakivi complexes as important carriers of tin (Haapala, 1974, 1977a). These granites are associated with greisen-, vein-, and skarn-type Sn (-W-Be-Zn-Cu-Pb) deposits and show the same petrographic, mineralogical, and geochemical peculiarities as tin granites in general (Haapala and Ojanperä, 1972; Haapala, 1995). These studies also led to a deeper understanding of the magmatic and postmagmatic processes in mineralized granites (Haapala, 1977b; see also Haapala, 1997). In his monograph on the Eurajoki rapakivi granite complex of southwestern Finland (Haapala, 1977b), Ilmari showed that the mineralogical and geochemical peculiarities of the peraluminous late-stage topaz-bearing intrusive phases of rapakivi suites (i.e., presence of rare accessory minerals such as monazite, cassiterite, bastnaesite, xenotime, columbite, and thorite; anomalously high contents of F, Li, Ga, Rb, Sn, and Nb; and low Mg, Fe, Ti, Ba, Sr, and Zr) can be ascribed to extreme magmatic fractionation and superimposed autometasomatic processes. For instance, the cassiterite in the topaz-bearing granite of the Eurajoki complex has clearly higher Nb and Ta contents than that in the associated pegmatites and greisen veins; this implies an early (magmatic) origin for the granite cassiterite (Haapala, 1974, 1997; see also Haapala and Rämö, 1990). Primarily magmatic origin of the topaz-bearing granites from Eurajoki is also indicated by the composition of melt inclusions (Haapala and Thomas, 1999).

In 1982, Ilmari was appointed as the Professor of Geology and Mineralogy at the Department of Geology, University of Helsinki. The post is the first chair in geology in Finland (founded in 1852) and had been previously held, among others, by Wilhelm Ramsay (1899–1928) and Pentti Eskola (1929–1953). Since the early 1980s, Ilmari's research activities have focused on the origin of granites, rapakivi granites in particular, with a special emphasis on bimodal igneous suites. In 1985, he explained the origin of the Finnish rapakivi granites by applying the magmatic underplate model (Haapala, 1985; see also Haapala and Rämö, 1990), and in the years that followed, this idea gained wide acceptance as deep seismic sounding studies (e.g., Luosto et al., 1990; Korja et al., 1993), and isotope geochemical data (e.g., Rämö, 1991) turned out to be compatible with the model. In 1992, Ilmari's work also led to a refined definition of rapakivi granite (Haapala and Rämö, 1992, p. 165; see also Best, 2003, p. 709) that takes into account the magmatic association, geochemical composition, and petrographic traits of these rocks but does not restrict their age:

Rapakivi granites are A-type granites characterized by the presence, at least in the larger batholiths, of granite varieties showing the rapakivi texture.

Ilmari Haapala has led numerous granite-oriented research projects in Finland and abroad. Those abroad include studies on the Paleoproterozoic Schachang rapakivi granite complex in the Sino-Korean craton near Beijing, China; the Miocene rapakivi-type granites in the Newberry Mountains of the Colorado River extensional corridor in southernmost Nevada; the Mesoproterozoic granites and associated anorthosites and minettes in the Big Burro Mountains of southwestern New Mexico; and the Cretaceous Spitzkoppe granite complex in the Damara orogenic belt in northwestern Namibia. The results of these studies are summarized in the lead paper of this volume (Haapala et al.). During his 21 years in office, Ilmari also trained a substantial number of PhD students (on average, two doctors a year).

Ilmari was the initiator and leader of the 1991–1996 rapakivi granite project of the International Geological Correlation Programme (IGCP-315 "Correlation of Rapakivi Granites and Related Rocks on a Global Scale"). This project stemmed from a comprehensive international symposium on Precambrian granitoids held at the University of Helsinki in 1989 (Haapala and Kähkönen, 1989; Haapala and

Fig. 1. Speakers of the January 12–14, 2003 symposium "Granitic Systems – State of the Art and Future Avenues" in the entrance hall of the Department of Geology, University of Helsinki. Key to the individuals: 1—Matti Vaasjoki; 2—Jorge Bettencourt; 3—Tapani Rämö; 4—Ed Stephens; 5—Jim Sears; 6—Anders Lindh; 7—Kent Condie; 8—Bob Linnen; 9—Jean-Louis Vigneresse; 10—Libby Anthony; 11—Lawford Anderson; 12—Karel Breiter; 13—Krister Sundblad; 14—Bernard Barbarin; 15—Bruce Ryan; 16—Bernard Bonin; 17—Hannu Huhma; 18—Olav Eklund; 19—Mikko Nironen; 20—Linata Sviridenko; 21—Raimo Lahtinen; 22—Sari Lukkari; 23—Eric Christiansen; 24—Roberto Dall'Agnol; 25—Tony Fallick; 26—Cal Barnes; 27—Hanna Nekvasil; 28—Ilmari Haapala. The table in front is a 1:400,000-scale bedrock map of Finland, designed by Ilmari Haapala and made of polished slabs of regionally representative rock types. Photo: Ari Aalto, University of Helsinki, IT Department, AV Unit.

Condie, 1991), had active participants from more than 20 countries, and established the rapakivi granites and related mafic rocks (gabbroids, anorthosites, tholeiitic diabase dikes) as a rock suite with profound global significance, not only in igneous petrology and metallogeny, but also in terms of Proterozoic crustal evolution (e.g., Rämö and Haapala, 1995; Haapala and Rämö, 1999a). Symposia and field trips were arranged in Finland, Russia, Japan, the United States, Canada, Italy, Brazil, and Sweden, and three proceedings volumes (Haapala and Rämö, 1994, 1999b; Dall'Agnol and Bettencourt, 1997) and a substantial number of original papers, abstract volumes, and field trip guides were published. The work of the project has been continued by IGCP-426 "Granite Systems and Proterozoic Lithospheric Processes" (cf. Rämö et al., 2002).

3. Contributions

The 20 papers that comprise this special issue fall under three general themes: (1) tectonics and source regions; (2) petrologic processes; and (3) fractionated granites and pegmatites. The papers by Haapala et al., Condie, Anderson and Morrison, and Anthony (theme 1), Barbarin (theme 2), and London (theme 3) are essentially reviews, and the remainder 14 are novel. Altogether, they cover a wide range of topics on granitic systems, dealing with general and experimental petrology, mineralogy, metallogeny, tectonics, and crustal evolution.

3.1. Tectonics and source regions

In a summary paper of rapakivi granite projects carried out by the Department of Geology, University of Helsinki since the late 1980s, Haapala et al. compare the tectonic setting, lithologic traits, geochemistry, and origin of the classic Proterozoic rapakivi granites of Finland and vicinity to the 1.70–1.68 Ga rapakivi granites of the Beijing region in China, the 130 Ma granites of western Namibia, and the 20–15 Ma granites of the Colorado River extensional corridor in the Basin and Range Province in southern Nevada. In all these cases, the granite magmatism was bimodal and related to extension. Rapakivi texture is often present in the granites, as are late intrusive phases

characterized by the presence of accessory topaz. Geochemically, the granites are aluminous A-type (cf. King et al., 1997) or ferrous alkali-calcic (cf. Frost et al., 2001). Magmatic underplating and resultant crustal anatexis is the favored petrogenetic model. The Cretaceous granites of Namibia are related to the Tristan plume and opening of South Atlantic, whereas a plume connection has not been positively established for the Fennoscandian, Chinese, or Nevadan suites.

Based on an extensive global-scale geochemical database (also accessible through the electronic data repository of Elsevier), Condie discusses the origin of TTG (tonalite–trondhjemite–granodiorite) and adakite suites. Both suites show fractionated REE and incompatible element patterns but differ in terms of compatible element contents and Nb/Ta. TTG melts are considered to have formed in the hornblende eclogite stability field from sources involving lower crust in convergent margins and, in the Archean, root zones of oceanic plateaus, whereas the origin of adakites is ascribed to slab melting. Compositions of the ~2.7 and ~1.9 Ga high-Al TTGs (unusually high La/Yb and compatible elements) are claimed to reflect catastrophic global mantle overturn events that involved accentuated mantle plume activity.

In a bipartitive review of mid-Proterozoic intracratonic granite magmatism, Anderson and Morrison present new oxygen isotope data on the granite suites of Laurentia and discuss the long-debated issue about the tectonic setting (anorogenic or orogenic?) of the pertinent granite suites in Laurentia and Baltica. The authors divide the Laurentian granites into (1) low-fO_2, ilmenite-series granites; (2) high-fO_2, magnetite-series granites; and (3) very high-fO_2, peraluminous two-mica granites. These compositional variations are claimed to reflect source variation involving a juvenile, shortly mantle-derived component (low-fO_2 group), a more oxidized source (high-fO_2 group), and a metasedimentary source component (very high-fO_2 group). Regarding Laurentia, the authors find no compelling evidence for an ~1.4-Ga orogeny and thus interpret the Laurentian mid-Proterozoic granites as anorogenic. On a global (Laurentia–Baltica) scale, the mid-Proterozoic event is considered unique and is related to the stabilization and breakup of Paleoproterozoic/Mesoproterozoic supercontinents.

Anthony presents a review focusing on the origin of three granite suites found in the south central

United States (west Texas, New Mexico, and Arizona). These range from Mesoproterozoic (~1.4 Ga) to Late Cretaceous and Early Tertiary in age and register various sources with older preexisting crust as a prominent source component. Studies dealing with the geochemical and mineral chemical composition of the suites are reviewed, and a correlation between source regions and tectonic setting is invoked. For further studies, Anthony calls for a detailed scrutiny of the cause of the varying melt regimes in relation to varying tectonic settings.

Tomascak et al. focus on the magmatic evolution of the U.S. Appalachians and present elemental geochemical and Nd, Pb isotope data on Middle and Late Devonian granitoid and monzodioritic plutons associated with a mid-Paleozoic suture zone in New Hampshire and Maine. Approximately 377-Ma quartz monzodiorite and biotite granodiorite have initial ϵ_{Nd} values of -2.8 to -0.7 and are related to a mafic-intermadiate lower crustal source. Approximately 370-Ma biotite granite, two-mica granite, and granodiorite show a wider range of ϵ_{Nd} values (-7.0 to -0.6) and are claimed to have derived from a crustal source with a metasedimentary component. Overall, the role of the continental crust as the sole source of these Devonian magmatic suites is stressed; a mantle melt component is considered unlikely. The authors also use the isotope data to refine the geographic position of Appalachian crustal belts at approximately 45°N latitude.

Dall'Agnol et al. discuss the petrography, mineral chemistry, elemental geochemistry, and isotope geology (Nd, Pb, O) of Paleoproterozoic intraplate granites of the Carajás Metallogenic Province, eastern Amazonian craton, Brazil. Three coeval (~1.88 Ga) A-type granite suites (Jamon, Serra do Carajás, Velho Guilherme) with different compositional traits are recognized. These suites register emplacement pressures in the 1–3 kbar range and vary from oxidized (Jamon) to reduced (Velho Guilherme) in character. Nd isotopes point to an Archean crustal source, and compositional variation in a lower crust regime is called for to explain the observed compositional diversity. A mantle superswell beneath a supercontinent (Trans-Amazonian) assembled at ~2.0 Ga is preferred as the trigger of the granite magmatism. The Carajás granites are ~100–200-Ma older than the corresponding magmatic suites of Laurentia and Baltica, reflecting the earlier assembly of the Trans-Amazonian supercontinent.

The current granite paradigm states that silicic material cannot be directly derived by partial melting of the mantle of the Earth. Bonin and Bébien challenge this paradigm by reviewing evidence from natural samples from the Earth, Moon, and Mars as well as from meteorites that show that silicic material (up to ~80 wt.% SiO_2) can be found in the mantle rocks of the terrestrial planets. On Earth, for instance, tiny amounts of silicic material are present as glass inclusions or as crystals (quartz, tridymite, alkali feldspar) in dominant mantle minerals (olivine, pyroxene, spinel). However small the volume of these silicic portions, Bonin and Bébien believe that their presence leads to at least an amendment of the paradigm and call for experimental studies to further explore the validity of the hypothesis.

Complying with the "platonistic" idea of Anderson (2002), Sears et al. question another paradigm—the role of mantle convection in lithospheric evolution. Rather than deep-mantle plume activity, the authors ascribe breakup of the Proterozoic Laurentia and Phanerozoic Gondwana supercontinents to passive rifting that occurred along linear zones that constitute a configuration consistent with truncated icosahedral tessellations of the lithosphere. These quasi-hexagonal patterns manifest a least-work configuration in which each seam forms a 23° great-circle arc, and the triple junctions join two hexagons and one pentagon each. For Laurentia and Gondwana, anorogenic magmatism is claimed to have formed by decompression melting along linear rupture zones that comply with the truncated-icosahedron configuration and were formed owing to increased surface-parallel tensile strain as the asthenosphere expanded underneath a stalled insulating supercontinent.

3.2. Petrologic processes

Barbarin presents a review of magma mingling and mixing processes in the calc-alkaline granitoid plutons in the central part of the Sierra Nevada batholith, California. The most common plutonic rocks of the batholith are hornblende-bearing tonalite, quartz diorite, granodiorite, and quartz monzonite that are associated with more mafic rocks and display a conspicuous series of mafic magmatic enclaves and other signs of mafic–silicic magma interaction. The mafic magmatic enclaves are, in general, dioritic to quartz dioritic in composition and show a wide range of textural and structural types. The plutons also include mafic dikes that may be undisturbed, composite, or thoroughly hybridized. Recent petrographical, mineralogical, geochemical, and isotope data presented on central Sierra Nevada plutons are reviewed. These are combined into a five-stage petrogenetic model for the granitoid and mafic rocks of the batholith, involving complex hybridization, differentiation, and segregation processes.

Using elemental geochemical and isotope (Nd, Sr, O, C) data, Barnes et al. discuss the origin of silica-saturated and silica-undersaturated monzodioritic, monzonitic, and syenitic rocks in the Caledonian Hortavær intrusive complex in west central Norway. The precursors of these were gabbroic and dioritic magmas that evolved along iron- and alkali-enrichment trends. Magma mixing and combined assimilation-fractional crystallization models are presented, the latter involving assimilation of carbonate rocks in a high-T regime and silicate rocks in a low-T regime. Substantial assimilation of carbonate rocks in the early stage is considered to have been possible because of the open-system nature of the complex—this allowed excess CO_2 to escape from the evolving magmas, enhancing incorporation of carbonate material into the evolving magmas.

Müller et al. discuss the textures and elemental geochemical features of quartz and feldspar phenocrysts recovered from Upper Carboniferous silicic volcanic rocks and their epizonal intrusive equivalents of the Erzgebirge Krušne Hory batholith in eastern Germany and northwestern Czech Republic. A texturally varying set of quartz phenocrysts (identified by SEM-CL imaging) bears evidence for detailed evolution of the Erzgebirge magma system and allows distinction of several magma reservoirs at different levels in the crust. Magma mixing is claimed to have been an important process and is evidenced by rounded quartz phenocrysts with dissolution surfaces and Ti-enriched internal zones, rapakivi textures, skeletal plagioclase crystals, and plagioclase with patchy growth zones. Alkali feldspar crystals are also texturally complex but have often reequilibrated and do not register the complexity of the magmatic system any more.

Eklund and Shebanov present new geochronological (U–Pb on zircon) and mineral chemical data on mineral separates recovered from a ~1.8-Ga ring complex (Åva) granite in the southwestern Finnish archipelago. U–Pb ages measured by ion microprobe are given for the granite, monzonite, and lamprophyre of the complex, and mineral chemistry of amphibole and mica is discussed in search of the intensive parameters of their crystallization. The authors also review existing geochemical data that imply a shoshonitic character for the magmatism. Amphibole and mica from different textural positions (from the cores of alkali feldspar megacrysts to the groundmass surrounding them) register middle and upper crust environments for the magmatic evolution of the complex. The ion microprobe U–Pb data are interpreted to suggest a ~30-Ma time frame for the magmatic evolution of the complex.

Lindh presents elemental geochemical and some Nd isotope data on three Paleoproterozoic late Svecofennian (~1.8 Ga) granites in the Bothnian basin of central Sweden on the western flank of the Fenoscandian shield. The three studied granites (Själevad, Härnö, Bergom) show differences in terms of silica and high field strength and rare earth element contents as well as initial Nd isotope composition, and their overall character ranges from I- to S-type. Approximate geochemical and thermal models are presented and are suggested to imply melting at different levels within the crust in the waning stages of the Svecofennian orogeny. An Archean source component is considered unlikely for these granites.

Linnen presents a pioneering systematic experimental study on the influence of water on the solubility of a wide range of accessory minerals in granitic melts. Solubility experiments were performed on columbite, tantalite, wolframite, rutile, zircon, and hafnon at 1035 °C and 2 kbar and water contents of ~0–6 wt.%. For peralkaline granitic melts, the results indicate that water content has no effect on accessory phase solubilities. For subaluminous melts with >2 wt.% H_2O, water has very little effect, whereas subaluminous melts with <2 wt.% H_2O show drastically decreased solubilities. As most, if not all, granitic melts contain more than 2 wt.% water, these results indicate that water content of the melt will not have a marked influence on the saturation of accessory phases in natural granitic melts in general. Linnen's results also call for caution while using solubilities of accessory phases (e.g., zircon) for temperature estimates.

3.3. Fractionated granites and pegmatites

In a thorough, interpretative synthesis of granitic pegmatites, London presents the current knowledge of these rocks with insights into the past and the future. The works of one of the pioneers in the field, Richard H. Jahns, are reviewed, and among several specific issues, the cooling history, pressure and temperature of crystallization and temperature gradients, importance of fluxes, significance of layered aplites, and viscosity of pegmatite-forming melts are discussed in detail. Rather than considering the characteristic textures and structures of granitic pegmatites as resulting from buoyant separation of aqueous vapor from silicate melt as Jahns did, the author favors constitutional zone refining in an undercooled silicate melt with or without the involvement of an aqueous vapor phase. Accordingly, pegmatites may represent crystal-free melts that emanated from plutons and had cooled substantially below their liquidus temperatures before the onset of crystallization. Future research should aim at relating granitic pegmatites to their sources and thus requires a shift of focus from individual pegmatite bodies to their surroundings and beyond.

Černý et al. discuss the origin of lepidolite-subtype pegmatites using the Archean Greer Lake leucogranite in the Superior Province in southern Manitoba, Canada as an example. The Greer Lake granite is an evolved, peraluminous, B-, P-, and S-poor leucogranite that was originally derived from a metatonalitic source. It contains barren, beryl–columbite-, and lepidolite-subtype pegmatitic segregations. Steep fractionation gradients mark the granite-to-pegmatite transition in the case of the lepidolite-subtype segregations in particular. The Greer Lake system is claimed to be rather unique in that the plutonic parent of the lepidolite-subtype pegmatites can be clearly discerned—normally such pagmatites crystallize far from their plutonic sources.

Breiter et al. present chemical and textural data on a highly fractionated peraluminous late Variscan granite stock (Podlesí) in western Czech Republic. The stock is characterized by magmatic layering and unidirectional solidification textures and includes quartz, K-feldspar, mica, and topaz crystals, showing evidence for multiple crystallization and resorption events. Geochemically, the Podlesí stock is particularly enriched in F, Li, and P. The textural evolution of the stock is explained by repeated opening of the magmatic system and resultant adiabatic pressure fluctuations and boundary–layer crystallization from undercooled melt.

Haapala and Lukkari provide field, petrographic, mineral chemical, and geochemical data on the Kymi stock, which represents one of the highly evolved late-stage intrusions of the classic Wiborg rapakivi granite batholith of southeastern Finland. The 3×6-km stock has a zoned structure (from the outer contact inwards: marginal pegmatite, equigranular alkali feldspar granite, porphyritic alkali feldspar granite) and is surrounded by a swarm of mineralized quartz veins as well as greisen veins and bodies. These veins and bodies are also found within the granites of the stock. A model involving convection and upward flow of evolved, fluid-enriched melt is favored—the equigranular granite and stockscheider pegmatite crystallized from this apical melt, whereas the porphyritic granite precipitated from the underlying less-enriched, denser magma. Autometasomatic processes in the stock were obviously substantial but not pervasive enough to mask the highly evolved pyrogenic nature of the stock.

Bettencourt et al. present a fluid inclusion and stable isotope (O, H) study on three tin-polymetallic deposits associated with late-stage intrusive phases in the Rondônia and Itu rapakivi provinces in western and southeastern Brazil, respectively. The Oriente Novo (Rondônia) and Correas (Itu) deposits are associated with stockwork and veins and are characterized by hydrothermal activity at ~240–450 °C and 1.2–2.6 kbar. The Santa Barbara deposit (Rondônia) is a greisenized cupola and registers higher temperatures for the hydrothermal activity: 500–570 °C for greisen, ~415 °C for cassiterite–quartz veins. The mineralizing fluids were aqueous with varying amounts of CO_2 as the dominant volatile. Three hydrothermal fluid regimes (orthomagmatic, mixed, meteoric) are identified, and precipitation of cassiterite, wolframite, and columbite-tantalite is claimed to have happened in the first two.

Using the Kymi topaz-bearing granite (cf. Haapala and Lukkari, this volume) and synthetic compositions as starting materials, Bhalla et al. have conducted experiments to study the influence of temperature, oxygen fugacity, and certain volatile elements on the solubility of SnO_2 in peraluminous granitic melt. The experiments were carried out at 700–800 °C, 2 kbar, and varying fO_2 (NNO to NNO +2 to +3) with ~2.5 to 3% normative corundum and H_2O close to the saturation level. The results show that the solubility of SnO_2 increases with temperature and decreases with the fO_2. Cl content also shows a slight positive correlation with SnO_2 solubility, whereas F content does not seem to be a controlling factor. A major finding of the paper is that the solubility of SnO_2 is dependent of the Al content of the melt, particularly in the case of a reduced melt that carries Sn in the divalent state. This agrees well with the observation that tin granites are characteristically peraluminous and reduced.

Acknowledgments

This special issue would not have been possible without generous help from devoted external reviewers: Don L. Anderson (Caltech), J. Lawford Anderson (Los Angeles), Ulf B. Andersson (Uppsala), Don Baker (Montreal), Bernard Barbarin (Clermont-Ferrand), Calvin G. Barnes (Lubbock), Sandra M. Barr (Wolfville), Bernard Bonin (Paris), James P. Calzia (Menlo Park), Eric H. Christiansen (Provo), Barrie Clarke (Dalhousie), Roberto Dall'Agnol (Belém), Olav Eklund (Turku), Brent A. Elliott (Florence), Francois Farges (Marne la Vallée), Steve Foley (Mainz), Carol Frost (Laramie), M. Charles Gilbert (Norman), Ilmari Haapala (Helsinki), Francois Holtz (Hanover), Hans Keppler (Tuebingen), Ilmo Kukkonen (Espoo), Seppo I. Lahti (Espoo), David London (Norman), Arto Luttinen (Helsinki), Gail A. Mahood (Stanford), Hervé Martin (Clermont-Ferrand), Robert F. Martin (Montreal), James McLelland (Saratoga Springs), Virginia T. McLemore (Socorro), Calvin F. Miller (Nashville), Hanna Nekvasil (Stony Brook), Mikko Nironen (Espoo), Øystein Nordgulen (Trondheim), Matti Poutiainen (Helsinki), Malcom P. Roberts (Grahamstown), Sergio Rocchi (Pisa), Hugh Rollinson (Al Khodh), Tapio Ruotoistenmäki (Espoo), Bruno Scaillet (Orleans), W. David Sinclair (Ottawa), the late Kjell P. Skjerlie (Tromsø), John Tarney (Leicester), Rainer Thomas (Potsdam), Peter Treloar (Kingston), Matti Vaasjoki (Espoo), W. Randy Van Schmus (Lawrence), Ron Vernon (Sydney), Karen Webber (New Orleans), and Jim Webster (New York). I would also like to thank Steve Foley, Editor in Chief of *Lithos*, for accepting this special issue, and Patricia Massar and Friso Veenstra for excellent collaboration in technical matters. This volume is a contribution to IGCP-426 (Granite Systems and Proterozoic Lithospheric Processes).

References

Anderson, D.L., 2002. How many plates? Geology 30, 411–414.

Best, M.G., 2003. Igneous and Metamorphic Petrology, Second Edition. Blackwell, Oxford.

Dall'Agnol, R., Bettencourt, J.S. (Eds.), 1997. Proceedings of the Symposium on Rapakivi Granites and Related Rocks, Belém (Brazil), August 2–5, 1995, Anais da Academia Brasileira de Ciências, vol. 69.

Frost, B.R., Barnes, C.G., Collins, W.J., Arculus, R.J., Ellis, D.J., Frost, C.D., 2001. A geochemical classification for granitic rocks. Journal of Petrology 42, 2033–2048.

Haapala, I., 1966. On the granitic pegmatites in the Peräseinäjoki–Alavus area, south Pohjanmaa, Finland. Bulletin de la Commission Géologique de Finlande 224, 98 pp.

Haapala, I., 1974. Some petrological and geochemical characteristics of rapakivi granite varieties associated with greisen-type Sn, Be, and W mineralization in the Eurajoki and Kymi areas, southern Finland. In: Štemprok, M. (Ed.), Metallization Associated with Acid Magmatism I. Ústredni ústav geologický, Praha, pp. 159–169.

Haapala, I., 1977a. The controls of tin and related mineralizations in the rapakivi-granite areas of south-eastern Fennoscandia. Geologiska Föreningens i Stockholm Förhandlingar 99, 130–142.

Haapala, I., 1977b. Petrography and geochemistry of the Eurajoki stock, a rapakivi-granite complex with greisen-type mineralization in southwestern Finland. Bulletin-Geological Survey of Finland 286, 128 pp.

Haapala, I., 1985. Metallogeny of the Proterozoic rapakivi granites of Finland. In: Taylor, R.P., Strong, D.F. (Eds.), Granite-Related Mineral Deposits; Geology, Petrogenesis and Tectonic Setting. Extended Abstracts of Papers Presented at the CIM Conference on Granite-Related Mineral Deposits, September 15th–17th, 1985. Canadian Institute of Mining and Metallurgy, Halifax, Canada, pp. 123–131.

Haapala, I., 1995. Metallogeny of the rapakivi granites. Mineralogy and Petrology 54, 141–160.

Haapala, I., 1997. Magmatic and postmagmatic processes in tin-mineralized granites: topaz-bearing leucogranite in the topaz-bearing leucogranite of the Eurajoki rapakivi stock, Finland. Journal of Petrology 38, 1645–1659.

Haapala, I., Ojanperä, P., 1972. Genthelvite-bearing greisens in southern Finland. Bulletin-Geological Survey of Finland 259, 22 pp.

Haapala, I., Kähkönen, Y. (Eds.), 1989. Symposium Precambrian Granitoids-Petrogenesis, Geochemistry and Metallogeny, Abstracts. Special Paper-Geological Survey of Finland, vol. 8.

Haapala, I., Rämö, O.T., 1990. Petrogenesis of the Proterozoic rapakivi granites of Finland. Special Paper-Geological Society of America 246, 275–286.

Haapala, I., Condie, K.C. (Eds.), 1991. Precambrian Granitoids—Petrogenesis, Geochemistry and Metallogeny, Special Issue, Precambrian Research, vol. 51.

Haapala, I., Rämö, O.T., 1992. Tectonic setting and origin of the Proterozoic rapakivi granites of the southeastern Fennoscandia. Transactions of the Royal Society of Edinburgh. Earth Sciences 83, 165–171.

Haapala, I., Rämö, O.T. (Eds.), 1994. IGCP Project 315 Publication No. 12. Mineralogy and Petrology 50 (1–3).

Haapala, I., Rämö, O.T., 1999a. Rapakivi granites and related rocks: an introduction. Precambrian Research 95, 1–7.

Haapala, I., Rämö, O.T. (Eds.), 1999b. Rapakivi Granites and Related Rocks, Special Issue. Precambrian Research 95 (1–2).

Haapala, I., Thomas, R., 1999. Melt inclusions in quartz and topaz of the topaz granite from Eurajoki, Finland. Journal of the Czech Geological Society 45, 149–154.

King, P.L., White, A.J.R., Chappell, B.W., Allen, C.M., 1997. Characterization and origin of aluminous A-type granites from the Lachlan Fold Belt, southeastern Australia. Journal of Petrology 38, 371–391.

Korja, K., Korja, T., Luosto, U., Heikkinen, P., 1993. Seismic and geoelectric evidence for collisional and extensional events in the Fennoscandian shield—implications for Precambrian crustal evolution. Tectonophysics 219, 129–152.

Luosto, U., Tiira, T., Korhonen, H., Azbel, I., Burmin, V., Buyanov, A., Kosminskaya, I., Ionkis, V., Sharov, N., 1990. Crust and upper mantle structure along the DSS Baltic profile in SE Finland. Geophysical Journal International 101, 89–110.

Rämö, O.T., 1991. Petrogenesis of the Proterozoic rapakivi granites and related basic rocks of the southeastern Fenoscandian: Nd and Pb isotopic and general geochemical constraints. Bulletin-Geological Survey of Finland 355, 161 pp.

Rämö, O.T., Haapala, I., 1995. One hundred years of rapakivi granite. Mineralogy and Petrology 52, 129–185.

Rämö, O.T., Van Schmus, W.R., Bettencourt, J.S., 2002. Preface: IGCP project 426-granite systems and proterozoic lithospheric processes. Precambrian Research 119, 1–7.

Rämö, O.T., Kosunen, P.J., Lauri, L.S., Karhu, J.A. (Eds.), 2003. Granitic Systems—State of the Art and Future Avenues. An International Symposium in Honor of Professor Ilmari Haapala, January 12–14, 2003, Abstract Volume. Helsinki University Press.

Tectonics and Source Regions

Available online at www.sciencedirect.com

Lithos 80 (2005) 1–32

www.elsevier.com/locate/lithos

Comparison of Proterozoic and Phanerozoic rift-related basaltic-granitic magmatism

Ilmari Haapala*, O. Tapani Rämö, Stephen Frindt[1]

Department of Geology, P.O. Box 64, FI-00014, University of Helsinki, Finland

Received 9 June 2003; accepted 9 September 2004
Available online 11 November 2004

Abstract

This paper compares the 1.67–1.47 Ga rapakivi granites of Finland and vicinity to the 1.70–1.68 Ga rapakivi granites of the Beijing area in China, the anorogenic ~130 Ma granites of western Namibia, and the 20–15 Ma granites of the Colorado River extensional corridor in the Basin and Range Province of southern Nevada. In Finland and China, the tectonic setting was incipient, aborted rifting of Paleoprotcrozoic or Archean continental crust, in Namibia it was continental rifting and mantle plume activity that led to the opening of southern Atlantic at ~130 Ma. The 20–15 Ma granites of southern Nevada were related to rifting that followed the Triassic–Paleogene subduction of the Farallon plate beneath the southwestern United States. In all cases, extension-related magmatism was bimodal and accompanied by swarms of diabase and rhyolite–quartz latite dikes. Rapakivi texture with plagioclase-mantled alkali feldspar megacrysts occurs in varying amounts in the granites, and the latest intrusive phases are commonly topaz-bearing granites or rhyolites that may host tin, tungsten, and beryllium mineralization. The granites are typically ferroan alkali-calcic metaluminous to slightly peraluminous rocks with A-type and within-plate geochemical and mineralogical characteristics. Isotope studies (Nd, Sr) suggest dominant crustal sources for the granites. The preferred genetic model is magmatic underplating involving dehydration melting of intermediate-felsic deep crust. Juvenile mafic magma was incorporated either via magma mingling and mixing, or by remelting of newly hybridized lower crust. In Namibia, partial melting of subcontinental lithospheric mantle was caused by the Tristan mantle plume, in the other cases the origin of the mantle magmatism is controversial. For the Fennoscandian suites, extensive long-time mantle upwelling associated with periodic, migrating melting of the subcontinental lithospheric mantle, governed by heat flow and deep crustal structures, is suggested.

Keywords: A-type granite; Bimodal association; Rifting; Diabase; Rapakivi granite

---

* Corresponding author. Tel.: +358 9 590400; fax: +358 9 19150826.

E-mail address: ilmari.haapala@helsinki.fi (I. Haapala).

[1] Present address: Geological Survey of Namibia, P.O. Box 2168, 1 Aviation Rd., Windhoek, Namibia.

1. Introduction

Continental rifting and resultant faulting, rapid erosion, and deposition of clastic sediments, as well as volcanic and plutonic activity are in several cases associated with hot mantle plumes. Rifting may vary from incipient (aborted) with only minor extension to the break-up of continents and opening of oceans (e.g., White and McKenzie, 1989; Windley, 1995, pp. 51–64, 98–104, 262–276; Peccerillo et al., 2003). The nature of the associated magmatism varies from alkaline to subalkaline, and bimodal suites are typical of most continental rifting environments.

This paper summarizes the results of studies led by the Department of Geology, University of Helsinki, on rapakivi-type granites and related mafic rocks associated with continental rifting in Finland, China, Namibia, and the western United States. The main aim of these studies has been to test whether the magmatic underplate model (see Bridgewater et al., 1974; Emslie, 1978), first applied to the Finnish rapakivi granites by Haapala (1985) and Haapala and Rämö (1990), is applicable to the rapakivi granites and related rocks of the other regions. The study areas were chosen to supplement each other and give as comprehensive information as possible on the tectonic setting, magmatic association, granite-related mineralization, and petrogenesis of the igneous suites. Southern Finland, the classic rapakivi area with 1.67–1.54 Ga rapakivi granites and associated diabase and quartz-porphyritic dike swarms enclosed in a 1.9 Ga cratonized crust, is taken as the reference area. The 1.7 Ga Shachang rapakivi granite complex near Beijing, China was emplaced, together with diabase dikes, in Neoarchean cratonized crust. In northwestern Namibia, Cretaceous granites resembling evolved rapakivi granites intruded, together with mafic and felsic dikes, into the cratonized Neoproterozoic–Paleozoic Damara orogenic crust. They were related to mantle plume activity and rifting which led to the opening of the South Atlantic. Miocene granites and associated diabase and rhyolite dikes of southernmost Nevada, Basin and Range Province, represent a modern example of an extensional environment with rapakivi-type mag-

matism. The studies have included field mapping as well as petrographic, geochronological, and geochemical (including isotopic) studies.

2. The 1.67–1.47 Ga rapakivi granites of Finland and vicinity

2.1. Geologic setting

The major Precambrian tectonic units of Fennoscandia comprise (1) the Archean 3.1–2.6 Ga granite gneiss–greenstone belt domain in the east, (2) the broad 1.9–1.8 Ga Svecofennian orogenic belt, and (3) the 1.7–1.5 Ga Gothian or Southwest Scandinavian domain in the southwest (see Fig. 1 inset). Major cratonic magmatic events are represented by 2.45 Ga layered mafic intrusions and associated granitoids in northern Finland (Alapieti, 1982; Rämö and Luukkonen, 2001; Lauri and Mänttäri, 2002) and the 1.67–1.47 Ga rapakivi granites and associated mafic rocks (diabase dikes, gabbros, anorthosites). Current syntheses of the evolution of the Fennoscandian lithosphere can be found in Nironen (1997), Korsman et al. (1999), and Pesonen et al. (2000). These studies utilized also extensive geophysical data including deep seismic soundings. As currently perceived, the Svecofennian crust was formed by complex accretion of island arcs and microcontinents from west and south to the Archean craton at 1.92–1.88 Ga. Subsequent thickening of the crust and remelting were related to magmatic underplating at 1.89–1.87 Ga (Korsman et al., 1999; Nironen et al., 2000). Collision-related melting of the felsic sedimentary-volcanic complex of southern Finland and central Sweden at 1.84–1.80 Ga produced a belt of migmatite-related potassic S-type granites. Emplacement of post-collisional shoshonitic and granitic plutons at ~1.80 Ga completed the Svecofennian orogenic evolution (Nironen, 1997; Eklund et al., 1998; Korsman et al., 1999).

Fig. 1. (a) Map showing the distribution and ages of the rapakivi granite complexes and diabase dikes as well as contours of crustal thickness in the south-central part of the Fennoscandian shield. The subvertical lines outline four rapakivi age zones (1.53–1.47, 1.59–1.54, 1.67–1.62, and 1.55–1.54 Ga). Sites for the five representative samples (Table 1) are indicated. The inset shows the area in relation to the major crustal domains in the shield. TIG is the Transcandinavian Igneous Belt. The map is from Rämö and Korja (2000) and Rämö et al. (2000), where the pertinent references are given. (b) Diagram showing variation of initial Nd isotope composition (ε_{Ndi}) in the rapakivi granites (open-circle pattern) and related mafic-intermediate rocks (black) in the four age zones. The composition of the depleted mantle (DePaolo, 1981), the ~1.9 Ga Svecofennian rocks (Huhma, 1986; Patchett and Kouvo, 1986), and the Archean crust of the Fennoscandian shield (Rämö et al., 1996) are shown at the time of interest. Adopted from Rämö et al. (2000), where also a more comprehensive list of references is given.

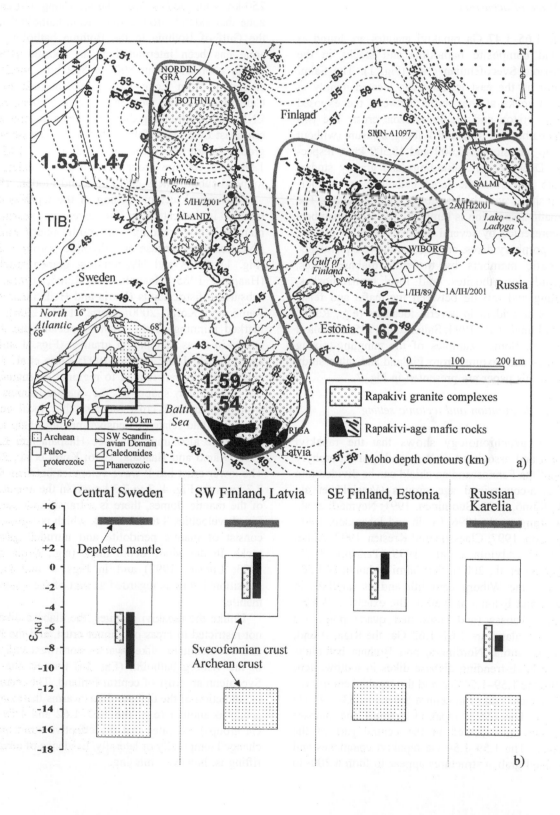

Fig.

2.2. Mode of occurrence

The 1.65–1.47 Ga rapakivi granites are found as epizonal composite batholiths and stocks in the Proterozoic Svecofennian crust (Fig. 1). The Salmi batholith in the east is, however, located mainly in Archean rocks at the border between the exposed Archean craton and the Proterozoic crust. The two largest batholiths, Riga and Wiborg, measure (without satellitic intrusions) 40,600 and 18,500 km^2, respectively. Deep seismic soundings and gravity studies indicate that even the large batholiths are subhorizontal sheets with thicknesses of about 5–10 km. The magmatic association is bimodal. The felsic members are represented by several different types of granite, quartz porphyry dikes, and rare extrusive equivalents, the mafic members by tholeiitic diabase dikes, gabbroids, anorthosites, and local monzodiorites. Mingling and mixing between the mafic and felsic magmas is evident in several localities (e.g., Rämö, 1991; Eklund et al., 1994; Salonsaari, 1995; Alviola et al., 1999). Some examples of the mingling-mixing structures and textures from Finland and other study areas of this paper are presented in Fig. 2.

2.3. Age distribution and tectonic setting

U–Pb geochronology shows that the rapakivi granites and associated diabase and quartz porphyry dikes of the Fennoscandian shield can be divided into four area-constrained age groups (Vaasjoki et al., 1991; Rämö, 1991; Suominen, 1991; Neymark et al., 1994; Rämö et al., 1996; Lindh and Johansson, 1996; Andersson, 1997; Claesson and Kresten, 1997; Ahl et al., 1997; Alviola et al., 1999; Persson, 1999; Andersson et al., 2002): the Salmi batholith is 1.56–1.53 Ga; the Wiborg batholith and its satellites in Finland and Estonia as well as the extensive WNW-trending diabase (and associated quartz porphyry) dikes in Finland are 1.67–1.62 Ga; the Riga, Åland, Vehmaa, Laitila, Nordingrå, and Bothnia batholiths and the NNE-trending diabase dikes in southwestern Finland are 1.59–1.54 Ga; and the rapakivi complexes and associated dikes in central Sweden 1.53–1.47 Ga (Fig. 1). The oldest rapakivi granites and diabase dikes are thus located in the central part of the province. The 1.59–1.54 Ga rapakivi complexes and associated graben structures appear to form a 200- to 250-km-wide and at least 800-km-long N-trending zone that extends from Latvia (Riga batholith) along the Gulf of Bothnia to the Bothnia batholith. This zone has been interpreted as an aborted paleorift within a broad (about 800 by 800 km) rifted area extending from Lake Ladoga in the east to the Caledonides in the west (Korja et al., 2001). Some ENE- and ~N-trending diabase dikes along the western coast of Finland follow this zone. The other age groups do not form clear belts, and the 1.67–1.62 and 1.56–1.53 Ga age groups can best be depicted as rounded subprovinces (see Puura and Flodén, 1999).

Extensional tectonic setting of the rapakivi magmatism is indicated by the extensive swarms of mainly WNW-trending (in southwestern Finland ENE-trending) diabase and rhyolitic porphyry dikes (Fig. 1), faults and NW-trending graben structures (Haapala, 1985; Haapala and Rämö, 1992), and subsurface listric faults (Korja and Heikkinen, 1995; Rämö and Korja, 2000; Korja et al., 2001). The marked thinning of the crust—especially the lower crust—in areas of rapakivi granites (Figs. 1 and 4a; Luosto et al., 1990; Luosto, 1997; Korja et al., 1993) is probably in part related to regional extension, but the subcircular shape of the mantle domes and overlying thinned crust suggest that local mantle upwelling and related extensive melting of the lower crust were also important (see Haapala and Rämö, 1992; Rämö and Haapala, 1996; Korja et al., 2001). The lower crust has P wave velocities between 7 and 7.7 km/s (in Fig. 4, 7 to 7.3 km/s). In the apical parts of the mantle domes, there is a transitional zone (in Fig. 4, velocities 8 to 8.2 km/s), which is supposed to consist of mantle peridotite and intruded gabbroic rocks. In the Moho depth maps (e.g., Korja et al., 1993; Luosto, 1997) and in Figs. 1 and 4a, the transitional zone is regarded as part of the uppermost mantle.

Unlike the rapakivi granites, the diabase dikes are not restricted to areas of thinner crust, and the Häme and Suomenniemi dike swarms northwest and north of the Wiborg batholith (Fig. 1a) transect the thick Svecofennian crust of central Finland. The change in the direction of the diabase dike swarms from one age group to another (e.g., the 1.67–1.62 and 1.59–1.54 Ga groups) indicates that the effective strain pattern changed temporally or laterally. Evidence of advanced rifting is, however, missing.

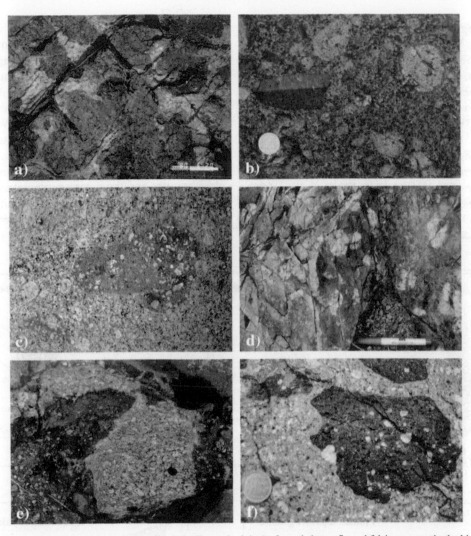

Fig. 2. Structures and textures related to the interaction (mingling and mixing) of coexisting mafic and felsic magmas in the bimodal magmatic suites of this study. (a) Pillows of monzodiorite in granite from the Ahvenisto rapakivi complex, southeastern Finland (Alviola et al., 1999). (b) Dark labradorite xenocryst (left) and andesine–oligoclase-mantled alkali feldspar ovoids in dark wiborgite on the Ristisaari Island in the southwestern part of the Wiborg batholith. (c) A dark hybrid enclave containing more mafic "enclaves in enclave" as well as alkali feldspar and quartz xenocrysts in a hybrid hornblende-bearing granite of the Jaala-Iitti complex, southeastern Finland (Salonsaari, 1995). (d) Alkali feldspar xenocrysts in a diabase dike cutting the biotite–hornblende granite of the Shachang pluton, China. (e) Comingled mafic pillows in the porphyritic granite at the northeastern margin of the Gross Spitzkoppe stock, Namibia (Frindt, 2002). In this case, close to the mafic pillows, the topaz-bearing granite has alkali feldspar megacrysts mantled by plagioclase (rapakivi texture sensu lato). (f) Mafic enclaves with alkali feldspar and quartz xenocrysts in rhyolite from a composite dike that cuts the Spirit Mountain granite in southernmost Nevada. Many of the alkali feldspar megacrysts in enclaves and rhyolite have plagioclase–quartz shells. Photos: (a), (c)–(f) Ilmari Haapala; (b) Tapani Rämö.

2.4. Petrography and geochemistry

The rapakivi granite batholiths and stocks are usually composed of several petrographically distinct granite types, and in some cases minor gabbroic, anorthositic, monzodioritic and syenitic bodies are also found (e.g., Vorma, 1976; Rämö, 1991; Alviola et al., 1999). The earlier granite phases are commonly hornblende–biotite (±fayalite) granites, followed by biotite granites and less common topaz-bearing alkali feldspar granites. The wiborgitic rapakivi granites with plagioclase-mantled alkali feldspar megacrysts are hornblende bearing. Greisen-type Sn–Be–W–Zn mineralization is associated with the topaz-bearing granites (Haapala, 1977a, 1995, 1997; Edén, 1991).

In geochemical and mineral chemical composition, the rapakivi granites are typical aluminous A-type granites (cf. King et al., 1997). Especial characteristics are high K, Fe/Mg, F, Ga, Rb, and Zr (except in the topaz-bearing granites) (e.g., Vorma, 1976; Haapala, 1977b; Rämö, 1991). The Fe/(Fe+Mg) number in biotite varies from 0.80 to 1.00 (Rieder et al., 1996), and in hornblende from 0.77 to 0.95 (Simonen and Vorma, 1969). In various differentiation diagrams, the

granites plot into the A-type fields of Whalen et al. (1987) and within-plate granite fields of Pearce et al. (1984) (Fig. 3a, b). The multielement diagram (Fig. 3c), with strong enrichments in Th and U and depletion in Sr, is typical of A-type granites (e.g., Whalen et al., 1996). The REE patterns (Fig. 3d) show enrichment in the LREEs and negative Eu anomalies. With increasing magmatic fractionation, the enrichment in LREE decreases and the Eu anomaly deepens. The fayalite–magnetite–quartz association in the more mafic rapakivi granites and the common prevalence of ilmenite over magnetite indicate reducing conditions (ilmenite series of Ishihara, 1977).

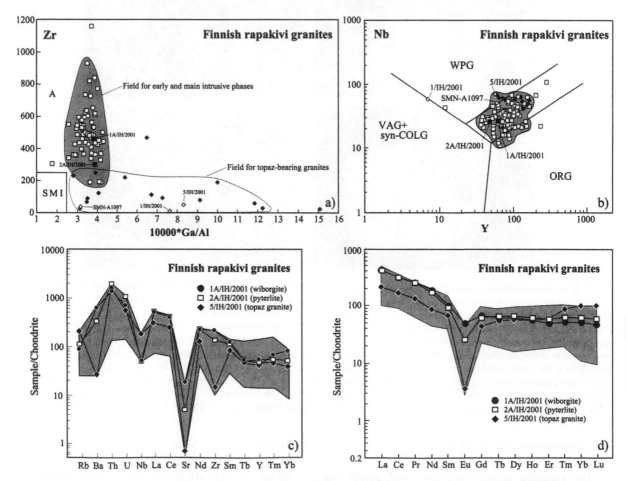

Fig. 3. Analyses of Finnish rapakivi granites plotted in (a) Zr versus 10,000 Ga/Al (Whalen et al., 1987), (b) Nb versus Y (Pearce et al., 1984), (c) a chondrite-normalized multielement, and (d) a chondrite-normalized REE diagrams (overall fields and type analyses marked). Data from Vorma (1976), Haapala (1977a,b), Rämö (1991), Lindberg and Bergman (1993), Salonsaari (1995), Rieder et al. (1996), Kosunen (1999), and Table 1.

2.5. Isotope geochemistry and petrogenesis

Since 1990, large amounts of new U–Pb, Nd, common Pb, and Sr isotope data have been published on the rapakivi complexes of Fennoscandia (Haapala and Rämö, 1990; Rämö, 1991, 2001; Vaasjoki et al., 1991; Neymark et al., 1994; Rämö et al., 1996; Amelin et al., 1997; Claesson and Kresten, 1997; Andersson, 1997; Persson, 1999; Alviola et al., 1999; Lindh, 2001; Andersson et al., 2002). The Sm–Nd system has proven to be an especially useful indicator of the overall age and nature of the sources of the rapakivi granites. Some results are summarized in Figs. 1 and 7.

The rapakivi granites of Finland have initial ε_{Nd} values between -3 and 0, suggesting derivation from the 1.9 Ga Svecofennian crust (Haapala and Rämö, 1990; Rämö, 1991). The Salmi rapakivi batholith, located at the Archean–Proterozoic boundary in Russian Karelia, shows more negative ε_{Nd} values of -9 to -5.5, suggesting that a major Archean source component mixed with Svecofennian or younger, rapakivi-age component (Rämö, 1991; Neymark et al., 1994). The rapakivi granites of central Sweden have also strongly negative ε_{Nd} values of -7.5 to -4.5, implying the presence of an Archean basement underneath the exposed Svecofennian crust (Andersson, 1997; Andersson et al., 2002). The clastic metasediments of the Svecofennian Bothnian Basin in central Sweden have ε_{Nd} (at 1.51 Ga) values (-10.2 to -5.8) that overlap those of the rapakivi granites (Welin et al., 1993; Andersson et al., 2002).

The mafic rocks (diabase dikes, gabbroids) have ε_{Nd} values that largely overlap with those of the associated rapakivi granites (Figs. 1 and 7). However, the diabase dikes of the Suomenniemi and Häme swarms (Wiborg rapakivi area) show ε_{Nd} values that are higher (up to $+1.6$) than those of the associated rapakivi granites. The mafic and hybrid intermediate rocks of central Sweden have low ε_{Nd} (at 1.51 Ma) values (-8.9 to -6.5), some of which are more negative than those of the rapakivi granites (Andersson et al., 2002). In the Salmi batholith of Russian Karelia, the ε_{Nd} values of the mafic and intermediate rocks (-6.5 to -7.5) overlap with those of the granites (Neymark et al., 1994).

The isotopic overlap of the mafic and felsic members of the rapakivi suites could suggest that the felsic magmas were formed by fractional crystallization of mantle-derived mafic parental magmas (see discussion in Frost et al., 2002). This is, however, regarded unlikely because granitic rocks generally greatly prevail over mafic and intermediate rocks in the rapakivi complexes. Numerous observations of mingling structures and textures indicate simultaneous presence of mafic and felsic magmas (Rämö, 1991; Eklund et al., 1994; Salonsaari, 1995). Further, the thinning of the lower crust above mantle domes and the transitional zones (Fig. 4b) is compatible with the interpretation that rapakivi magmas are of deep crustal origin rather than mantle fractionates. Therefore, different main sources, mantle and crust, are preferred for the mafic and felsic magmas, respectively.

Rämö (1991) suggested that the rapakivi granites of Finland (the Suomenniemi batholith) were originated by about 20% partial melting of intermediate-felsic (tonalitic-granodioritic) Svecofennian lower crust; for the rapakivis of the Salmi batholith, a mixed (Archean–Svecofennian) source was invoked. Kosunen et al. (2004) proposed that the parent magmas of the Obbnäs rapakivi granites in southern Finland originated by partial melting of ferrodioritic/ferromonzodioritic or tonalitic source, with contribution through mixing from a juvenile mafic (diabase) melt, whereas the adjacent Bodom granite had a granodioritic source. Kosunen et al. (2004) concluded that stepwise fractionation of tholeiitic basalt is an unlikely process for producing the trace element distribution of the granites. For the rapakivi granites of central Sweden, Andersson et al. (2002) preferred a mixed source composed of dominantly Svecofennian and 30–40% Archean metaigneous rocks.

Neymark et al. (1994) suggested an interesting model for the rocks of the Salmi batholith. This involves Svecofennian magmatic underplating with interaction between the mantle-derived magmas and the Archean crust, which led to the formation of a hybrid, melt-depleted lower crust. When the mafic residues of these melting and differentiation processes returned to the mantle, carrying with them material from the lower continental crust, a zone of hybrid uppermost mantle was formed. Both the hybrid zones had mixed isotopic character. Subsequent mantle upwelling caused partial melting of the lower crust and formation of the rapakivi granite magmas, whereas partial melting of the enriched upper mantle

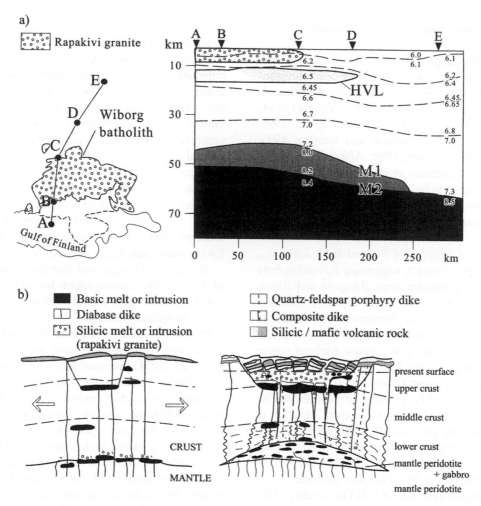

Fig. 4. (a) A vertical section across the Wiborg rapakivi granite batholith showing the structure of the crust and upper mantle according to Luosto et al. (1990) and Korja et al. (1993). The broken lines show changes in P wave velocities. HVL is a high-velocity layer (probably gabbroic rocks), and M1 and M2 mark a transitional zone at the crust–mantle boundary (from Rämö and Haapala, 1996). (b) A model for the origin of the bimodal rapakivi granite complexes of Finland according to Haapala (1989) and Rämö and Haapala (1996). Mantle-derived mafic magmas are intruded to the crust–mantle boundary, where they form a magmatic underplate, a hybrid of mantle peridotite, gabbro, and possible crustal material. Parts of the mafic magmas intrude to higher crustal levels producing gabbroic-anorthositic intrusions, diabase dikes, and basaltic lava fields. The magmatic underplate causes extensive partial melting of the lower crust producing A-type granitic magmas—parents of the rapakivi granites, quartz porphyry dikes, and rhyolitic volcanic rocks. Mingling and hybridization of the granitic and mafic magmas occurs locally in all crustal levels. Partial melting and transit of the granitic magmas to higher crustal levels, possibly coupled with extensional stretching, causes the thinning of the lower crust.

produced the parental magmas of the mafic rocks. This interpretation has similarities with the model of Emslie et al. (1994) for the Nain plutonic suite of Labrador. The model of Neymark et al. (1994) is supported by the discoveries of granulitic lower crustal xenoliths in a Devonian lamprophyre diatreme on Elovy island in southern Kola Peninsula (Kempton et al., 2001, and references therein) and in a 0.6 Ga

kimberlite diatreme in Lahtojoki near Kuopio, eastern Finland, in the area of >58 km thick crust (Hölttä et al., 2000). The xenoliths are mainly mafic garnet-bearing granulites, with some intermediate-felsic granulites. It appears that below the Archean crust, close to the Paleoproterozoic (Svecofennian) domain, is a predominantly mafic lower crust, which shows petrographic and isotopic evidence of at least two episodes of magmatism and reworking including K-metasomatism. U–Pb zircon age determinations of the lower crustal xenoliths suggest that in the Kola area the main magmatism and reworking episodes took place at 2.5–2.4 Ga (time of basaltic magmatism that produced the large layered mafic intrusions of the northern Fennoscandian shield) and at ~1.7 Ga (time of granitic magmatism in the upper crust) (Downes et al., 2002). For the lower crust of the Kuopio area (eastern Finland), U–Pb ages of zircon in xenoliths and zircon xenocrysts in the Lahtojoki kimberlite imply main growth and reworking episodes at 2.7 and 1.8 Ga (Hölttä et al., 2000; Peltonen and Mänttäri, 2001). The granulite xenoliths of Elovy island show a wide variation in isotopic composition; ε_{Nd} (at 1.54 Ga) values vary from +6.8 to −11.1 (commonly from −5 to −9), covering the field of the Salmi rapakivi complex (Neymark et al., 1994; Kempton et al., 2001). Available three whole-rock Nd isotope analyses of the Lahtojoki xenoliths (Hölttä et al., 2000) give ε_{Nd} (at 1.64 Ga) values of −3.5, −2.5, and +2.8, grossly matching the field of the granites of the Wiborg batholith. It is unlikely, however, that partial melting of this type of mafic lower crust could have produced the typical REE distribution of the rapakivi granites (see Elliott, 2003).

A problem with a metaigneous tonalitic-granodioritic source is that melting of such a source would generally produce high-fO_2 (magnetite-type) granites rather than reduced ilmenite-type rocks typical of the rapakivi association. Frost and Frost (1997) suggested melting of a tholeiitic mafic-intermediate igneous source for the formation of the reduced rapakivi granites of Wyoming. Taking into account the metaluminous-peraluminous character of the Fennoscandian rapakivi granites and the overlapping Nd isotope compositions of the Svecofennian metasediments and rapakivi granites of the same areas (e.g., Welin et al., 1993; Andersson et al., 2002), fusion of some metasedimentary rocks (metagreywackes) is also possible. In the Archean and early Paleoproterozoic, sediments and their derivatives would generally be in reduced state because of the common presence of biogenic carbon in the sedimentary rocks and the low oxygen content of the atmosphere (see Karhu and Holland, 1996). However, low boron contents of the rapakivi granites do not support any higher incorporation of sedimentary material.

The diabases, gabbros, and anorthosites of Finland and Estonia (age groups 1.67–1.62 and 1.59–1.54 Ga) with ε_{Nd} +1.6 to −1.7 were obviously fractionated from mafic magmas that had been derived from a mantle source characterized by long-term LREE depletion, and were variously contaminated with crustal materials (Rämö, 1991). The origin of the mafic rocks of the Salmi batholith and central Sweden is more problematic. They may have been formed from mafic magmas generated by partial melting of LREE-enriched subcontinental lithospheric mantle, with little crustal assimilation (Neymark et al., 1994; Rämö et al., 2000), or from non-enriched mafic magmas that had assimilated and melted Archean/Svecofennian crustal rocks. The crustal assimilants may have been intermediate-felsic metaigneous rocks (Rämö, 1991; Andersson et al., 2002), mafic-intermediate metaigneous rocks formed by earlier (Svecofennian or later) magmatic underplatings (see Frost and Frost, 1997; Frost et al., 1999), or plagioclase–pyroxene granulite residue after extraction of granitic magmas (see Emslie et al., 1994). Advanced assimilation of plagioclase-rich granulite could help to explain the presence of anorthositic rocks in some rapakivi complexes.

The evidence for the source of the mafic and granitic magmas is not unequivocal. If the mantle-derived magmas were able to cause large-scale remelting of the lower crust, then also effective contamination with older crustal components and change in isotope composition of the mantle magmas would be expected. Processes involving low-degree remelting and assimilation of tholeiitic mafic-intermediate lower crust by mantle magmas could explain chemical characteristics of the mafic and felsic members of the bimodal association as well as the reduced (ilmenite-series) character of the rapakivi granites (see Frost and Frost, 1997; Frost et al., 1999). Another question is could such low-degree partial melting have produced the huge amount of granitic magma of the rapakivi batholiths of Fennoscandia and thinning of the lower crust below the rapakivi batholiths.

We think that the source of large rapakivi granite complexes of Fennoscandia did not consist of one rock type only, but comprised a suite of rocks, mainly intermediate-felsic but not necessarily only metaigneous rocks, that were fused successively and in varying amounts. The input of mantle magmas left its signatures also in the granites, either by hybridization of felsic crustal magmas and mafic mantle-derived magmas (for which there is convincing field evidence), or by anatectic melting of crustal rocks hybridized in earlier (Svecofennian and/or later, rapakivi-age) magmatic underplating(s) (cf. Hildreth et al., 1991). The differences in the average chemical composition of the rapakivi complexes in southeast Fennoscandia (Vorma, 1976; Rämö and Haapala, 1995) probably reflect variations in the relative amounts of the fused source rocks, although also depth of erosion level may have caused significant differences in this respect. The overall source of the different plutons has probably varied from ferrodioritic to granodioritic in composition.

2.6. Petrogenetic model

The model including magmatic underplating and crustal anatexis in extensional tectonic environment was first applied to Finnish rapakivi granites by Haapala (1985, 1989) and Nurmi and Haapala (1986), and has been further developed in several papers (Rämö, 1991; Haapala and Rämö, 1992; Elo and Korja, 1993; Rämö and Haapala, 1996). A simplified two-stage model, based on the model of Huppert and Sparks (1988), is presented in Fig. 4b. The mafic rocks are derivatives of mantle magmas that evolved through variable crustal assimilation, local hybridization, and fractional crystallization. The felsic rocks are essentially crust derived, but probably have a mantle component, and evolved through minor assimilation, hybridization, and fractional crystallization. The erupted lavas formed volcanoes and lava and tuff plateaus above the plutonic rapakivi complexes, and were subsequently eroded away. This simple model explains the origin of the rapakivi granite magmas, bimodal character of the magmatism, thinning of the lower crust, and extensional setting, but does not explain the reason for the partial melting of the mantle. In this regard, possible mechanisms include passive or active rifting, deep mantle plumes, extensional collapse of the Paleoproterozoic orogen, mantle superswells under a Paleoproterozoic supercontinent, and melting of subcontinental mantle owing to instabilities related to distant subduction zones on the flanks of accretionally growing continents (see Rämö and Haapala, 1995; Haapala and Rämö, 1999; Åhäll et al., 2000).

In his discussion on the origin of Laurentia's Mesoproterozoic anorogenic granitic magmatism, Hoffman (1989) suggested that the mid-Proterozoic supercontinent thermally insulated the mantle beneath its interior causing convective large-scale upwelling— thousands of kilometers in diameter—of the heated mantle, which led to anorogenic magmatism across Laurentia. According to Hoffman (1989), inward mantle flow at the periphery of the growing supercontinent may contribute to the development of mantle upwelling beneath the interior of the supercontinent, to balance the downward mantle flow at the periphery. This model can be applied to the Fennoscandian rapakivi magmatism as well as, at 1.6 Ga, Fennoscandia was probably part of a supercontinent that included Laurentia and Amazonia (Pesonen and Mertanen, 2002). The thick continental crust provided a radiogenic thermal blanket below which the subcontinental mantle became heated, started to upwell and melt locally, eventually bulging up into the lower crust in areas where the heat flow was especially high or the crust weak (e.g., along deep fracture zones). The substantial thinning of the lower crust above subcircular mantle domes (e.g., the Wiborg batholith area; Figs. 1 and 4a, b) suggests that, in such cases, thinning was mainly caused by upwelling mantle magmas (magmatic underplating) and extensive melting of the lower crust. The latter was promoted by the fertile (juvenile) character of the newly formed Paleoproterozoic crust (cf. Anderson and Bender, 1989). Mantle upwelling and related reorganization of the crust stabilized lithosphere in one area, and mantle diapirism shifted to other areas, producing the area-constrained age groups (Fig. 1a). Possible subduction-related downward mantle flows at the margin of the accretionally growing continent at 1.65–1.5 Ga (see Åhäll et al., 2000) may have contributed to the onset of the mantle upwellings beneath the interior of the continent (cf. Hoffman, 1989).

3. The 1.7 Ga Shachang rapakivi pluton

3.1. Geologic setting

Several rift-related 1.70–1.68 Ga rapakivi granite plutons and related mafic rocks intruded Archean gneisses and migmatites in the northern part of the

North China (or Sino-Korean) craton. This Archean craton records continental growth from 3.85–3.55 to ~2.5 Ga (Liu et al., 1990; Wang et al., 1990). During the Paleoproterozoic, the craton was reworked by several tectonic events and became surrounded by orogenic belts in the south and east. Between 1.85 and 1.60 Ga, an E-trending 50- to 100-km-wide intracratonic rift developed and was the site of extensive

continental sedimentation and igneous activity, including rapakivi granite.

The emplacement of the rapakivi granites and associated quartz syenites and mafic rocks (anorthosites, gabbros, diabase dikes) was controlled by three fault zones within the rift (Yu et al., 1994, 1996a; Fig. 5a). The U–Pb zircon ages of the rapakivi granites and quartz syenites show little variation: the Lanying

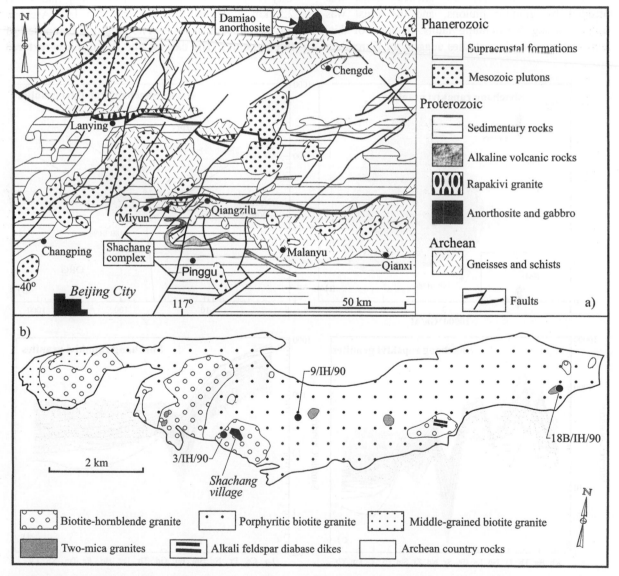

Fig. 5. (a) Geological map northeast of Beijing City showing the location of Proterozoic rapakivi granite–anorthosite–syenite complexes in relation to roughly E-trending fault zones. (b) The Shachang rapakivi granite pluton. The locations of representative type specimens (Table 1) are marked. Modified from Rämö et al. (1995).

pluton is 1697.3±4 Ma, the Chiecheng pluton (located west of the area shown in Fig. 4a) 1696.7±2.4 Ma, and the Shachang pluton 1690.5±2.4 to 1683.8±2.2 Ma (Yu et al., 1996a,b). The age of the Damio anorthosite–quartz mangerite complex is less accurate with an Sm–Nd isochron age of 1735±239 Ma for anorthosite and an Rb–Sr isochron age of 1686±193 Ma for quartz mangerite (Yu et al., 1996a).

During the Mesozoic, the Archean crust was intruded by granitic plutons, and the crust was ruptured along NE-trending faults. In spite of this, deep seismic sounding studies suggest an E-trending

mantle dome under the Proterozoic rift where the rapakivi granites are found. In the area of Shachang complex, the crust is ~37 km thick (Yu et al., 1996a).

3.2. Petrography

The Shachang rapakivi granite pluton is located in the southern rift fault zone, 75 km northeast of Beijing City (Fig. 5). The complex is 13 km long and 0.5 to 2.5 km wide and cuts sharply the late Archean gneisses and migmatites. The pluton is composed of three granites. The earliest one is a rapakivi-textured

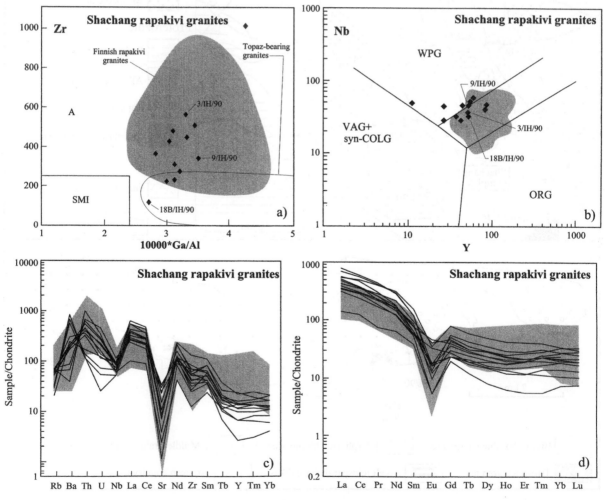

Fig. 6. Analyses of the Shachang rapakivi granite pluton (Rämö et al., 1995; Yu et al., 1996a,b) plotted in (a) Zr versus 10,000 Ga/Al (Whalen et al., 1987), (b) Nb versus Y (Pearce et al., 1984); (c) chondrite-normalized multielement, and (d) chondrite-normalized REE diagrams. The representative type samples (Table 1) are indicated in (a) and (b). The fields of the Finnish rapakivi granites (Fig. 3, shadowed) are indicated.

hornblende–biotite granite with rounded alkali feld-spar megacrysts (6–8 cm in diameter) consisting of several sectorially intergrown crystal units; about one third of the ovoids have plagioclase shells. The mafic minerals are enriched in iron: Fe/(Fe+Mg) is 0.80–0.82 in hornblende and 0.76–0.92 in biotite. The prevailing granite type is a porphyritic biotite granite, and the westernmost part of the complex consists of a medium-grained biotite granite. The youngest intrusive phase is a medium- to fine-grained two-mica granite that occasionally also contains topaz. Wolframite-bearing quartz veins are found along the southern margin of the complex, and have been mined in the past.

Associated with the Shachang granites are felsic porphyry and diabase dikes. Two diabase dikes that cut the porphyritic hornblende–biotite granite contain abundant corroded alkali feldspar xenocrysts. These must have been derived from partially crystallized granitic magma and thus demonstrate the bimodal character of the magmatism.

3.3. Geochemistry and isotopes

In terms of their whole-rock and mineral chemistry, the Shachang granites resemble the rapakivi granites of Finland. They are metaluminous to peraluminous and plot in the discrimination diagrams in the fields of A-type and within-plate granites, thus overlapping with the Finnish rapakivi granites (Fig. 6a,b). In multielement and REE diagrams, the Shachang granites are nearly identical to the Finnish rapakivis (Fig. 6c,d).

Nd isotope data on 10 granite and 3 country rock samples are presented, together with data of the Finnish rapakivi granites, in Fig. 7 (cf. Rämö et al., 1995). The Shachang granites show uniform ε_{Nd} (at 1685 Ma) values that average at -5.7 ± 0.3 (1σ), and depleted mantle model ages (T_{DM}) of 2220 to 2410 Ma. The country rocks show ε_{Nd} (at 2500 Ma) values of +0.6 to +1.5, and T_{DM} ages of 2560 to 2630 Ma (Rämö et al., 1995). The data suggest that the late Archean crust was the main source of the Shachang

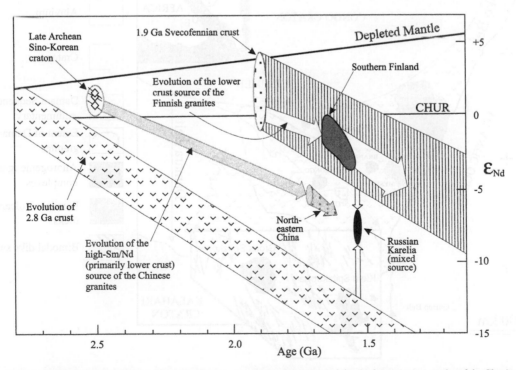

Fig. 7. ε_{Nd} versus age diagram showing the initial Nd isotope composition of granites and the Archean country rocks of the Shachang rapakivi granite pluton (Rämö et al., 1995) as well as the rapakivi granites (Rämö, 1991) and the earlier Svecofennian igneous rocks (Huhma, 1986; Patchett and Kouvo, 1986) of Finland. The diagram also shows the evolution of the rapakivi granites and their probable sources in the Shachang area and in Finland, evolution of the undifferentiated Earth (CHUR; DePaolo and Wasserburg, 1976) and depleted mantle (DM; DePaolo, 1981).

rapakivi granites, possibly with minor mantle magma input. The granites are thus at least 800 Ma younger than their principal source. Compared to the Finnish rapakivis, the striking similarities regarding the magmatic association and chemical composition further suggest similar genetic processes (magmatic underplate) and approximately similar source regions.

4. The Cretaceous Gross Spitzkoppe granite, northwestern Namibia

4.1. Geologic setting

The Gross Spitzkoppe granite stock is one of about 20 anorogenic Mesozoic Damaraland plutonic com-

plexes in the Pan-African (Neoproterozoic–Paleozoic) Damara orogenic belt of Namibia (Fig. 8). The Damara orogen can be divided into three branches: the northern (Kaoko belt) and southern (Gariep belt) branches that follow the Atlantic coast, and the NE-trending 400-km-wide inland branch (Damara belt). These mobile belts developed by amalgamation of three Precambrian cratons: the Congo craton in the north, the Kalahari craton in the south, and the Rio de la Plata craton in the west (Miller, 1983; Prave, 1996; Seth et al., 1998). The Damara orogenic belt was intensively metamorphosed and intruded by syn-orogenic to post-orogenic granitic-dioritic plutons at ~650–450 Ma (Miller, 1983).

The anorogenic Damaraland complexes include subvolcanic-epizonal complexes of three categories

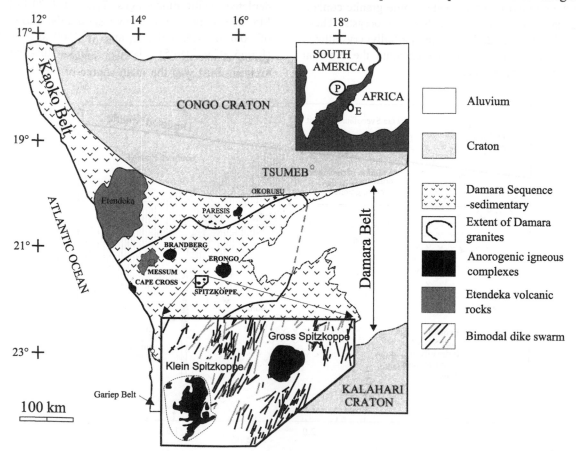

Fig. 8. Geological sketch map of northwestern Namibia showing the distribution of the Cretaceous 137–123 Ma Etendeka volcanic rocks and Damaraland volcanic-plutonic complexes. The lower inset shows the Gross Spitzkoppe and Klein Spitzkoppe granite stocks and bimodal dike swarms. In the upper inset map, P marks the bimodal Parana volcanic rocks in Brazil and E the corresponding Etendeka volcanic rocks in Namibia. Simplified from Miller (1983) and Milner et al. (1995).

(Martin et al., 1960; Milner et al., 1995): (1) granitic complexes (Brandberg, Erongo, Gross and Kleine Spitzkoppe); (2) differentiated basic complexes (e.g., Cape Cross, Messum, Okenjeje); and (3) peralkaline and carbonatitic complexes (e.g., Paresis, Kalkfeld, Okorusu). They extend in the Damara belt from the Cape Cross complex on the coast to the Okoruso complex about 350 km inland. Their isotopic Rb–Sr and Ar–Ar ages range from 124 to 137 Ma (Milner et al., 1995). Thus they are ~50–60 Ma younger than the Karoo basalts and more than 300 Ma younger than the post-orogenic (500–450 Ma) Damara granites. Associated with the Damaraland complexes are large amounts of diabase and quartz latite–rhyolite dikes and lavas (Fig. 8). The bimodal Etendeka lava field represents the eastern part of the wide 135–132 Ma Etendeka–Parana volcanic province.

The origin and emplacement of the Damaraland complexes and associated volcanic rocks and their counterparts along the Atlantic margin in South America has been related to the Tristan mantle plume, as well as to continental rifting that led to the opening of the southern Atlantic (O'Conner and le Roex, 1992; Milner and le Roex, 1996). The intracratonic rifting developed to seafloor spreading at 130 Ma, as can be judged from the oldest magnetic anomaly (M4) off the coast of Namibia

(Rabinowitz and LaBrecque, 1979). Although the Damaraland–Parana lavas and intrusive complexes are found over a wide area, they were formed within a relatively short time period of ~15 Ma. The emplacement of the Damaraland intrusive complexes was controlled by old zones of weakness within the Damara orogenic belt between the Precambrian cratons (e.g., Milner et al., 1995). As in the Fennoscandian shield, the metasedimentary-metavolcanic mobile belt with old zones of weakness was more prone to heat flow and mantle diapirism than the Archean cratons.

Deep seismic reflection soundings suggest that, at the southern margin of the Damara belt, the crust is layered and more than 50 km thick. It is much thinner than this (with strongly reduced lower crust) in the central parts of the belt where the Damara and Damaraland granitic plutons are found (Green, 1983). This may be related to voluminous partial melting of the lower crust and extraction of granitic melts during the Damara orogen and Damaraland anorogenic event (Green, 1983; Ewart et al., 1998b).

4.2. Petrography

The Gross Spitzkoppe stock is an inselberg of an epizonal, rounded 30 km^2 granite complex that

Fig. 9. An areal view from the northwest showing Gross Spitzkoppe (right) and Pondok (centre) mountains of the Gross Spitzkoppe stock. The outer limit of Gross Spitzkoppe stock in foreground is indicated by dashed line. Photo: Ilmari Haapala.

intrudes the Damara metasediments and granites (Figs. 9 and 10). The Damara crust is also cut by mainly NE- and NW-trending diabase and quartz latite dikes, which obviously represent the same magmatic event as the bimodal Etendeka volcanic rocks and are slightly older than the Gross Spitzkoppe and Klein Spitzkoppe granites. Bimodal character of the Spitzkoppe granite magmatism is indicated by synplutonic mafic dikes and magmatic mafic-intermediate inclusions, which contain quartz and alkali feldspar xenocrysts (Fig. 2e; Frindt, 2002; Frindt et al., 2004a). Locally, lamprophyric dikes cut the Spitzkoppe granites.

The Gross Spitzkoppe stock is composed of three zonally arranged granite types: medium-grained biotite granite at the margins, and a coarse-grained biotite granite and porphyritic biotite granite at the center. The outward dipping contact of the pluton as well as some inner contacts are marked by layered pegmatites and aplites, which show textures indicating crystallization from under-cooled magma (Frindt and Haapala, 2002, 2004). The granites are cut by subhorizontal aplite dikes up to 20 m in width and by a number of variously dipping thinner pegmatite and aplite dikes. Miarolitic cavities and pegmatite pockets with topaz and beryl are common, indicating local fluid saturation. The porphyritic granite is occasionally greisenized and mineralized with wolframite; otherwise the granites are well preserved (Frindt and Poutiainen, 2002). The three granites are all topaz-bearing monzogranites that contain a few percent of extremely iron-rich biotite [(Fe/(Fe+Mg) 0.90 to 1.00]. Common accessory minerals include fluorite, columbite, monazite, zircon, magnetite, and niobian rutile.

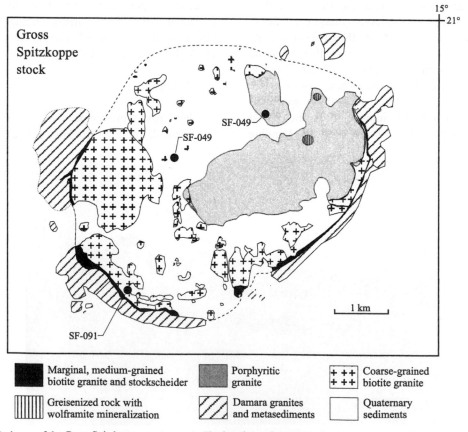

Fig. 10. Geological map of the Gross Spitzkoppe granite stock. The locations of representative type specimens (Table 1) are marked. Adopted from Frindt and Haapala (2002) and Frindt (2002).

4.3. Geochemistry, isotopes, and petrogenesis

The granites are marginally peraluminous high-silica monzogranites and, in all discrimination diagrams, plot into the fields of A-type and within plate granites, just like the rapakivi granites of Finland (Fig. 11). There are only minor differences in the average composition of the different granite types (Frindt et al., 2004a): SiO_2 75.7–76.6, Al_2O_3 12.22–12.48, Fe_2O_3 1.45–1.71, FeO 0.80–1.07, MgO 0–0.01, CaO 0.37–0.76, Na_2O 3.19–3.46, K_2O 5.04–5.53, and F 0.47–0.51 wt.%. Na_2O+K_2O is high, on average 8.45 wt.%. The granites belong to the "low-P" subtype ($P_2O_5<0.1$, $Al_2O_3<14.5$, $SiO_2>73$ wt.%) of F-rich granites by Taylor (1992). In the normative Ab–Or–Q diagram, they plot near the ternary minimum with 0.5 wt.% F at 1 kbar water pressure. High contents of incompatible trace elements (Rb ranges from 446 to 831, Nb from 60 to 176, Ta from 3.0 to 15, and Ga from 24 to 42 ppm) as well as low contents of Sr (5 to 48 ppm), Ba (0 to 164 ppm), and Eu/Eu* (0.23 to 0.00) suggest a high degree of fractionation (Frindt et al., 2004a). The granites show many geochemical similarities with the topaz-bearing rapakivi granites of Finland (Fig. 11a–d) but also some differences. The Sn contents are elevated (11 to 14 ppm), but not so high as in typical tin granites. The sum of alkalies is higher than in the Finnish rapakivi granites with corresponding silica contents. Accessory magnetite is ubiquitous in the Spitzkoppe granite but ilmenite is very rare. This indicates more oxidizing conditions in the Spitzkoppe granites than in the Finnish rapakivis.

Isotopic studies (Rb–Sr, Ar–Ar, Sm–Nd, Rb–Sr) have recently been published from the Etendeka and

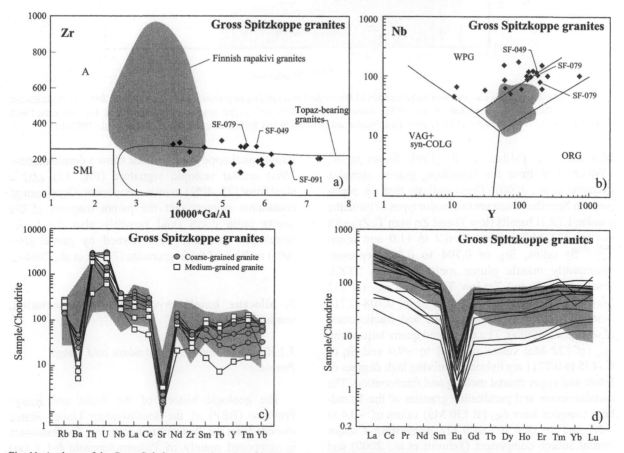

Fig. 11. Analyses of the Gross Spitzkoppe granites plotted in (a) Zr versus 10,000 Ga/Al (Whalen et al., 1987), (b) Nb versus Y (Pearce et al., 1984), (c) chondrite-normalized multielement, and (d) chondrite-normalized REE diagrams. The representative type samples (Table 1) are indicated in (a) and (b). The fields of the Finnish rapakivi granites (Fig. 3, shadowed) are shown.

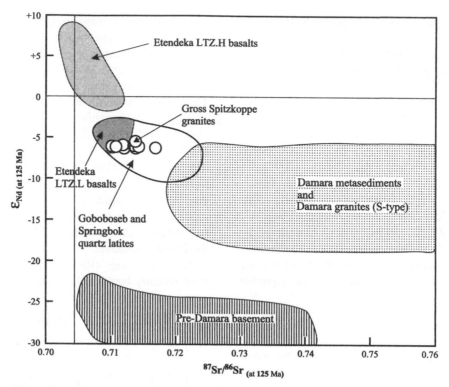

Fig. 12. Initial Nd–Sr isotope compositions (calculated at 125 Ma) of the Gross Spitzkoppe granites (open circles; Frindt, 2002), compared to the fields of pre-Damara basement (Seth et al., 1998), Damara metasediments and S-type granites (McDermott and Hawkesworth, 1990; McDermott et al., 1996), Etendeka LTZ.H and LTZ.L basalts and Goboboseb and Springbok quartz latites (Ewart et al., 1998a,b).

Messum lavas (Milner et al., 1995; Ewart et al., 1998a,b) and from the Brandberg granite complex (Schmitt et al., 2000). These indicate that, in north-western Namibia, there are two major types of tholeiitic basalts. LTZ.H basalts (low Ti and Zr; high Ti/Zr) with ε_{Nd} (at 132 Ma) values of +6.2 to +1.0 and initial $^{87}Sr/^{86}Sr$ ratios, Sr_i, of 0.704 to 0.705 represent dominantly mantle plume melts, whereas LTZ.L basalts (low Ti and Zr; low Ti/Zr) with ε_{Nd} (at 132 Ma) values of −8.0 to −9.5 and Sr_i of 0.708–0.717 represent plume–crust interactions and fractionation (Ewart et al., 1998a). The associated quartz latites with ε_{Nd} (at 132 Ma) values of −7.3 to −9.0 and Sr_i of 0.7175 to 0.7211 are hybrids involving high degrees of lower and upper crustal melting and fractionation. The metaluminous and peralkaline granites of the Brand-berg complex have ε_{Nd} (at 130 Ma) values of −0.4 to −5.1 and Sr_i of 0.707 to 0.727, suggesting a major crustal source component (Schmitt et al., 2000) and significant mantle input. Preliminary isotope studies by Trumbull et al. (2000) and Frindt et al. (2004b) on the

Gross Spitzkoppe granitic stock show a dominant (60–75%) crustal isotopic signature (Fig. 12), and a significant (25–40%) mantle component. Geochemical constraints suggest that the parent magmas of the quartz latite dikes (and possibly also the Gross Spitzkoppe granites) were formed by partial (10–20%) melting of felsic granulite (Frindt et al., 2004b).

5. Miocene basalt–rhyolite–granite association, southern Nevada

5.1. Geologic outline of the Basin and Range Province

The geologic history of the Basin and Range Province (BRP) of the southwestern United States extends from the Precambrian to today. The basement is composed mainly of Paleoproterozoic (~1.7 Ga) high- to medium-grade gneisses, which were intruded by mid-Proterozoic ~1.4 Ga granites. In the northern part of the BRP, the basement consists of the Archean rocks of the Wyoming Province. During late Paleozoic, the coast of southwest America became an active continental margin. Eastward subduction continued through the Mesozoic leading to a succession of orogenic events and igneous activity and accretionary growth of the continent. The Cretaceous magmatic arc is represented by the 135 to 80 Ma granitic batholiths of Baja California, Sierra Nevada, and Idaho. From the mid-Cretaceous to early Paleogene, the rocks of the continental margin were thrust eastward, giving rise to a foreland fold-and-thrust belt (Hamilton, 1987). The rapid eastward accretion and thrusting during the late Cretaceous–Paleogene Sevier-Laramide orogenies led to major crustal shortening and thickening of the orogenic crust.

The rapid subduction related to the Sevier-Laramide orogenies led to the arrival of the East Pacific spreading axis at the trench at about 30 Ma. Consumption of the Farallon plate was accompanied by steepening and cessation of the subduction, development of the San Andreas transform fault zone, and strong extension (BRP; Coney, 1987; Wernicke, 1985; Hodges and Walker, 1992). Active extension and rifting was accompanied by magmatism, which changed gradually and with some overlapping from intermediate-felsic calc-alkaline to bimodal basaltic-rhyolitic A-type magmatism. Although active debate continues about the details, the general interpretation is that this change in composition of the Tertiary magmatism reflects the change from plate subduction magmatism to continental rift magmatism.

The Tertiary extension was quite pronounced in the 50- to 100-km-wide NW-trending zone known as the Colorado River extensional corridor (CREC; Fig 13). This zone is characterized by extensive normal and listric faulting, large-scale subhorizontal detachment faults, and a central zone of tectonically uplifted domal metamorphic core complexes (Howard and John, 1987; McCarthy et al., 1991; Howard et al., 1994). Bimodal volcanism overlapped with extension, and several granitic plutons of this age are now exposed—one of them is the Spirit Mountain pluton (SMP). Strong extension led to thinning of the crust from about 60 to 26–30 km (McCarthy et al., 1991; Zandt et al., 1995). This thinning was especially strong in the lower crust. Although the extension and magmatism overlapped in time, there are indications

that in some cases the magmatism started first and extension followed thereafter (e.g., Gans et al., 1989; Scott et al., 1995). Debate continues on whether the extension (decompression) was the cause or effect of mantle magmatism.

The Miocene bimodal extension-related magmatism in BRP is mainly represented by basaltic and rhyolitic volcanites and dikes, but in places also high-level plutonic members, including rapakivi-textured granites, are known (Volborth, 1973; Calzia, 1994; Calzia and Rämö, 2000). Some of the rhyolites are topaz-bearing and may host Sn–Be–W mineralization (Christiansen et al., 1986; Congdon and Nash, 1991).

5.2. Petrography

The Newberry Mountains in the CREC host three Tertiary granite plutons surrounded by Proterozoic granites and gneisses (Fig. 13): the White Rock Wash pluton (WRWP), the Spirit Mountain pluton (SMP), and the Mirage pluton (MP) (Volborth, 1973; Howard et al., 1994; Haapala et al., 1995; Rämö et al., 1999).

WRWP consists of a medium- to coarse-grained two-mica granite, which shows deformation and recrystallization, and gives an ion probe U–Pb age of 68.5 Ma (Miller et al., 1997) and a whole-rock Rb–Sr age of about 65 Ma (Rämö et al., 1999). Petrographically, it is similar to the Cretaceous two-mica granites in the Eldorado Mountains (D'Andrea Kapp et al., 2002) and elsewhere in the Cordillerian interior (Miller and Barton, 1990). Plagioclase is more abundant than alkali feldspar. Magnetite, apatite, and zircon are common accessory minerals, and garnet is locally present. In places, the granite shows gradual migmatitic contacts with surrounding Proterozoic mica gneisses. Its origin and emplacement is related to thrusting and crustal thickening at the convergent plate boundaries during late stages of the Sevier orogeny.

SMP is mainly composed of a medium- to coarse-grained weakly porphyritic biotite±hornblende granite (Spirit Mountain granite) and gives a whole-rock Rb–Sr age of ~20 Ma (Rämö et al., 1999). Part of the alkali feldspar megacrysts are mantled by plagioclase. Magnetite and titanite are characteristic minor constituents and apatite, zircon, and allanite are present in small amounts. At the western margin of the pluton, the Spirit Mountain granite grades into a fine- to

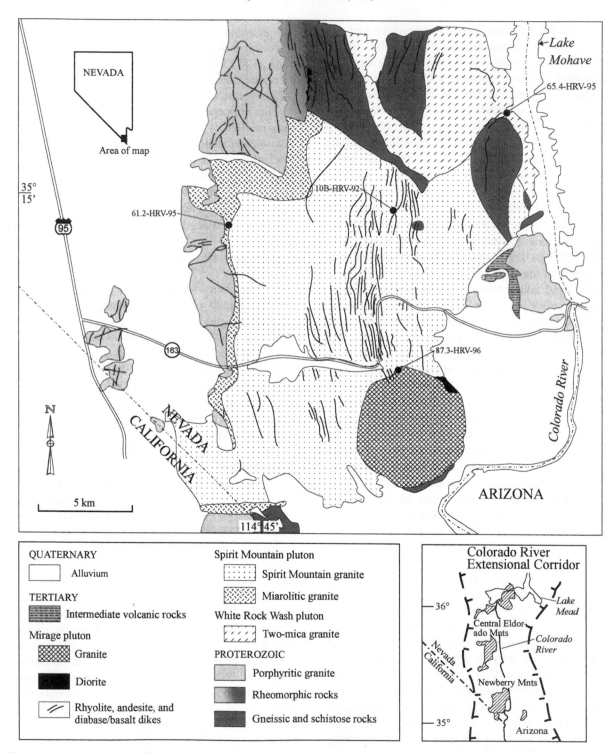

Fig. 13. Geological map of the Newberry Mountains area in southern Nevada, based on Volborth (1973) and 1992–1996 field studies by I. Haapala, O.T. Rämö, and A. Volborth. The locations of representative type specimens (Table 1) are marked. The inset shows location of the study area (Newberry Mountains) relative to the Colorado River extensional corridor.

medium-grained, granophyric and miarolitic alkali feldspar granite (Volborth, 1973). Perthitic alkali feldspar is the prevailing or sole feldspar and muscovite and biotite are locally present. Minor fluorite is also commonly present. The miarolitic granite probably represents the apical parts of the magma chamber and suggests westward tilting of the pluton during rifting.

MP consists of fine- to medium-grained biotite granite, which usually is porphyritic but may grade into an even-grained microgranite; dioritic rocks are found in the northeastern and southeastern margin of the pluton. The texture is hypidiomorphic with euhedral andesine and biotite crystals, biotite is often chloritized. Magnetite, apatite, and zircon are typical accessory minerals. According to Howard et al. (1994), all the Miocene dike swarms and plutons in the southern Newberry Mountains are probably 18–14 Ma in age.

The Tertiary granite plutons and their Precambrian country rocks are cut by numerous, generally NS-striking mafic and rhyolite dikes. Some dikes yield spectacular examples of mingling and hybridization of the two magmas (Volborth, 1973).

5.3. Geochemistry

The three granite plutons are distinct from each other in chemical composition (Table 1; Fig. 14a–f). The WRWP granites are peraluminous with molecular $Al_2O_3/(CaO+Na_2O+K_2O)$ (A/CNK) varying from 1.0 to 1.15, whereas the granites of SMP and MP straddle the metaluminous-peraluminous boundary (SMP 0.93–1.04, MP 0.98–1.08). In the Spirit Mountain granite, the one with mantled alkali feldspar megacrysts, SiO_2 varies from 66.6 to 76.4 wt.%, and Na_2O+K_2O from 8.2 to 9.6 wt.%. SMP has the highest K/Ca and Rb/Ba, MP the highest Ca, Mg and Fe, and WRWP the highest Na/K and lowest Rb/Ba. The miarolitic granite of SMP is geochemically the most evolved with lowest Ti, Mg, Ba, and Sr, and highest Rb and F. SMP has a clear negative Eu/Eu* anomaly, which is deepest in the miarolitic granite (Fig. 14d,f).

In the discrimination diagrams of Whalen et al. (1987), the MP granite plots into the field of I, S, M-type granites, whereas the SMP and WWP plot in both A- and I, S, M-fields (Fig. 14a). In the diagrams of Pearce et al. (1984), the SMP granites plot into the within plate granite fields, and those of WWP and MP into the fields of volcanic arc granites (Fig. 14b). The multielement spidergram and REE diagram of the SMP granites show similarities with the rapakivi granites of Finland (Fig. 14c,d), whereas the corresponding diagrams of WRWP and MP differ markedly from them (Fig. 14e,f).

The granite plutons of the Newberry Mountains provide an example of the flaws involved in the use of tectonomagmatic discrimination diagrams. Although both SMP and MP appear to be synextensional and related to continental rifting, only the SMP granites plot into the within-plate granite field. Similar observations have been made by several authors (Falkner et al., 1995, and references therein). Extensional processes were intimately associated with magmatic underplating, which hybridized the lithosphere shortly after cessation of convergence (see Falkner et al., 1995; Rämö et al., 1999; Miller and Miller, 2002). In such rapidly changing tectonic regime, sources generated at pre-, syn-, and post-hybridization stages can yield typologically different granites.

5.4. Isotope studies and petrogenesis

Preliminary Nd, Pb, and Rb–Sr isotopic studies on the plutons and diabase dikes of the Newberry Mountain area have been presented by Haapala et al. (1995, 1996) and Rämö et al. (1999). The Nd isotopic results are summarized in Table 2 and Fig. 15.

The WRWP has ε_{Nd} (at 65 Ma) values of -16.2 ± 0.3 Ma and Nd model ages averaging 1766 ± 100 Ma (Rämö et al., 1999). This indicates a major Proterozoic source. SMP has ε_{Nd} (at 20 Ma) values of -10.1 ± 0.8 Ma and Nd model ages averaging 1222 ± 74 Ma, which indicates clearly younger overall source. MP has isotopic composition between SMP and WRWP: ε_{Nd} (at 15 Ma) ranges from -11.4 to -14.2 and Nd model ages from 1338 to 1552 Ma. The corresponding Sr isotope initial values are for WRWP 0.7119 ± 0.006, for SMP 0.7098 ± 0.0002, and for MP 0.7109 to 0.7115 (Rämö et al., 1999). Three diabase dikes from SMP have ε_{Nd} (at 15 Ma) values of 0, -4.1, and -8.4, suggesting variable crustal contamination or a variably enriched mantle source.

Table 1

Representative chemical composition of the granites studied: wiborgite batholith and topaz granite stocks (Finland); Shachang (China); Gross Spitzkoppe (Namibia); Newberry Mountains (southern Nevada): Spirit Mountain (SM) pluton, Mirage (M) pluton, White Rock Wash (WRW) pluton

	Finnish rapakivi granites					Shachang rapakivi			Gross Spitzkoppe			Newberry Mountains			
	Wiborgite	Topaz granites				Wiborgite	Porphyritic granite	Fine-grained two-mica granite	Topaz granites			SM pluton		M pluton	WRW pluton
		Pyterlite	Eurajoki	Kymi	Suomenniemi				Medium-grained granite	Coarse-grained granite	Porphyritic granite	Main granite	Miarolitic granite	Main granite	Two-mica granite
Sample #	1A/IH/2001	2A/IH/2001	5/IH/2001	1/IH/89	SMN-A1097	3/IH/90	9/IH/90	18B/IH/90	SF-091	SF-049	SF-079	10B-HRV-92	61.2-HRV-95	87.3-HRV-96	65.4-HRV-95
SiO_2 (wt.%)	69.96[2]	76.70[2]	75.78[2]	73.5	74.1	68.00	72.90	74.80	76.9	74.90	76.20	73.20	76.50	70.00	73.30
TiO_2	0.42[2]	0.18[2]	0.02[2]	0.02	0.04	0.60	0.24	0.09	0.04	0.13	0.10	0.29	0.07	0.36	0.18
Al_2O_3	13.41[2]	11.70[2]	13.62[2]	15.1	14.4	14.20	12.90	13.00	12.6	12.40	11.90	14.00	12.20	14.80	15.20
FeO^w	3.35	1.53	0.65	0.6	0.1	2.20	0.80	0.10	0.7	1.20	1.00	0.7	0.10	1.1	0.60
Fe_2O_3	0.45[2]	0.31[2]	0.19[2]	0.2	0.81	1.72	1.44	0.96	1.24	1.81	1.70	1.06	0.63	1.35	0.69
MnO	0.06[2]	0.02[2]	0.05[2]	0.02	0.02	0.00	0.04	0.02	0.01	0.02	0.01	0.05	0.07	0.04	0.02
MgO	0.35[2]	0.09[2]	0.00[2]	0.1	0.01	0.49	0.15	0.08	0.00	0.00	0.00	0.51	0.01	0.74	0.31
CaO	2.16[2]	0.80[2]	0.64[2]	0.76	0.99	1.85	1.11	0.79	0.36	0.82	0.77	1.08	0.47	2.21	1.85
Na_2O	3.09[2]	2.49[2]	3.70[2]	4.05	4.32	3.59	3.43	3.80	3.71	3.20	3.18	3.98	4.06	3.67	4.40
K_2O	5.46[2]	5.59[2]	4.35[2]	4.55	4.14	5.23	5.29	5.37	4.93	5.48	5.09	4.87	4.54	4.35	3.10
P_2O_5	0.11[2]	0.02[2]	0.02[2]	0.01	0.02	0.14	0.04	0.0	0.02	0.02	0.00	0.06	0.01	0.13	0.06
H_2O^+ w	0.31	0.17	0.24	0.4	na	0.90	0.50	0.4	0.2	0.30	0.20	0.4	0.20	0.50	0.30
F^w	0.24	0.43	1.19	1.45	0.88	0.26	0.59	0.49	0.49	0.58	0.50	0.08	0.15	0.05	0.04
Total	99.37	100.03	100.43	100.76	99.83	99.18	99.43	99.91	101.20	100.86	100.65	100.28	99.01	99.30	100.05
$-O=F_2$	0.10	0.18	0.50	0.61	0.37	0.11	0.25	0.21	0.21	0.24	0.21	0.03	0.06	0.02	0.02
Total	99.27	99.85	99.93	100.15	99.46	99.07	99.18	99.70	100.99	100.62	100.44	100.25	98.95	99.28	100.03
Cl^w (ppm)	800	300	100	110	na	563[1]	317[1]	200[1]	131	218	218	na	na	na	na
Rb[1]	271[4]	349[4]	1050[4]	978	714	160	335	346	831	470	455	139	316	123	89
Sr[2]	155	69	8	22	16[1]	558[1]	171[1]	53[1]	15	23	19	159	12	338	527
Ba[1]	1144[2]	541[2]	28[2]	151	118	2890	864	203	30	112	90	398	68	1240	965
Ga[4]	27	24	60	61[2]	24	25[2]	24[2]	19[2]	44[1]	38[1]	35[1]	20	22	19	20
Be[2]	5	5	16	17	na	7	7	32	7	5	5	5	<1	2	<1
Zn[2]	125	86	197	67.2	28[1]	87.6	39.5	21	55.6	44.9	40.9	6.7	7.4	5.6	39
Sc[2]	8[3]	3[3]	12[3]	8	3.3[3]	6.5	2.4	0.7	na	na	na	23	12	4	2
Zr[1]	459[4]	304[4]	51[4]	12	39	560	340	116	179	271	280	184	119	135	119
Hf[3]	12.7	10	5.7	7.9	2.6	15	11	6.7	8.9	8.3	8.2	6.7	7.4	5.6	3.9
Sn[4]	8	9	110	15[1]	35[1]	<2[1]	27[1]	28[1]	18	14	12	23	12	2	15

W[3]	3	3	9	23	na	4	4	4	26	5	4	2	10	3	na	2
Nb[1]	26.4[4]	25.8[4]	69.8[4]	58	50.7	37	46	33	82	120	107	39	51	14	10
Ta[3]	2.3[4]	2.41[4]	22.6[4]	45	na	1	3	1.0	13	7.7	7.7	1.5	4	<0.5	1.1
Th[3]	24.7[4]	38.6[4]	27.6[4]	31	na	8	23	51.0	61	93	88	24	50	19	5.4
U[3]	8.08[4]	14.7[4]	6.34[4]	35	na	1	3	5.4	7	27	28	2.3	4.6	1.5	1.5
Pb[2]	38	55	88	130	na	21	41	38	26	26	23	17	26	17	15
Y[1]	63[2]	75[2]	55[2]	7[2]	59	51[2]	52[2]	52[2]	199	174	182	24	19	19	11
La[4]	95.9	99	47.4	65	28.1	155.0	176.0	52.7	39	131	125.0	51.6	27	44.5	26.3
Ce[4]	182	182	97.2	154	73.2	338.0	345.0	121.0	105	253	233.0	100	45	85.9	52.1
Pr[4]	21.5	21.2	11.1	15	8.6	39.4	35.8	10.2	13.9	26.7	25.5	9.9	3.4	9.1	5.7
Nd[4]	80.6	76.2	37.3	34.9	30.95	154.0	123.0	45.6	53.2	90.6	87.9	31.5	7.4	29.4	20.4
Sm[4]	14.6	13.7	9.26	8.1	8.55	22.2	18.2	8.7	18.6	20.8	21.9	5.8	1.4	5.3	3.7
Eu[4]	2.56	1.26	1.18	<0.05	0.1	3.6	1.4	0.5	0.1	0.4	0.3	0.77	0.1	1.2	1.18
Gd[4]	12.6	12.2	7.77	3.2	7.4	14.0	10.9	7.1	16.6	19	19.5	4.8	1.5	4.5	3
Tb[4]	2.14	2.2	1.84	1.4	1.6	1.8	1.6	1.1	3.5	3.7	3.8	0.7	0.2	0.6	0.3
Dy[4]	12.7	13.8	12.7	9.7	13.6	10.7	8.8	7.2	24.9	25.8	25.6	4.1	1.9	3.3	2.1
Ho[4]	2.67	2.98	2.71	1.98	3.19	1.9	1.8	1.5	5.7	5.9	5.9	0.84	0.5	0.6	0.38
Er[4]	7.26	8.51	8.6	8.6	10.7	5.3	5.4	4.9	19.5	19.7	19.0	2.8	1.9	1.7	1
Tm[4]	1.07	1.3	1.77	2.5	1.8	0.8	0.8	0.9	3.7	3.2	3.1	0.5	0.4	0.3	0.1
Yb[4]	7.31	8.56	15	27.9	14.8	4.8	5.8	5.8	4.6	20.8	21.4	3.2	3.1	1.6	0.8
Lu[4]	1.05	1.23	2.3	4.43	2.19	0.7	0.9	1.1	4.4	3	3.1	0.5	0.5	0.2	0.13

Samples 1A/IH/2001, 2A/IH/2001, and 5/IH/2001 analyzed by Activation Laboratories (Canada), all other by X-ray Assay Laboratories (Canada). Analytical methods denoted by superscript codes as follows: [1]XRF, [2]ICP, [3]NA, [4]ICP-MS, [w]wet chemical; na—not analyzed. SMN–A1097 from Rämö (1991), analyses of Shachang granites from Yu et al. (1996a), analyses of Gross Spitzkoppe granites from Frindt et al. (2004a). Abbreviations for the plutons of the Newberry Mountains: SM–Spirit Mountain; M–Mirage; WRW–White Rock Wash.

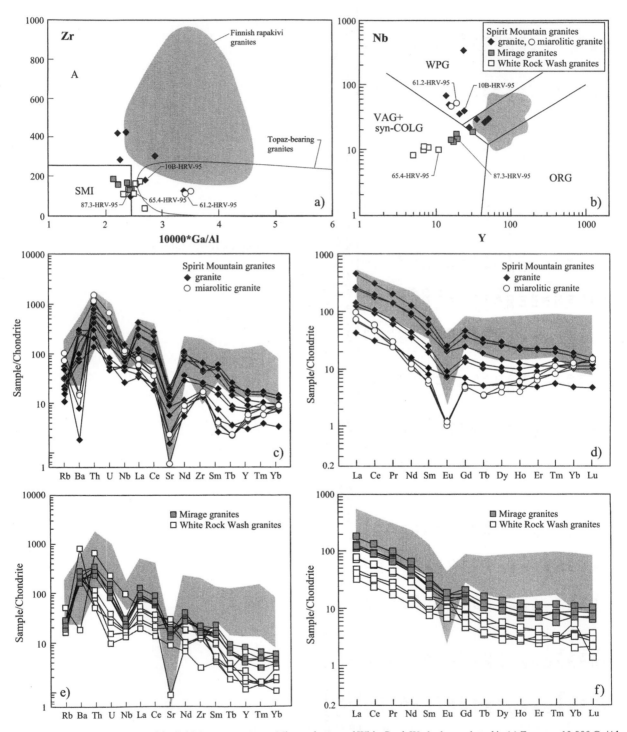

Fig. 14. Analyses of the granites of the Spirit Mountain pluton, Mirage pluton, and White Rock Wash pluton plotted in (a) Zr versus 10,000 Ga/Al (Whalen et al., 1987) and (b) Nb versus Y diagrams (Pearce et al., 1984); representative samples (Table 1) are indicated. Granites of the Spirit Mountain pluton (c and d) and granites of the Mirage and White Rock Wash plutons (e and f) are plotted separately on chondrite-normalized multielement spidergrams (c and e) and REE diagrams (d and f), respectively. The fields of the Finnish rapakivi granites (Fig. 3, shadowed) are indicated.

Table 2
Summary of Nd isotope data on the Spirit Mountain pluton, Newberry Mountains, southern Nevada (unpublished data by the authors)

Sample	Rock type	ε_{Nd} (at 15 Ma)	T_{DM}[a]
Spirit mountain pluton			
3-HRV-92	Granite	−9.6	1138
10B-HRV-92	Granite	−10.1	1228
14-HRV-93	Leucogranite	−9.6	1178
15A-HRV-93	Granite	−10.2	1231
15D-HRV-93	Rhyolite	−9.8	1034
Mirage pluton			
4.2-HRV-95	Granite	−14.2	1491
14.3-IIRV-95	Granite	−14.2	1512
87.2-HRV-96	Granite	−11.4	1338
28.1-HRV-95	Diorite	−8.1	1317
White Rock Wash pluton			
65.4-HRV-95	Two-mica granite	−16.1	1698
65.5-HRV-95	Two-mica granite	−16.5	1884
73.1-HRV-95	Two-mica granite	−16.3	1812
73.4-HRV-95	Two-mica granite	−15.7	1670
Mafic dikes			
10-HRV-93	Diabase	−4.1	946
11D-HRV-93	Diabase	−8.4	1118
86.1-HRV-96	Diabase	0.0	992
Country rocks			
1-HRV-92	Precambrian granite	−19.3 (−3.1)[b]	2028
11-HRV-92	Precambrian gneiss	−22.4 (−2.9)[b]	1924

[a] Calculated according to the model of DePaolo (1981).
[b] Values in parenthesis calculated at 1450 Ma.

The isotopic and other geochemical data indicate different sources for the two Miocene extension-related granite plutons. SMP may have had both Mesozoic and Proterozoic felsic-intermediate crustal source components with some juvenile input, whereas MP has older (mainly Proterozoic) and probably more calcic average source. Falkner et al. (1995) reported for the granite of the Miocene Aztec Wash pluton from the adjacent Eldorado Mountains isotopic values, which are nearly identical with the SMP granites: ε_{Nd} (at 15.7 Ma) of −10.1, Sr initial ratios of 0.7101 and 0.7102. The two granite plutons (SMP and Aztec Wash) obviously had remarkably similar sources and petrogenetic histories. Magmatic underplating or intraplating and partial melting of lower or middle crust is the preferred genetic model. The isotopic composition of the SMP granites suggests a

mixed source—either the melted crust was previously hybridized or the mantle-derived magma and crustal magma interacted to form the hybrid magma. For the granites of the Eldorado Mountains, a similar conclusion was presented by Miller and Miller (2002).

6. Concluding remarks

The four described granitic associations from Finland and adjacent Fennoscandia, China, Namibia, and Nevada represent intracontinental magmatism that ranges in age from 1700 to 15 Ma. They are spatially and temporally far from each other but show many striking similarities, allowing some common petrogenetic interpretations.

1. In all four cases, the tectonic setting during granite emplacement was extensional, but the rocks represent different geotectonic environments. In the case of the rapakivi granites of Finland and adjacent Fennoscandia (age range 1670–1470 Ma, four area-constrained age groups), the magmatism was probably associated with long-lasting, large-scale mantle upwelling and incipient, aborted rifting. In China, the rapakivi magmatism had a short age span (1700–1680 Ma) and was associated with aborted rifting in the Archean North China craton. In Namibia, the anorogenic magmatism (137–123 Ma) was related to a mantle plume and "early" stages of continental rifting, which has continued as oceanic rifting. In the Basin and Range Province of Nevada and vicinity, 20–15 Ma rifting and related magmatism followed Mesozoic–early Tertiary subduction-related orogenic processes.

2. The magmatic association is in all cases bimodal. The silicic members are represented by granites and rhyolitic-quartz latitic dikes, the mafic members by gabbroic rocks and diabase dikes, and locally (Namibia) alkaline rocks. Intermediate rocks are relatively rare. Local magma mingling and hybridization are present in all case areas.

3. The granites show the chemical and mineralogical characteristics of within-plate granites and A-type granites. In the Newberry Mountains of southern

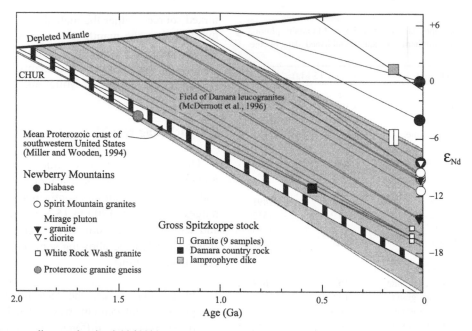

Fig. 15. ε_{Nd} versus age diagram showing initial Nd isotope composition of the granites and associated rocks of the 125 Ma Gross Spitzkoppe stock in Namibia (Frindt, 2002) and the Newberry Mountains in southern Nevada (Haapala et al., 1996; Rämö et al., 1999). The ε_{Nd} values were calculated for the Gross Spitzkoppe granites and a lamprophyre dike at 125 Ma, for the country rock (Damara granite) at 450 Ma, and for the rocks of the Newberry Mountains as follows: Spirit Mountain granites at 20 Ma, Mirage and diorite at 15 Ma, White Rock Wash granites at 65 Ma, diabase dikes at 15 Ma, Proterozoic granite gneiss at 1400 Ma. The shaded band shows the evolution of the Damara leucogranites in Namibia according to McDermott et al. (1996). The striped path shows the evolution of mean Proterozoic crust of the southwestern United States according to Miller and Wooden (1994). The thin lines show the evolution trends of individual samples. Evolution line of undifferentiated Earth (CHUR) is from DePaolo and Wasserburg (1976) and that of depleted mantle from DePaolo (1981).

Nevada, only the Spirit Mountain pluton shows clear within-plate character, however. In all four regions, the most evolved silicic members are topaz-bearing granites and rhyolites, and Sn–Be–W mineralization is in many cases associated with them.

4. Isotopic and geochemical studies suggest a prevailingly intermediate-felsic crustal source for the granites. Associated mafic rocks of all regions frequently have Nd isotope signatures largely overlapping with those of the associated granites. In some cases, however, some of the diabase dikes or basalts show chondritic or depleted mantle compositions [in Namibia ε_{Nd} (at 125 Ma) up to +6], suggesting an asthenospheric mantle source. The composition of the mafic magmas was changed by assimilation of crustal rocks.

5. Magmatic underplating with voluminous dehydration partial melting of the deep crust is the most plausible mechanism for the bimodal magmatic association in all four cases. It explains the observed structural features of the lithosphere (thinning of the lower crust above mantle diapirs) as well as the geochemical and isotopic characteristics of the bimodal suites. The observed mantle component in the granites is probably related to mixing of the mantle-derived mafic magmas and crust-derived felsic magmas, or to anatectic melting of hybridized lower crust.

6. The reason for the mantle melting is not known in detail, and may also vary from one area to another. There is good evidence (e.g., isotopic and geochemical studies indicating depleted or near-chondritic mantle source) that in Namibia the Cretaceous anorogenic bimodal magmatism was caused by a deep mantle plume; it is not impossible that mantle plumes played a role in other areas as well. In the case of the Fennoscandian rapakivi granites, partial melting of the

mantle material was possibly caused by thick crust that thermally insulated the underlying subcontinental mantle. This caused heating and large-scale upwelling of the solid mantle, eventually leading to magmatic underplating and partial melting. Mantle diapirism and associated extensive partial melting and reorganization of the crust stabilized the lithosphere in one area, and subsequent mantle diapirism appeared in neighboring area, being controlled by the heat flow in the mantle and the strength of the crust. In the case of the Colorado River extensional corridor (Basin and Range Province), the bimodal magmatism was largely synextensional, but opinions differ whether the strong extension (decompression) was the cause or the effect of the mantle magmatism.

Acknowledgments

This report is based on long-term petrologic, geochemical, and isotopic research made by the rapakivi granite group of the University of Helsinki since the mid-1980s on anorogenic granitic magmatism in Finland, China, Namibia, and Nevada. Throughout the research, the authors have received support from several quarters, which is gratefully acknowledged. The results from the different areas have already been largely published, and this article has the character of a synthesizing review. During the fieldwork in China (1990, 1992), I.H. had fruitful cooperation with Professor Yu, Jianhua, Ms. Fu, Huiqin, Mr. Zhang, Fenglan, and Mr. Wan, Fangxiao, all from the Geological Institute of Beijing. In 1994–2002, S.F. and I.H. received important logistic and personal support from the Geological Survey of Namibia and the Geological Survey of Finland that had a joint mapping project in Damaraland in the1990s. The Master's theses of S.F. (Gross Spitzkoppe) and John Kandara (Klein Spitzkoppe) were done in the framework of this project. In southern Nevada, Professor Alex Volborth who had mapped the Newberry Mountains area in the 1960s, guided I.H. and O.T.R. in the field in southern Nevada in 1992. His later visits to the field, as well as his and Nadja von Volborth's hospitality, made our field research periods as pleasant as possible, the latest unforgettable one was in 2001. Isotope geological research has been an essential part of our studies. O.T.R. has had the privilege to use the isotope laboratory facilities of the Geological Survey of Finland since 1987—support and help from Drs. Hannu Huhma and Matti Vaasjoki in particular are warmly acknowledged. The manuscript was reviewed by Professors Eric H. Christiansen and Roberto Dall'Agnol, their suggestions and corrections improved the text markedly. Our studies have acquired financial support from Academy of Finland (Projects 36002 and 4067), the University of Helsinki, and the Centre of International Mobility (CIMO).

References

Åhäll, K.I., Connelly, J.N., Brewer, T.S., 2000. Episodic rapakivi magmatism due to distal orogenesis? Correlation of 1.69–1.50 Ga orogenic and inboard, "anorogenic" events in the Baltic Shield. Geology 28, 823–826.

Ahl, M., Andersson, U.B., Lundqvist, T., Sundblad, K. (Eds.), 1997. Rapakivi Granites and Related Rocks in Central Sweden. Ser. Ca, vol. 87. pp. 1–99.

Alapieti, T., 1982. The Koillismaa layered igneous complex, Finland—its structure, mineralogy and geochemistry, with emphasis on the distribution of chromium. Bulletin-Geological Survey of Finland 319. 116 pp.

Alviola, R., Johanson, B.S., Rämö, O.T., Vaasjoki, M., 1999. The Proterozoic Ahvenisto rapakivi granite-massif-type anorthosite complex, southeastern Finland; Petrography and U–Pb chronology. In: Haapala, I., Rämö, O.T. (Eds.), Rapakivi Granites and Related Rocks. Precambrian Research, vol. 95. pp. 89–107.

Amelin, Yu., Larin, A.M., Tucker, R.D., 1997. Chronology of multiphase emplacement of the Salmi rapakivi granite–anorthosite complex, Baltic Shield: implications for magmatic evolution. Contributions to Mineralogy and Petrology 127, 348–368.

Anderson, J.L., Bender, E.E., 1989. Nature and origin of Proterozoic A-type granitic magmatism in the southwestern United States of America. Lithos 23, 19–52.

Andersson, U.B., 1997. Petrogenesis of some Proterozoic granitoid suites and associated basic rocks in Sweden geochemistry and isotope geology. Rapporter och Meddelanden-Sveriges Geologiska Undersökning 9. 213 pp.

Andersson, U.B., Neymark, L.A., Billström, K., 2002. Petrogenesis of the Mesoproterozoic (Subjotnian) rapakivi complexes of central Sweden: implications from U–Pb zircon ages, Nd, Sr and Pb isotopes. Transactions of the Royal Society of Edinburgh. Earth Sciences 92, 201–228.

Bridgewater, D., Sutton, J., Watterson, J., 1974. Crustal downfolding associated with igneous activity. Tectonophysics 21, 57–77.

Calzia, J.P., 1994. Petrogenesis of the middle Miocene rapakivi granite, Death Valley, California, USA. International Mineralogical Association 16th General Meeting, Pisa. p. 62. Abstract volume.

Calzia, J.P., Rämö, O.T., 2000. Late Cenozoic crustal extension and magmatism, southern Death Valley region, California. In: Lageson, D.R., Peters, S.G., Lahren, M.M. (Eds.), Great Basin and Sierra Nevada. Field Guide-Geological Society of America, vol. 2. pp. 135–164.

Christiansen, E.H., Sheridan, M.F., Burt, D.M., 1986. The geology and geochemistry of the Cenozoic topaz rhyolites from the western United States. Special Paper-Geological Society of America 205. 82 pp.

Claesson, S., Kresten, P., 1997. The anorogenic Noran intrusion—a Mesoproterozoic rapakivi massif in south-central Sweden. Geologiska Foreningens i Stockholm Forhandlingar 119, 115–122.

Coney, P.J., 1987. The regional tectonic setting and possible causes of Cenozoic extension in North America Cordillera. In: Coward, M.P., Dewy, J.F., Hancock, P.L. (Eds.), Continental Extensional Tectonics. Special Publication-Geological Society of London, vol. 28. pp. 177–186.

Congdon, R.D., Nash, W.P., 1991. Eruptive pegmatite magma: rhyolite of the Honeycomb Hills, Utah. American Mineralogist 76, 1261–1278.

D'Andrea Kapp, J., Miller, C.F., Miller, J.S., 2002. Ireteba Pluton, Eldorado Mountains, Nevada: late, deep-source, peraluminous magmatism in the Cordilleran Interior. Journal of Geology 2002, 649–669.

DePaolo, D.J., 1981. Neodymium isotopes in the Colorado Front Range and crust–mantle evolution in the Proterozoic. Nature 291, 193–196.

DePaolo, D.J., Wasserburg, G.J., 1976. Nd isotopic variations and petrogenetic models. Geophysical Research Letters 3, 249–252.

Downes, H., Peltonen, P., Mänttäri, I., Sharkov, E.V., 2002. Proterozoic zircon ages from lower crustal granulite xenoliths, Kola Peninsula, Russia: evidence for crustal growth and reworking. Journal of the Geological Society (London) 159, 485–488.

Edén, P., 1991. A specialized topaz-bearing rapakivi granite and associated mineralized greisen in the Ahvenisto complex, SE Finland. Bulletin of the Geological Society of Finland 63, 25–40.

Eklund, O., Fröjdö, S., Lindberg, B., 1994. Magma mixing, the petrogenetic link between anorthositic suites and rapakivi granites, Åland, SW Finland. Mineralogy and Petrology 50, 3–19.

Eklund, O., Konopelko, D., Rutanen, H., Fröjdö, S., Shebanov, A., 1998. 1.8 Ga Svecofennian post-collisional shoshonitic magmatism in the Fennoscandian shield. Lithos 45, 87–108.

Elliott, B.A., 2003. Petrogenesis of the postkinematic magmatism of the Central Finland Granitoid Complex II; sources and magmatic evolution. Journal of Petrology 44, 1681–1701.

Elo, S., Korja, A., 1993. Geophysical interpretation of the crustal and upper mantle structure in the Wiborg rapakivi granite area, southeastern Finland. In: Gorbatschev, R. (Ed.), Baltic Shield, Special Volume. Precambrian Research, vol. 64. pp. 273–288.

Emslie, R.F., 1978. Anorthosite massifs, rapakivi granites, and late Proterozoic rifting of North America. Precambrian Research 7, 61–98.

Emslie, R.F., Hamilton, M.A., Theriault, R.J., 1994. Petrogenesis of a Mid-Proterozoic anorthosite–mangerite–charnockite–granite (AMCG) complex: isotopic and chemical evidence from the Nain Plutonic Suite. Journal of Geology 102, 539–558.

Ewart, A., Milner, S.C., Armstrong, R.A., Duncan, A.R., 1998a. Etendeka volcanism of the Goboboseb Mountains and Messum Igneous Complex, Namibia: Part I. Geochemical evidence of early Cretaceous Tristan Plume melts and the role of crustal contamination in the Paraná-Etendeka CFB. Journal of Petrology 39, 191–225.

Ewart, A., Milner, S.C., Armstrong, R.A., Duncan, A.R., 1998b. Etendeka volcanism of the Goboboseb Mountains and Messum Igneous Complex, Namibia: Part II. Voluminous quartz latite volcanism of the Awahab magma system. Journal of Petrology 39, 227–253.

Falkner, C.M., Miller, C.F., Wooden, J.L., Heizler, M.T., 1995. Petrogenesis and tectonic significance of the calc-alkaline, bimodal Aztec Wash pluton, Eldorado Mountains, Colorado River extensional corridor. Journal of Geophysical Research 100, 10453–10476.

Frindt, S., 2002. Petrology of the Cretaceous anorogenic Gross Spitzkoppe granite stock, Namibia. PhD thesis, Department of Geology, University of Helsinki, Finland.

Frindt, S., Haapala, I., 2002. The Cretaceous Gross Spizkoppe granite stock in Namibia, a highly evolved A-type granite with structures and textures demonstrating magma flow, undercooling and vapor saturation during crystallization. 11th Quadrennial IAGOD Symposium and Geocongress 2002. Windhoek Namibia, 22–26 July. Extended Abstract volume [CD-ROM].

Frindt, S., Haapala, I., 2004. Anorogenic Gross Spitzkoppe granite stock in central western Namibia: Part II. Structures and textures indicating crystallization from undercooled melt. American Mineralogist 89, 857–866.

Frindt, S., Poutiainen, M., 2002. P–T path fluid evolution in the Gross Spitzkoppe granite stock, Namibia. Bulletin of the Geological Society of Finland 74, 103–114.

Frindt, S., Haapala, I., Pakkanen, L., 2004a. Anorogenic Gross Spitzkoppe granite stock in central western Namibia: Part I. Petrology and geochemistry. American Mineralogist 89, 841–856.

Frindt, S., Trumbull, R.B., Romer, R.L., 2004. Petrogenesis of the Gross Spitzkoppe topaz granite, central western Namibia: a geochemical and Nd–Sr–Pb isotope study. Chemical Geology 206, 43–71.

Frost, C.D., Frost, B.R., 1997. Reduced rapakivi-type granites: the tholeiite connection. Geology 25, 647–650.

Frost, C.D., Frost, B.R., Chamberlain, K.R., Edwards, B.R., 1999. Petrogenesis of the 1.43 Ga Sherman batholith, SE Wyoming, USA: a reduced, rapakivi-type anorogenic granite. Journal of Petrology 40, 1771–1802.

Frost, C.D., Frost, B.R., Bell, J.M., Chamberlain, K.R., 2002. The relationship between A-type granites and residual magmas from anortosites: evidence from the northern Sherman batholith, Laramie Mountains, Wyoming, USA. Precambrian Research 119, 45–71.

Gans, P.B., Mahood, G.A., Schermer, E., 1989. Synextensional magmatism in the Basin and Range Province; a case study from the eastern Great Basin. Special Paper-Geological Society of America 233. 53 pp.

Green, R.W.E., 1983. Seismic refraction observations in the Damara orogen and flanking craton and their bearing on deep seated processes in the orogen. Special Publication-Geological Society of South Africa 11, 267–355.

Haapala, I., 1977a. The controls of tin and related mineralizations in the rapakivi granite areas of southeastern Fennoscandia. Geologiska Foreningens i Stockholm Forhandlingar 99, 130–142.

Haapala, I., 1977b. Petrography and geochemistry of the Eurajoki stock, a rapakivi–granite complex with greisen-type mineralization in southwestern Finland. Bulletin-Geological Survey of Finland 286. 128 pp.

Haapala, I., 1985. Metallogeny of the Proterozoic rapakivi granites of Finland. In: Taylor, R.P., Strong, D.F. (Eds.), Granite-Related Mineral Deposits: Geology, Petrogenesis, and Tectonic Setting. Extended abstracts of Papers Presented at the Canadian Institute of Mining and Metallurgy Conference on Granite-Related Mineral Deposits. September 15–17, 1985, Halifax, Canada. pp. 123–131.

Haapala, I., 1989. Suomen rapakivigraniitcista. Rapakivi granites of Finland. Academica Scientiarum Fennica. Year Book 1988–1989, 135–140 (in Finnish with an English summary).

Haapala, I., 1995. Metallogeny of the rapakivi granites. Mineralogy and Petrology 54, 149–160.

Haapala, I., 1997. Magmatic and postmagmatic processes in tin-mineralized granites: topaz-bearing leucogranites in the Eurajoki rapakivi granite stock, Finland. Journal of Petrology 38, 1645–1659.

Haapala, I., Rämö, O.T., 1990. Petrogenesis of the rapakivi granites of Finland. In: Stein, H.J., Hannah, J.L. (Eds.), Ore-bearing Granite Systems: Petrogenesis and Mineralizing Processes. Special Paper-Geological Society of America, vol. 246. pp. 275–286.

Haapala, I., Rämö, O.T., 1992. Tectonic setting and origin of the Proterozoic rapakivi granites of southeastern Fennoscandia. Transactions of the Royal Society of Edinburgh. Earth Sciences 83, 165–171.

Haapala, I., Rämö, O.T., 1999. Rapakivi granites and related rocks: an introduction. Precambrian Research 95, 1–7.

Haapala, I., Rämö, O.T., Volborth, A., 1995. Miocene rapakivi-like granites of southern Newberry Mountains, Nevada, U.S.A: Comparison to the Proterozoic rapakivi granites of Finland. In: Brown, M., Niccoli, P.M. (Eds.), The origin of granites and related rocks; Third Hutton Symposium, Abstracts. U.S. Geological Survey Circular, vol. 1129. pp. 61–62.

Haapala, I., Rämö, O.T., Volborth, A., 1996. Petrogenesis of the Miocene granites of Newberry Mountains, Colorado River extensional corridor, southern Nevada. Abstracts with Programs-Geological Society of America 28, 458.

Hamilton, W., 1987. Crustal extension in the Basin and Range province, southwestern United States. In: Coward, M.P., Dewey, J.F., Hancock, P.L. (Eds.), Continental Extensional Tectonics. Special Publication-Geological Society of London, vol. 28. pp. 155–176.

Hildreth, W., Halliday, A.N., Christiansen, R.L., 1991. Isotopic and chemical evidence concerning the genesis and contamination of

basaltic and rhyolitic magma beneath the Yellowstone Plateau volcanic field. Journal of Geology 32, 63–138.

Hodges, K.V., Walker, J.D., 1992. Extension in the Cretaceous Sevier orogen, North American Cordillera. Geological Society of America Bulletin 104, 560–569.

Hoffman, P.J., 1989. Speculations on Laurentia's first giga-year (2.0–1.0 Ga). Geology 17, 135–138.

Hölttä, P., Huhma, H., Mänttäri, I., Peltonen, P., Juhanoja, J., 2000. Petrology and geochemistry of mafic granulite xenoliths from the Lahtojoki kimberlite pipe, eastern Finland. Lithos 51, 109–133.

Howard, K.A., John, B.E., 1987. Crustal extension along rooted systems of imbricate low-angle faults: Colorado River extensional corridor, California and Arizona. In: Coward, M.P., Dewey, J.F., Hancock, P.L. (Eds.), Continental Tectonics. Special Publication-Geological Society of London, vol. 28. pp. 299–311.

Howard, K.A., John, B.E., Davis, G.A., Anderson, J.L., Gans, P.B., 1994. A guide to Miocene extension in the lower Colorado River region, Nevada, Arizona, and California. Open-File Report (United States Geological Survey), 94–246.

Huhma, H., 1986. Sm–Nd, U–Pb and Pb–Pb isotopic evidence for the origin of Early Proterozoic Svecokarelian crust in Finland. Bulletin-Geological Survey of Finland 337. 48 pp.

Huppert, H.E., Sparks, S.J., 1988. The generation of granitic magmas by intrusion of basalt into continental crust. Journal of Petrology 29, 599–624.

Ishihara, S., 1977. The magnetite-series and ilmenite-series granite rocks. Mining Geology 27, 293–305.

Karhu, J.A., Holland, H.D., 1996. Carbon isotopes and the rise of atmospheric oxygen. Geology 24, 867–870.

Kempton, P.D., Downes, H., Neymark, L.A., Wartho, J.A., Zartman, R.E., Sharkov, E.V., 2001. Garnet granulite xenoliths from the northern Baltic shield—0the underplated lower crust of a Palaeoproterozoic large igneous province? Journal of Petrology 42, 731–763.

King, P.L., White, A.J.R., Chappell, B.W., Allen, C.M., 1997. Characterization and origin of aluminous A-type granites from the Lachlan Fold Belt, southeastern Australia. Journal of Petrology 38, 371–391.

Korja, A., Heikkinen, P., 1995. Proterozoic extensional tectonics of the central Fennoscandian Shield: results of the Baltic and Bothnian Echoes from the Lithosphere experiment. Tectonics 14, 504–517.

Korja, K., Korja, T., Luosto, U., Heikkinen, P., 1993. Seismic and geoelectric evidence for collisional and extensional events in the Fennoscandian Shield—implications for Precambrian crustal evolution. Tectonophysics 219, 129–152.

Korja, A., Heikkinen, P., Aaro, S., 2001. Crustal structure of the northern Baltic Sea paleorift. Tectonophysics 331, 341–358.

Korsman, K., Korja, T., Pajunen, M., Virransalo, P., 1999. The GGT/SVEKA Transect: Structure and evolution of continental crust in the Paleoproterozoic Svecofennian Orogen in Finland. International Geology Review 41, 287–333.

Kosunen, P., 1999. The rapakivi granite plutons of Bodom and Obbnäs, southern Finland. Bulletin of the Geological Society of Finland 71, 275–304.

Kosunen, P.J., Rämö, O.T., Vaasjoki, M., 2004. Petrogenesis of the Bodom and Obbnäs rapakivi plutons, southern Finland:

reduced, aluminous A-type granites from variable sources. Journal of Petrology (in review).

Lauri, L., Mänttäri, I., 2002. The Kynsijärvi quartz alkali feldspar syenite, Koillismaa, eastern Finland—silicic magmatism associated with 2.44 Ga continental rifting. Precambrian Research 119, 121–140.

Lindberg, B., Bergman, L., 1993. Pre-Quaternary rocks of the Vehmaa map-sheet area. Geological map of Finland 1:100,000. Explanation to the maps of pre-Quaternary rocks, sheet 1042. Geological Survey of Finland, Espoo, Finland.

Lindh, A., 2001. Microgranular enclaves in the Nordingrå rapakivi granite, east central Sweden—the early rapakivi development. Neues Jahrbuch für Mineralogie Abhandlungen 176, 299–322.

Lindh, A., Johansson, I., 1996. Rapakivi granites of the Baltic Shield: the Nordingrå granite, its chemical variation and Sm–Nd isotope composition. Neues Jahrbuch für Mineralogie. Abhandlungen 170, 291–312.

Liu, D.Y., Shen, Q.H., Zhang, Z.Q., Jahn, B.M., Auvray, B., 1990. Archean crustal evolution in China: U–Pb geochronology of the Qianxi Complex. Precambrian Research 14, 185–202.

Luosto, U., 1997. Structure of the Earth's crust in Fennoscandia as revealed from refractive and wide-angle reflection studies. Geophysica 33, 3–16.

Luosto, U., Tiira, T., Korhonen, H., Azel, I., Burmin, V., Buyanov, A., Kosminskaya, I., Ionkis, V., Sharov, N., 1990. Crust and upper mantle structures along the DSS Baltic profile in SE Finland. Geophysical Journal International 10, 89–110.

Martin, H., Mathias, M., Simpson, E.S.W., 1960. The Damaraland sub-volcanic ring complexes in South West Africa. Report of the International Geological Congress XXI Session, vol. 13, pp. 156–174.

McCarthy, I., Larkin, S.P., Fuis, G.S., Simpson, R.W., Howard, K.A., 1991. Anatomy of a metamorphic core complex: seismic refraction/wide-angle reflection profiling in southeastern California and western Arizona. Journal of Geophysical Research 26, 12259–12291.

McDermott, F., Hawkesworth, C.J., 1990. Intracrustal recycling and upper-crustal evolution: a case study from the Pan-African Damara mobile belt, central Namibia. Chemical Geology 83, 263–280.

McDermott, F., Harris, N.B.W., Hawkesworth, C.J., 1996. Geochemical constraints on crustal anatexis: a case study from the Pan-African granitoids of Namibia. Contributions to Mineralogy and Petrology 123, 406–423.

Miller, R.M., 1983. The Pan-African Damara Orogen of South West Africa/Namibia. Special Publication-Geological Society of South Africa 11, 431–515.

Miller, C.F., Barton, M.D., 1990. Phanerozoic plutonism in the Cordilleran interior, USA. Special Paper-Geological Society of America 241, 213–231.

Miller, C.F., Miller, J.S., 2002. Contrasting stratified plutons exposed in tilt blocks, Eldorado Mountains, Colorado River rift, NV, USA. Lithos 61, 209–224.

Miller, C.F., Wooden, J.L., 1994. Anatexis, hybridization and modification of ancient crust: Mesozoic plutonism in the Old Woman Mountains area, California. Lithos 32, 135–156.

Miller, C.F., D'Andrea, J.L., Ayers, J.C., Coath, C.D., Harrison, T.M., 1997. BSE imaging and ion probe geochronology of zircon and monazite from plutons of the Eldorado and Newberry Mountains, Nevada: age, inheritance, and subsolidus modification. EOS 78, F783.

Milner, S.C., le Roex, A.P., 1996. Isotope characteristics of the Okenyenya igneous complex northwestern Namibia: constraints on the composition of the early Tristan Plume and the origin of the Emi mantle component. Earth and Planetary Science Letters 141, 277–291.

Milner, S.C., le Roex, A.P., O'Conner, J.M., 1995. Age of the Mesozoic igneous rocks in northwestern Namibia and their relationship to continental breakup. Journal of the Geological Society (London) 152, 97–104.

Neymark, L.A., Yu, V., Larin, A.M., 1994. Pb–Nd–Sr isotopic constraints on the origin on the 1.54–1.56 Ga Salmi rapakivi granite–anorthosite batholith (Karelia, Russia). Mineralogy and Petrology 50, 173–194.

Nironen, M., 1997. The Svecofennian orogen: a tectonic model. Precambrian Research 86, 21–44.

Nironen, M., Elliott, B.A., Rämö, O.T., 2000. 1.88–1.87 Ga post kinematic intrusions of the Central Finland Granitoid Complex: a shift from C-type to A-type magmatism during lithospheric convergence. Lithos 53, 37–58.

Nurmi, P.A., Haapala, I., 1986. The Proterozoic granitoids of Finland: granite types, metallogeny and relation to crustal evolution. Bulletin of the Geological Society of Finland 58, 203–233.

O'Conner, J.M., le Roex, A.P., 1992. South Atlantic hotspot–plume systems: 1. Distribution of volcanism in time and space. Earth and Planetary Science Letters 113, 343–364.

Patchett, J., Kouvo, O., 1986. Origin of continental crust of 1.9–1.7 Ga age: Nd isotopes and U–Pb zircon ages in the Svecokarelian terrain of South Finland. Contributions to Mineralogy and Petrology 92, 1–12.

Pearce, J.A., Harris, N.B.W., Tindle, A.G., 1984. Trace element discrimination diagrams for the tectonic interpretation of granitic rocks. Journal of Petrology 25, 956–983.

Peccerillo, A., Barberio, M.R., Yirgu, G., Ayalev, D., Barbieri, M., Wu, T.W., 2003. Relationships between mafic and peralkaline silicic magmatism in continental rift settings: a petrological, geochemical and isotopic study of Gedemsa Volcano, Central Ethiopian Rift. Journal of Petrology 44, 2003–2032.

Peltonen, P., Mänttäri, I., 2001. An ion microprobe U–Th–Pb study of zircon xenocrysts from the Lahtojoki kimberlite pipe, eastern Finland. Bulletin of the Geological Society of Finland 73, 47–58.

Persson, A.I., 1999. Absolute (U–Pb) and relative age determinations of intrusive rocks in the Ragunda rapakivi complex, central Sweden. Precambrian Research 95, 109–127.

Pesonen, L.J., Mertanen, S., 2002. Paleomagnetic configuration of continents during the Proterozoic. Institute of Seismology, University of Helsinki, Report S-42, pp. 103–109.

Pesonen L.J., Korja, A., Hjelt, S.-E., 2000. Lithosphere 2000, a symposium on structure, composition and evolution of the lithosphere in Finland. Programme and extended abstracts. Institute of Seismology Report S-41.

Prave, A.R., 1996. Tale of three cratons: tectonostratiographic anatomy of the Damara orogen in northwestern Namibia and the assembly of Gondwana. Geology 24, 1115–1118.

Puura, V., Flodén, T., 1999. Rapakivi–granite–anorthosite magmatism—a way of thinning and stabilisation of the Svecofennian crust, Baltic Sea Basin. Tectonophysics 305, 75–92.

Rabinowitz, P.D., LaBrecque, J.L., 1979. The Mesozoic South Atlantic Ocean and evolution of its continental margins. Journal of Geophysical Research 84, 5973–6002.

Rämö, O.T., 1991. Petrogenesis of the Proterozoic rapakivi granites and related basic rocks of southeastern Fennoscandia: Nd and Pb isotopic and general geochemical constraints. Bulletin-Geological Survey of Finland 355. 161 pp.

Rämö, O.T., 2001. Isotopic composition of pyterlite in Vyborg (Viipuri), Wiborg batholith, Russia. Bulletin of the Geological Society of Finland 73, 111–115.

Rämö, O.T., Haapala, I., 1995. One hundred years of rapakivi granite. Mineralogy and Petrology 52, 129–185.

Rämö, O.T., Haapala, I., 1996. Rapakivi granite magmatism: a global review with emphasis on petrogenesis. In: Demaiffe, D. (Ed.), Petrology and Geochemistry of Magmatic Suites of Rocks in the Continental and Oceanic Crust. A volume dedicated to Jean Michot. Universite Libre de Bruxelles, Royal Museum for Central Africa, pp. 177–200.

Rämö, O.T., Korja, A., 2000. Mid-Proterozoic evolution of the Fennoscandian Shield. Extended abstract, 31st IGC, Rio de Janciro, August 2000. Abstract volume [CD-ROM].

Rämö, O.T., Luukkonen, E.J., 2001. 2.4 Ga A-type granites of the Kainuu region, eastern Finland: characterization and tectonic significance. EUG XI. European Union of Geosciences, April 8th–12th, 2001, Strasbourg, France: abstract volume. European Union of Geosciences, Strasbourg, pp. 771.

Rämö, O.T., Haapala, I., Vaasjoki, M., Yu, J., Fu, H., 1995. The 1700 Ma Shachang complex, northeast China: Proterozoic rapakivi granite not associated with Paleoproterozoic orogenic crust. Geology 23, 815–818.

Rämö, O.T., Huhma, H., Kirs, J., 1996. Radiogenic isotopes of the Estonian and Latvian rapakivi granite suites: new data from the concealed Precambrian of the East European Craton. Precambrian Research 79, 209–226.

Rämö, O.T., Haapala, I.J., Volborth, A., 1999. Isotopic and general geochemical constraints on the origin of Tertiary granitic plutonism in the Newberry Mountains, Colorado River Extensional Corridor, Nevada. Abstracts with Programs-Geological Society of America 31 (6), A86.

Rämö, O.T, Korja, A., Haapala, I., Eklund, O., Fröjdö, S., Vaasjoki, M., 2000. Evolution of the Fennoscandian lithosphere in the mid-Proterozoic: the rapakivi magmatism. Institute of Seismology, University of Helsinki, Report S-41, pp. 129–136.

Rieder, M., Haapala, I., Povondra, P., 1996. Mineralogy of dark mica from the Wiborg rapakivi batholith, southeastern Finland. European Journal of Mineralogy 8, 593–605.

Salonsaari, P.T., 1995. Hybridization in the bimodal Jaala-Iitti complex and its petrogenetic relationship to rapakivi granites and associated mafic rocks of southeastern Finland. Bulletin of the Geological Society of Finland 67 (Part 1b). 104 pp.

Schmitt, A.K., Emmermann, R., Trumbull, R.B., Bühn, B., Henjes-Kunst, F., 2000. Petrogenesis and $^{40}Ar/^{39}Ar$ geochronology of the Brandberg complex, Namibia: evidense for major mantle contribution in metaluminous and peralkaline granites. Journal of Petrology 41, 1207–1239.

Scott, R.B., Unruh, D.M., Snee, L.W., Harding, A.E., Nealey, L.D., Blank Jr., H.R., Budahn, J.R., Mehnert, H.H., 1995. Relation of peralkaline magmatism to heterogeneous extension during the middle Miocene, southeastern Nevada. Journal of Geophysical Research 100, 10381–10401.

Seth, B., Kröner, A., Mezger, K., Nemchin, A., Pidgeon, R.T., Okrusch, M., 1998. Archean to Neoproterozoic magmatic events in the Kaoko belt of NW Namibia and their geodynamic significance. Precambrian Research 92, 341–363.

Simonen, A., Vorma, A., 1969. Amphibole and biotite from rapakivi. Bulletin de la Commission Geologique de Finlande 238. 28 pp.

Suominen, V., 1991. The chronostratigraphy of southwestern Finland with special reference to Postjotnian and Subjotnian diabases. Bulletin-Geological Survey of Finland 356. 100 pp.

Taylor, R.P., 1992. Petrological and geochemical characteristics of the Pleasant Ridge zinnwaldite–topaz granite, south New Brunswick, and comparisons with other topaz-bearing felsic rocks. Canadian Mineralogist 30, 895–921.

Trumbull, R.B., Emmermann, R., Bühn, B., Gerstenberger, H., Mingram, B., Schmitt, A., Volker, F., 2000. Insight on the genesis of the Cretaceous Damaraland igneous complexes in Namibia from a Nd- and Sr-isotopic perspective. Communications of the Geological Survey of Namibia 12, 313–325.

Vaasjoki, M., Rämö, O.T., Sakko, M., 1991. New U–Pb ages from the Wiborg rapakivi area: constraints on the temporal evolution of the rapakivi granite–anorthosite–diabase dyke association of southeastern Finland. Precambrian Research 51, 227–243.

Volborth, A., 1973. Geology of the granite complex of the Eldorado, Newberry, and northern Dead Mountains, Clark County, Nevada. Bulletin-Nevada Bureau of Mines and Geology 80. 40 pp.

Vorma, A., 1976. On the petrochemistry of rapakivi granites with special reference to the Laitila massif, southwestern Finland. Bulletin-Geological Survey of Finland 285. 98 pp.

Wang, K., Windley, B.F., Sills, J.D., Yan, Y., 1990. The Archean gneiss complex in E. Hebei Province, north China: geochemistry and evolution. Precambrian Research 48, 245–265.

Welin, E., Christansson, K., Kähr, A.-M., 1993. Isotopic investigations of metasedimentary and igneous rocks in the Palaeoproterozoic Bothnian Basin, central Sweden. Geologiska Foreningens i Stockholm Forhandlingar 115, 285–296.

Wernicke, B., 1985. Uniform-sense normal simple shear of the continental lithosphere. Canadian Journal of Earth Sciences 22, 108–125.

Whalen, J.B., Currie, K.L., Chappell, B.W., 1987. A-type granites; Geochemical characteristics, discrimination, and petrogenesis. Contributions to Mineralogy and Petrology 95, 407–419.

Whalen, J.B., Jenner, G.A., Longstaffe, F., Robert, F., Gariéty, C., 1996. Geochemical and isotopic (O, Nd, Pd and Sr) constraints on A-type granite petrogenesis based on the Topsails igneous suit, Newfoundland Appalachians. Journal of Petrology 37, 1463–1489.

White, R., McKenzie, D., 1989. Magmatism at rift zones: the generation of volcanic continental margins and flood basalts. Journal of Geophysical Research 94, 7685–7729.

Windley, B.F., 1995. The Evolving Continents, 3rd edition. John Wiley & Sons, Chichester.

Yu, I., Fu, H., Wan, F., 1994. Petrogenesis of potassic alkaline volcanics associated with rapakivi granites in the Proterozoic rift of Beijing, China. Mineralogy and Petrology 50, 83–96.

Yu, J.H., Fu, H.Q., Zhang, F.L., Wan, F.X., Haapala, I., Ramo, O.T., Vaasjoki, M., 1996a. Anorogenic Rapakivi Granites and Related Rocks in the Northern of the North China Craton. China Science and Tecnology Press, Beijing. In Chinese with English summary on pp. 137–182.

Yu, I., Fu, H., Haapala, I., Rämö, O.T., Vaasjoki, M., Mortenson, J.K., 1996b. Paleoproterozoic anorogenic rapakivi granite–anorthosite suite in the north part of the North China Craton. Progress in Geology of China. (1993–1996)—Papers to 30th IGC. pp. 104–108.

Zandt, G., Myers, S.C., Wallace, T.C., 1995. Crust and mantle structure across the Basin and Range-Colorado Plateau boundary at 37°N latitude and implications for Cenozoic extensional mechanism. Journal of Geophysical Research 100, 10529–10548.

Available online at www.sciencedirect.com

Lithos 80 (2005) 33–44

www.elsevier.com/locate/lithos

TTGs and adakites: are they both slab melts?

Kent C. Condie*

Department of Earth and Environmental Science, New Mexico Institute of Mining and Technology, Socorro, NM 87801, USA

Received 25 March 2003; accepted 9 September 2003
Available online 5 November 2004

Abstract

Although both high-Al TTG (tonalite–trondhjemite–granodiorite) and adakite show strongly fractionated REE and incompatible element patterns, TTGs have lower Sr, Mg, Ni, Cr, and Nb/Ta than most adakites. These compositional differences cannot be easily related by shallow fractional crystallization. While adakites are probably slab melts, TTGs may be produced by partial melting of hydrous mafic rocks in the lower crust in arc systems or in the Archean, perhaps in the root zones of oceanic plateaus. It is important to emphasize that geochemical data can be used to help constrain tectonic settings, but it cannot be used alone to reconstruct ancient tectonic settings.

Depletion in heavy REE and low Nb/Ta ratios in high-Al TTGs require both garnet and low-Mg amphibole in the restite, whereas moderate to high Sr values allow little, if any, plagioclase in the restite. To meet these requirements requires melting in the hornblende eclogite stability field between 40- and 80-km deep and between 700 and 800 °C.

Some high-Al TTGs produced at 2.7 Ga and perhaps again at about 1.9 Ga show unusually high La/Yb, Sr, Cr, and Ni. These TTGs may reflect catastrophic mantle overturn events at 2.7 and 1.9 Ga, during which a large number of mantle plumes bombarded the base of the lithosphere, producing thick oceanic plateaus that partially melted at depth.
© 2004 Elsevier B.V. All rights reserved.

Keywords: TTG; Adakite; Archean tectonics; Arc systems; Mantle plume event

1. Introduction

Since the classic book edited by Fred Barker on trondhjemites and related rocks (Barker, 1979), this suite of rocks, now known as the TTG (tonalite–trondhjemite–granodiorite) suite, has received considerable attention. Because they are the most volumi-

nous rock type in the preserved Archean crust, it is critical to understand the source and origin of TTGs to better understand the origin and early evolution of continents. Even in the Tertiary, TTGs are an important juvenile component added to the continental crust in the Andes and other continental-margin arcs. Because of geochemical similarities of TTGs to modern adakites, they are commonly assumed to have similar origins (Drummond and Defant, 1990; Martin, 1999), and in some cases, the two terms are used interchangeably. For instance, on a primitive mantle

---

* Tel.: +1 505 835 5531; fax: +1 505 835 6436.
E-mail address: kcondie@nmt.edu.

0024-4937/$ - see front matter © 2004 Elsevier B.V. All rights reserved.
doi:10.1016/j.lithos.2003.11.001

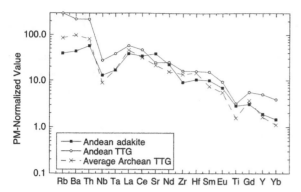

Fig. 1. Primitive-mantle normalized incompatible element distributions in adakites and TTGs. Data from Kay et al. (1993, 1999) and Table 1 and Appendix A. Primitive mantle values from Sun and McDonough (1989).

normalized element distribution diagram, both rock types exhibit strong enrichment in the most incompatible elements, with notable negative anomalies at Nb–Ta and Ti and strong depletion in heavy REE and Y (Fig. 1). Hence, as with adakites, TTGs are commonly assumed to represent melts of descending slabs, a conclusion with important implications for tectonic settings in the Archean. However, some investigators have pointed to dissimilarities between adakites and TTGs and hence to different tectonic settings (Smithies, 2000; Kamber et al., 2002). One of the issues in this controversy is that of how to define TTG and whether or not TTGs are strictly an Archean phenomena or if they occur throughout geologic time.

In this study, these questions are addressed by comparing published chemical analyses of TTGs and adakites of varying ages and locations. Results show that TTGs and adakites are not the same thing, and consequently, they may have quite different origins. Furthermore, TTGs are not an Archean phenomenon but have contributed to the growth of continental crust throughout geologic time.

2. Data collection and definitions

To compare adakites and TTGs, I have compiled chemical analyses from the literature together with unpublished analyses from our laboratory. Included are major elements and a group of both compatible and incompatible trace elements (see Appendix A). We have used a definition for TTG similar to that of

Barker (1979) and Defant and Drummond (1990): Al_2O_3 contents >15% at 70% SiO_2, Sr >300 ppm, Y<20 ppm, Yb<1.8 ppm, and Nb≤10 ppm. Although adakites commonly show these same chemical characteristics, they are more mafic and can be distinguished from TTGs by their relatively high Mg, Ni, Cr, and Sr contents. TTGs also have been referred to as "low-Mg adakites" (Rapp et al., 1999). For a rock to be called an adakite, in addition to the above geochemical characters, it should have a Mg number greater than 50, Sr >500 ppm (often >1000 ppm), Cr ≥50 ppm, and Ni ≥20 ppm (Figs. 2–5; Appendix A). Also, as pointed out by Kamber et al. (2002), Archean TTGs do not resemble adakites in terms of fluid-mobile trace elements such as Ba, K, and Rb (Table 1). The relatively high contents of these elements in TTGs compared to adakites may reflect a subduction zone component in TTG sources, which is not present in the descending slab source of adakites. The relatively high Mg, Cr, and Ni contents of adakites are generally ascribed to interaction of the slab melts with the overlying mantle wedge, a process that has been verified by experimental studies (Drummond and Defant, 1990; Rapp et al., 1999). Another relatively minor group of plutonic rocks known a sanukitoids is also recognized in the Archean (Shirey and Hanson, 1984). These rocks, which show the similar distributions of incompatible elements as TTGs, contain larger amounts of K, Rb, Th, and U and are generally thought to be the products of melting metasomatized mantle (Stern and Hanson, 1991). They are not included in the present investigation.

Average values of selected element concentrations and element ratios are give in Appendix A, and a summary of average TTG compositions of various ages is given in Table 1 compared with average adakite. Although the geochemical definitional screens given in the previous paragraph are used to distinguish TTG from adakite, in some cases, screens give conflicting results. In these cases, the rock is classified according the majority of the screens. One thing is very clear from Table 1, TTGs are not an Archean phenomenon but occur throughout geologic time. Some of the youngest examples are found in the Miocene batholiths of the Andes and comprise a large proportion of the Andean Cordillera (Petford and Atherton, 1996; Kay et al., 1999). On average,

Table 1
Average chemical compositions of TTGs and adakites

	TTGs								Adakites
	Early Archean		Late Archean		Proterozoic		Phanerozoic		
	Mean	1σ	Mean	1σ	Mean	1σ	Mean	1σ	Mean
SiO_2	70.4	2.9	68.3	2.6	67.3	3.6	65.9	3.5	62.43
TiO_2	0.31	0.14	0.42	0.12	0.47	0.2	0.47	0.16	0.67
Al_2O_3	15.2	1.1	15.5	0.75	15.8	0.78	16.5	0.94	17.05
Fe_2O_3T	2.79	1.2	3.42	1.1	4.04	1.4	4.11	1.3	3.99
MgO	0.96	0.62	1.39	0.59	1.48	0.70	1.67	0.69	3.31
CaO	2.74	0.88	3.26	0.73	3.42	1.10	4.36	1.10	6.53
Na_2O	4.71	0.61	4.51	0.48	4.33	0.66	4.00	0.50	4.25
K_2O	2.22	0.68	2.20	0.66	2.30	0.82	2.14	0.63	1.42
P_2O_5	0.1	0.06	0.14	0.06	0.14	0.06	0.12	0.05	0.26
MnO	0.06	0.02	0.07	0.03	0.08	0.05	0.09	0.02	0.08
	99.49		99.21		99.36		99.36		99.99
Th	4.1	2.4	8.1	5.3	6.1	3.1	7.6	4.4	3.9
U	1.2	0.49	1.5	0.99	2.1	1.1	1.9	1.1	1.2
Ni	17	19	22	13	23	12	12	5.0	64
Cr	45	48	35	26	55	31	32	15	82
Y	8.5	6.4	9.1	5.4	17.3	8.9	14.5	4.7	9.7
Zr	152	44	154	49	152	63	122	30	117
Nb	6.1	3.5	6.2	2.4	7.1	4.6	6.7	1.7	9.7
Hf	3.8	1.6	4.7	1.9	4.3	1.6	3.4	0.73	3.3
Ta	0.41	0.29	0.84	0.3	0.72	0.54	0.75	0.32	0.60
La	22	9.4	36	16	26	12	17	6.2	24
Ce	40	16	65	29	45	26	34	11	65
Nd	16	8.2	25	9.1	18	12	16	4.5	26
Sm	2.9	1.7	4.2	1.7	3.5	2	3.1	0.89	4.7
Eu	0.82	0.31	1.07	0.32	0.95	0.44	0.84	0.23	1.37
Gd	2.2	1.4	2.9	1.2	3	1.9	2.8	0.65	2.30
Tb	0.31	0.23	0.38	0.17	0.49	0.27	0.40	0.11	0.40
Yb	0.82	0.7	0.71	0.41	1.33	0.86	1.16	0.39	0.81
Lu	0.14	0.1	0.11	0.07	0.23	0.14	0.18	0.06	0.09
Rb	76	26	67	24	63	25	63	26	15
Ba	500	258	769	288	717	239	716	251	309
Sr	362	117	515	127	473	159	493	105	1550
(La/Yb)n	25	15	36	24	14.2	8.3	11.3	5.3	18.2
Sr/Y	72	24	89	49	37	30	56	12	160
Nb/Ta	12.0	2.8	13.0	6.3	9.9	6.7	12.2	9.2	16.1
Mg#	40.8	8.3	46.2	6.7	43.2	7.5	45.4	6.9	60.4
K_2O/Na_2O	0.51	0.18	0.51	0.24	0.56	0.25	0.68	0.5	0.33
n	212		831		752		698		221

Major elements as oxides in wt.%, trace elements in ppm; n, number of samples; 1σ, one standard deviation of mean; Mg#=MgO/ $(MgO+0.79Fe_2O_3T)$, molecular ratio; chondrite normalizing values from Haskin et al. (1968); Early Archean, ≥3.5 Ga; Late Archean, 3.5–2.5 Ga; Proterozoic, 2.5–0.54 Ga; Phanerozoic, <0.54 Ga. TTG references are given in Appendix A; Adakite references (and references cited therein): Li and Li (2003), Percival et al. (2003), Kay et al. (1993), Martin (1999), Stern and Kilian (1996), Samaniego et al. (2002), Rapp et al. (1999), Polat and Kerrich (2002), Drummond and Defant (1990), Kay and Kay (2002), Peacock et al. (1994), Smithies and Champion (2000), Bourdon et al. (2002), Drummond et al. (1996), Defant and Drummond (1993), Kepezhinskas et al. (1997), Myers et al. (1985), Yogodzinski et al. (1995), Xu et al. (2002), Gutscher et al. (2000).

however, Archean TTGs are more depleted in Y and heavy REE and have higher K_2O/Na_2O ratios than post-Archean TTGs (Table 1). In terms of tectonic setting, all young TTGs share in common an arc-related setting, often, but not always a continental-margin arc. Archean TTGs usually are intruded into

greenstone belts with arc or oceanic plateau geochemical affinities. Until recently, adakites were
described only from Phanerozoic successions, but
they are now recognized in the Proterozoic and a few
rare occurrences in Archean terranes (Polat and
Kerrich, 2002). As with TTGs, adakites are found in
arc-related tectonic settings, both oceanic arcs and
continental-margin arcs. Unlike TTGs, Archean adakites are not found associated with greenstones with
oceanic plateau geochemical affinities.

Data shown on the graphs in this study are mean
values of TTGs from different geographic locations
and different ages (Appendix A). Over 150 sites and
nearly 2500 samples are represented in the database,
although not all trace elements are available for each
site. In evaluating similarities and differences, not
only mean values, but also median values and ranges
for each site are considered.

3. Adakites and TTGs

3.1. Andean examples

Perhaps the best described comparison of adakites
and TTGs is from the southern Andes (Kay et al.,
1993, 1999; Stern and Kilian, 1996; Petford and
Atherton, 1996). Well-documented adakites are associated with the Chilean triple junction where the Chile
rise is being subducted beneath the Andes in

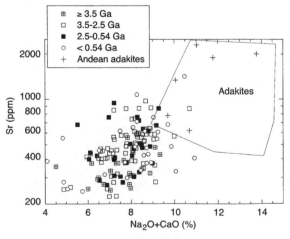

Fig. 2. Sr vs. Na₂O+CaO graph showing the distribution of TTGs in
comparison to the adakite field. Data sources given in Table 1 and
Appendix A. Andean adakites from Kay et al. (1993).

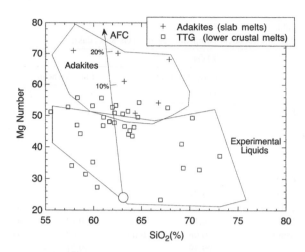

Fig. 3. Mg number vs. SiO₂ showing a comparison of Andean TTGs
and adakites. Data sources given in Table 1 and Appendix A.
Experimental liquids from data published in Rapp et al. (1991),
Wylie et al. (1997) and Winther (1996). AFC, assimilation fractional
crystallization trajectory from Stern and Kilian (1996); percents are
amounts of assimilated ultramafic rock.

Patagonia (Kay et al., 1993). The Andean adakites
were erupted 12 Ma and appear to represent melts
derived from the hot descending plate. These adakitic
magmas share many geochemical features with the
type adakites from Adak Island in the Aleutians, such
as similar incompatible element distributions and
similar high Sr contents (Figs. 1 and 2). Another
feature of the Andean adakites is relatively high Mg
number (50–70) as well as high Ni (>40 ppm), Cr
(>80 ppm), and Sr (1800 ppm; Figs. 2–4), all of which
are generally interpreted to reflect reaction of slab
melts with overlying ultramafic rocks of the mantle
wedge. The AFC line in Fig. 3 is one possible
trajectory of assimilation-fractional crystallization of
ultramafic rocks in the mantle wedge.

In contrast to the Patagonia Andean adakites, a
second group of felsic volcanic and related plutonic
rocks is found in the Central Volcanic Zone (CVZ) of
the Andes. These rocks, which have similar incompatible element distributions to the Patagonian adakites (Fig. 1), have relatively low Mg numbers (<55)
and also relatively low Sr (300 ppm), Ni (5 ppm), and
Cr (50 ppm; Kay et al., 1999). Kay et al. (1999)
proposed a model for CVZ volcanic rocks involving
thickening of the crust due to tectonic shortening and
lithosphere delamination. The rapidly delaminated
lithosphere was replaced by hot asthenosphere, which
partially melted producing basaltic magmas. These

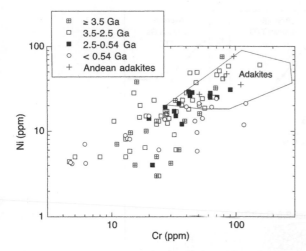

Fig. 4. Mg number vs. SiO_2 showing the distribution of TTGs in comparison to the adakite field. Data sources given in Table 1 and Appendix A. Experimental liquid field references given in Fig. 3.

magmas, in turn, heated the thickened lower crust to produce TTG magmas, leaving eclogitic residues in the lower crust. Unlike the adakitic magmas, the CVZ volcanic rocks are similar in composition to experimental melts of wet mafic sources (Fig. 3). The low Mg numbers, Ni, and Cr in these rocks mean that unlike the Patagonian adakites, the CVZ magmas did not react with the mantle wedge. Employing the definitions used in this study, the CVZ volcanics are classic TTGs.

Hence, it would appear that, in the Andes, both adakites and TTGs were produced during the Tertiary depending on whether a hot descending slab or a thickened/delaminated lower crust was melted. The key differences in the derivative magmas are not the incompatible element distributions but compatible elements such Mg, Sr, Ni, and Cr.

3.2. Some geochemical contrasts

For comparative purposes, TTGs are grouped into five age groups: ≥3.5, 3.5–2.5, 2.7, 2.5–0.54, and <0.54 Ga. On a Mg#–SiO_2 graph, the majority of TTGs fall in the hydrous experimental liquid field with only a small number plotting in the adakite field (Fig. 4). Most TTGs have lower Mg numbers than adakites, a feature recently emphasized by Smithies (2000). Again, this suggests that most TTGs, although partial melts of hydrous mafic sources, have not

reacted with the mantle wedge. Although there is more overlap between TTGs and adakites in terms of Cr and Ni distributions (Fig. 5), a large number of the TTGs, regardless of age, have average Ni and Cr contents less than most adakites (Appendix A). Sr is a compatible element in plagioclase, and hence, the Sr distribution in adakite and TTG magmas reflects, at least in part, the role of plagioclase fractionation (Ellam and Hawkesworth, 1988; Tarney and Jones, 1994; Martin and Moyen, 2002). The partitioning of Sr into the melt is also related to the An content of the plagioclase (Foley et al., 2002). Most adakites have higher Sr and Na_2O+CaO values than most TTGs, although there is minor overlap in Sr contents (Fig. 2). TTGs with <500 ppm Sr probably reflect either or both plagioclase in the restite or plagioclase fractional crystallization. Calcalkaline TTGs (low-Al TTGs) contain even less Sr (<300 ppm), a feature commonly ascribed to plagioclase fractional crystallization.

As described by numerous investigators and especially by Martin (1993, 1999), Archean high-Al TTGs and adakites both have steep REE patterns, resulting in relatively high La/Yb ratios at low Yb values (Fig. 6; Appendix A). In contrast, most post-Archean TTGs belong to the calcalkaline suite and often show a fractional crystallization sequence extending from diorite (or gabbro) to granite. These rocks typically have very low La/Yb ratios and exhibit a large range of Yb values (Fig. 6). The reason that

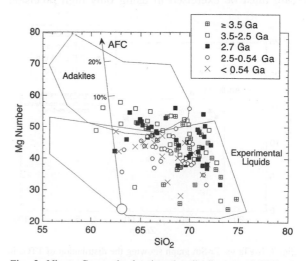

Fig. 5. Ni vs. Cr graph showing the distribution of TTGs in comparison to the adakite field. Data sources given in Table 1 and Appendix A.

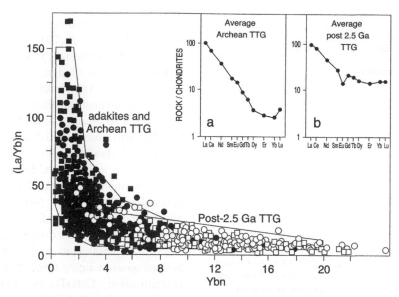

Fig. 6. $(La/Yb)_n$ vs. Yb_n graph showing distribution of adakites and TTGs. Modified after Martin (1993). Post-Archean TTG=calcalkaline plutonic suites. n—normalized to primitive mantle using values from Sun and McDonough (1989).

adakites and high-Al TTGs are indistinguishable on the La/Yb–Yb plot is that reaction with the mantle wedge does not appreciably affect REE distributions or incompatible element ratios (Rapp et al., 1999). In both cases, the highest La/Yb ratios require that garnet occur in the restite (Martin, 1993).

An important incompatible element ratio that distinguishes adakite from high-Al TTG is the Nb/Ta ratio (Kamber et al., 2002). In making this contrast, care must be exercised in using only high precision

analyses of Nb and Ta, and in this respect, most Nb analyses by XRF must be rejected. Using only high precision ICPMS Nb and Ta data, TTGs typically have Nb/Ta ratios less than the primitive mantle value of 16.7 and sometimes much lower (with averages ranging down to about 5; Fig. 7; Appendix A). In contrast, adakites have Nb/Ta ratios near the primitive mantle value (generally between 15 and 20). In addition, TTGs generally show Zr/Sm ratios greater than the primitive mantle value of 25.2 while adakites scatter on both sides of this value. Because rutile is probably not a residual phase during melting of eclogite (Rapp et al., 1991), eclogite-derived TTGs will have high Nb/Ta ratios, reflecting the high ratios in melted rutile (Foley et al., 2002; Klemme et al., 2002). For this reason, it is unlikely that TTGs are derived from the partial melting of rutile-bearing eclogites.

4. Discussion

4.1. Fractional crystallization or partial melting?

Is it possible that adakites and TTGs are both slab melts that reacted with the mantle wedge but that TTGs later underwent fractional crystallization, which reduced their compatible element contents? For

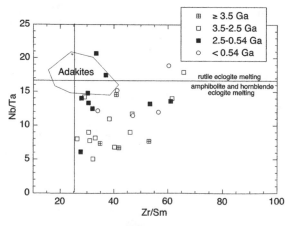

Fig. 7. Nb/Ta vs. Zr/Sm graph showing the distribution of TTGs in comparison to the adakite field. Lines represent primitive mantle values from Sun and McDonough (1989) and melting fields from Foley et al. (2002).

instance, partial melting of Sr-rich eclogite in a descending slab produces adakitic magmas with high Sr contents (as there is no plagioclase in the restite) and steep REE patterns (reflecting restite garnet or/ and amphibole). Fractional crystallization of these magmas at shallower depths could reduce the Sr contents by plagioclase removal and reduce the Mg number and Cr and Ni contents by amphibole/biotite removal without appreciably affecting the REE patterns (Kamber et al., 2002; Samaniego et al., 2002). Amphibole removal during fractional crystallization could also lower the Nb/Ta ratios, as observed in TTGs relative to adakites (Fig. 7).

However, there are at least three lines of evidence that suggest that TTGs are not simply the fractional crystallization products of adakitic magmas. As pointed out by several investigators, the apparent absence of thick sequences of cumulates does not favor a fractional crystallization connection (Smithies, 2000). To produce felsic TTGs from intermediate to mafic adakites requires more than 50% fractional crystallization, leaving behind large volumes of cumulates of mafic to intermediate composition. Where are these cumulates? This is especially a problem in the Archean crust where large volumes of preserved TTGs demand enormous volumes of early cumulates. One cannot hide these in the deep crust because there are many areas of deep Archean crust exposed at the surface today, such as those in southern India and Antarctica (Condie and Allen, 1984; Black et al., 1986; Sheraton et al., 1987). Even in these cratons with Archean high-grade rocks exposed at the surface, few cumulates of mafic to intermediate composition are found.

Another argument against a fractional crystallization connection between adakites and TTGs is the similarity in composition of TTGs to that of experimental melts, as revealed by numerous melting studies of amphibolites and eclogites (Fig. 3; Rapp et al., 1991; Winther, 1996; Wylie et al., 1997; Prouteau et al., 2001). Is it coincidental that so many TTGs fall in the experimental melt field if they are the products of extensive fractional crystallization?

Also not favoring a fractional crystallization connection between adakites and TTGs is the fact that trace element distributions define trajectories similar to calculated batch melting trajectories and not to shallow fractional crystallization trajectories.

This is especially apparent on the $(La/Yb)_n$–Yb_n graph (Martin, 1993; Samaniego et al., 2002). Although amphibole fractional crystallization produces residual melts that follow batch-melting trends, it cannot produce the high La/Yb values characteristic of many Archean TTGs.

4.2. P–T regimes

Are P–T regimes the same for production of adakite and TTG magmas? Numerous experimental studies provide a robust framework upon which the pressures and temperatures of melting of these magmas can be estimated (references above). Experimental results agree that both adakites and TTGs are produced from partial melting of a wet mafic protolith. Although experiments have covered a wide range of water contents, only those performed at relatively high water contents ($\geq 5\%$) produce felsic magmas similar to TTG at relatively low melting temperatures, leaving amphibole in the restite (Rapp et al., 1991; Prouteau et al., 2001). Also, because amphibole has a Kd for Ti (~3) greater than neighboring elements Eu and Gd (~1.6) on an incompatible element diagram, restite amphibole can explain the ubiquitous negative Ti anomaly in TTGs (Fig. 1). The high La/Yb and Sr/Y ratios in both adakite and TTG magmas seem to require that garnet also be present in the restite. In the case of TTG magmas, the low Nb/Ta ratios indicate the presence of a low-Mg amphibole in the restite, such that Nb is retained in the restite compared to Ta (Foley et al., 2002). In contrast, adakitic magmas, with their comparatively high Nb/Ta ratios, cannot have low-Mg amphibole in the restite (Appendix A; Fig. 7). These results also show that rutile-bearing eclogites cannot serve as sources for most TTG magmas, as partial melting of such eclogites produces melts with high Nb/Ta ratios (Foley et al., 2002).

So what P–T regimes are allowed for production of TTG magmas? To have both low-Mg amphibole and garnet in the restite but little, if any, plagioclase requires melting at depths >40 km but generally less than about 80 km over a temperature range of 700–800 °C (Martin, 1999). This is in the stability field of hornblende eclogite rather than garnet amphibolite, which by definition contains significant amounts of plagioclase.

4.3. Tectonic setting

If adakitic magmas have reacted with mantle wedges, they reflect not only an arc setting but the subduction of young relatively hot slabs such as the Chile rise beneath the southern Andes. Martin (1993, 1999) and Foley et al. (2003) have suggested similar tectonic scenarios for TTGs in the Archean. However, if the Andes Central Volcanic Zone is a better analogue for Archean TTG, perhaps it is the Archean lower crust that is melting rather than the descending slab (for an Archean example, see Whalen et al., 2002). As pointed out by Andean investigators, this crust may be thickened by magmatic underplating of basaltic magmas coming from the mantle wedge or by tectonic shortening in the overriding plate along a convergent plate boundary (Petford and Atherton, 1996; Kay et al., 1999). Subduction erosion, which brings mafic lower crust down the subducting plate to higher pressure and temperature, can lead to partial melting. Of course, this model requires a considerable amount of water in Archean subduction zones to obtain the necessary hydrous melting. If plates were recycled faster in the Archean due to hotter mantle, perhaps water was also recycled faster in subduction zones, providing a rich supply at Archean convergent margins. Regardless of the specific tectonic scenario, it is the lower crust of the arc that is melting to produce the Andean TTG and not the descending slab. If the high Mg, Cr, and Ni in adakites require reaction with the mantle wedge, the low contents of these elements in TTGs mean a lack of such reaction and does not favor a slab origin for TTGs (Smithies, 2000). The fact that true adakites are rare in the Archean may be due to comparatively hot subducted plates, which ride high displacing all or most of the mantle wedge. Hence, there is little, if any, opportunity for Archean slab melts to react with the mantle wedge.

Are there other acceptable tectonic settings for TTGs and especially for Archean TTGs when plate tectonics may not have quite the same as it is today? Perhaps the root zones of Archean oceanic plateaus were buried deep enough so the mafic component underwent partial melting. Such a model was suggested as a possible origin for the Aruba batholith in the Caribbean oceanic plateau during the Cretaceous (White et al., 1999). Of course, these deep roots must be hydrated to form amphibole, a requirement that would seem probable in the Archean mantle, as previously discussed. Hydrothermal alteration of plateau basalts followed by deep burial would also introduce water into the system. In fact, any tectonic setting in which hornblende eclogite is stable and the temperature reaches 700–800 °C could produce TTG magmas.

At this point, it is important to emphasize a fact that is commonly overlooked in using geochemical data to assign tectonic settings; that is, geochemical data can be used to help constrain tectonic settings, but it cannot be used alone to reconstruct ancient tectonic settings.

4.4. Secular changes in TTGs

Martin and Moyen (2002) have recently suggested secular changes in the composition of TTGs and that these changes, if real, can be used to constrain evolving tectonic settings from the Early to the Late Archean. To overcome the problem of fractional crystallization, the authors suggested that the upper limit in the range of variation in compatible elements (Mg, Sr, Cr, Ni) be used as most representative of parental magma compositions. From their study, they concluded that, between 4 and 2.5 Ga, the Mg number, Sr, Ni, and Cr contents of TTGs increased, reflecting the cooling of the mantle and increased interactions of TTG magmas (generated in descending slabs) with the mantle wedge.

To explore this idea further, both the mean value and the maximum values of TTGs from the database (Appendix A) for Mg number, Ni, Sr, and $(La/Yb)_n$ are plotted as a function of age from 4 Ga to the present (Figs. 8–11). Unlike the conclusions of Martin and Moyen (2002), the Mg number shows no evidence of a trend with time either between 4 and 2.5 Ga or afterwards (Fig. 8; Appendix A). This observation is valid for both the mean and maximum Mg number. The TTG Ni distribution with age suggests a maximum around 2.7 Ga with decreasing Ni values from 2.7 Ga to the present (Fig. 9; Appendix A). As pointed out by Martin and Moyen (2002), there is also a suggestion of increasing Ni between 4 and 2.7 Ga. However, this change may reflect a rather sudden increase in Ni around 2.7 Ga rather than a gradual increase from 4 to 2.7 Ga.

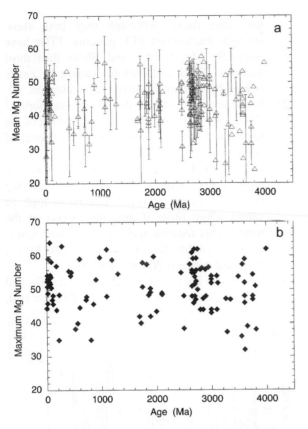

Fig. 8. Mg number in TTGs vs. age. (a) Mean values with one standard deviation, and (b) maximum values. Data sources given in Table 1 and Appendix A.

Although not shown, the same patterns are evident for Cr. Sr shows an irregular pattern with time (Fig. 10; Appendix A). In both the mean and maximum value plots, there is a suggestion of a maximum around 2.7 Ga and perhaps again at about 1.9 Ga. Although Sr may increase from 4 to 2.7 Ga, as pointed out by Martin and Moyen, the data in Fig. 10 are also consistent with little change between 4 and 3 Ga and a relatively sudden increase around 2.7 Ga. These results also suggest that Sr in TTGs decreases after 1.9 Ga, reaching a minimum around 400–800 Ma and then shows an upward swing in the last 100 Ma. The time distribution of $(La/Yb)_n$ (Fig. 11) and Sr/Y (Table 1) shows patterns much like those of Sr and Ni, with possible maxima around 2.7 and 1.9 Ga. It is also notable that the K_2O/Na_2O ratio remains approximately constant in TTGs with time (at about 0.5) until the Phanerozoic when it increases (Table 1).

If these secular trends are real, they may be important in tracking the cooling history of the Earth and in tracking changing tectonic regimes with time. The apparent maxima in Sr, Ni, and $(La/Yb)_n$ at 2.7 Ga and possibly at 1.9 Ga may record catastrophic mantle overturn events, as suggested by Condie (1998). Increased mantle plume activity at these times could lead to increased production rates of oceanic plateaus with thick mafic roots. These mafic roots may have been buried deep enough so that plagioclase was not stable and garnet was present throughout. Partial melting of such sources would produce TTG magmas with higher than normal La/Yb and Sr/Y ratios due to widespread restitic garnet. Although smaller degrees of melting could also yield higher La/Yb ratios and lower Yb contents, this is unlikely in the Archean when the mantle was hotter by a factor of 3 or 4. It is also possible that TTG magmas may have been contaminated with komatiites or ultramafic

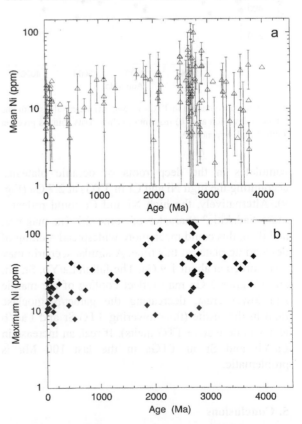

Fig. 9. Ni concentration in TTGs vs. age. (a) Mean values with one standard deviation and (b) maximum values. Data sources given in Table 1 and Appendix A.

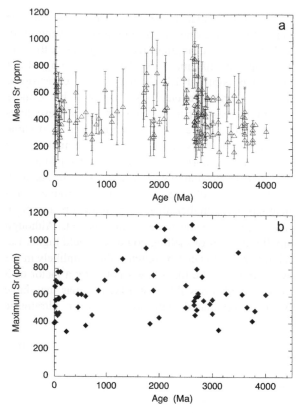

Fig. 10. Sr concentration in TTGs vs. age. (a) Mean values with one standard deviation and (b) maximum values. Data sources given in Table 1 and Appendix A.

cumulates in the deep roots of oceanic plateaus, accounting the high Ni and Cr in TTGs at 2.7 Ga (Fig. 9). Alternatively, the high Ni and Cr could reflect a more "adakitic" character of some 2.7 Ga magmas, and if so, this could mean more widespread melting of descending plates at this time. A similar scenario may be recorded at about 1.9 Ga. The fall in La/Yb, Sr, Ni, and Cr after 2 Ga may reflect cooling of the mantle and lower crust, decreasing the garnet/plagioclase ratio in the restite (thus lowering TTG Sr and La/Yb levels in derivative TTG melts). If real, an increase in La/Yb and Sr in TTGs in the last 100 Ma is problematic.

5. Conclusions

(1) For the most part, high-Al TTG (tonalite–trondhjemite–granodiorite) and adakite are not

the same rocks. Although they both show strongly fractionated REE patterns, TTGs have lower Sr, Mg, Ni, Cr, and Nb/Ta than most adakites. TTGs and adakites cannot be easily related by shallow fractional crystallization.

(2) While adakites are probably slab melts, high-Al TTGs may be produced by partial melting of the lower crust in arc systems or in the root zones of oceanic plateaus.

(3) Depletion in heavy REE and low Nb/Ta ratios in high-Al TTGs requires both garnet and low-Mg amphibole in the restite. Moderate to high Sr values allow little, if any, plagioclase in the restite. This requires melting in the hornblende eclogite stability field between 40- and 80-km deep and between 700 and 800 °C.

(4) If peaks in La/Yb, Sr, Cr, and Ni at 2.7 Ga and perhaps also at 1.9 Ga in high-Al TTG are real,

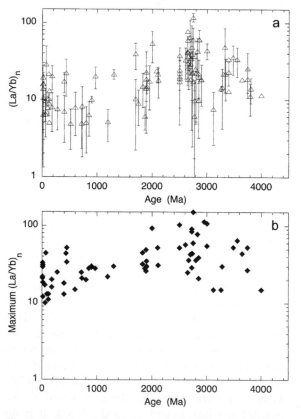

Fig. 11. $(La/Yb)_n$ ratio in TTGs vs. age. (a) Mean values with one standard deviation and (b) maximum values. Data sources given in Table 1 and Appendix A. n—normalized to primitive mantle using values from Sun and McDonough (1989).

they may reflect catastrophic mantle overturn events at these times.

Acknowledgments

This paper was substantially improved by in-depth reviews by Stephen Foley, Herve Martin, John Tarney, and Hugh Rollinson. It should be pointed out, however, that some of the reviewers do not agree with the geochemical distinctions between TTG and adakite as proposed in the paper. I am happy to dedicate the paper to Professor Ilmari Haapala on the occasion of his retirement from the University of Helsinki.

Appendix A. Supplementary data

Supplementary data associated with this article can be found, in the online version, at doi:10.1016/j.lithos.2003.11.001.

References

Barker, F., 1979. Trondhjemites, Dacites and Related Rocks. Elsevier, New York.

Black, L.P., Sheraton, J.W., James, P.R., 1986. Late Archean granites of the Napier complex, Enderby Land, Antarctica: a comparison of Rb–Sr, Sm–Nd and U–Pb systematics in a complex terrain. Precambrian Research 32, 343–368.

Bourdon, E., Eissen, J.P., Monzier, M., Robin, C., Martin, H., Cotton, J., Hall, M.L., 2002. Adakite-like lavas from Antisana volcano: evidence for slab melt metasomatism beneath the Andean northern volcanic zone. Journal of Petrology 43, 199–217.

Condie, K.C., 1998. Episodic continental growth and supercontinents: a mantle avalanche connection? Earth and Planetary Science Letters 163, 97–108.

Condie, K.C., Allen, P., 1984. Origin of charnockites from southern India. In: Kröner, A., Hanson, G.N., Goodwin, A.M. (Eds.), Archean Geochemistry. Springer-Verlag, New York, pp. 182–203.

Defant, M.J., Drummond, M.S., 1990. Derivation of some modern arc magmas by melting of young subducted lithosphere. Nature 347, 662–665.

Defant, M.J., Drummond, M.S., 1993. Mount St. Helens: potential example of the partial melting of the subducted lithosphere in a volcanic arc. Geology 21, 547–550.

Drummond, M.S., Defant, M.J., 1990. A model for trondhjemite–tonalite–dacite genesis and crustal growth via slab melting: Archean to modern comparisons. Journal of Geophysical Research 95, 21503–21521.

Drummond, M.S., Defant, M.J., Kepezhinskas, P.K., 1996. Petrogenesis of slab-derived trondhjemites–tonalite–dacite/adakite magmas. Transactions of the Royal Society of Edinburgh. Earth Sciences 87, 205–215.

Ellam, R.M., Hawkesworth, C.J., 1988. Is average continental crust generated at subduction zones? Geology 16, 314–317.

Foley, S., Tiepolo, M., Vannucci, R., 2002. Growth of early continental crust controlled by melting of amphibolite in subduction zones. Nature 417, 837–840.

Foley, S., Buhre, S., Jacob, D.E., 2003. Evolution of the Archean crust by delamination and shallow subduction. Nature 421, 249–252.

Gutscher, M.A., Maury, R., Eissen, J.P., Bourdon, E., 2000. Can slab melting be caused by flat subduction? Geology 28, 535–538.

Haskin, L.A., Haskin, M.A., Frey, F.A., Wildeman, T.R., 1968. Relative and absolute terrestrial abundances of the REE. Origin and Distribution of the Elements. Pergamon Press, New York, pp. 889–912.

Kamber, B.S., Ewart, A., Collerson, K.D., Bruce, M.C., McDonald, G.D., 2002. Fluid-mobile trace element constraints on the role of slab melting and implications for Archean crustal growth models. Contributions to Mineralogy and Petrology 144, 38–56.

Kay, R.W., Kay, S.M., 2002. Andean adakites: three ways to make them. Acta Petrologica Sinica 18, 303–311.

Kay, S.M., Ramos, V.A., Marquez, M., 1993. Evidence in Cerro Pampa volcanic rocks for slab-melting prior to ridge–trench collision in southern south America. Journal of Geology 101, 703–714.

Kay, S.M., Mpodozis, C., Coira, A.B., 1999. Neogene magmatism, tectonism, and mineral deposits of the central Andes. In: Skinner, B.J. (Ed.), Geology and Ore Deposits of the Central Andes. Society of Economic Geology, Special Publication, vol. 7, pp. 27–59.

Kepezhinskas, P., McDermott, F., Defant, M.J., Hochstaedter, A., Drummond, M.S., Hawkesworth, C., Koloskov, A., Maury, R.C., Bellon, H., 1997. Trace element and Sr–Nd–Pb isotopic constraints on a three-component model of Kamchatka arc petrogenesis. Geochimica et Cosmochimica Acta 61, 577–600.

Klemme, S., Blundy, J.D., Wood, B.J., 2002. Experimental constraints on major and trace element partitioning during partial melting of eclogite. Geochimica et Cosmochimica Acta 66, 3109–3123.

Li, W-X., Li, X-H., 2003. Adakitic granites within the NE Jiangxi ophiolites, south China: geochemical and Nd isotopic evidence. Precambrian Research 122, 29–44.

Martin, H., 1993. The mechanisms of petrogenesis of the Archean continental crust—comparison with modern processes. Lithos 30, 373–388.

Martin, H., 1999. Adakitic magmas: modern analogues of Archean granitoids. Lithos 46, 411–429.

Martin, H., Moyen, J-F., 2002. Secular changes in tonalite–trondhjemite–granodiorite composition as markers of the progressive cooling of Earth. Geology 30, 319–322.

Myers, J.D., Marsh, B.D., Sinha, K., 1985. Srontium isotopic and selected trace element variations between two aleutian volcanic centers: implications for the development of arc volcanic plumbing systems. Contributions to Mineralogy and Petrology 91, 221–234.

Peacock, S.M., Rushmer, T., Thompson, A.B., 1994. Partial melting of subducting oceanic crust. Earth and Planetary Science Letters 121, 227–244.

Percival, J.A., Stern, R.A., Rayner, N., 2003. Archean adakites from the Ashuanipi complex, eastern superior province, Canada: geochemistry, geochronology and tectonic significance. Contributions to Mineralogy and Petrology 145, 265–280.

Petford, N., Atherton, M., 1996. Na-rich partial melts from newly underplated basaltic crust: the Cordillera Blanca batholith, Peru. Journal of Petrology 37, 1491–1521.

Polat, A., Kerrich, R., 2002. Nd-isotope systematics of 2.7 Ga adakites, magnesian andesites, and arc basalts, superior province: evidence for shallow crustal recycling at Archean subduction zones. Earth and Planetary Science Letters 202, 345–360.

Prouteau, G., Scaillet, B., Pichavant, M., Maury, R.C., 2001. Evidence for mantle metasomatism by hydrous silicic melts derived from subducted oceanic crust. Nature 410, 197–200.

Rapp, R.P., Watson, E.B., Miller, C.F., 1991. Partial melting of amphibolite/eclogite and the origin of Archean trondhjemites and tonalites. Precambrian Research 51, 1–25.

Rapp, R.P., Shimizu, N., Norman, M.D., Applegate, G.S., 1999. Reaction between slab-derived melts and peridotite in the mantle wedge: experimental constraints at 3.8 GPa. Chemical Geology 160, 335–356.

Samaniego, P., Martin, H., Robin, C., Monzier, M., 2002. Transition from calc-alkalic to adakitic magmatism at Cayambe volcano, Ecuador: insights into slab melts and mantle wedge interactions. Geology 30, 967–970.

Sheraton, J.W., Tingey, R.J., Black, L.P., Offe, L.A., Ellis, D.J., 1987. Geology of an unusual Precambrian high-grade metamorphic terrane-Enderby Land and western Kemp Land, Antarctica. Australian Bureau of Mineral Resources Geology and Geophysics Bulletin, p. 223.

Shirey, S.B., Hanson, G.H., 1984. Mantle-derived Archean monzodiorites and trachyandesites. Nature 310, 222–224.

Smithies, R.H., 2000. The Archean tonalite–tondhjemite–granodiorite (TTG) series is not an analogue of cenozoic adakite. Earth and Planetary Science Letters 182, 115–125.

Smithies, R.H., Champion, D.C., 2000. The Archean high-Mg diorite suite: links to tonalite–trondhjemite–granodiorite mag-

matism and implications for early Archean crustal growth. Journal of Petrology 41, 1653–1671.

Stern, R.A., Hanson, G.N., 1991. Archean high-Mg granodiorites: a derivative of light REE enriched monzodiorite of mantle origin. Journal of Petrology 32, 201–238.

Stern, C.R., Kilian, R., 1996. Role of the subducted slab, mantle wedge and continental crust in the generation of adakites from the Andean Austral volcanic zone. Contributions to Mineralogy and Petrology 123, 263–281.

Sun, S.S., McDonough, W.F., 1989. Chemical and isotopic systematics of oceanic basalts: implications for mantle composition and processes. In: Saunders, A.S., Norry, M.J. (Eds.), Magmatism in Ocean Basins. Geological Society of London, Special Publication, vol. 42, pp. 313–345.

Tarney, J., Jones, C.E., 1994. Trace element geochemistry of orogenic igneous rocks and crustal growth models. Journal of the Geological Society 151, 855–868.

Whalen, J.B., Percival, J.A., McNicoll, V.J., Longstaffe, F.J., 2002. A mainly crustal origin for tonalitic granitoid rocks, superior province, Canada: implications for Late Archean tectonomagmatic processes. Journal of Petrology 43, 1551–1570.

White, R.V., Tarney, J., Kerr, A.C., Saunders, A.D., Kempton, P.D., Pringle, M.S., Klaver, G.T., 1999. Modification of an oceanic plateau, Aruba, Dutch Caribbean: implications for the generation of continental crust. Lithos 46, 43–68.

Winther, K.T., 1996. An experimentally based model for the origin of tonalitic and trondhjemitic melts. Chemical Geology 127, 43–59.

Wylie, P.J., Wolf, M.B., van der Laan, S.R., 1997. Conditions for formation of tonalites and trondhjemites: magmatic sources and products. In: De Wit, M., Ashwal, L.D. (Eds.), Greenstone Belts. Oxford University Press, Oxford, pp. 256–266.

Yogodzinski, G.M., Volynets, R.W., Koloskov, O.N., Kay, A.V., Kay, S.M., 1995. Magnesian andesite in the western Aleutian Komandorsky region: implications for slab melting and processes in the mantle wedge. Geological Society of America Bulletin 107, 505–519.

Xu, J.F., Shinjo, R., Defant, M.J., Wang, Q., Rapp, R.P., 2002. Origin of Mesozoic adakitic intrusive rocks in the Ningzhen area of east China: partial melting of delaminated lower continental crust. Geology 30, 1111–1114.

Available online at www.sciencedirect.com

SCIENCE DIRECT®

Lithos 80 (2005) 45–60

ELSEVIER

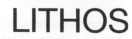

LITHOS

www.elsevier.com/locate/lithos

Ilmenite, magnetite, and peraluminous Mesoproterozoic anorogenic granites of Laurentia and Baltica

J. Lawford Anderson*, Jean Morrison

Department of Earth Sciences, University of Southern California, Los Angeles, CA 90089-0040, USA

Received 25 June 2003; accepted 9 September 2004
Available online 8 December 2004

Abstract

Emplacement of 1.6 to 1.3 Ga Mesoproterozoic plutons in Baltica and Laurentia formed an immense belt of A-type granite batholiths that include (1) low-fO_2, ilmenite-series granite intrusions from the Baltic region to Wyoming, (2) high-fO_2, magnetite-series granite intrusions of the central to southwestern U.S., and (3) peraluminous, two-mica granite intrusions from Colorado to central Arizona. These mineralogic divisions are mirrored by substantial elemental and oxygen isotopic differences. The ilmenite-series granites, which often contain classic rapakivi textures, have the highest Fe/Mg ratios and are highest in LIL element enrichment. They also have the lowest whole-rock $\delta^{18}O$ values at 5.7‰ to 7.7‰. The magnetite-series granites are less potassic, less LILE-enriched, and have higher whole-rock $\delta^{18}O$ values, ranging from 7.6‰ to 10.8‰. Although they retain A-type characteristics, the peraluminous granites are the least LILE-enriched and have the lowest Fe/Mg ratios. They also have the highest whole-rock $\delta^{18}O$ values ranging from 8.8‰ to 12.0‰. Feldspar, where strongly reddened, can exhibit elevated $\delta^{18}O$ values, which is interpreted to indicate subsolidus exchange with surface-derived aqueous fluids. Quartz $\delta^{18}O$ values are interpreted to generally retain their magmatic values. The transcontinental mineralogic, chemical, and oxygen isotopic variations are interpreted as indicative of broad changes in the composition of a lower crustal source, which is compatible with a reduced mantle-derived crustal source for the ilmenite-series granites and a more oxidized crustal source for the others, including a metasedimentary component in the source for the two-mica granite subprovince.

Widespread thermal metamorphism at ~1.4 Ga is present throughout much of the magmatic province and is viewed as a consequence of this immense event. Compressional deformation associated with several western 1.4 Ga Laurentia granite batholiths, alternatively interpreted as the distal expressions of a presumed 1.4 Ga orogeny, have at least in part been shown to be localized on preexisting Paleoproterozoic zones of deformation. Thus, we do not find compelling evidence for a ~1.4 Ga orogeny related to the formation of most of these granites. Renewed intrusions at ~1.0–1.1 Ga between and immediately following phases of the Grenville orogeny indicate that situations leading to their formation need to be more broadly considered.

The origin of this red granite-forming event in Laurentia and Baltica is considered as part of a global magmatic event that was coeval with intrusion of massif anorthosites and associated charnockites. Most are viewed as anorogenic, but it is

---

* Corresponding author. Tel.: +1 213 740 6727; fax: +1 213 740 8801.
E-mail address: anderson@usc.edu (J.L. Anderson).

0024-4937/$ - see front matter © 2004 Elsevier B.V. All rights reserved.
doi:10.1016/j.lithos.2004.05.008

recognized that the same conditions leading to their formation may have occurred during extensional phases of orogens. The immense volumes of red granites produced are also essentially unique to the Mesoproterozoic and appear to be tied to the stabilization and eventual break up of supercontinents of both Paleoproterozoic and Mesoproterozoic age.

Keywords: Mesoproterozoic; Granite; Anorogenic; Oxygen fugacity; Oxygen isotope

1. Introduction

The interval of 1.6–1.3 Ga follows an era of orogenic growth throughout much of Laurentia and Baltica (terms for Proterozoic continental assemblies as reviewed by Torsvik, 2003) and is characterized by a distinct range of Mesoproterozoic igneous activity including anorthosite massifs, mangerite and charnockite intrusions, and batholiths of A-type, red granite of rapakivi affinity. The origin of this magmatism (termed AMCG for anorthosite, mangerite, charnockite, and granite) that swept across Laurentia/Baltica remains one of the more enigmatic igneous episodes of the Mesoproterozoic Era. Moreover, similar Mesoproterozoic igneous activity has been documented on nearly every other continent and appears to occur between periods of orogenic growth. The amount of crustal readjustment is profound, and in many areas, up to 40% of exposed Proterozoic crust consists of these intrusions.

Recently, some workers have considered the 1.6–1.3 Ga magmatism to represent an in-board manifestation of continued orogenesis along the margin of Laurentia (Kirby et al., 1995; Nyman and Karlstrom, 1997; Rivers, 1997; Karlstrom et al., 2001; Gower and Krogh, 2002; Menuge et al., 2002). However, others have maintained the view that the intrusions are anorogenic, related to either extension or mantle upwelling, and further suggest that the granite batholiths were derived from low degrees of partial melting of a crustal source dominantly of Paleoproterozoic or older age (Anderson and Bender, 1989; Emslie, 1991; Rämö, 1991; Anderson and Morrison, 1992; Creaser et al., 1991; Bickford and Anderson, 1993; Frost and Frost, 1997; Amelin et al., 1997; Rämö et al., 2003). Anderson (1987), Anderson and Bender (1989), and Hoffman (1989) have called upon a model involving mantle diapirism, coining the phrases "mantle and crustal overturn" and "mantle superswells", respectively, leading to a succession of crustal melting events within fertile, yet relatively dry, crust principally formed during preceding Proterozoic orogenies.

This paper has two goals: (1) to offer a trans-supercontinental threefold mineralogical and chemical characterization of these granites supported by new oxygen isotope data and (2) to reexamine the orogenic and plate tectonic setting leading to their formation. In short, we view most of these granites, with some exceptions, as anorogenic and uniquely part of the well-known Mesoproterozoic "anorthosite event", during which massif anorthosites were emplaced together with mangerites, charnockites, and granites. It is further suggested that these voluminous granites formed in response to stabilization and eventual breakup of a Proterozoic supercontinent, consistent with conclusions offered by Windley (1993) and Rogers and Santosh (2002). Similar models have been offered for post-Grenville magmatism suites by Mukherjee and Das (2002) and post-Pan African anorogenic suites by Doblas et al. (2002).

2. Compositional variation among Mesoproterozoic A-type Laurentian granites

Mesoproterozoic Laurentian and Baltica granites characteristically have A-type chemistry due to their high concentrations of alkali and other lithophile and high field strength elements, F, and high Fe/Mg ratios. Alkali–lime indices typically fall in the alkali–calcic or alkalic field due to low CaO and high K_2O. Building on the findings of Anderson and Morrison (1992, 1998), we have found that the principle differences among the Laurentia/Baltica granites are the degree of alumina saturation and the Fe–Ti oxide mineralogy, leading to identification of three trans-continental petrographic provinces, as depicted in Fig. 1. Most are metaluminous, contain biotite+titanite±hornblende, and can be divided into either the

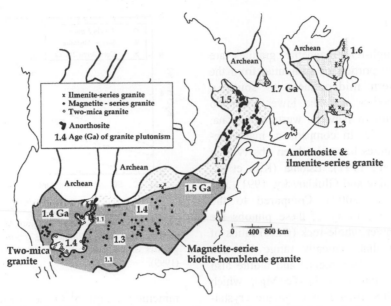

Fig. 1. Regional distribution of mid-Proterozoic granite and anorthosite complexes. Modified from Anderson and Morrison (1998).

ilmenite- or magnetite-series, which form two extensive and separate regions of the former supercontinents. A third subprovince in a major western Laurentian region is composed of peraluminous granites that contain biotite or two micas±other aluminous phases. Although all of these granites are uniquely enriched in megacrystic or phenocystic K-feldspar and many have rapakivi textures, only the granites of the ilmenite-series subprovince include the classic wiborgitic or pyterlitic variety of rapakivi granite. Currently, the ilmenite-series granites are also the only ones known to be exposed with near coeval anorthosite or charnockite.

2.1. Ilmenite granites

Low-fO_2, ilmenite granites that occupy much of northeastern Laurentia and most of Baltica (Haapala and Rämö, 1990; Fig. 1), including most of the classic rapakivi massifs of Finland, large biotite–hornblende rapakivi granite intrusions in Labrador, the Wolf River batholith of Wisconsin, and the Sherman granite of Wyoming. Rapakivi granites of South Africa (Kerr and Thomas, 1991) and Brazil (Dall'Agnol et al., 1991, 1999) are examples of other ilmenite Mesoproterozoic granites outside of Laurentia/Baltica. The ilmenite granites are uniformly metaluminous and contain iron-rich hornblende and biotite as mafic

phases as a result of crystallization under reducing conditions. Ilmenite is the sole or dominant Fe–Ti oxide, and typically hastingsitic hornblende and annitic biotite have atomic ratios of Fe/(Fe+Mg) ranging form 0.80 to 0.99. The granites are potassic and have the highest observed weight-based Fe/Mg ratios (Fig. 2). Frost and Frost (1997) have presented a model for the origin of the low-fO_2, ilmenite-series rapakivi granites though partial melting of a tholeiitic, mafic lower crustal source derived by fractionation from anorthositic magmas.

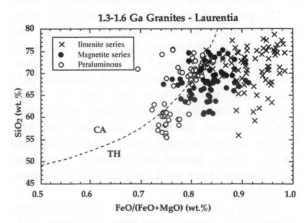

Fig. 2. SiO_2 versus $FeO^*/(FeO^*+MgO)$ diagram for Mesoproterozoic granites of North America. Modified from Anderson and Cullers (1999).

2.2. Magnetite granites

Metaluminous high-fO$_2$ magnetite granites are exposed in a sizable portion of Laurentia, from the central and southern midcontinent to southern Colorado, New Mexico, and the lower Colorado River region of southern Nevada, western Arizona, and southern California. In comparison, magnetite-bearing rapakivi granites have also been observed in China (Jianhua et al., 1991), Estonia (Kirs et al., 1991), Siberia (Moralev and Glukhovsky, 1991), and Brazil (Rämö et al., 2002). Compared to the ilmenite-series granites, many of these plutons are less potassic with lower whole-rock (wt.%) ratios of FeO*/(FeO*+MgO) that typically range between 0.80 and 0.88 (Fig. 2). Hornblende and biotite also have lower atomic ratios of Fe/(Fe+Mg), which range from 0.3 to 0.8, owing to magmatic crystallization under more oxidizing conditions (Anderson and Bender, 1989).

2.3. Peraluminous granites

Peraluminous Mesoproterozoic granites form a distinct subprovince in Laurentia from central Colorado to New Mexico and central Arizona contain biotite or two micas. They often lack titanite and have monazite as an accessory phase. Magnetite is the principle Fe–Ti oxide, and ilmenite is less abundant or absent. Anderson and Cullers (1999) noted two petrographic suites in this subprovince. One was termed Silver Plume-type, after the Silver Plume granite (Colorado), and is relatively anhydrous based on the occurrence of magmatic sillimanite, late crystallization of muscovite, biotite and fluorite, and calculations of crystallization conditions including low fH$_2$O (Anderson and Thomas, 1985). The second, termed Oracle-type after the Oracle granite (Arizona), were emplaced under more hydrous solidus conditions and commonly lack fluorite as an accessory phase (Anderson and Bender, 1989). Several are strikingly pegmatitic. These granites have the lowest level of iron enrichment (Fig. 2).

Although the Silver Plume- and Oracle-type peraluminous granite suites have similarities to S-type and similar granites of orogenic belts (Thompson and Barnes, 1999), including their aluminous

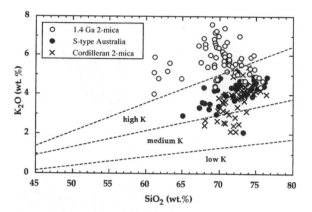

Fig. 3. K$_2$O versus SiO$_2$ diagram showing a comparison of peraluminous Silver Plume-type granites of Colorado to S-type granites of orogenic belts. Modified from Anderson and Cullers (1999).

mineralogy, high δ^{18}O (see below), and low Fe/Mg ratios, they are also much different. As depicted in Fig. 3, these granites have much higher abundances of K, Rb, Nb, and Y, typical of "within-plate" granites (Anderson and Cullers, 1999) and also La, Th, and Zr. Fundamentally, these chemical attributes reinforce the interpretation that they are simply a subset of A-type granites, like the other 1.4 Ga granites of this region, and Laurentia/Baltica in general.

3. Oxygen isotopic compositions

Oxygen isotope systematics have been studied in several Mesoproterozoic intrusions (Wenner and Taylor, 1976; Shieh, 1983; Shieh et al., 1976; Heaman et al., 1982; Kim, 1989; Wu and Kerrich, 1986; Emslie and Taylor, 1993). Anderson and Morrison (1992) also reported whole-rock δ^{18}O values for a range of Laurentia Mesoproterozoic granites. Additional oxygen isotope analyses of quartz and feldspar were conducted at the University of Southern California, using a CO$_2$ laser extraction system and represent the first attempt to look at the oxygen isotope compositions of coexisting minerals in these granites at the trans-supercontinent scale. Oxygen was extracted from handpicked quartz and feldspar separates, using a CO$_2$ laser probe system (Sharp, 1990). Mineral separates weighing ~0.2–10 mg were loaded into individual wells in a solid Ni sample holder and were heated in the presence of

Table 1
Compilation of oxygen isotope data for Mesoproterozoic Laurentian granites

Pluton	Sample	SiO$_2$ (wt.%) whole-rock	δ^{18}O whole-rock	δ^{18}O quartz	δ^{18}O feldspar	Δquartz–feldspar	Comments
A. Ilmenite series granites, Wolf River batholith, WI[a]							
Wolf River	DR8	72.32	6.8	7.4	6.6	0.8	
granite	GR36A	73.36	7.1				
	8L	75.18	6.9				
	VSGR6	73.23	6.9				
	DR15	69.94	7.2				
	GR-6	66.92	7.2	8.1	7.4	0.7	
	GR24A	61.63	7.2	8.0			
	GR34	68.48	7.0	7.2	6.0	1.2	
	HT30	71.67	7.0	7.5	6.7	0.8	
Wiborgite	GR36B	73.54	7.0	7.6	6.2	1.4	
porphyry	XS1	72.57	7.1				
Red River	GR3	72.58		8.8	7.1(g), 8.6(p), 9.4(rd)	1.7 (g), 0.2 (p), −1.6 (rd)	
granite	TG11b	72.28	6.6	7.2	6.0	1.2	
	SP24	73.07	7.6				
	XW1	72.76	6.8				
Waupaca	TG38	69.29	7.0	6.7	8.4(p), 9.4(rd)	−1.7 (p), −2.7 (rd)	
granite	WP5A	65.40	5.7	6.4	6.5(g), 6.7 (p), 8.1(rd)	−0.1 (g), −0.3 (p), −1.7 (rd)	
	XW4B	68.39	6.4	7.0	5.6(g), 5.8 (p)	1.4 (g), 1.2 (p)	
Belongia fine	12ATC	77.36	6.5	7.9	6.5	1.4	
granite	73M	76.59					
Belongia coarse	58M	75.53		8.9	7.4	1.5	
granite	83AT	75.60	7.1	7.4			
Hager granite	50M	77.32	7.7				
	72M	70.59	6.9				
Peshtigo	92ATC	62.05	6.8				
mangerite	GR17B	63.44	5.8				
	GR26A	59.20	6.3				
	TM3	55.86	7.5				
B. Magnetite-series granites[b]							
Gold Butte	GB3	66.06	9.1				
granite, NV	GB6	66.76	9.2				
Newberry	NY6a	65.20	10.8				
granite, CA	NY6b	64.03	8.9				Chloritized
Beer Bottle	LG83-1A	65.99	8.1				
granite, NV	LG83-10	67.77	9.6				
San Isabel	TG14	56.05	8.1	8.71, 8.90	8.02, 7.79	0.9	
granite, CO	TG91	64.50	9.0	9.81, 10.08	8.14, 8.23	1.8	
Bowmans	JP44	60.85	8.2				
Wash, CA	JP-410	61.32	6.5				Chloritized
Parker Dam, CA	JP-280	66.33	10.0				
	JP-281	67.42	8.4				
	JP-282	69.41	9.1				
	JP-412	69.88	8.9				
Holy Moses	HP1A2	67.05	7.6	8.36, 8.58	6.83, 6.84	1.6	
granite	HPIB2	68.28	8.4	8.63			
Hualapai granite	HP-7	68.44	6.1				Chloritized
	HP-3	66.67	9.1				
	HP-2	71.00	9.1				

(continued on next page)

Table 1 (*continued*)

Pluton	Sample	SiO$_2$ (wt.%) whole-rock	$\delta^{18}O$ whole-rock	$\delta^{18}O$ quartz	$\delta^{18}O$ feldspar	Δquartz–feldspar	Comments
C. Peraluminous biotite–sphene granites[b]							
Marble Mountain	MB2A	67.84	10.4	10.04	7.90	2.1	
granite, CA	MB1B	68.35	8.8				
	MB5B	69.50	9.7	10.03	8.58, 8.49	1.5	
Fort Huachuca	AHU83-1	67.55	11.1	10.52, 10.64	9.32, 8.01, 8.96	1.8	
granite, AZ	AHU83-2	63.52	12.0	10.70, 10.81	8.73, 8.24, 8.40, 8.98, 9.20	2.0	
Davis Dam	NY-11A	70.36	10.1				
granite, CA	NY-13A	66.49	11.4				
D. Peraluminous two-mica granites[b]							
Ruin granite, AZ	ARU83-1b	71.27	10.3				
	ARU83-1a	70.27	10.3				
Sierra Estrella granite, AZ	ASE83-1	70.15	10.2	10.71, 10.73			
Ak–Chin granite, AZ	AMC83-1	73.43	11.2	11.50, 11.66	7.53	4.1	
Oracle granite, AZ	AOR83-2	67.30	10.5	11.81, 11.80	9.14, 9.17	2.7	
	AOR83-1	67.89	11.0	12.16, 12.12	7.26	4.9	
St. Vrain	SVL-1	72.18	11.6	11.71, 12.17	11.35	0.6	
granite, CO	SVL-3	71.24	10.8				
	SVL-6	69.49	10.6				
Silver Plume	SP-9	72.49	6.8				Chloritized
granite, CO	SP-1	70.55	10.4				
	SP-3	67.47	11.0				
Oak Creek granite, CO	js77	60.70	10.0	11.28	8.73, 8.87	2.5	

[a] Wolf River batholith oxygen isotope data from the PhD dissertation of Kim (1989) on samples collected by Anderson. Feldspar color noted as follows: g—grey; p—pink; rd—red. Elemental chemistry of these samples can be found in Anderson and Cullers (1978), Anderson and Thomas (1985), and Anderson and Bender (1989).

[b] Whole-rock oxygen isotope data from Anderson and Morrison (1992). Quartz and feldspar (this study) analyses from single or multiple separates using a laser probe system.

BrF5 by a 20W CO$_2$ laser beam. The O$_2$ gas was converted to CO$_2$ by reaction with hot graphite, then analyzed in a VG PRISM gas ratio mass spectrometer. Accuracy and precision were monitored by repeated analyses of Gore Mountain garnet (Valley et al., 1996). Analysis of 26 aliquots yielded an average $\delta^{18}O$ of 5.81‰±0.20‰. The data, along with whole-rock data from Anderson and Morrison (1992) and additional data from Kim (1989), are reported in Table 1. In most cases, more than one aliquot of quartz and feldspar were analyzed from each sample. Thus, multiple $\delta^{18}O$ values reported in Table 1 provide an indication of the within sample variability of quartz and feldspar $\delta^{18}O$. Four samples contained petrographic evidence of extensive hydrothermal alteration, and those are referred as "chloritized" in Table 1.

3.1. Whole-rock data

The whole-rock data reveal clear distinctions among the three mineralogic divisions. The highest whole-rock $\delta^{18}O$ values (8.8–12.0‰) are from the peraluminous province, intermediate values (7.6–10.8‰) are from the magnetite-series subprovince of Laurentia, and the lowest values (5.7–7.7‰) are from the ilmenite-series province, specifically, the granites of the Wolf River batholith of Wisconsin. The whole-rock $\delta^{18}O$ values versus wt.% SiO$_2$ are shown in Fig. 4. The latter are similar in range to those reported for elsewhere in northeast Laurentia and Baltica, including the Nain province plutons of central Labrador ($\delta^{18}O$=6.9 to 8.4; Emslie and Taylor, 1993), granite and charnockite of the Roga-land complex, Norway ($\delta^{18}O$=6.1 to 7.4; Demaiffe

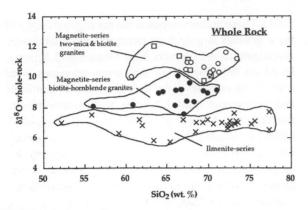

Fig. 4. Whole-rock $\delta^{18}O$ versus SiO_2 diagram for Mesoproterozoic granites. Data for two-mica, and biotite granites are from Colorado and central Arizona (Anderson and Morrison, 1992), data for biotite-hornblende granites are from western Arizona, southern Nevada, and southern California (Anderson and Morrison, 1992), and data for ilmenite-series granites are from the Wolf River batholith of Wisconsin (Kim, 1989, from samples supplied by Anderson). The designation of ilmenite- and magnetite-series refers to the dominant Fe–Ti oxide.

and Javoy, 1980), and the Wiborg massif, Finland (Dempster et al., 1994).

Because many of these granites have epizonal features or are otherwise susceptible to postcrystallization hydrothermal alteration, it is crucial to assess whether postcrystallization processes have altered magmatic $\delta^{18}O$ values. For example, Wenner and Taylor (1976) demonstrated that regionally extensive hydrothermal alteration in the St. Francois Mountains of Missouri led to elevated $\delta^{18}O$ values. Thus, in the absence of $\delta^{18}O$ values for coexisting quartz, feldspar, and other minerals, caution must be exercised in interpreting whole-rock $\delta^{18}O$ values.

The granites in question are commonly red in color, owing to an abundance of alkali feldspar and the tendency of this phase to alter in the presence of low-temperature, oxidizing solutions. Kim (1989) completed a detailed study of feldspars in the Wolf River batholith and found an increase in feldspar $\delta^{18}O$ with reddening (Table 1). Kim further concluded that the reddening was due to exchange with sedimentary formation waters at temperatures between 260 and 350 °C. The same increase in $\delta^{18}O$ of reddened feldspar has been reported by Wenner and Taylor (1976) for granites of the St. Francois Mountains and by Dempster et al. (1994) for the Wiborg massif. These observations stand in contrast to the common

lowering of feldspar $\delta^{18}O$ values associated with alteration of many Phanerozoic granites, which results from either higher temperature alteration and/or preferential involvement of low $\delta^{18}O$ meteoric fluids (Criss and Taylor, 1983; Solomon and Taylor, 1991; Morrison, 1994; Wickham et al., 1996; Morrison and Anderson, 1998).

3.2. Quartz and feldspar data

We have analyzed 24 quartz and feldspar separates from 14 samples from these granites. The quartz $\delta^{18}O$ values are plotted versus wt.% SiO_2 in Fig. 5. Many of the $\Delta Qtz–Fsp$ values ($=\delta^{18}O_{Qtz} - \delta^{18}O_{Fsp}$) are between $+1‰$ and $+2‰$ (Table 1), which is appropriate for broadly preserved magmatic compositions. Notably, some granites, including the Oracle and Ak–Chin granites of Arizona and the Oak Creek batholith of Colorado, have $\Delta Qtz–Fsp$ values significantly in excess of $2‰$ and thus indicative of subsolidus exchange with hydrothermal fluids. Otherwise, the strong correlation between quartz $\delta^{18}O$ and the other above described chemical and mineralogical characteristics for granites from these three Laurentian/Baltica subprovinces is interpreted to indicate that quartz $\delta^{18}O$ values reflect magmatic compositions, including that of their source region.

The present data support the model of Frost and Frost (1997) of a tholeiitic, mantle-derived crustal source for the ilmenite-series granites, as such can be expected to derive a low-$\delta^{18}O$ and low-fO_2 granitic magma. The magnetite-series and the two-mica

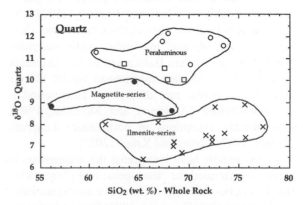

Fig. 5. $\delta^{18}O$ (in quartz) versus SiO_2 (in whole-rock) diagram for Mesoproterozoic peraluminous, magnetite-series, and ilmenite-series granites from Laurentia.

granites require a more oxidized crustal source with considerable sedimentary component for the latter. The data further indicate significant and fundamental regional changes in lower crustal compositions across Laurentia and Baltica at 1.6–1.3 by ago.

4. Anorogenic or orogenic?

As part of the well known Mesoproterozoic "anorthosite event" (Herz, 1969; Emslie, 1978; Anderson, 1983; Emslie, 1991), these A-type granite batholiths were emplaced worldwide and formed significant portions of all present-day continents, with the possible exception of Antarctica. Between 2.0 and 1.7 Ga, the Paleoproterozoic portions of the Laurentian/Baltica cratons, which included North America, Greenland, and the Fennoscandian regions of Sweden, Finland, and northwest Russia, underwent considerable crustal growth through punctuated orogenic episodes. Much of the new crust generated was juvenile mantle-derived material with minimal reworking of Archean cratons (DePaolo, 1981; Nelson and DePaolo, 1985; Bennett and DePaolo, 1987; Anderson and Cullers, 1987; Orrell et al., 1999; Patchett and Bridgwater, 1984; Patchett and Arndt, 1986; Patchett and Ruiz, 1987; Condie, 1990; Rämö, 1991; Rämö and Haapala, 1995). Following this period of orogenic growth, the time period of 1.6–1.3 Ga age is characterized by a very different spectrum of igneous activity including the remarkably voluminous emplacement of anorthosite massifs, charnockite and related rock units, and batholiths of potassic red granite of rapakivi affinity. Åhäll and Connelly (1998), for example, have utilized ages of these intrusions, which they term "interorganic", to correlate similar magmatic events between western Baltica and eastern Laurentia. In contrast, calcalkaline magmatism typical of orogenic belts became rare during this time period in Laurentia/Baltica (Anderson and Morrison, 1992), with notable exceptions in northeastern Laurentia (Rivers, 1997; Gower and Krogh, 2002).

Many authors have interpreted this magmatism as anorogenic (termed the "anorogenic trinity" by Anderson, 1983), but others, as noted above, have questioned this interpretation (Graubard, 1991; Duebendorfer and Christensen, 1995; Nyman et al., 1994; Kirby et al., 1995; Nyman and Karlstrom, 1997; Karlstrom et al.,

2001). These authors opposed the anorogenic origin of these granites largely from structural and cooling-age investigations in Colorado, New Mexico, and Arizona, based on the observation that some intrusions were coeval with compressional deformation and significant and widespread crustal heating and metamorphism. For example, Åhäll et al. (2000) and Karlstrom et al. (2001) concluded that this Mesoproterozoic granite-forming event was part of a long-lived convergent orogen on the margin of Laurentia, leading to the formation of the Grenvillian (~1 Ga) Rodinia supercontinent. In contrast, Selverstone et al. (2000), while finding that 1.4 Ga granites of Colorado were emplaced synchronously with development of contractional fabrics, determined that the contractional deformation is localized along preexisting shear zones that were simply reactivated during 1.4 Ga plutonism, a conclusion that mirrors that of Reed et al. (1993) and Tweto (1987). Selverstone et al. (2000) further found little evidence for 1.4 Ga penetrative deformation outside of the shear zones and considered the event a "far-field" effect of compressional stresses on the southern margin of Laurentia.

It is clear that some western Laurentian granitic batholiths of ~1.5–1.4 Ga age have foliations indicative of emplacement synchronous with regional tectonic deformation (Graubard, 1991; Kirby et al., 1995; Nyman and Karlstrom, 1997), and many Paleoproterozoic metamorphic rocks yield cooling ages close to 1.4 Ga (Shaw et al., 1999; Selverstone et al., 2000). However, most Laurentia/Baltica 1.5–1.4 Ga plutons crosscut their host rocks and are generally not metamorphosed and are only locally deformed. Given the vast amount of Mesoproterozoic granite that has passed into and through Laurentian crust, that considerable resetting of isotopic systems that occurred at 1.4 Ga should not be surprising. Metamorphic terranes lacking these plutons but with widespread Mesoproterozoic metamorphic ages (e.g., northern New Mexico; Pedrick et al., 1998) could simply be due to the fact that they represent crustal sections pervaded by these granite masses in their common passage to more shallow levels of emplacement. This was clearly an immense event and one not just on a transcontinental but rather on a nearly global scale.

Although several Mesoproterozoic Laurentian/Baltica plutons contain areas of foliation, and some of this foliation may be tectonic in origin, clear evidence of orogeny in the regions hosting these intrusions is lacking. Absent from these host regions are important orogenic attributes such as oceanic terrane accretion and production of juvenile calcalkaline magmas typical of either modern or ancient subduction zones. In regard to the latter, such rock types should be intermediate or dioritic to tonalitic in composition, and such rock types have long been recognized to be conspicuously absent among these specific Mesoproterozoic intrusions. This may be a problem, as an example, for the proposed 1.52–1.46 Ga Pinwarian orogeny of the eastern Grenville Province (Gower and Krogh, 2002), given that most of the dated plutons are K-rich granites or of the AMCG suite. Gower and Krogh (2002) also consider flat-slab subduction to explain Elsonian age (1.46–1.29 Ga) anorthosites and other AMCG intrusions in the Grenville Province, including that of the Michikamau, Harp Lake, Mistastin, and Nain, for which we find little support given the lack of contractional deformation and the abundance of mafic dike swarms, in addition to the absence of juvenile calcalkaline magmatism.

Nd–Sr isotopic studies indicate that most of the ~1.4 Ga granites have compositions consistent with derivation from 1.9 to 1.7 Ga Paleoproterozoic crust with a variable, unusually minor, Archean contribution (as reviewed by Anderson and Morrison, 1992; Anderson and Cullers, 1999). However, a sizeable region of the southern midcontinent have yielded Nd T_{DM} ages close to 1.4–1.5 Ga, implying derivation from young mantle-derived crust (Nelson and DePaolo, 1985; Van Schmus et al., 1996; Menuge et al., 2002). Menuge et al. (2002) further interpret the data as indicating that the source was of calcalkaline rocks and that the melting occurred in an extensional setting or back-arc, possibly related to the Pinwarian orogen. Such is certainly a possible scenario but not unique. The data do require a young crustal source, but it need not be subduction-related or calcalkaline in composition. In contrast, the inferred lower crustal source could as well have been derived from derivative magmas of the AMCG suite emplaced along or above the mantle–crust boundary. Such would be in agreement with the model of Frost and Frost (1997).

Regardless of presumed tectonic setting, it must be acknowledged that this Mesoproterozoic magmatism is an expression of a huge global event that, on a magma volume scale, has no clearly defined analog in younger geologic time. It was a profound red granite-forming event affecting Paleoproterozoic terranes worldwide. An emerging view of the Late Archean and Proterozoic is that orogenic growth was globally episodic and at times quite rapid. Originally proposed by Gastil (1960) and Sutton (1963), the idea continues to receive renewed attention (Condie, 1989; Hoffman, 1989; Anderson and Bender, 1989; Anderson and Morrison, 1992; Stein and Hofmann, 1994; Condie, 1998). Condie (1998) regards the times of 2.7, 1.9, and 1.2 Ga as coincident with orogenic "superevents", each of 50–80 my duration, which he suggested are tied to the assembly of supercontinents. Notably, the geologic record shows minimal evidence for preserved orogenies between these times, although there are exceptions. The 2.6–2.0 Ga era is better known for its large mafic intrusions and the initial significant appearance of iron formations and geoclinal sedimentation. Likewise, the global appearance of anorthosite–charnockite–rapakivi granite first appears subsequent to the 2.1–1.8 Ga orogenies, of which Zhao et al. (2002) presented a comprehensive summary. Their findings are critically important for they demonstrated the global nature of this orogenic "superevent" and the likewise near restriction of significant Paleoproterozoic orogenic deformation/plutonism to this time frame. It is remarkable that so few orogenies have been recognized after ~1.7 Ga and before the onset of Grenvillian orogenies beginning at 1.3 Ga, including that of the Elzevirian (McLelland et al., 1996). Exceptions are geographically limited but may include the 1.5–1.45 Ga Pinwaran orogeny of the eastern Grenville Province, Canada (Krogh et al., 1996; Rivers, 1997; Gower and Krogh, 2002), the 1.59–1.50 Isa orogeny of northeastern Australia (Spikings et al., 2002), the 1.57–1.54 Ga Kararan orogeny of southern Australia (Betts et al., 2003), and a 1.45 Ga orogeny recently recognized in the southwestern part of the Amazonian craton in Brazil (Geraldes et al., 1998).

5. Mesoproterozoic crustal formation and ties to supercontinents

Although this work is focused on the A-type granites of Laurentia and Baltica, comparable intrusions occurred globally during much of the Mesoproterozoic. Almost all are hosted in earlier formed Paleoproterozoic orogenic crustal sections. An exception is that noted by Rämö et al. (2002), who have recently documented 1.88 Ga A-type granites hosted by, and largely derived from, the Archean Amazonian craton in Brazil. The greatest volumes of these granites appear in the geologic record subsequent to Paleoproterozoic orogenies. As noted above, many occur in the 1.6–1.3 Ga age range. However, a significant number of younger intrusions occur in the Grenville Province of Canada and the Adirondacks of New York, all associated with anorthosite and charnockite and appearing at 1.16–1.14 Ga between the 1.3–1.2 Ga Elzevirian and 1.1 Ga Ottawan phases of the Grenvillian orogeny or closely post-dating the Grenvillian orogeny at 1.08–1.05 Ga (McLelland et al., 1988, 1996; Corrigan and Hanmer, 1997). Several other examples postdate or occur outboard of the Grenvillian orogen, including the 1.1–1.0 Ga Pikes Peak batholith of Colorado (Barker et al., 1975; Smith et al., 1999), the Enchanted Rock batholith and related intrusions of the Llano region of central Texas (Smith et al., 1997), and the Red Bluff and Burro Mountain granites of west Texas and southwestern New Mexico (Shannon et al., 1997; Smith et al., 1997; Rämö et al., 2003).

Sederholm (1891) offered one of the very earliest characterizations of the changing character of plutonism within Proterozoic orogens as a function of timing of intrusion relative to peak deformation. Likewise, Eskola (1932), Marmo (1971), and Vorma (1971) contributed similar classifications leading to a widely accepted three-part classification of syn-, late-, and postkinematic for batholiths within orogenic belts. Although the details of their respective classifications differed, all noted the increasing abundance of K and other incompatible elements with decreasing age of granitic batholiths within orogenic belts. The classic rapakivi granites of Finland and nearby regions are notable in the postkinematic or anorogenic suites along with near coeval emplacement of anorthosite and charnockite. The chemical features of rapakivi and other presumed anorogenic granites subsequently led to the definition of A-type granite (Loiselle and Wones, 1979; Collins et al., 1982; Whalen et al., 1987).

However, it should not be presumed that all A-type granites or other members of the AMCG suite are anorogenic. Certainly rift and/or plume models for their origin (Anderson, 1983; Emslie, 1991; Rämö and Haapala, 1995) are appropriate for many, but the case warrants reexamination. Based on the close timing of AMCG intrusions to phases of the Grenvillian orogeny, McLelland et al. (1996) and Corrigan and Hanmer (1997) have proposed models involving lithospheric delamination and convective thinning of continental lithosphere, respectively, to achieve an extensional regime in a broadly convergent setting, implying that the tectonic setting for the origin of A-type granites needs to be broadened. Kay and Mahlburg-Kay (1993) have likewise presented cogent arguments for potassic magmatism as a result of delamination. The ultimate cause must relate to those settings that lead to coupled melting of fertile mantle and relatively dry and fertile crust. We also find that the appearance of A-type granites, and thus, the conditions predicted for their formation, strongly follows the assembly of supercontinents, as detailed below.

There has been considerable new research dealing with the movement of pre-Pangaea supercontinents, including (1) Rodinia, which formed following the ~1.1 Ga Grenville-age orogenies (Moores, 1991; Dalziel, 1991; Hoffman, 1991; Rogers, 1996; Dalziel, 1997; Karlstrom et al., 1999), and (2) Columbia, which is concluded as being assembled following the 2.0 to 1.6 Ga Paleoproterozoic orogenies (Rogers and Santosh, 2002; Condie, 2002; Sears and Price, 2002; Zhao et al., 2002).

Windley (1993), Rogers and Santosh (2002), and Mukherjee and Das (2002) have suggested that the Mesoproterozoic AMCG association is intimately tied to the breakup of the Earth's earlier supercontinents, such as Columbia and Rodinia. Outside of Laurentia/Baltica, less volumes of these granites appear in west Africa and elsewhere following the 0.7 to 0.5 Ga Pan African orogeny (Küster and Harms, 1991; Doblas et al., 2002). Phanerozoic supercontinents also led, during their breakup, to similar but much lesser volumes of A-type granite and rhyolite, including those of southern Gondwana (Kay et al., 1989) and west Africa (Doblas et al., 2002).

A fundamental question is why should the world's greatest volume of anorthosite, charnockite, and rapakivi or A-type granite be restricted to these periods of time of the Mesoproterozoic? Anderson and Bender (1989) and Hoffman (1989) broke new ground on this question by suggesting that it was tied to the rapid growth of continents during the Proterozoic above a mantle that was subsequently isolated from ongoing processes of ocean crust production. They further argued that this portion of the Earth's mantle, then existing in a chondritic to more enriched composition and in a subcontinental setting, underwent radiogenic warming and thermal expansion, leading to the ascent of adiabatic mantle plumes, subsequent mantle melting, and transfer of heat to and melting of the overlying newly formed continents.

A parallel question is why should the time of ~1.4 Ga, which is centered at the peak of A-type granite production worldwide, also be the time when few orogenies are recognized? As indicated earlier, Karlstrom et al. (2001) have suggested an orogenic origin for tectonic deformation at ~1.4 Ga in the western and southwestern U.S., which we, and others, consider not plausible (Selverstone et al., 2000; Rämö et al., 2003). In the Phanerozoic, areas of orogeny coincide in time with intraplate occurrences of rifting and mantle upwelling (Condie, 1989). This may not be the case for the Proterozoic or Archean, hence much of the Earth's history, thus directing attention to the apparent episodic orogenic growth of continents, as reviewed earlier, and to proposals for the construction of supercontinents. Clearly, this issue needs to be further examined. Perhaps, plate consumption during the times following supercontinent assembly became largely intraoceanic, or alternatively, plate motions slowed in response to global changes in the convective cycles of the deep mantle (Condie, 1998). Regardless, this time was a time of several key events of the Earth's history, including (1) global production of A-type granite as part of the AMCG suite and what has been termed the "anorogenic trinity", (2) the formation of one, and likely two, supercontinents, (3) the restricted occurrence of orogenies as recognized by juvenile crust addition and ocean terrane accretion, and (4) the restriction of most intrusions of the AMCG suite in regions underlain by Paleoproterozoic crust.

Thus, we suggest that the A-type granites of the Mesoproterozoic have an origin very much related to the episodic nature of crust formation during this time and are fundamentally, but not always, anorogenic. Moreover, we attribute the succession of mantle and crustal melting to mantle instability, which at times led to rifting but otherwise to "hot spot" upwelling. Lithospheric delamination and other forms of thinning of the continental lithosphere could have also played a key role. We further suggest that the prior rapid construction of continental crust and eventual collisions leading to the formation of supercontinental masses were ultimately responsible for the mantle instability.

6. Concluding remarks

Prior petrologic and radiogenic isotopic studies of 1.6–1.3 Ga magmatism in Laurentia and Baltica have revealed important and fundamental changes attributable to source age or mixture of source ages of the many granite batholiths that pervaded Proterozoic terranes during this interval of the Earth history. Our current work has led to the recognition of transcontinental mineralogic, elemental, and oxygen isotopic signatures of these granites, which we view as reflective of changes in the composition and state of oxidation of the lower crust and of a magmatic episode that affected not only Laurentia and Baltica but also the entire planet 1–2 billion years ago. We further suggest that the composition of these red granites relates not to their variable tectonic setting but to the composition of their lower crustal source, including its state of oxidation and hydration, at the stage of melting.

The ilmenite-series granites, which often contain classic rapakivi textures, are the most potassic and LIL-enriched and have the highest Fe/Mg ratios of the granites emplaced across Laurentia/Baltica during this time. They also exhibit quartz $\delta^{18}O$ values that are the lowest and that are indicative of a crust-inherited mantle signature. We suggest that (1) the principle source of these granites was an oxygen-reduced quartz dioritic, tonalitic, or similar mantle-derived lower crustal rock type of Paleoproterozoic age and that (2) mantle-derived intrusions led to partial melting of the lower crust of Laurentia and Baltica, the extent of which was limited by the stability of crustal hydrous phases under vapor-absent conditions. At low fO_2, such as with fractionated phases of the anorthosite suite (Frost and Frost, 1997), the hydrous phases remained stable until fairly high temperatures in excess of 900 °C. Subsequent low fractions of melting led to the extreme A-type compositions.

The magnetite-series granites are often less potassic, less LILE-enriched, and have higher $\delta^{18}O$ values. We infer that they were derived from lower crustal sections consisting of oxidized and more hydrous calcalkaline plutons. Such a crustal source would melt at a lower temperature, and resultant higher fractions of melting would dilute the abundance of incompatible elements and the Fe/Mg ratio of derived melts. Although they retain A-type characteristics, the peraluminous granites are the least LILE-enriched and have the lowest Fe/Mg ratios and the highest $\delta^{18}O$ values. They are also nearly exclusively hosted in country rock terranes dominated by metasedimentary gneisses, and, assuming such material exists at depth, partial melting would proceed at even lower temperatures and to a greater extent. The abundance of K and Th could be expected to remain high due to the effect of source composition. The metasedimentary component in the crustal source would lead to higher degrees of melting and would further lower the abundances of most incompatible elements.

Thus, we believe that the transcontinental variations in these granites are indicative of broad regional changes in the composition of the lower crust of Laurentia and Baltica, including a reduced mantle-derived crustal source for the ilmenite-series granites, a more oxidized magmatic crustal source for the magnetite-series granites, and a likewise oxidized but more metasedimentary-rich source for the two-mica granites. Although discussion regarding their anorogenic versus orogenic origin continues, the fact that most of these intrusions postdate prior orogenic events by 100s of millions of years is significant. That some intrusions have been shown to be coeval with regional deformation is important but does not necessarily provide evidence for an orogenic setting. Otherwise, this would be akin to having the Yellowstone volcanic province of western North America as a far-field effect of the Cascadia subduction zone. Yellowstone's bimodal volcanism, the result of a long-term "hot spot" dating back to the origins of the Snake River and Columbia flood basalts, is coeval with, but essentially unrelated to, subduction beneath the western margin of the northwest Cordilleran orogen. Current continental hot spots, such as Yellowstone, serve as contemporary analogues to this Mesoproterozoic form of igneous activity, although we also note that no Phanerozoic example compares to the magnitude of A-type granite magma production during this stage of the Earth's history.

Most of A-type granite production in Laurentia/ Baltica predated orogenic construction of Rodinia, but they serve as an important tie to test models for its reconstruction. For example, the current truncated western margin of Laurentia is bordered by the 1.4 Ga magnetite-series granite province, and granites of their age, composition, and state of oxidation should be an important tie in linking this portion of Laurentia to other, now separated, continental mass.

Acknowledgments

We wish to thank Jim McLelland and Calvin Miller for their detailed and very helpful reviews, and our editor, Tapani Rämö, for his considerable patience with our delayed submission to this volume.

References

Åhäll, K.-I., Connelly, J., 1998. Intermittent 1.53–1.13 Ga magmatism in western Baltica: age constraints and correlations within a postulated supercontinent. Precambrian Research 92, 1–20.

Åhäll, K.-I., Connelly, J.N., Brewer, T.S., 2000. Episodic rapakivi magmatism due to distal orogenesis? Correlation of 1.69–1.50 Ga orogenic and inboard 'anorogenic' events in the Baltic shield. Geology 28, 823–826.

Amelin, Y.V., Larin, A.M., Tucker, R.D., 1997. Chronology of multiphase emplacement of the Salmi rapakivi granite–anorthosite complex, Baltic Shield: implications for magmatic evolution. Contributions to Mineralogy and Petrology 127, 353–368.

Anderson, J.L., 1983. Proterozoic anorogenic granite plutonism of North America. In: Medaris Jr., L.G., Byers, C.W., Mickelson, D.M., Shanks, W.C. (Eds.), Proterozoic Geology; Selected Papers from an International Proterozoic Symposium, Geological Society of America Memoir, vol. 161, pp. 133–152.

Anderson, J.L., 1987. The origin of A-type Proterozoic magmatism: a model of mantle and crustal overturn. Abstracts with Programs - Geological Society of America 19, 571.

Anderson, J.L., Bender, E.E., 1989. Nature and origin of Proterozoic A-type granitic magmatism in the southwestern United States. Lithos 23, 19–52.

Anderson, J.L., Cullers, R.L., 1978. Geochemistry and evolution of the Wolf River batholith, a late Precambrian rapakivi massif in north Wisconsin, U.S.A. Precambrian Research 7, 287–324.

Anderson, J.L., Cullers, R.L., 1987. Crust-enriched, mantle-derived tonalites of the Early Proterozoic Penokean orogen of Wisconsin. Journal of Geology 95, 139–154.

Anderson, J.L., Cullers, R.L., 1999. Paleo- and Mesoproterozoic granite plutonism of Colorado and Wyoming. Rocky Mountain Geology 34, 149–164.

Anderson, J.L., Morrison, J., 1992. The role of anorogenic granites in the Proterozoic crustal development of North America. In: Condie, K.C. (Ed.), Proterozoic Crustal Evolution. Elsevier, Amsterdam, pp. 263–299.

Anderson, J.L., Morrison, J., 1998. Oxygen isotope systematics of 1.3 to 1.6 Ga granites of Laurentia. Abstracts with Programs - Geological Society of America 30, A89.

Anderson, J.L., Thomas, W.M., 1985. Proterozoic anorogenic two-mica granite plutonism. Geology 13, 177–180.

Barker, F., Wones, D.R., Sharp, W.N., Desborough, G.A., 1975. The Pikes Peak batholith, Colorado Front Range, and a model for the origin of the gabbro–anorthosite–syenite–potassic granite. Precambrian Research 2, 97–160.

Bennett, V.C., DePaolo, D.J., 1987. Proterozoic crustal history of the western United States as determined by neodymium isotopic mapping. Geological Society of America Bulletin 99, 674–685.

Betts, P.G., Valenta, R.K., Finlay, J., 2003. Evolution of the Mount Woods Inlier, northern Gawler Craton, southern Australia: an integrated structural and aeromagnetic analysis. Tectonophysics 366, 83–111.

Bickford, M.E., Anderson, J.L., 1993. Middle Proterozoic magmatism. In: Reed, J.C., Bickford, M.E., Houston, R.S., Link, P.K., Rankin, D.W., Sims, P.K., Van Schmus, W.R. (Eds.), Geology of North America, Geological Society of America DNAG Volume C-2, Precambrian, Conterminous U.S.A. pp. 281–292.

Collins, W.J., Beams, S.D., White, A.J.R., Chappell, B.W., 1982. Nature and origin of A-type granites with particular reference to southeastern Australia. Contributions to Mineralogy and Petrology 80, 189–200.

Condie, K.C., 1989. Plate Tectonics and Crustal Evolution, 3rd edition. Pergamon Press, New York.

Condie, K.C., 1990. Growth and accretion of continental crust: inferences based on Laurentia. Chemical Geology 83, 183–194.

Condie, K.C., 1998. Episodic continental growth and supercontinents: a mantle avalanche connection? Earth and Planetary Science Letters 163, 97–108.

Condie, K.C., 2002. Breakup of a Paleoproterozoic supercontinent. Gondwana Research 5, 41–43.

Corrigan, D., Hanmer, S., 1997. Anorthosites and related granitoids in the Grenville orogen: a product of convective thinning of the lithosphere? Geology 25, 61–64.

Creaser, R.A., Price, R.C., Wormald, R.J., 1991. A-type granites revisited: assessment of a residual source model. Geology 19, 163–166.

Criss, R.E., Taylor, H.P., 1983. An $^{18}O/^{16}O$ study of the Tertiary hydrothermal systems of the southern half of the Idaho batholith. Geological Society of America Bulletin 94, 640–663.

Dall'Agnol, R., Macambira, M., Lafon, J.M., 1991. Petrological and geochemical characteristics of the lower and middle anorogenic granites of the central Amazonian province, Amazonian craton. In: Haapala, I., Rämö, O.T. (Eds.), Abstract Volume, Symposium on Rapakivi Granites and Related Rocks, IGCP Project 315, Geological Survey of Finland Guide, vol. 34, pp. 12–13.

Dall'Agnol, R., Costi, H.T., Leite, A.A. da S., Magalháes, M.S. de, Teixeira, N.P., 1999. Rapakivi granites from Brazil and adjacent areas. Precambrian Research 95, 9–39.

Dalziel, I.W.D., 1991. Pacific margins of Laurentia and East Antarctica–Australia as a conjugate rift pair: evidence and implications for an Eocambrian supercontinent. Geology 19, 598–601.

Dalziel, I.W.D., 1997. Overview: Neoproterozoic–Palaeozoic geography and tectonics: review, hypothesis, environmental speculations. Geological Society of America Bulletin 109, 16–42.

Demaiffe, D., Javoy, M., 1980. $^{18}O/^{16}O$ ratios of anorthosites and related rocks from the Rogaland complex (SW Norway). Contributions to Mineralogy and Petrology 72, 311–317.

Dempster, T.J., Jenkin, G.R.T., Rogers, G., 1994. The origin of rapakivi texture. Journal of Petrology 35, 963–981.

DePaolo, D.J., 1981. Neodymium isotopes in the Colorado Front Range and crust–mantle evolution in the Proterozoic. Nature 291, 193–196.

Doblas, M., Lopez-Ruiz, J., Cebria, J.M., Youbi, N., Degroote, E., 2002. Mantle insulation beneath the West African craton during the Precambrian–Cambrian transition. Geology 30, 839–842.

Duebendorfer, E.M., Christensen, C., 1995. Synkinematic (?) intrusion of the "anorogenic" 1425 Ma Beer Bottle Pass pluton, southern Nevada. Tectonics 14, 168–184.

Emslie, R.F., 1978. Anorthosite massifs, rapakivi granites, and late Proterozoic rifting of North America. Precambrian Research 7, 61–98.

Emslie, R.F., 1991. Granitoids of rapakivi granite–anorthosite and related associations. Precambrian Research 51, 173–192.

Emslie, R.F., Taylor, B.E., 1993. Rapakivi granitoids, central Labrador: fluids and stable isotopes. Abstracts with Programs - Geological Society of America 25, 31.

Eskola, P., 1932. On the origin of granite magmas. Mineralogie Petrographie 42, 455–481.

Frost, C.D., Frost, B.R., 1997. Reduced rapakivi-type granites: the tholeiitic connection. Geology 25, 647–650.

Gastil, G., 1960. The distribution of mineral dates in time and space. American Journal of Science 258, 1–35.

Geraldes, M.C., Van Schmus, W.R., Teixeira, W., 1998. Age of Proterozoic crust in SW Mato Grosso, Brazil: evidence for a 1450 Ma magmatic arc in SW Amazonia. Abstracts with Programs - Geological Society of America 30, 96–97.

Gower, C.F., Krogh, T.E., 2002. A U–Pb geochronological review of the Proterozoic history of the eastern Grenville Province. Canadian Journal of Earth Sciences 39, 795–829.

Graubard, C.M., 1991. Extension in a transpressional setting: emplacement of the mid-Proterozoic Mt. Evans batholith, central Front Range, Colorado. Abstracts with Programs - Geological Society of America 23, 27.

Haapala, I., Rämö, O.T., 1990. Petrogenesis of the Proterozoic rapakivi granites of Finland. In: Stein, H.J., Hannah, J.L. (Eds.), Ore-bearing Granite Systems: Petrogenesis and Mineralizing Processes, Geological Society of America Special Paper, vol. 246, pp. 275–286.

Heaman, L.M., Shieh, Y.-N., McNutt, R.H., Shaw, D.M., 1982. Isotopic and trace element study of the Loon Lake pluton, Grenville Province, Ontario. Canadian Journal of Earth Sciences 19, 1045–1054.

Herz, N., 1969. Anorthosite belts, continental drift, and the anorthosite event. Science 164, 944–947.

Hoffman, P.F., 1989. Speculations on Laurentia's first gigayear (2.0 to 1.0 Ga). Geology 17, 135–138.

Hoffman, P.F., 1991. Did the breakout of Laurentia turn Gondwanaland inside out? Science 252, 1409–1412.

Jianhua, Y., Huiqin, F., Fenglan, Z., Meisheng, G., 1991. Geochemistry of the rapakivi granite suite in a Proterozoic rift trough in Beijing and its vicinity. Acta Geologica Sinica 4, 169–186.

Karlstrom, K.E., Harlan, S.S., Williams, M.L., McLelland, J., Geissman, J.W., Åhäll, K.-I., 1999. Refining Rodinia: geologic evidence for the Australia–western U.S. connection. GSA Today 9, 1–7.

Karlstrom, K.E., Åhäll, K.-I., Harlan, S.S., Williams, M.L., McLelland, J., Geissman, J.W., 2001. Long-lived (1.8–1.0 Ga) convergent orogen in southern Laurentia, its extensions to Australia and Baltica, and implications for refining Rodinia. Precambrian Research 111, 5–30.

Kay, R.W., Mahlburg-Kay, S., 1993. Delamination and delamination magmatism. Tectonophysics 219, 177–189.

Kay, S.M., Ramos, V.A., Mpodozis, C., Sruoga, P., 1989. Late Paleozoic to Jurassic silicic magmatism at the Gondwana margin: analogy to the Middle Proterozoic of North America? Geology 17, 324–328.

Kerr, A., Thomas, R.J., 1991. Rapakivi granites and fayalite-bearing charnockites from the Proterozoic Mobile Belt in Natal, South Africa. In: Haapala, I., Rämö, O.T. (Eds.), Abstract Volume, Symposium on Rapakivi Granites and Related Rocks, IGCP Project 315, Geological Survey of Finland Guide, vol. 34, pp. 26–27.

Kim, S.-J., 1989. Oxygen and sulfur isotope studies of the Wolf River Batholith in Wisconsin and related Precambrian anorogenic granitic rocks in the Mid-continent of North America. PhD thesis, Purdue University, Lafayette, Indiana, U.S.A.

Kirby, E., Karlstrom, K.E., Andronicos, C.L., 1995. Tectonic setting of the Sandia pluton: an orogenic 1.4 Ga granite in New Mexico. Tectonics 14, 185–201.

Kirs, J., Huhma, H., Haapala, I., 1991. Petrological–chemical features and age of Estonian anorogenic potassium granites. In: Haapala, I., Rämö, O.T. (Eds.), Abstract Volume, Symposium on Rapakivi Granites and Related Rocks, IGCP Project 315, Geological Survey of Finland Guide, vol. 34, pp. 28–29.

Krogh, T.E., Gower, C.F., Wardle, R.J., 1996. Pre-Labradorian crust and later Labradorian, Pinwarian and Grenvillian metamorphism in the Mealy Mountains terrane, Grenville Province, eastern Labrador. In: Gower, C.F. (Ed.), Program and Abstracts, Proterozoic Evolution in the North Atlantic Realm, COPENA-ECSOOT-IBTA Conference, Goose Bay, Labrador, 29 July-2 August 1996. Newfoundland Department of Mines and Energy, St. John's, Newfoundland, Canada, pp. 106–107. (Comp.).

Küster, D., Harms, U., 1991. Late Proterozoic/Early Paleozoic rapakivi–granitoids in NE-Africa—evidence for Pan African crustal consolidation. In: Haapala, I., Rämö, O.T. (Eds.), Abstract Volume, Symposium on Rapakivi Granites and Related Rocks, IGCP Project 315, Geological Survey of Finland Guide, vol. 34, pp. 31.

Loiselle, M.C., Wones, D.R., 1979. Characteristics and origin of anorogenic granites. Abstracts with Programs - Geological Society of America 11, 468.

Marmo, V., 1971. Granite Petrology and the Granite Problem. Elsevier, New York.

McLelland, J.M., Chiarenzelli, J., Whitney, P., Isachsen, Y., 1988. U–Pb zircon geochronology of the Adirondack Mountains and implications for their geologic evolution. Geology 16, 920–924.

McLelland, J.M., Daly, S., McLelland, J., 1996. The Grenville orogenic cycle: an Adirondack perspective. Tectonophysics 265, 1–29.

Menuge, J.F., Brewer, T.S., Seeger, C.M., 2002. Petrogenesis of metaluminous A-type rhyolites from the St. Francois Mountains, Missouri and the Mesoproterozoic evolution of the southern Laurentian margin. Precambrian Research 113, 269–291.

Moores, E.M., 1991. Southwest U.S.–East Antarctica (SWEAT) connection: a hypothesis. Geology 19, 425–428.

Moralev, V.M., Glukhovsky, M.Z., 1991. Tectonic setting of the rapakivi granites of the Aldan Shield, Siberia. In: Haapala, I., Rämö, O.T. (Eds.), Abstract Volume, Symposium on Rapakivi Granites and Related Rocks, IGCP Project 315, Geological Survey of Finland Guide, vol. 34, pp. 34–35.

Morrison, J., 1994. Meteoric water–rock interaction in the lower plate of the Whipple Mountain metamorphic core complex, California. Journal of Metamorphic Geology 12, 827–840.

Morrison, J., Anderson, J.L., 1998. Footwall refrigeration along a detachment fault: implications for thermal evolution of core complexes. Science 279, 63–66.

Mukherjee, A., Das, S., 2002. Anorthosites, granulites and the supercontinent cycle. Gondwana Research 5, 133–146.

Nelson, B.K., DePaolo, D.J., 1985. Rapid production of continental crust 1.7–1.9 b.y. ago: Nd and Sr isotopic evidence from the basement of the North American midcontinent. Geological Society of America Bulletin 96, 746–754.

Nyman, M.W., Karlstrom, K.E., 1997. Pluton emplacement processes and tectonic setting of the 1.42 Ga Signal batholith, SW USA: important role of crustal anisotropy during regional shortening. Precambrian Research 82, 237–263.

Nyman, M.W., Karlstrom, K.E., Kirby, E., Graubard, C.M., 1994. Mesoproterozoic contractional orogeny in western North America: evidence from ca. 1.4 Ga plutons. Geology 22, 901–904.

Orrell, S.E., Bickford, M.E., Lewry, J.F., 1999. Crustal evolution and age of thermotectonic reworking in the western hinterland of the trans-Hudson orogen, northern Saskatchewan. Precambrian Research 95, 187–223.

Patchett, P.J., Bridgwater, D., 1984. Origin of continental crust of 1.9–1.7 Ga defined by Nd isotopes in the Ketilidian terrain of South Greenland. Contributions to Mineralogy and Petrology 87, 311–318.

Patchett, P.J., Arndt, P.J., 1986. Nd isotopes and tectonics of 1.9–1.7 Ga crustal genesis. Earth Planetary Science Letters 78, 329–338.

Patchett, P.J., Ruiz, J., 1987. Nd isotopic ages of crust formation and metamorphism in the Precambrian of eastern and southern Mexico. Contributions to Mineralogy and Petrology 96, 523–528.

Pedrick, J.N., Karlstrom, K.E., Bowring, S.A., 1998. Reconciliation of conflicting tectonic models for Proterozoic rocks of northern New Mexico. Journal of Metamorphic Geology 16, 687–707.

Rämö, O.T., 1991. Petrogenesis of the Proterozoic rapakivi granites and related basic rocks of southeastern Fennoscandia: Nd and Pb isotopic and general geochemical constraints. Geological Survey of Finland, 355.

Rämö, O.T., Haapala, I., 1995. One hundred years of rapakivi granite. Mineralogy and Petrology 52, 129–185.

Rämö, O.T., Dall'Agnol, R., Macambira, M.J.B., Leite, A.A.S., de Oliveira, D.C., 2002. 1.88 Ga oxidized A-Type granites of the Rio Maria Region, eastern Amazonian craton, Brazil: positively anorogenic!. Journal of Geology 110, 603–610.

Rämö, O.T., McLemore, V.T., Hamilton, M.A., Kosunen, P.J., Heizler, M., Haapala, I., 2003. Intermittent 1630–1220 Ma magmatism in central Mazatzal province: new geochronologic piercing points and some tectonic implications. Geology 31, 335–338.

Reed, J.C., Bickford, M.E., Tweto, O., 1993. Proterozoic accretionary terranes of Colorado and southern Wyoming. In: Van Schmus, W.R., Bickford, M.E. (Eds.), Transcontinental Proterozoic Provinces, Geology of North America, Geological Society of America DNAG Volume C-2, Precambrian, Conterminous U.S.A. pp. 211–228.

Rivers, T., 1997. Lithotectonic elements of the Grenville province: review and tectonic implication. Precambrian Research 86, 117–154.

Rogers, J.J.W., 1996. A history of continents in the past three billion years. Journal of Geology 104, 91–107.

Rogers, J.J.W., Santosh, M., 2002. Configuration of Columbia, a Mesoproterozoic supercontinent. Gondwana Research 5, 5–22.

Sears, J.W., Price, R.A., 2002. The hypothetical supercontinent Columbia: implications for the Siberian–West Laurentian connection. Gondwana Research 5, 35–39.

Sederholm, J.J., 1891. Studien über archäische Eruptivgesteine aus dem südwestlichen Finnland. Tschermaks Mineralogische und Petrographische Mitteilungen 12, 97–142.

Selverstone, J., Hodgins, M., Aleinikoff, J.N., Fanning, C.M., 2000. Mesoproterozoic reactivation of a Paleoproterozoic transcurrent boundary in the northern Colorado Front Range: implications for ~1.7- and 1.4-Ga tectonism. Rocky Mountain Geology 35, 139–162.

Shannon, W.M., Barnes, C.G., Bickford, M.E., 1997. Grenville magmatism in west Texas: petrology and geochemistry of the Red Bluff granitic suite. Journal of Petrology 38, 1279–1305.

Sharp, Z.D., 1990. A laser-based microanalytical method for the in situ determination of oxygen isotope ratios of silicates and oxides. Geochimica et Cosmochimica Acta 54, 1353–1357.

Shaw, C.A., Snee, L.W., Selverstone, J., Reed, J.C., 1999. $^{40}Ar/^{39}Ar$ thermochronology of Mesoproterozoic metamorphism in the Colorado Front Range. Journal of Geology 107, 49–67.

Shieh, Y.-N., 1983. Oxygen isotope study of Precambrian granites from the Illinois Deep Hole Project. Journal of Geophysical Research 88, 7300–7304.

Shieh, Y.-N., Schwarcz, H.P., Shaw, D.M., 1976. An oxygen isotope study of the Loon Lake Pluton and the Apsley Gneiss, Ontario. Contributions to Mineralogy and Petrology 54, 1–16.

Smith, D.R., Barnes, C.G., Shannon, W., Roback, R.C., James, E., 1997. Petrogenesis of Mid-Proterozoic granitic magmas: examples from central and west Texas. Precambrian Research 85, 53–79.

Smith, D.R., Noblett, J., Wobus, R.A., Unruh, D., Douglass, J., Beane, R., Davis, C., Goldman, S., Kay, G., Gustavson, B., Saltoun, B., Stewart, J., 1999. Petrology and geochemistry of late-stage intrusions of the A-type, mid-Proterozoic Pikes Peak batholith (central Colorado, USA): implications for genetic models. Precambrian Research 98, 271–305.

Solomon, G.C., Taylor, H.P., 1991. Oxygen isotope studies of Jurassic fossil hydrothermal systems, Mojave Desert, southeastern California. In: Taylor, H.P., O'Neil, J.R., Kaplan, I.R.Special Publication, vol. 3. The Geochemical Society, pp. 449–462.

Spikings, R.A., Foster, D.A., Kohn, B.P., Lister, G.S., 2002. Post-orogenic (<1500 Ma) thermal history of the Palaeo-Mesoproterozoic, Mt. Isa province, NE Australia. Tectonophysics 349, 327–365.

Stein, M., Hofmann, A.W., 1994. Mantle plumes and episodic crustal growth. Nature 372, 63–68.

Sutton, J., 1963. Long-term cycles in the evolution of continents. Nature 198, 731–735.

Thompson, A.G., Barnes, C.G., 1999. Petrology and geochemistry of the 1.4 Ga Priest pluton. Rocky Mountain Geology 34, 223–244.

Torsvik, T.H., 2003. The Rodinia Jigsaw puzzle. Science 300, 1379–1381.

Tweto, O., 1987. Rock units of the Precambrian basement in Colorado. U.S. Geological Survey Professional Paper 1321-A.

Valley, J.W., Kitchen, N., Kohn, M.J., Niendorf, C.R., Spicuzza, M.J., 1996. UWG-2, a garnet standard for oxygen isotope ratios: strategies for high precision and accuracy with laser heating. Geochimica et Cosmochimica Acta 59, 5223–5231.

Van Schmus, W.R., Bickford, M.E., Turek, A., 1996. Proterozoic geology of the east-central Midcontinent basement. In: van der Pluijm, B.A., Catacosinos, P.A. (Eds.), Basement and Basins of Eastern North America, Geological Society of America Special Paper, vol. 308, pp. 7–32.

Vorma, A., 1971. Alkali feldspars of the Wiborg rapakivi massif in southeastern Finland. Bulletin de la Commission Geologique de Finlande, 246.

Wenner, D.B., Taylor, H.P., 1976. Oxygen and hydrogen isotope studies of a Precambrian granite–rhyolite terrane, St. Francois Mountains, southeastern Missouri. Geological Society of America Bulletin 87, 1587–1598.

Whalen, J.B., Currie, K.L., Chappell, B.W., 1987. A-type granites: geochemical characteristics, discrimination, and petrogenesis. Contributions to Mineralogy and Petrology 95, 407–419.

Wickham, S.M., Alberts, A.D., Zanvilevich, A.N., Litvinovsky, B.A., Bindeman, I.N., Schauble, E.A., 1996. A stable isotopic study of anorogenic magmatism in east central Asia. Journal of Petrology 37, 1063–1095.

Windley, B.F., 1993. Proterozoic anorogenic magmatism and its orogenic connections. Journal of the Geological Society (London) 150, 39–50.

Wu, T.-W., Kerrich, R., 1986. Combined oxygen isotope-compositional studies of some granitoids from the Grenville Province of Ontario, Canada: implications for source regions. Canadian Journal of Earth Sciences 23, 1412–1432.

Zhao, G., Cawood, P.A., Wilde, S.A., Sun, M., 2002. Review of global 2.1–1.8 Ga orogens: implications for a pre-Rodinia supercontinent. Earth-Science Reviews 59, 125–162.

Available online at www.sciencedirect.com

Lithos 80 (2005) 61–74

www.elsevier.com/locate/lithos

Source regions of granites and their links to tectonic environment: examples from the western United States

Elizabeth Y. Anthony[*]

Department of Geological Sciences, The University of Texas at El Paso, El Paso, TX 7968-0555, USA

Received 9 May 2003; accepted 9 September 2004
Available online 10 November 2004

Abstract

This review, in honor of Ilmari Haapala's retirement, reflects on lessons learned from studies of three granitic systems in western North America: (1) Mesoproterozoic samples from west Texas and east New Mexico; (2) Laramide granitic systems associated with porphyry-copper deposits in Arizona; and (3) granites of the Colorado Mineral Belt. The studies elucidate relationships amongst tectonic setting, source material, and magma chemistry.

Mesoproterozoic basement samples are from two different felsic suites with distinct elemental and isotopic compositions. The first suite, the "plutonic province", is dominantly magnesian, calc-alkalic to alkali-calcic, and metaluminous. It has low K_2O/Na_2O and Rb/Sr, and Nd model ages of 1.56 to 1.40 Ga. The second suite, the "Panhandle igneous complex", is magnesian, metaluminous, alkalic, and is part of the Mesoproterozoic belt of magmatism that extends from Finland to southwestern United States. Samples from the Panhandle igneous complex demonstrate three episodes of magmatism: the first pulse was intrusion of quartz monzonite at 1380 to 1370 Ma; the second was comagmatic epizonal granite and rhyolite at 1360 to 1350 Ma. Both of these rock types are high-K to slightly ultra-high-K. The third pulse at 1338 to 1330 Ma was intrusion of ultra-high-K quartz syenite. Nd model ages (1.94 to 1.52 Ga) are distinct from those of the "plutonic province" and systematically older than crystallization ages, implying a substantial crustal input to the magmas.

At the Sierrita porphyry-copper deposit in the Mazatzal Province of southeastern Arizona, trace element, Sr, and Nd isotopic compositions were determined for a suite of andesitic and rhyolitic rocks (67 Ma) intruded by granodiorite and granite. Isotopic composition and chemical evolution are well correlated throughout the suite. Andesite has the least negative initial ε_{Nd} (−4.3) and lowest $^{87}Sr/^{86}Sr_i$ (0.7069). It is also the oldest and chemically most primitive, having low concentrations of Rb, SiO_2, and high concentrations of transition elements. These parameters change through the system to the youngest unit (granite), which has the most negative ε_{Nd} (−8.5), the highest $^{87}Sr/^{86}Sr_i$ (0.7092), and is chemically most evolved. Correlation between chemical and Nd isotopic evolution probably resulted from a continuous process of progressive assimilation, in which mafic magmas invade and incorporate continental crust. Deposits in Arizona with ε_{Nd} values more negative than the −8.5 of Sierrita lie in the older Yavapai province in the northwestern part of the state. The difference in the most negative epsilon Nd implies that Nd isotopic signature is sensitive to the age of the Precambrian domain.

---

[*] Tel.: +1 915 747 5483; fax: +1 915 747 5073.
  E-mail address: eanthony@geo.utep.edu.

The granites from the Colorado Mineral Belt were emplaced during the transition from Laramide convergence to mid-Tertiary extension. Three different groups of granites are recognized. The first is Laramide and was formed during assimilation-fractional crystallization involving lower crustal mafic source materials; the second and third groups are mid-Tertiary and represent intracrustal melting of heterogeneous sources. This change in source regions and melt regimes in transition from convergence to extension is fundamental to the Mesozoic and Cenozoic evolution of western North America.

Keywords: Granite; Mesoproterozoic; Laramide; Colorado Mineral Belt; North America

1. Introduction

This article draws on studies of three granite systems of diverse ages in the western United States. These are Mesozoic-aged granites related to porphyry-copper deposits in Arizona (Anthony, 1986; Anthony and Titley, 1988, 1994; Titley and Anthony, 1989), granites of the Colorado Mineral Belt that span the transition from Mesozoic compressional tectonics to mid-Tertiary extension in Colorado (Ouimette, 1993, 1995), and Mesoproterozoic granites of the Southern Granite-Rhyolite Province of Texas and New Mexico (Barnes et al., 1999a, 2002; Amarante et al., 2004). A recurring theme in these studies is the importance of magmatic source materials in controlling fundamental characteristics of granites. This idea is not new; it has been either implicit or explicit in many studies through the decades and was elegantly summarized by Chappell (1979) in the statement "...granites image their sources...". Recently, Frost et al. (2001) advocated a return to a descriptive classification of granitoids in terms of Fe-number, alkali-lime index, and aluminum saturation, and abandonment of the alphabetical typologic nomenclature. One reason the alphabetical nomenclature is so enduring is that it addresses tectonic setting, which is a primary justification for the study of granite petrogenesis. This review presents examples of the first-order correlation between the chemical and isotopic signature of granites and their source regions and argues that the link between granites and tectonics is via the source region. Confusion and obfuscation occur in the alphabetical or any other classification when a direct link is attempted between granite chemistry and tectonics that neglects the important intermediate step of evaluating the role of source materials.

2. The Mesoproterozoic of west Texas and New Mexico

2.1. Observations

This study used core samples from the basement of west Texas and eastern New Mexico (Fig. 1) retrieved from drilling into Precambrian basement during oil exploration. Therefore, the majority consisted of one sample per well resulting in a reconnaissance charac-

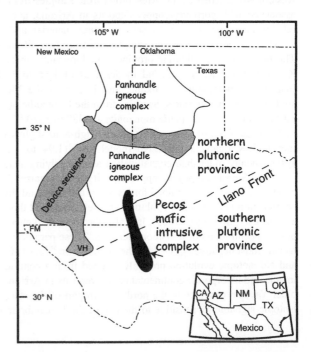

Fig. 1. Location map for Mesoproterozoic geological features of the southern United States; FM—Franklin Mountains, VH—Van Horn Mountains. Other features as described in text. Inset shows location relative to the United States and Mexico; CA—California, AZ—Arizona, NM—New Mexico, TX—Texas, OK—Oklahoma. Modified from Amarante et al. (2004).

terization of the distribution of igneous rock types in the area (Barnes et al., 1999a). Petrography of the samples was originally described by Flawn (1956), Muehlberger et al. (1967), and Thomas et al. (1984) and was used to draft the Decade of North American Geology (DNAG) basement map for the area. A second set of cuttings came from 41 closely spaced wells that penetrated the Southern Granite-Rhyolite Province in the Panhandle area of west Texas (Barnes et al., 2002). Approximately half of the wells penetrate the basement at least 90 m, another quarter as much as 190 m, and some more than 210 m. Detailed logs of all wells are in Barnes (2001), and a summary figure of stratigraphic columns and interpretative cross-sections are in Barnes et al. (2002). Finally, samples were studied from a well, the Mescalero #1, that penetrated approximately 2652 m of basement in eastern New Mexico. The well includes from top to bottom Mesoproterozoic Debaca sequence (a cover sequence of metasedimentary and metavolcanic rocks intruded by gabbros), which is in depositional contact with underlying rocks from the Southern Granite-Rhyolite Province. The latter are analogous in age and lithology to samples from the Panhandle area. Geophysical logs for the well correlate to lithology and seismic reflection profiles, yielding a three-dimensional interpretation of this basement area (Amarante et al., 2004).

Petrography, whole-rock chemistry, mineral chemistry, and Nd isotopes were used to characterize the samples. Whole-rock chemistry was used to classify the samples according to the scheme of Frost et al. (2001) and to explore the influence of source material on melt composition (e.g., Patiño Douce, 1996, 1999). Mineral chemistry was used primarily to estimate the oxidation state of the magma. A traditional method for determining oxidation state is to determine the composition of iron-oxide minerals. The studies included here used, however, the composition of biotite in equilibrium with quartz, K-feldspar, and magnetite (Wones and Eugster, 1965; Wones, 1972). For this assemblage, high Mg numbers correlate with high oxidation states, because Fe^{3+} partitions into magnetite. Nd isotopic signature was used to discriminate between crustal provinces of different ages and to evaluate mixing of reservoirs during magma genesis.

Two felsic suites were recognized on the basis of the criteria described above (Table 1; Barnes et al., 1999a, 2002). The first suite, referred to as the "plutonic province" (cf., Fig. 1), is dominantly magnesian, metaluminous, calc-alkalic to alkalic-calcic, with a few ferroan, metaluminous, and alkalic samples. Biotite chemistry indicates that the magnesian samples are moderately to highly oxidized, and the ferroan samples are strongly reduced. Nd model ages (calculated according to Nelson and DePaolo, 1985) are 1.35 to 1.34 Ga for the ferroan samples and 1.56 to 1.40 Ga for the magnesian samples. No crystallization ages exist for the plutonic province. The second suite is from the Panhandle area of the Southern Granite-Rhyolite Province and is referred to as the "Panhandle igneous complex". It consists of undeformed epizonal granites and ignimbritic rhyolites intruded by two generations of mafic sills and dikes. It is distinct from the plutonic province in that it is magnesian, metaluminous, but alkalic rather than calc-alkalic or alkali-calcic, and is characterized by $K_2O/Na_2O>1$ and $Rb/Sr>1$. Biotite chemistry indicates that all samples are highly oxidized, and Nd model ages, which range from 1.94 to 1.52 Ga, are older than in the plutonic province. U–Pb dates on zircon yield crystallization ages of 1.38 to 1.34 Ga for the Panhandle igneous complex (Barnes et al., 2002, and references therein).

The samples from the 41 closely spaced wells yield a detailed picture of the Panhandle igneous complex that clarifies its characteristics and serves as an important contribution to understanding Mesoproterozoic magmatism in southern Laurentia. Petrography, geochemistry, and geochronology define three pulses of magmatism. The first pulse, intrusion of quartz monzonite, occurred at 1380 to 1370 Ma; the second pulse was contemporaneous emplacement of comagmatic, high-K epizonal granite and rhyolite at 1360 to 1350 Ma. The third pulse of magmatism, at 1338 to 1330 Ma, was intrusion of ultra-high-K quartz syenite. Biotite chemistry implies that all samples are highly oxidized. In fact, these samples are among the highest Mg numbers for biotite from ~1.4 Ga magmatism in western North America (Barnes et al., 2002, Fig. 13). F/Cl ratios in the biotites are uniform and low for the granites and rhyolite, and higher and more variable for the monzonite and syenite.

Table 1
Characteristics of Mesoproterozoic granites from western Texas and eastern New Mexico

Lithology	Classification	Chemistry	Oxidation state	Nd isotopes	Comments
Panhandle igneous complex					
• Quartz monzonite (1380 to 1370 Ma) granite/rhyolite (1360 to 1350 Ma)	Magnesian-metaluminous to peraluminous-alkalic	High-K to borderline ultra-high-K $K_2O/Na_2O>1$ $Rb/Sr>1$	Oxidized, amongst most oxidized of ~1.4 Ga granitoids of North America	t_{DM}=1.94 to 1.52 Ga	t_{DM} greater than crystallization age
• Quartz syenite (1338 to 1330 Ma)		Ultra-high-K	High F/Cl relative to granite/rhyolite		
Plutonic province					
• Crosbyton gavity and magnetic high	Ferroan-metaluminous-alkalic		Reducing	t_{DM}=1.35 to 1.34 Ga	
• Deformed granitoids	Magnesian-metaluminous-calc-alkalic to alkali-calcic	$K_2O/Na_2O<1$ $Rb/Sr<1$	Intermediate to high	t_{DM}=1.56 to 1.40 Ga	
Mafic rocks					
• Alkaline	Low Mg number	No Nb anomaly in majority of samples	Not applicable	PMIC: t_{DM}=1.53 Ga panhandle: t_{DM}=1.56 to 1.44 Ga Debaca sequence: t_{DM}=1.26 Ga	Detrital zircon from base of Debaca sequence: 1690 and 1320 Ma populations
• subalkaline	High Mg number				

t_{DM} is depleted mantle model age calculated according to DePaolo (1988).

Mafic rocks are ubiquitous (Table 1) and include samples from the Debaca sequence and Pecos mafic intrusive complex, a voluminous layered mafic intrusion (Kargi and Barnes, 1995; Barnes et al., 1999b) with a clear gravity and seismic reflection signature (Adams and Miller, 1995). Mafic rocks also are found as sills and dikes in the 41 wells of the Panhandle igneous complex. They constitute a substantial portion (~20% to 30%) of the cuttings, creating a bimodal character to the rock types. In the Mescalero #1 well, mafic rocks are found as diorite associated with the Mesoproterozoic quartz syenite and as gabbro intruding the overlying Debaca sequence. The mafic rocks are in part alkaline, in part subalkaline, and the majority lacks a negative Nb anomaly. The subalkaline rocks tend to have lower Fe numbers, whereas the alkaline rocks have high Fe numbers and are more evolved. The oldest Nd model ages are from the Panhandle igneous complex (1.56 to 1.44 Ga) and the Pecos mafic intrusive complex (1.53 Ga). A model age for one sample from the Debaca sequence is distinctly younger at 1.26 Ga. SHRIMP

analysis of detrital zircon from an arkose at the base of Debaca sequence in the Mescalero #1 well yields populations at 1708 and 1308 Ma (Barnes, 2001; Amarante et al., 2004). The immediately underlying quartz syenite is 1332 Ma. $^{40}Ar/^{39}Ar$ spectra for hornblende and biotite from the gabbros that intrude the upper portion of the Debaca sequence in this well yield apparent ages of ~1090 Ma.

2.2. Interpretation

2.2.1. Ages of mafic magmatism

The Nd model ages for mafic samples indicate that the rocks are Mesoproterozoic; because, these are only model ages, no more detailed interpretation is warranted. This point is made clear in the Franklin Mountains of west Texas, where mafic dikes and sills with Nd model ages of 1.47 to 1.31 Ga (Patchett and Ruiz, 1989; Barnes et al., 1999b) crosscut the Red Bluff granitic suite, which has a U–Pb zircon age of 1.12 Ga (Shannon et al., 1997). A second example is the Pecos mafic intrusive complex with its Nd model age of 1.53 Ga compared to U–Pb crystallization ages of 1163 to 1072 Ma (Keller et al., 1989). Crystallization ages, on the other hand, do support at least two cycles of mafic magmatism. The first one is associated with the Debaca sequence and is bimodal. Ages for this magmatism are from zircons in a felsic tuff in the Casner Marble of the Franklin Mountains (Pittenger et al., 1994) and a series of zircon ages for the Allamore and Tumbledown formations in the Van Horn Mountains (Bickford et al., 2000). The second cycle of mafic magmatism is Grenvillian in age (~1.1 Ga) and is represented by the gabbros crosscutting the Debaca sequence in the Mescalero #1 well and the Pecos mafic intrusive complex. The Grenville was a time of pervasive mafic magmatism in this southern part of Laurentia. Examples include diabase sills in the Apache Group of Arizona (Wrucke, 1989) and the Cardenas basalt in the Unkar Group (Larson et al., 1994; Timmons et al., 2001).

2.2.2. Petrogenesis of the basement rocks and tectonic implications

Source indicators summarized in Table 1 argue strongly that the majority of the felsic rocks are crustal melts. These indicators include the oxidized nature of the magmas as deduced from biotite chemistry, with the argument being that mantle derived magmas are generally reduced, and the observation that Nd model ages are older than crystallization ages for the Panhandle igneous complex. The Nd model ages are similar to the ages of the Paleoproterozoic crust from this portion of Laurentia, while the crystallization ages of the Panhandle igneous complex are Mesoproterozoic. Given this geologic context for the Nd model ages, it is most reasonable to hypothesize that Paleoproterozoic crust was involved in magma genesis. The possible exceptions to this conclusion are the ferroan samples from the plutonic province. Insufficient detail exists for the ferroan samples to discuss their petrogenesis other than to observe that their ferroan chemistry and reduced oxidation state make them candidates to be either (1) melts of mafic crustal underplate, as has been interpreted for the 1.43 Ga Sherman batholith of southeast Wyoming (Frost et al., 1999) or (2) end stage crystallization of a mafic magmatic system, as is the case for the 1.12 Ga Red Bluff granitic suite of the Franklin Mountains (Shannon et al., 1997). The samples come from the

basement in the vicinity of the "Crosbyton high" of west Texas, an ovoid gravity and magnetic high, which would be consistent with involvement of mafic rocks in their genesis.

The chemistry of the Panhandle igneous complex is consistent with their source being intermediate composition meta-igneous rock (Patiño Douce, 1996, 1999). This type of source material has been suggested for rapakivi granites of similar age in Finland (Rämö and Haapala, 1995) and is in contradistinction to the ferroan, alkali-calcic, metaluminous Sherman batholith, whose source is considered to be mafic (Frost et al., 1999). An active controversy exists concerning the tectonic setting for the 1.4 Ga magmatic event and what, if any, Phanerozoic analog is applicable. Several petrogenetic studies (e.g., Rämö and Haapala, 1995; Anderson and Cullers, 1999; Frost et al., 1999; Haapala and Rämö, 1999; Barnes et al., 2002) have presented evidence for a dominantly crustal origin of the magmas, probably triggered by heat from contemporaneous mafic magmatism. In the Phanerozoic, such bimodal magmatism is most often associated with hot spots or extension. An important observation for the theme of this paper is that the tectonic setting can generate magmas with contrasting characteristics, examples being the Panhandle igneous complex and the Sherman batholith, with the contrast being a function of the source material. Finally, with respect to tectonic evolution in the Mesoproterozoic, a number of studies indicate that there were three cycles of magmatism at ~1.4, 1.2, and 1.1 Ga (e.g., Rämö et al., 2003; Amarante et al., 2004). These suites tend to be magnesian or ferroan and calc-alkalic to alkalic, and are distinct from the magnesian, calcic to calc-alkalic suites. The latter suites are common in, for example, the Mesozoic Cordilleran batholiths of western North America and thus are typical of the plutonic portions of continental magmatic arcs. It is significant for the interpretation of the tectonic setting of Mesoproterozoic magmatism in southern Laurentia that the three cycles of magmatism are not characterized by Cordilleran-style igneous activity. Furthermore, the southwestern United States has no record of Cordilleran-style magmatism during intervals between these cycles, as has been suggested for Sweden by Åhäll et al. (2000).

3. Porphyry-copper mineralized granites of Arizona

3.1. Observations

The western United States is composed of an amalgamation of Archean and Proterozoic provinces (Karlstrom and CD-ROM Group, 2002), with, for example, the Archean Wyoming Province, as well as the Paleoproterozoic Yavapai and Mazatzal provinces. In Arizona, the Yavapai province occupies the northwestern part of the state and is 1.80 to 1.75 Ga in age. The Mazatzal Province in the southeastern part of the state is 1.70 to 1.68 Ga in age (Fig. 2). Superposed on this Proterozoic grain is Laramide-aged granitic magmatism associated with porphyry-copper deposits. A number of studies have been conducted to explore

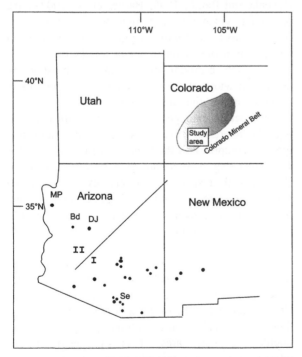

Fig. 2. Map of Proterozoic terranes of Arizona and the Colorado Mineral Belt. Terrane I, Mazatzal (1.70 to 1.68 Ga); Terrane II, Yavapai (1.80 to 1.75 Ga). Dots indicate locations of porphyry-copper deposits in Arizona and New Mexico with deposits referred to in text indicated; Se—Sierrita, MP—Mineral Park, Bd—Bagdad, and DJ—Diamond Joe. Location of the Laramide and mid-Tertiary samples discussed in the text is indicated by the rectangle in the southwestern section of the Colorado Mineral Belt. Modified from Titley and Anthony (1989) and Titley (2001).

petrogenesis of the igneous rocks associated with porphyry-copper deposits. Among the first that included Nd isotopic data was a reconnaissance study by Farmer and DePaolo (1984), followed by a more detailed study of petrologic evolution for a single porphyry-copper deposit, the Sierrita deposit south of Tucson, Arizona (Anthony and Titley, 1988, 1994). This research was continued by the University of Arizona research group with additional chemical and Nd, Sr isotopic studies (Asmerom et al., 1991; Lang and Titley, 1998), and Pb isotopic studies of igneous rocks and ore minerals (Bouse et al., 1999). All these demonstrated the fundamental importance of the Proterozoic domains or provinces on the nature of magmatism and metallogeny (Titley, 2001). The theme of close linkages between characteristics of granites and type of mineralization was pioneered for tin-mineralized rapakivi granites by Ilmari Haapala (e.g., Haapala, 1977a,b, 1995, 1997).

At the Sierrita porphyry-copper deposit in the Mazatzal Province of southeastern Arizona, elemental and Sr and Nd isotopic data for whole-rock samples, biotite compositions, and trace-element analyses of mineral separates were determined for a suite of magnesian, calc-alkalic, metaluminous rocks. The suite consists of 67-Ma volcanic rocks (andesite, rhyolite) intruded by granodiorite and granite. Isotopic composition and chemical evolution are well correlated throughout the suite. The andesite has the least negative initial ε_{Nd} (−4.3) and lowest $^{87}Sr/^{86}Sr_i$ (0.7069). The andesite is also the oldest and chemically most primitive rock type, having low concentrations of Rb, SiO_2, and high concentrations of the transition elements. These parameters change through the system to the youngest unit (granite), which has the most negative ε_{Nd} (−8.5), highest $^{87}Sr/^{86}Sr_i$ (0.7092), and is chemically the most evolved. The only deposits with ε_{Nd} values more negative than the −8.5 of Sierrita are Mineral Park (−11), Bagdad (−11), and Diamond Joe (−14). These deposits lie in the older Yavapai province in the northwestern part of the state.

Magmas associated with porphyry-copper deposits tend to be quite oxidized, resulting in the biotite coexisting with magnetite, potassium feldspar, and quartz being phlogopitic (Anthony and Titley, 1988, and references therein). A rigorous calculation of oxidation state from this equilibrium assemblage requires knowledge of the Fe^{2+}/Fe^{3+} ratio for the biotites, which was determined for the Sierrita samples from a combination of Mössbauer spectroscopy and wet chemical methods. Values of log fO_2 from Wones (1972) for Sierrita samples for all intrusive phases (the diorite, granodiorite, and granite) varied from approximately -15 at 700 °C to -17 at 600 °C. These values are approximately at the NNO buffer at the higher temperatures and approach the hematite–magnetite buffer at lower temperatures. The values of oxygen fugacity agree with estimates made from the magnitude of the positive Eu anomaly for feldspar separates from the granodiorite. Finally, whole-rock REE patterns suggest early crystallization of amphibole, which is corroborated by petrography.

3.2. Interpretation

The correlation between chemical and Nd isotopic evolution is interpreted to have resulted from a continuous process of progressive assimilation in which mafic magmas invade and incorporate continental crust. Inverse modeling, using geophysical matrix inversion code and combined assimilation-fractional crystallization (AFC) equations, yields a number of insights. Isotopic and elemental data fit best a value of r (the rate of assimilation relative to crystallization) between 0.2 and 0.7 (Anthony and Titley, 1988). The goodness of fit of the model in matrix inversion is measured by the diagonal of the model resolution matrix, with a value of unity corresponding to small standard deviations. The values of the diagonal for r values between 0.7 and 0.2 were usually greater than 0.69 and as high as 0.98 (Anthony and Titley, 1988). Second, the inverse modeling converged best at isotopic values for the assimilant very similar to the observed values for the granite. This strengthened the hypothesis, based on the observation of consistent differences in most negative ε_{Nd} in Yavapai versus Mazatzal, that the granites were in isotopic equilibrium with the crustal reservoir. Finally, the solution for f, the fraction of melt remaining during AFC, is sensitive to whether the initial isotopic values, i.e., the input parameters, for the mantle reservoir were values appropriate to depleted mantle or to enriched, lithospheric mantle. For depleted mantle isotopic values, the first magma-

tism (the andesite) represented an f of 0.8, and the final granites a value of 0.2. When chondritic values were used to represent a lithospheric signature, the values of f were 0.8 and 0.5 for the andesite and the granite, respectively.

Using the ε_{Nd} value of -8.5 of the granites and the age of the Paleoproterozoic crust in southeastern Arizona (1.7 Ga), one can calculate that the crustal source materials had a $^{147}Sm/^{144}Nd$ of approximately 0.135. Similar calculations for the Sr data yield a $^{87}Rb/^{86}Sr$ of 0.30. These ratios are quite different from those of felsic (upper crustal) rocks and are consistent with assimilation of intermediate to mafic, mid- to lower crustal rocks. This interpretation agrees with data from Esperança et al. (1988), who report lower crustal xenoliths of amphibolite in latites from central Arizona with Nd model ages of 2.3 to 1.5 Ga, ε_{Nd} of -9, and $^{87}Sr/^{86}Sr_i$ of 0.7081. Corroborative evidence for an amphibolite source is found in the petrography, chemistry, and oxidation state of the Sierrita rock suite. Early crystallization of amphibole requires water contents of 3% to 4% in the magma, which suggests a crustal source that has not yet undergone dehydration. Mueller (1971) argues that this same magmatic water content would be an effective buffer and oxidant, thus explaining the high oxidation states of the Sierrita magmas.

A final implication of the Sierrita study is that the Nd isotopic signature is sensitive to the age of the Precambrian domain. The incremental change in ε_{Nd} is approximately one epsilon unit per 100 my, and thus the more negative epsilon values for deposits from the Yavapai domain can be attributed to the greater age of that crustal province rather than to a source with a different long-term Sm/Nd ratio. The importance of the contrasting Precambrian domains has been reinforced in subsequent investigations. These include Pb isotopic studies of igneous rocks and ore minerals (Bouse et al., 1999), Ag/Au ratios (Titley, 2001), and most recently Os isotopes (Barra et al., 2002) for Arizona. Recent studies in the Death Valley region of California (Rämö et al., 2002) also document that the isotopic signature of Mesozoic Cordilleran plutons reflects diverse Precambrian lithospheric terranes. The study also finds isotopic evidence for mixing between mantle-derived magmas and Precambrian crust in the magma genesis process.

4. Laramide to Tertiary magmatism in the Colorado Mineral Belt

4.1. Observations

In western North America, the transition from Mesozoic to Cenozoic is represented by two periods of magmatism: the first, associated with Laramide tectonism, is from 75 to 45 Ma, and the second is mid-Tertiary, from 45 to 29 Ma (Christiansen and Yeats, 1992). These time intervals reflect a change in tectonic regime from subduction to extension, as the North American plate overrode the Farallon plate (Severinghaus and Atwater, 1990). Our study of this transition (Ouimette, 1993, 1995) focused on the southern portion of the Colorado Mineral Belt, a northeast-striking Mesoproterozoic (~1.4 Ga) shear zone (Tweto and Sims, 1963), which was reactivated in the Mesozoic and Cenozoic by plutonism and related ore deposits (Stein and Crock, 1990; Cunningham et al., 1994). The Colorado Mineral Belt is associated with a major gravity low (Isaacson and Smithson, 1976). The study characterized two stocks from the Laramide and nine from the mid-Tertiary. The study area (Fig. 2) lies within the transition from the central Colorado Mineral Belt to the southern Colorado Mineral Belt of Stein and Crock (1990).

Table 2 and Figs. 3 and 4 summarize the characteristics of the samples studied. They are divided into three groups, Laramide (Group I) and two from the Mid-Tertiary, based on major and trace element composition. The Laramide suite is magnesian, calc-

alkalic, and metaluminous, which is a signature shared by the majority of Cordilleran granitoids (Frost et al., 2001). It is characterized by moderate REE slopes (Fig. 3) and negative Nb, P, and Ti anomalies (Fig. 4). A notable feature on the multielement diagram is that Ta is decoupled from Nb, resulting in a Nb/Ta ratio of 5. These rocks correspond to the Laramide suite of Stein and Crock (1990), for which ε_{Nd} values of -1 to -9 have been reported. The range in ε_{Nd} suggests that mixing of source reservoirs may have been significant. The mid-Tertiary rocks are divided into two groups on the basis of major and trace element chemistry. The first (Group II) is magnesian, alkali-calcic, and peraluminous. It is characterized by a steep REE slope and a multielement diagram with patterns similar to the Laramide suite, but with deeper anomalies at the same SiO_2 content. The steep REE patterns suggest residual garnet in the source, and the peraluminous nature of the rocks implies metasedimentary source material. These Group II samples are chemically equivalent to the mid-Tertiary granite suite of the southern Colorado Mineral Belt (Stein and Crock, 1990) that has ε_{Nd} values of -7 to -9. Stein and Crock argue that the limited range in isotopic values implies a homogeneous source. The third group is magnesian, alkali-calcic, and borderline metaluminous/peraluminous. It has a third, distinctive REE pattern with an inflection at Eu. The multielement pattern is also distinctive, with negative anomalies at Sr, P, and Ti. For this suite, Nb is coupled with Ta, resulting in a Nb/Ta ratio of 14 to 16, within the range reported as typical for crustal rocks (Barth et al.,

Table 2
Characteristics of granites associated with the Colorado Mineral Belt

	Age	Classification	Trace element chemistry	Comments
Laramide (Group I)	72–71 Ma	Magnesian Calc-alkalic Metaluminous	Moderate REE slope $(La/Yb)_n$=11–12 Nb decoupled from Ta	Source material lower crustal and mafic
Mid-Tertiary (Group II)	41–40 Ma	Magnesian Alkali-calcic Peraluminous	Steep linear REE slope $(La/Yb)_n$=13–45 Nb decoupled from Ta	Steep REE and peraluminosity imply metasedimentary source
Mid-Tertiary (Group III)	37–29 Ma	Magnesian Alkali-calcic Metaluminous/ peraluminous	Break in REE slope at Eu $(La/Yb)_n$=5–16 Nb coupled to Ta	Correlates with Stein and Crock (1990) granodiorite/quartz monzonite suite

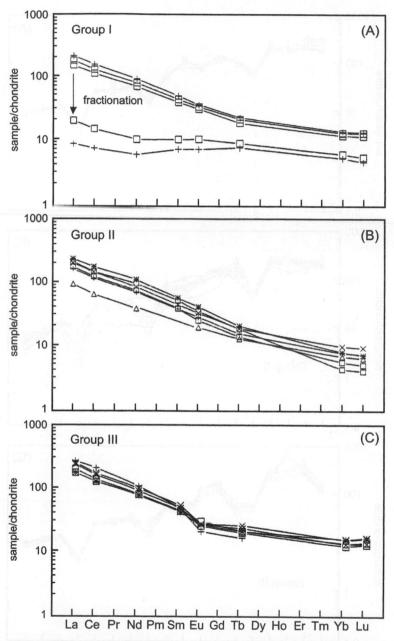

Fig. 3. Chondrite-normalized rare-earth element compositions for Laramide and mid-Tertiary granites of the Colorado Mineral Belt (modified from Ouimette, 1995). (A) Group I, Laramide-aged plutons: the arrow shows the change in REE pattern that accompanies increase in SiO$_2$; (B) Group II, mid-Tertiary granites (41–40 Ma) with steep REE patterns; (C) Group III, mid-Tertiary granites (37–29 Ma) with inflection of REE pattern at Eu. Other characteristics of the groups are summarized in Table 2 and Fig. 4. Normalizing values from Anders and Ebihara (1982).

2000). These samples are chemically equivalent to the mid-Tertiary granodiorites and quartz monzonites of the central Colorado Mineral Belt of Stein and Crock, and have ε_{Nd} values of −8 to −11.

4.2. Interpretation

The most significant observation from this study of the Colorado Mineral Belt is that changes in the

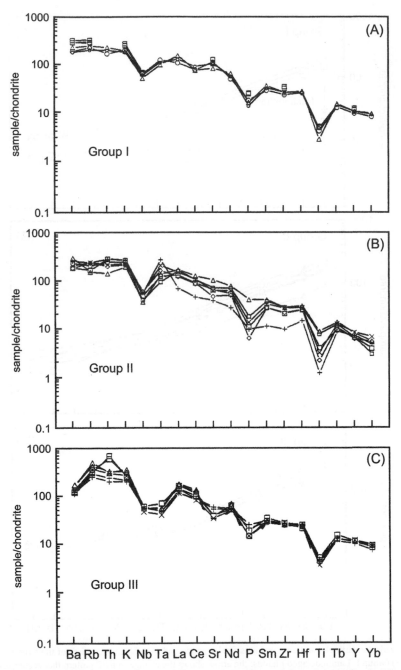

Fig. 4. Chondrite-normalized multielement diagrams for Laramide and mid-Tertiary granites of the Colorado Mineral Belt (modified from Ouimette, 1995). (A) Group I, Laramide-aged plutons characterized by negative Nb, P, and Ti anomalies. (B) Group II, mid-Tertiary granites (41–40 Ma) with patterns similar to Group I except for lower HREE concentrations. (C) Group III, mid-Tertiary granites (37–29 Ma) with coupling of Nb and Ta. Other characteristics of the groups are summarized in Table 2 and Fig. 3. Chondrite normalizing values from Anders and Ebihara (1982).

chemical signature of the granites correlate to the transition in tectonic style from Laramide convergence to mid-Tertiary extension. Based on the combination of elemental and isotopic data, Stein and Crock (1990) proposed a model of assimilation and fractional crystallization involving mantle and mafic, amphibolitic to granulitic lower crust to generate the Laramide granitoids. This combination of source and process is very similar to the porphyry-copper-related granitoids described in the previous section. Both Group II and Group III samples, because of their negative ε_{Nd} values, are interpreted to originate from crustal sources with little or no mantle chemical input. Group II mid-Tertiary granites, with their REE patterns, suggesting garnet in the source and their peraluminosity, were probably derived from metasedimentary sources. Group III mid-Tertiary samples are isotopically and chemically distinct from the Group II samples, requiring that heterogeneous crustal sources were tapped during the mid-Tertiary time interval (Ouimette, 1995).

This shift in source regions as a function of the changing tectonic style has recently been documented for New Mexico by McMillan et al. (2000). They found that at ~36 Ma, volcanism changed abruptly from an arc-like, convergent signature to a bimodal suite. The mafic rocks in this latter suite have elemental and isotopic characteristics that imply they are partial melts of lithosphere with small amount of contamination of lower crustal material. The coeval rhyolitic magmas on the other hand reflect the involvement of upper-crustal components. Barton (1996), in his overview of Mesozoic and Cenozoic magmatism in southwestern North America, reached a similar conclusion. He found that there are two distinctive types of igneous suites: the first consists of calc-alkalic to alkalic rocks formed during periods of convergence and compression. These suites vary from early intermediate and mafic rocks to late felsic rocks over intervals lasting 20 to 50 Ma. The second suite is formed during periods of quiescence or extensional tectonics and is characterized by contemporaneous igneous rocks having widely different compositions.

5. Discussion and conclusions

The three studies described here provide examples of the importance of source materials in determining the chemical and petrological characteristics of granites. The role of source was seen in the differences between the plutonic province and the Panhandle igneous complex in the Mesoproterozoic basement, in the fundamentally different isotopic and metallogenic character of Laramide granites from the Yavapai and Mazatzal provinces in Arizona, and in the changes in elemental character for Laramide versus mid-Tertiary granites in the Colorado Mineral Belt.

The three characteristics most useful are oxidation state, as revealed in these studies through mafic mineral chemistry, isotopic signature, and elemental chemistry. Many previous studies have discussed one or more of these indicators. The following examples are not meant to be an exhaustive review, but rather to highlight some of the pioneering studies. These include, for the oxidation state of the magma, a study by Mueller (1971), which demonstrated the importance of iron content and water fugacity in buffering magmatic oxidation states. Mueller argued that, given the difficulty of changing either parameter during the main stage of differentiation, these variables are inherited from the source. Oxidation state can move away from established buffers during late-stage processes as documented by Czamanske and Wones (1973) and Pichavant et al. (1996). These studies provide examples where magmatic processes do modify the initial oxidation state. They perhaps best illustrate how unusual modification is for the main stage of magmatism and, therefore, that the majority of magmas are buffered by their sources. This point has also been argued by Carmichael (1991) and Blevin and Chappell (1992). A pioneering study on the inheritance of radiogenic and stable isotopic signatures in granites was by McCulloch and Chappell (1982), in which they showed consistent difference in Nd, Sr, and O isotopic values for I- and S-type granites. Numerous others that use isotopic signatures to image source region of magmas followed their study. Important contributions pertinent to Mesozoic and Cenozoic granites of the western United States include Farmer and DePaolo (1983, 1984), Anthony and Titley (1988), Barton (1996), and Lang and Titley (1998). Finally, inherited elemental composition was the original motivation for the Australian typological classification and nomenclature of granites. This is implicit in the original use of "infra" and "supra" for "I" and "S". A follow-on study by Burnham (1992) showed the consistent variations in normative mineralogy for the different granite types of Australia. Elemental inheritance has been brought to a very sophisticated state by Patino Douce and others (e.g., Patiño Douce, 1999), who have provided, through experimental studies, a rigorous evaluation of the chemical composition of magmas as a function of source lithology and chemistry.

The studies reviewed here also demonstrate a strong correlation between the source regions tapped during magma genesis and tectonic setting. For the Laramide of both Arizona and Colorado, AFC processes involving mantle-derived magmas and mafic lower crust characterize convergent tectonic settings. For nonconvergent settings in both the Mesoproterozoic and the mid-Tertiary, the style of magmatism shifted to intracrustal melting of heterogeneous sources. Implicit in this conclusion is that fundamentally different melt regimes may exist for convergent vs. nonconvergent tectonism. The observations presented in this paper leave a twofold challenge for future avenues of study. The first is to continue to refine chemical and petrological characteristics, source materials, and tectonic setting of magmatic suites. The second is to conduct studies that document melt regimes to elucidate the reasons for different melt regimes in individual tectonic settings. A recent example of such a study is from the Mesozoic of New Zealand (Klepeis et al., 2003). A principal justification for continued study is that unraveling the answers to these questions will enhance the role that granites play in interpreting past tectonic environments.

Acknowledgments

I would like to thank Professors Tapani Rämö and Ilmari Haapala for the invitation to the retirement symposium for Professor Haapala. This review includes themes for which Professor Haapala has consistently made important contributions over the years: the petrogenesis of Mesoproterozoic granites, their geologic and tectonic setting, and metallogeny related to granites. I have benefited tremendously from studying his contributions, and it was a pleasure to have the opportunity to synthesize my ideas on these topics. I thank my mentor Professor Spencer Titley, who has played a pivotal role in my understanding of geology, and my students, particularly Mark Ouimette and Melanie Barnes whose excellent work is reflected in this review. The manuscript benefited from reviews by M.C. Gilbert, J.P. Calzia, an anonymous reviewer, and Tapani Rämö.

References

Adams, D.C., Miller, K.C., 1995. Evidence of late Middle Proterozoic extension in the Precambrian basement beneath the Permian Basin. Tectonics 14, 1263–1272.

Åhäll, K.-I., Connelly, J.N., Brewer, T.S., 2000. Episodic rapakivi magmatism due to distal orogenesis? Correlation of 1.69–1.50 Ga orogenic and inboard, "anorogenic" events in the Baltic Shield. Geology 28, 823–826.

Amarante, J.F.A., Kelley, S.A., Heizler, M.T., Barnes, M.A., Miller, K.C., Anthony, E.Y., 2004. Characterization and age of the Mesoproterozoic Debaca sequence in the Tucumcari Basin, New Mexico. In: Karlstrom, K., Keller, G.R. (Eds.), The Rocky Mountain Region–An Evolving Lithosphere: Tectonics, Geochemistry, and Geophysics. American Geophysical Union Monograph CD-ROM Special Volume (in press).

Anders, E., Ebihara, M., 1982. Solar abundances of the elements. Geochimica et Cosmochimica Acta 46, 2363–2380.

Anderson, J.L., Cullers, R.L., 1999. Paleo- and Mesoproterozoic granitic plutonism of Colorado and Wyoming. Rocky Mountain Geology 34, 149–164.

Anthony, E.Y., 1986. Geochemical evidence for crustal melting in the origin of the igneous suite at the Sierrita porphyry copper deposit, southeastern Arizona. PhD thesis, University of Arizona, Tucson, U.S.A.

Anthony, E.Y., Titley, S.R., 1988. Progressive mixing of isotopic reservoirs during magma genesis at the Sierrita porphyry copper deposit: inverse solutions. Geochimica et Cosmochimica Acta 52, 2235–2249.

Anthony, E.Y., Titley, S.R., 1994. Patterns of element mobility during hydrothermal alteration of the Sierrita porphyry copper deposit, Arizona. Economic Geology 89, 186–192.

Asmerom, Y., Patchett, J.P., Damon, P.E., 1991. Crust–mantle interactions in continental arcs: inferences from the Mesozoic arc in the southwestern United States. Contributions to Mineralogy and Petrology 107, 124–134.

Barnes, M.A., 2001. The petrology and tectonics of the Mesoproterozoic margin of southern Laurentia. PhD thesis, Texas Tech University, Lubbock, U.S.A.

Barnes, M.A., Rohs, R.C., Anthony, E.Y., Van Schmus, R.W., Denison, R.E., 1999a. Isotopic and elemental chemistry of subsurface Precambrian igneous rocks, west Texas and eastern New Mexico. Rocky Mountain Geology 34, 245–262.

Barnes, C.G., Shannon, W.M., Kargi, H., 1999b. Diverse Mesoproterozoic basaltic magmatism in west Texas. Rocky Mountain Geology 34, 263–273.

Barnes, M.A., Anthony, E.Y., Williams, I., Asquith, G.B., 2002. Architecture of a 1.38 Ga granite–rhyolite complex as revealed by geochronology and isotopic and elemental geochemistry of subsurface samples from west Texas, USA. Precambrian Research 119, 9–43.

Barra, F., Ruiz, J., Mathur, R., Titley, S.R., 2002. A Re–Os study of sulfide minerals from the Bagdad porphyry Cu–Mo deposit, northern Arizona, USA. Mineralium Deposita 38, 585–596.

Barth, M.G., McDonough, W.F., Rudnick, R.L., 2000. Tracking the budget of Nb and Ta in the continental crust. Chemical Geology 165, 197–213.

Barton, M.D., 1996. Granitic magmatism and metallogeny of southwestern North America. Transactions of the Royal Society of Edinburgh. Earth sciences 87, 261–280.

Bickford, M.E., Kristian, S., Nielsen, K.C., McLelland, J.M., 2000. Geology and geochronology of Grenville-age rocks in the Van Horn and Franklin Mountains area, west Texas: implications for the tectonic evolution of Laurentia during the Grenville. Geological Society of America Bulletin 112, 1134–1148.

Blevin, P.L., Chappell, B.W., 1992. The role of magma sources, oxidation states, and fractionation in determining the granite metallogeny of eastern Australia. Transactions of the Royal Society of Edinburgh. Earth Sciences 83, 305–316.

Bouse, R.M., Ruiz, J., Titley, S.R., Tosdal, R.M., Wooden, J.L., 1999. Pb isotopic compositions of late Cretaceous–early Tertiary igneous rocks and sulfide mineralization in Arizona: implications for the source of plutons and metals in porphyry copper deposits. Economic Geology 94, 211–244.

Burnham, C.W., 1992. Calculated melt and restite compositions of some Australian granites. Transactions of the Royal Society of Edinburgh. Earth Sciences 83, 387–397.

Carmichael, I.S.E., 1991. The redox state of basic and silicic magmas: a reflection of their source regions? Contributions to Mineralogy and Petrology 106, 129–141.

Chappell, B.W., 1979. Granites as images of their source rocks. Abstracts - Geological Society of America 11, 400.

Christiansen, R.L., Yeats, R.S., 1992. Post-Laramide geology of the U.S. Cordilleran region. In: Burchfiel, B.C., Lipman, P.W., Zoback, M.L. (Eds.), The Cordilleran Orogen: Conterminous U.S., DNAG G-3. The Geological Society of America, Boulder, pp. 261–406.

Cunningham, C.G., Naeser, C.W., Marvin, R.F., Luedke, R.G., Wallace, A.R., 1994. Ages of selected intrusive rocks and associated ore deposits in the Colorado Mineral Belt. United States Geological Survey Bulletin 2109, 31.

Czamanske, G.K., Wones, D.R., 1973. Oxidation during magmatic differentiation, Finnmarka complex, Oslo area, Norway: Part 2. The mafic silicates. Journal of Petrology 14, 349–380.

DePaolo, D.J., 1988. Neodymium Isotope Geochemistry. In: Wyllie, P.J. (Ed.), Monograph Series Rocks and Minerals. Monograph Number 20. Springer, Berlin. 187 pp.

Esperança, S., Carlson, R.W., Shirey, S.B., 1988. Lower crustal evolution under central Arizona: Sr, Nd, and Pb isotopic and geochemical evidence from the mafic xenoliths of Camp Creek. Earth and Planetary Science Letters 90, 26–40.

Farmer, L.G., DePaolo, D.J., 1983. Origin of Mesozoic and Tertiary granite in the western United States and implications for Pre-Mesozoic crustal structure Nd and Sr: isotopic studies in the geocline of the northern Great Basin. Journal of Geophysical Research 88, 3379–3401.

Farmer, L.G., DePaolo, D.J., 1984. Origin of Mesozoic and Tertiary granite in the western United States and implications for Pre-Mesozoic crustal structure: Nd and Sr isotopic studies of unmineralized and Cu- and Mo-mineralized granite in the Precambrian craton. Journal of Geophysical Research 89, 10141–10160.

Flawn, P.T., 1956. Basement Rocks of Texas and Southeast New Mexico. The University of Texas Press, p. 5605.

Frost, C.D., Frost, B.R., Chamberlain, K.R., Edwards, B.R., 1999. Petrogenesis of the 1–43 Ga Sherman batholith, SE Wyoming, USA: a reduced, rapakivi-type anorogenic granite. Journal of Petrology 40, 1771–1802.

Frost, R.B., Barnes, C.G., Collins, W.J., Arculus, R.J., Ellis, D.J., Frost, C.D., 2001. A geochemical classification for granitic rocks. Journal of Petrology 42, 2033–2048.

Haapala, I., 1977a. The controls of tin and related mineralizations in the rapakivi-granite areas of south-eastern Fennoscandia. Geologiska Föreningens i Stockholm Förhandlingar 999, 130–142.

Haapala, I., 1977b. Petrography and geochemistry of the Eurajoki stock, a rapakivi-granite complex with greisen-type mineralization in southwestern Finland. Geological Survey of Finland, 286.

Haapala, I., 1995. Metallogeny of the rapakivi granites. Mineralogy and Petrology 54, 149–160.

Haapala, I., 1997. Magmatic and postmagmatic processes in tin-mineralized granites: topaz-bearing leucogranite in the Eurajoki Rapakivi Granite Stock, Finland. Journal of Petrology 38, 1645–1659.

Haapala, I., Rämö, O.T., 1999. Rapakivi granites and related rocks: an introduction. Precambrian Research 95, 1–7.

Isaacson, L.B., Smithson, S.B., 1976. Gravity anomalies and granite emplacement in west-central Colorado. Geological Society of America Bulletin 87, 22–28.

Kargi, H., Barnes, C.G., 1995. A Grenville-age layered intrusion in the subsurface of west Texas: petrology, petrography, and possible tectonic setting. Canadian Journal of Earth Sciences 32, 2159–2166.

Karlstrom, K., CD-ROM Group, 2002. Structure and evolution of the lithosphere beneath the Rocky Mountains: initial results from the CD-ROM experiment. GSA Today 12, 4–10.

Keller, G.R., Hills, J.M., Baker, M.R., Wallin, E.T., 1989. Geophysical and geochronological constraints on the extent and age of mafic intrusions in the basement of west Texas and eastern New Mexico. Geology 11, 1049–1052.

Klepeis, K.A., Clarke, G.L., Rushmer, T., 2003. Magma transport and coupling between deformation and magmatism in the continental lithosphere. GSA Today 13, 4–11.

Lang, J.R., Titley, S.R., 1998. Isotopic and geochemical characteristics of Laramide magmatic systems in Arizona and implications for the genesis of porphyry copper deposits. Economic Geology 93, 138–170.

Larson, E.E., Patterson, P.E., Mutschler, F.E., 1994. Lithology, chemistry, age, and origin of the Proterozoic Cardenas Basalt, Grand Canyon, Arizona. Precambrian Research 65, 255–276.

McCulloch, M.T., Chappell, B.W., 1982. Nd isotopic characteristics of S- and I-type granites. Earth and Planetary Science Letters 58, 51–64.

McMillan, N.J., Dickin, A.P., Haag, D., 2000. Evolution of magma source regions in the Rio Grande rift, southern New Mexico. Geological Society of America Bulletin 112, 1582–1593.

Muehlberger, W.R., Denison, R.E., Lidiak, E.G., 1967. Basement rocks in the continental interior of the United States. American Association of Petroleum Geologists Bulletin 51, 2351–2380.

Mueller, R.F., 1971. Oxidative capacity of magmatic components. American Journal of Science 270, 236–243.

Nelson, B.K., DePaolo, D.J., 1985. Rapid production of continental crust 1.7 to 1.9 b.y. ago: Nd isotopic evidence from the basement of the North American mid-continent. Geological Society of America Bulletin 96, 746–754.

Ouimette, M.A., 1993. Trace element compositions of late Cretaceous to late Tertiary epizonal plutons from west-central Colorado. Abstracts - Geological Society of America, A-42.

Ouimette, M.A., 1995. Petrology and geochemistry of Laramide and Tertiary igneous rocks in the Elk Mountains, Gunnison County, Colorado. PhD thesis, University of Texas at El Paso, El Paso, U.S.A.

Patchett, J.P., Ruiz, J., 1989. Nd isotopes and the origin of Grenville-age rocks in Texas: implications for Proterozoic evolution of the United States mid-continent region. Journal of Geology 97, 685–695.

Patiño Douce, A.E., 1996. Effects of pressure and H_2O content on the compositions of primary crustal melts. Special Paper - Geological Society of America 315, 11–22.

Patiño Douce, A.E., 1999. What do experiments tell us about the relative contributions of crust and mantle to the origin of granitic magmas? Special Publication - Geological Society of London 168, 55–75.

Pichavant, M., Hammouda, T., Scaillet, B., 1996. Control of redox state and Sr isotopic composition of granitic magmas: a critical evaluation of the role of source rocks. Transactions of the Royal Society of Edinburgh. Earth Sciences 87, 321–330.

Pittenger, M.A., Marsaglia, K.M., Bickford, M.E., 1994. Depositional history of the middle Proterozoic Casner Marble and basal Mundy Breccia, Franklin Mountains, west Texas. Journal of Sedimentary Research. Section B, Stratigraphy and Global Studies 64, 282–297.

Rämö, O.T., Haapala, I., 1995. One hundred years of Rapakivi Granite. Mineralogy and Petrology 52, 129–185.

Rämö, O.T., Calzia, J.P., Kosunen, P.J., 2002. Geochemistry of Mesozoic plutons, southern Death Valley region, California: insights into the origin of Cordilleran interior magmatism. Contributions to Mineralogy and Petrology 143, 416–437.

Rämö, O.T., McLemore, V.T., Hamilton, M.A., Kosunen, P.J., Heizler, M.T., Haapala, I., 2003. Intermittent 1630–1220 Ma magmatism in central Mazatzal province: new geochronological piercing points and some tectonic implications. Geology 31, 335–338.

Severinghaus, J., Atwater, T., 1990. Cenozoic geometry and thermal state of the subducting slabs beneath western North America. Memoir - Geological Society of America 176, 1–22.

Shannon, W.M., Barnes, C.G., Bickford, M.E., 1997. Grenville magmatism in west Texas: petrology and geochemistry of the Red Bluff granitic suite. Journal of Petrology 38, 1279–1305.

Stein, H.J., Crock, J.G., 1990. Late Cretaceous–Tertiary magmatism in the Colorado Mineral Belt: rare earth element and samarium–neodymium isotopic studies. Memoir - Geological Society of America 174, 195–224.

Thomas, J.J., Shuster, R.D., Bickford, M.E., 1984. A terrane of 1350–1400 million year-old silicic volcanic and plutonic rocks in the buried Proterozoic of the mid-continent and in the Wet Mountains. Geological Society of America Bulletin 95, 1150–1157.

Timmons, J.M., Karlstrom, K.E., Dehler, C.M., Geissman, J.W., Heizler, M.T., 2001. Proterozoic multistage (1.1 and 0.8 Ga) extension in the Grand Canyon Supergroup and establishment of northwest and north–south tectonic grains in the southwestern United States. Geological Society of America Bulletin 113, 163–180.

Titley, S.R., 2001. Crustal affinities of metallogenesis in the American Southwest. Economic Geology 96, 1323–1342.

Titley, S.R., Anthony, E.Y., 1989. Laramide mineral deposits in Arizona. Arizona Geological Society Digest 17, 485–514.

Tweto, O., Sims, P.K., 1963. Precambrian ancestry of the Colorado Mineral Belt. Geological Society of America Bulletin 74, 991–1014.

Wones, D.R., 1972. Stability of biotite: a reply. American Mineralogist 57, 316–317.

Wones, D.R., Eugster, H.P., 1965. Stability of biotite: experiment, theory, and application. American Mineralogist 50, 1228–1272.

Wrucke, C.T., 1989. The Middle Proterozoic Apache Group, Troy Quartzite, and associated diabase of Arizona. In: Jenney, J.P., Reynolds, S.J. (Eds.), Arizona Geological Society Digest, vol. 117, pp. 239–258.

Available online at www.sciencedirect.com

Lithos 80 (2005) 75–99

www.elsevier.com/locate/lithos

Source contributions to Devonian granite magmatism near the Laurentian border, New Hampshire and Western Maine, USA

Paul B. Tomascak[a,*], Michael Brown[b], Gary S. Solar[b,c], Harry J. Becker[a],
Tracey L. Centorbi[a,b], Jinmei Tian[b]

[a]*Isotope Geochemistry Laboratory, Department of Geology, University of Maryland, College Park, MD 20742, USA*
[b]*Laboratory for Crustal Petrology, Department of Geology, University of Maryland, College Park, MD 20742, USA*
[c]*Department of Earth Sciences, SUNY, College at Buffalo, Buffalo, NY 14222, USA*

Received 9 May 2003; accepted 9 September 2004
Available online 8 December 2004

Abstract

Radiogenic isotope data (initial Nd, Pb) and elemental concentrations for the Mooselookmeguntic igneous complex, a suite of mainly granitic intrusions in New Hampshire and western Maine, are used to evaluate petrogenesis and crustal variations across a mid-Paleozoic suture zone. The complex comprises an areally subordinate monzodiorite suite [377±2 Ma; ε_{Nd} (at 370 Ma)=−2.7 to −0.7; initial $^{207}Pb/^{204}Pb$=15.56–15.58] and an areally dominant granite [370±2 Ma; ε_{Nd} (at 370 Ma)=−7.0 to −0.6; initial $^{207}Pb/^{204}Pb$=15.55–15.63]. The granite contains meter-scale enclaves of monzodiorite, petrographically similar to but older than that of the rest of the complex [389±2 Ma; ε_{Nd} (at 370 Ma)=−2.6 to +0.3; initial $^{207}Pb/^{204}Pb$ c. 15.58, with one exception]. Other granite complexes in western Maine and New Hampshire are c. 30 Ma older than the Mooselookmeguntic igneous complex granite, but possess similar isotopic signatures.

Derivation of the monzodioritic rocks of the Mooselookmeguntic igneous complex most likely occurred by melting of Bronson Hill belt crust of mafic to intermediate composition. The Mooselookmeguntic igneous complex granites show limited correlation of isotopic variations with elemental concentrations, precluding any significant presence of mafic source components. Given overlap of initial Nd and Pb isotopic compositions with data for Central Maine belt metasedimentary rocks, the isotopic heterogeneity of the granites may have been produced by melting of rocks in this crustal package or through a mixture of metasedimentary rocks with magmas derived from Bronson Hill belt crust.

New data from other granites in western Maine include Pb isotope data for the Phillips pluton, which permit a previous interpretation that leucogranites were derived from melting heterogeneous metasedimentary rocks of the Central Maine belt, but suggest that granodiorites were extracted from sources more similar to Bronson Hill belt crust. Data for the Redington pluton are best satisfied by generation from sources in either the Bronson Hill belt or Laurentian basement. Based on these data, we infer that Bronson Hill belt crust was more extensive beneath the Central Maine belt

---

* Corresponding author. Present address: Department of Earth Sciences, Piez Hall, SUNY-Oswego, Oswego, NY 13126, USA. Tel.: +1 315 312 2786; fax: +1 315 312 3059.

E-mail address: tomascak@oswego.edu (P.B. Tomascak).

than previously recognized and that mafic melts from the mantle were not important to genesis of Devonian granite magma.

Keywords: Acadian; Appalachians; Bronson Hill belt; Central Maine belt; Granite; Geochemistry; Pb isotopes; Nd isotopes

1. Introduction

In the study of the histories of orogenic belts, granites are indispensable sources of information. Age relations, elemental and isotopic compositions, fabrics and three-dimensional forms provide evidence as to the dynamics of exhumed convergent plate margins and give pivotal clues to the identity of unexposed basement terranes. Thus, as long as geochemical contrasts between source materials are clear, regional-scale geochemical mapping may illuminate basement relations (Bennett and DePaolo, 1987; Ayuso and Bevier, 1991; Dorais and Paige, 2000). However, in areas like the northern Appalachians of New Hampshire and Maine, where end-member source compositions may bear considerable similarity to one another, detailed studies of individual precisely-dated plutonic complexes may be a more appropriate approach to provide the kind of information required.

In the northern Appalachians, the principal exposed basement source candidates are Laurentian (North American) and Avalonian (non-North American) continental crust, which exhibit strong contrasts in their Pb and Nd isotopic signatures (Ayuso and Bevier, 1991; Barr and Hegner, 1992; Whalen et al., 1994; Kerr et al., 1995). However, the nature of the basement beneath New Hampshire and western Maine is likely to be much more complex (Stewart et al., 1992), requiring assessment of additional source components, including various mantle reservoirs, Taconic arc crust, and mid-crustal metasedimentary rocks. The specific nature of the basement near the edge of the geophysically defined margin of North American continental crust is still rather poorly known. For this reason, our study of granite complexes in New Hampshire and western Maine is important because the magmas from which they crystallized potentially record information that directly addresses this uncertainty

in basement terranes. There is abundant evidence for lower crust of circum-Gondwanan affinity across strike to the south of this boundary (Whalen et al., 1998; Tomascak et al., 1996), but whether the non-North American crust extends fully to the boundary with North American crust is a point of speculation.

In addition to questions about the details of tectonic assembly of a continental margin that was active for more than 100 Ma, the northern Appalachians present longstanding problems regarding the heat sources for the generation of abundant peraluminous granite bodies (DeYoreo et al., 1989; Chamberlain and Sonder, 1990; Brown and Solar, 1999). Furthermore, orogens control the mechanics of interactions between converging plates; thus, it is important to understand both weakening and hardening mechanisms. Syntectonic pervasive melt flow, episodic melt expulsion, ascent and emplacement, and crystallization of melt all affect the rheology and control the mechanical response to imposed stresses. For these reasons it is important to evaluate crustal melting and the relative contributions of mantle-derived and crustal-derived melts to upper crustal plutons. Geochemical tracers, particularly isotopic systems, yield important petrogenetic information.

The plutons studied are in part petrologically primitive (with potentially important mantle contributions). Confirming or ruling out mantle source components for these rocks, or indeed placing constraints on the volume of mantle contributions, will allow development of better-constrained tectonic models. In this study we use precise geochronology, isotope tracers and elemental geochemistry of plutonic rocks in order to more clearly define the nature of the crust beneath the boundary of the Bronson Hill and Central Maine belts of the northern Appalachians during the Devonian.

2. Geology and samples

The New Hampshire and Maine part of the northern Appalachians is divided into several tecto-nostratigraphic units defined by discrete northeast–southwest oriented tracts of deformed and metamorphosed rock (Fig. 1). The Central Maine belt (CMB) underlies much of the area; it is composed of a Lower Paleozoic sedimentary succession, deformed and metamorphosed at greenschist to upper amphibolite facies conditions, and intruded by Devonian plutons. The CMB is located between Ordovician rocks of the Bronson Hill belt (BHB) to the northwest and Neoproterozoic to Silurian rocks of the Avalon Composite terrane (ACT) to the southeast, from which it is separated by the dextral-transcurrent Norumbega shear zone system (NSZS).

The area of study covers part of the western side of the Central Maine belt where it is in contact with the Bronson Hill belt to the northwest (Figs. 1 and 2). The CMB comprises Siluro-Devonian metasedimentary rocks (metaturbidite) of the "Rangeley stratigraphic sequence" (Moench, 1971; Moench et al., 1995; Solar and Brown, 2001a), migmatites (Brown and Solar, 1998a, 1999; Solar and Brown, 2001b) and predominantly granitic plutons of various volumes and shapes (e.g., Moench et al., 1995; Brown and Solar, 1998b, 1999; Pressley and Brown, 1999). To the northwest, the CMB is bounded by a Devonian tectonite zone (Fig. 1; Solar and Brown, 2001a) across which the belt is juxtaposed against the BHB. The main deformation and metamorphism within the CMB were the result of Devonian Acadian orogenesis, whereas the main deformation and metamorphism within the BHB were the result of Ordovician Taconic orogenesis. However, both belts record Devonian metamorphism and deformation in the area of their contact, recorded by structures such as refolded folds found in the BHB rocks, but not in the adjacent CMB rocks, and the tectonite zone that includes the contact. Plutonism across the area was Devonian and late syntectonic (Solar et al., 1998), and may be concordant with, or discordant to, the regional metamorphic mineral fabrics (Brown and Solar, 1998b; Solar and Brown, 1999).

The penetrative deformation within the CMB is recorded by metamorphic mineral fabrics and regional-scale folds of the metasedimentary rock units consistent with bulk transpression during the Acadian orogeny (Solar and Brown, 2001a). Oblique (dextral) southeast-side up contraction of the belt was largely accommodated within the broad Central Maine belt shear zone system (Fig. 1; Brown and Solar, 1998a). Solar and Brown (1999, 2001a) and Brown and Solar (1998a,b, 1999) subdivided the region into kilometer-scale alternating NE–SW-trending structural zones of apparent flattening (AFZ in Fig. 2) and apparent constrictional finite strain as defined by the bulk rock fabrics, supported by the style and intensity of folds of the metasedimentary rock layers (tighter in the apparent flattening zones). The grade of Devonian regional metamorphism within the CMB ranges from greenschist to upper amphibolite facies; granulite facies assemblages occur in Massachusetts (Chamberlain and Robinson, 1989). In Maine and New Hampshire regional metamorphism occurred at low pressure; andalusite is abundant in rocks of suitable grade and composition. At the highest grade (upper sillimanite zone; e.g., Guidotti and Holdaway, 1993), anatectic migmatites are variably developed in metapelitic rocks (Solar and Brown, 2001b; Johnson et al., 2003). Contact metamorphic effects of pluton emplacement are local and have modified and variably overprinted the regional metamorphism (e.g., Solar and Brown, 1999, 2000; Guidotti and Johnson, 2002; Johnson et al., 2003).

Peraluminous granite bodies are widespread (Fig. 1), ranging from millimeter- to centimeter-scale leucosomes and meter-scale tabular- and cylinder-shaped bodies in migmatites (e.g., Brown and Solar, 1999; Solar and Brown, 2001b), to kilometer-scale composite plutons (Brown and Solar, 1998b). The Phillips and Redington plutons and the central lobe of the Lexington composite pluton (Fig. 1) have yielded ages of c. 404 Ma (U–Pb zircon, monazite), which are interpreted to record the age of crystallization (Solar et al., 1998; Brown and Pressley, 1999; see Appendix B). Schlieric granite and decimeter- to meter-scale tabular granite bodies within the migmatites crystallized synchronously with the c. 404 Ma plutons (Solar et al., 1998). The three-dimensional geometry of the granites, their relationship to the regional structure and the close association of smaller bodies, such as the Phillips pluton, with heterogeneous migmatite support a genetic relationship between deformation, granite ascent and pluton emplacement (Brown and

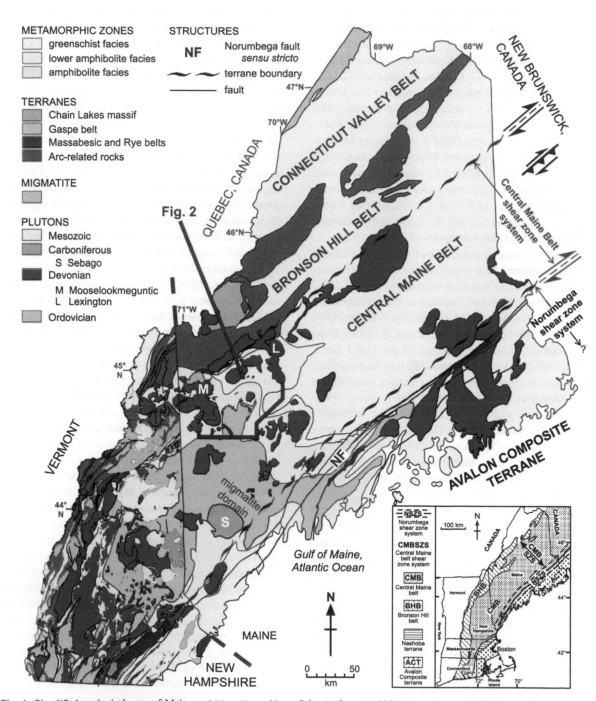

Fig. 1. Simplified geological map of Maine and New Hampshire, of the northeastern U.S.A. (see inset map for location), to illustrate the distribution of plutons, migmatite, metamorphic zones and principal terranes. Area shown in Fig. 2 is indicated. Modified after Lyons et al. (1997) and Solar and Brown (1999).

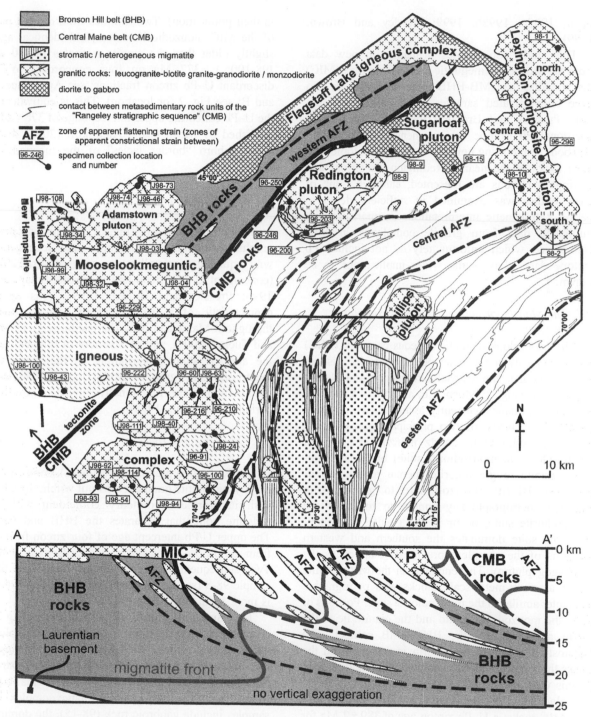

Fig. 2. Simplified geological map and schematic structure section of the area of study (see Fig. 1 for location) based upon the mapping of Solar and Brown (2001a), and Tian (2000). Abbreviations: MIC—Mooselookmeguntic igneous complex; P—Phillips pluton; BHB—Bronson Hill belt; CMB—Central Maine belt. The regional structure is illustrated as the alternating zones of apparent flattening strain (AFZ) and zones of apparent constrictional strain (ACZ). The tectonite zone in the northwest is coincident with the BHB-CMB boundary at the surface which is cross-cut by the MIC.

Solar, 1998a, 1998b, 1999; Pressley and Brown, 1999).

This study concentrates primarily on new data from the Mooselookmeguntic igneous complex (MIC) that straddles the CMB-BHB contact (Fig. 2). We also have reinvestigated samples of the Phillips pluton from Pressley and Brown (1999), and have made additional Pb isotope measurements on rocks from this body. Supporting new elemental and isotopic data from rocks of the nearby Redington, Sugarloaf and Lexington plutons are presented, although none of these bodies was examined in sufficient detail to permit petrogenetic interpretation. Comparison with published data from granite complexes in New Hampshire, which unfortunately do not include Pb isotope data, allow some wider implications to be drawn.

2.1. The Mooselookmeguntic igneous complex

The Mooselookmeguntic igneous complex (MIC) was previously mapped as petrographically distinct plutonic bodies, the Mooselookmeguntic and Umbagog plutons (Fig. 2; Moench et al., 1995). Considering the information currently available, we consider the MIC to consist of two principal types of rock: biotite granodiorite to quartz monzodiorite (herein referred to as the monzodiorite suite), biotite and two-mica granite to granodiorite (herein referred to as granite). Biotite monzodiorite to granodiorite enclaves, petrographically similar to rocks of the monzodiorite suite, occur in the granite. The monzodiorite suite dominates the southern and western portions of the complex (previously mapped as the Umbagog pluton). The enclaves in the MIC granite occur as multi-meter-scale blocks with petrographic character similar to the monzodiorite suite. Portions of both the monzodiorite suite and the granite lie to the north and south of the BHB-CMB contact. The Adamstown pluton in the north is distinct from the rocks that comprise the MIC and will not be discussed in this work.

Solar et al. (1998) reported crystallization ages for two MIC rocks: a U–Pb zircon age of 389 ± 2 Ma for sample 96-216 (see Fig. 2), a granodiorite enclave ("biotite granite" in their publication), and a concordant U–Pb monazite age of 370 ± 1 Ma for sample 96-210 of the MIC granite ("two-mica leucogranite"

in their publication). Two samples from distinct parts of the MIC monzodiorite suite yield identical ages, slightly older than the granite (samples J98-40 and J98-100: c. 377 Ma on four concordant to 3% discordant U–Pb zircon fractions; see Appendices A and B; see Fig. 2 for location). This is equivalent to the U–Pb concordia upper intercept age of 378 ± 2 Ma published by Moench and Aleinikoff (2002) for an alkali gabbro "border facies" of the monzodiorite suite.

2.2. The Phillips pluton

The Phillips pluton (Fig. 2) comprises petrographically distinct granodiorite and leucogranite with equivalent crystallization ages of 404 ± 2 Ma (concordant U–Pb zircon and monazite; Solar et al., 1998; Pressley and Brown, 1999). This age is characteristic of a wide band of igneous bodies throughout Maine and New Hampshire (e.g., Bradley and Tucker, 2002), and of a variety of migmatites in the area of study, contributing to the thesis that metamorphism, granite production and regional deformation were functionally linked during this period (Solar et al., 1998).

2.3. Other granitic rocks

The Redington pluton (5 specimens), dominantly a porphyritic biotite granite, crops out within the CMB, with its northern boundary coincident with the tectonite zone that separates the BHB and CMB. The upper U/Pb intercept age of four zircon fractions from sample 96-246 is 408 ± 5 Ma (Solar et al., 1998). Two subsequent concordant zircon fractions from sample 98-8 (see Fig. 2) suggest an age of c. 406 Ma is more accurate (see Appendix B).

The Sugarloaf pluton (2 specimens; Fig. 2) occupies a similar position to the Redington pluton with respect to the BHB-CMB contact. Mingled felsic-mafic rocks in outcrops of this body bear strong resemblance to large portions of the more mafic part of the adjacent Flagstaff Lake intrusive complex. The samples include gabbroic rock (98-15), the dominant rock type in outcrop, and subordinate material from a felsic pod (sample 98-9).

The Lexington composite pluton (5 specimens) comprises northern, central and southern portions, based partly on interpretation of the three-dimensional character of the pluton (Brown and Solar, 1998b), occurring east along strike from the main body of the MIC (Fig. 2). Rock types are primarily granite and granodiorite, based on normative composition, although one sample (98-10-2) is a gabbro enclave. All granite samples contain biotite and some contain two igneous micas (e.g., 98-2). The U–Pb zircon age determined by Solar et al. (1998; 404±2 Ma) is from sample 96-296-6, from the central portion of the pluton, and an identical zircon age has been measured from the southern lobe (sample 98-2: c. 403 Ma on two concordant fractions; see Appendix B). Interestingly, the northern part of the pluton has yielded a younger age (sample 98-1: c. 365 Ma on two concordant U–Pb zircon fractions; see Appendix B).

3. Analytical procedures

Major and trace element analyses (except Nd, Sm) were produced by XRF at Washington University following their standard procedures (R. Couture, analyst). Most samples for Sm–Nd isotope dilution analyses were prepared by flux fusion, Fe-coprecipitation, and cation exchange identically to Tomascak et al. (1996). Despite a Nd blank of c. 3 ng during the course of the study, the ratio of sample/blank for Nd was always greater than 500. Samples from the Redington, Lexington and Sugarloaf pluton were digested for two days in concentrated HF+HNO$_3$ in Paar Teflon bombs at 210 °C. Blanks for these preparations were <400 pg for Nd and <100 pg for Sm, and corrections were insignificant. Samples of alkali feldspar were isolated, cleaned and leached as detailed in Tomascak et al. (1996). The total Pb blank was 130–370 pg, of which 1–4 pg was introduced during loading. Using whole-rock Pb concentrations as a proxy for concentrations in alkali feldspar, a minimum of 200 ng Pb was processed for each sample, indicating that no significant blank correction was necessary.

Isotope ratio measurements carried out at the Isotope Geochemistry Laboratory, Department of Geology, University of Maryland, used a VG Sector 54 mass spectrometer, and methods described in Tomascak et al. (1996). Measured Nd isotope ratios were normalized to $^{146}Nd/^{144}Nd$=0.72190. Analyses of the La Jolla Nd standard during the study yielded a mean $^{143}Nd/^{144}Nd$ of 0.511844±16 (2σ; n=10), translating to an uncertainty of ±0.35 ε unit. To facilitate discussion of rocks with distinct crystallization ages, Nd isotope data in this work are calculated for the age of the youngest rocks, 370 Ma. Given the Sm/Nd of the samples, there is at most a 0.3 unit difference between ε_{Nd} (at 370 Ma) and the initial ε_{Nd} for the oldest samples.

A fractionation correction of 0.079±0.008% amu^{-1} was applied to the Pb isotope data, referenced relative to the updated recommended isotopic values of NBS Pb standard SRM-981 (Thirlwall, 2000). The Pb standard analyses collected throughout the study yielded a mean reproducibility of ±0.032% amu^{-1} (2σ; n=16) for ratios involving measurement of ^{204}Pb. To assess potential long-term, user-to-user reproducibility, a K-feldspar sample from Tomascak et al. (1996; SG-3-14: $^{206}Pb/^{204}Pb$=18.467, $^{207}Pb/^{204}Pb$=15.622, $^{208}Pb/^{204}Pb$= 38.256) was hand-picked, cleaned and leached. The mean offset between results is ±0.046% amu^{-1} ($^{206}Pb/^{204}Pb$=18.439, $^{207}Pb/^{204}Pb$=15.612, $^{208}Pb/^{204}Pb$=38.192), similar to the reproducibility of replicate preparations of individual feldspar samples from this study (±0.025% amu^{-1}).

4. Geochemical results

4.1. Elemental data

The MIC monzodiorite suite and enclaves range from metaluminous to peraluminous, whereas all of the MIC granites are peraluminous (Table 1). Despite their c. 12 Ma age difference, compositions of samples from the monzodiorite suite and of the enclaves overlap almost completely, with the exception that the enclaves have lower Mg# (average of 48.6 versus 57.6). Contents of Fe$_2$O$_3^T$, MgO and CaO decrease with increasing SiO$_2$, and correlations are absent or poor between SiO$_2$ and Al$_2$O$_3$, Na$_2$O, or K$_2$O (Fig. 3). All of these samples show light rare earth element enrichment typical of crustal rocks ($^{147}Sm/^{144}Nd$=0.1002 to 0.1273; Table 2).

Table 1
Major (wt.%) and trace (ppm) elements in samples from this study

Pluton group	MIC enclave					MIC monzodiorite			
Sample number	96-216	J9832-2	J98-54	J98-93	J98-111-3	96-222	J98-43	J98-100	J98-63-2
Rock type	gdr	gdr	qtz mdr	mzdr	mdr	qtz mdr	qtz mdr	mzn	gdr
SiO_2	65.0	67.2	63.8	51.7	52.1	57.9	57.7	57.2	64.6
TiO_2	0.85	0.51	0.91	1.74	2.03	1.28	1.35	1.36	0.90
Al_2O_3	15.97	15.89	16.25	14.87	15.77	16.10	15.00	15.34	15.95
Fe_2O_3*	4.69	4.54	4.58	8.62	9.14	7.02	6.70	6.03	4.43
MnO	0.10	0.10	0.10	0.24	0.17	0.12	0.10	0.11	0.08
MgO	2.83	1.75	2.04	3.75	4.99	4.57	5.20	4.20	3.09
CaO	3.66	4.12	3.78	6.26	7.39	5.32	5.70	4.73	3.98
Na_2O	3.64	3.11	4.03	3.94	3.42	3.45	3.55	3.27	3.48
K_2O	2.43	1.90	3.28	2.86	2.18	2.56	3.50	5.46	2.58
P_2O_5	0.24	0.09	0.34	1.08	0.65	0.45	0.50	0.69	0.26
L.O.I.	0.47	0.39	0.35	0.43	0.48	0.47	0.37	0.38	0.42
Sum	99.90	99.59	99.45	98.50	98.31	99.22	99.68	98.77	99.77
Rb	196	145	105	117	79	193	177	234	120
Sr	437	185	480	994	1253	623	798	910	510
Ba	528	484	872	1187	1093	667	888	1297	730
Y	16.0	15.9	22.8	30.3	26.5	22.7	19.8	27.1	16.9
Zr	181	162	323	419	185	218	161	342	185
Nb	10.7	11.2	17.9	28.1	17.5	12.4	15.2	20.6	10.7
Sn	5.5	<4.2	3.8	6.3	<7.8	<4.1	<6.7	<6.8	<4.8
Pb	12.0	12.1	19.6	12.4	22.6	12.1	29.1	36.1	15.3
V	81.4	75.1	70.7	136.1	151.7	112.6	123.4	131.5	84.7
Cr	93.7	<21.7	22.9	<26.8	59.3	152.5	177.5	105.4	93.4
Ni	48.3	<4.2	13.3	13.1	25.9	73.3	113.6	57.4	45.4
Co	17.3	8.0	12.8	22.7	27.8	23.0	24.2	22.6	14.1
Zn	71.9	58.6	93.2	147.3	105.5	76.7	70.5	79.8	65.2
Ga	19.3	16.6	21.5	26.4	19.3	20.0	21.5	20.0	19.1
Nd	25.20	28.30	55.30	81.45	75.89	41.35	45.18	72.57	30.27
Sm	5.305	5.067	9.441	13.68	12.67	7.433	8.489	12.03	5.751
Mg#	54.5	43.3	46.9	46.3	52.0	56.3	60.6	58.0	58.0
A/CNK	1.05	1.08	0.95	0.71	0.74	0.89	0.75	0.77	1.01

Pluton group	MIC monzodiorite	MIC granite							
Sample number	J98-40	96-60	96-91	96-100	96-100-2	96-210	96-229	J98-03	J98-04
Rock type	gdr	gdr	gdr	grn	grn	grn	grn	grn	grn
SiO_2	67.0	71.8	69.6	71.6	73.8	72.5	71.9	74.8	72.7
TiO_2	0.81	0.27	0.32	0.30	0.14	0.18	0.30	0.19	0.27
Al_2O_3	15.35	15.23	15.98	15.10	14.12	14.95	14.81	13.29	14.49
Fe_2O_3*	3.86	1.64	2.03	2.11	0.92	1.29	1.82	1.55	1.77
MnO	0.07	0.04	0.04	0.08	0.04	0.06	0.04	0.07	0.05
MgO	2.38	0.55	0.88	0.43	0.23	0.44	0.52	0.47	0.49
CaO	3.02	2.36	3.13	1.51	0.77	1.48	1.67	1.30	1.36
Na_2O	3.25	4.42	3.87	4.21	3.98	3.89	3.66	3.24	3.46
K_2O	2.64	3.22	2.99	3.91	4.98	4.17	4.22	4.09	4.78
P_2O_5	0.21	0.10	0.13	0.28	0.13	0.12	0.17	0.04	0.17
L.O.I.	0.86	0.29	0.28	0.37	0.29	0.42	0.39	0.41	0.31
Sum	99.34	99.86	99.46	100.08	99.47	99.64	99.70	99.44	99.84

Table 1 (*continued*)

Pluton group	MIC monzodiorite	MIC granite							
Sample number	J98-40	96-60	96-91	96-100	96-100-2	96-210	96-229	J98-03	J98-04
Rock type	gdr	gdr	gdr	grn	grn	grn	grn	grn	grn
Rb	136	125	89	194	212	228	200	173	194
Sr	441	266	387	129	63	135	166	202	151
Ba	770	695	861	484	253	317	487	483	448
Y	17.8	6.7	6.3	21.3	9.2	11.8	12.1	14.2	9.2
Zr	147	123	154	187	81	82	156	111	138
Nb	11.4	8.2	5.7	22.4	10.4	12.8	11.1	13.2	9.6
Sn	4.3	3.5	3.4	9.2	5.5	6.7	7.9	6.1	<6.4
Pb	19.8	27.6	22.1	26.1	33.2	32.0	28.5	23.2	31.5
V	74.9	19.8	28.2	16.4	<5.1	13.3	18.2	14.1	12.8
Cr	69.5	21.9	29.1	16.4	17.3	19.7	<22.8	<14.4	<22.3
Ni	32.3	<5.6	<3.6	<4.4	<3.5	6.2	<4.0	<3.6	<3.6
Co	15.0	4.0	5.3	3.9	4.5	4.3	<5.0	4.6	<6.1
Zn	57.0	33.8	38.8	64.2	25.1	34.3	62.8	19.0	50.4
Ga	18.4	16.4	18.9	21.3	16.9	18.4	20.5	14.0	16.9
Nd	29.90	18.56	14.64	37.57	–	21.40	–	14.55	39.66
Sm	6.204	3.093	2.602	6.466	–	3.975	–	2.613	7.081
Mg#	55.0	39.9	46.2	28.8	33.1	40.3	36.1	37.5	35.4
A/CNK	1.12	1.01	1.04	1.09	1.06	1.10	1.09	1.10	1.09

Pluton group	MIC granite								
Sample number	J98-24	J98-32	J98-63-1	J98-88	J98-92-2	J98-94	J98-99	J98-108	J98-111-2
Rock type	grn	grn	gdr	gdr	gdr	gdr	grn	grn	gdr
SiO_2	73.4	74.2	75.3	74.8	71.5	72.6	74.3	73.6	72.2
TiO_2	0.12	0.11	0.08	0.06	0.47	0.28	0.11	0.21	0.41
Al_2O_3	14.90	14.42	14.02	14.38	14.60	15.21	14.47	14.14	14.95
Fe_2O_3*	0.96	0.80	0.55	0.47	2.90	1.84	0.96	1.15	2.51
MnO	0.02	0.03	0.02	0.01	0.05	0.06	0.04	0.05	0.04
MgO	0.32	0.29	0.13	0.04	0.80	0.56	0.26	0.29	0.85
CaO	1.14	1.06	1.15	0.89	1.65	2.30	0.53	0.91	1.99
Na_2O	3.46	4.09	4.49	4.90	4.06	4.68	3.51	3.31	4.09
K_2O	4.98	4.57	3.62	3.91	3.23	1.50	4.85	4.93	2.45
P_2O_5	0.15	0.14	0.13	0.09	0.12	0.18	0.20	0.14	0.04
L.O.I.	0.52	0.23	0.30	0.23	0.47	0.50	0.69	0.69	0.49
Sum	99.97	99.93	99.79	99.78	99.85	99.69	99.92	99.40	100.03
Rb	202	220	169	179	145	85	298	285	84
Sr	143	78	59	95	182	125	51	101	263
Ba	368	196	56	172	452	79	190	297	421
Y	12.0	8.2	10.3	6.4	41.5	10.7	9.8	11.7	14.2
Zr	60	73	44	27	231	133	56	97	242
Nb	9.5	8.4	9.9	6.9	21.2	7.5	14.3	10.5	13.7
Sn	6.7	5.8	7.6	6.9	6.7	<6.8	11.4	8.1	4.7
Pb	50.0	31.8	31.4	48.7	37.6	19.6	33.2	29.2	36.9
V	6.9	<5.1	4.8	<4.7	33.4	15.4	<6.3	6.4	25.3
Cr	<14.6	<16.4	<23.7	<14.2	<14.8	<14.6	<25.6	<14.8	<14.4
Ni	<3.4	<3.4	<3.2	<3.2	<7.2	<6.6	<3.4	<5.0	<4.9

(continued on next page)

Table 1 (continued)

Pluton group	MIC granite								
Sample number	J98-24	J98-32	J98-63-1	J98-88	J98-92-2	J98-94	J98-99	J98-108	J98-111-2
Rock type	grn	grn	gdr	gdr	gdr	gdr	grn	grn	gdr
Co	4.2	<3.7	<3.7	<5.6	4.2	<6.0	<6.2	<3.5	5.9
Zn	28.5	26.2	16.9	24.8	62.7	42.2	37.3	51.2	46.2
Ga	19.1	16.8	21.3	16.1	18.7	16.1	19.5	21.7	18.1
Nd	17.54	16.65	31.28	4.650	53.37	24.94	6.349	36.71	70.78
Sm	4.038	3.623	6.806	1.177	9.624	4.656	1.698	6.734	10.59
Mg#	39.8	41.8	31.9	14.4	35.3	38.1	34.9	33.3	40.1
A/CNK	1.13	1.06	1.05	1.03	1.11	1.13	1.21	1.14	1.15

Pluton group	MIC granite	Redington					Lexington north	Lexington south	Lexington central
Sample number	J98-114	96-200	96-203	96-246	96-250	98-8	98-1	98-2	96-296-2
Rock type	grn	grn	grn	grn	grn	grn	gdr	grn	grn
SiO_2	74.6	73.8	67.2	68.7	64.4	72.8	71.0	73.3	72.9
TiO_2	0.09	0.36	0.68	0.56	0.72	0.10	0.38	0.19	0.40
Al_2O_3	14.41	13.78	16.12	16.02	17.77	16.17	15.01	14.25	13.56
Fe_2O_3*	0.81	2.24	4.41	3.69	4.65	1.24	2.57	1.46	2.77
MnO	0.02	0.02	0.04	0.03	0.04	0.05	0.08	0.04	0.05
MgO	0.18	0.64	1.22	1.06	1.30	0.21	0.69	0.32	0.82
CaO	0.62	1.51	2.12	1.84	2.34	0.70	2.41	0.94	1.55
Na_2O	3.39	2.60	2.78	3.08	3.50	3.74	3.95	3.24	3.05
K_2O	5.13	3.84	4.12	3.86	4.26	4.04	3.71	5.51	3.97
P_2O_5	0.22	0.15	0.14	0.15	0.11	0.24	0.08	0.21	0.22
L.O.I.	0.56	0.74	0.78	0.87	0.66	0.66	0.37	0.51	0.48
Sum	100.04	99.71	99.57	99.90	99.74	99.91	100.2	99.96	99.80
Rb	243	123	139	144	138	–	143	287	172
Sr	54	124	168	149	196	–	215	70.5	156
Ba	104	580	750	590	990	–	495	197	272
Y	11.8	9.1	15.0	12.1	16.2	–	17.8	12.2	15.3
Zr	44	140	254	212	262	–	171	99.7	153
Nb	11.7	8.9	13.9	12.2	13.9	–	12.2	15.6	17.4
Sn	7.4	4.4	<5.1	<6.3	<6.3	–	<5.7	10	<5.7
Pb	38.0	26.2	30.6	33.0	32.3	–	23.5	37.1	36.3
V	<3.4	30	41.6	43.7	46.0	–	27.6	5.1	28.3
Cr	<17.7	23.9	18.1	23.9	32.1	–	<15.0	<14.8	15.5
Ni	<3.4	5.4	12.0	12.4	11.3	–	<3.8	<3.6	3.8
Co	<3.3	3.6	7.7	6.6	6.2	–	4.0	<4.8	8.3
Zn	34.7	44.8	80.2	70.1	77.5	–	44.3	40.5	55.5
Ga	20.2	16.3	20.7	18.6	22.3	–	18.3	16.1	15.4
Nd	15.73	32.61	52.83	52.34	66.71	8.943	29.42	22.44	33.70
Sm	3.930	5.659	9.357	9.077	11.47	2.460	5.138	4.708	6.423
Mg#	30.6	36.1	35.4	36.3	35.6	25.1	32.9	30.3	37.0
A/CNK	1.18	1.23	1.25	1.27	1.22	1.37	1.01	1.10	1.12

Fe_2O_3*=total Fe as Fe_2O_3; L.O.I.=loss on ignition at 600 °C; Mg#=100* molar MgO/(MgO+FeO), where FeO is estimated to be 0.9* total Fe; A/CNK= molar Al_2O_3/(Na_2O+K_2O+CaO). Rock type (as defined by CIPW normative compositions): grn—granite; gdr—granodiorite; qtz mdr—quartz monzodiorite; mdr—monzodiorite; mzn—monzonite. <x=samples where element was below limit of accurate quantification. –=not determined.

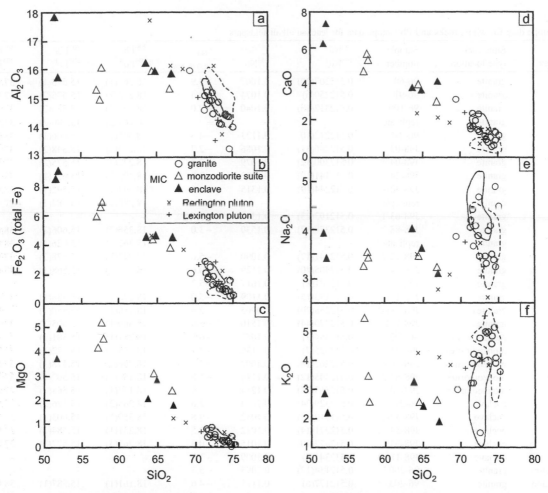

Fig. 3. SiO$_2$ versus major element variation diagrams: (a) Al$_2$O$_3$; (b) Fe$_2$O$_3$ (total Fe); (c) MgO; (d) CaO; (e) Na$_2$O; (f) K$_2$O (all in wt.%). Fields denote compositions of Phillips pluton granodiorites (solid line) and leucogranites (dashed line), after Pressley and Brown (1999).

The MIC granites have high SiO$_2$ concentrations (69–78 wt.%) and show internally coherent decreases in Al$_2$O$_3$, TiO$_2$, Fe$_2$O$_3^T$, MgO, CaO with increasing SiO$_2$, despite minimal spread (Fig. 3). The MIC granites show a substantial range in trace element content, particularly the trace alkaline earth Ba and the high field strength elements Y (Fig. 4), Nb and Zr. The MIC granites are depleted in transition metals such as Ni, Cr, V and Co, which reach moderate concentrations in the MIC monzodiorite suite. Although the majority of the MIC granites have Sm/Nd close to average upper crustal values, there is significant variability in the degree of fractionation (^{147}Sm/^{144}Nd=0.0905–0.1617; Table 2).

The Redington pluton samples have major element compositions distinct from the MIC granites, generally spanning the gap on Harker plots between the monzodioritic and granitic MIC samples (Fig. 3). They tend to have lower CaO and Na$_2$O contents and higher Al$_2$O$_3$ and Fe$_2$O$_3^T$ than rocks with comparable SiO$_2$. The samples are uniformly weakly evolved with respect to Rb/Sr and are enriched in Ba and compatible transition metals (V, Cr, Ni, Co) relative to MIC granites (Fig. 4).

The samples from each of the three lobes of the Lexington composite pluton have major and trace element concentrations that overlap the MIC granites, although with slightly lower Na$_2$O (Fig. 3). Together

Table 2
Nd isotope data for whole rocks and Pb isotope data for leached alkali feldspars

Pluton/ complex	Suite/rock type/location	Sample number	$^{143}Nd/$ $^{144}Nd^a$	$^{147}Sm/$ ^{144}Nd	ε_{Nd} (at 370 Ma)	$^{206}Pb/$ $^{204}Pb^b$	$^{207}Pb/$ $^{204}Pb^b$	$^{208}Pb/$ $^{204}Pb^b$
MIC	granite	96-60	0.512207(20)	0.1007	−3.9	18.265(1)	15.573(1)	37.932(2)
MIC	granite	96-91	0.512250(6)	0.1074	−3.4	18.273(2)	15.575(2)	37.921(5)
MIC	granite	96-100	0.512312(18)	0.1040	−2.0	18.269(2)	15.586(2)	37.962(6)
MIC	granite	replicate	–	–	–	18.263(2)	15.580(2)	37.937(5)
MIC	granite	96-210	0.512212(10)	0.1123	−4.3	18.395(1)	15.606(1)	38.085(2)
MIC	granite	J98-03	0.512286(31)	0.1086	−2.7	18.239(3)	15.548(5)	37.746(5)
MIC	granite	J98-04	0.512300(12)	0.1079	−2.4	18.222(2)	15.575(2)	37.917(5)
MIC	granite	J98-24	0.512141(10)	0.1392	−7.0	18.405(1)	15.621(1)	38.101(2)
MIC	granite	J98-32-1	0.512244(19)	0.1315	−4.6	18.293(1)	15.592(1)	38.014(2)
MIC	granite	replicate	–	–	–	18.289(1)	15.584(1)	37.999(3)
MIC	granite	J98-63-1	0.512192(23)	0.1315	−5.6	–	–	–
MIC	granite	J98-88	0.512378(17)	0.1530	−3.0	18.350(2)	15.606(2)	38.032(4)
MIC	granite	replicate	–	–	–	18.367(1)	15.626(1)	38.092(4)
MIC	granite	J98-92-2	0.512141(17)	0.1090	−5.6	18.242(3)	15.578(2)	37.927(6)
MIC	granite	J98-94	0.512405(40)	0.1129	−0.6	18.282(1)	15.568(1)	37.881(2)
MIC	granite	J98-99	0.512281(8)	0.1617	−5.3	–	–	–
MIC	granite	J98-108	0.512173(15)	0.1109	−5.0	18.278(1)	15.588(1)	38.016(3)
MIC	granite	J98-111-2	0.512258(20)	0.0905	−2.4	18.180(1)	15.550(1)	37.865(4)
MIC	granite	J98-114	0.512270(44)	0.1510	−5.0	18.360(1)	15.605(1)	38.011(2)
MIC	monzodiorites	96-222	0.512384(7)	0.1087	−0.7	18.281(1)	15.581(1)	37.907(3)
MIC	monzodiorites	J98-43	0.512325(7)	0.1136	−2.1	18.253(1)	15.572(1)	37.896(3)
MIC	monzodiorites	J98-100	0.512262(18)	0.1002	−2.7	18.296(1)	15.578(1)	37.958(2)
MIC	monzodiorites	J98-63-2	0.512345(12)	0.1149	−1.8	18.159(2)	15.562(1)	37.821(4)
MIC	monzodiorites	J98-40	0.512363(13)	0.1254	−1.9	18.175(1)	15.564(1)	37.845(2)
MIC	enclave	96-216	0.512329(24)	0.1273	−2.6	18.265(2)	15.576(2)	37.905(4)
MIC	enclave	J98-32-2	0.511910(15)	0.1082	−9.8	18.359(1)	15.603(1)	38.100(5)
MIC	enclave	J98-54	0.512271(14)	0.1032	−2.5	18.231(1)	15.584(1)	37.944(3)
MIC	enclave	J98-93	0.512412(10)	0.1015	0.3	18.267(3)	15.577(3)	37.920(7)
MIC	enclave	J98-111-3	0.512341(7)	0.1009	−1.0	–	–	–
Redington	granite	96-200	0.512134(17)	0.1093	−5.3	–	–	–
Redington	granite	96-203	0.512177(6)	0.1115	−4.6	18.161(1)	15.587(1)	38.024(3)
Redington	granite	96-250	0.512184(13)	0.1083	−4.3	18.150(1)	15.576(1)	37.989(3)
Redington	granite	98-8	0.512346(14)	0.1732	−4.5	18.162(6)	15.581(6)	38.008(13)
Redington	granite	96-246	0.512161(9)	0.1092	−4.8	18.159(1)	15.586(1)	38.015(2)
Lexington	north portion	98-1	0.512286(13)	0.1100	−2.8	18.066(1)	15.545(1)	37.925(2)
Lexington	central portion	98-10-1	0.512247(8)	0.1202	−4.0	18.201(1)	15.594(1)	38.051(2)
Lexington	central portion	96-296-2	0.512213(13)	0.1200	−4.7	18.185(1)	15.585(1)	38.042(2)
Lexington	central portion	98-10-2	0.512690(62)	0.1670	+2.4	–	–	–
Lexington	south portion	98-2	0.512225(16)	0.1321	−4.7	–	–	–
Sugarloaf	felsic pod	98-9	–	0.1994	–	–	–	–
Sugarloaf	gabbro	98-15	0.512709(15)	0.1941	+1.5	–	–	–
Phillips	leucogranite	PH-03	0.512224(12)^c	0.1695^c	−6.8^c	18.177(1)	15.583(1)	38.016(3)
Phillips	leucogranite	PH-04	0.512253(11)^c	0.1751^c	−6.5^c	18.194(1)	15.591(1)	38.029(2)
Phillips	leucogranite	PH-05	0.512222(10)^c	0.1789^c	−7.3^c	18.170(1)	15.567(1)	37.961(3)
Phillips	granodiorite	P-74a	0.512386(6)^c	0.1257^c	−1.6^c	18.213(2)	15.605(2)	38.050(4)
Phillips	granodiorite	PH-06	0.512431(9)^c	0.1242^c	−0.6^c	18.230(1)	15.609(1)	38.040(2)
Phillips	granodiorite	P-92	0.512333(6)^c	0.1104^c	−1.9^c	18.185(3)	15.562(2)	37.926(6)

MIC denotes the Mooselookmeguntic igneous complex.

—=not determined.

[a] Measured value, corrected for fractionation and blank; absolute 2σ within-run uncertainty on last digit(s) of individual analysis in parentheses.

[b] Same as a, but also corrected for blank.

[c] Data from Pressley and Brown (1999).

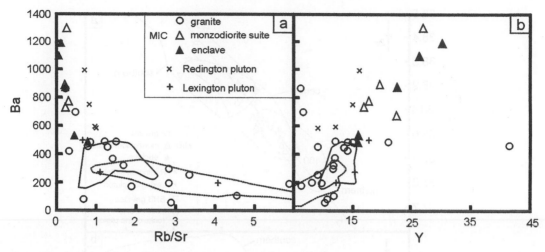

Fig. 4. (a) Rb/Sr versus Ba and (b) Y versus Ba variation diagrams (all concentrations in ppm). Fields denote compositions of Phillips pluton granodiorites (solid line) and leucogranites (dashed line), after Pressley and Brown (1999).

with the MIC granites, these are the only rocks studied that have Rb/Sr significantly greater than unity (Fig. 4).

4.2. Isotope data

Samples of the MIC monzodiorite suite ($\varepsilon_{Nd}=-2.8$ to -0.7) and enclaves ($\varepsilon_{Nd}=-2.8$ to 0.1, except for J98-32-2) span a similar, restricted range in initial Nd isotopic composition, whereas the granites are highly variable ($\varepsilon_{Nd}=-7.0$ to -0.6) (Table 2). Enclave sample J98-32-2 contains the least radiogenic Nd ($\varepsilon_{Nd}=-10.0$) of all of the samples analyzed. The Redington pluton samples have homogeneous initial Nd isotopic compositions, in the middle of the MIC granite range ($\varepsilon_{Nd}=-5.3$ to -4.3). Samples from the felsic portions of the Lexington composite pluton have initial ε_{Nd} that overlap the MIC granite values (-5.0 to -2.8). The mafic sample from the central lobe of the Lexington pluton and the single Sugarloaf sample that was analyzed (a gabbro) yield ε_{Nd} values of $+2.4$ and $+1.5$, respectively.

Alkali feldspars from the MIC monzodiorite suite span a narrow Pb isotopic range, all within analytical uncertainty and overlapping the ratios of the enclaves, except sample J98-32-2, which has higher $^{206}Pb/^{204}Pb$, $^{207}Pb/^{204}Pb$ and $^{208}Pb/^{204}Pb$ (Fig. 5). Save for that one sample, the monzodiorite suite and enclaves have a mean $^{207}Pb/^{204}Pb$ of 15.574. The Pb

isotopic compositions of the MIC granites span a large range that overlaps the data for the less evolved rocks and extends to higher $^{206}Pb/^{204}Pb$, $^{207}Pb/^{204}Pb$ and $^{208}Pb/^{204}Pb$, and with a slightly higher mean $^{207}Pb/^{204}Pb$ of 15.583 (Fig. 5). The granite data occupy all three of the fields in $^{207}Pb/^{206}Pb$ and $^{208}Pb/^{206}Pb$ space from Ayuso and Bevier (1991), who distinguished northern, central and southern plutons. The MIC data show similar $^{206}Pb/^{204}Pb$ but slightly lower $^{207}Pb/^{204}Pb$ and $^{208}Pb/^{204}Pb$, though still within uncertainty, to the Mooselookmeguntic granite feldspar reported by Ayuso and Bevier (1991).

The new Pb isotope data from the Phillips pluton are consistent with the three analyses from Pressley and Brown (1999): average compositions of granodiorites are slightly more radiogenic in uranogenic Pb than the leucogranites, although the samples overlap within uncertainty, with values equivalent to those of the MIC (Fig. 5). The Redington pluton initial Pb data are, like the whole-rock Nd initial isotopic compositions, tightly clustered, with a mean $^{207}Pb/^{204}Pb$ of 15.582, identical to the mean for the MIC granites (Fig. 5).

All samples from this study have initial Pb isotopic compositions that lie close to or below the two-stage growth curve of Stacey and Kramers (1975), advanced relative to their model ages. The rocks of this study all have $^{208}Pb/^{204}Pb$ typical of upper crustal

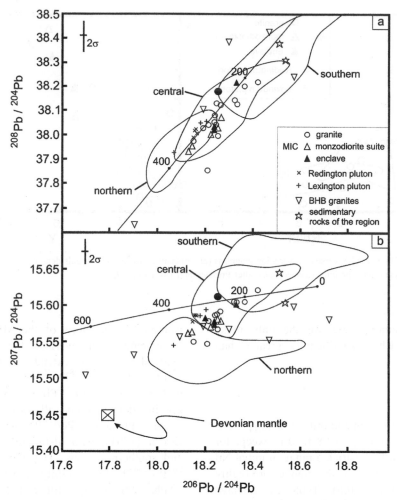

Fig. 5. Plots of Pb isotopic compositions of leached alkali feldspars from rocks of this study: (a) $^{206}Pb/^{204}Pb$ versus $^{208}Pb/^{204}Pb$ with the model two-stage curve of Stacey and Kramers (1975); (b) $^{206}Pb/^{204}Pb$ versus $^{207}Pb/^{204}Pb$ with the Stacey and Kramers curve. Tics are all 200 Ma. Fields (northern, central, southern) from Ayuso and Bevier (1991) are based on geographic variability of Pb isotopes in the northern Appalachians. Bronson Hill belt granite data (inverted triangles) are from Moench and Aleinikoff (2002), on samples of pre-Devonian plutons. The filled circle is the Mooselookmeguntic granite feldspar datum from Ayuso and Bevier (1991). Age-corrected whole-rock data of Siluro-Devonian sedimentary rocks from Maine (star symbols) are from Krogstad (1993). The estimate of Devonian oceanic mantle is from Zartman and Doe (1981).

rocks with sources that have not experienced significant Th/U changes in their histories.

5. Discussion

5.1. Constraints on sources

Ayuso (1986) defined three distinct regions in Pb isotope space based on feldspar data from granites occupying different geographic areas in Maine (Fig. 5), interpreted to reflect influences of (primarily) Laurentian (with low $^{207}Pb/^{204}Pb$ at a given $^{206}Pb/^{204}Pb$) and non-North American (Avalon-like or circum-Gondwanan, with high $^{207}Pb/^{204}Pb$ at a given $^{206}Pb/^{204}Pb$) crustal sources. This approach has been fundamental to the use of Pb isotopes in probing the unexposed crust in Maine, but has limitations, particularly in assessing crust–mantle mixtures and in the specific identity of sources for samples with isotopic compositions that are neither clearly Laurentian nor clearly non-North American.

Traditionally, Nd and Pb data are viewed separately and give potentially complementary results based on differences in time-integrated parent/daughter ratios of source materials. Combining the two systems on one diagram (using $^{207}Pb/^{204}Pb$) permits more comprehensive assessment of possible sources of the plutonic rocks. Owing to the shorter half life of ^{235}U compared to ^{238}U, the most significant production of ^{207}Pb in the Earth took place in the Precambrian; since then the most significant changes in Pb isotopic compositions have been the steady increase in $^{206}Pb/^{204}Pb$ and $^{208}Pb/^{204}Pb$. As such, in areas where magmatic source components have significant time integrated differences in U/Pb, $^{207}Pb/^{204}Pb$ is a more powerful tracer of these components than $^{206}Pb/^{204}Pb$, which will vary considerably in mantle and crustal materials in response to small changes in U/Pb—changes that may be unrelated to the nature of source materials.

Of the known, potential, or anticipated sources of the magmas parental to plutons of this study, we discriminate mantle of two isotopically distinct types, three basement terranes, and a broad mid-crustal source that takes into account metasedimentary rocks that overlie the basement framework. These can be distinguished to varying extents based on their respective elemental and isotopic characteristics. We summarize the general distinguishing features of these components below.

Material source components from the mantle that have been called into prominence in the petrogenesis of other plutonic complexes in the northern Appalachians range from chemically homogeneous convecting upper mantle reflected in modern mid-ocean ridge basalts (MORB; e.g., Waight et al., 2001; Coish and Rogers, 1987) to subduction-modified lithospheric mantle (e.g., Arth and Ayuso, 1997; Whalen et al., 1996). Sources within the latter reservoir are characterized by specific trace element enrichment (large-ion lithophile elements) and depletion (high field strength elements), and by long-term decrease in Sm/Nd, yielding initial ε_{Nd} lower than the MORB range (<+8 to +10, present-day). The Pb isotopic composition of Paleozoic MORB has been estimated (Zartman and Doe, 1981), although this does not take into account the significant variability expected from analogy with modern MORB (e.g., White, 1993).

An estimate of the variability of lithospheric mantle Pb was made by Pegram (1990), through analysis of Mesozoic tholeiites in eastern North America. The Mesozoic tholeiites show considerable, positively correlated variability in initial ε_{Nd} and $^{207}Pb/^{204}Pb$, with samples from in and near New England having the highest ratios for both systems, with average ε_{Nd} (at 370 Ma) c. +1.2 and average $^{207}Pb/^{204}Pb$ (corrected to 370 Ma) c. 15.63. The Nd initial isotopic compositions are consistent with data from mafic rocks in the Mount Ascutney complex in eastern Vermont (Foland et al., 1988).

The isotopic compositions of both Nd and Pb in Laurentian crustal materials are now well-established from studies in Maine (Ayuso et al., 1988) and in 'classic' Laurentian crustal massifs, such as at the Grenville front and Adirondacks (Fletcher and Farquhar, 1977; Daly and McLelland, 1991; McLelland et al., 1993; DeWolf and Mezger, 1994). In these rocks, ε_{Nd} (at 370 Ma) ranges from−10 to −4 and $^{207}Pb/^{204}Pb$ is characteristically <15.60, with the majority <15.56.

Continental crustal fragments accreted to North America throughout Paleozoic closure of Iapetus are thought to have circum-Gondwanan origins (Williams and Hatcher, 1983). Although it is not clear whether multiple accreted crustal basement terranes with slightly different initial isotopic characteristics can be defined (Whalen et al., 1996; Tomascak et al., 1999), the general nature of these terranes is well constrained, primarily from basement exposures in Atlantic Canada, but also from granitic plutons in the coastal region throughout the northern Appalachians. These non-North American ("Avalon-like") sources are characterized by high (juvenile) ε_{Nd} (at 370 Ma) (>−4) and radiogenic $^{207}Pb/^{204}Pb$ (>15.6) (Ayuso and Bevier, 1991; Barr and Hegner, 1992; Whalen et al., 1994; Kerr et al., 1995; Samson et al., 2000; Fig 6).

Studies of Bronson Hill belt (BHB) rocks in which both Nd and Pb isotopes were analyzed are lacking. Whalen et al. (1998) reported initial Nd and Pb isotope data for arc-related Mid- to Late Ordovician volcanic and plutonic rocks in the Miramichi belt in western New Brunswick. The granites in this part of the Taconic orogen (the age of which is perhaps as much as 15 Ma older than Taconic orogenesis in New Hampshire and Maine), have $^{207}Pb/^{204}Pb$ in the range

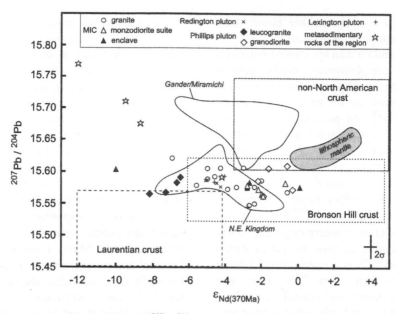

Fig. 6. Plot of whole-rock ε_{Nd} (at 370 Ma) versus initial ^{207}Pb/^{204}Pb. Metasedimentary supracrustal rocks of the CMB (star symbols) are from Ayuso and Schulz (2003) and Arth and Ayuso (1997). White fields with italic labels define ranges of paired Nd, Pb data for granites in the northern Appalachians (Silurian to Devonian Gander zone and Ordovician Miramichi highlands plutons from New Brunswick: Whalen et al., 1996, 1998; Northeast Kingdom batholith, Vermont, granites: Arth and Ayuso, 1997). Shaded field represents compositions of mid-Atlantic tholeiites from the central and northern Appalachians (Pegram, 1990), an estimate of the lithospheric mantle at 370 Ma. Dashed rectangular fields show estimated ranges for basement terranes based on literature data from unpaired Nd, Pb samples (see text).

15.67 to 15.71, and ε_{Nd} (470 Ma) of -5.4 to $+0.2$. The coincidence of Nd and Pb isotope data from these rocks and Silurian to Devonian plutons in the same region (Whalen et al., 1994) suggests that the Pb in Ordovician granites may in large part not reflect arc basement sources, but rather significant contributions from lower- to mid-crustal metasedimentary rocks with strongly non-North American Pb.

Samson and Tremblay (1996) reported ε_{Nd} (at 450 Ma) in the range -5 to $+5$ for Ordovician volcanic rocks from Québec correlative with the BHB in Maine. Hingston (1992) presented Nd isotope data from mafic to felsic metavolcanic rocks in New Hampshire of estimated Ordovician age. These Nd data combined with initial Pb isotope data from 435 to 469 Ma granite plutons that intrude the BHB in New Hampshire (Moench and Aleinikoff, 2002), allow an estimate to be made for the nature of BHB crust. The Pb in these granites has low ^{207}Pb/^{204}Pb (<15.58). It should be stressed that the estimate of BHB isotopic composition used herein is based on data not from the same or even spatially equivalent samples. Further-

more, the data chosen from Moench and Aleinikoff (2002) show a large range in ^{206}Pb/^{204}Pb and ^{208}Pb/^{204}Pb (although not ^{207}Pb/^{204}Pb). Considering that these are single samples from large plutonic bodies, caution must be applied in their absolute interpretation, as they may not be uncompromised representatives of BHB sources.

Ordovician to Devonian metasedimentary rocks in the northern Appalachians have not been extensively studied with Pb isotopes, although there is a growing body of Nd isotope data. In general these rocks, which crop out in the CMB and BHB in the region between central Maine and eastern Vermont, show relatively unradiogenic initial Nd isotopic compositions, the majority are in the range ε_{Nd} (at 370 Ma) -12 to -4 (Ayuso and Schulz, 2003; Arth and Ayuso, 1997; Cullers et al., 1997; Pressley and Brown, 1999; Lathrop et al., 1996; Foland et al., 1988), which completely overlaps the Nd range for Laurentian crust at this time. Hence, Nd isotopes alone are not an effective discriminator between these two potential sources.

The small number of extant Pb isotope data for these kinds of metasedimentary rocks show an inverse correlation with Nd initial ratios, with low values of $^{207}Pb/^{204}Pb$ being present in samples with the highest ε_{Nd}. The Pb isotopic compositions have to be viewed carefully, because they were obtained from whole rocks and are calculated initial compositions which integrate measured U/Pb assumed to be applicable to these rocks. If the sources of the original sediments contained Laurentian (low ε_{Nd}, low $^{207}Pb/^{204}Pb$) and non-North American components (high ε_{Nd}, high $^{207}Pb/^{204}Pb$), a positive correlation would be expected in Pb–Nd space; however, the opposite is apparent based on the few literature data available for which both isotope systems were measured. This intriguing feature cannot be examined further without a greater abundance of Pb isotope data for samples where Nd isotopes have been determined.

5.1.1. Sources of the MIC monzodiorite suite and enclaves

Samples from the MIC monzodiorite suite have a combination of radiogenic initial Nd and unradiogenic initial Pb (low $^{207}Pb/^{204}Pb$), but both are homogeneous, especially compared to the scale of major element variation in these rocks (Fig. 6). Comparing the Nd–Pb relations of the MIC monzodiorite suite to lithospherically derived tholeiites from elsewhere in New England allows a comparison with a plausible enriched mantle source. Samples from the MIC monzodiorite suite consistently have both lower $^{207}Pb/^{204}Pb$ and lower ε_{Nd} than the tholeiites of Pegram (1990). The monzodiorite samples have significantly less radiogenic initial Nd and more radiogenic initial Pb than the Devonian MORB mantle. The absence of mafic rocks in the area with characteristic high ε_{Nd} (>+8) also does not support a depleted mantle source component for these magmas.

Samples from the MIC monzodiorite suite have fractionated Sm/Nd relative to normal mantle rocks and high Nd concentrations, resembling ordinary upper crustal rocks in this regard. However, this is not unusual among orogenic diorites (e.g., van de Flierdt et al., 2003; McCulloch and Woodhead, 1993; Liew et al., 1989), where source rocks have significant lower crustal components. Samples from the MIC monzodiorite suite have compatible element concen-

trations (V, Cr, Ni, Co) equivalent to diorites of similar SiO_2 content interpreted to have been generated exclusively by lower crustal melting (van de Flierdt et al., 2003), suggesting that a mantle component is not necessary in their petrogenesis. The estimated composition of BHB crust overlaps the Nd–Pb initial ratios of the MIC monzodiorite suite. Laurentian crust of mafic composition (e.g., Shaw and Wasserburg, 1984) is plausible as a basement source for the monzodiorite suite as well, but the scarcity of paired Nd–Pb data from them make it unfeasible to test this alternative.

Given the isotopic and elemental data, two models for the petrogenesis of the monzodiorite suite appear most probable: assimilation of Laurentian crustal material by melts derived from the enriched lithospheric mantle, or derivation from arc crust of the BHB. The isotopic compositions of the monzodiorite suite can be reproduced using either two component mixing calculations or assimilation-fractional crystallization (Fig. 7; details in Appendix C). However, the elemental contents of these rocks, in particular the compatible elements, are not consistent with significant input of mantle material. Additionally, although ε_{Nd} and SiO_2 are weakly anticorrelated (Fig. 8), as would be expected with a mantle-Laurentian crust mixture, the absence of trends between compatible element contents and Nd isotopes in samples of the MIC monzodiorite suite makes a mantle-mixing scenario less likely than derivation from a single source in the lower crust, specifically BHB crust. It is likely that mantle heat was required to produce large-scale melting of the mafic to intermediate, metaigneous lower crust, even if compelling evidence for mantle material involvement is lacking.

With the exception of sample J98-32-2, the MIC enclaves display isotopic and elemental characteristics indistinguishable from the monzodiorite suite, suggesting the groups may be related by a single petrogenetic process. The geochronology of neither the monzodiorite suite nor the enclaves has been investigated in detail. However, the apparent potential c. 12 Ma age difference between these two groups of rocks is consistent with variations in age populations seen in microgranitoid enclaves in of other granitic suites (e.g., Elburg, 1996). Compositions of the enclaves do not argue for significant interaction with their host granite, which is reasonable considering their age difference (c. 20 Ma) and likely thermal

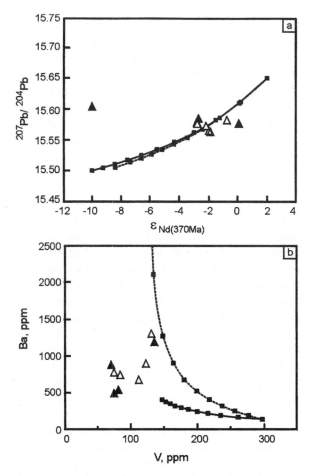

Fig. 7. Compositions of MIC monzodiorite suite and enclave samples: (a) whole-rock ε_{Nd} (at 370 Ma) versus initial $^{207}Pb/^{204}Pb$; (b) V versus Ba (ppm). Symbols as in Fig. 6. Also plotted are calculated two-component mixing (solid line; Langmuir et al., 1978) and assimilation-fractional crystallization (dashed line; DePaolo, 1988) curves, with 10% increments. In the mixing calculations (for details see Appendix C) enriched mantle is mixed with or assimilates felsic crust of Laurentian origin. Although the isotopic compositions of the MIC monzodiorite suite are generally consistent with derivation from a mantle-crust mixture, the compatible element signatures of these rocks are far lower than predicted using any plausible end members.

difference between the enclaves and their host magma. The enclave sample with much less radiogenic Nd (J98-32-2; $\varepsilon_{Nd}=-10.0$) most plausibly assimilated material with isotopic characteristics similar to CMB metasedimentary rocks.

5.1.2. Sources of the MIC granites

The MIC granites form an array in Pb–Nd space, such that samples with lower initial ε_{Nd} have slightly

less radiogenic $^{207}Pb/^{204}Pb$ (Fig. 6). Despite coherent major and trace element trends in the MIC granite data, the significant variation in initial isotopic compositions of Nd without a strong deviation in Pb does not allow the granites to be derived from a single homogeneous source. The isotopic variation of these samples could be produced by magma of more mafic composition mixing with or assimilating evolved crust, or by melting of a heterogeneous package of crustal material.

If the spread in initial ε_{Nd} in the MIC granites was due to an assimilation process between a mantle/mafic lower crust component and a more siliceous mid-crustal (metasedimentary) component, the data should describe trends in ε_{Nd} versus SiO_2 (or other indicative major element) space, which are not evident (Fig. 8). Furthermore, the best estimate of an incompatible element enriched mantle component is unlike the high-ε_{Nd} end member of the granites.

The spread in initial ε_{Nd} could be produced by mixing compositionally similar yet isotopically distinct components. Evolved Laurentian crustal rocks have significantly lower ε_{Nd} than the MIC granites that have the most Laurentian-like Pb (i.e., lowest $^{207}Pb/^{204}Pb$). Our estimate of the Nd–Pb characteristics of BHB crust provides a possible lower-$^{207}Pb/^{204}Pb$ source component for the MIC granites, but does not appear to satisfy the lower initial ε_{Nd} of some samples.

Elemental and isotopic parameters of magma compositions can be approximated by an assimilation-fractional crystallization model wherein melts of BHB crust assimilate CMB metasedimentary material and undergo variable extents of crystallization (Fig. 9). Nevertheless, the data for certain elements (notably Ba and Nd; Fig. 9c,d) are sufficiently dispersed that they require either source heterogeneity or non-ideal behavior in the melt extraction process (e.g., retained residual K-feldspar and monazite).

These data do not rule out derivation of magmas from heterogeneous mid-crustal sources. The initial Nd and Pb isotopic compositions of the MIC granites form an array that is subparallel to that for metasedimentary rocks from the CMB in Maine and the Connecticut Valley belt in Vermont (Ayuso and Schulz, 2003; Arth and Ayuso, 1997) (Fig. 6). The granite data extend to lower $^{207}Pb/^{204}Pb$ and higher ε_{Nd} than the metasedimentary rock data. This could

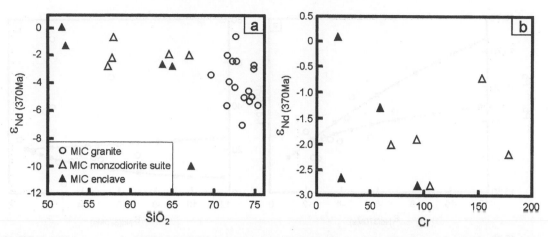

Fig. 8. (a) SiO$_2$ (wt.%) versus ε_{Nd} (at 370 Ma) for MIC samples. (b) Cr (ppm) versus $\varepsilon_{Nd(370\ Ma)}$ for MIC monzodiorite suite and enclave data. Lack of correlations among the samples of either the granite or monzodiorite suites argue against significant mixing between mafic and felsic components.

be a shortcoming of an inadequate data set for these potential source rocks. Indeed, variation in Nd and Pb isotopes with stratigraphic age has been observed in Appalachian sedimentary rocks and metamorphic equivalents, indicating higher ε_{Nd} and lower $^{207}Pb/^{204}Pb$ in the oldest rocks (Bock et al., 1996; Krogstad, 1993).

The low-$^{207}Pb/^{204}Pb$ MIC granite data overlap the MIC monzodiorite suite in Nd–Pb space. Although the geochronologic data suggest a >4 Ma gap in the timing of crystallization of these two groups of rocks, it is possible that their emplacement occurred as part of a continuous process that might be adequately defined only by a much larger number of age determinations. As discussed above, it would also be fortuitous to develop the minimal trends of the granite elemental compositions from a mixture of a more evolved crustal component and something equivalent to the MIC monzodiorite suite. Although no petrogenetic link between these two groups is apparent, it is worthwhile considering a thermal link.

The Nd–Pb isotope trend of the MIC granites overlaps that of the Northeast Kingdom batholith in Vermont (Arth and Ayuso, 1997; Fig. 6), a suite of Devonian plutons that intrude west across strike, within the Connecticut Valley belt, where the underlying crust is Laurentian. Arth and Ayuso interpreted the Northeast Kingdom batholith to have been derived substantively from igneous lower crust, consistent with our interpretation for the monzodiorite suite of

the MIC. However, primitive segments of the Northeast Kingdom batholith have been interpreted as possessing mantle source components (Ayuso and Arth, 1992; Arth and Ayuso, 1997). Although a mantle source is not explicitly ruled out in our analysis of the MIC data, we favor the simpler interpretation, that BHB crust forms the low-$^{207}Pb/^{204}Pb$, high-ε_{Nd} source component to these magmas and that most of the Pb and Nd isotopic heterogeneity has been introduced through additions of metasedimentary material prior to final emplacement.

5.1.3. Sources of the other western Maine granitic rocks

The additional Pb isotope data from the Phillips pluton allow reinterpretation of the source for these rocks (Fig. 6). The interpretation that the leucogranites derive from melting of isotopically heterogeneous CMB metasedimentary rocks (Pressley and Brown, 1999) remains acceptable, given the limits on our understanding of Pb in these rocks.

The Phillips pluton granodiorites were interpreted to come from melting of Avalon-like crustal basement by Pressley and Brown (1999). Whereas these sources are consistent with ε_{Nd} of −2.0 to 0, the mean $^{207}Pb/^{204}Pb$ of 15.59 is not simply accounted for by such an end member. The data are more consistent with the granodiorites originating from a BHB source, which requires a modification of the crustal assembly

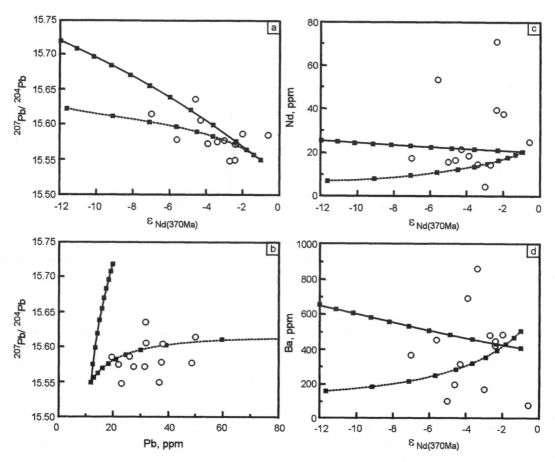

Fig. 9. Compositions of MIC granite samples: (a) whole-rock ε_{Nd} (at 370 Ma) versus initial $^{207}Pb/^{204}Pb$; (b) Pb (ppm) versus initial $^{207}Pb/^{204}Pb$; (c) ε_{Nd} (at 370 Ma) versus Nd (ppm); (d) $\varepsilon_{Nd(370\ Ma)}$ versus Ba (ppm). Also plotted are calculated two-component mixing (solid line; Langmuir et al., 1978) and assimilation-fractional crystallization (dashed line; DePaolo, 1988) curves, with 10% increments. In the mixing calculations (for details see Appendix B) Bronson Hill belt crust partial melts are mixed with or assimilate Central Maine belt metasedimentary rock. The isotopic data permit either mechanism to have produced the samples, at moderate degrees of crystallization or mixing. The dispersion of Ba and Nd elemental data may reflect mineralogical heterogeneity of source materials or crustal assimilant, or entrained residual K-feldspar and monazite.

at depth as shown in interpretive cross-sections (Fig. 2). Viewed as a whole, we interpret the MIC and Phillips pluton data to require the extension of BHB crust farther into central Maine than previously interpreted. This suggests that examination of granites between the Phillips and Sebago plutons should delineate the transition between BHB and non-North American crust at depth.

The source of the Redington pluton was apparently more thoroughly homogenized than that of the MIC granites, or alternatively the magmas that make up the Redington pluton were homogenized after melt generation and possibly during emplacement

(Fig. 6). The Nd–Pb relations of Redington pluton samples are satisfied either by derivation from the estimated BHB crust, Laurentian basement, or CMB metasedimentary rocks. Plutons interpreted to have a significant CMB-supracrustal source tend to show significant initial Nd isotope heterogeneity (e.g., Sebago pluton group 2 granites, Tomascak et al., 1996; Phillips pluton leucogranites, Pressley and Brown, 1999), and thus such sources are not favored for the Redington pluton. Although $^{207}Pb/^{204}Pb$ of the Redington pluton samples are high relative to most Laurentian samples, they are within the range, and hence potentially derived from this crustal source.

The significant differences in major and trace element composition between MIC granites and the Redington samples suggests a fundamental difference in the nature of the sources, although they share isotopic character.

Without paired Nd and Pb isotope data from the Lexington and Sugarloaf plutons detailed interpretation of their sources is precluded. Textures suggestive of magma mingling in the Sugarloaf pluton and the central lobe of the Lexington plutonic complex suggest petrogenetic processes that may be different from those for the other plutons of this study. Nevertheless, the positive ε_{Nd} values of the mafic portions of these mingled zones do not require mantle components, considering the nature of BHB crust. More detailed study of these rocks, as well as the Flagstaff Lake intrusive complex (Fig. 2), is an important next step toward investigating the potential for material input from mantle sources in this area during the Devonian.

The volume of melt generated to form the plutonic complexes in this study, the sources of which are interpreted to have been dominantly lower crustal metaigneous rocks, requires significant thermal input that is best satisfied by advection of thermal energy during lithospheric thinning and upwelling of the convecting asthenospheric mantle. Although material and thermal input from the mantle are commonly assumed to go hand in hand, certain areas of exposed lower crust bearing mantle intrusions show evidence for an absence of appreciable material contributions to melting overlying crust by these melts (e.g., Barboza et al., 1999).

6. Summary

Data presented for the Mooselookmeguntic igneous complex (MIC) in western Maine support the following inferences about melting in the crust in this area between c. 389 and 370 Ma. Earliest magmas, represented by the monzodiorite suite and enclaves in the MIC, were generated by melting Bronson Hill belt (BHB) crust of mafic to intermediate composition. The compositions are most consistent with derivation of these magmas from a single crustal source rather than a mixture of mantle and crust. Heterogeneity in initial Nd and Pb isotopic compositions of the MIC

granites are most plausibly attributable to either melting of rocks in the Central Maine belt stratigraphic package or through assimilation of this material by magmas from intermediate to felsic BHB crust. The absence of isotope-element correlations indicates that metasedimentary-mafic lower crust (or crust–mantle) mixtures are not tenable.

Data from granites emplaced 10–15 Ma before the MIC also yield new insight on crustal architecture. New data for the Phillips pluton are consistent with derivation of the leucogranites from CMB sources whereas the granodiorites are best considered to have originated in BHB crust. The geochemistry of the Redington pluton suggests sources in either BHB crust or Laurentian crust. Collectively, we interpret the data to indicate that BHB crust extended southeastward to significant distance beyond the exposed limit of BHB exposure and that, at least in this part of the orogen, direct material contributions from mantle sources were not important (see Fig. 2, structure section).

Acknowledgments

This work was supported by the NSF (EAR-9705858) and the University of Maryland, and by a University of Maryland Honors Research grant to Centorbi. Much of the laboratory data collected here formed part of the MS and BS theses of Tian and Centorbi, respectively. U–Pb data were collected by Solar and Tian in the Geochronological Laboratory at Washington University; we thank R.D. Tucker for access to his laboratory and NSF for funding the analytical work (EAR-0003531). We thank E. Troy Rasbury-Holt for generously sharing high-purity silica gel, and A. Gale, J.H. Long and E. Moseley-Stewart for assistance in sample preparation. The manuscript was significantly improved after considering the reviews of Sandra M. Barr and Calvin G. Barnes and the editorial comments of O. Tapani Rämö.

Appendix A. Supplementary data

Supplementary data associated with this article can be found, in the online version, at doi:10.1016/j.lithos.2004.04.059.

Appendix B

U and Pb isotope data for zircon samples from this study

Sample number	Sample weight (mg)	Pb (ppm)	U (ppm)	Pb_{com} (pg)	$^{206}Pb/^{204}Pb^{a}$	$^{206}Pb/^{238}U^{b}$	$^{207}Pb/^{235}U^{b}$	$^{207}Pb/^{206}Pb^{b}$	$^{206}Pb/^{238}U$ age (Ma)	$^{207}Pb/^{235}U$ age (Ma)	$^{207}Pb/^{206}Pb$ age (Ma)
J98-40	0.015	14.82	235.8	2.1	6200	0.05919 (14)	0.4418 (16)	0.05413 (14)	371	372	377
J98-40	0.029	14.14	227.1	2.4	10,060	0.05811 (16)	0.4338 (14)	0.05414 (8)	364	366	378
J98-100	0.028	11.49	150.8	2.6	6351	0.06077 (16)	0.4536 (15)	0.05414 (12)	380	380	377
J98-100	0.018	43.47	583.8	1.0	37,470	0.06009 (14)	0.4485 (10)	0.05413 (4)	376	376	377
98-8	0.003	48.66	810.1	0.8	10,480	0.06446 (16)	0.4870 (14)	0.05479 (10)	403	403	404
98-8	0.002	120.4	2028	1.5	11,090	0.06418 (14)	0.4855 (12)	0.05486 (10)	401	402	407
98-1	0.004	21.65	353.8	2.0	2514	0.05782 (18)	0.4285 (30)	0.05375 (36)	362	362	361
98-1	0.006	47.49	786.7	1.2	14,260	0.05852 (14)	0.4346 (12)	0.05385 (8)	367	366	365
98-2	0.013	85.08	1446	2.7	26,960	0.06395 (16)	0.4832 (12)	0.05480 (6)	400	400	404
98-2	0.002	79.40	1247	5.1	2125	0.06401 (28)	0.4831 (26)	0.05474 (18)	400	400	402

Analyses of 1–21 clear, colorless to pale brown, prismatic to needle-like grains for each sample. Blanks: Pb=2 pg; U=0.2 pg.

[a] Measured value, uncorrected.

[b] Corrected for spike contribution, instrumental mass fractionation, blank and estimated initial Pb content; absolute 2σ within-run uncertainty on last digit(s) of individual analysis in parentheses.

Appendix C. Details of calculations

The fractional crystallization model for the MIC monzodiorite suite and enclaves assumes a residual mineralogy of 60% plagioclase, 30% clinopyroxene, and 10% amphibole, and uses a mass ratio of assimilated to fractionated material (r) of 0.5. The starting material is enriched lithospheric mantle with V=300 ppm, Ba=140 ppm, Nd=8 ppm, and Pb=6 ppm. These parameters are estimated from the Mesozoic tholeiite data of Pegram (1990) and from the compilations of McDonough and Sun (1995) and Sun and McDonough (1989). This starting magma assimilates Laurentian crust with V=150 ppm, Ba=400 ppm, Nd=11 ppm, and Pb=20 ppm. This composition is estimated from lower crustal averages of Shaw et al. (1986) and Rudnick and Fountain (1995). Isotopic characteristics of the end members are discussed in the text.

The fractional crystallization model for MIC granites assumes a residual mineralogy of 45% plagioclase, 25% amphibole, 15% K-feldspar, 15% biotite, and 0.1% monazite, and uses r=0.2. The composition of the starting material, average Bronson Hill belt crust, has Ba=500 ppm, Nd=20 ppm, and Pb=12 ppm. These parameters are estimated from data in Leo (1985), Hingston (1992), and Kim and Jacobi

(1996). This starting magma assimilates Central Maine belt crust, with Ba=660 ppm, Nd=25 ppm, and Pb=20 ppm. Compositional features are consistent with data from western Maine Siluro-Devonian metasedimentary rocks (Solar and Brown, 2001b). Isotopic characteristics of the end members are discussed in the text.

Mineral distribution coefficients for the calculations come from a variety of sources in order to attempt to best match the estimated magma compositions. These sources include Bacon and Druitt (1988), Bea et al. (1994), Brenan et al. (1995), Ewart and Griffin (1994), Luhr and Carmichael (1980), Mahood and Hildreth (1983), Nash and Crecraft (1985), and Sisson (1994). Bulk distribution coefficients calculated from these references, using the assemblages above, were V=1.1; Nd=0.46 (using basalt/andesite coefficients), 1.6 (using rhyolite coefficients); Ba=0.26 (mafic), 1.7 (felsic); Pb=0.61 (mafic), 0.40 (felsic).

References

Arth, J.G., Ayuso, R.A., 1997. The Northeast Kingdom batholith, Vermont: Geochronology and Nd, O, Pb, Sr isotopic constraints on the origin of Acadian granitic rocks. In: Sinha, A.K., Whalen, J.B., Hogan, J.P. (Eds.), The Nature of Magmatism in the

Appalachian Orogen. Geological Society of America, Memoir, vol. 191, pp. 1–17.

Ayuso, R.A., 1986. Lead-isotopic evidence for distinct sources of granite and for distinct basements in the northern Appalachians, Maine. Geology 14, 322–325.

Ayuso, R.A., Arth, J.G., 1992. The Northeast Kingdom batholith, Vermont: magmatic evolution and geochemical constraints on the origin of Acadian granitic rocks. Contributions to Mineralogy and Petrology 111, 1–23.

Ayuso, R.A., Bevier, M.L., 1991. Regional differences in Pb isotopic compositions of feldspars in plutonic rocks of the northern Appalachian mountains, U.S.A. and Canada: a geochemical method of terrane correlation. Tectonics 10, 191–212.

Ayuso, R.A., Schulz, K.J., 2003. Pb–Nd–Sr isotope geochemistry and origin of the Ordovician Bald Mountain and Mount Chase massive sulfide deposits, northern Maine. Economic Geology Monographs 11, 611–630.

Ayuso, R.A., Horan, M.F., Criss, R.E., 1988. Pb and O isotope geochemistry of granitic plutons in northern Maine. American Journal of Science 288-A, 421–460.

Bacon, C.R., Druitt, T.H., 1988. Compositional evolution of the zoned calcalkaline magma chamber of Mt. Mazama, Crater Lake, Oregon. Contributions to Mineralogy and Petrology 98, 224–256.

Barboza, S.A., Bergantz, G.W., Brown, M., 1999. Regional granulite facies metamorphism in the Ivrea zone: is the Mafic Complex the smoking gun or a red herring? Geology 27, 447–450.

Barr, S.M., Hegner, E., 1992. Nd isotopic compositions of felsic igneous rocks in Cape Breton Island, Nova Scotia. Canadian Journal of Earth Sciences 29, 650–657.

Bea, F., Pereira, M.D., Stroh, A., 1994. Mineral/leucosome trace-element partitioning in a peraluminous migmatite (a laser ablation-ICP-MS study). Chemical Geology 117, 291–312.

Bennett, V.C., DePaolo, D.J., 1987. Proterozoic crustal history of the western United States as determined by neodymium isotopic mapping. Geological Society of America Bulletin 99, 674–685.

Bock, B., McLennan, S.M., Hanson, G.N., 1996. The Taconic orogeny in southern New England: Nd-isotope evidence against addition of juvenile components. Canadian Journal of Earth Sciences 33, 1612–1627.

Bradley, D., Tucker, R.D., 2002. Emsian synorogenic paleogeography of the Maine Appalachians. Journal of Geology 110, 483–492.

Brenan, J.M., Shaw, H.F., Ryerson, F.J., Phinney, D.L., 1995. Experimental determination of trace-element partitioning between pargasite and synthetic hydrous andesitic melt. Earth and Planetary Science Letters 135, 1–11.

Brown, M., Pressley, R.A., 1999. Crustal melting in nature: prosecuting source processes. Physics and Chemistry of the Earth. Part A: Solid Earth and Geodesy 24, 305–316.

Brown, M., Solar, G.S., 1998a. Shear zone systems and melts: feedback relations and self organization in orogenic belts. Journal of Structural Geology 20, 211–227.

Brown, M., Solar, G.S., 1998b. Granite ascent and emplacement during contractional deformation in convergent orogens. Journal of Structural Geology 20, 1365–1393.

Brown, M., Solar, G.S., 1999. The mechanism of ascent and emplacement of granite magma during transpression: a syntectonic granite paradigm. Tectonophysics 312, 1–33.

Chamberlain, C.P., Robinson, P., 1989. Styles of metamorphism with depth in the central Acadian high, New England. Contribution, vol. 63. Department of Geology and Geography, University of Massachusetts, Amherst.

Chamberlain, C.P., Sonder, L.J., 1990. Heat-producing elements and the thermal and baric patterns of metamorphic belts. Science 250, 1365–1393.

Coish, R.A., Rogers, N.W., 1987. Geochemistry of the Boil Mountain ophiolitic complex, northwestern Maine, and tectonic implications. Contributions to Mineralogy and Petrology 97, 51–65.

Cullers, R.L., Bock, B., Guidotti, C.V., 1997. Elemental distributions and neodymium isotopic compositions of Silurian metasediments, western Maine, USA: redistribution of the rare earth elements. Geochimica et Cosmochimica Acta 61, 1847–1861.

Daly, J.S., McLelland, J.M., 1991. Juvenile mid-Proterozoic crust in the Adirondack highlands, Grenville Province, northeastern North America. Geology 19, 119–122.

DePaolo, D.J., 1988. Neodymium Isotope Geochemistry: An Introduction. Springer-Verlag, New York.

DeWolf, C.P., Mezger, K., 1994. Lead-isotope analyses of leached feldspars—constraints on the early crustal history of the Grenville orogen. Geochimica et Cosmochimica Acta 58, 5537–5550.

DeYoreo, J.J., Lux, D.R., Guidotti, C.V., Decker, E.R., Osberg, P.H., 1989. The Acadian thermal history of western Maine. Journal of Metamorphic Geology 7, 169–190.

Dorais, M.J., Paige, M.L., 2000. Regional geochemical and isotopic variations of northern New England plutons: implications for magma sources and for Grenville and Avalon basement-terrane boundaries. Geological Society of America Bulletin 112, 900–914.

Elburg, M.A., 1996. U–Pb ages and morphologies of zircon in microgranitoid enclaves and peraluminous host granite: evidence for magma mingling. Contributions to Mineralogy and Petrology 123, 177–189.

Ewart, A., Griffin, W.L., 1994. Application of proton-microprobe to trace-element partitioning in volcanic rocks. Chemical Geology 117, 251–284.

Fletcher, I.R., Farquhar, R.M., 1977. Lead isotopes in Grenville and adjacent Paleozoic formations. Canadian Journal of Earth Sciences 14, 56–66.

Foland, K.A., Raczek, I., Henderson, C.M.B., Hofmann, A.W., 1988. Petrogenesis of the magmatic complex at Mount Ascutney, Vermont, USA. Contributions to Mineralogy and Petrology 98, 408–416.

Guidotti, C.V., Holdaway, M.J., 1993. Petrology and field relations of successive metamorphic events in pelites of west-central Maine. In: Cheney, J.T., Hepburn, J.C. (Eds.), Field Trip Guidebook for the Northeastern United States, vol. 1. Geological Society of America, pp. L1–L23.

Guidotti, C.V., Johnson, S.E., 2002. Pseudomorphs and associated microstructures of western Maine, USA. Journal of Structural Geology 24, 1139–1156.

Hingston, M., 1992. The geochemistry of the Ammonoosuc Volcanics and Oliverian granites, west-central New Hampshire. MSc thesis, Dartmouth College, Hanover, New Hampshire, U.S.A.

Johnson, T.E., Brown, M., Solar, G.S., 2003. Low-pressure sub-solidus and suprasolidus phase equilibria in the MnNCKFMASH system: constraints on conditions of regional metamorphism in western Maine, northern Appalachians. American Mineralogist 88, 624–638.

Kerr, A., Jenner, G.A., Fryer, B.J., 1995. Sm–Nd isotope geo-chemistry of Precambrian to Paleozoic granitoid suites and the deep-crustal structure of the southeast margin of the Newfound-land Appalachians. Canadian Journal of Earth Sciences 32, 224–245.

Kim, J., Jacobi, R.D., 1996. Geochemistry and tectonic implications of the Hawley Formation meta-igneous units, northern Massa-chusetts. American Journal of Science 296, 1126–1174.

Krogstad, E.J., 1993. Pb isotopic composition of Paleozoic sedi-ments derived from the Appalachian orogen. Abstracts with Programs-Geological Society of America 25, 30–31.

Langmuir, C.H., Vocke, R.D., Hanson, G.N., Hart, S.R., 1978. General mixing equation with applications to Iceland basalts. Earth and Planetary Science Letters 37, 380–392.

Lathrop, A.S., Blum, J.D., Chamberlain, C.P., 1996. Nd, Sr and O isotopic study of the petrogenesis of two syntectonic members of the New Hampshire Plutonic Series. Contributions to Mineralogy and Petrology 124, 126–138.

Leo, G.W., 1985. Trondhjemite and metamorphosed quartz kerato-phyre tuff of the Ammonoosuc Volcanics (Ordovician), western New Hampshire and adjacent Vermont and Massachusetts. Geological Society of America Bulletin 96, 1493–1507.

Liew, T.C., Finger, F., Hock, V., 1989. The Moldanubian granitoid plutons of Austria—chemical and isotopic studies bearing on their environmental setting. Chemical Geology 76, 41–55.

Luhr, J.F., Carmichael, I.S., 1980. The Colima volcanic complex, Mexico: I. Post-caldera andesites from Volcan Colima. Con-tributions to Mineralogy and Petrology 71, 343–372.

Lyons, J.B., Bothner, W.S., Moench, R.H., Thompson, J.B., 1997. Bedrock geologic map of New Hampshire. U.S. Geological Survey State Geologic Map, 1:250000.

Mahood, G.A., Hildreth, E.W., 1983. Large partition coefficients for trace elements in high-silica rhyolites. Geochimica et Cosmo-chimica Acta 47, 11–30.

McCulloch, M.T., Woodhead, J.D., 1993. Lead isotopic evidence for deep crustal-scale fluid transport during granite petrogenesis. Geochimica et Cosmochimica Acta 57, 659–674.

McDonough, W.F., Sun, S.S., 1995. Composition of the Earth. Chemical Geology 120, 223–253.

McLelland, J.M., Daly, J.S., Chiarenzelli, J., 1993. Sm–Nd and U–Pb isotopic evidence of juvenile crust in the Adirondack lowlands and implications for the evolution of the Adirondack Mts. Journal of Geology 101, 97–105.

Moench, R.H., 1971. Geologic map of the Rangeley and Phillips quadrangles, Franklin and Oxford Counties, Maine. U.S. Geo-logical Survey Map I-605.

Moench, R.H., Aleinikoff, J.N., 2002. Stratigraphy, geochronology, and accretionary terrane settings of two Bronson Hill arc sequences, northern New England. Physics and Chemistry of the Earth 27, 47–95.

Moench, R.H., Boone, G.M., Bothner, W.A., Boudette, E.L., Hatch Jr., N.L., Hussey II, A.M., Marvinney, R.G., Aleinikoff, J.N., 1995. Geologic map of the Sherbrooke-Lewiston area, Maine, New Hampshire, and Vermont, United States, and Quebec Canada. U.S. Geological Survey Map I-1898-D.

Nash, W.P., Crecraft, H.R., 1985. Partition coefficients for trace elements in silicic magmas. Geochimica et Cosmochimica Acta 49, 309–322.

Pegram, W.J., 1990. Development of continental lithospheric mantle as reflected in the chemistry of the Mesozoic Appalachian Tholeiites, U.S.A. Earth and Planetary Science Letters 97, 316–331.

Pressley, R.A., Brown, M., 1999. The Phillips pluton, Maine, USA: evidence of heterogeneous crustal sources and implications for granite ascent and emplacement mechanisms in convergent orogens. Lithos 46, 335–366.

Rudnick, R.L., Fountain, D.M., 1995. Nature and composition of the continental crust—a lower crustal perspective. Reviews of Geophysics 33, 267–309.

Samson, S.D., Tremblay, A., 1996. Nd isotopic composition of volcanic rocks in the Ascot complex, Québec: comparison with other Ordovician terranes. Abstracts with Programs-Geological Society of America 28, 96.

Samson, S.D., Barr, S.M., White, C.E., 2000. Nd isotopic character-istics of terranes within the Avalon Zone, southern New Brunswick. Canadian Journal of Earth Sciences 37, 1039–1052.

Shaw, H.F., Wasserburg, G.J., 1984. Isotopic constraints on the origin of Appalachian mafic complexes. American Journal of Science 284, 319–349.

Shaw, D.M., Cramer, J.J., Higgins, M.D., Truscott, M.G., 1986. Composition of the Canadian Precambrian shield and the continental crust of the Earth. In: Dawson, J.D., et al., (Eds.), The Nature of the Lower Continental Crust, Special Publication-Geological Society of London, vol. 24, pp. 275–282.

Sisson, T.W., 1994. Hornblende-melt trace-element partition parti-tioning measured by ion microprobe. Chemical Geology 117, 331–344.

Solar, G.S., Brown, M., 1999. The classic high-T–low-P meta-morphism of west-central Maine, USA: is it post-tectonic or syn-tectonic? Evidence from porphyroblast-matrix relations. Canadian Mineralogist 37, 311–333.

Solar, G.S., Brown, M., 2000. The classic high-T–low-P meta-morphism of west-central Maine, USA: is it post-tectonic or syn-tectonic? Evidence from porphyroblast-matrix relations: reply. Canadian Mineralogist 38, 1007–1026.

Solar, G.S., Brown, M., 2001a. Deformation partitioning during transpression in response to Early Devonian oblique conver-gence, northern Appalachian orogen, USA. Journal of Structural Geology 22, 1043–1065.

Solar, G.S., Brown, M., 2001b. Petrogenesis of migmatites in Maine, USA: possible source of peraluminous granite in plutons. Journal of Petrology 42, 789–823.

Solar, G.S., Pressley, R.A., Brown, M., Tucker, R.D., 1998. Granite ascent in contractional orogenic belts: testing a model. Geology 26, 7111–7714.

Stacey, J.S., Kramers, J.D., 1975. Approximation of terrestrial Pb isotope evolution by a two-stage model. Earth and Planetary Science Letters 26, 207–221.

Stewart, D.B., Wright, B.E., Unger, J.D., Phillips, J.D., Hutchinson, D.R., 1992. Global Geoscience Transect 8: Quebec-Maine-Gulf of Maine Transect, Southeastern Canada, Northeastern United States of America. U. S. Geological Survey Map I-2329.

Sun, S.S., McDonough, W.F., 1989. Chemical and isotopic systematics of oceanic basalts: implications for mantle composition and processes. In: Saunders, A.D., Norry, M.J. (Eds.), Magmatism in the Ocean Basins, Special Publication-Geological Society of London, vol. 42, pp. 313–345.

Thirlwall, M.F., 2000. Inter-laboratory and other errors in Pb isotope analyses investigated using a Pb-207-Pb-204 double spike. Chemical Geology. Isotope Geoscience Section 163, 299–322.

Tian, J. 2000. A geologic and geochemical study of the Mooselookmeguntic composite pluton, west central Maine and east central New Hampshire. MSc thesis, University of Maryland, College Park, U.S.A.

Tomascak, P.B., Krogstad, E.J., Walker, R.J., 1996. Nature of the crust in Maine, USA: evidence from the Sebago batholith. Contributions to Mineralogy and Petrology 125, 45–59.

Tomascak, P.B., Krogstad, E.J., Walker, R.J., 1999. The significance of the Norumbega Fault Zone in southwestern Maine: clues from the geochemistry of granitic rocks. In: Ludman, A., West Jr., D.P. (Eds.), The Norumbega Fault System of the Northern Appalachians, Special Paper-Geological Society of America, vol. 331, pp. 105–119.

van de Flierdt, T., Hoernes, S., Jung, S., Masberg, P., Hoffer, E., Schaltegger, U., Friedrichsen, H., 2003. Lower crustal melting and the role of open-system processes in the genesis of synorogenic quartz diorite-granite-leucogranite associations: constraints from Sr–Nd–O isotopes from the Bandombaai Complex, Namibia. Lithos 67, 205–226.

Waight, T.E., Wiebe, R.E., Krogstad, E.J., Walker, R.J., 2001. Isotopic responses to basaltic injections into silicic magma chambers: a whole-rock and microsampling study of macrorhythmic units in the Pleasant Bay layered gabbro-diorite complex, Maine, USA. Contributions to Mineralogy and Petrology 142, 323–335.

Whalen, J.B., Jenner, G.A., Currie, K.L., Barr, S.M., Longstaffe, F.J., Hegner, E., 1994. Geochemical and isotopic characteristics of granitoids of the Avalon Zone, southern New Brunswick: possible evidence for repeated delamination events. Journal of Geology 102, 269–282.

Whalen, J.B., Fyffe, L.R., Longstaffe, F.J., Jenner, G.A., 1996. The position of the Gander-Avalon boundary, southern New Brunswick, based on geochemical and isotopic data from granitoid rocks. Canadian Journal of Earth Sciences 33, 129–139.

Whalen, J.B., Roger, N., van Staal, C.R., Longstaffe, F.J., Jenner, G.A., Winchester, J.A., 1998. Geochemical and isotopic (Nd, O) data for Ordovician felsic plutonic and volcanic rocks of the Miramichi Highlands: petrogenetic and metallogenic implications for the Bathurst Mining Camp. Canadian Journal of Earth Sciences 35, 237–252.

White, W.M., 1993. U-238/Pb-204 in MORB and open system evolution of the depleted mantle. Earth and Planetary Science Letters 115, 211–226.

Williams, H., Hatcher, R.D., 1983. Appalachian suspect terranes. In: Hatcher, R.D., Williams, H., Zietz, I. (Eds.), Contributions to the Tectonics and Geophysics of Mountain Belts, Geological Society of America, Memoir, vol. 158, pp. 33–53.

Zartman, R.E., Doe, B.R., 1981. Plumbotectonics—the model. Tectonophysics 75, 135–162.

Available online at www.sciencedirect.com

SCIENCE @ DIRECT®

Lithos 80 (2005) 101–129

LITHOS

www.elsevier.com/locate/lithos

Petrogenesis of the Paleoproterozoic rapakivi A-type granites of the Archean Carajás metallogenic province, Brazil

Roberto Dall'Agnol[a,*], Nilson P. Teixeira[a], O. Tapani Rämö[b], Candido A.V. Moura[c], Moacir J.B. Macambira[c], Davis C. de Oliveira[a]

[a]Group of Research on Granite Petrology, Centro de Geociências, Universidade Federal do Pará, Caixa Postal 1611, 66075-100 Belém, PA, Brazil
[b]Department of Geology, P.O. Box 64, FI-00014 University of Helsinki, Finland
[c]Isotope Geology Laboratory, Centro de Geociências, Universidade Federal do Pará, Caixa Postal 1611, 66075-100 Belém, PA, Brazil

Received 2 May 2003; accepted 9 September 2004
Available online 26 November 2004

Abstract

Three Paleoproterozoic A-type rapakivi granite suites (Jamon, Serra dos Carajás, and Velho Guilherme) are found in the Carajás metallogenic province, eastern Amazonian craton. Liquidus temperatures in the 900–870 °C range characterize the Jamon suite, those for Serra dos Carajás and Velho Guilherme are somewhat lower. Pressures of emplacement decrease from Jamon (3.2 ± 0.7 kbar) through Serra dos Carajás (2.0 ± 1.0 kbar) to Velho Guilherme (1.0 ± 0.5 kbar). Oxidizing conditions (NNO+0.5) characterized the crystallization of the Jamon magma, the Velho Guilherme magmas were reducing (marginally below FMQ), and the Serra dos Carajás magmas were intermediate between the two in this respect. The three granite suites have Archean T_{DM} model ages and strongly negative ε_{Nd} values (-12 to -8 at 1880 Ma), and they were derived from Archean crust. The Jamon granite suite may have been derived from a quartz dioritic source, and the Velho Guilherme granites from K-feldspar-bearing granitoid rocks with some sedimentary input. The Serra dos Carajás granites either had a somewhat more mafic source than Velho Guilherme or were derived by a larger degree of melting. Underplating of mafic magma was probably the heat source for the melting. The petrological and geochemical characteristics of the Carajás granite suites imply considerable compositional variation in the Archean of the eastern Amazonian craton. The oxidized Jamon suite granites are similar to the Mesoproterozoic magnetite-series granites of Laurentia, and they were derived from Archean igneous sources that were more oxidized than the sources of the Fennoscandian rapakivi granites. The Serra dos Carajás and Velho Guilherme granites approach the classic reduced rapakivi series of Fennoscandia and Laurentia. No counterparts of the Mesoproterozoic two-mica granites of Laurentia have been found, however. Following the model of Hoffman [Hoffman, P., 1989. Speculations on Laurentia's first gigayear (2.0 to 1.0 Ga). Geology 17, 135–138], the origin of the ~1.88 Ga Carajás granites is related to a mantle superswell beneath the Trans-Amazonian supercontinent. This caused breakup of the continent and was associated with magmatic underplating and resultant crustal melting and generation of A-type granite magmas. The Paleoproterozoic continent that included the

* Corresponding author. Tel.: +55 91 211 1477; fax: +55 91 211 1609.
  E-mail address: robdal@ufpa.br (R. Dall'Agnol).

Archean and Trans-Amazonian domains of the Amazonian craton was assembled at ~2.0 Ga; its disruption was initiated at ~1.88 Ga, at least 200 Ma earlier than in Laurentia and Fennoscandia. The Carajás granites were related to the breakup of the supercontinent, not to subduction processes.

Keywords: Carajás; Amazonian craton; Nd isotopes; Oxygen fugacity; Paleoproterozoic; A-type granite; Archean

---

1. Introduction

Rapakivi granites are A-type granites characterized by the presence of the rapakivi texture (Haapala and Rämö, 1992), and their origin is generally associated with crustal anatexis promoted by magmatic underplating (Huppert and Sparks, 1988; Rämö and Haapala, 1995; Dall'Agnol et al., 1999a). Proposed sources vary from granodiorite, tonalite (Anderson and Cullers, 1978; Anderson, 1983; Creaser et al., 1991), and quartz diorite (Dall'Agnol et al., 1999c) to tholeiites and their differentiates (Frost and Frost, 1997; Frost et al., 1999). Fractional crystallization from alkaline basalts (Eby, 1992) or other mantle-derived magmas (Bonin, 1986) and melting of residual granulitic sources (Collins et al., 1982; Clemens et al., 1986) have also been proposed for the origin of A-type granites. A-type granites were originally defined as anorogenic granites characterized by low water and oxygen fugacity (Loiselle and Wones, 1979). However, these rocks are also found in postcollisional settings (Whalen et al., 1987; King et al., 1997), and there is increasing evidence of oxidized A-type granites (Anderson, 1983; Anderson and Bender, 1989; Anderson and Smith, 1995; Dall'Agnol et al., 1997a, 1999b; Anderson and Morrison, 2003). Moreover, studies on natural (King et al., 2001) and experimental (Clemens et al., 1986; Dall'Agnol et al., 1999b; Holtz et al., 2001; Klimm et al., 2003) systems point to significant water contents (>2 to 6.5 wt.%) in A-type granite magmas.

One important recently discovered rapakivi granite province is located in the Amazonian craton (Rämö and Haapala, 1995; Haapala and Rämö, 1999; Dall'Agnol et al., 1999a). In age, these granites range from 1.88 Ga in the southeastern part of the craton (Dall'Agnol et al., 1994, 1999a) to ~1.0 Ga in southwestern provinces (Bettencourt et al., 1999). Granitic suites are found in the southern (~1.82 Ga)

(Costi et al., 2000) and northwestern (~1.55 Ga) (Fraga et al., 2003) Guiana shield.

Three 1.88–1.86-Ga granite suites are present in the Carajás metallogenic province of the eastern Amazonian craton: Jamon, Velho Guilherme, and Serra dos Carajás (Dall'Agnol et al., 1994, 1999c; Barros et al., 1995; Javier Rios et al., 1995; Teixeira et al., 2002). These are composed of batholiths and stocks intruded into different domains of the Archean Carajás province. We consider these granites as rapakivi granites because they are A-type granites that have commonly, or in places, more scarcely, K-feldspar megacrysts rimmed by plagioclase (Dall'Agnol et al., 1999a). They are characterized by high $FeO_t/(FeO_t+MgO)$, and HFSE, and comply with the ferroan alkali-calcic to calc-alkalic metaluminous to peraluminous granites of Frost et al. (2001a).

Neodymium isotope data indicate that the Carajás Paleoproterozoic granites were derived from Archean sources (Dall'Agnol et al., 1999c; Rämö et al., 2002; Teixeira et al., 2002). Most of the rapakivi granites are Mesoproterozoic and located in older Proterozoic terranes (Rämö and Haapala, 1995). Only few suites are demonstrably related to Archean crust (cf. Rämö et al., 1995; Amelin et al., 1997; Frost et al., 1999). Thus, the Carajás province is a key area for understanding of the influence of Archean crust in the origin of A-type rapakivi granites. The studied granite suites are also the initial markers at ~1.88 Ga of rapakivi magmatism that spread over most of the Amazonian craton and continued until ~1.0 Ga. Moreover, the Carajás granite suites display strong contrasts in the degree of oxidation. One aim of this paper is to study the reasons for these differences and to discuss the origin of the A-type granites from this perspective. To do this, the studied granites will be compared to the mid-Proterozoic A-type granites of Laurentia and Fennoscandia.

2. Geologic setting

The Carajás metallogenic province is now recognized as one of the main Archean tectonic provinces of the Amazonian craton (Machado et al., 1991; Macambira and Lafon, 1995; Dall'Agnol et al., 1997b; Rämö et al., 2002). It has been included into the Central Amazonian province (Fig. 1a) by some authors (Tassinari and Macambira, 1999) and considered an independent tectonic province (Carajás) by others (Santos et al., 2000). The Central Amazonian and Imataca provinces started to form in the Archean and are the oldest provinces of the craton.

The eastern part of the Central Amazonian province (e.g., the Carajás metallogenic province) comprises mostly Archean rocks (Dall'Agnol et al., 1999a; Tassinari and Macambira, 1999). In the more poorly studied central part (e.g., the western Xingu region; Fig. 1b), Archean basement is not exposed, and Paleoproterozoic granitoids and volcanic rocks are dominant; Archean age of the domain is, however, indicated by Nd model ages (T_{DM}; Cordani and Sato, 1999; Teixeira et al., 2002). The eastern domain is limited in the north by the Maroni-Itacaiúnas province (Fig. 1a) that was formed in the 2.2–2.1-Ga Trans-Amazonian event. In the east, it is limited by the Neoproterozoic Araguaia belt (Fig. 1b) related to the Brasiliano (Pan-African) cycle that did not significantly affect the Amazonian craton.

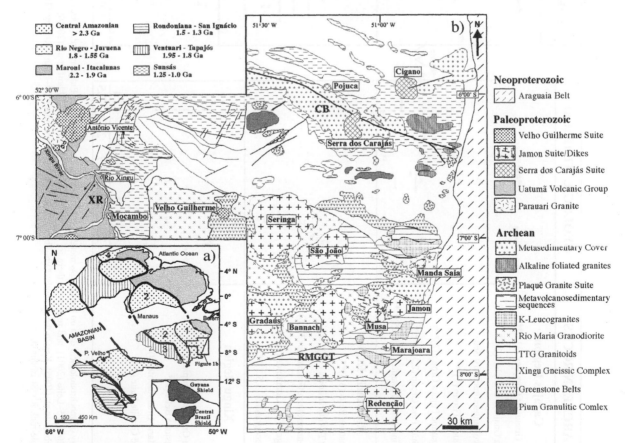

Fig. 1. (a) Sketch map of the Amazonian craton (modified from Tassinari and Macambira, 1999). 1–Carajas metallogenic province; 2–Xingu–Iricoumé domain; 3–Tapajós region; 4–Imataca. (b) Simplified geological map of the Carajás Mining Province showing the distribution of Paleoproterozoic A-type granites (modified from Teixeira, 1999; Leite, 2001). RMGGT–Rio Maria Granite-Greenstone Terrane; CB–Carajás Basin; XR–Xingu Region. Inset in panel (a) shows the two exposed parts of the Amazonian craton–the Guiana shield in the north and the Central Brazil shield in the south.

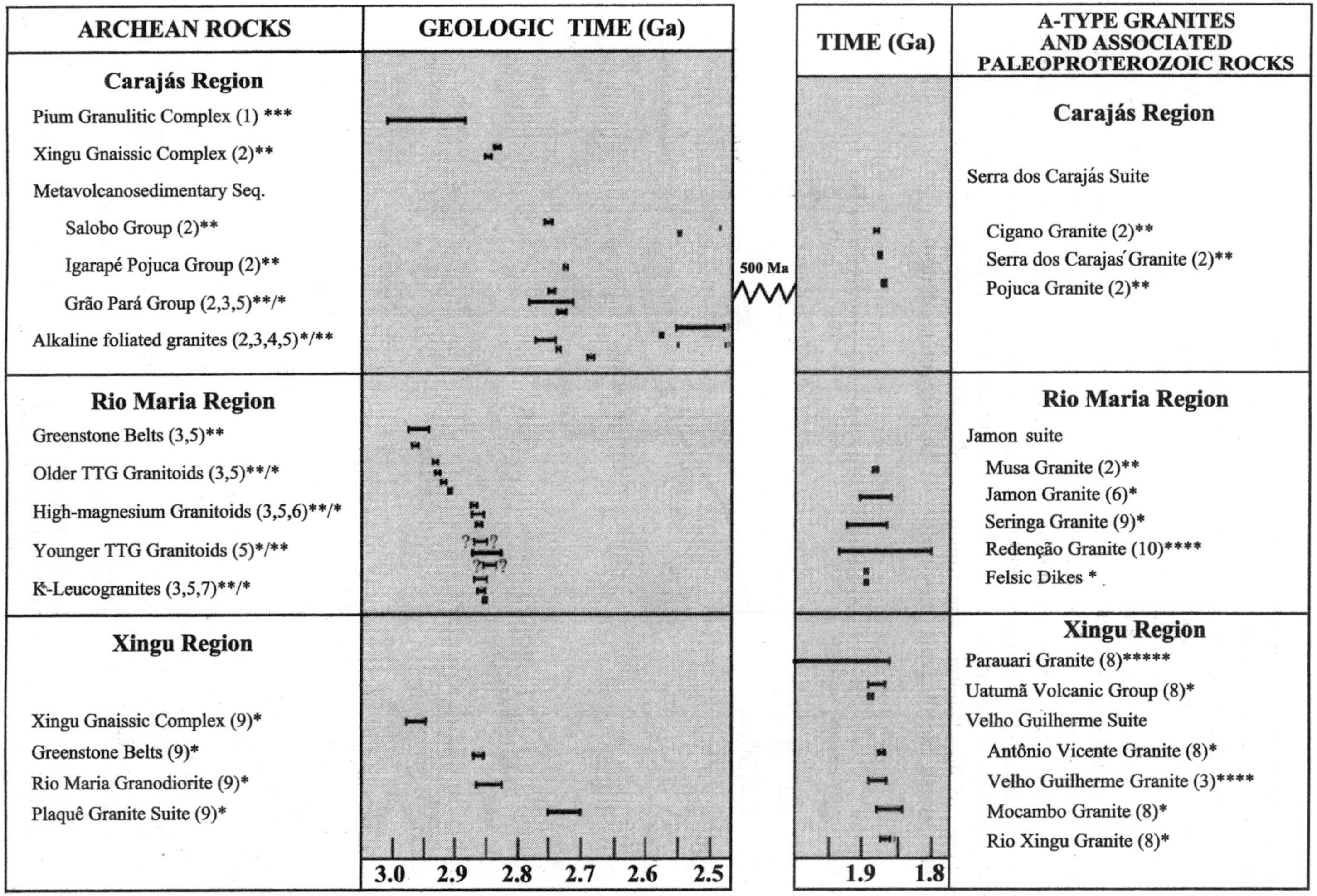

Fig. 2. Summary of geochronological data on the Carajás metallogenic province. Data sources: 1–Pidgeon et al. (2000); 2–Machado et al. (1991); 3–Macambira and Lafon (1995); 4–Barros et al. (2001); 5–Leite (2001), and references there in; 6–Dall'Agnol et al. (1999c); 7–Althoff et al. (2000); 8–Teixeira et al. (2002) and references therein; 9–Avelar et al. (1999); 10–Barbosa et al. (1995). Key to the methods: *Pb–Pb in zircon, evaporation method; **U–Pb in zircon by TIMS; ***U–Pb in zircon by SHRIMP; ****whole-rock Pb–Pb age; *****whole-rock Rb–Sr isochron.

The Carajás metallogenic province was cratonised at the end of the Archean. The A-type granite suites were formed at ~1.88 Ga in an extensional tectonic regime. In the Rio Maria terrane, this is indicated by dike swarms that are coeval with the granites (cf. Fig. 2). In the adjacent provinces, orogenic events are significantly older (the Trans-Amazonian event in the north) or younger (the Brasiliano event in the east) than these granites. The 1.88-Ga event is also identified in other areas of the Central Amazonian (Xingu and Iricoumé domains) and Tapajós provinces (Fig. 1a), where various plutonic and volcanic rocks have that age. The 1.88-Ga event has been interpreted to mark the beginning of the breakup of the Paleoproterozoic continent that was formed at the end of the Trans-Amazonian event (Lamarão et al., 2002). However, Santos et al. (2001) considered that a younger magmatic arc, following older Trans-Amazonian arcs, was formed in the Tapajós province (west of the studied area; Fig. 1a) at that time. Thus, a link between this hypothetical arc and the A-type 'anorogenic' granites of Carajás should be examined (cf. Åhäll et al., 2000).

The eastern Central Amazonian province is divided into two Archean tectonic domains (Figs. 1b and 2), the 3.0–2.86-Ga Rio Maria Granite-Greenstone Terrane (Macambira and Lafon, 1995; Dall'Agnol et al., 1997b) and the rift-related Carajás Basin dominantly composed of 2.76–2.55-Ga metavolcanic rocks, banded iron formations, and granitoids (Machado et al., 1991; Macambira and Lafon, 1995; Barros et al., 2001). In the Xingu region in the west, the Archean sequences of the Carajás metallogenic province are bordered by Paleoproterozoic terranes; these are dominated by calc-alkaline granitoids (Parauari suite), volcanic sequences (Uatumã Supergroup), and A-type granites (Figs. 1, 2; cf. Teixeira et al., 2002).

The Archean crust of the three domains was intruded by ~1.88-Ga rapakivi-type granite suites (Dall'Agnol et al., 1994, 1999a; Figs. 1 and 2). Petrological, geochemical, and isotope data (Dall'Agnol et al., 1999b,c; Rämö et al., 2002; Teixeira et al., 2002) indicate some contrasts between the granite suites—Jamon, Serra dos Carajás, and Velho Guilherme in the Rio Maria Granite-Greenstone Terrane, Carajás Basin, and Xingu region, respectively. Unlike the Carajás Basin, the Rio Maria Terrane was not affected by events younger than 2.77 Ga (Fig. 2). The eastern domains of the Xingu region constitute Archean rocks similar to those of Rio Maria. In the west, only Paleoproterozoic units are exposed, and their role in the evolution of the Velho Guilherme granite suite should be considered (Figs. 1b and 2).

Differences between the three Archean domains are also reflected in their metallogenic traits. The most important gold, copper, manganese, and iron deposits of the province are located in the Carajás Basin. In the Rio Maria Granite-Greenstone Terrane, only small gold and tungsten deposits have been found. The Xingu region is characterized by tin deposits associated with the Velho Guilherme suite.

3. Characterization of the Paleoproterozoic granite suites

3.1. Geologic and geochronologic aspects

All three granitic suites are composed of ~1.88–1.86-Ga granite batholiths and stocks (~5 to 50 km in diameter; Fig. 1) that were emplaced at shallow levels (~1 to 3 kbar; Table 1). Contacts are sharp, and angular enclaves of country rocks are commonly observed in the granites, indicating a high viscosity contrast between the magmas and the Archean bedrock. The granites are not foliated, they cut discordantly the E–W or NW–SE trend of the Archean country rocks and have caused hornblende hornfels contact metamorphism (Dall'Agnol et al., 1994, 1999c). Swarms of mafic, intermediate, and felsic dikes are associated with the Jamon suite. Composite mafic–felsic dikes cutting Archean high-Mg granodiorites have been locally found. The felsic dikes have yielded Pb–Pb zircon ages of 1885 ± 4 and 1885 ± 2 Ma (Oliveira D.C., unpublished data). One dated rhyolite porphyry shows evidence of mingling with an associated mafic dike, demonstrating that the mafic and felsic magmas were coeval. There are no anorthosites or charnockites associated with the Paleoproterozoic granite suites of the Carajás province possibly because of the relatively shallow level of erosion.

3.2. Petrography

Isotropic, equigranular, coarse- or medium-grained and seriated, coarse- to medium-grained, or medium- to fine-grained rocks are dominant in most of the

Table 1
Geochronology of the Paleoproterozoic A-type granites of the Carajás region

Pluton	Method	Analyzed material	Age
Serra Dos Carajás granite suite			
Cigano	U–Pb	Zircon	1883±2 Ma (1)
Serra dos Carajás	U–Pb	Zircon	1880±2 Ma (1)
Pojuca	U–Pb	Zircon	1874±2 Ma (1)
Jamon granite suite			
Musa	U–Pb	Zircon	1883+5/−2 Ma (1)
Jamon	Pb–Pb	Zircon	1885±32 Ma (2)
Seringa	Pb–Pb	Zircon	1893±15 Ma (3)
Redenção	Pb–Pb	Whole rock	1870±68 Ma (4)
Felsic dikes	Pb–Pb	Zircon	1885±4 Ma (7)
		Zircon	1885±2 Ma (7)
Velho Guilherme granite suite			
Velho Guilherme	Pb–Pb	Whole rock	1874±30 Ma (5)
Rio Xingu	Pb–Pb	Zircon	1866±3 Ma (6)
	Pb–Pb	Whole rock	1906±29 Ma (6)
Mocambo	Pb–Pb	Zircon	1862±32 Ma (6)
Antonio Vicente	Pb–Pb	Zircon	1867±4 Ma (6)
	Pb–Pb	Whole rock+feldspar	1896±9 Ma (6)

Data sources: (1) Machado et al. (1991); (2) Dall'Agnol et al. (1999c); (3) Avelar et al. (1999); (4) Barbosa et al. (1995); (5) Macambira and Lafon (1995); (6) Teixeira et al. (2002); (7) Oliveira, D.C., unpublished data.

plutons. Microgranites are also present but subordinate in volume. Coarse porphyritic rocks are less abundant, and typical wiborgitic and pyterlitic textures have not been found. However, plagioclase mantled K-feldspar megacrysts are observed in every pluton, and they are particularly common in the Redenção and Bannach batholiths of the Jamon suite, as well as in the Serra dos Carajás batholith. Only local occurrence of rapakivi texture is a feature described in many A-type rapakivi plutons [e.g., the Bodom and Obbnäs plutons of Finland (Kosunen, 1999) and the Sherman batholith of Wyoming (Frost et al., 1999)]. Mantled K-feldspar megacrysts are more common in the coarse porphyritic facies displaying evidence of interaction between different felsic magmas. In this regard, the porphyritic granites of the Redenção and Bannach plutons of the Jamon suite look similar to the porphyritic granite of the Sherman batholith (Frost et al., 1999). Unlike in the Sherman batholith, comprehensive evidence for mingling between mafic and felsic magmas has not been found.

In all three suites, the rocks are essentially granites sensu stricto, but the modal composition shows some contrasts (Fig. 3). The Jamon, Redenção, and Musa plutons of the Jamon suite are composed of monzogranite with subordinate syenogranite. Contents of mafic minerals are normally between 15% and 5%; in the less evolved facies, they reach >20% and are <5% in the differentiated leucogranites. In the Serra dos Carajás suite, monzogranite and syenogranite are dominant, and the amount of mafic minerals is generally <15%. In the Velho Guilherme suite, syenogranite is dominant over monzogranite and alkali-feldspar granites. In general, the Velho Guilherme granites are more leucocratic than the granites of the other suites. In all three suites, biotite or, in the less evolved facies, biotite and hornblende are the major mafic phases. Muscovite is present only in evolved hydrothermally altered leucogranites.

In the Jamon suite, typical primary accessory mineral assemblage includes zircon, apatite, magnetite, ilmenite, allanite, and titanite (Dall'Agnol et al., 1999c; Oliveira, 2001). Fluorite is significant only in the more evolved facies. Subsolidus processes were limited to alteration of plagioclase and mafic phases. In the Serra dos Carajás suite, accessory minerals are similar, but primary titanite is rare or absent, fluorite is more common, and tourmaline is sometimes present (Javier Rios et al., 1995; Barros et al., 1995). Secondary K-feldspar replacing primary K-feldspar and plagioclase is common as is local greisenisation and intense oxidation. Sulfide mineralization is frequent in the altered granites. In the Velho Guilherme suite, the dominant syenogranites do not contain titanite or significant magnetite, are enriched in fluorite, and have sporadic monazite and xenotime. They may also contain topaz and siderophyllite and generally show intense alteration (Teixeira, 1999). Plagioclase is deanorthized and partially replaced by sericite, fluorite, and topaz; K-feldspar is albitized; primary biotite is replaced by chlorite and/or muscovite and oxide minerals. Greisen is common in the more advanced alteration zones.

3.3. Mineralogy

3.3.1. Jamon suite

In the Jamon suite, biotite is the dominant mafic mineral, amphibole is abundant only in the less

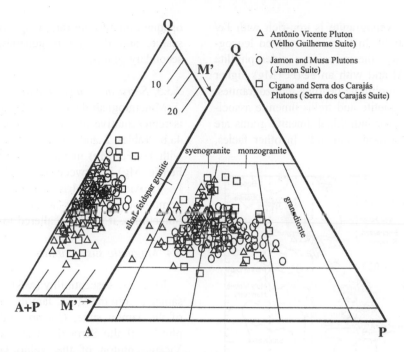

Fig. 3. QAP and Q–(A+P)–M' plots for the A-type granites of Carajás. Data sources: Jamon Suite–Jamon and Musa plutons (Dall'Agnol et al., 1999c and references therein); Velho Guilherme Suite–Antonio Vicente pluton (Teixeira, 1999); Serra dos Carajás Suite–Serra dos Carajás and Cigano plutons (Dall'Agnol et al., 1994; Barros et al., 1995).

evolved facies. The composition of these minerals in the Jamon pluton was discussed by Dall'Agnol et al. (1999b). They compared natural compositions with the results of experimental studies (reduced and oxidized conditions) on hornblende–biotite monzogranite. Chemical and experimental evidences demonstrate that the Jamon pluton magma evolved in relatively oxidizing conditions above the NNO buffer. The amphiboles are ferroedenitic or edenitic hornblende with Fe/(Fe+Mg) in the 0.47–0.65 range in the hornblende–biotite monzogranite and 0.6–0.73 in the biotite monzogranite (Fig. 4a; Table 2). In the rhyolite porphyry, these ratios are 0.35–0.40, indicating a more oxidizing character. Fluorine contents in the amphiboles are relatively high and vary generally between 1 and 2 wt.%. Aluminum in amphibole suggests crystallization around 3.2±0.7 kbar (Dall'Agnol et al., 1999b). Fe/(Fe+Mg) of biotite ranges from 0.6–0.63 (hornblende–biotite monzogranite) to 0.65–0.70 (biotite monzogranite). In the rhyolite porphyry dikes and micromonzogranite, this ratio range is 0.4–0.45 (Fig. 4b; Table 3). Fluorine content in biotite is also relatively high and similar to that in the amphiboles. A

study of Fe–Ti oxide minerals indicated that the ilmenite in the granite has been reequilibrated (Dall'Agnol et al., 1997a). However, primary compositions (78.5–90% ilmenite) have been preserved in the porphyry dikes. These and reconstructed primary titano–magnetite compositions (46–28% ulvospinel) suggest (at a fO_2 of NNO+0.5) temperatures of 775 and 1065 °C, respectively, for the rhyolite and dacite porphyry dikes (Dall'Agnol et al., 1997a, 1999b).

3.3.2. Velho Guilherme suite

Most of the available mineralogical data on the Velho Guilherme suite come from the Antonio Vicente complex (Teixeira, 1999; Dall'Agnol, R., unpublished data). The amphibole of the less evolved monzogranites is hastingsite with Fe/(Fe+Mg) between 0.74 and 0.85. Associated biotite displays similar Fe/(Fe+Mg) ratios. In the alkali-feldspar granite, the amphibole is transitional between hastingsite and ferroedenite. Fe/(Fe+Mg) in the amphibole and associated biotite are higher (0.95–0.98) than in the monzogranites. Both amphiboles are enriched in Al[IV] compared to those of the Jamon suite (Fig. 4a).

The biotite of the syenogranite is iron-rich with Fe/(Fe+Mg) of 0.85–0.89. In the more evolved leucogranites of this suite, the mica is a siderophyllite enriched in total Al and with an Fe/(Fe+Mg) higher than 0.9. In the amphibole-bearing monzogranites, intergrowths of magnetite and trellis ilmenite associated with composite or individual ilmenite grains are the dominant Fe–Ti oxide minerals. In other facies,

opaque minerals are rare; magnetite is absent or very scarce, and ilmenite is generally present as small secondary grains.

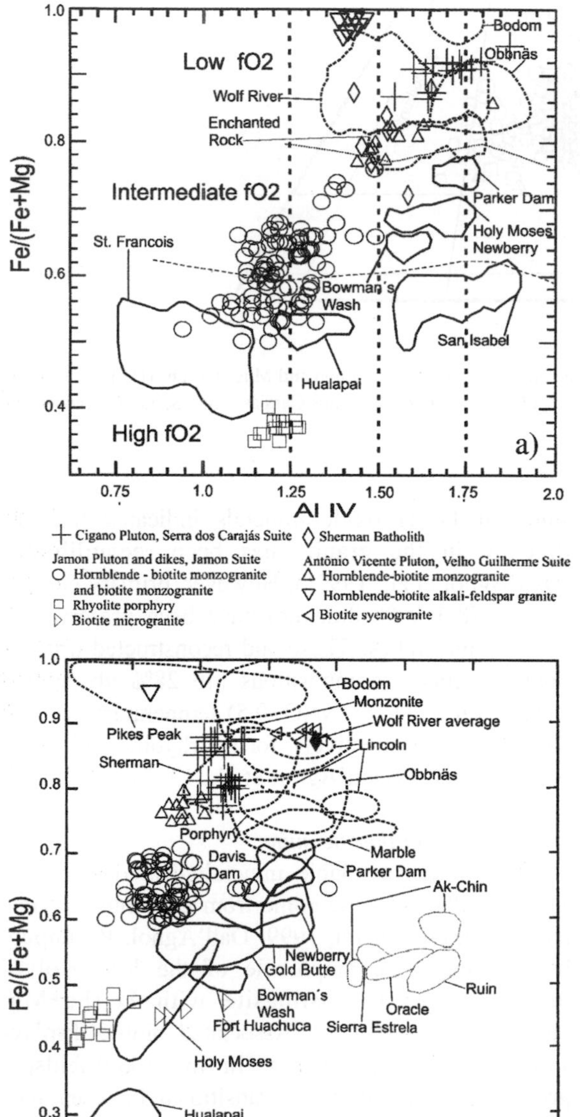

3.3.3. Serra dos Carajás suite

Mineralogical data of the Cigano pluton (Table 2) is representative of the Serra dos Carajás suite (Fig. 4a,b; Tables 2 and 3). The amphibole is hastingsite with Fe/(Fe+Mg) in the 0.85–0.94 range. In biotite, Fe/(Fe+Mg) is between 0.78 and 0.88. The dominant Fe–Ti oxide mineral is magnetite, which is found as homogeneous grains without associated ilmenite. Sometimes, it has been altered to hematite.

3.4. Magnetic susceptibility

Magnetic susceptibility (MS) data of representative samples from the Musa, Jamon, and Redenção plutons of the Jamon suite, Cigano, and Serra dos Carajás plutons of the Serra dos Carajás suite and Antonio Vicente pluton of the Velho Guilherme suite are available (Magalhães et al., 1994; Oliveira, 2001; see Fig. 5). These indicate that (1) MS values have an unimodal distribution and are higher in the Jamon suite (1.05×10^{-3} to 54.73×10^{-3}, most values $>5.0 \times 10^{-3}$; MS in SI volume); (2) MS values of Cigano and Serra dos Carajás plutons are more variable with dominant moderate MS values (1.0×10^{-3} to 5.0×10^{-3}); (3) in Velho Guilherme (Antonio Vicente pluton), the range of MS values is similar to that of Serra dos Carajás, but relatively low MS values ($<1.0 \times 10^{-3}$) dominate; furthermore, the MS distribution is bimodal with the syenogranites and evolved leucogranites having lower and the amphib-

Fig. 4. (a) Fe/(Fe+Mg) versus Al^{IV} diagram showing composition of amphibole in the A-type granites of Carajás (this paper and Rämö et al., 2002) and A-type mid-Proterozoic granites of the United States (Anderson and Smith, 1995; Frost et al., 1999) and Fennoscandia (Obbnäs and Bodom plutons; Kosunen, 2004). Continuous fields—oxidized magnetite-series granites; dashed fields—reduced granites. (b) Fe/(Fe+Mg) versus $Al^{IV}+Al^{VI}$ diagram showing composition of biotite in the A-type granites of Carajás (this paper) and A-type mid-Proterozoic granites of the United States (Anderson and Bender, 1989: continuous fields–magnetite-series granites; dashed fields–ilmenite-series granites; dotted fields–two-mica granite; Barker et al., 1975: Pikes Peak; Anderson and Cullers, 1978: Wolf River; Frost et al., 1999: Sherman batholith) and Fennoscandia (Obbnäs and Bodom plutons; Kosunen, 2004).

Table 2
Electron microprobe analyses of amphiboles from the Paleoproterozoic granites of the Carajás region

Pluton	Cigano			Jamon					Antônio Vicente				BHSG	
Rock type	HBMzG	HBPMzG	HBPMzG	BMzG	BMzG	HBMzG	HBMzG	HBMzG	R. Porph	R. Porph	HBMzG	HBMzG		
Sample	CIG 14A	CIG 34A	CIG 34C	AU 375	AU 382	AU 390	AU 390	AU 391	CRE 106A		IE-32		MUT 21	
SiO_2 (wt.%)	37.061	38.489	39.339	43.710	41.990	44.520	43.630	43.280	46.350	46.120	38.006	40.007	40.169	39.956
TiO_2	0.464	1.276	1.556	1.550	0.960	1.870	1.630	1.390	1.680	1.610	0.790	1.436	1.583	1.413
Al_2O_3	10.503	9.482	9.104	6.610	8.080	6.960	7.100	7.730	7.110	7.400	10.349	8.885	7.471	7.404
FeO	32.241	30.371	29.448	23.940	26.050	18.190	21.290	23.570	13.290	14.930	28.904	28.548	34.850	33.860
MnO	0.281	0.594	0.497	1.050	0.950	0.680	0.560	1.030	0.470	0.630	0.484	0.471	0.639	0.478
MgO	1.133	1.975	2.803	6.620	5.330	10.100	8.400	7.140	13.700	12.740	3.111	3.954	0.461	0.796
CaO	10.897	10.531	11.061	10.670	11.150	11.260	11.210	11.130	11.330	11.510	10.939	10.321	9.791	9.932
Na_2O	1.498	1.830	1.701	2.030	1.790	1.860	1.830	1.720	2.000	2.030	1.633	1.764	2.006	1.899
K_2O	2.168	1.704	1.626	1.080	1.340	1.090	1.090	1.130	1.060	1.130	1.804	1.342	1.109	1.345
H_2O	1.245	1.330	3.539	1.140	1.220	1.370	1.310	1.350	0.890	0.920	1.041	1.083	1.425	1.144
F	0.265	0.633	0.582	1.420	1.280	1.100	1.120	1.080	2.210	2.140	1.122	1.340	0.567	1.67
Cl	1.537	0.618	0.000	0.240	0.130	0.130	0.210	0.160	0.100	0.150	0.825	0.373	0.424	1.311
O–F–Cl	0.460	0.410	0.250	0.650	0.570	0.490	0.520	0.490	0.950	0.930	0.760	0.650	0.330	0.560
Total	98.840	98.420	101.01	99.140	99.700	98.640	98.860	100.220	99.240	100.380	99.050	98.780	100.170	99.140
Si	6.124	6.297	6.356	6.810	6.595	6.816	6.751	6.718	6.853	6.809	6.175	6.370	6.484	6.522
Al	1.876	1.703	1.644	1.190	1.405	1.184	1.249	1.282	1.147	1.191	1.825	1.630	1.420	1.423
Fe_3	0.000	0.000	0.000	0.000	0.000	0.000	0.000	0.000	0.000	0.000	0.000	0.000	0.095	0.055
Al^{VI}	0.168	0.124	0.089	0.023	0.089	0.069	0.044	0.106	0.091	0.095	0.155	0.036	0.000	0.000
Fe_3	0.796	0.637	0.479	0.414	0.521	0.232	0.344	0.287	0.319	0.303	0.780	0.934	0.888	0.777
Ti	0.058	0.157	0.189	0.182	0.113	0.215	0.190	0.203	0.187	0.179	0.097	0.161	0.192	0.173
Mg	0.279	0.482	0.675	1.538	1.248	2.302	1.938	1.697	3.020	2.804	0.753	0.938	0.111	0.194
Fe_2	3.659	3.518	3.500	2.706	2.900	2.094	2.410	2.622	1.324	1.540	3.148	2.867	3.721	3.790
Mn	0.039	0.082	0.068	0.139	0.126	0.088	0.073	0.085	0.059	0.064	0.059	0.064	0.087	0.066
Ca	1.929	1.846	1.915	1.781	1.876	1.845	1.858	1.845	1.795	1.821	1.904	1.761	1.693	1.737
Na	0.071	0.154	0.085	0.219	0.124	0.155	0.142	0.155	0.205	0.179	0.096	0.239	0.307	0.263
ANa	0.409	0.426	0.448	0.394	0.421	0.396	0.407	0.405	0.368	0.402	0.419	0.305	0.321	0.338
AK	0.457	0.356	0.335	0.215	0.268	0.213	0.215	0.233	0.200	0.213	0.374	0.273	0.228	0.280
SumA	0.866	0.782	0.783	0.609	0.690	0.609	0.623	0.638	0.568	0.615	0.792	0.578	0.550	0.618
Sumcat	15.866	15.782	15.783	15.609	15.690	15.609	15.623	15.638	15.568	15.615	15.792	15.578	15.550	15.618
Fe/(Fe+Mg)	0.940	0.900	0.850	0.670	0.730	0.500	0.590	0.630	0.350	0.400	0.840	0.800	0.980	0.960

Total Fe reported as FeO. Structural formulae as 13-CNK. Key to abbreviations: H–hornblende; B–biotite; MzG–monzogranite; SG–syenogranite; P–porphyritic; R. Porph–rhyolite porphyry.

Table 3
Electron microprobe analyses of biotites from the Paleoproterozoic granites of the Carajás region

Pluton	Cigano				Jamon						Antonio Vicente					
	HBMzG		HBPMzG		BMzG		HBMzG	BMcG	R. Porph		HBMzG		HBSG	BMzG	BSG	IABSG
	CIG 14A	CIG 3	CIG 34A	CIG 34C	AU 375	AU 375	AU 390	HZR 751	CRE 106A		IE-02		MUT 21	GAM 37	SL 07	NR 48
SiO$_2$ (wt.%)	33.936	34.916	34.330	34.676	36.980	36.560	37.180	38.580	39.970	38.760	35.075	34.692	33.574	33.508	34.401	36.287
TiO$_2$	4.268	3.621	3.369	3.571	2.160	2.640	2.050	2.460	1.280	1.840	3.653	4.069	4.025	2.923	3.111	1.361
Al$_2$O$_3$	12.549	12.236	12.115	12.359	11.920	11.910	12.130	12.390	10.870	11.400	12.193	11.981	11.114	13.719	13.694	19.597
FeO	32.698	30.202	32.707	31.004	26.140	26.890	24.450	18.570	17.870	19.890	31.296	28.803	36.134	33.013	32.035	25.837
MnO	0.202	0.180	0.331	0.274	0.540	0.580	0.500	0.430	0.290	0.410	0.155	0.360	0.236	0.172	0.242	0.545
MgO	2.642	4.753	3.160	4.152	7.820	7.170	9.380	12.590	14.220	12.240	4.763	5.491	1.008	2.871	2.154	1.549
CaO	0.000	0.000	0.000	0.000	0.000	0.000	0.010	0.000	0.000	0.000	0.016	0.000	0.000	0.010	0.000	0.001
Na$_2$O	0.060	0.073	0.048	0.060	0.000	0.040	0.040	0.060	0.060	0.010	0.107	0.070	0.033	0.060	0.001	0.128
K$_2$O	8.813	9.145	8.853	9.108	9.040	9.090	9.080	9.610	9.650	9.000	8.593	9.340	8.582	8.640	9.048	9.247
F	0.362	0.609	0.304	0.460	1.570	0.830	0.570	2.120	3.190	2.240	0.541	0.509	0.269	0.642	0.271	2.532
H$_2$O	3.455	3.373	3.473	3.436	3.020	3.360	3.530	2.910	2.400	2.790	3.433	3.424	3.415	3.302	3.506	2.499
O–F	0.150	0.260	0.130	0.910	0.660	0.350	0.240	0.890	1.340	0.940	0.230	0.210	0.110	0.270	0.110	1.070
Total	98.830	98.780	98.560	98.910	98.530	98.720	98.680	98.830	98.460	97.640	99.600	98.530	98.280	98.590	98.350	98.510
Si	5.698	5.782	5.757	5.761	5.992	5.894	5.895	6.039	6.324	6.171	5.738	5.752	5.760	5.592	5.798	5.993
Al^{IV}	2.302	2.214	2.243	2.239	2.008	2.106	2.105	1.961	1.676	1.829	2.262	2.248	2.240	2.408	2.202	2.007
Al^{VI}	0.179	0.174	0.150	0.179	0.267	0.155	0.160	0.323	0.349	0.309	0.087	0.091	0.005	0.288	0.516	1.804
Ti	0.539	0.451	0.425	0.446	0.263	0.320	0.244	0.290	0.153	0.220	0.450	0.507	0.519	0.367	0.394	0.169
Fe$_2$	4.591	4.186	4.587	4.308	3.542	3.625	3.242	2.431	2.364	2.648	4.282	3.994	5.184	4.607	4.516	3.568
Mn	0.029	0.015	0.047	0.039	0.074	0.079	0.067	0.057	0.039	0.055	0.021	0.051	0.034	0.024	0.035	0.076
Mg	0.661	1.174	0.790	1.028	1.889	1.723	2.170	2.938	3.354	2.905	1.162	1.357	0.258	0.714	0.541	0.381
Na	0.020	0.023	0.016	0.019	0.000	0.013	0.012	0.018	0.018	0.003	0.034	0.023	0.011	0.019	0.000	0.041
K	1.888	1.933	1.894	1.931	1.869	1.869	1.837	1.919	1.948	1.828	1.793	1.975	1.878	1.839	1.946	1.948
Cátion	23.907	23.956	23.909	23.950	23.904	23.784	23.781	23.976	24.224	23.968	23.832	23.998	23.889	23.860	23.948	23.987
Fe/Fe+Mg	0.870	0.780	0.850	0.810	0.65	0.680	0.590	0.450	0.410	0.480	0.790	0.750	0.950	0.870	0.890	0.900

Total Fe reported as FeO. Key to abbreviations: H–hornblende; B–biotite; MzG–monzogranite; McG–microgranite; SG–syenogranite; P–porphyritic; R. Porph–rhyolite porphyry; I–intensively; A–altered.

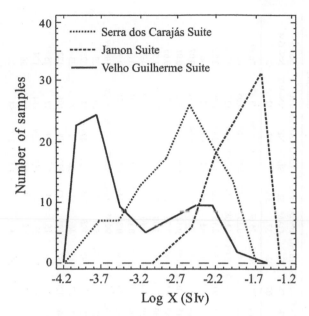

Fig. 5. Volume magnetic susceptibility frequency histogram of the Jamon (Musa granite), Serra dos Carajás (Cigano granite), and Velho Guilherme (Antonio Vicente granite) suites of Carajás (data from Magalhães et al., 1994).

ole-bearing monzogranites and syenogranites having higher MS values.

In the Jamon suite, contents of Fe–Ti oxide minerals vary generally between 0.5% and 2%, and magnetite is dominant over ilmenite. These granites are thus typical magnetite series (Ishihara, 1981). In Serra dos Carajás, the contents of Fe–Ti oxides are lower (<1%) than in the Jamon suite, and magnetite is generally not accompanied by ilmenite. Following the classification of Ishihara (1981), the Serra dos Carajás granites are also magnetite series. In the syenogranites of the Velho Guilherme suite, opaque mineral contents are generally <0.1%, and magnetite is absent in most of the studied samples, indicating that the syenogranites are ilmenite series. In the amphibole-bearing facies of Velho Guilherme (the Antonio Vicente batholith), magnetite is a common accessory mineral. They thus differ in this respect from the syenogranites of the suite.

3.5. Elemental geochemistry

Representative chemical composition of the three granite suites are given in Table 4 and have been discussed previously (Jamon: Dall'Agnol et al., 1999c; Oliveira, 2001; Velho Guilherme: Dall'Agnol

et al., 1994; Teixeira, 1999; Serra dos Carajás: Barros et al., 1995; Javier Rios et al., 1995). All granites display the general characteristics of the A-type (see Collins et al., 1982; Whalen et al., 1987; Eby, 1992; King et al., 1997) and within-plate (Pearce et al., 1984) granites. K_2O/Na_2O ratios are generally between 1.0 and 2.0 and increase from the Jamon suite to the Velho Guilherme and Serra dos Carajás suites (Fig. 6a). The silica content is >65 wt.%, generally >70 wt.% (Table 4; Fig. 6b). The studied granites are ferroan, transitional between alkali-calcic and calc-alkalic, metaluminous to slightly peraluminous granites (Fig. 6b–d) as most A-type granites are (cf. Frost et al., 2001a). They plot in the field of A-type granites in the $FeOt/(FeOt+MgO)$ versus SiO_2 diagram, and their $FeOt/(FeOt+MgO)$ ratios are over ~0.80, increasing from the Jamon suite to the Velho Guilherme and Serra dos Carajás suites. The Jamon and Velho Guilherme suites show a positive correlation between $FeOt/(FeOt+MgO)$ and SiO_2, but rocks with lower silica contents also have relatively high $FeOt/(FeOt+MgO)$ (Table 4; Fig. 6b). The trend observed in the $FeOt/(FeOt+MgO)$ versus SiO_2 diagram indicates that the silica-rich rocks of the Jamon suite are not evolved members of a magnesian or Cordilleran granite series. The more oxidized character of the Jamon suite is also evident in the $FeOt/(FeOt+MgO)$ versus SiO_2 plot (Fig. 6b).

REE patterns (Fig. 7) display low $(La/Lu)_N$, moderate to pronounced negative Eu anomalies, and flat HREE (Table 3). The Eu anomaly generally increases from the less evolved to the more evolved facies in each pluton. Total REE contents is highest in the Serra dos Carajás suite, moderate in the Jamon suite, and lowest in the Velho Guilherme suite. The most pronounced negative Eu anomalies are found in the tin-specialized syenogranites of the Velho Guilherme suite and in the leucogranites (e.g., Pojuca) of the Serra dos Carajás suite. LREE decrease and HREE increase in the tin–leucogranites of the Velho Guilherme suite, resulting in typical gullwing-shaped patterns (Fig. 7e,f).

3.6. Intensive parameters: temperature, pressure, and oxygen fugacity

Relatively high temperatures of crystallization (≥900 °C) characterize A-type granite magmas

Table 4
Chemical composition of the Paleoproterozoic A-type granites of the Carajás region

Pluton	Serra dos Carajás (1)				Cigano (2)				Redenção (3)						Antonio Vicente (4)					Velho Guilherme (4)	Mocambo (4)
	HBMzG	BSG	BSG	BAFG	HBMzG	HBPMzG	BMzG	ABMzG	CBAMzG	BAMzG	ABMzG	BMzG	BMzP	LMzG	HBMG	BMzG	BSG	ABSG	IABSG	BSG	SGMv
	CJ	CJ	CJ	CJ	ECR	ECR	ECR	CIG	DCR	DCR	JCR	JCR	JCR	DCR	IE	GAM	SL	SL	NE	NN-VG	NN-GM
	19	29B	32B	32	59C	34A	91B	3	34	63A	09	01D	07	07	02	37	3A	4	83	32	24
SiO$_2$ (wt.%)	71.90	73.89	74.50	73.41	72.80	70.47	69.47	72.56	66.10	71.10	70.70	74.20	74.40	76.00	70.42	75.60	74.40	75.20	76.04	75.67	76.19
TiO$_2$	0.25	0.16	0.20	0.13	0.36	0.35	0.39	0.20	1.20	0.60	0.44	0.30	0.22	0.12	0.68	0.15	0.17	0.07	0.05	0.05	0.07
Al$_2$O$_3$	13.03	12.68	11.85	12.33	12.08	12.36	13.07	13.01	13.10	13.50	14.20	13.30	13.10	13.10	12.08	12.08	12.42	12.19	12.42	12.29	12.76
Fe$_2$O$_3$	1.79	1.10	1.03	0.83	4.30*	5.89*	5.19*	2.66*	3.47	2.19	1.24	1.31	1.01	0.65	5.34*	2.15*	2.25*	1.76*	2.13*	1.39*	2.56*
FeO	1.85	1.34	1.92	1.35	nd	nd	nd	nd	3.36	1.54	2.03	0.62	0.53	0.31	nd	nd	nd	nd	nd	nd	nd
MnO	0.03	nd	0.01	nd	0.05	0.08	0.01	Tr	0.13	0.07	0.06	0.05	0.04	0.03	0.07	Tr	0.02	Tr	0.04	0.01	0.07
MgO	0.10	0.03	0.04	nd	0.20	0.26	0.50	0.32	1.10	0.60	0.46	0.22	0.18	<0.10	0.50	0.12	0.14	0.03	Tr	0.02	0.01
CaO	1.23	0.68	1.11	0.89	1.98	1.37	1.22	1.45	2.80	2.10	1.60	1.10	0.67	0.55	1.95	0.76	1.02	0.81	0.81	0.58	0.65
Na$_2$O	3.02	2.74	2.82	2.52	2.95	2.63	3.00	3.22	3.70	3.70	3.50	3.50	3.40	3.10	2.83	3.00	3.29	3.52	3.42	3.79	3.50
K$_2$O	5.64	6.03	5.57	7.26	3.94	5.75	5.25	5.41	3.90	4.20	4.80	5.10	5.30	5.40	4.74	5.10	5.22	4.95	4.17	4.57	4.22
P$_2$O$_5$	0.16	0.13	0.14	0.12	0.19	0.20	0.24	0.17	0.42	0.22	0.14	0.05	0.04	0.01	0.34	0.12	0.12	0.10	0.10	0.01	0.01
LOI	0.99	0.55	0.22	0.41	0.60	0.91	1.25	0.74	0.32	0.38	0.35	0.31	0.73	0.17	0.88	0.80	0.86	0.84	1.18	0.71	0.95
Total	99.99	99.33	99.41	99.26	99.45	100.27	99.59	99.74	99.60	100.2	99.52	100.1	99.62	99.44	99.83	99.88	99.91	99.47	100.36	99.09	100.99
Ba (ppm)	nd	243	253	362	2838	1444	1460	780	919	1498	1310	909	489	32	1454	539	386	95	8	20	29
Rb	209	243	nd	nd	79	185	213	256	151	139	193	204	281	396	192	196	339	524	477	533	628
Sr	96	69	62	29	223	176	166	111	241	332	271	196	122	27	168	46	51	20	11	9	23
Zr	403	244	301	62	547	428	362	168	686	377	353	258	240	126	397	193	163	132	88	106	172
Nb	39	28	75	37	22	31	27	30	29	16	20	21	24	21	16	15	34	50	49	37	75.81
Y	241	115	78	44	45	70	52	38	80	50	43	71	62	30	70	36	63	105	159	114	174
Ga	nd	nd	nd	nd	30	43	45	22	26	26	25	23	27	27	5	13	<5	17	19	29	36
Sc	nd	nd	nd	nd	9	5.3	6.09	2	14	<10	15	10	<10	<10	10	3	3.6	1.5	2.7	<1	2

Th	nd	nd	nd	nd	19	221	126	68	<5	<5	<5	<5	<5	<5	23	75	56	56	47	48	70
U	nd	nd	nd	nd	nd	nd	nd	nd	<10	<10	<10	<10	<10	<10	10	11	30	38	42	12	19
V	nd	nd	nd	nd	5	6	29	13	53	22	14	<10	<10	<10	39	5	6	<5	<5	<5	<5
Cr	nd	nd	nd	nd	24	23	15	7	nd	nd	nd	nd	nd	nd	9	7	57	6	22	nd	nd
Co	nd	nd	nd	nd	39	10	16	5	nd	nd	nd	nd	nd	nd	23	11	26	10	5	28	22
Ni	nd	nd	nd	nd	14	14	14	<5	nd	nd	nd	nd	nd	nd	7	13	21	7	8	nd	nd
Cu	nd	nd	nd	nd	12	12	111	26	nd	nd	nd	nd	nd	nd	42	13	9	9	8	8	37
Zn	nd	nd	nd	nd	38	59	22	22	nd	nd	nd	nd	nd	nd	79	30	24	19	39	27	111
La	334	268	160	65.5	214	529	215	83	102.9	60.5	82.0	71.8	56.8	11.9	81	110.1	80.90	57.2	28.3	26.38	77.29
Ce	449	259	331	55.8	329	707	330	143	196.5	120.9	146.7	139.7	114.4	31.9	153.8	194	157.80	120.7	76.9	61.58	152.91
Nd	226	168	107	32.5	105	187	106	50	84.5	48.4	46.6	51.7	36.3	9.2	59.4	57.3	49.4	39.5	33.2	32.31	56.10
Sm	37.3	27	16.5	5.14	16.5	26.7	16.7	10.1	14.9	8.1	8.4	9.3	6.6	1.8	12.6	10.1	10.7	11.3	13.8	9.13	12.51
Eu	3.08	2.33	1.24	0.71	1.97	2.89	1.94	1.26	1.6	1.5	1.4	1.4	1.8	0.2	2.15	0.7	0.7	0.4	0.2	0.10	0.096
Gd	30.8	18.9	11.1	3.58	12.4	20.9	12.7	10.0	11.8	6.3	6.0	7.6	5.0	1.3	11.7	7.2	8.7	10.7	15	10.49	13.62
Dy	30.4	17.2	8.91	3.77	8.82	11.7	8.95	5.54	7.9	4.4	3.8	5.0	3.9	0.8	11.5	5.7	9.8	14.5	23.8	14.35	20.36
Er	17.8	9.78	4.82	2.52	5.07	6.64	5.09	3.47	3.7	2.3	1.7	3.6	2.1	0.4	6.9	3.3	6.3	10.5	15.6	10.69	16.17
Yb	15.3	8.44	4.05	2.63	5.16	6.36	5.18	3.39	2.9	1.8	1.2	2.8	1.9	0.4	7	3.6	7.6	13.4	17.5	11.56	19.47
Lu	2.01	1.06	0.56	0.37	0.89	1.23	0.91	0.74	0.4	0.3	0.2	0.4	0.3	0.1	0.9	0.5	1.0	1.7	2.1	1.79	2.95
$FeO_t/(FeO_t+MgO)$	0.97	0.98	0.98	1.00	0.95	0.95	0.91	0.89	0.87	0.87	0.88	0.90	0.90	0.92	0.91	0.94	0.94	0.98	1.00	0.98	0.99
K_2O/Na_2O	1.87	2.20	1.98	2.88	1.34	2.20	1.75	1.68	1.05	1.14	1.37	1.46	1.56	1.74	1.67	1.70	1.59	1.41	1.22	1.22	1.21
Eu/Eu^*	0.27	0.30	0.26	0.48	0.40	0.36	0.39	0.38	0.35	0.61	0.57	0.49	0.37	0.29	0.53	0.24	0.21	0.11	0.04	0.03	0.22
$(La/Lu)_N$	17.02	25.90	28.57	17.70	24.04	43.00	23.62	11.21	26.70	20.94	42.59	18.63	19.66	12.37	9.00	22.02	8.09	3.36	1.34	1.47	2.62

nd—not determined, *—Fe$_2$O$_3$ as total. Key to abbreviations: C—clinopyroxene; H—hornblende; B—biotite; MzG—monzogranite; SG—syenogranite; P—porphyritic; L—leuco; I—intensively; A—altered; AFG—alkali feldspar granite; DP—dacite porphyry; RP—rhyolite porphyry; Mv—muscovite. Data sources: (1) Barros et al. (1995); (2) Dall'Agnol et al. (1994); (3) Oliveira (2001); (4) Teixeira (1999).

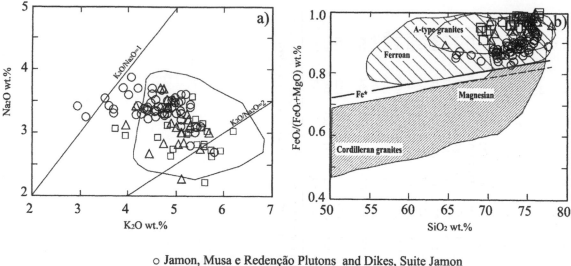

o Jamon, Musa e Redenção Plutons and Dikes, Suite Jamon
△ Antônio Vicente Pluton ,Velho Guilherme Suite
□ Cigano and Serra dos Carajás Plutons, Serra dos Carajás Suite
◯ Finnish Rapakivi

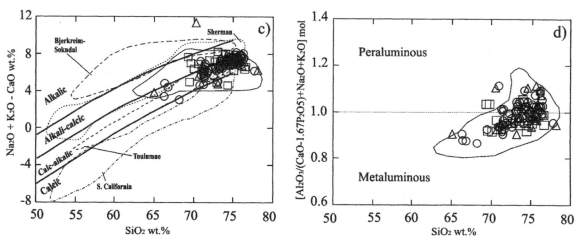

Fig. 6. Whole-rock (a) Na_2O versus K_2O, (b) $FeOt/(FeOt+MgO)$ versus SiO_2, (c) Na_2O+K_2O-CaO versus SiO_2, and (d) $[Al_2O_3/(CaO-1.67*P_2O_5+Na_2O+K_2O)]$ versus SiO_2 diagrams showing composition of A-type granites of Carajás. Fe* line, A-type, and Cordilleran granite fields in b and alkalic, alkalic-calcic, calc-alkalic, and calcic fields in c are from Frost et al. (2001a). Data sources: Finnish rapakivi–plutons from the Wiborg area (Rämö and Haapala, 1995); Serra dos Carajás suite–Serra dos Carajás granite (Barros et al., 1995); Cigano Granite (Dall'Agnol et al., 1994, and Dall'Agnol, R., unpublished); Jamon suite–Jamon and Musa granites (Dall'Agnol et al., 1999c); Velho Guilherme suite–Antonio Vicente granite (Teixeira, 1999).

(Collins et al., 1982; Anderson, 1983; Clemens et al., 1986; Rämö and Haapala, 1995; King et al., 1997; Frost et al., 1999; Dall'Agnol et al., 1999b; Holtz et al., 2001; Klimm et al., 2003). For the studied granites, temperature constraints were obtained in an experimental study of the Jamon hornblende–biotite monzogranite (Dall'Agnol et al., 1999b). Comparison

of the sequence of crystallization and the composition of minerals in natural samples and experimental results indicated that the dacite porphyry dikes associated with the Jamon granites had liquidus temperatures near 900 °C. A temperature of ~870 °C was estimated for the beginning of crystallization of the hornblende–biotite monzogranite, with gradu-

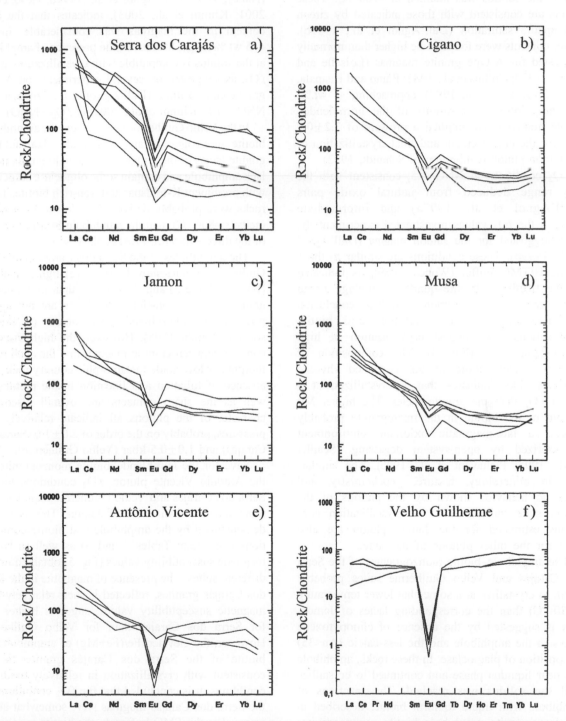

Fig. 7. Chondrite-normalized (Evensen et al., 1978) REE patterns for representative granites of the Serra dos Carajás (a,b), Jamon (c,d), and Velho Guilherme (e, f) suites. Data source as in Fig. 6.

ally decreasing temperatures for the more evolved facies; the solidus was attained at ~700 °C. These results are consistent with those indicated by zircon and apatite saturation (Dall'Agnol et al., 1999b). Water contents were found to be higher than normally suggested for A-type granite magmas (Loiselle and Wones, 1979; Collins et al., 1982; Rämö and Haapala, 1995; Frost and Frost, 1997), approaching ~6 wt.%. The amphibole compositions of the hornblende–biotite monzogranite implied a pressure of 3.2±0.7 kbar for the emplacement and final crystallization of the Jamon pluton (calibration of Schmidt, 1992).

fO_2 conditions at NNO+0.5, consistent with the fO_2 range deduced from natural oxide pairs (Dall'Agnol et al., 1997a) and intermediate between the fO_2 values investigated experimentally, were suggested for the Jamon magma (Dall'Agnol et al., 1999b). These conditions are similar to those of the HITMQ buffer (Wones, 1989), as indicated by the titanite–magnetite–quartz assemblage found in the granites of the Jamon suite. This conclusion is consistent with the amphibole and biotite compositions that suggest crystallization at high to moderate fO_2 (Fig. 4). The composition of amphibole and biotite in an associated rhyolite porphyry dike indicates that it crystallized at a higher fO_2 than the granite facies. The higher Mg content of the biotite in the microgranite probably reflects a late-magmatic oxidizing environment characterized by open-system degassing (Scaillet et al., 1995; Pichavant et al., 1996). The similarities in mineralogy, texture, geochemistry, and magnetic behavior shown by the plutons of the Jamon suite indicate that the crystallization conditions estimated for the Jamon pluton are also valid for the other plutons of the suite.

The amphibole–biotite monzogranites of the Serra dos Carajás and Velho Guilherme suites probably began to crystallize at a somewhat lower temperature (~850 °C) than the corresponding facies of Jamon. This is suggested by the absence of clinopyroxene relicts in the amphibole and the less calcic (An<35) composition of plagioclase. In these rocks, amphibole is a near liquidus phase and continued to crystallize until the solidus was reached. The reactions of amphibole with liquid to form biotite, described in the Jamon suite amphibole–biotite monzogranites (Dall'Agnol et al., 1999b,c), are not common. This,

together with experimental results in granite systems (Naney, 1983; Dall'Agnol et al., 1999b; Holtz et al., 2001; Klimm et al., 2003), indicates that the H_2O content in the magmas was considerable, in the 4–5 wt.% range. Moreover, the presence of amphibole at the solidus is compatible with crystallization at low fO_2, as suggested by several experiments on A-type granite compositions (Dall'Agnol et al., 1999b, at fO_2 NNO −1.5; Klimm et al., 2003, at fO_2 ~NNO).

In the Velho Guilherme suite, except the amphibole–biotite monzogranite and amphibole alkali-feldspar granite facies of the Antonio Vicente pluton, the rocks lack amphibole and contain sodic oligoclase associated with prominent K-feldspar and iron-rich biotite. These rocks were probably derived from lower temperature melts (<~800 °C) than the amphibole-bearing granites of the three suites (cf. Frost et al., 1999).

The amphiboles of the Serra dos Carajás and Velho Guilherme granites show Fe/(Fe+Mg) higher than 0.75 (Fig. 4a; Table 2) and thus higher than in the Jamon granites. These iron-rich amphiboles are not appropriate for Al-in-hornblende geobarometry (cf. Anderson and Smith, 1995). However, the high viscosity contrast between granitic magmas and the anchimetamorphic or low-grade metamorphic country rocks, the absence of foliation and lineation in the plutons, as well as the sharp contacts and overall discordant character of the plutons all indicate relatively low pressures, probably on the order of 2.0±1.0 (Serra dos Carajás) and 1.0±0.5 kbar (Velho Guilherme).

Save for the hornblende–biotite monzogranites of the Antonio Vicente pluton, fO_2 conditions for the Serra dos Carajás and Velho Guilherme magmas were more reduced than those of Jamon. This is clearly demonstrated by the amphibole and biotite compositions (Fig. 4a,b; Tables 2 and 3) as well as by the magnetic susceptibility values (Fig. 5) obtained for the different suites. The presence of magnetite in the Serra dos Carajás granites, reflected in their relatively high magnetic susceptibility values, suggests higher fO_2 for Serra dos Carajás than for Velho Guilherme. However, the elevated Fe/(Fe+Mg) of amphibole and biotite of the Serra dos Carajás granites is not consistent with crystallization in relatively oxidizing conditions. It is concluded that the fO_2 conditions for the Serra dos Carajás magmas were somewhat above or equal to the FMQ buffer. In the biotite syenogranites of the Velho Guilherme suite, the absence of significant magnetite and the low magnetic susceptibility values indicate more reducing conditions, probably a little below FMQ.

3.7. Isotope data

3.7.1. Neodymium

Whole-rock Nd isotope data for the Jamon suite can be found in Dall'Agnol et al. (1999b) and Rämö et al. (2002) and for the Velho Guilherme suite in Teixeira et al. (2002). New data for the Serra dos Carajás suite are presented in Table 5. The Carajás plutons are strongly enriched in LREE and show a wide range in $^{147}Sm/^{144}Nd$ (0.068 to 0.140). Their ε_{Nd} (at 1880 Ma) values (Fig. 8a,b) are remarkably similar, ranging from −10.5 to −7.9 (mean value −9.5), and the T_{DM} (DePaolo, 1981) show considerable variation (~3.35 to 2.60 Ga). The ε_{Nd} (at 1880 Ma) values for the Jamon Suite (−10.5 to −8.1) are similar to those of the Serra dos Carajás suite (−9.7 to −7.9). In the Velho Guilherme suite, the ε_{Nd} (at 1880 Ma) values for the Antonio Vicente and Rio Xingu plutons are somewhat lower (−12.1 to −12.2), while that of the Mocambo pluton (−7.9) is similar compared to those of the other two suites. Samples from the Jamon, Musa, and Bannach

plutons of the Jamon suite fall on an isochron with an age of 1879±52 Ma (MSWD of 1.46, initial ε_{Nd} of −9.4; Rämö et al., 2002). This is compatible with the available U–Pb and Pb–Pb zircon ages (~1.88 Ga; cf. Table 1, Fig. 2) and shows that the plutons have retained their initial magmatic Nd isotope composition.

The Nd evolution lines of the granitoids of the ~3-Ga Rio Maria Granite-Greenstone Terrane define a distinct evolution path for the Archean crust (Fig. 8a). Save for the two Velho Guilherme granites, all the granites of the three suites plot into the more radiogenic part of this path at 1.88 Ga. This suggests that the granites were derived from deeper parts of the Archean crust with slightly higher long-term Sm/Nd than that of the samples from the exposed parts of the crust. Thus, the 1.88-Ga granites of Serra dos Carajás and Jamon were derived from broadly similar Archean crustal sources. This also suggests that the deep crust beneath the supracrustal sequences of the Carajás basin—that are younger than the Archean upper crustal sequences of Rio Maria—has a similar overall age as the crust beneath the exposed parts of the Rio Maria Granite-Greenstone Terrane. This further implies that the evolution of the Carajás basin was intracratonic (Gibbs et al., 1986) and not

Table 5
Nd isotope data for the Paleoproterozoic granites of the Carajás region

Sample	Rock type	Sm (ppm)	Nd (ppm)	$^{147}Sm/^{144}Nd$[a]	$^{143}Nd/^{144}Nd$[b]	ε_{Nd}[c]	T_{DM}[d] (Ma)
Serra dos Carajás pluton							
CJ-29b	Granite	23.62	160.1	0.08917	0.510909±7	−7.9	2611
CJ-38	Granite	16.99	109.9	0.09349	0.510894±8	−9.2	2727
Pojuca pluton							
F14/395	Granite	9.08	39.30	0.1397	0.511440±11	−9.7	3353
Cigano pluton							
ECR-CG-59C	Granite	11.97	63.88	0.1133	0.511115±13	−9.7	2939
ECR-CG-96	Granite	23.53	173.9	0.08177	0.510733±10	−9.5	2668

Analyses were performed at the Unit for Isotope Geology, Geological Survey of Finland in 2001. Rock powders (150–200 mg) were dissolved in Teflon bombs at 180 °C in a mixture of HNO_3 and HF, dissolved in HCl, and spiked with a $^{149}Sm-^{150}Nd$ tracer. Light REE were separated using standard cation exchange chromatography, and Sm and Nd were purified by and on quartz columns (Richard et al., 1976). The total procedural blank was <300 pg for Nd. Isotope ratios of Sm and Nd were measured on a VG Sector 54 mass spectrometer. Repeated analyses of La Jolla Nd standard gave $^{143}Nd/^{144}Nd$ of 0.511848±0.000016 (mean and external 2σ error of 10 measurements); external error in the reported $^{143}Nd/^{144}Nd$ is better than 0.003%.
[a] Estimated error for $^{147}Sm/^{144}Nd$ is less than 0.5%.
[b] $^{143}Nd/^{144}Nd$ normalized to $^{146}Nd/^{144}Nd$=0.7219. Within-run error expressed as $2\sigma_m$ in the least significant digits.
[c] Initial ε_{Nd} (at 1880 Ma) values calculated using $^{143}Nd/^{144}Nd$=0.512638 and $^{147}Sm/^{144}Nd$=0.1966. Maximum error is ±0.35 ε units.
[d] Depleted mantle model ages according to the model of DePaolo (1981).

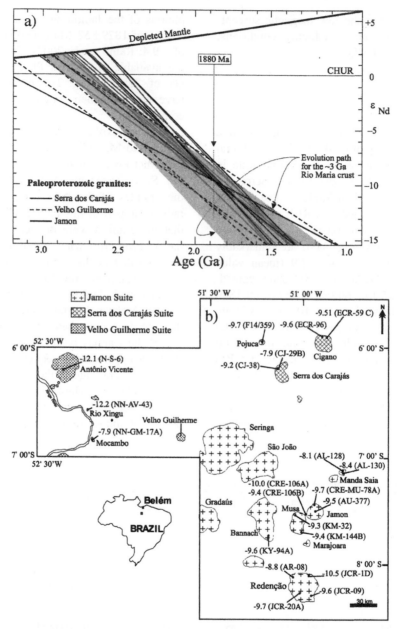

Fig. 8. (a) ε_{Nd} versus age diagram showing Nd isotope composition of Paleoproterozoic ~1880-Ma A-type granites of the Serra dos Carajás (data from this study), Velho Guilherme (Teixeira et al., 2002), and Jamon (Rämö et al., 2002) suites. Evolution path for the Archean Rio Maria crust is from Rämö et al. (2002). CHUR is undifferentiated Earth (DePaolo and Wasserburg, 1976), depleted mantle evolution is according to DePaolo (1981). Descending lines represent isotopic evolution of individual samples. (b) Simplified geological map of the Carajás metallogenic province showing the geographic distribution of the ε_{Nd} (at 1880 Ma) values of the analyzed A-type granites.

directly related to continental margin magmatic activity, as proposed by Teixeira and Egler (1994).

The Velho Guilherme Suite granites were probably also derived from an Archean crust (Teixeira et al., 2002). However, as two of the analyzed granites have clearly less radiogenic initial Nd isotopic compositions (~−12 versus ~−9; Fig. 8a), it is possible that the sources of their magmas were characterized by larger mean crustal residence time compared to that of Jamon and Serra dos Carajás (Fig. 8b).

3.7.2. Lead

Pb isotope data on representative alkali-feldspar fractions from the three granitic suites (Souza, 1996; Barbosa et al., 1995; this study) are presented in Table 6 and Fig. 9. Samples from the Antonio Vicente (Velho Guilherme suite) and Pojuca (Serra dos Carajás suite) plutons, in particular, are highly radiogenic and plot above the conventional curves used to characterize overall tectonic environments (e.g., Doe and Zartman, 1979). The Archean rocks that host the Pojuca pluton have radiogenic Pb isotopic compositions, with $^{206}Pb/^{204}Pb$ and $^{207}Pb/^{204}Pb$ ranging between 35 to 340 and 19 to 65, respectively (Macambira et al., 2001), and could have contaminated the Pojuca magma. However, owing to the leaching technique used to analyze the feldspar fractions (cf. Table 6), the measured isotopic ratios may not quite represent the actual initial ratios (that is, they could be somewhat less radiogenic).

The feldspars from the Musa and Jamon plutons (Jamon suite) plot around the Doe and Zartman (1979) upper crust and orogen curves and those from the Redenção pluton plot between mantle and orogen. In view of the crustal origin of the magmas as suggested by the Nd isotope data, this can be interpreted to reflect a Pb isotope composition

intermediate between an unradiogenic lower crust and a radiogenic upper crust. In the two-stage model of Stacey and Kramers (1975), the Jamon suite falls along the growth curve of average crustal Pb.

The Carajás samples are plotted in the $^{208}Pb/^{204}Pb$ versus $^{206}Pb/^{204}Pb$ diagram in Fig. 9b. The samples of the Jamon and Musa plutons are more radiogenic and plot close to the orogen, upper crust, and mantle, whereas those from the Redenção pluton plot closer to the lower crust and indicate a higher overall Th/U. Considering the similarities in age, tectonic setting, petrography, magnetic susceptibility, and geochemistry of the plutons, this may suggest that the Redenção pluton was derived from a less radiogenic source (lower crust). The samples from the Antonio Vicente pluton plot between the upper crust and lower crust, and they are more radiogenic than the samples from the other plutons. Sample CRE-106B from a dacite porphyry dike associated with the Jamon and Musa plutons is also more radiogenic compared to the granite samples of these plutons. This can be explained by isotopic contamination of the dike magma by its Archean country rocks.

The feldspar fractions from the granites of the Velho Guilherme suite (Antonio Vicente pluton) and Serra dos Carajás suite (Pojuca pluton) are radiogenic probably because of radiogenic lead contamination from the highly radiogenic host rocks. The Pb isotope data of the samples from the Jamon suite suggest

Table 6
Pb isotopic data for the alkali feldspar fractions of the Paleoproterozoic granites of the Carajás region

Pluton	Sample	$^{206}Pb/^{204}Pb$	Error	$^{207}Pb/^{204}Pb$	Error	$^{208}Pb/^{204}Pb$	Error
Pojuca (1)	PSM-10	24.460	0.090	17.600	0.076	42.290	0.070
Jamon	AU-391	15.644	0.010	15.387	0.014	35.274	0.044
	AU-377	15.674	0.009	15.292	0.014	35.534	0.043
Musa	KM-06	15.721	0.010	15.350	0.014	35.563	0.043
	CR-106	16.442	0.010	15.396	0.014	36.044	0.043
	KM-77AI	15.690	0.018	15.479	0.018	35.597	0.051
Redenção (2)	AVR-51B	15.026	0.010	15.103	0.015	35.425	0.045
	AVR-53A	15.109	0.009	15.146	0.014	35.510	0.043
Antonio Vicente	IE-02	17.116	0.010	15.935	0.014	37.267	0.045
	IE-05	17.273	0.010	15.981	0.050	37.503	0.045

Analyses were performed at the Isotope Geology Laboratory of the Federal University of Pará. Approximately 200 mg of handpicked feldspar from crushed rock were leached two times at 100 °C with HCl 6N and once with 1 N HF (20 min each leaching step) in teflon bombs, and, finally, washed with distilled-water. Then, the samples were digested in HF (48%, 8 h-coll+12 h at 100 °C). After evaporation, the samples were dissolved in HBr (8 N), evaporated, then dissolved in HBr (0.5 N), centrifugated, and the Pb was extracted by DOWEX 1×8 chromatrography. The total procedure blank was less than 1 ng. Isotope ratios were measured on a VG54E mass spectrometer. The isotope ratios were corrected for mass fractionation (0.00120±0.00030 amu), and the uncertainties are presented at 1σ. External data were from (1) Souza (1996) and (2) Barbosa et al. (1995).

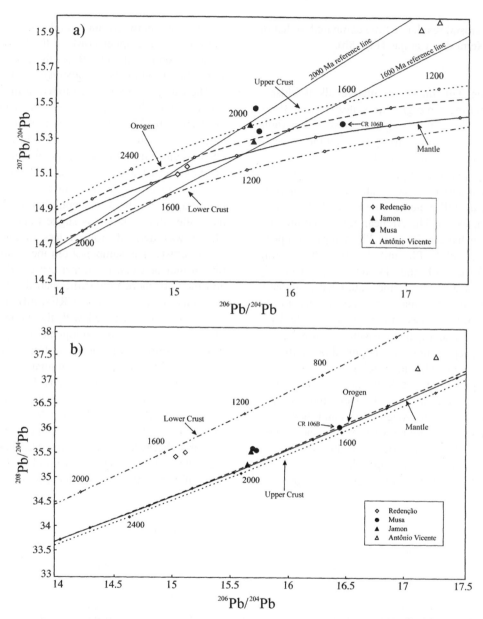

Fig. 9. Pb isotope composition of K-feldspar fractions from the A-type granites of Carajás plotted in (a) $^{207}Pb/^{204}Pb$ versus $^{206}Pb/^{204}Pb$ and (b) $^{208}Pb/^{204}Pb$ versus $^{206}Pb/^{204}Pb$ diagrams. Growth curves for mantle, orogen, upper crust, and lower crust are from Doe and Zartman (1979).

derivation from sources intermediate between upper and lower crust. The Jamon and Musa plutons are more radiogenic than Redenção—this implies a less radiogenic source for the latter.

3.7.3. Oxygen

Whole-rock oxygen isotope composition of samples of Jamon pluton and associated dikes (Jamon suite), Cigano and Serra dos Carajás plutons (Serra dos Carajás suite), and Antonio Vicente pluton (Velho Guilherme suite) are shown in Table 7. Oxygen isotope data on quartz are also quoted for the Serra dos Carajás pluton (Javier Rios et al., 1995), the Antonio Vicente pluton (Teixeira, 1999), and the Musa pluton (Javier Rios, 1995). The whole-rock $\delta^{18}O$ values show a moderate range from 6.5‰ to

Table 7
Oxygen isotope data and SiO_2 contents of the Paleoproterozoic granites of the Carajás region

Granite Pluton	Sample	SiO_2 (wt.%)	$\delta^{18}O$ in whole rock (‰ SMOW)	$\delta^{18}O$ in quartz (‰ SMOW)
Serra dos Carajás granite suite				
Cigano	59C	72.8	+8.1[a]	
	34A	70.47	+7.4[a]	
	96	69.09	+8.1[a]	
	CIG1	75.12	+8.1[a]	
	CIG3	72.56	+8.3[a]	
Serra dos Carajás	CJ19	71.90	+8.5[a]	
	CJ38	75.00	+10.1[a]	
	CJ32B	74.5	+6.6[a]	
	CJ35	76.5	+8.4[a]	
	CJ32	73.41	+9.0[a]	
	CJ29	75.53	+7.8[a]	
	GC-27	nd	nd	+7.06 (3)
Jamon granite suite				
Jamon	374	73.02	+6.5[a]	
	385	73.00	+7.3 [a]	
	393	71.20	+8.1[a]	
	CRE-78A	70.62	+7.7[b]	
	AU-377	74.60	+8.5[b]	
	AU-391	71.12	+12.2[b]	
	AU-375	73.71	+7.7[b]	
	AU-397	75.47	+8.6[b]	
	HZR-752	71.09	+8.5[b]	
Musa	PP72	nd	nd	+7.61 (2)
	CRE-25A	nd	nd	+7.16 (2)
Felsic Dykes	CRE-106A	71.29	+8.2[b]	
	CRE-106B	67.93	+7.7[b]	
	CRE-82	72.34	+6.6[a]	
	CRE-103	73.30	+6.9[a]	
Velho Guilherme granite suite				
Velho Guilherme	NN-VG-77	76.78		+8.2 (1)
Mocambo	NN-GM56	76.05		
	NN-GM-20	nd		+9.0 (1)
	NN-GM-28	nd		+8.9 (1)
Antonio Vicente	SL-9B-DT	nd	nd	+8.0 (1)
	IG-SN-11	nd	nd	+8.5 (1)
	IE-02	70.42	+6.7[b]	+8.3 (1)
	GAM-CS-37	75.60	+7.7[b]	
	SL-03A-DT	74.40	+7.0[b]	
	SL-07C-DT	78.06	+7.1[b]	
	SL-04-DT	75.20	+7.3[b]	
	NE-B-83	76.04	+7.4[b]	

Analyses were performed at the Scottish Universities Research & Reactor Centre (a) and at Washington State University (b). nd–not determined. Quartz data were from (1) Teixeira (1999), (2) Javier Rios (1995), and (3) Javier Rios et al. (1995). SiO_2 contents were from Dall'Agnol et al. (1999c) (Jamon granite suite), Teixeira (1999) (Velho Guilherme granite suíte), Barros et al. (1995) (Serra dos Carajás granite), and Dall'Agnol et al. (1994) and Dall'Agnol, R., unpublished (Cigano granite).

10.1‰ and are mostly between 6.5‰ and 8.5‰ (Fig. 10); this is considered typical for granitic rocks (cf. Taylor, 1978; Sheppard, 1986). The $\delta^{18}O$ values of the Antonio Vicente pluton range from 6.7‰ to 7.7‰ and are, on average, lower than those of the other suites. $\delta^{18}O$ values for the Jamon suite and associated dikes range from 6.5‰ to 8.6‰, but values higher than 7.5‰ are dominant; the values of the Serra dos Carajás plutons vary between 6.6‰ and 10.1‰, with most values between 7.8‰ and 9.0‰.

Quartz from the Serra dos Carajás granite (Javier Rios et al., 1995) has a $\delta^{18}O$ value of 7.06‰, which is lower than the $\delta^{18}O$ values of the whole rocks. Quartz and calcite from hydrothermal veins gave $\delta^{18}O$ values of 8.3‰ to 8.7‰ (Javier Rios et al., 1995), similar to those obtained for the whole rock. However, $\delta^{18}O$ values for quartz of samples of the Antonio Vicente, Velho Guilherme, and Mocambo plutons (Velho Guilherme suite) range from 8.0‰ to 9.0‰ and tend to be significantly higher than the whole-rock $\delta^{18}O$ values (Fig. 10). $\delta^{18}O$ values for quartz of granites of the Musa pluton (Jamon suite) are 7.16‰ and 7.61‰,

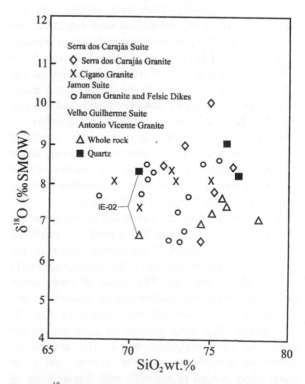

Fig. 10. $\delta^{18}O$ (in whole rock/quartz) versus SiO_2 (in whole rock) diagram for the A-type granites of Carajás. For data sources, see Table 7.

while, in the quartz of hydrothermal veins, they range from 7.67‰ to 9.67‰ (Javier Rios, 1995). Save for values exceeding 9.0‰, these are similar to those obtained for whole-rock of the Jamon pluton.

Of the studied granites, the reduced tin-bearing granites of the Velho Guilherme suite have the lowest $\delta^{18}O$ values. In this regard, they approach the ilmenite-series granites of the midcontinental and western United States (Anderson and Morrison, 2003). On the other hand, the $\delta^{18}O$ whole-rock values of the granites of the Jamon and Serra dos Carajás suites, all containing magnetite as a significant accessory mineral, are similar to those obtained for the magnetite-series granites of the western United States.

4. Discussion

4.1. Petrogenesis of A-type granites

The three studied granite suites are composed of A-type syenogranites to monzogranites with subordinate alkali-feldspar granite. They have similar ages and strongly negative ε_{Nd} values and were emplaced into Archean crust. They show some variation in liquidus temperature and H_2O content, but their most striking difference is the degree of oxidation. Except for the dike swarm associated with the Jamon suite, mafic and intermediate rocks associated with the Carajás granites have not been found. The absence of charnockitic and anorthositic rocks is also worth of note. Evidence of mingling processes involving mafic and felsic magmas is restricted to a composite dike associated with the Jamon suite. A plausible petrogenetic model should explain all these aspects.

The studied suites are composed of granite sensu stricto, and their magmas had a relatively restricted compositional variation. If we consider the presence of mafic magmas in the dikes, the magmatism could be considered bimodal. The rarity of other rocks besides granite in the studied plutons (more than 200 modal analyses, not all represented in Fig. 3) demonstrates that these granites do not represent the felsic end of a magmatic series. The absence of significant granodiorite, tonalite, syenite, and more mafic rocks is hard to reconcile with fractionation of mafic or intermediate mantle-derived magmas. The

considerable size of the plutons would also imply significantly larger volumes of mafic or intermediate precursors (cf. Frost et al., 1999).

Regarding anatectic origin (cf. Rämö and Haapala, 1995), three principal sources for A-type granites can be envisioned: (1) a melt-depleted granulite, proposed originally to explain the origin of the A-type granites of the Lachlan Fold Belt (Collins et al., 1982; Clemens et al., 1986; Whalen et al., 1987); (2) a quartz–feldspathic quartz–dioritic–tonalitic–granodioritic source (Anderson, 1983; Anderson and Bender, 1989; Emslie, 1991; Creaser et al., 1991; King et al., 1997, 2001; Dall'Agnol et al., 1999a,b,c); and (3) underplated basalts and their differentiated equivalents (Frost and Frost, 1997; Frost et al., 1999).

The first hypothesis was criticized by Creaser et al. (1991) because the residual rocks will not have the adequate composition to generate A-type granite magmas. This has been accepted by most authors, including those that have worked on the A-type granites of the Lachlan Fold Belt more recently (cf. King et al., 1997, 2001). Besides the general mineralogical and geochemical restrictions involved in the depleted granulitic source model (cf. Creaser et al., 1991), the magmas of the Paleoproterozoic suites of Carajás were relatively enriched in water (4 to 6 wt.%), rendering their derivation from such sources even more unlikely. Thus, for Carajás, the two latter models should be scrutinized.

The tholeiitic model is not able to explain the origin of the Jamon suite and similar oxidized A-type granites of Laurentia (cf. Anderson and Bender, 1989; Anderson and Smith, 1995; Anderson and Morrison, 2003) because partial melting of tholeiites and their differentiates will lead to reduced magmas (Frost and Lindsley, 1991; Frost and Frost, 1997; Frost et al., 1999). It is theoretically possible, however, to explain the low oxygen fugacity of the Velho Guilherme and, in some extent, the Serra dos Carajás suites by the tholeiite model. Nevertheless, the granites of these suites differ in several aspects from the Sherman batholith for which the tholeiitic model has been proposed (Frost et al., 1999); except for the dikes of the Jamon suite, mafic and intermediate rocks have not been described in Carajás. They also lack fayalite and clinopyroxene and biotite (accompanied by amphibole in the less evolved granites) is the principal mafic phase. Moreover, significant water content in the least evolved magmas seems plausible, and the temperatures of crystallization of the Velho Guilherme suite are lower than those of the Sherman batholith granites.

The Nd isotope data on the Jamon, Velho Guilherme, and Serra dos Carajás suites require an Archean source for the granites (Fig. 8a). The ε_{Nd} (at 1880 Ma) values range from -10.5 to -7.9 and are thus consistent with an Archean crustal source (Fig. 8a,b). The Pb isotope data are more limited and less conclusive, but they are not contradictory with upper and lower crustal sources. The magmas of the three granite suites were presumably derived from Archean sources isotopically quite similar in terms of Nd in the case of the Jamon and Serra dos Carajás granites and somewhat less radiogenic in the case of the Velho Guilherme granites.

Proterozoic A-type rapakivi granites have been generally considered to have ben derived from Paleoproterozoic or hybrid Archean–Paleoproterozoic sources (Rämö and Haapala, 1995). Archean sources have been identified for the Shachang rapakivi complex near Beijing, China (Rämö et al., 1995). Our Nd isotope data indicate that the three Carajás granite suites were also derived from Archean crustal sources. This, combined with the petrological traits, gives low credibility to the hypothesis of their derivation by partial melting of underplated Paleoproterozoic tholeiites and their differentiates. The Mule Creek lobe of the Sherman batholith was intruded into the border zone between Archean and Proterozoic provinces, but it does not show isotopic evidence in favor of derivation from ancient crust, and a mantle or mantle-like isotopic reservoir has been assumed (Frost et al., 2001b). However, the Nd isotope composition of the Mule Creek rocks is entirely distinct from that of the Carajás granites. We conclude that the intraplate suites of Carajás are rapakivi A-type granites derived from Archean quartz–feldspathic crustal sources and that the main compositional differences between the suites probably relate to compositional variation in their sources.

For the Jamon pluton and associated dacite and rhyolite porphyry dikes, geochemical modeling and Nd isotope data are consistent with a silica-poor quartz dioritic source (Dall'Agnol et al., 1999c). Twenty-five to thirty percent of melting of such a source would be able to generate, respectively, the hornblende–biotite monzogranite and dacite porphyry magmas. The Velho Guilherme magma sources probably contained more K-feldspar and were more radiogenic than those of the Jamon suite. Granitoid rocks containing significant modal K-feldspar may include a sedimentary rock component. This is suggested by the dominantly peraluminous reduced character of the magmas and associated tin mineralization. The prominent negative Eu anomalies of most of the granites of the Velho Guilherme suite (Fig. 7e, f) are indicative of important feldspar fractionation. This suggests that feldspar could have been retained in the magma residue, but it is not contradictory with feldspar fractionation during magma evolution. The flat HREE patterns are also indicative of tin-specialized character and reflect the influence of fluorine. In this respect, the Velho Guilherme granites differ from those of Jamon and Serra dos Carajás.

The Serra dos Carajás granites are petrographically and geochemically rather similar to the Velho Guilherme granites. However, Nd isotope data suggest that they were derived from Archean sources, with a crustal residence time similar to that of Rio Maria Archean granitoids. The REE patterns of Serra dos Carajás are also different from those of Velho Guilherme. In this respect, the Cigano pluton is similar to the Musa pluton of the Jamon suite, both showing evidence of amphibole fractionation (Fig. 7b, d). The REE patterns of the Serra dos Carajás granite approach the Jamon patterns (Fig. 7a,c) as well as those of the Bodom pluton in the Wiborg batholith area of Finland (Kosunen, 1999). Moderate to large negative Eu anomalies are ubiquous, suggesting an important role for feldspar fractionation. Accordingly, the source of the Serra dos Carajás magma could have been quartz–feldspathic rocks geochemically intermediate between the sources of Jamon and Velho Guilherme or similar to the source of the Velho Guilherme suite but somewhat more mafic or the result of larger degree of partial melting.

Magmatic underplating has been proposed as the heat source for the melting that formed the Carajás granite magmas (Dall'Agnol et al., 1994, 1999a,b,c). The presence of mafic dike swarms and composite mafic and felsic dikes, temporally and spatially associated with the Jamon and Musa plutons and similar to those described in the rapakivi complexes of

Finland (Rämö, 1991), is consistent with this. The limited occurrence of mafic rocks and the absence of anorthosites and charnockitic rocks can be explained by a relatively high erosional level in Carajás (cf. Rämö and Haapala, 1995).

4.2. Tectonic setting of the Carajás granites: implications for A-type granite origin

The eastern part of the Central Amazonian province corresponds to the Carajás metallogenic province and is limited in the north by the >2.0-Ga Trans-Amazonian Maroni-Itacaiúnas province and in the east by the ~0.6-Ga (Brasiliano) Araguaia belt (Fig. 1). The Xingu–Iricoumé domain of the Central Amazonian province was probably formed during the Archean and was subsequently affected by Paleopro-terozoic volcano-plutonic magmatism (Tassinari and Macambira, 1999; Santos et al., 2000; Teixeira et al., 2002). Thus, there is no geological or geochronolog-ical evidence for the 1.88-Ga granites of Carajás being related to remote subduction processes. Furthermore, these granites are not foliated, they were emplaced at a shallow level in a rigid Archean crust, and they are associated with dike swarms, implying an extensional tectonic setting. These traits do not favor the Carajás granites being related to extension following a subduction event, such as in the Basin and Range province of the western United States (Hawkesworth et al., 1995; Hooper et al., 1995).

Hoffman (1989) and Windley (1993) proposed that the mid-Proterozoic anorogenic granites of Laurentia–Baltica were related to the formation of a super-continent (cf. discussion in Frost et al., 1999). Hoffman (1989) postulated that the upwelling of a mantle superswell developed beneath a stationary supercontinent aggregated in the late Paleoprotero-zoic. He stated that rifting occurred as a consequence, rather than being the cause, of mantle upwelling. The greater abundance of anorogenic magmatism in the mid-Proterozoic could be explained by the secular cooling of the mantle. Windley (1993) stated that the rapakivi granites of Fennoscandia began to form about 100 Ma after the last major (Svecofennian) compres-sional event. Crustal and lithospheric thinning led to upwelling of the asthenosphere and decompression melting of the upper mantle. The resulting lithospheric thickening was followed by lithospheric detachment

and may have led to the extension and collapse of the orogen. According to Windley (1993), the detachment may have extended to at least 100 Ma after the end of thickening, and the Mesoproterozoic anorogenic magmatism was not really anorogenic, having been formed in extensional regimes during the formation and breakup of a 1.5-Ga supercontinent.

The model proposed by Windley (1993) is not applicable to the evolution of the Carajás province because the Paleoproterozoic A-type granites of the province were emplaced ~1.0 to ~0.65 Ga after the stabilization of the Archean crust. Windley's time scale for the tectonic evolution for the Proterozoic provinces of Laurentia and Fennoscandia was thus quite different. Crustal thickening of the Carajás province could have occurred in the end of the Archean, following episodes of crustal accretion (Macambira and Lafon, 1995; Althoff et al., 2000; Leite, 2001; Rämö et al., 2002). After this, the crust remained stable until the Paleoproterozoic granites were formed.

Hoffman's model is able to explain the different aspects of the evolution of the A-type granites of Carajás. There is strong evidence for a supercontinent in the Paleoproterozoic (~2.0 Ga) that was the result of the accretion related to the Trans-Amazonian event (Brito Neves, 1999). This supercontinent could have included the Central Amazonian and Maroni-Itacaiu-nas provinces of the Amazonian craton (Tassinari and Macambira, 1999; cf. Fig. 1a) and possibly also the West African and São Francisco/Congo cratons (Neves, 2003). Lamarão et al. (2002) proposed that the beginning of the breakup of the supercontinent was related to the 1.88-Ga event. This event is registered in several parts of the Amazonian craton and is particularly remarkable in the Central Ama-zonian province (Tassinari and Macambira; cf. Fig. 1a). Hoffman's model implies that a mantle super-swell was formed beneath the Trans-Amazonian supercontinent. It would have caused uplift of the Amazonian craton and was followed by magmatic underplating that was responsible for melting of the Archean crust and resulting A-type magmas.

An important difference between the model of Hoffman (1989) and our hypothesis is that the Paleoproterozoic supercontinent, including the Archean and Trans-Amazonian provinces of the Ama-zonian craton was formed at ~2.0 Ga (cf. Brito Neves,

1999; Lamarão et al., 2002; Neves, 2003), whereas Laurentia was formed at ~1.8 Ga (Hoffman, 1989) and Laurentia–Baltica at ~1.5 Ga (Windley, 1993). We believe that this ~200-Ma age difference is significant. It is reflected not only in the age of the orogenic events but also in the ages of the rapakivi granites (cf. Rämö and Haapala, 1995; Dall'Agnol et al., 1999a)—the A-type granites of Carajás are ~100- to 200-Ma older than the oldest rapakivi granites of Laurentia and Fennoscandia.

Santos et al. (2001) explained the evolution of the Tapajós province (Fig. 1a) by the formation of four Paleoproterozoic (2.01 to 1.88 Ga) magmatic arcs. These arcs were followed by continental breakup responsible for the origin of the Maloquinha A-type granites and the Iriri Group volcanic rocks (Uatumã Supergroup). On the other hand, Lamarão et al. (2002) considered that the evolution of the Tapajós province at 1.88 Ga probably involved reworking of an older orogenic terrane by continental extension. In their model, the Tapajós province was affected at 1.88 Ga by the incipient breakup of the Trans-Amazonian supercontinent, resulting a distensional tectonic setting probably similar to that of Basin and Range province. The magmatism associated with the Basin and Range province displayed high-K calc-alkaline and A-type signatures (Hawkesworth et al., 1995; Hooper et al., 1995) corresponding to the Parauari suite and Maloquinha suite and Iriri Group in the Tapajós province. At same time, in the Archean craton (Carajás), extensive underplating and extension generated the A-type granite suites. In this model, the origin of the Carajás granites is related to the disruption of the Trans-Amazonian supercontinent, not to subduction processes.

4.3. Comparison to Laurentia and Fennoscandia

Three distinct groups of Mesoproterozoic granites with rapakivi affinity have been distinguished in Laurentia: ilmenite-series granite, magnetite-series granite, and two-mica granite (Anderson and Bender, 1989; Anderson and Morrison, 2003). The ilmenite-series granites correspond to the granite members of the anorthosite–mangerite–charnockite–rapakivi granite series (AMCG series; Emslie, 1991) and are correlated with the rapakivi granites of the Fenno-scandian shield. Representative data from the Wiborg rapakivi area (Rämö, 1991; Rämö and Haapala, 1995; Kosunen, 1999, 2004) have been selected for comparison.

Anderson and Morrison (2003) stated that, compared to the ilmenite-series granites, the magnetite-series granites of Laurentia are less potassic and display lower FeOt/(FeOt+MgO) in the 0.80–0.88 range. Hornblende and biotite Fe/(Fe+Mg) ratios range from 0.3 to 0.8 and are thus also lower. The Jamon suite granites are similar to the magnetite-series granites of Laurentia, as indicated by the Fe/(Fe+Mg) of amphibole, biotite, and whole-rock (Table 4; Figs. 4, 6a and b). They are magnetite-bearing oxidized granites that crystallized at fO_2 conditions near the NNO and HITMQ buffers. They differ in this respect from the more typical reduced (~FMQ) rapakivi granites of Fennoscandia (cf. Rämö and Haapala, 1995). Thus, the Serra dos Carajás and Velho Guilherme granites have affinities with the classic rapakivi series. In spite of their relatively reduced character (Figs. 4 and 6b), the Serra dos Carajás granites are also magnetite-series (Ishihara, 1981). On the other hand, most of the Velho Guilherme granites correspond to the ilmenite-series granites. The presence or absence of magnetite and the significant variation in the Fe/(Fe+Mg) ratios of biotite and amphiboles in the Serra dos Carajás and Velho Guilherme granites (Fig. 4) indicate that the fO_2 conditions were not uniform in the plutons considered. A similar picture is shown by the Bodom and Obbnäs plutons of the Wiborg rapakivi area in southeastern Finland—the Obbnäs pluton was formed in more oxidizing conditions than the Bodom pluton (Fig. 4a and b; Kosunen, 1999, 2004). Nevertheless, the amphiboles of both plutons plot in the low-fO_2 field (Fig. 4a) and differ in this respect from the magnetite-series granites of the Jamon suite and the western United States. The composition of amphiboles in the Obbnäs pluton is similar in terms of Fe/(Fe+Mg) but differs by its higher Al^{IV} content from the amphibole in the hornblende–biotite monzogranite of Velho Guilherme.

The general picture emerging from the Paleoproterozoic rapakivi-type granites of Carajás contrasts in some aspects with the Laurentian granites (cf. Anderson and Morrison, 2003). In Carajás, the country rocks and probably also the sources of the

A-type magmas were all Archean, and the three studied granite suites are pratically coeval. They evolved in variable fO_2 conditions, but they do not correspond exactly to the three main granite groups identified in Laurentia. The Jamon suite granites are typical magnetite-series and similar to the oxidized granites of Laurentia. The dominant syenogranites of the Velho Guilherme Suite are ilmenite series. The Serra dos Carajás granites are magnetite-bearing granites but are also relatively reduced. Both evolved in fO_2 conditions similar to those estimated for the rapakivi granites of Laurentia and Fennoscandia (Figs. 4 and 6b). Peraluminous two-mica granites, similar to those in Laurentia, are not found in Carajás.

The tin-bearing granites of the Velho Guilherme suite have the lowest $\delta^{18}O$ values of the Carajás granite suites and approach the ilmenite-series granites of the midcontinental and western United States (Anderson and Morrison, 2003). On the other hand, the $\delta^{18}O$ whole-rock values of the Jamon and Serra dos Carajás granites are similar to those of the magnetite-series granites of the United States. This further stresses the similarities between the Laurentian and Carajás suites.

5. Conclusion

Proterozoic-oxidized A-type granites are not restricted to the central and southwestern United States, but they are abundant in the Amazonian craton. They represent a variety of granite with rapakivi affinity, featuring magmas derived from oxidized (Dall'Agnol et al., 1999b), not reduced, sources (Frost and Frost, 1997), as commonly observed in other areas (Rämö and Haapala, 1995). A-type granites are thus not derived from reduced sources only. The Carajás granites magmas tapped Archean granitoid metaigneous rocks, in some cases, with a contribution of metasedimentary rocks. The oxidation state of the magmas was essentially dependent on the nature of their sources.

The origin of A-type magmas in the Amazonian craton is related to a mantle superswell that was formed beneath the Trans-Amazonian supercontinent (Hoffman, 1989). It caused uplift of the Amazonian craton and was followed by magmatic underplating that was responsible for extensive melting of the

Archean crust. The Trans-Amazonian supercontinent is older than the Laurentia–Baltica supercontinent, and the A-type granites of Carajás are ~100- to 200-Ma older than the oldest rapakivi granites of Laurentia and Fennoscandia.

Acknowledgments

C.E.M. Barros, M.S. Magalhães, M.A.B.M. Figueiredo, M.C. Gastal, R.O. Silva Jr., A.A.S. Leite, F. Javier Rios, R.N.N. Villas, M.G.B. Gonçalez, M.A. Horbe, M.A. Oliveira, and J.A.C. Almeida contributed to the research on the Paleoproterozoic granites. DOCEGEO (CVRD Group) gave support in the field, helped in sampling, and transmitted unpublished information. The stimulating reviews of Carol Frost and W. Randy Van Schmus greatly contributed for improvement of the paper. The authors express special thanks to the Analytical Department and the direction of the CRPG (Vandoeuvre, France) for whole-rock chemical analyses, the Microanalyses Laboratory of the University of Nancy I (RD data) and the Microprobe laboratory of São Paulo University (NPT data) for microprobe analyses, A.E. Fallick of the Scottish Universities Environmental Research Centre and P.B. Larson of the Washington State University (USA) for oxygen isotopic analyses, P. Kosunen for unpublished data on the Bodom and Obbnäs plutons, C.N. Lamarão for support on the treatment of mineralogical and chemical data, and C.M.D. Fernandes for help in the preparation of illustrations. O.T.R. acknowledges help from the staff of the Unit for Isotope Geology, Geological Survey of Finland while making the Nd isotope analyses. This research received support from CNPq (000400038/99, 463196/00-7, and 550739/01-7) and UFPA. This paper is a contribution to PRONEX/CNPq (Proj. 103/98–Proc. 66.2103/1998-0) and IGCP-426.

References

Åhäll, K.I., Conelly, J.N., Brewer, T.S., 2000. Episodic rapakivi magmatism due to distal orogenesis: correlation of 1.69–1.50 Ga orogenic and inboard, "anorogenic" events in the Baltic Shield. Geology 28, 823–826.

Althoff, F.J., Barbey, P., Boullier, A.-M., 2000. 2.8–3.0 Ga plutonism and deformation in the SE Amazonian craton: the

Archean granitoids of Marajoara (Carajás mineral province). Precambrian Research 104, 187–206.

Amelin, Y.V., Larin, A.M, Tucker, R.D., 1997. Chronology of multiphase emplacement of the Salmi rapakivi granite–anorthosite complex, Baltic shield: implications for magmatic evolution. Contributions to Mineralogy and Petrology 127, 353–368.

Anderson, J.L., 1983. Proterozoic anorogenic granite plutonism of the North America. In: Medaris, L.G., Mickelson, D.M., Byers, C.W., Shanks, W.C. (Eds.), Proterozoic Geology, Memoir-Geological Society of America, vol. 161, pp. 133–154.

Anderson, J.L., Bender, E.E., 1989. Nature and origin of Proterozoic A-type granitic magmatism in the southwestern United States of America. Lithos 23, 19–52.

Anderson, J.L, Cullers, R.L., 1978. Geochemistry and evolution of the Wolf River Batholith, a Late Precambrian rapakivi massif in North Wisconsin, U.S.A. Precambrian Research 7, 287–324.

Anderson, J.L., Morrison, J., 2003. Ilmenite, magnetite, and Mesoproterozoic granites of Laurentia. In: Rämö, O.T., Kosunen, P.J., Lauri, L.S., Karhu, J.A. (Eds.), Granitic Systems—State of the art and Future Avenues An International Symposium in Honor of Professor Ilmari Haapala, January 12–14, 2003, Abstract volume. Helsinki University Press, pp. 7–12.

Anderson, J.L., Smith, D.R., 1995. The effects of temperature and fO_2 on the Al-in-hornblende barometer. American Mineralogist 80, 549–559.

Avelar, V.G., Lafon, J.-M., Correia Jr., F.C., Macambira, E.M.B., 1999. O magmatismo Arqueano da região de Tucumã-Província Mineral de Carajás: novos resultados geocronológicos. Revista Brasileira de Geociências 29, 453–460 (in Portuguese).

Barbosa, A.A., Lafon, J.M., Neves, A.P., Vale, A.G., 1995. Geocronologia Rb–Sr e Pb–Pb do Granito Redenção, SE do Pará: implicações para a evolução do magmatismo proterozóico da região de Redenção. Boletim do Museu Paraense Emílio Goeldi. Ciências da Terra 7, 147–164 (in Portuguese).

Barker, F., Wones, D.R., Sharp, W.N., Desborough, G.A., 1975. The Pikes Peak batholith, Colorado Front Range, and a model for the origin of the gabbro–anorthosite–syenite–potassic granite suite. Precambrian Research 2, 97–160.

Barros, C.E.M., Dall'Agnol, R., Vieira, E.A.P., Magalhães, M.S., 1995. Granito Central da Serra dos Carajás: avaliação do potencial metalogenético para estanho com base em estudos da borda oeste do corpo. Boletim do Museu Paraense Emílio Goeldi. Série Ciências da Terra 7, 93–123 (in Portuguese).

Barros, C.E.M., Sardinha, A.S., Barbosa, J.P.O., Krimski, R., Macambira, M.J.B., 2001. Pb–Pb and U–Pb zircon ages of Archean syntectonic granites of the Carajás metallogenic province, Northern Brazil. 3th Simposio Sudamericano de Geologia Isotopica, 3, Pucon, Chile, Resumos Expandidos. Servicio Nacional de Geologia Y Mineria. CD-ROM.

Bettencourt, J.S., Tosdal, R.M., Leite Jr., W.B., Payolla, B.L., 1999. Mesoproterozoic rapakivi granites of the Rondônia tin province, southwestern border of the Amazonian craton, Brazil: I. Reconnaissance U–Pb geochronology and regional implications. Precambrian Research 95, 41–67.

Bonin, B., 1986. Ring Complex Granites and Anorogenic Magmatism. Bureau des Recherches Géologiques et Minières, Orléans.

Brito Neves, B.B., 1999. América do Sul: quatro fusões, quatro fissões e o processo acrescionário andino. Revista Brasileira de Geociências 29, 379–392 (in Portuguese).

Clemens, J.D., Holloway, J.R., White, A.J.R., 1986. Origin of the A-type granite: experimental constraints. American Mineralogist 71, 317–324.

Collins, W.J., Beams, S.D., White, A.J., Chappell, B.W., 1982. Nature and origin of A-type Granites with particular reference to southeastern Australia. Contributions to Mineralogy and Petrology 80, 189–200.

Cordani, U.G., Sato, K., 1999. Crustal evolution of the South American Platform, based on Sr and Nd systematics on granitoid rocks. Episodes 22, 167–173.

Costi, H.T., Dall'Agnol, R., Moura, C.A.V., 2000. Geology and Pb–Pb geochronology of Paleoproterozoic volcanic and granitic rocks of Pitinga province, Amazonian craton, Northern Brazil. International Geology Review 42, 832–849.

Creaser, R.A., Price, R.C., Wormald, R.J., 1991. A-type granites revisited: assessment of a residual-source model. Geology 19, 163–166.

Dall'Agnol, R., Lafon, J.M., Macambira, M.J.B., 1994. Proterozoic anorogenic magmatism in the Central Amazonian province: geochronological, petrological and geochemical aspects. Mineralogy and Petrology 50, 113–138.

Dall'Agnol, R., Pichavant, M., Champenois, M., 1997a. Iron–titanium oxide minerals of the Jamon granite, Eastern Amazonian region, Brazil: implications for the oxygen fugacity in Proterozoic A-type granites. Anais da Academia Brasileira de Ciências 69, 325–347.

Dall'Agnol, R., Souza, Z.S., Althoff, F.J., Barros, C.E.M., Leite, A.A.S., Jorge João, X.S., 1997b. General aspects of the granitogenesis of the Carajás metallogenetic province. Second International Symposium on Granites and Associated Mineralizations, Excursions Guide, Salvador, Companhia Baiana de Pesquisa Mineral, Superintendência de Geologia e Recursos Minerais, pp. 135–161.

Dall'Agnol, R., Costi, H.T., Leite, A.A., Magalhães, M.S., Teixeira, N.P., 1999a. Rapakivi granites from Brazil and adjacent areas. Precambrian Research 95, 9–39.

Dall'Agnol, R., Scaillet, B., Pichavant, M., 1999b. An experimental study of a lower Proterozoic A-type granite from the eastern Amazonian craton, Brazil. Journal of Petrology 40, 1673–1698.

Dall'Agnol, R., Rämö, O.T., Magalhães, M.S., Macambira, M.J.B., 1999c. Petrology of the anorogenic, oxidised Jamon and Musa granites, Amazonian craton: implications for the genesis of Proterozoic A-type granites. Lithos 46, 431–462.

DePaolo, D.J., 1981. Neodymium isotope in the Colorado Front Range and crust–mantle evolution in the Proterozoic. Nature 291, 193–196.

DePaolo, D.J., Wasserburg, G.J., 1976. Nd isotopic variations and petrogenetic models. Geophysical Research Letters 3, 249–252.

Doe, B.R., Zartman, R.E., 1979. Plumbotectonics. In: Barnes, H. (Ed.), Geochemistry of Hydrotermal Ore Deposits. Wiley, New York, pp. 22–70.

Eby, G.N., 1992. Chemical subdivision of the A-type granitoids: petrogenesis and tectonic implications. Geology 20, 641–644.

Emslie, R.F., 1991. Granitoids of rapakivi granite–anorthosite and related associations. Precambrian Research 51, 173–192.

Evensen, N.M., Hamilton, P.J., O'Nions, R.K., 1978. Rare earth abundances in chondritic meteorites. Geochimica et Cosmochimica Acta 42, 1199–1212.

Fraga, L.M., Dall'Agnol, R., Macambira, M.J.B., 2003. The Mucajai anorthosite–mangerite rapakivi granite (AMG) complex, north Amazonian craton, Brazil. EGS–AGU–EUG Joint Assembly, Nice, France. VGP6 Granite Systems and Proterozoic Lithospheric Processes, EAE03-A-14489.

Frost, C.D., Frost, B.R., 1997. Reduced rapakivi type granites: the tholeiitic connection. Geology 25, 647–650.

Frost, B.R., Lindsley, D.H., 1991. Occurrence of iron–titanium oxides in igneous rocks. In: Lindsley, D.H. (Ed.), Oxide Minerals: Petrologic and Magnetic Significance, Mineralogical Society of America Reviews in Mineralogy, vol. 25, pp. 1–9.

Frost, C.D., Frost, B.R., Chamberlain, K.R., Edwards, B., 1999. Petrogenesis of the 1.43 Ga Sherman batholith, SE Wyoming, USA: a reduced, rapakivi-type anorogenic granite. Journal of Petrology 40, 1771–1802.

Frost, B.R., Barnes, C.G., Collins, W.J., Arculus, R.J., Ellis, D.J, Frost, C.D., 2001a. A geochemical classification for granitic rocks. Journal of Petrology 42, 2033–2048.

Frost, C.D., Bell, J.M., Frost, B.R., Chamberlain, K.R., 2001b. Crustal growth by magmatic underplating: isotopic evidence from the northern Sherman batholith. Geology 29, 515–518.

Gibbs, A.K., Wirth, K.R., Hirata, W.K., Olszewski Jr., W.J., 1986. Age and composition of the Grão Pará Group volcanics, Serra dos Carajás. Revista Brasileira de Geociências 16, 201–211.

Haapala, I., Rämö, O.T., 1992. Tectonic setting and origin of the Proterozoic rapakivi granites of the southeastern Fennoscandia. Transactions of the Royal Society of Edinburgh. Earth Sciences 83, 165–171.

Haapala, I., Rämö, O.T., 1999. Rapakivi granites and related rocks: an introduction. Precambrian Research 95, 1–7.

Hawkesworth, C., Turner, S., Gallagher, K., Hunter, A., Bradshaw, T., Rogers, N., 1995. Calc-alkaline magmatism, lithospheric thinning and extension in the Basin and Range. Journal of Geophysical Research 100, 10,271–10,276.

Hoffman, P., 1989. Speculations on Laurentia's first gigayear (2.0 to 1.0 Ga). Geology 17, 135–138.

Holtz, F., Johannes, W., Tamic, N., Behrens, H., 2001. Maximum and minimum water contents of granitic melts generated in the crust: a reevaluation and implications. Lithos 56, 1–14.

Hooper, P.R., Bailey, D.G., McCarley Holder, G.A., 1995. Tertiary calc-alkaline magmatism associated with lithospheric extension in the Pacific Northwest. Journal of Geophysical Research 100, 10,303–10,319.

Huppert, H.E., Sparks, R.S., 1988. The generation of granitic magmas by intrusion of basalt into continental crust. Journal of Petrology 29, 599–624.

Ishihara, S., 1981. The granitoid series and mineralization. Economic Geology, 75th Anniversary volume, 458–484.

Javier Rios, F., 1995. A jazida de wolframita de Pedra Preta, Granito Musa, Amazônia Oriental (PA): Estudo dos fluidos mineralizantes e isótopos estáveis de oxigênio em veios hidrotermais.

PhD thesis, Federal University of Pará, Belém, Brazil (in Portuguese).

Javier Rios, F., Villas, R.N., Dall'Agnol, R., 1995. O Granito Serra dos Carajás: fácies petrográficas e avaliação do potencial metalogenético para estanho no setor norte. Revista Brasileira de Geociências 25, 20–31 (in Portuguese).

King, P.L., White, A.J.R., Chappell, B.W., Allen, C.M., 1997. Characterization and origin of aluminous A-type granites from the Lachlan Fold Belt, southeastern Australia. Journal of Petrology 38, 371–391.

King, P.L., Chappell, B.W., Allen, C.M., White, A.J.R., 2001. Are A-type granites the high-temperature felsic granites? Evidence from fractionated granites of the Wangrah Suite. Australian Journal of Earth Sciences 48, 501–514.

Klimm, K., Holtz, F., Johannes, W., King, P.L., 2003. Fractionation of metaluminous A-type granites: an experimental study of the Wangrah Suite, Lachlan Fold Belt, Australia. Precambrian Research 124, 327–341.

Kosunen, P., 1999. The rapakivi granite plutons of the Bodom and Obbnäs, Southern Finland: petrography and geochemistry. Bulletin of the Geological Society of Finland 71, 275–304.

Kosunen, P.J., 2004. Petrogenesis of mid-Proterozoic A-type granite: Case Studies from Fennoscandia (Finland) and Laurentia (New Mexico). PhD thesis, Department of Geology, University of Helsinki, Finland.

Lamarão, C.N., Dall'Agnol, R., Lafon, J.-M., Lima, E.F., 2002. Geology, geochemistry, and Pb–Pb zircon geochronology of the Paleoproterozoic magmatism of Vila Riozinho, Tapajós Gold Province, Amazonian craton, Brasil. Precambrian Research 119, 189–223.

Leite, A.A.S., 2001. Geoquímica, petrogênese e evolução estrutural dos granitóides arqueanos da região de Xinguara, SE do Cráton Amazônico. Belém, Universidade Federal do Pará, Centro de Geociências. PhD thesis, Federal University of Pará, Belém, Brazil (in Portuguese).

Loiselle, M.C., Wones, D.R., 1979. Characteristics and origin of anorogenic granites. Abstracts with programs-Geological Society of America 11, 468.

Macambira, M.J.B., Lafon, J.-M., 1995. Geocronologia da Província mineral de Carajás: síntese dos dados e novos desafios. Boletim do Museu Paraense Emílio Goeldi. Ciências da Terra 7, 263–288 (in Portuguese).

Macambira, M.J.B., Galarza-Toro, M.A., Souza, S.R.B., Silva, C.M.G., 2001. Pb isotope investigations on Cu–Au deposits from Carajás province, Amazonian craton, Brazil. Simposio Sudamericano de Geologia Isotopica, 3, Pucon, Chile, Resumos Expandidos. Servicio Nacional de Geologia Y Mineria. CD-ROM.

Machado, N., Lindenmayer, Z., Krogh, T.E., Lindenmayer, D., 1991. U–Pb geochronology of Archean magmatism and basement reactivation in the Carajás área, Amazon Shield, Brazil. Precambrian Research 49, 329–354.

Magalhães, M.S., Dall'Agnol, R., Sauck, W.A., Luiz, J.G., 1994. Suscetibilidade magnética: um indicador da evolução petrológica de granitóides da Amazônia. Revista Brasileira de Geociências 24, 139–149 (in Portuguese).

Naney, M.T., 1983. Phase equilibria of rock-forming ferromagnesian silicates in granitic systems. American Journal of Science 283, 993–1033.

Neves, S.P., 2003. Proterozoic history of the Borborema province (NE Brazil): correlations with neighboring cratons and Pan-African belts and implications for the evolution of western Gondwana. Tectonics 22.

Oliveira, D.C., 2001.Geologia, geoquímica e petrologia magnética do granito paleoproterozóico Redenção, SE do Cráton Amazônico. MSc Thesis, Federal University of Pará, Belém, Brazil (in Portuguese).

Pearce, J.A., Harris, N.B.W, Tindle, A.C., 1984. Trace element discrimination diagrams for the tectonic interpretation of granitic rocks. Journal of Petrology 25, 956–983.

Pichavant, M., Hammouda, T., Scaillet, B., 1996. Control of redox state and Sr isotopic composition of granitic magmas: a critical evaluation of the role of source rocks. Transactions of the Royal Society of Edinburgh. Earth Sciences 88, 321–329.

Pidgeon, R., Macambira, M.J.B., Lafon, J.-M., 2000. Th–U–Pb isotopic systems and internal structures of complex zircons from an enderbite from the Pium Complex, Carajás Province, Brazil: evidence for the ages of granulite facies metamorphism and the source of the enderbite. Chemical Geology 166, 159–171.

Rämö, O.T., 1991. Petrogenesis of the Proterozoic rapakivi granites and related basic rocks of the southeastern Fennoscandia: Nd and Pb isotopic and general geochemical constraints. Geological Survey of Finland Bulletin, 355.

Rämö, O.T., Haapala, I., 1995. One hundred years of rapakivi granite. Mineralogy and Petrology 52, 129–185.

Rämö, O.T., Haapala, I., Vaasjoki, M., Yu, J.H., Fu, H.Q., 1995. 1700 Ma Sachang complex, northeast China: Proterozoic rapakivi granite not associated with Paleoproterozoic orogenic crust. Geology 23, 815–818.

Rämö, O.T., Dall'Agnol, R., Macambira, M.J.B., Leite, A.A.S., de Oliveira, D.C., 2002. 1.88 Ga oxidized A-type granites of the Rio Maria region, eastern Amazonian craton, Brazil: positively anorogenic! Journal of Geology 110, 603–610.

Richard, P., Shimizu, N., Allègre, C.J., 1976. 143Nd/146Nd, a natural tracer: an application to oceanic basalts. Earth and Planetary Science Letters 31, 269–278.

Santos, J.O.S., Hartmann, L.A., Gaudette, H.E., Groves, D.I., McNaughton, N.J., Fletcher, I., 2000. A new understanding of the provinces of the Amazonian craton based on integration of field mapping and U–Pb and Sm–Nd geochronology. Gondwana Research 3, 453–488.

Santos, J.O.S., Groves, D.I., Hartmann, L.A., Moura, M.A., McNaughton, N.J., 2001. Gold deposits of the Tapajós and

Alta Floresta domains, Tapajós-Parima orogenic belt, Amazonian craton, Brazil. Mineralium Deposita 36, 278–299.

Scaillet, B., Pichavant, M., Roux, J., 1995. Experimental crystallization of leucogranite magmas. Journal of Petrology 36, 663–705.

Schmidt, M.W., 1992. Amphibole composition in tonalite as a function of pressure: an experimental calibration of the Al-in-hornblende barometer. Contributions to Mineralogy and Petrology 110, 304–310.

Sheppard, S.M.F., 1986. Igneous rocks: III. Isotopic studies of magmatism in Africa, Eurasia, and Oceanic Islands. In: Valley, J.W., Taylor, H.P., O'Neil, J.R. (Eds.), Stable Isotopes in High Temperature Geological Processes, Mineralogical Society of America Reviews in Mineralogy, vol. 16, pp. 319–371.

Souza, S.R.B., 1996. Estudo Geocronológico e de Geoquímica Isotópica da Área Pojuca (Província Mineral de Carajás). MSc thesis, Federal University of Pará, Belém, Brazil (in Portuguese).

Stacey, J.S., Kramers, J.D., 1975. Approximation of terrestrial lead isotope evolution by a two-stage model. Earth and Planetary Science Letters 26, 207–221.

Tassinari, C.C.G., Macambira, M.J.B., 1999. Geochronological provinces of the Amazonian craton. Episodes 22, 174–182.

Taylor Jr., H.P., 1978. Oxygen and hydrogen isotope studies of plutonic granitic rocks. Earth and Planetary Science Letters 38, 177–210.

Teixeira, N.P., 1999. Contribuição ao estudo das rochas granitóides e mineralizações associadas da Suíte Intrusiva Velho Guilherme, Província Estanífera do Sul do Pará. PhD thesis, University of São Paulo, Brazil (in Portuguese).

Teixeira, J.B., Egler, D.H., 1994. Petrology, geochemistry, and tectonic setting of Archean basaltic and dioritic rocks from the N4 iron deposit, Serra dos Carajás, Pará, Brazil. Acta Geologica Leopoldensia XVII, 71–114.

Teixeira, N.P., Bettencourt, J.S., Moura, C.A.V., Dall'Agnol, R., Macambira, E.M.B., 2002. Archean crustal sources for Paleoproterozoic tin-mineralized granites in the Carajás province, SSE Pará, Brazil: Pb–Pb geochronology and Nd isotope geochemistry. Precambrian Research 119, 257–275.

Whalen, J.B., Currie, K.L., Chappell, B.W., 1987. A-type granite: geochemical characteristics, discrimination and petrogenesis. Contributions to Mineralogy and Petrology 95, 407–419.

Windley, B.F., 1993. Proterozoic anorogenic magmatism and its orogenic connections. Journal of the Geological Society (London) 150, 39–50.

Wones, D.R., 1989. Significance of the assemblage titanite+magnetite+quartz in granitic rocks. American Mineralogist 74, 744–774.

Available online at www.sciencedirect.com

ELSEVIER

Lithos 80 (2005) 131–145

www.elsevier.com/locate/lithos

The granite-upper mantle connection in terrestrial planetary bodies: an anomaly to the current granite paradigm?

Bernard Bonin*, Jean Bébien

Orsayterre, UPS-CNRS FRE 2566, Département des Sciences de la Terre, Université de Paris-Sud, F-91405 Orsay CEDEX, France

Received 1 April 2003; accepted 9 September 2004
Available online 10 November 2004

Abstract

Granite formed in the terrestrial planets very soon after their accretion. The oldest granite-forming minerals (4.4 Ga zircon) and granite (4.0 Ga granodiorite) indicate conditions resembling the present-day ones, with the presence of oceans and external processes related to liquid water. As a result, the current granite paradigm states that granite is not issued directly from the melting of the mantle. However, a granite-upper mantle connection is well established from several pieces of evidence. Tiny micrometre- to millimetre-sized enclaves of granite-like glassy and crystalline materials in Earth's mantle rocks are known in oceanic and continental areas. Earth's mantle-forming minerals, such as olivine, pyroxene, and chromite, can contain silicic materials, either as glass inclusions or as crystallised products (quartz or tridymite, sanidine, K-feldspar, and/or plagioclase close to albite end-member). Importantly, the same evidence is amply found in some types of meteorites, whether they are primitive, such as ordinary chondrites, or differentiated, such as IIE irons, howardite–eucrite–diogenite (HED), and Martian shergottite–nakhlite–chassignite (SNC) achondrites. Although constituting apparently an anomaly, the granite-upper mantle connection can be reconciled with the current granite paradigm by recognising that the conditions prevailing in the formation of granite are not only necessarily crustal but can occur also at depths in mantle rocks. Unresolved problems to be explored further include whether tiny amounts of granitic material within terrestrial mantles may be hints of greater abundances and more direct mantle involvement, and what role can be played by granite trapped within the upper mantle in lithosphere buoyancy.
© 2004 Elsevier B.V. All rights reserved.

Keywords: Granite; Silicic glass; Mantle; Terrestrial planets; Meteorites; Paradigm

* Corresponding author. Tel.: +33 1 69 15 67 66; fax: +33 1 69 15 67 72.
  E-mail address: bbonin@geol.u.-psud.fr (B. Bonin).

0024-4937/$ - see front matter © 2004 Elsevier B.V. All rights reserved.
doi:10.1016/j.lithos.2004.03.059

1. Introduction

Even now, the proportion of granitoid and associated volcanic rocks present in the Earth is low, about 0.001 of the bulk Earth (Clarke, 1996). That small proportion corresponds nevertheless to a total mass of at least 10^{22} kg and a volume of about 3.74×10^9 km^3 (Bonin et al., 2002). Roughly 86 vol.% of the upper continental crust is granitic in composition (Wedepohl, 1991). Granite occurs also, albeit in smaller amounts, within lower continental crust (charnockite series), oceanic crust, and upper mantle. The granite-upper mantle connection constitutes an important issue that will be addressed here.

That all the granitic material exposed on the Earth came ultimately from upper mantle through a protracted sequence of magmatic processes and events was never questioned. However, there is currently no consensus about the volume of granite that can be generated directly from mantle. Vein and dyke swarms of granitic composition cross-cutting peridotite exposed on the walls of slow-spreading midoceanic ridges have been observed, thus providing fuel to the hypothesis that magma chambers in which granitic liquids are elaborated may occur in upper mantle.

As granitic material is known in a number of extraterrestrial occurrences (for a review, see Bonin et al., 2002), the meteoritic record should also be considered. The Earth accretes currently 15,000 to 50,000 t a^{-1} of extraterrestrial matter, but most of it (~95 wt.%) is dust, within the 40–1500 μm particle size-range, corresponding to micrometeorites. Pristine micrometeorites recovered in polar ice sheets consist of matter similar to carbonaceous chondrites of CM group and possibly CI and CR groups. Presumably, a nonchondritic component is present but has yet to be identified (Maurette et al., 1991; Kurat et al., 1994). The probability to detect extraterrestrial granitic material in the dust collection and to distinguish it from terrestrial particles is exceedingly low.

Silicate meteorites are mostly peridotite and basalt, resembling the Earth's mantle and mafic crust. Some iron meteorites have small amounts of silicate materials. To what extent the current meteoritic record constitutes a representative sampling of extraterrestrial bodies, or is biased by secondary effects occurring during the entry in the Earth's atmosphere, should be discussed.

The first point to examine is whether reasonably large granitic fragments can survive impact and ejection from the Earth or other planets. The Earth's surface is constituted by ~72% oceans and ~28% continents, where granite is exposed on large areas. The probability of ejecting granitic material from the Earth's surface is, therefore, fairly high, considering that emerged parts of the globe are more likely to be scavenged by impacts than oceanic areas covered by 4 to 5 km of liquid water. The occurrence of terrestrial meteorites is known since the 1920s, although their origin was not yet fully understood. Tektites are silicic ($SiO_2 > 65$ wt.%) glasses ejected from terrestrial impact craters over large distances; their compositions evidence that they formed by melting of upper crust. Their mass can reach up to 24 kg for exceptional samples, but their common size is about 2–4 cm. Some samples exhibit evidence of aerodynamic ablation resulting from partial remelting of the glass during reentry into the Earth's atmosphere. In their review, Dressler and Reimold (2001) reported no crystallised silicic material, suggesting either that they did not survive impact and were completely melted during ejection or that they could not be distinguished from other granitic blocks or gravels in the sedimentary record. No mafic material was reported so far, probably owing to the fact that the currently identified tektite-strewn fields were issued from impacts based on continental areas.

The second point to examine relates to the meteoritic record itself. Shergottite–nakhlite–chassignite (SNC) meteorites, thought to be Martian igneous rocks, are mostly cumulates (McSween, 1994). The discovery at the Mars Pathfinder site of hard rocks with more silicic compositions (McSween et al., 1999; Foley et al., 2001) was unexpected. The possibility that sedimentary silica may be an important phase on the surface of Mars is explored (McLennan, 2003), although no material of such composition was ever recovered, in evidence of that the SNC collection is constituted only by very hard rocks and that softer rocks were destroyed during their ejection and travel in the space. Ejection of all SNC meteorites could have occurred during no more than three discrete events, suggesting that only limited areas of the Martian surface have been scavenged.

Thus, meteorites should not be regarded as a representative sampling of the asteroid belt and terrestrial planets. Taking account of the Martian evidence, we may speculate that fragments ejected from the Earth and arriving on Venus, or Mercury, could be less granitic than usually supposed. At this point, we contend that the proportion of rock types, particularly the apparent paucity of extraterrestrial granitic material, is meaningless in terms of abundances, but provides qualitative insights on extraterrestrial matter.

2. When did granite form first?

Condensation in the space of the major granite-forming minerals, e.g., plagioclase and alkali feldspar, could take place at 1400–1000 K and 10^2 Pa in a solar-type nebula yielding $C/O=0.55$, thus pre-dating the Earth's birth (Encrenaz and Bibring, 1987). However, preservation of these early grains during further accretionary episodes building the entire Solar System is completely unlikely.

The Sun and the associated accretionary disk formed 4566 ± 2 Ma ago, as determined from the Allende refractory inclusions (Zhao, 2002). Accretion of terrestrial planets involved three major stages: (i) sticking and friction between dust grains in the solar nebula, ultimately producing kilometre-scale planetesimals within 1 to 3 Ma of the birth of the Solar System (Wetherill, 1994); (ii) runaway gravitational growth in $\sim10^5$ years, resulting in $\sim10^{23}$ kg bodies (lunar to Mercury sizes) of (carbonaceous) chondritic compositions (Wetherill and Stewart, 1993); and (iii) collisions produced by neighbouring ~1000-km-sized bodies (Melosh, 1990) and planetary differentiation starting less than 8 to 10 Ma after the onset of the Solar System.

The use of extinct nuclides, which have half-lives two orders of magnitude shorter than U, shows that terrestrial planets completed their accretion at various periods of time, ranging from as early as 4.56 Ga for the howardite–eucrite–diogenite (HED) parent body, possibly the asteroid 4Vesta (Yamaguchi et al., 2001), to 4.54 Ga for Mars to 4.52–4.50 Ga for Moon (Lee et al., 1997, 2002), the origin of which is still disputed (O'Neill, 1991; Ruzicka et al., 2001; Dreibus and Wänke, 2002).

The Earth's growth was more protracted and dominated by impacts and planetary collisions, which induced in particular the formation of Moon. The Earth's present mass, with its metallic core and water abundance ($\sim5\times10^{-4}$ Earth's mass, Morbidelli et al., 2000), was reached at 4.47–4.45 Ga, i.e., 100–120 Ma after the beginning of the Solar System (Halliday, 2000; Zhao, 2002). A late period of heavy bombardment occurred at 4.4 to 3.8 Ga, with each catastrophic asteroidal or cometary collision inducing for a few millennia resurfacing, water evaporation, and incubation at an effective temperature of 2000 K, thus potentially frustrating the emergence of life (Wells et al., 2003).

Until recently, it was commonly considered that, at the end of its accretion, the Earth almost certainly contained no granite sensu lato. This assumption became questionable when a zircon grain was discovered, yielding the very ancient age of 4404 ± 8 Ma, only ~160 Ma younger than the beginning of the Solar System (Wilde et al., 2001). Younger 4.3 to 3.8 Ga zircon grains, coeval with the late heavy bombardment, are detected as detrital grains in 3 Ga metasediments from Western Australia. They are characterised by high LREE contents, consistent with growth in evolved granitic magmas (Peck et al., 2001). This conclusion is currently questioned, as the zircon crystals collected in the lunar highland anorthositic crust have about the same age and document the same type of LREE overabundance (Whitehouse and Kamber, 2002).

A second outstanding feature of the 4.4 to 3.8 Ga zircon crystals is the range of oxygen isotope ratios from 5.4 to 15‰, corresponding to magmatic $\delta^{18}O$ values from about 7 up to 11‰ (Mojzsis et al., 2001). There is no known primitive or mantle reservoir of this composition (Wilde et al., 2001). High $\delta^{18}O$ values in terrestrial zircon show that liquid hydrosphere was present and low-temperature surficial processes, such as weathering, alteration, and diagenesis, have occurred, shortly following the giant impact that formed the Moon (Peck et al., 2001). Fairly constant $\delta^{18}O$ values throughout the Archean, above the mean value of mantle zircon (5.3‰ in xenocrysts hosted by kimberlite), suggest uniform processes and conditions conducive to the presence of liquid-water oceans (Valley et al., 2002).

Albeit unexpected because of continuing heavy bombardment between 4.4 and 3.8 Ga, ancient continental crust materials, having interacted with a liquid hydrosphere, could escape volatilisation and are still preserved at the Earth's surface. The earliest surviving continental crust, the Acasta gneisses in the western Slave Province of Canada, resembles other Archean gneiss complexes and is not unusual in any way. It includes a 4031 ± 3 Ma metagranodiorite and slightly younger, yet older than 4.0 Ga, metatonalites (Bowring and Williams, 1999). Zircon-inherited cores imply that even older gneisses can await discovery. Bulk rock $\epsilon_{Nd(t)}$ ranging from +4 to −4 suggest that the parental magmas were formed by interaction of mantle-derived magmas with a preexisting zircon-bearing crust. Such evidence means that mantle–crust differentiation started very soon after crystallisation of the magma ocean.

3. The current granite paradigm

Although seemingly chaotic, the general progress of any branch of science is defined by three stages of development (Kuhn, 1970; summary in Clarke, 1996):

(i) Preconsensus, preparadigm, or prenormal stage. No single theory is capable of explaining the observations. In this type of world, each observation can have an explanation entirely independent of that for any other observation.

(ii) Normal or paradigm-governed stage. A branch of science operates for relatively long periods of time within an accepted framework called a 'paradigm'. A good paradigm should be simple, internally consistent, broad in scope. and capable of successful applications and predictions. Any unsolved problem can be set aside, either because the solution exists and the researcher has not found it or because the solution does not exist within the current paradigm and, therefore, it becomes an 'anomaly'. Such anomalies accumulate with time.

(iii) Revolution. After a relatively short period of crisis, the old paradigm is overthrown in favour of a new one which not only accounts for the anomalies under the old paradigm, but also has a better explanatory and predictive power. The new, postrevolution, paradigm incorporates much of the vocabulary and apparatus of the old one, but rarely in the same way.

Two main paradigms were successively powerful in earth sciences: uniformitarism and plate tectonics. In this general frame, each discipline developed its own accepted framework and vocabulary that are not always easy to grasp for the other disciplines. Hutton, in 1795, brought up the first major revolution in granite geology by offering the plutonist paradigm which lasted for about 150 years, until the accumulating anomalies forced a major change in thinking from plutonism to magmatism.

Since the middle of the 20th century, the magmatic paradigm has satisfactorily governed the granite scientific community. Granite compositions correspond to thermal minima in experimental silicate systems, which means that they can be produced through different ways, the main processes capable of generating the different granite types being partial melting and magma differentiation (Fig. 1).

The 'Current Granite Paradigm', discussed during the 3rd Hutton Symposium on Granites held in Maryland in 1995 states first: "Members of the granite family... are magmatic rocks that form indirectly by partial melting of preexisting rocks such as mantle peridotite (thereby requiring subsequent compositional modification) or directly by partial melting of mantle-derived mafic igneous rocks, felsic igneous rocks, and crustal siliciclastic metasedimentary rocks... (wackes and pelites), or combinations of these types of source materials" (Clarke, 1996, p. 356). Direct generation of granite by partial melting of mantle peridotite is clearly discarded. In Fig. 1, the different processes yielding primary, secondary, and tertiary granitoids (Clarke, 1992) are marked by dashed lines, with direct derivation from mantle topped by a question mark. The major reason generally advocated is that such a process would require an unrealistically low degree of melting of peridotite, especially if depleted, and preclude any large-scale mobility for the supposedly viscous magma.

However, as shown hereafter, in a large range of terrestrial planetary bodies, granite-like liquids, now

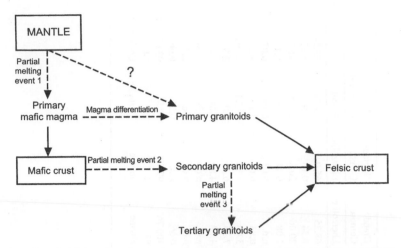

Fig. 1. Genetic classification of granitoids based on the number of partial melting events from an original peridotite mantle source (modified from Clarke, 1992). Solid lines connect magmas and their solid products in terms of crustal types. Dashed lines indicate partial melting and magma differentiation processes. Direct connection of mantle with granitoids by a single partial melting event is indicated by the dashed line topped by the question mark.

converted into glass or silica mineral–feldspar intergrowths, did occur in equilibrium with and trapped within upper mantle minerals. The upper mantle-granite connection constitutes an 'anomaly' to the first statement of the paradigm. As such, it deserves further examination.

4. Silicic glasses in Earth's mantle and mantle-derived magmas

Terrestrial planetary mantles have compositions derived from the primitive solar nebula, imaged by CI carbonaceous chondrites (McDonough and Sun, 1995). Modal compositions of the upper mantle are determined by olivine±orthopyroxene±clinopyroxene± Al-bearing accessory mineral assemblages. Experimental data show that liquids capable of being extracted from static and isotropic peridotite could correspond to melting amounts as low as ~0.1 vol.% (McKenzie, 1989). By contrast, in deformation experiments at high stress and strain rates, the liquid distribution is largely stress controlled and results predict retention up to a few vol.% in parts of the mantle without lattice preferred orientation or where the preferred orientation tend to trap liquid (Faul, 2000). The picritic to silicic liquids can be trapped either inside the pores in the crystalline framework, or within the

crystals growing during deformation. Cooling of silicic liquid inclusions during ascent and exhumation of upper mantle materials results in development of glass, or of [quartz/tridymite+feldspar] intergrowths.

4.1. Silicic glasses in natural samples from the mantle of the Earth

The Earth's upper mantle xenoliths in volcanic provinces are widespread, especially in the within-plate areas, where mantle metasomatism is portrayed by secondary melt and fluid inclusions forming trails along healed fracture planes in minerals. Melt inclusions from ultramafic peridotite from both oceanic and continental regions show that volatile-rich, silicic metasomatic glasses are present throughout the lithosphere (Table 1). The compositions of silicic glasses and their derived crystalline products differ strikingly from the erupted, mantle-derived mafic magmas and are more akin to the continental than to the oceanic crust (Schiano and Clocchiatti, 1994).

In oceanic areas, e.g., Tahaa Island (Schiano et al., 1992), Kerguelen Archipelago (Schiano et al., 1994), Canary Islands (Neumann and Wulff-Pedersen, 1997), and Grande Comore (Schiano et al., 1998), glasses trapped by mantle olivine contain in average 60–64 wt.% SiO_2 and ~10 wt.% total alkali oxides, with some compositions up to 71 wt.% SiO_2. They

Table 1
Selected compositions of silicic glass inclusions within Earth's mantle rocks and mantle-derived rocks

Sample locality	Tahaa[a]	Kerguelen Archipelago[b]	Canary Islands La Palma[c]				Lanzarote	Grande Comore[d]	Nógrád-Gömör[e]	Skien (Oslo Rift)[f]	
Host rock	Lherzolite	Harzburgite	Harzburgite	Harzburgite	Harzburgite	Lherzolite	Harzburgite	Lherzolite	Lherzolite	Basalt	
SiO$_2$	62.84	60.69	68.19	63.94	64.51	67.48	70.51	63.25	65.0	45.54	55.54
TiO$_2$	0.69	0.81	0.24	0.18	0.47	0.30	0.07	0.34	0.16	3.97	1.22
Al$_2$O$_3$	16.33	20.37	15.68	17.14	16.91	14.67	17.59	20.06	20.8	12.85	11.10
FeO*[g]	3.47	0.89	1.68	1.76	2.32	1.61	1.28	1.08	0.41	12.80	5.47
MnO	0.02	0.03	0.07	0.03	0.06	0.04	n.d.	n.d.	0.03	0.19	0.07
MgO	3.04	1.71	1.59	1.15	1.60	1.49	0.87	0.83	0.01	7.19	8.68
CaO	1.61	4.56	1.21	1.02	1.85	1.33	4.52	3.00	1.90	13.24	11.93
Na$_2$O	4.17	5.15	6.62	6.91	6.88	6.61	3.90	7.35	6.23	2.07	2.88
K$_2$O	6.81	4.35	4.35	5.10	4.64	4.43	1.58	4.19	4.89	1.64	1.41
P$_2$O$_5$	0.02	0.50	0.08	0.14	0.22	0.68	0.03	n.d.	n.d.	0.39	0.06
H$_2$O	n.d.	1.20	n.d.	n.d.	n.d.	n.d.	n.d.	n.d.	n.d.	n.d.	n.d.
Cl	0.15	0.12	0.08	0.12	0.15	0.12	0.11	0.66	n.d.	0.01	0.06
Total	99.15	100.38	99.79	97.49	99.62	98.18	100.46	100.76	99.43	99.81	98.68
mg#[h]	61	77	63	41	55	62	48	58	4	50	74
Host mineral	olivine	olivine	interstitial	ortho-pyroxene	olivine	ortho-pyroxene	ortho-pyroxene	olivine	olivine	Ti-augite	Cr-diopside

n.d.—not determined.

[a] Tahaa Island (French Polynesia): data from Schiano et al. (1992).
[b] Kerguelen Archipelago: data from Schiano et al. (1994).
[c] Canary Islands: data from Neumann and Wulff-Pedersen (1997).
[d] Grande Comore: data from Schiano et al. (1998).
[e] Nógrád-Gömör: data from Szabó et al. (1996).
[f] Skien (Oslo Rift): data from Kirstein et al. (2002).
[g] Compositions expressed in wt.% FeO*=total iron as FeO (no correction for the presence of Fe$_2$O$_3$).
[h] mg#=100 MgO/(MgO+FeO*) (M).

commonly coexist with carbonates, as glass and minerals, and/or aqueous brines associated with CO_2 (e.g., Tenerife, Frezzotti et al., 2002), as well as with mafic glass of basaltic composition, related to the coeval xenolith-bearing mantle-derived magmas.

In continental areas, evidence for silicic trends is represented by intermediate glass trapped by mantle olivine crystals (e.g., Hungary, Szabó et al., 1996), monzodiorite to syenite either disrupted in mantle-derived magmas (e.g., Scotland, Aspen et al., 1990) or occurring as veinlets cross-cutting peridotite (Romania, Chalot-Prat and Arnold, 1999). In most cases, silicic products coexist with more mafic compositions and crystalline søvite.

4.2. Silicic glasses in rock-forming minerals of mantle-derived magmas

On the Earth, rock-forming phenocrysts (e.g., olivine, clinopyroxene) crystallising from mantle-derived magmas contain melt inclusions. Their compositions range from ultrabasic to basic to intermediate, some are similar to the enclosing magma, the others evidence liquids coming from other sources and replenishing deep-seated evolving magma chambers (e.g., Kirstein et al., 2002). The paucity of more silicic compositions is especially noteworthy, compared with the mantle xenoliths evidence (Table 1).

5. Silicic glasses in terrestrial planetary mantles and mantle-derived magmas

Meteorites are subdivided into undifferentiated and differentiated families (Bischoff, 2001). The undifferentiated family comprises four distinct classes of chondrite, defining 12 groups and may include three classes of primitive achondrite. The differentiated family is made up of five classes of metal-poor achondrite, two classes of stony-iron meteorite, and seven main classes of iron meteorite. The Martian SNC and the lunar meteorites form two supplementary classes that are generally included within the achondrites.

During the 1960s and 1970s, pioneering isotope studies have been performed on meteorites to cast light on the time sequences in the formation of planetary bodies and their evolution. Owing to the specificity of Rb–Sr and K–Ar isotope systematics, alkali-bearing material was systematically searched in meteorites, even the iron ones, to obtain reliable ages. As a result, extraterrestrial "granites" (for a review, see Bonin et al., 2002) were known well before their discovery on the Moon (Shih et al., 1993). They make up clasts in ordinary chondrites and, more surprisingly, enclaves in iron meteorites. Whatever their ultimate origin, and despite their scarcity and small sizes, the granitic material observed in contrasting types of meteorites shed light on the feasibility of highly evolved liquids of alkali feldspar granite composition in the asteroid belt.

5.1. Silicate enclaves within iron meteorites

Alkali-rich, silicate inclusions in the IIE irons were the first constituents in iron meteorites for which reliable ages were obtained during the 1960s. They show diversity in mineralogy, among which Colomera, Kodaikanal, and Weekeroo Station form a group (Takeda et al., 2003). In the model of "rhyolitic plums in the pudding" proposed by Wasserburg in the 1960s, silicates (plums) crystallised from strongly differentiated silicate liquids trapped in the cooling metal (pudding) at a relatively shallow depth on the parent body after a shock event.

The Colomera IIE iron displays two types of silicate inclusions, K-rich in the surface and K-poor in the interior. A Rb–Sr isochron using silicates, phosphates, and glasses yields an age of 4.51 ± 0.04 Ga, recalculated with the decay constant of 1.42×10^{-11} a^{-1} (Sanz et al., 1970). The Colomera data are consistent with a simple evolutionary history during the formation of the Solar System. An upper limit of 47 ± 7 Ma can be set for the time interval between dispersion of the silicate crystal mush within the partly molten metal phase and final Sr isotopic equilibration. In the K-rich surface inclusion, K-feldspar (sanidine) phenocrysts up to several centimetres in length, orthopyroxene and relict olivine coexist with silica and subrounded Cr-diopside. Textural relationships suggest that a K-rich silicate liquid was present and segregated as a fluid phase that leached K from surrounding materials. The interior silicate inclusions (Table 2) are chiefly composed of

Table 2
Selected compositions of silicic inclusions within iron meteorite, terrestrial planetary mantle rocks, and mantle-derived rocks

Meteorite	Colomera[a]	Adzhi-Bogdo[b]		Bovedy[c]	Dar al Gani 489[d]		Shergotty[e]		Chassigny[f]	
Host rock	IIE iron	LL chondrite		L3 chondrite	Lherzolitic shergottite		Basaltic shergottite		Chassignite	
SiO_2	60.08	72.2	78.6	60.5	63.15	80.49	69.1	72.4	78.6	71.1
TiO_2	0.51	0.4	0.1	n.d.	0.28	0.21	0.2	0.2	0.24	0.12
Al_2O_3	10.53	14.0	11.2	10.2	21.62	10.71	17.1	17.4	11.9	16.4
FeO^{*g}	3.82	1.0	0.2	5.08	3.23	1.48	1.6	1.5	0.76	2.08
MnO	0.22	<0.1	<0.1	n.d.	n.d.	n.d.	n.d.	n.d.	0.00	0.04
MgO	7.26	0.1	<0.1	10.8	0.6	0.27	0.1	0.1	0.06	0.51
CaO	9.50	0.3	<0.1	7.16	5.40	2.56	1.4	1.6	0.14	1.24
Na_2O	5.95	2.0	0.6	4.79	3.52	1.89	2.8	4.7	2.77	3.8
K_2O	0.40	9.8	9.2	1.30	0.43	0.39	6.9	2.1	3.8	4.1
P_2O_5	n.d.	0.1	<0.1	n.d.	n.d.	n.d.	0.4	0.3	n.d.	n.d.
Total	99.91	100.0	100.3	99.8	98.22	97.96	99.6	100.3	98.3	99.4
mg#[h]	77	15	<47	79	25	25	10	11	12	30
Occurrence	inclusions in Fe-Ni alloy	millimeter-sized clasts		mesostasis in a clast	immiscible blebs in the same inclusion in olivine		inclusions in pyroxene		inclusion in chromite	inclusion in olivine

n.d.—not determined.

[a] Colomera: average of calculated compositions based on modal composition and mineral analyses (Takeda et al., 2003).
[b] Adzhi-Bogdo (Mongolia): calculated bulk composition based on modal composition and mineral analyses (Bischoff, 1993).
[c] Bovedy: data from Ruzicka et al. (1995); note the similarity of composition with the average composition of Colomera interior inclusions.
[d] Dar al Gani 489 (Libya): data from Folco et al. (2000).
[e] Shergotty: data from Shih et al. (1982).
[f] Chassigny: data from Varela et al. (2000).
[g] Compositions expressed in wt.%. FeO*=total iron as FeO (no correction for the presence of Fe_2O_3).
[h] mg#=100 MgO/(MgO+FeO*) (M).

Cr-diopside and albite, coexisting with chromite, Cl-apatite, whitlockite and albitic glasses with excess silica, and minor Fe-Mg components (Takeda et al., 2003). By comparison with the Caddo County IAB iron, the mineralogical and chemical data are consistent with a two-stage model. The original partial melt derived from an H chondrite-like source had already crystallised significant amounts of Cr-diopside and albite before it was dispersed and encapsulated by the partly molten Fe–Ni alloy.

The Weekeroo Station iron meteorite yields K-feldspar crystals and is geochemically similar (Burnett and Wasserburg, 1967a). The Kodaikanal iron meteorite is characterised by abundant up to 2-cm-long silicate enclaves containing diopside, orthopyroxene, two species of alkali feldspar, and glass of alkali feldspar composition. A well-defined Rb–Sr isochron yields a recalculated age of 3.72 ± 0.10 Ga, with an initial ratio of 0.713 ± 0.020. Kodaikanal is interpreted as a 'secondary' body produced by ill-defined differentiation process about 0.78 Ga after the formation of the Solar System (Burnett and Wasserburg, 1967b).

5.2. Silicic glasses within ordinary chondrites

The ordinary chondrite class is subdivided into H (high-iron), L (low-iron), and LL (very low-iron) groups. At least two groups yield silicic materials as clasts or mesostases of clasts.

The Adzhi-Bogdo meteorite is a breccia belonging to the LL group. The breccia contains various types of clasts, including impact glasses, some of which are K-rich, and alkali feldspar granite fragments. The alkali feldspar granite consists chiefly of K-feldspar plus either quartz, or tridymite, and minor albite, Cl-apatite, whitlockite, zircon, ilmenite, Ca-poor Fe-rich pyroxene, and a Na, Ti-bearing silicate (Bischoff et al., 1993). Calculated bulk compositions (Bischoff, 1993) are Na-poor and Si- and K-rich (Table 2).

So far, K-rich clasts are known in 14 LL chondritic breccias, but only Vishnupur and Adzhi-Bogdo yield alkali feldspar granite and associated pyroxenite clasts (Jäckel and Bischoff, 1997). Two types of K-rich clasts are identified: (i) the "coarse-grained" (~100 μm grain size) granitic clasts (up to 700 μm in length) may result from slow cooling of the chondritic parent magma differentiating ultimately into alkali feldspar

granite residual liquid and pyroxenite cumulates; and (ii) the fine-grained (~1 μm grain size) clasts (several mm in length), composed of euhedral olivine and skeletal pyroxene set in a K-rich groundmass, crystallised from liquids formed by impact melting of various mixtures of already crystallised granite and LL chondritic parent-body.

The Bovedy L3 chondrite contains a large $(4.5 \times 7 \times 4$ mm) igneous-textured clast of silica-rich orthopyroxenite, which consists of normally zoned orthopyroxene, tridymite, an intergrowth of feldspar and sodic glass, pigeonite, and small amounts of a silicic glass, chromite, augite, and Fe–Ni metal. The silicic glass is remarkably similar in composition with the average interior inclusions of the Colomera IIE iron (Table 2). The clast is interpreted as an igneous differentiate from a siliceous liquid that probably cooled in the near-surface region of the parent object, an asteroid or planetesimal formed in the vicinity of ordinary chondrites. Although the age of the clast is not determined, it was likely produced from a magma within 100 Ma of solar nebula formation because small planetesimals cool extremely quickly. Fractionated magmas can also account for most silica polymorph-bearing objects in ordinary chondrites (Ruzicka et al., 1995).

5.3. Silica minerals in HED achondrites

The howardite–eucrite–diogenite (HED) family of achondrites is probably issued from the asteroid 4Vesta (for the identification of the parent body, see references in Barrat et al., 2000; Yamaguchi et al., 2001). Eucrite meteorites are currently classified into cumulate and noncumulate (or ordinary) groups, based on their TiO_2 versus $FeO*/MgO$ contents. Noncumulate eucrites are likely to have erupted and crystallised in situ during the differentiation of a magma ocean, while cumulate eucrites could correspond to solid+trapped liquid at the bottom of the magma ocean (Barrat et al., 2000). In the case of EET90020, thermal metamorphism resulted into a hot crust (~870 °C) and was followed by a short reheating event up to 1060 °C, causing partial melting and rapid cooling, 4.51 ± 0.04 Ga ago. New minerals, such as tridymite, Cr-ulvöspinel, some pyroxene and plagioclase, crystallised from a silica-oversaturated liquid. The other noncumulative eucrites were completely reset during impact bombardment of the HED parent body less than 4 Ga ago (Yamaguchi et al., 2001).

Diogenites are orthopyroxenite cumulates believed to have originated during an extensive magmatic event at about 4.6 Ga. In the Tatahouine diogenite, orthopyroxene crystals form a mosaic texture resulting from shock effects. Inclusions within orthopyroxene of α-cristobalite, a low-pressure, high-temperature silica polymorph, locally in symplectic association with chromite, or associated with metal, can be explained by pyroxene breaking down under reducing conditions to give SiO_2+chromite+iron metal+aluminous pyroxene. Their occurrence is consistent with an early (roughly 100 Ma after the differentiation of the HED parent body) shock history occurring within a still hot orthopyroxenite (Benzerara et al., 2002). In that case, there is no need to invoke silicic liquids to generate cristobalite.

5.4. Silicic glasses in SNC Martian meteorites

The shergottite–nakhlite–chassignite (SNC) family of achondrites, with the addition of ALH84001, is thought to be made up of igneous Martian rocks based on young crystallisation ages and a close match between the gases implanted in them during the shock event and the Martian atmosphere (McSween, 1994). Shergottites are subdivided into basaltic and lherzolitic or picritic. Martian mantle samples include lherzolitic or picritic shergottites (lherzolite, harzburgite), nakhlites (clinopyroxenite, wehrlite), chassignite (dunite), and ALH84001 (orthopyroxenite).

Nakhlites and chassignite yield ages of about 1.3 Ga. Chassignite is currently represented by only one meteorite, fallen in 1815 at Chassigny (France). It consists of Fe-rich (Fo_{68}) olivine that contains glassy and partially crystallised inclusions. Glasses yield granitic compositions (Table 2), that are not modified by heating experiments at 1200 °C, suggesting that the glass cannot be a residual melt but rather is an independent component that was trapped with or without mineral phases (Varela et al., 2000).

Lherzolitic shergottites yield ages around 180 Ma. Their isotopic heterogeneity, higher than Earth's mantle, suggests assimilation of an evolved component, strongly enriched in incompatible elements and in LREE relatively to HREE, that formed at 4.5 Ga and was an oxidant, possibly reflecting hydrous alteration of a Martian crustal component or hydrous metasomatism within the Martian mantle (Borg et al., 2002). Dar al Gani 489, from Libya, offers an occurrence of silicic glasses within mantle minerals (Folco et al., 2000). Although its texture and modal composition are close to basalt, its bulk chemistry and occurrence of olivine+chromite+enstatite suggest close relationships with lherzolitic shergottites. Olivine crystals and one large chromite grain display inclusions composed of pyroxene and two compositionally different glasses showing a liquid immiscibility fabric. Blebs of alkali-poor, highly silicic glass are set within an intermediate dacitic glass (Table 2). The dacitic glass resembles the calculated sulphur-free rock of the Mars Pathfinder landing site (McSween et al., 1999, but see Foley et al., 2001) by its silica, lime, and alkali contents, but strongly differs by higher Al and lower Fe abundances. Silicic glass immiscibility appears to be common in olivine megacrysts of lherzolitic shergottites. Inclusions in LEW88516 consist of augite and two discrete glasses (65 and 95 wt.% SiO_2; Kring and Gleason, 1999). In NWA 1068, the largest olivine crystals contain inclusions of two discrete glasses (71.5 and 93 wt.% SiO_2) embedding pyroxene and sulphide grains (Barrat et al., 2002). These inclusions indicate that protracted crystallisation could produce siliceous magmas from mafic shergottite-type magmas on Mars.

The timing of crystallisation of the basaltic shergottites has long been a very controversial subject. Three meteorites provide a Sm–Nd isotope alignment at ~1.34 Ga (Shih et al., 1982), while individual meteorites yield younger dates around 180 Ma, which were interpreted either as the time of shock metamorphism (Shih et al., 1982) or as true crystallisation ages (e.g., Borg et al., 2002). The basaltic shergottites yield silicic glass (Table 2) included within pyroxene and olivine (Shih et al., 1982). In Zagami shergottite, mesostases containing fayalite, phosphates, and opaque minerals that crystallised at the very end of crystallisation provide evidence for fractional crystallisation of a single magma unit on Mars (McCoy et al., 1999). In this respect, they resemble the late-stage rhyolitic glass collected in the Alae lava lake of Hawaii (Wright and Peck, 1978).

In ALH84001, silica is included within orthopyroxene, suggesting that the parental magma that produced the rock was silica-oversaturated (Kring and Gleason, 1999). The respective crystallisation ages of ALH84001 and shergottites, 4.5 Ga versus 180 Ma, suggest that silicic igneous rocks could have been produced on Mars throughout its entire geologic history and can occur in the ancient cratered highlands and in the much younger terranes, as well—a challenge to future Martian geological studies.

6. Discussion: towards a consensus?

The review of the silicic magmas occurring within terrestrial planetary mantles reveals two contrasting groups: (i) residual liquids formed as end-products of differentiation, mainly through fractional crystallisation and/or immiscibility of mantle-derived mafic magmas (Bonin, 1996; Bonin et al., 2002); and (ii) liquids compositionally (e.g., mg#, Ti/Al ratios) in equilibrium with their mantle host-rocks. As silicic magmas are rich in incompatible elements, they could correspond at best to extremely low degrees of melting under near-solidus conditions.

6.1. Current explanatory models

Various models are offered to explain the coexistence in equilibrium of silicic liquids and mantle rocks (Neumann and Wulff-Pedersen, 1997). Some have no bearing on mantle processes, they include: (i) breakdown of hydrous minerals (e.g., amphibole) in response to decompression during transport of the xenoliths to the surface and heating by the lava; and (ii) partial melting of mantle xenoliths during short residence times (up to a few years) in magma chambers during ascent in the host magma.

The other models imply formation of silicic liquids at mantle depths. They include:

(A)   small degrees of partial melting of subducted crust, followed by percolation into the overlying depleted mantle wedge;

(B)   equilibrium or disequilibrium partial melting involving breakdown of various anhydrous phases;

(C)   partial melting of a metasomatised, hydrated, and/or carbonated, mantle;

(D)   immiscible separation of a single mantle-derived magma into coexisting silicic and carbonate liquids and/or two silicate liquids; and

(E)   infiltration by migrating metasomatic liquid/fluid phases, or basaltic magmas, genetically unrelated with the mantle wall rock, with which they react.

6.2. Future avenues to explore

Model A implies active plate tectonics. As such, it is probably irrelevant to the terrestrial planets other than the Earth, with the possible exceptions of Venus (Vorder Bruegge and Head, 1989) and early Mars (Sleep, 1994; Fairén et al., 2002).

Anhydrous experiments on fertile and depleted MORB-type anhydrous peridotite at 1.5 GPa fail to reproduce highly silicic liquid. As the degree of melting approaches 0% in fertile peridotite, the very first droplet of liquid yields an iron-rich nepheline-normative composition with no more than 53 wt.% SiO_2 and up to 8 wt.% Na_2O (Robinson et al., 1998). Thus, Model B should be discarded.

In addition to the experimental data obtained on Earth's mantle fluids (Schneider and Eggler, 1986), equilibrium and dynamic crystallisation experiments under low-pressure, water-saturated conditions on basaltic shergottite (McCoy and Lofgren, 1999; Dann et al., 2001) produce andesitic liquids resembling the Mars Pathfinder rock compositions (McSween et al., 1999), thus validating in some ways Model C.

Experiments conducted at pressures down to less than 0.1 MPa and melt inclusion studies substantiate that numerous variants of Model D can operate in the terrestrial planets (Earth, Mars, Moon, and asteroids; for a review, see Bonin et al., 2002).

The variant of Model E involving metasomatic phases has a bearing on Models C and D, and can operate at high pressures, while the variant of Model E involving only basaltic magmas seem to be progressively abandoned (Frezzotti et al., 2002).

6.3. Problems to be solved

The Earth seems to remain a special place, because granite is a significant part of its crust and that major and protracted processes of differentiation of its upper mantle must have operated to produce so voluminous silicic material. The meteoritic record does not provide evidence for so large amounts of silicic products in the other terrestrial planets and the asteroid belt.

The apparent relationship between planet size and time duration of igneous activity is explained on the basis of insulating capacity of large bodies that permits retention of heat produced by the decay of long-lived radionuclides (McSween, 1994). Like the Earth, Venus (Namiki and Solomon, 1994) and Mars (Greeley and Schneid, 1991; Hartmann et al., 1999) are capable of igneous activity today. On the contrary, duration of igneous activity on Mercury, Moon, and the HED parent body should have been much shorter. In all cases except Mercury, silicic material was observed, implying very short time scales to produce felsic liquids by magma differentiation and/or partial melting.

The problem consists of how representative the occurrence of granite-like inclusions is in terms of abundances and direct mantle involvement. We agree with the point raised by Clarke (written communication, 2003) that a silicic inclusion does not make a batholith or a continental crust. As far as scaling is concerned, micrometre- to millimetre-sized silicic inclusions within cm-sized crystals correspond to volume ratios of 10^{-9}–10^{-3} and 1 dm^3 of granite-like material can be included within 1 m^3 of mafic to ultramafic material. In 100-km-thick lithosphere, 100-m-thick silicic sheets and layers may be envisaged, although large pieces of felsic rocks have not yet been identified in the meteoritic record.

The Earth exsolved large volumes of granite, mostly as a constituent of continental crust which probably began to form only ~160 Ma after the beginning of the Solar System (Wilde et al., 2001). To the present knowledge, the other terrestrial planets seem to have retained most of their silicic matter as inclusions and enclaves, although the occurrence on the surface of large-scale silicic formations cannot be ruled out, at least in the case of Mars (Scott and Tanaka, 1982) and Venus (Petford, 2001).

Silicic material plays a significant role in lithosphere buoyancy, so that subduction of continental crust thicker than ~12–25-km- into ~100–200-km-thick lithosphere is rare (Cloos, 1993). Plate tectonics are commonly said to be currently lacking in terrestrial planets other than the Earth (e.g., Turcotte, 1996), although it is not ruled out that they could have

operated in older times. To what extent that feature is due to protracted accumulation at depths of silicic pockets and sheets within planetary lithospheres remains a matter of speculation.

7. Conclusions

Silicic material formed very early in the history of the terrestrial planets, probably within 100 Ma of the solar nebula formation. However, the first granites to have crystallised in the Solar System elsewhere than the Earth still await discovery. The oldest granite-forming minerals (~4.4 Ga) and granites (~4.0 Ga) detected on Earth suggest that mantle-derived magmas interacted with a preexisting crust, of probable tonalitic composition. Accordingly, the current granite paradigm states that granite is not a direct product from the mantle, but needs a crustal history (reservoirs, sources, etc.).

However, apparently conflicting with the paradigm, silicic material does occur within the mantle of the terrestrial planets, even in the smallest asteroidal bodies, as shown by natural samples analysed from the Earth, Moon, and Mars, and by ample meteoritic evidence. Various explanations are offered, mostly based on melt inclusion studies. As most models are not fully substantiated by experimental data, they remain elusive and speculative.

The volume of granite enclosed in a volume unit of solid mantle is lower by several orders of magnitude than that issued from mantle-derived magmas (juvenile crust) or extracted from the recycled crust. A challenge is to determine the actual amounts of granitic material that can be trapped in the mantle of the terrestrial planets, and the role they can play in lithosphere buoyancy.

The granite-upper mantle connection can be regarded as an anomaly to the current granite paradigm. Does it mean that the paradigm should be abandoned now? As long as no experimental evidence for direct derivation of silicic liquids from mantle peridotite is offered, there are no other alternatives based on firm grounds. It is therefore suggested to amend slightly the paradigm by stating that most conditions prevailing in the formation of silicic liquids are likely to occur also at mantle depths. The future will tell whether the paradigm will prevail.

Acknowledgements

By launching in 1995 the granite-research list network, Barrie Clarke did the wonderful job of electronically connecting granitologists throughout the world and we owe a great debt to him. His publication of the Current Granite Paradigm prompted us to reexamine occurrences of granitic material in the Earth, the terrestrial planets, and the asteroid belt. We benefited from careful reviews by Barrie Clarke and Sergio Rocchi and useful editorial remarks by Tapani Rämö. However, they should not be held responsible for the ideas and prejudices expressed here, which are our own. Last, we would like to express our admiration for the half-century work performed by Ilmari Haapala in the granitic realm, especially in that fascinating part dedicated to granite-mantle connections.

References

Aspen, P., Upton, B.G.J., Dickin, A.P., 1990. Anorthoclase, sanidine and associated megacrysts in Scottish alkali basalts: high-pressure syenitic debris from upper mantle sources? European Journal of Mineralogy 2, 503–517.

Barrat, J.A., Blichert-Toft, J., Gillet, P., Keller, F., 2000. The differentiation of eucrites: the role of in situ crystallization. Meteoritics and Planetary Science 35, 1087–1100.

Barrat, J.A., Jambon, A., Bohn, M., Gillet, P., Sautter, V., Göpel, C., Lesourd, M., Keller, F., 2002. Petrology and chemistry of the picritic shergottite North West Africa 1068 (NWA 1068). Geochimica et Cosmochimica Acta 66, 3505–3518.

Benzerara, K., Guyot, F., Barrat, J.A., Gillet, P., Lesourd, M., 2002. Cristobalite inclusions in the Tatahouine achondrite: implications for shock conditions. American Mineralogist 87, 1250–1256.

Bischoff, A., 1993. Alkali–granitoids as fragments within the ordinary chondrite Adzhi-Bogdo: evidence for highly fractionated, alkali–granitic liquids on asteroids. Lunar and Planetary Science Conference XXIV, Abstracts, pp. 113–114.

Bischoff, A., 2001. Meteorite classification and the definition of new chondrite classes as a result of successful meteorite search in hot and cold deserts. Planetary and Space Science 49, 769–776.

Bischoff, A., Geiger, T., Palme, H., Spettel, B., Schultz, L., Scherer, P., Schlüter, J., Lkhamsuren, J., 1993. Mineralogy, chemistry, and noble gas contents of Adzhi-Bogdo—an LL3-6 chondritic breccia with L-chondritic and granitoidal clasts. Meteoritics 28, 570–578.

Bonin, B., 1996. A-type granite ring complexes: mantle origin through crustal filters and the anorthosite–rapakivi magmatism connection. In: Demaiffe, D. (Ed.), Petrology and Geochemistry of Magmatic Suites of Rocks in the Continental and Oceanic Crusts. ULB-MRAC, Bruxelles, pp. 201–217. A volume dedicated to Professor J. Michot.

Bonin, B., Bébien, J., Masson, P., 2002. Granite: a planetary point of view. Gondwana Research 5, 261–273.

Borg, L.E., Nyquist, L.E., Wiesmann, H., Reese, Y., 2002. Constraints on the petrogenesis of Martian meteorites from the Rb–Sr and Sm–Nd isotopic systematics of the lherzolitic shergottites ALH77005 and LEW88516. Geochimica et Cosmochimica Acta 66, 2037–2053.

Bowring, S.A., Williams, I.S., 1999. Priscoan (4.00–4.03 Ga) orthogneisses from northwestern Canada. Contributions to Mineralogy and Petrology 134, 3–16.

Burnett, D.S., Wasserburg, G.J., 1967a. $^{87}Rb–^{87}Sr$ ages of silicate inclusions in iron meteorites. Earth and Planetary Science Letters 2, 397–408.

Burnett, D.S., Wasserburg, G.J., 1967b. Evidence for the formation of an iron meteorite at 3.8×10^9 years. Earth and Planetary Science Letters 2, 137–147.

Chalot-Prat, F., Arnold, M., 1999. Immiscibility between calcio-carbonatitic and silicate melts and related wall rock reactions in the upper mantle: a natural case study from Romanian mantle xenoliths. Lithos 46, 627–659.

Clarke, D.B., 1992. Granitoid Rocks, Topics in the Earth Sciences, vol. 7. Chapman & Hall, London. 283 pages.

Clarke, D.B., 1996. Two centuries after Hutton's 'Theory of the Earth': the status of granite science. Transactions of the Royal Society of Edinburgh. Earth Sciences 87, 353–359.

Cloos, M., 1993. Lithospheric buoyancy and collisional orogenesis: subduction of oceanic plateaus, continental margins, island arcs, spreading ridges, and seamounts. Geological Society of America Bulletin 105, 715–737.

Dann, J.C., Holzheid, A.H., Grove, T.L., McSween Jr., H.Y., 2001. Phase equilibria of the Shergotty meteorite: constraints on pre-eruptive water contents of Martian magmas and fractional crystallization under hydrous conditions. Meteoritics and Planetary Science 36, 793–806.

Dreibus, G., Wänke, H., 2002. Comment on: "Comparative geochemistry of basalts from the Moon, Earth, HED asteroid, and Mars: implications for the origin of the Moon", by Ruzicka, A., Snyder, G.A., Taylor, L.A., 2001, Geochimica et Cosmochimica Acta 65, 979–997. Geochimica et Cosmochimica Acta 66, 2631–2632.

Dressler, B.O., Reimold, W.U., 2001. Terrestrial impact melt rocks and glasses. Earth-Science Reviews 56, 205–284.

Encrenaz, T., Bibring, J.P., 1987. Le Système Solaire. Savoirs Actuels. InterEditions/Editions du CNRS, Paris.

Fairén, A., Ruiz, J., Anguita, F., 2002. An origin for the linear magnetic anomalies on Mars through accretion of terranes: implications for dynamo timing. Icarus 160, 220–223.

Faul, U., 2000. Melt distribution and grain growth in polycrystalline olivine. In: Bagdassarov, N., Laporte, D., Thompson, A.B. (Eds.), Physics and Chemistry of Partially Molten Rocks. Kluwer Academic Publishers, Dordrecht, pp. 67–92.

Folco, L., Franchi, I.A., D'Orazio, M., Rocchi, S., Schultz, L., 2000. A new Martian meteorite from the Sahara: the shergottite Dar al Gani 489. Meteoritics and Planetary Science 35, 827–839.

Foley, C.N., Economou, T.E., Clayton, R.N., 2001. Chemistry of Mars Pathfinder samples determined by the APXS. Lunar and Planetary Science Conference XXXII, Abstract, pp. 1979. pdf [CD-ROM].

Frezzotti, M.L., Andersen, T., Neumann, E.-R., Simonsen, S.L., 2002. Carbonatite melt-CO_2 fluid inclusions in mantle xenoliths from Tenerife, Canary Islands: a story of trapping, immiscibility and fluid–rock interaction in the upper mantle. Lithos 64, 77–96.

Greeley, R., Schneid, B.D., 1991. Magma generation on Mars: amounts, rates, and comparisons with Earth, Moon, and Venus. Science 254, 996–998.

Halliday, A.N., 2000. Terrestrial accretion rates and the origin of the Moon. Earth and Planetary Science Letters 176, 17–30.

Hartmann, W.K., Malin, M., McEwen, A., Carr, M., Soderblom, L., Thomas, P., Danielson, E., James, P., Veverka, J., 1999. Evidence for recent volcanism on Mars from crater counts. Nature 397, 586–589.

Jäckel, A., Bischoff, A., 1997. Potassium-rich fragments in LL-chondritic breccias. Meteoritics and Planetary Science 32, A66.

Kirstein, L.A., Dunworth, E.A., Nikogosian, I.K., Touret, J.L.R., Lustenhouwer, W.J., 2002. Initiation of melting beneath the Oslo Rift: a melt inclusion perspective. Chemical Geology 183, 221–236.

Kring, D.A., Gleason, J.D., 1999. Siliceous igneous rocks on Mars. Lunar and Planetary Science Conference XXX, Abstract, pp. 1611. pdf [CD-ROM].

Kuhn, T., 1970. The Structure of Scientific Revolutions, 2nd ed. International Encyclopedia of Unified Science 2, no. 2. 210 pp.

Kurat, G., Koeberl, C., Presper, T., Brandstätter, F., Maurette, M., 1994. Petrology and geochemistry of Antarctic meteorites. Geochimica et Cosmochimica Acta 58, 3879–3904.

Lee, D.-C., Halliday, A.N., Snyder, G.A., Taylor, L.A., 1997. Age and origin of the Moon. Science 278, 1098–1103.

Lee, D.-C., Halliday, A.N., Leya, I., Wieler, R., Wiechert, U., 2002. Cosmogenic tungsten and the origin and earliest differentiation of the Moon. Earth and Planetary Science Letters 198, 267–274.

Maurette, M., Olinger, C., Christophe Michel-Lévy, M., Kurat, G., Pourchet, M., Brandstätter, F., Bourot-Denise, M., 1991. A collection of diverse micrometeorites recovered from 100 tonnes of Antarctic blue ice. Nature 351, 44–47.

McCoy, T.J., Lofgren, G.E., 1999. Crystallization of the Zagami shergottite: an experimental study. Earth and Planetary Science Letters 173, 397–411.

McCoy, T.J., Wadhwa, M., Keil, K., 1999. New lithologies in the Zagami meteorite: evidence for fractional crystallization of a single magma unit on Mars. Geochimica et Cosmochimica Acta 63, 1249–1262.

McDonough, W.F., Sun, S.-S., 1995. The composition of the Earth. Chemical Geology 120, 223–253.

McKenzie, D., 1989. Some remarks on the movement of small melt fractions in the mantle. Earth and Planetary Science Letters 95, 53–72.

McLennan, S.M., 2003. Sedimentary silica on Mars. Geology 31, 315–318.

McSween Jr., H.Y., 1994. What we have learned about Mars from SNC meteorites. Meteoritics 29, 757–779.

McSween Jr., H.Y., et al., 1999. Chemical, multispectral, and textural constraints on the composition and origin of rocks at the Mars Pathfinder landing site (19 authors). Journal of Geophysical Research 104, 8679–8715.

Melosh, H.J., 1990. Giant impacts and the thermal state of the early earth. In: Newsom, H.E., Jones, J.H. (Eds.), Origin of the Earth. Oxford University Press, Oxford, pp. 69–83.

Mojzsis, S.J., Harrison, T.M., Pidgeon, R.T., 2001. Oxygen-isotope evidence from ancient zircons for liquid water at the Earth's surface 4,300 Myr ago. Nature 409, 178–181.

Morbidelli, A., Chambers, J., Lunine, J.I., Petit, J.M., Robert, F., Valsecchi, G.B., Cyr, K.E., 2000. Source regions and timescales for the delivery of water to Earth. Meteoritics and Planetary Science 36, 371–380.

Namiki, N., Solomon, S.C., 1994. Impact crater densities on volcanoes and coronae on Venus: implications for volcanic resurfacing. Science 265, 929–933.

Neumann, E.-R., Wulff-Pedersen, E., 1997. The origin of highly silicic glass in mantle xenoliths from the Canary Islands. Journal of Petrology 38, 1513–1539.

O'Neill, H.S.C., 1991. The origin of the Moon and the early history of the Earth—a chemical model: Part 1. The Moon. Geochimica et Cosmochimica Acta 55, 1135–1157.

Peck, W.H., Valley, J.W., Wilde, S.A., Graham, C.M., 2001. Oxygen isotope ratios and rare earth elements in 3.3 to 4.4 Ga zircons: ion microprobe evidence for high $\delta^{18}O$ continental crust and oceans in the Early Archean. Geochimica et Cosmochimica Acta 65, 4215–4229.

Petford, N., 2001. Dyke widths and ascent rates of silicic magmas on Venus. Transactions of the Royal Society of Edinburgh. Earth Sciences 91, 87–95.

Robinson, J.A.C., Wood, B.J., Blundy, J.D., 1998. The beginning of melting of fertile and depleted peridotite at 1.5 GPa. Earth and Planetary Science Letters 155, 97–111.

Ruzicka, A., Kring, D.A., Hill, D.H., Boynton, W.V., Clayton, R.N., Mayeda, T.K., 1995. Silica-rich orthopyroxenite in the Bovedy chondrite. Meteoritics 30, 57–70.

Ruzicka, A., Snyder, G.A., Taylor, L.A., 2001. Comparative geochemistry of basalts from the Moon, Earth, HED asteroid, and Mars: implications for the origin of the Moon. Geochimica et Cosmochimica Acta 65, 979–997.

Sanz, H.G., Burnett, D.S., Wasserburg, G.J., 1970. A precise $^{87}Rb/^{87}Sr$ age and initial $^{87}Sr/^{86}Sr$ for the Colomera iron meteorite. Geochimica et Cosmochimica Acta 34, 1227–1239.

Schiano, P., Clocchiatti, R., 1994. Worldwide occurrence of silica-rich melts in sub-continental and sub-oceanic mantle minerals. Nature 368, 621–624.

Schiano, P., Clocchiatti, R., Joron, J.L., 1992. Melt and fluid inclusions in basalts and xenoliths from Tahaa Island, Society Archipelago: evidence for a metasomatized upper mantle. Earth and Planetary Science Letters 111, 69–82.

Schiano, P., Clocchiatti, R., Shimizu, N., Weis, D., Mattielli, N., 1994. Cogenetic silica-rich and carbonate-rich melts trapped in mantle minerals in Kerguelen ultramafic xenoliths: implications for metasomatism in the oceanic upper mantle. Earth and Planetary Science Letters 123, 167–178.

Schiano, P., Bourdon, B., Clocchiatti, R., Massare, D., Varela, M.E., Bottinga, Y., 1998. Low-degree partial melting trends recorded in upper mantle minerals. Earth and Planetary Science Letters 160, 537–550.

Schneider, M.E., Eggler, D.H., 1986. Fluids in equilibrium with peridotite minerals: implications for mantle metasomatism. Geochimica et Cosmochimica Acta 50, 711–724.

Scott, D.H., Tanaka, K.L., 1982. Ignimbrites of Amazonis Planitia region of Mars. Journal of Geophysical Research 87, 9839–9851.

Shih, C.-Y., Nyquist, L.E., Bogard, D.D., McKay, G.A., Wooden, J.L., Bansal, B.M., Wiesmann, H., 1982. Chronology and petrogenesis of young achondrites, Shergotty, Zagami, and ALHA77005: late magmatism on a geologically active planet. Geochimica et Cosmochimica Acta 46, 2323–2344.

Shih, C.-Y., Nyquist, L.E., Wiesmann, H., 1993. K–Ca chronology of lunar granites. Geochimica et Cosmochimica Acta 57, 4827–4841.

Sleep, N.H., 1994. Martian plate tectonics. Journal of Geophysical Research 99, 5639–5655.

Szabó, C., Bodnar, R.J., Sobolev, A.V., 1996. Metasomatism associated with subduction-related, volatile-rich silicate melt in the upper mantle beneath the Nógrád-Gömör field, northern Hungary/southern Slovakia: evidence from silicate melt inclusions. European Journal of Mineralogy 8, 881–899.

Takeda, T., Hsu, W., Huss, G.E., 2003. Mineralogy of silicate inclusions of Colomera IIE iron and crystallization of Cr–diopside and alkali feldspar from a partial melt. Geochimica et Cosmochimica Acta 67, 2269–2288.

Turcotte, D.L., 1996. Magellan and comparative planetology. Journal of Geophysical Research 101, 4765–4773.

Valley, J.W., Peck, W.H., King, E.M., Wilde, S.A., 2002. A cool early Earth. Geology 30, 351–354.

Varela, M.E., Kurat, G., Bonnin-Mosbah, M., Clocchiatti, R., Massare, D., 2000. Glass-bearing inclusions in olivine of the Chassigny achondrite: heterogeneous trapping at sub-igneous temperatures. Meteoritics and Planetary Science 35, 39–52.

Vorder Bruegge, R.W., Head, J.W., 1989. Fortuna Tessera, Venus: evidence of horizontal convergence and crustal thickening. Geophysical Research Letters 16, 699–702.

Wedepohl, K.H., 1991. Chemical composition and fractionation of the continental crust. Geologische Rundschau 80, 207–223.

Wells, L.E., Armstrong, J.C., Gonzalez, G., 2003. Reseeding of early Earth by impacts of returning ejecta during the late heavy bombardment. Icarus 162, 38–46.

Wetherill, G.W., 1994. Provenance of the terrestrial planets. Geochimica et Cosmochimica Acta 58, 4513–4520.

Wetherill, G.W., Stewart, G.R., 1993. Formation of planetary embryos: effects of fragmentation, low relative velocity, and independent variation of eccentricity and inclination. Icarus 106, 190–209.

Whitehouse, M.J., Kamber, B.S., 2002. On the overabundance of light rare earth elements in terrestrial zircons and its implication for Earth's earliest magmatic differentiation. Earth and Planetary Science Letters 204, 333–346.

Wilde, S.A., Valley, J.W., Peck, W.H., Graham, C.M., 2001. Evidence from detrital zircons for the existence of continental crust and oceans on the Earth 4.4 Gyr ago. Nature 409, 175–178.

Wright, T.L., Peck, D.L., 1978. Crystallization and differentiation of the Alae magma, Alae lava lake, Hawaii. Solidification of Alae Lava Lake, Hawaii, U.S. Geological Survey Professional Paper, vol. 935-C, pp. 1–20.

Yamaguchi, A., Taylor, G.J., Keil, K., Floss, C., Crozaz, G., Nyquist, L.E., Bogard, D.D., Garrison, D.H., Reese, Y.D., Wiesmann, H., Shih, C.-Y., 2001. Post-crystallization reheating and partial melting of eucrite EET90020 by impact into the hot crust of asteroid 4 Vesta ~4.50 Ga ago. Geochimica et Cosmochimica Acta 65, 3577–3599.

Zhao, Y., 2002. The age and accretion of the Earth. Earth-Science Reviews 59, 235–263.

Available online at www.sciencedirect.com

SCIENCE DIRECT®

Lithos 80 (2005) 147–154

ELSEVIER

LITHOS

www.elsevier.com/locate/lithos

Continental rift systems and anorogenic magmatism

James W. Sears*, Gregory M. St. George, J. Chris Winne

University of Montana, Missoula MT 59812, USA

Received 2 April 2003; accepted 9 September 2004
Available online 24 November 2004

Abstract

Precambrian Laurentia and Mesozoic Gondwana both rifted along geometric patterns that closely approximate truncated-icosahedral tessellations of the lithosphere. These large-scale, quasi-hexagonal rift patterns manifest a least-work configuration. For both Laurentia and Gondwana, continental rifting coincided with drift stagnation, and may have been driven by lithospheric extension above an insulated and thermally expanded mantle. Anorogenic magmatism, including flood basalts, dike swarms, anorthosite massifs and granite-rhyolite provinces, originated along the Laurentian and Gondwanan rift tessellations. Long-lived volcanic regions of the Atlantic and Indian Oceans, sometimes called hotspots, originated near triple junctions of the Gondwanan tessellation as the supercontinent broke apart. We suggest that some anorogenic magmatism results from decompression melting of asthenosphere beneath opening fractures, rather than from random impingement of hypothetical deep-mantle plumes.
© 2004 Elsevier B.V. All rights reserved.

Keywords: Rift; Gondwana; Laurentia; Anorogenic magma; Plume

1. Introduction

The ultimate causes and controls of continental rifting and anorogenic magmatism remain controversial questions in geotectonics. Geodynamicists commonly model anorogenic magmatism as the result of impingement of hypothetical deep-mantle plumes on the lithosphere, with continental rifting as a secondary consequence of lithospheric doming above the plume head (cf. Wilson, 1963; Morgan, 1981; Bijwaard and Spakman, 1999; Campbell, 2001; Condie, 2001; Storey et al., 2001; Turcotte and Schubert, 2002). However, in a serious challenge to the deep-mantle plume paradigm, Hamilton (2002, 2003) asserts that the endothermic phase boundary at 660-km depth isolates lower and upper mantle convection, and that propagation of lithospheric cracks triggers anorogenic magmatism through decompression melting of ordinary asthenosphere. Similarly, Anderson (1982, 2001, 2002) takes a top-down tectonic viewpoint, in which a stagnant supercontinent insulates the underlying mantle, leading to thermal expansion, partial melting and a geoid bulge. The lithosphere then fractures over

* Corresponding author. Tel.: +1 406 243 2341; fax: +1 406 243 4028.
E-mail address: jwsears@selway.umt.edu (J.W. Sears).

0024-4937/$ - see front matter © 2004 Elsevier B.V. All rights reserved.
doi:10.1016/j.lithos.2004.05.009

this bulge and anorogenic magma injects propagating rift zones.

Here, we show that continental rift zones and associated anorogenic magmatic provinces in some cases approximate a mathematically precise, large-scale tessellation, the truncated icosahedron. This quasi-hexagonal fracture pattern manifests a least-work configuration that minimizes perimeter, area and energy (cf. Anderson, 2002). The tessellation appears to be a brittle phenomenon organized within the lithosphere, and so favors the top-down origin for some rifts and anorogenic magmatic provinces.

2. Truncated icosahedron

The icosahedron has 20 identical triangular faces that meet in fives at each of 12 vertices. Truncation of the icosahedron creates pentagonal and hexagonal faces with equal edge-lengths (Fig. 1). Spherical truncation creates the familiar soccer-ball tessellation of regularly arranged hexagons and pentagons, for

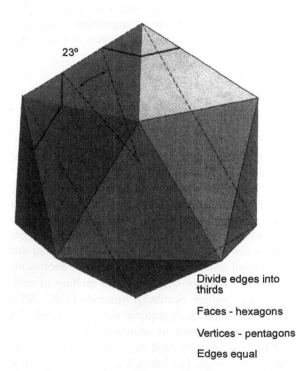

23°

Divide edges into thirds

Faces - hexagons

Vertices - pentagons

Edges equal

Fig. 1. Truncation of icosahedron divides edges into thirds and produces hexagonal and pentagonal faces with central angle of 23.28°.

which each seam forms a 23° great-circle arc, and each triple junction joins two hexagons and one pentagon. The distribution of polygons is mathematically precise; establishment of a single pentagon fixes the entire tessellation. The truncated icosahedron incorporates the largest regular hexagonal tiles permissible on a tessellation of a sphere.

Icosahedral structures typify many natural spherical equilibrium systems. Examples include fullerene molecules, wart and herpes viruses, radiolaria, tortoise shells, zeolites, quasi-crystals, clathrates, boron hydrates and foams (cf. Anderson, 2002). We propose that, under conditions of uniform horizontal extension, the pattern also characterizes fractures of spherical shells, such as the lithosphere, and associated emplacement of anorogenic magma.

3. Laurentia

Laurentia is a relic of a Paleoproterozoic supercontinent that likely included parts of Baltica, Australia, Siberia, China and other cratons (Sears and Price, 2003; Karlstrom et al., 2001; Hoffman, 1989). Mesoproterozoic and Neoproterozoic rift zones truncated Paleoproterozoic and Archean basement trends. Upon restoration of Cenozoic drift, the rifts approximate a hexagon and part of a pentagon of a precise truncated-icosahedral tessellation (Fig. 2). As detailed elsewhere (Sears, 2001), the edges of these tiles follow broad zones of Mesoproterozoic and Neoproterozoic anorogenic magmatism that include dike swarms, volcanic-sedimentary rift basins, and anorthosite, granite and rhyolite suites. The anorogenic suites are characteristically bimodal, A-type magmatic rocks having within-plate trace-element signatures (Van Schmus and Bickford, 1993). The rifts evolved into early Paleozoic passive-margin miogeoclines (Bond et al., 1984). Phanerozoic orogenic belts that compressed these miogeoclines now frame the craton and highlight the hexagon.

The Greenland, Arctic and western Canadian sides of Laurentia have arc-lengths of 23° and define three hexagonal edges. A fourth edge of this hexagon follows the Montana–Tennessee structural corridor, a >600-km wide zone of rifts, dike swarms, and anorogenic magmatism in the subsurface of the central United States (Hatcher et al., 1987; Paulsen

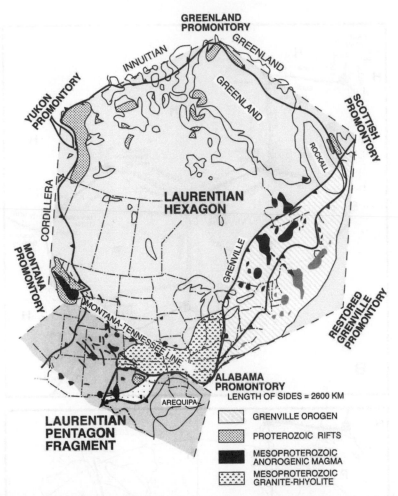

Fig. 2. Truncated-icosahedral fracture tessellation of Laurentia, after Sears (2001). Each side of hexagon subtends 23° of arc. Mesoproterozoic and Neoproterozoic anorogenic provinces occur along edges. Restored for Cenozoic continental drift.

and Marshak, 1994; Marshak et al., 2000; Sears, 2003). The Alabama, Montana, Yukon, Greenland and Scottish promontories are congruent with five vertices of the hexagon. This congruence satisfies two independent, precisely defined parameters of the truncated-icosahedral hexagon, the arc-length of edges (2600 km) and the turning-angles between edges (120°). The late Mesoproterozoic Grenville orogen may have tectonically collapsed the sixth vertex, as reconstructed in Fig. 2.

The tessellated rift zones initiated approximately 200 Ma after completion of tectonic amalgamation of Laurentia (Hoffman, 1989), when the supercontinent stalled in its apparent polar wander path (Elston et al., 2002).

4. Gondwana

Gondwana provides a Mesozoic example of truncated-icosahedral rifting and anorogenic magmatism. Fig. 3 restores Gondwana into its tight 200-Ma structural configuration, after Lawver et al. (1999), and illustrates the close congruence of Gondwanan rift zones and anorogenic magmatic provinces to the tessellation. The truncated-icosahedral tessellation is rigorous; Antarctica approximates a pentagon, the other polygons are dependent. More than 20,000 km of rifts followed 15 edges of this singular tessellation during breakup of the supercontinent. Triple-rift junctions are well established for Gondwana (Burke and Dewey, 1973). However, it has not previously

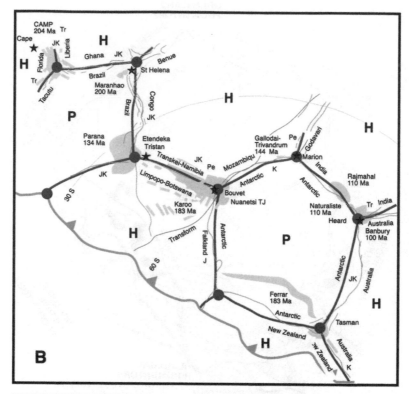

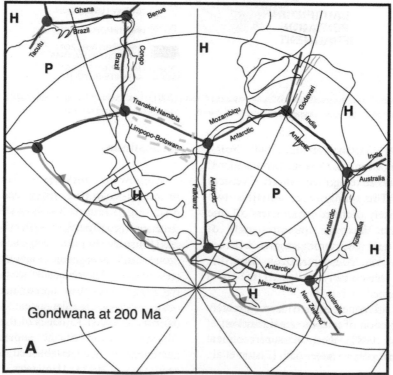

been noted that many of the triple junctions are separated by 23° of arc and fit the unique arrangement of the truncated icosahedron.

The following conjugate coasts correspond to tessellation edges (Fig. 3B): Florida–Liberia, Brazil–Ghana, Brazil–Congo, Falkland–Antarctica, Mozambique–Antarctica, India–Antarctica, Australia–Antarctica, New Zealand–Antarctica, Australia–India and Australia–New Zealand. The following branching rift zones, or failed arms, match edges of the same tessellation: Tacutu, Benue and Godavari (cf. Şengör and Natal'in, 2001). The discontinuity across central South America coincides with a tessellation edge (Fig. 3A). The Limpopo–Botswana and Transkei–Namibia mafic and kimberlite dike swarms (Hunter and Reid, 1987) follow a tessellation edge across southern Africa. The tessellated rifts mostly fractured relatively coherent Precambrian cratons within Gondwana. Gondwanan rifts that were discordant to the tessellation were those that reactivated older Gondwanan sutures. According to the compilation of Şengör and Natal'in (2001), the tessellated Gondwanan rifts mostly evolved in late Jurassic to early Cretaceous time, but some have Permian or Triassic ancestry (Fig. 3B).

Large anorogenic igneous outbreaks occurred along tessellation edges, near vertices. These comprise the CAMP (Central Atlantic mafic province), Maranhão, Paraña-Etendeka, Karoo, Gallodai-Trivandrum, Rajmahal, Naturaliste, Banbury and Tasman provinces. The igneous outbreaks were episodic from 204 to 100 Ma and are commonly considered to be plume eruptions (Ernst and Buchan, 2001).

Several volcanic "hotspots" are thought to represent continuing anorogenic magmatism at the approximate sites of the initial outbreaks of the Gondwanan igneous provinces; well-documented tracks of extinct volcanoes on the sea bed connect some of these active volcanic sea mounts to the provinces (Morgan, 1981). The following hotspot-igneous province links are possible: Cape Verde (or Fernando?) hotspot—CAMP, St Helena hotspot—

Maranhao province, Tristan hotspot—Parana-Etendeka province, Bouvet hotspot—Karoo province, Marion hotspot—Gallodai-Trivandrum province and Heard hotspot—Naturaliste province. With the exception of Cape Verde, these hotspots conform rather closely to the vertices of a single truncated-icosahedral tessellation. Excluding Cape Verde, we determined a 1.5° mean angular distance between these hotspots and the triple junctions of a best-fit truncated-icosahedral tessellation (St. George, 2003). As shown in Fig. 3, the hotspot and rift tessellations are approximately congruent when Gondwana is restored onto the hotspot framework.

We suggest that the fracture tessellation propagated across Gondwana before late Jurassic seafloor spreading began to separate its parts. Fracture propagation may have begun in Permian or Triassic, because some of the rifts that fall on the tessellation are known to date from that time. The fractures relieved tensile strain within the lithosphere, but may not have erupted magma until the resulting tiles separated sufficiently to drive decompression melting. That is, although organization of the fractures may have been a passive, within-Gondwana response to tension, separation of the tiles and magmatic outbreaks may have depended on global plate-dynamics.

As observed by A. Luttinen, one of the reviewers of this manuscript, if the hotspots were indeed generated top-down, drifting of Gondwana during propagation of the fractures and opening of the triple junctions should have left behind a framework of asthenospheric hotspots that exhibit distorted rather than ideal truncated-icosahedral geometry. We suggest that initial opening of the Central Atlantic may have moved Gondwana and its fracture tessellation off the outbreak point of CAMP, so that the hotspot tessellation is displaced with respect to Cape Verde (or Fernando) hotspot. Furthermore, initial opening of the Mozambique Channel may have split the fracture tessellation between the halves of Gondwana, so that

Fig. 3. (A) Gondwana configuration at 200 Ma, after Lawver et al. (1999), showing congruence of Gondwanan rifts with truncated-icosahedral tessellation. Circles at truncated-icosahedral vertices (triple junctions) are separated by 23° of arc. (B) Simplified map of Gondwana on same base as (A), except Earth coordinates rotated slightly from Lawver's placement to demonstrate congruence of active hotspots associated with Gondwanan large igneous provinces and tessellation triple-junctions. Stars: active hotspots, names italicised. Yellow: large igneous provinces and dikes. Blue: Gondwanan rifts. Red: ideal truncated-icosahedral tessellation. P=pentagon, H=hexagon. Ages of rifts from Şengör and Natal'in (2001): Pe=Permian, Tr=Triassic, J=Jurassic, K=Cretaceous.

the modern hotspots are slightly dispersed from an ideal tessellation.

Despite these minor distortions, the approximate overall congruence of the hotspots to the tessellation implies that Gondwana was largely stagnant during late Jurassic–early Cretaceous magmatic outbreaks. This stagnation may be reflected in the late Jurassic–early Cretaceous loop in the Gondwana apparent-polar-wander path (De Wit et al., 1988) and is consistent with the De Wit et al. (1988) restoration of Jurassic and early Cretaceous Gondwana. It differs slightly from models of Golonka et al. (1994), Lawver et al. (1999) and Şengör (2001), however, in the position of southern Gondwana with respect to the hotspots.

Because the symmetry of sea-floor spreading is incompatible with that of the truncated icosahedron, Gondwanan plate boundaries evolved new ridge-transform configurations as the continental fragments dispersed. The incongruent southeast coast of Africa is one such transform boundary.

5. Discussion

The truncated-icosahedral fractures of Laurentia and Gondwana argue against a deep-mantle plume origin for some continental rifts, associated anorogenic igneous provinces and volcanic hotspots. The tessellated fractures better characterize brittle lithosphere than a chaotically convecting mantle. Although both Rayleigh-Bénard and Rayleigh-Taylor convection can adapt polygonal cells such as those of the truncated icosahedron under stagnant-lid conditions, such cells are unlikely to remain stable because high Rayleigh numbers (10^6–10^7) expected in the upper mantle lead to time-dependent convection and shifting cell patterns (Talbot et al., 1991). Worm-like ascents of hypothetical plumes from the mantle (Steinberger, 2000) argue against a geometrically rigorous hotspot distribution. The hotspot tessellation may, however, be consistent with the top-down tectonic models of Anderson (2001) and Hamilton (2003).

Anderson (1982) and De Wit et al. (1988) determined that a stalled supercontinent should insulate the mantle, and that the resulting accumulation of heat should partially melt and expand the asthenosphere. The increased radius of curvature of

the lifted plate should then impose homogenous surface-parallel tensile strain on the lithosphere that should eventually fracture and disperse the supercontinent (Anderson, 1982). Anderson (1982) proposed that thermal uplift of Gondwana began in Permian, and that rift response began in Triassic–Jurassic; he hypothesized that the Atlantic–African geoid anomaly is a residual feature that is genetically related to the growth, stagnation and breakup of the supercontinent. We argue that the truncated-icosahedral fractures record the lithospheric response predicted by Anderson's (1982) model.

Jagla and Rojo (2002) demonstrated that cracks spontaneously self-organize into quasi-hexagonal tessellations in isotropic elastic media under conditions of uniform, layer-parallel tension, such as that proposed for Anderson's (1982) stagnant supercontinent. This is because the hexagonal pattern relieves the greatest strain energy for the least work invested in nucleation and propagation of fractures. Quasi-hexagonal fracture patterns emerge at several scales in natural materials; mud cracks and columnar basalt provide familiar examples. The scale of the polygons may be a proxy for the strength of the material; weaker and thinner materials, such as cornstarch, break into small polygons. The truncated icosahedron includes the largest hexagonal tiles permissible within a spherical tessellation; it may provide the most efficient configuration of fracture sets to relieve tension within large tracts of stiff Precambrian lithosphere. Geometry requires pentagons to alternate with hexagons on a sphere, leading to the truncated icosahedral tessellation.

Once a fracture tessellation has been established, the resulting lithospheric tiles are then free to passively separate above an expanding asthenosphere, much like sutures between hexagonal plates of a tortoise shell. Separation of the lithospheric tiles may then draw curtains of asthenosphere upward to grout the fractures, as a secondary process. Decompression-melting may then lead to injection of dike swarms along tessellation edges. Diapirs may rise to fill the larger openings that form at triple-rift junctions and erupt large igneous provinces. Injection of mafic magma into the lower crust may lead to secondary melting and formation of other anorogenic magmas, resulting, for example, in anorthosite massifs and granite-rhyolite provinces. Silicic anorogenic magmatism may be more typical where the fractures cross relatively young continental lithosphere, as for example in Laurentia, where large silicic provinces are more typical in Proterozoic orogenic belts than in Archean terranes. In Gondwana, mafic anorogenic magmatism is more common in Archean regions, but the large Chon Akie felsic province (cf. Ernst and Buchan, 2001) occupies the Phanerozoic Gondwanides. Perhaps fracture zones are more diffuse in younger, weaker continental lithosphere, leading to broader zones with more silicic magmatism. At the surface, grabens may evolve as lithospheric attenuation zones widen to 600 km or more, perhaps due to melting, eruption, sapping and separation of the tiles to accommodate continued expansion of the asthenosphere.

Resumption of plate motion may remove a supercontinent from the thermal bulge and arrest anorogenic magmatism, except for lingering asthenospheric hotspot plumes. However, should plate motion again stagnate and thermal expansion renew beneath the continent, established fractures may experience a resurgence of anorogenic magmatism, because significantly less energy is required to open existing fractures than to create new ones. Thus, as in Laurentia, several generations of magma may intrude the same fractures (Sears, 2001). Conversely, a second, discordant rift tessellation may begin to propagate. The East African rift system, which is discordant with the Gondwanan tessellation, may be such an example.

6. Conclusions

We conclude that a stalled supercontinent may fracture into a least-work, truncated-icosahedral configuration. Dike swarms may be intimately associated with the fracturing process, and magma injection may lead to melting of lower continental crust and further anorogenic magmatism. Continental drift may arrest anorogenic magmatism, but periods of stagnation may lead to recurrent anorogenic magmatism along established truncated-icosahedral fractures. Tessellated hotspots appear to herald lingering diapiric activity at the asthenospheric sites of the Gondwanan triple junctions on the eve of dispersal. They remain disposed within the Atlantic–African geoid anomaly,

like hobnails in the footprint of Gondwana (cf. Anderson, 1982).

Acknowledgments

This manuscript was much improved following reviews by D.L. Anderson, A. Luttinen, T. Ruotoistenmäki and editorial comments by O.T. Rämö. Sears thanks Professor Haapala and the Department of Geology at the University of Helsinki for their hospitality during the preparation of this manuscript.

References

Anderson, D.L., 1982. Hotspots, polar wander, Mesozoic convection and the geoid. Nature 297, 391–393.

Anderson, D.L., 2001. Top-down tectonics? Science 293, 2017–2018.

Anderson, D.L., 2002. How many plates? Geology 30, 411–414.

Bijwaard, H., Spakman, W., 1999. Tomographic evidence for a narrow whole mantle plume below Iceland. Earth and Planetary Science Letters 166, 121–126.

Bond, G.C., Nickeson, P.A., Kominz, M.A., 1984. Breakup of a supercontinent between 625 Ma and 555 Ma: new evidence and implications for continental histories. Earth and Planetary Science Letters 70, 325–345.

Burke, K., Dewey, J.F., 1973. Plume-generated triple junctions: key indicators in applying plate tectonics to old rocks. Journal of Geology 81, 406–433.

Campbell, I.H., 2001. Identification of ancient mantle plumes. In: Ernst, R.E., Buchan, K.L. (Eds.), Mantle Plumes: Their Identification Through Time. Special Paper-Geological Society of America, vol. 352, pp. 5–21.

Condie, K.C., 2001. Mantle Plumes and their Record in Earth History. Cambridge University Press, Cambridge.

De Wit, M., Jeffery, M., Bergh, H., Nicolaysen, L., 1988. Geological map of sectors of Gondwana reconstructed to their disposition ~150 Ma. America Association of Petroleum Geologists, Tulsa, OK 74101, scale 1:10,000,000.

Elston, D.P., Enkin, R.J., Baker, J., Kisilevsky, D.K., 2002. Tightening the belt: paleomagnetic-stratigraphic constraints on deposition, correlation, and deformation of the Middle Proterozoic (ca. 1.4 Ga) Belt-Purcell Supergroup, United States and Canada. Geological Society of America Bulletin 114, 619–638.

Ernst, R.E., Buchan, K.L., 2001. Large mafic magmatic events through time and links to mantle-plume heads. In: Ernst, R.E., Buchan, K.L. (Eds.), Mantle Plumes: Their Identification Through Time. Special Paper-Geological Society of America, vol. 352, pp. 483–566.

Golonka, J., Ross, M.I., Scotese, C.R., 1994. Phanerozoic paleogeographic and paleoclimatic modeling maps. In: Embry, A.F, Beauchamp, B., Glass, D.J. (Eds.), Pangaea: Global Environments and Resources. Canadian Society of Petroleum Geologists, Calgary, Canada, pp. 1–47.

Hamilton, W.B., 2002. Plate-tectonic circulation is driven by cooling from the top and is closed within the upper mantle. Saskatoon, Saskatchewan Abstracts Geological Association of Canada-Mineralogical Association of Canada Annual Meeting, Abstracts, vol. 27, p. 45.

Hamilton, W.B., 2003. An alternative earth. GSA Today 13 (11), 4–12.

Hatcher Jr., R.D., Zietz, I., Litehiser, J.J., 1987. Crustal subdivisions of the eastern and central United States and a seismic boundary hypothesis for eastern seismicity. Geology 15, 528–532.

Hoffman, P.F., 1989. Speculations on Laurentia's first gigayear (2.0 to 1.0 Ga). Geology 17, 135–139.

Hunter, D.R., Reid, D.L., 1987. Mafic dyke swarms in southern Africa. In: Halls, H.C., Fahrig, W.F. (Eds.), Mafic Dyke Swarms. Special Paper-Geological Association of Canada, vol. 34, pp. 445–456.

Jagla, E.A., Rojo, A.G., 2002. Sequential fragmentation: the origin of columnar quasihexagonal patterns. Physical Review. E 65, 026203.

Karlstrom, K.E., Åhäll, K.-I., Harlan, S.S., Williams, M.L., McLelland, J., Geissman, J.W., 2001. Long-lived (1.8–1.0 Ga) convergent orogen in southern Laurentia, its extensions to Australia and Baltica, and implications for refining Rodinia. Precambrian Research 111, 5–30.

Lawver, L.A., Gahagan, L.M., Dalziel, I.W.D., 1999. A tight fit-early Mesozoic Gondwana, a plate reconstruction perspective. National Institute of Polar Research. Special Issue 53, 214–229.

Marshak, S., Karlstrom, K.E., Timmons, J.M., 2000. Inversion of Proterozoic extensional faults: an explanation for the pattern of Laramide and ancestral Rockies intracratonic deformation, United States. Geology 28, 735–738.

Morgan, W.J., 1981. Hot spot tracks and the opening of the Atlantic and Indian Oceans. In: Emiliani, C. (Ed.), The Sea. Wiley, New York, pp. 443–487.

Paulsen, T., Marshak, S., 1994. Cratonic weak zone in the U.S. continental interior: the Dakota–Carolina corridor. Geology 22, 15–18.

Sears, J.W., 2001. Icosahedral fracture tessellation of early Mesoproterozoic Laurentia. Geology 29, 327–330.

Sears, J.W., 2003. On the edge of the icosahedron: anatomy of the Montana–Tennessee structural corridor. Abstracts with Programs-Geological Society of America 35 (5), 41.

Sears, J.W., Price, R.A., 2003. Tightening the Siberian connection to western Laurentia. Geological Society of America Bulletin 115, 943–953.

Şengör, A.M.C., 2001. Elevation as indicator of mantle-plume activity. In: Ernst, R.E., Buchan, K.L. (Eds.), Mantle Plumes: Their Identification Through Time. Special Paper-Geological Society of America, vol. 352, pp. 183–225.

Şengör, A.M.C., Natal'in, B.A., 2001. Rifts of the world. In: Ernst, R.E., Buchan, K.L. (Eds.), Mantle Plumes: Their Identification Through Time. Special Paper-Geological Society of America, vol. 352, pp. 389–482.

St. George, G.M., 2003. Rotate TI. Computer program, available upon request from Department of Mathematical Sciences, University of Montana, Missoula MT 59812, USA.

Steinberger, B., 2000. Plumes in a convecting mantle: models and observations for individual hotspots. Journal of Geophysical Research 105, 11127–11152.

Storey, B.C., Leat, P.T., Ferris, J.K., 2001. The location of mantle-plume centers during the initial stages of Gondwana breakup. In: Ernst, R.E., Buchan, K.L. (Eds.), Mantle Plumes: Their Identification Through Time. Special Paper-Geological Society of America, vol. 352, pp. 71–80.

Talbot, C.J., Ronnlund, P., Schmeling, H., Koyi, H., Jackson, M.P.A., 1991. Diapiric spoke patterns. Tectonophysics 188, 187–201.

Turcotte, D.L., Schubert, G., 2002. Geodynamics. Cambridge University Press, Cambridge.

Van Schmus, W.R., Bickford, M.E., 1993. Transcontinental Proterozoic provinces. In: Reed, J.C., Bickford, M.E., Houston, R.S., Link, P.K., Rankin, D.W., Sims, P.K., Van Schmus, W.R. (Eds.), Geology of North America. Geological Society of America DNAG, vol. C-2, pp. 171–334. Precambrian: Conterminous U.S.

Wilson, J.T., 1963. A possible origin of the Hawaiian Islands. Canadian Journal of Physics 41, 863–870.

Petrologic Processes

Available online at www.sciencedirect.com

Lithos 80 (2005) 155–177

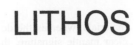

www.elsevier.com/locate/lithos

ELSEVIER

Mafic magmatic enclaves and mafic rocks associated with some granitoids of the central Sierra Nevada batholith, California: nature, origin, and relations with the hosts

Bernard Barbarin[*]

Université Blaise Pascal and CNRS, Département des Sciences de la Terre and UMR 6524 Magmas et Volcans,
5 rue Kessler, F-63038 Clermont-Ferrand cedex, France

Received 14 April 2003; accepted 9 September 2004
Available online 25 November 2004

Abstract

The calc-alkaline granitoids of the central Sierra Nevada batholith are associated with abundant mafic rocks. These include both country-rock xenoliths and mafic magmatic enclaves (MME) that commonly have fine-grained and, less commonly, cumulate textures. Scarce composite enclaves consist of either xenoliths enclosed in MME, or of MME enclosed in other MME with different grain size and texture. Enclaves are often enclosed in mafic aggregates and form meter-size polygenic swarms, mostly in the margins of normally zoned plutons. Enclaves may locally divert schlieren layering. Mafic dikes, which also occur in swarms, are undisturbed, composite, or largely hybridized. In central Sierra Nevada, with the exception of xenoliths that completely differ from the other rocks, host granitoids, mafic aggregates, MME, and some composite dikes exhibit a bulk compositional diversity and, at the same time, important mineralogical and geochemical (including isotopic) similarities. MME and host granitoids display distinct major and trace element compositions. However, strong correlations between MME–host granitoid pairs indicate interactions and parallel evolution of MME and enclosing granitoid in each pluton. Identical mafic mineral compositions and isotopic features are the result of these interactions and parallel evolution. Mafic dikes have broadly the same major and trace element compositions as the MME although variations are large between the different dikes that are at distinctly different stages of hybridization and digestion by the host granitoids. The composition of the granitoids and various mafic rocks reflects three distinct stages of hybridization that occurred, respectively, at depth, during ascent and emplacement, and after emplacement. The occurrence and succession of hybridization processes were tightly controlled by the physical properties of the magmas. The sequential thorough or partial mixing and mingling were commonly followed by differentiation and segregation processes. Unusual MME that contain abundant large crystals of hornblende resulted from disruption of early cumulates at depth, whereas those richer in large crystals of biotite were formed by disruption of late mafic aggregates or schlieren layerings at the level of emplacement. MME and host granitoids are considered cogenetic, because both are hybrid rocks that were produced by the mixing of the same two components in different proportions. The felsic component was produced by partial melting of preexisting crustal

* Tel.: +33 4 73 34 67 30; fax: +33 4 73 34 67 44.
E-mail address: B.Barbarin@opgc.univ-pbclermont.fr.

0024-4937/$ - see front matter © 2004 Elsevier B.V. All rights reserved.
doi:10.1016/j.lithos.2004.05.010

materials, whereas the dominant mafic component was probably derived from the upper mantle. However, in the lack of a clear mantle signature, the origin of the mafic component remains questionable.

Keywords: Mafic enclaves; Magma mixing; Calc-alkaline granitoids; Sierra Nevada batholith; California

1. Introduction

Mafic magmatic enclaves (MME) are common in calc-alkaline granitoids (Didier and Barbarin, 1991) and are abundant in most of the Cordilleran granitoids (e.g., Pitcher, 1983; Barbarin, 1999). They provide evidence of the role of mafic magmas in the initiation and evolution of calc-alkaline granitoid magmas and thus their origin is of fundamental significance in interpreting the history of batholiths.

Field and petrographic features of MME have been determined in detail by many investigators. The main references are an early paper of Phillips (1880), a descriptive survey of "autoliths" of the Sierra Nevada by Pabst (1928), comprehensive books on the subject by Didier (1973) and Didier and Barbarin (1991), which contain many examples and a review of the various characteristics of MME, and a discussion of microstructures by Vernon (1991). Geochemical and isotopic studies of MME (e.g., Fourcade and Allègre, 1981; Domenick et al., 1983; Tindle and Pearce, 1983; Reid et al., 1983; Noyes et al., 1983a; Barbarin et al., 1985; Kistler et al., 1986; Hill et al., 1988; Dorais et al., 1990; Dodge and Kistler, 1990) emphasize their significance to the genesis of granitoid magmas and their relations to associated mafic rocks.

MME are particularly abundant in the calc-alkaline granitoids of the Sierra Nevada batholith (Fig. 1) (e.g., Pabst, 1928; Bateman, 1992). They have been described in many plutons (Bateman and Nokleberg, 1978; Bateman and Chappell, 1979; Noyes et al., 1983a,b; Bateman, 1992) and represent the main topic of several other investigations (Pabst, 1928; Link, 1969; Reid et al., 1983; Domenick et al., 1983; Furman and Spera, 1985; Frost and Mahood, 1987; Dodge and Kistler, 1990; Dorais et al., 1990; Barbarin, 1991).

This report concerns the characteristics and origin of MME and associated mafic rocks such as mafic aggregates, schlieren layering, undisturbed or compo-site mafic dikes, mainly from selected plutons in the central Sierra Nevada batholith (Bateman, 1992). The localities described or mentioned in this report are identified in Fig. 1 and Table 1. Petrographic, modal, mineralogical, and chemical data on these rocks (Barbarin et al., 1989) and their host granitoids (Bateman et al., 1984a; Bateman, 1992), impose constraints to a general model for their origin and evolution.

2. The Sierra Nevada granitoids

The Sierra Nevada batholith represents a portion of the nearly-continuous chain of Circum-Pacific granitoids (Pitcher, 1983). Like the Coastal batholith of Peru (Pitcher et al., 1985), the Sierra Nevada batholith is an example of a batholith emplaced in a continental magmatic arc (references in Bateman, 1992). The plutonic rocks of the Sierra Nevada batholith range in composition from gabbro to granite. Tonalite, quartz diorite, granodiorite, and quartz monzonite are the most common types and hornblende is the characteristic mafic silicate. They form normally zoned plutons (e.g., Tuolumne Intrusive Suite: Bateman and Chappell, 1979) that commonly are associated in suites of various ages. In general, the plutons are rounded or elongated ~NE–SW, parallel to the long axis of the batholith (Bateman, 1992).

The axial part of the batholith consists of Cretaceous granitoid suites that progressively young eastward (from ~125 to ~88 Ma) (Stern et al., 1981; Chen and Moore, 1982). Jurassic plutons and suites (from 186 to 155 Ma) are exposed along both margins and, locally, in the interior of the batholith. A single Triassic suite is present in the east side of the batholith. The central part between 37°N and 38°N has been mapped at a scale of 1:62,500 (e.g., Bateman, 1992), and relations between the different plutons and suites are clearly defined (Fig. 1). Many plutons and suites have been studied in detail (Mount

Givens Granodiorite: Bateman and Nokleberg, 1978, Gilder and McNulty, 1999, McNulty et al., 2000; Tolumne Intrusive Suite: Bateman and Chappell, 1979; Red Lake and Eagle Peak plutons: Noyes et al., 1983a,b; Palisade Crest Intrusive Suite: Sawka et al., 1990; Dinkey Creek pluton: Dorais et al., 1990, Cruden et al., 1999; Jackass Lakes pluton: McNulty et al., 1996). Samples from various plutons and suites have been compared mineralogically (Dodge et al., 1968, 1969; Piwinskii, 1968; Barbarin, 1986, 1990),

chemically (Kistler and Peterman, 1973; Bateman et al., 1984a,b), and experimentally (Piwinskii, 1973).

3. Petrography and field relations of mafic rocks

The Sierra Nevada batholith granitoids, especially the intermediate rocks that range in composition from tonalite to quartz monzonite, contain abundant MME (e.g., Pabst, 1928; Bateman, 1992) and associated

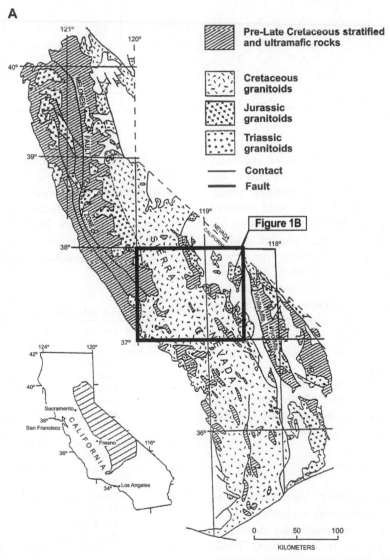

Fig. 1. (A) Simplified geological map of the Sierra Nevada batholith. (B) Simplified geological map of the central part of the Sierra Nevada batholith between latitudes 37°N and 38°N showing sample locations described in the paper (both from Bateman, 1992). See Table 1 for key to sample locations.

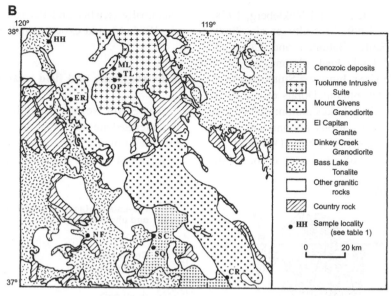

Fig. 1 (*continued*).

mafic rocks. The amount of mafic rocks, including MME, generally increases toward the margins of the plutons, especially those that are normally zoned. MME and other mafic rocks are also found scattered throughout hornblende-bearing plutons and suites but are scarce or absent in biotite granites and in intrusions that contain large K-feldspar megacrysts. The distribution and abundance of MME are mainly related to the composition of the host rock rather than to contacts and magmatic movements as suggested by Pabst (1928). In the normally zoned plutons of the Sierra Nevada batholith, MME and other mafic rocks are concentrated near contacts. A striking connection exists among the abundance of MME, the color index of the host granitoid, and compositional zoning of plutons or suites (e.g., Barbarin, 1989). The absence of biotite-rich and sillimanite-bearing restites is also noticeable in the Sierra Nevada granitoids.

3.1. Mafic magmatic enclaves

Because the term "inclusion" has been used to describe fluids or mineral grains enclosed in other minerals, "enclave" is more suitable for describing rock enclosed in another rock (Lacroix, 1893; Holmes, 1928; Read, 1957; Didier, 1973; Vernon, 1983, 1984; Didier and Barbarin, 1991). In this report, the term "mafic magmatic enclave" (Barbarin, 1988)

is preferred to "mafic microgranular enclave" (Didier, 1973) or "microgranitoid enclave" (Vernon, 1983, 1984) because descriptions of enclaves from calc-alkaline granitoid plutons throughout the world indicate that they are invariably finer grained than the enclosing granitoids, but not necessarily microgranular or microgranitoid. On the other hand, the use

Table 1
Location and rock unit of the selected localities

Map symbol (cf. Fig. 1)	Locality	Lat. (°N)	Long. (°W)	Rock unit
TL	Tenaya lake	37°49.8′	119°28.0′	Half Dome granodiorite
OP	Olmstead Point	37°48.9′	119°29.2′	Half Dome granodiorite
ML	May Lake	37°50.6′	119°29.9′	Half Dome granodiorite
CR	Courtright Reservoir	37°04.4′	118°58.0′	Mount Givens granodiorite
ER	Elephant Rock	37°43.2′	119°42.8′	El Capitan granite
SQ	Shaver Lake Quarry	37°09.3′	119°17.7′	Dinkey Creek granodiorite
SC	San Joaquin River Canyon	37°11.2′	119°20.9′	Dinkey Creek granodiorite
NF	North Fork	37°11.6′	119°35.8′	Bass Lake tonalite
HH	Hetch Hetchy Reservoir	37°56.8′	119°47.2′	Bass Lake tonalite

of "magmatic" emphasizes the crystallization of these enclaves from magmas, and consequently their typical igneous texture (Fig. 2). In addition, "mafic" indicates that these enclaves are darker-colored than their enclosing granitoid. "Mafic" also distinguishes dark-colored magmatic enclaves from felsic microgranular enclaves, which have a different origin (Didier, 1973).

Detailed descriptions of MME from the northern and southern parts of the Sierra Nevada batholith are given in Pabst's (1928) benchmark paper and identical observations have been made in the central part. Different types of MME can be distinguished by their grain size, texture, structure, mineral content, nature and abundance of phenocrysts, composition, external

morphology, and contacts with host granitoids. Some MME are composite and consist of one MME enclosed in another MME of different grain size or texture (see Fig. 2 in Barbarin, 1991). Composite MME with fine-grained and porphyritic or xenocryst-rich parts are particularly common. However, there is no rule concerning which type of MME is enclosed in another. More than 50% of the MME are fine-grained and dark, contain abundant plagioclase phenocrysts, scarce small hornblende phenocrysts, and are usually surrounded by a fine-grained margin of variable width (Fig. 2). In most comparisons of enclave–host granitoid pairs, these MME are used as a reference. The fine-grained margins are commonly characterized by a trachytic texture where strong flow foliation, parallel to the contact, is shown by plagioclase laths and is locally diverted by plagioclase phenocrysts (Fig. 2B). Close relationships also appear between the grain size of MME and the nature of contacts with hosts, and between aspect ratios of MME and intensity of flow foliation in the hosts. Because contacts correspond to interlocking of crystals of the felsic and mafic components, the larger the grain size of the MME, the more diffuse the contacts.

The same minerals are present in MME as in their granitoid hosts but in different proportions (Barbarin, 1986; Barbarin et al., 1989). MME range from diorite to quartz diorite and are composed of plagioclase (45–55%) and mafic minerals (40–50%). The host granitoids contain less plagioclase (35–45%) and mafic minerals (5–20%), and much more quartz and K-feldspar (35–50%) (Fig. 3). The proportions of hornblende and biotite and color index vary widely from one enclave to another (Fig. 4); in general, the ratio of hornblende to the total amount of mafic minerals increases with color index. The abundance of hornblende in MME also correlates with its abundance in the host: MME enclosed in tonalites generally contain 5–10% more hornblende than MME enclosed in granodiorites. In contrast, the mafic mineral content of the granitoids is relatively constant, and biotite is almost always more abundant than hornblende.

Although quartz and K-feldspar are scarce or absent in most MME, they are present in some MME that are enclosed in felsic granitoids (Fig. 3) and are then interstitial. Quartz and scarce megacrystic K-feldspar xenocrysts are also found in some MME.

A

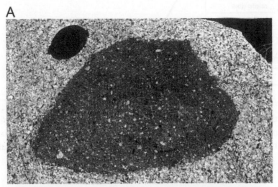

B

Fig. 2. (A) A typical mafic magmatic enclave in the Cretaceous granodiorite of the central Sierra Nevada batholith (Mount Givens granodiorite at Courtright Reservoir). (B) Photomicrograph of a contact zone of a typical mafic magmatic enclave (right) against the host granodiorite (left). Within the fine-grained margin, flow alignment of elongate plagioclase laths is diverted by the euhedral phenocrysts of hornblende and zoned plagioclase. Crossed polars, base of photomicrograph corresponds to 20 mm.

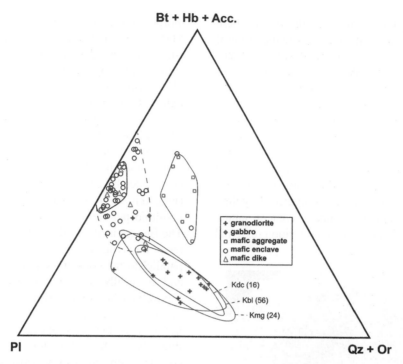

Fig. 3. Modal composition of granitoids and associated mafic rocks plotted in a plagioclase (Pl)-quartz plus K-feldspar (Qz+Or)-mafic plus accessory minerals (Bt+Hb+Acc.) diagram. Granitoids, mafic aggregates, and MME plot in three distinct zones. Mafic dikes plot in the same area as MME. Modes of granitoids (from Bateman et al., 1984a) are for the Dinkey Creek Granodiorite (Kdc: 16 analyses), the Bass Lake Tonalite (Kbl: 56 analyses), and the Mount Givens Granodiorite (Kmg: 24 analyses).

Globular quartz grains about the same size as quartz crystals in the host (a few millimeters in diameter), exhibit a typical ocellar texture with spheric shape, corrosion embayments, and hornblende mantle. However, ocellar texture (e.g., Palivcová, 1978; Hibbard, 1991) is scarce in the MME of the Sierra Nevada batholith granitoids, except for the El Capitan Granite. K-feldspar megacrysts are rare to absent in MME, even where the host is megacrystic such as the Mount Givens Granodiorite. Titanite, where present in the MME, is also present in the host granitoid and has similar composition, crystal habit, and opaque inclusions (Barbarin and Bateman, 1986). Apatite occurs as large stubby crystals in the granitoids, whereas in MME it occurs as needles enclosed in all other minerals. MME contain rare large zircon grains with pitted surfaces, whereas host granitoids contain abundant small zircon crystals with clear surfaces. Apatite needles, acicular and hollow hornblende prisms, plagioclase laths, and locally thin and elongate biotite crystals are common in MME, mainly in fine-grained margins.

Clots of mafic and accessory minerals are also a distinctive feature of the MME. They consist predominantly of euhedral hornblende crystals intergrown with biotite, titanite, and opaque minerals. They may represent concentrations of dense, early-formed phases as proposed by Reid and Hamilton (1987), but are more probably pseudomorphs of amphibole or pyroxene phenocrysts, as suggested by their commonly geometric shapes.

A rare type of MME occurs only once or twice per hundreds or thousands of regular MME. These MME are 10–30 cm in diameter, contain up to 70% of mafic minerals, and exhibit a cumulate texture (see Fig. 1g in Barbarin, 1991). Detailed microscopic descriptions and analyses of such enclaves are given in Dorais et al. (1990).

3.2. Mafic aggregates and schlieren layering

Mafic aggregates are especially common in the margins of intrusions, adjacent to contacts with other granitoids or with country rocks. They are charac-

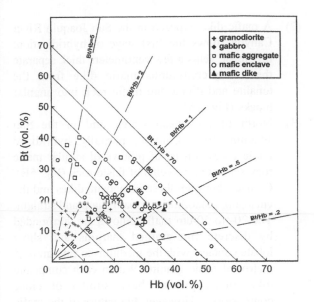

Fig. 4. Biotite versus hornblende contents for the granitoids and associated mafic rocks of the central Sierra Nevada batholith. MME are mostly scattered. Most MME contain broadly the same amount of mica and amphibole, but a few are either unusually hornblende-rich or exceptionally biotite-rich. Hornblende is clearly dominant in mafic dikes, whereas biotite is dominant in mafic aggregates and most granitoids.

terized by abundant large euhedral crystals of biotite and hornblende. Like MME, mafic aggregates contain from 40% to 50% of mafic minerals (Fig. 4). Mafic aggregates, however, contain more quartz and K-feldspar than MME, and the proportions of plagioclase to quartz plus K-feldspar are identical to those in the granitoid (Fig. 3). In some mafic aggregates, the modal abundances of titanite and allanite reach 5%. Mafic aggregates form schlieren layerings, huge masses, or are associated with MME in swarms, many of which are dike-like. Their contacts with the various types of MME and country rock xenoliths are mostly sharp, whereas contacts with the granitoids can be either sharp or diffuse.

Schlieren layering consists of rhythmic successions of layers with graded bedding. In some granitoids, such as the Half Dome Granodiorite at Olmstead Point or Tenaya Lake (cf. Fig. 1B), graded bedding is clearly noticeable within each layer. A 10-cm-thick layer begins with a thin seam of titanite overlain by small crystals of mafic minerals. Upward, hornblende and biotite crystals are progressively replaced by felsic minerals with regularly

increasing grain size. These layers were described as "schlieren" by Cloos (1936) and "layered mafic–felsic magma system" by Coleman et al. (1995). The schlieren layerings may locally be several meters thick and extend over more than one hundred meters. At Courtright Reservoir in the Mount Givens Granodiorite, crossbedding is commonly observed (Bateman et al., 1984a,b). These layers were deposited from and partly eroded by flowing magma (Gilbert, 1906). Graded bedding, crossbedding, unconformities, and channel structures are caused by differential magmatic movements along shear planes separating layers that were accompanied by settling of minerals within each layer. The middle and upper parts of the layers actually include large euhedral crystals of hornblende and biotite that are ~10 mm long. These crystals of hornblende and biotite probably represent phenocrysts that were stopped by schlieren layerings while they were settling in the granitoid magma. When the MME were almost solid, they were incorporated into the mafic aggregates. Rare xenoliths of host metamorphic rocks are also enclosed in the mafic aggregates. Contacts between lower parts of the layers and enclosing granitoids are mostly sharp, whereas contacts between their upper parts and granitoids are mostly diffuse. At Courtright Reservoir, some MME are locally found above the upper contact; others deform or divert the layers. Schlieren layerings that include abundant MME of varying nature and size resemble successive lag deposits of solid crystals, MME, and xenoliths, concentrated along pluton margins by flowing magma.

Mechanical concentrations of accessory minerals, euhedral mafic minerals, and MME can also form broadly spherical masses of mafic aggregates from a few centimeters to several meters across and these mafic aggregates may grade to pipes or dikes. Some dikes formed with MME of various type and size enclosed in a mafic aggregate, are described as composite dikes. However, their contacts with the enclosing granitoid are mostly diffuse. At Courtright Reservoir and Shaver Lake Quarry (Fig. 1B), the shapes of the masses of mafic aggregates and their contacts with the host suggest that, after their formation by concentration of solid particles in the granitic magma, they were injected in local early fractures. The mafic aggregates seem to initiate near the margins of the plutons and then move on short distances toward the center by such processes as filter pressing.

MME swarms are abundant in the Sierra Nevada granitoids (e.g., Tobisch et al., 1997). Two types of MME swarms can be distinguished (see Fig. 3 in Barbarin, 1991):

(1) "Polygenic" swarms consist of MME of various types and sizes enclosed in mafic aggregates mainly formed of large crystals of mafic minerals. They may also contain some xenoliths. Their contacts with host granitoids are distinct, but neither sharp nor straight, even where they occur as composite dikes or pipes. They are common especially in the margins of the plutons where MME and other mafic rocks are abundant. The best examples of these MME swarms are exposed at Courtright Reservoir (Bateman et al., 1984b).

(2) "Monogenic" swarms consist of relatively similar MME enclosed in medium to fine-grained granitoids. Country rock xenoliths have not been observed in these swarms. In contrast to the other type of MME swarms, monogenic swarms are common in areas where mafic or composite dikes crosscut the plutons.

3.3. Mafic dikes

The distribution of mafic dikes appears to be independent of the composition of their hosts. Mafic dikes commonly form dense swarms. The thickness of the dikes varies from a few tens of centimeters to 5 m. All the intermediate stages between undisturbed mafic dikes and dike-like monogenic MME swarms are found in the granitoids of the central Sierra Nevada batholith, and can be summarized as follows:

(1) Scarce undisturbed mafic dikes are associated with dense swarms of composite dikes (e.g., San Joaquin River Canyon in the Dinkey Creek pluton; Hetch Hetchy Reservoir and North Fork in the Bass Lake pluton; Fig. 1B). The grain size commonly increases from fine-grained margins towards the center. Locally, undisturbed mafic dikes contain angular xenoliths of the host granodiorite.

(2) A mafic dike exposed in the San Joaquin River Canyon shows the first stage of hybridization: tiny felsic veins a few centimeters thick separate the nearly-continuous mafic dike from the tonalite and divide the mafic rock into angular blocks (Fig. 5A).

(3) North of Stevenson Creek and in the San Joaquin River Canyon, partially hybridized mafic dikes consist of pillows of similar mafic rock separated by veins of felsic rock (Fig. 5B). Contacts between these composite dikes and the enclosing granodiorite remain sharp and straight. The 20- to 50-cm mafic pillows are surrounded by discrete fine-grained margins and show lobate contacts with the medium- to fine-grained matrix of felsic granite. Within each composite dike, mafic pillows have similar or close compositions. However, the nature of the mafic pillows may vary markedly from one locality to another: they display either a basaltic appearance

Fig. 5. (A and B) Two types of composite dikes from the Dinkey Creek pluton (San Joaquin River Canyon): the felsic component is much less abundant in A than in B.

with dark color and fine-grained texture (e.g., Highway 140 in the Guadalupe Igneous Complex: Best, 1963), or a more granitic appearance with gray color and fine-grained texture (e.g., Elephant Rock in the El Capitan pluton; Fig. 1B). Intermediate types exist between these two extremes (e.g., Hetch Hetchy Reservoir in the Bass Lake pluton; San Joaquin River Canyon in the Dinkey Creek pluton; Fig. 1B). Whatever their appearance, these mafic pillows have relatively fine-grained, equigranular texture, and phenocrysts or xenocrysts have not been observed. They have fine-grained lobate margins that are well developed where compositional, chemical, and thermal contrasts between the two magmas were strong (dark fine-grained mafic pillows enclosed in a granitic host). Uncommon angular fragments of mafic rocks locally occur in some zones of the composite dikes: they may result from the local disruption of pillows. Mafic pillows are commonly flattened and slightly elongated parallel to the main direction of the dikes. In the bottom of the San Joaquin River Canyon, several composite dikes sharply crosscut each other. A dike of this type, more than 100 m wide, was described and studied in detail in the Eagle Lake quartz monzodiorite pluton, Sequoia National Park, south-central Sierra Nevada batholith (Furman and Spera, 1985).

(4)   At Hetch Hetchy Reservoir, in addition to undisturbed or composite dikes, there is a type of more hybridized composite dikes in which mafic enclaves and the felsic host display many textural heterogeneities. The mafic enclaves vary in size, and are commonly elongated. Contacts with the abundant aplitic granitoid host are locally diffuse. The composition varies from one mafic enclave to another both within a dike and within large enclaves. The color, mineralogy, as well as major and trace elements and isotopes of these enclaves are intermediate between those of the undisturbed dikes and the host Bass Lake tonalite (Barbarin et al., 1989). Boundaries between these composite dikes and the host tonalite are relatively diffuse.

(5)   The next stage involves relatively rare monogenic MME swarms that may represent fragments of composite dikes. Although these swarms do not exhibit a dike shape with parallel, straight, and sharp contacts with the host, they contain compositionally and texturally identical mafic enclaves enclosed in a medium- to fine-grained granitoid similar to the most hybridized composite dikes.

The suggested evolution from undisturbed mafic dikes through composite dikes to the monogenic MME swarms parallels an increase in the proportion of the felsic component. Although undisturbed mafic dikes, various types of composite dikes, and monogenic MME swarms are commonly found in the same area, neither close structural nor chronological relationships were found between them. At Hetch Hetchy Reservoir, dikes that parallel the foliation of the granodiorite or are close to it, are undisturbed, whereas dikes that cut across the foliation are composite. Unfortunately, contacts between these dikes were not found.

The most mafic enclaves have a dioritic composition and mainly consist of plagioclase and mafic minerals, like the MME (Fig. 3). Relatively more felsic blobs are enriched in quartz and K-feldspar, and their composition tends toward those of the granitoid hosts. The mafic blobs contain more hornblende than biotite (Fig. 4) and the ratio of hornblende to the total amount of mafic minerals is higher than in the MME. Compositions of the mafic dikes (Barbarin et al., 1989) resemble those of the dikes described in detail at Hell Hole Meadow in the upper part of San Joaquin River Canyon (Reid and Hamilton, 1987). The composite dikes that consist of a single type of mafic pillows or fragments enclosed in fine-grained and relatively felsic granitoid differ from the dike-like polygenic MME swarms that consist of various types of MME enclosed in coarse-grained and mafic granitoids.

In the Sierra Nevada batholith, undisturbed mafic dikes or composite dikes with various stages of hybridization have not been traced to trains of enclaves, as for example in Donegal (Pitcher, 1991), the Coastal Batholith of Peru (Cobbing and Pitcher, 1972), or the Klamath Mountains of California (Barnes, 1983). In the Sierra Nevada batholith, the genesis of MME and mafic dikes may correspond to two distinct events; this is implied by mafic dikes that sharply crosscut MME (e.g., North Fork in the Bass

Lake pluton; Highway 168 in the Big Sandy Bluffs granite) and may define the foliation in the host granitoids (e.g., Hetch Hetchy Reservoir in the Bass Lake pluton). This suggests that MME were as crystalline as their granitoid host at the time mafic or composite dike intrusion. However, as will be shown in a model developed later, these observations do not exclude the possibility that MME were formed by the disruption of earlier mafic dikes. The modes and evolution of mafic dikes and MME are quite similar, and they might share the same origin, although they were formed at two distinct periods of the history of the pluton (e.g., Barbarin and Didier, 1992).

3.4. Xenoliths

The margins of plutons that are bounded by metamorphic screens or roof pendants commonly contain country-rock xenoliths ranging from some millimeters to some meters in diameter. These xenoliths are commonly mixed with the other enclave types in polygenic enclave swarms (e.g., Courtright Reservoir: Bateman et al., 1984b). However, they are readily distinguished from other enclaves by their angular shapes and rusty halos. Pabst (1928) listed several characteristics, which distinguish country-rock

xenoliths from MME even when they are chemically and mineralogically very similar. At Courtright Reservoir, scattered strongly foliated xenoliths enclosed in MME form composite enclaves. Contacts between xenoliths and MME are always sharp. Although xenoliths are relatively common in the margins of the plutons, neither residues of melting nor thoroughly digested xenoliths, commonly referred to as restites (e.g., Didier and Barbarin, 1991), were observed in the surveyed granitoid plutons of the central Sierra Nevada batholith.

4. Whole-rock and mineral chemistry

Whole-rock major, trace, and rare-earth element and isotope data on the MME and various mafic rocks associated with granitoids in four plutons of central Sierra Nevada were compiled by Barbarin et al. (1989). Major and minor element variation diagrams for MME and host granitoids are given in Dodge and Kistler (1990). The diversity in modal composition of MME and mafic rocks associated with the granitoids (Fig. 3) also shows in chemical composition (Fig. 6). In the granitoids, MgO ranges from <1% to 3.5% and CaO from <1% to 7%, whereas in MME MgO ranges from 2% to 7% and CaO from 4% to 8.5%. The

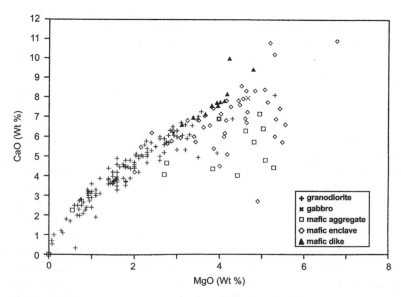

Fig. 6. MgO versus CaO diagram for the granitoids and associated mafic rocks of the central Sierra Nevada batholith. Analyses are given in Bateman et al. (1984a) for most of the granitoids, and in Barbarin et al. (1989) for the other rocks.

composition of the MME spreads over a large area and about half the MME follow the trend of the host granitoids. The rest are enriched in MgO (or depleted in CaO) and plot below the trend for granitoids. These compositional differences are also apparent in the El Capitan granite and the Tuolumne Intrusive Suite, although MME are not so widely scattered here (Reid et al., 1983). MME are distinctly enriched in CaO, MgO, FeOt, and less strongly enriched in TiO_2, MnO, and P_2O_5 compared to their hosts, reflecting their greater abundance of plagioclase, mafic minerals, and apatite. Mafic aggregates and schlieren layerings commonly contain less CaO (about 4%) than MgO (2.5–5%). Where MgO content is about 5%, CaO content may vary from about 4% to 7%. Differences between MME and mafic aggregates are related to the relative abundance of andesine in MME and of mafic minerals and especially of biotite in mafic aggregates. The MME that plot below the main granitoid trend are enriched in mafic minerals, especially biotite. Mafic dikes form a trend parallel to that of the granitoid hosts, with the exception of one mafic dike and one MME. Xenoliths (not shown in Fig. 6) are poor in MgO (1.5–2.5%). One xenolith plots close to the trend and two others that contain 15% and 17% CaO plot far above all the other rocks.

An even more striking relation shown by the major elements is that MgO and FeOt in MME and their host granitoids vary sympathetically within the same pluton and from one pluton to another (Barbarin, 1991). Where MME are relatively poor in MgO and FeOt, the host granitoid is also poor in them. Furthermore, the FeOt/MgO ratios are almost constant in the various rocks. These relations reflect similar chemical compositions of the mafic minerals in each enclave–granitoid pair.

The relative abundance of REE, especially of heavy REE, and their slight fractionation in MME results in flatter REE patterns, whereas the relatively greater fractionation of REE in host granitoids results in steeper REE patterns (Fig. 7). There is a close relationship between the slope of REE patterns in MME and granitoids (Barbarin et al., 1985). It is noticeable that granitoids with less fractionated REE (tonalites) contain MME with less fractionated REE, whereas granitoids with relatively more fractionated REE (granodiorites) contain MME with relatively more fractionated REE. Like the MgO and FeOt

contents, the differences in REE contents and their fractionation between enclave–host granitoid pairs are broadly constant. In contrast, the relative Eu anomaly characteristic of the MME is commonly absent in the host granitoids. Enrichment of REE in MME and especially of HREE can be explained either by concentration of hornblende and possibly apatite in the MME, or more probably, by preferential partitioning of REE between mafic and felsic liquids (Watson, 1976; Ryerson and Hess, 1978).

The trace element contents of MME and host granitoids are distinct (Barbarin et al., 1989; Dodge and Kistler, 1990). Generally, MME contain larger amounts of the transition elements (Sc, Cr, Co, Zn, Sb, Nb) and smaller amounts of Ba and Th than the granitoids (Fig. 8). The contents of Sr, Zr, Hf, and Ta are broadly the same in MME and their hosts. Cs and Th vary differently in the various pairs. However, enrichment patterns are similar for all MME–granitoid pairs, and a strong correlation exists between the trace element content of MME and the nature of their granitoid hosts (Fig. 8). The variations are mainly related to the degree of differentiation of the granitoids and the abundance of minerals such as K-feldspar. Fig. 8 shows that the Mount Givens granodiotite is particularly enriched in K, Rb, Cs, and Ba compared to its MME, because it is enriched in K-feldspar (3- to 5-cm-long megacrysts). The trace element contents of the mafic dikes are similar to those of the MME and distinct from those of the granitoids. Large compositional variations exist between the different types of dikes (undisturbed, composite, or largely hybridized).

Sr and Nd isotope compositions have been determined for MME, mafic aggregates, mafic dikes, and host granitoids (Barbarin et al., 1989). For each pluton, MME and some of the mafic aggregates plot on Rb–Sr isochrons previously defined using granitoid samples only (Dodge and Kistler, 1990). In addition, there are no significant differences in the ε_{Nd} and ε_{Sr} values between MME and the host granitoids. Similar results have been obtained from many other MME–granitoid pairs worldwide (e.g., Holden et al., 1987; Pin et al., 1990). In contrast, mafic dikes may plot above or below the isochrons when they are undisturbed or slightly hybridized (e.g., San Joaquin River Canyon in the Dinkey Creek pluton), and close

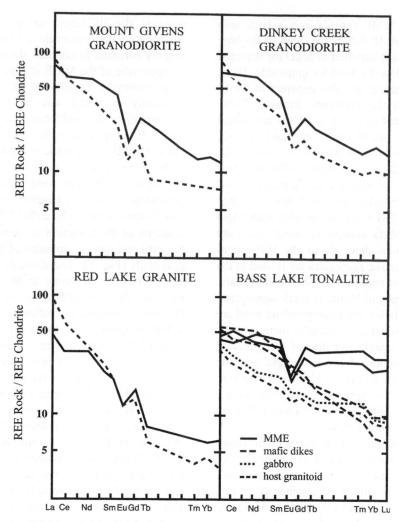

Fig. 7. Selected and typical chondrite-normalized REE patterns of host granitoids, mafic magmatic enclaves and mafic dikes from some plutons of the central Sierra Nevada batholith (data from Barbarin et al., 1989).

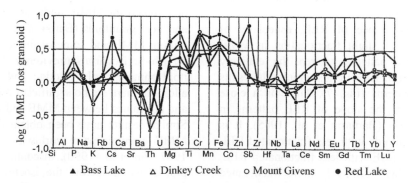

Fig. 8. Host granitoid-normalized major and trace element patterns for selected mafic magmatic enclaves from some plutons of the central Sierra Nevada batholith (data from Barbarin et al., 1989).

to or on the isochrons when they are composite and partially digested (e.g., Hetch Hetchy Reservoir in the Bass Lake pluton).

Electron microprobe analyses of biotite, hornblende, and plagioclase from MME, host granitoids, and mafic aggregates are given in Barbarin et al. (1989). The mafic minerals from the cores and the fine-grained, darker margins of a few enclaves were analyzed separately. Despite differences in whole-rock composition, biotite and hornblende in MME have the same chemical compositions as in contiguous mafic aggregates or host granitoid, regardless of differences in grain size, morphology, crystal habit, or location within the MME, mafic aggregates, or granitoids (Fig. 9). Atomic Fe/(Fe+Mg) ratios clearly show this relation. When compositions of mafic minerals change within a pluton, parallel changes are observed in the MME and mafic aggregates (Fig. 9). These compositional variations are also found in biotite and hornblende from closely spaced enclave–granitoid or enclave–mafic aggregate pairs from the same locality. Identical chemical compositions of the mafic minerals in each enclave–granitoid pair may explain the sympathetic variations of MgO and FeOt in MME and their host granitoids (Barbarin, 1991) and the almost constant FeOt/MgO ratios. Plagioclase phenocrysts in enclaves and large crystals of plagioclase in host granitoids are also compositionally similar and

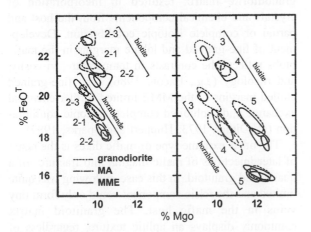

Fig. 9. MgO and FeOt contents in biotite and hornblende in various host granitoid or mafic aggregate–MME pairs from either the same locality (2-1, 2-2, and 2-3: quarry near the Shaver Lake Dam) or from different localities of the Dinkey Creek pluton (data from Barbarin et al., 1989). Each zone is defined by analyses from about five different grains.

show similar zoning (Barbarin, 1990). However, small crystals from the matrix of enclaves are more albitic and generally unzoned.

5. The role of hybridization processes in granitoid magmas

Morphology, magmatic textures, mineralogy, and chemical features of MME indicate that they crystallized from a relatively mafic magma of intermediate composition. Varying textures and chemical compositions (Barbarin et al., 1989), and exceptional composite enclaves that consist of xenolith(s) enclosed in a MME (see Fig. 2a in Barbarin, 1991), clearly show that MME cannot have formed by partial or total transformation and recrystallization of wall rocks xenoliths such as plagioclase–diopside hornfelses (e.g., Link, 1969). The presence of phenocrysts of hornblende and oscillatory zoned andesine in the fine-grained margins and the magmatic foliation (Fig. 2B) favor magmatic origin. MME are not restites and cannot have formed by transformation of the residues of partial melting of crustal rocks (as proposed by Bateman and Chappell, 1979): they do not contain peraluminous minerals (like most restites do in which at least sillimanite is present), and their metaluminous composition also excludes indirect derivation by melting of peraluminous restites.

It is commonly accepted that mafic magma was associated with granitic magma in most plutons and that MME represent modified blobs of this mafic magma. Furthermore, observations of xenocrysts in MME (Hibbard, 1981, 1991; Reid and Hamilton, 1987; Barbarin, 1990; Didier and Barbarin, 1991) and the development of felsic haloes in the granitoid hosts near the contact with the enclaves (see Fig. 1d in Barbarin, 1991) infer interaction between mafic and granitic magmas. Although some investigators have questioned the importance of magma mixing or mingling (e.g., Furman and Spera, 1985; Frost and Mahood, 1987), no one denies their significance in the calc-alkaline granitoid plutons of the Sierra Nevada batholith.

The different granitoids and associated mafic rocks in the Sierra Nevada batholith may result from hybridization processes that acted at three distinct but continuous periods and at different levels in the crust: (1) thorough mixing at depth produced homogeneous magmas that crystallized to granitoids, possibly with some MME; (2) mingling and local mixing during ascent and emplacement produced the many types of MME; (3) mingling and limited mixing at the emplacement level produced the MME of the composite and strongly hybridized dikes. Differentiation or local segregation commonly followed mixing and mingling (Barbarin and Bateman, 1986).

Granitoids represent homogeneous hybrid rocks in which the two original mafic and felsic components are obscure. Sr, Nd, and O isotopic data indicate that most of the Sierran calc-alkaline granitoid magmas were initially hybrids produced by mixing of mafic mantle-derived melts and felsic crustal materials (Kistler and Peterman, 1973; DePaolo, 1981; Domenick et al., 1983; Kistler et al., 1986; Hill et al., 1988).

The many different types of MME were produced by mingling and local mixing of mafic magma with granitoid magma during ascent and emplacement. Local mixing occurred between each blob of mafic magma and the enclosing granitoid magma, and differed from the thorough mixing at depth. Mafic magma could melt surrounding crustal rocks in the lower crust only if it was trapped long enough to permit thermal exchange (Marsh, 1982; Huppert and Sparks, 1988; Annen and Sparks, 2002). Thus, mafic magma that was injected into an open system in which granitoid magma was already moving upward did not have enough time to thoroughly mix with the felsic host and was disrupted into small scattered blobs in the moving granitoid magma. During ascent and emplacement, mingling continued and interaction between granitoid and MME involved thermal, mineral, and chemical transfers. The many types of MME (Figs. 3 and 6) crystallized from distinct magmas related to distinct mixing events that involved different proportions of the end members. These local mixings explain the similar mineralogical and chemical traits of each MME–granitoid pair and the differences between the various pairs in the same pluton. The various types of MME may then have aggregated into swarms. Other relatively solid particles (crystals, xenoliths) that were present in the granitoid magma were also concentrated in the swarms. These polygenic swarms typically consist of many different types of MME and scattered xenoliths, both enclosed in a matrix rich in biotite and hornblende phenocrysts. They are concentrated near the margins of the plutons, because, in a flowing magma, they formed lag deposits between domains of contrasting temperatures and rheologies. Polygenic swarms that include many different types of MME should be distinguished from monogenic swarms in which MME are similar and formed at the same time.

At the emplacement level, incompletely crystallized granitoid magmas were affected by early fractures (cf. Hibbard and Watters, 1985) that channelled the residual liquid and fluids. Mafic magmas that were injected into the fractures fragmented into many blobs and interacted with the residual granitoid magma (e.g., Furman and Spera, 1985). Each fracture represented a particular mixing system with characteristic proportions of the end members and physical conditions, and then produced a certain type of MME. Enclaves followed their host, moved within the fracture, became slightly flattened or elongated, but did not leave the fracture. The great variety of dikes indicates that they formed and evolved differently. Major controlling factors were the relative physical properties of the two components and, consequently, depended on the amount of felsic magma available and the time when mafic magma was injected into the early fractures (e.g., Barbarin and Didier, 1992). Hybridization of MME scattered in a tonalite or granodiorite matrix resulted in incorporation of crystals and chemical components from the host and partial or complete isotopic equilibration. Development of fine-grained and lobate margins in the mafic blobs reflects the contrasts in temperature, viscosity, and rheology (e.g., Bacon, 1986). The fine-grained chilled margins of the MME limited further chemical exchange and prohibited complete isotopic equilibration (cf. Didier, 1973; Huppert and Sparks, 1989).

The other extreme type of mafic dikes is the result of later injections of mafic magma into fractures of a quasi-solid granitoid. In this case, pillowing was quite limited and the felsic component could only form tiny veins in the mafic host. The granitoid matrix commonly displays an aplitic texture, regardless of its composition. The composite dikes that sharply crosscut each other in the bottom of San Joaquin River Canyon show that intervals between injections of mafic magma were relatively short. Immediately following the emplacement of a composite dike, a new fracture formed and allowed injection of another batch of mafic magma and crystallization of a new composite dike.

The distribution of MME and different types of mafic dikes in the granitoid plutons of the Sierra Nevada batholith vary a great deal. The abundance of MME decreases from the most mafic to the most felsic facies of the intrusive suites. Both the abundance of MME and the relatively mafic composition of the host granitoids are explained by the higher proportion of mafic component involved in the hybridization processes (Kistler et al., 1986; Barbarin, 1989). If the addition of the mafic component to the mixing system was continuous, the entire pluton has a relatively mafic composition and contain many MME (e.g., the western Cretaceous tonalite pluton of the Sierra Nevada batholith such as the Bass Lake Tonalite; Bateman, 1992). If a large batch of mafic magma was introduced into the mixing system at depth only once, the first pulse of granitoid would

consist of mafic granitoids that contain many MME, whereas the next pulses would consist of more felsic granitoids and fewer MME (e.g., the Tuolumne Intrusive Suite; Bateman and Chappell, 1979). As the amount of the mafic component decreased and that of the anatectic magma increased by enchanged melting of crustal rocks (cf. Huppert and Sparks, 1988), involvement of the mafic component decreased progressively and possibly was nil during the latest granitoid pulses (Barbarin, 1989). Consequently, MME are scarce or absent in the latest suites such as Tuolumne. In contrast, the distribution of mafic dikes is independent of the nature of the host and is not controlled by the relative proportion of the end members in a mixing system at depth, but mainly depends on the presence of fracture systems in the partially crystallized plutons and availability of mafic magma.

In granitoid plutons, different types of interactions between felsic and mafic components are constrained

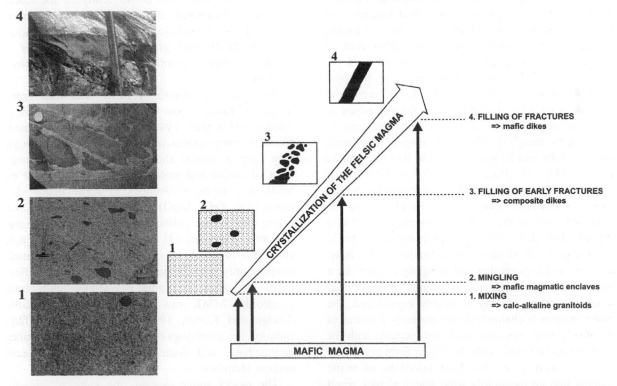

Fig. 10. Sketch showing the various types of hybridization processes resulting from injection of mafic magma into a felsic magma at different stages of crystallization of the felsic magma (from Barbarin and Didier, 1992), illustrated with cases from the Dinkey Creek pluton: granodiorite (photograph 1) and MME (photograph 2) from Shaver Lake Quarry; composite dike (photograph 3) and undisrupted dike (photograph 4) from San Joaquin River Canyon. See text for detailed explanation.

not only by the relative proportions of the two components but also by the physical properties of the contrasted magmas (e.g., Furman and Spera, 1985; Sparks and Marshall, 1986; Frost and Mahood, 1987). The physical features of these magmas and especially the relative rheology of the felsic and mafic magmas are closely related to their compositions, temperatures, and rates of crystallization. As these factors do not change simultaneously during the evolution of the system, several types of interactions may appear sequentially, and yield various types of hybrid rocks from the same parent magmas (e.g., Barbarin, 1988). Barbarin and Didier (1992) proposed that the various types of mafic rocks and enclaves are produced by injection of mafic magma into granitoid magma at different stages of crystallization of the latter. Accordingly, a fourfold model can be applied to the Sierra Nevada granitoids and associated mafic rocks (Fig. 10). (1) If mafic magma is introduced before the beginning of the crystallization of the felsic magma, thorough mixing results in homogeneous hybrid magmas and calc-alkaline granitoids. This type of mixing occurs at depth and is usually favored by convection. If crystals are already present in one or both of the components, they may become xenocrysts in apparently-homogeneous hybrid rocks. (2) If the mafic magma is introduced slightly later, the viscosities of the two magmas can be sufficiently different to permit only mingling. The mafic magma may break up into blobs and be scattered in the felsic magma to form MME. Mingling results in an increase of contact surfaces between the two components and promotes chemical transfer between MME and the granitoid host. In the granitoid plutons of the Sierra Nevada batholith, the large proportions of mafic magma probably delayed crystallization of the felsic magmas and kept mixing and mingling efficient for a long time. (3) If the mafic magma is introduced when the felsic magma is largely crystallized, the mafic magma is channelled into the early fractures of the nearly-solid granitoid rock and interacts with the last magmatic liquids only locally to form composite or fragmented dikes. (4) Late injections of mafic magma into an essentially solid granitoid rock result in undisturbed mafic dikes. Rheologies of the two components are so different that most exchanges are inhibited.

6. Origin of the mafic magma

In the Sierra Nevada batholith, as in many other plutons worldwide, the problem of the origin of the mafic magma remains. Most MME probably represent modified blobs of mafic magma and their study might give information on the nature and origin of this magma. The mafic magma is broadly coeval with the felsic magma and interactions of these magmas are evident particularly at the emplacement level. In the many models proposed (e.g., Kistler and Peterman, 1973; Brown, 1977; DePaolo, 1981; Hill et al., 1988; Kistler et al., 1986; Dodge and Kistler, 1990), the mafic magma either originates in the upper mantle or represents non-differentiated, cumulative parts of the granitoid magma. In the central part of the Sierra Nevada batholith, one of the goals of the petrographic, mineralogical, geochemical, and isotopic survey of the MME and host granitoids from several selected plutons (Barbarin et al., 1989) was to determine the origin of the MME and consequently of the mafic magma. Unfortunately, the data do not clearly favor either of the hypotheses. The isotopic correlation between MME and granitoids requires either a common source or complete equilibration of dissimilar materials.

The cogenetic nature of MME and host granitoids is based upon the following arguments (Dodge and Kistler, 1990): (1) mafic minerals have identical compositions in MME and host, although they vary in modal abundance; (2) although contents of major and trace elements are different in MME and hosts, differences in some major elements (e.g., FeOt, MgO) are relatively constant, and strong correlations exists between variation of major and trace elements; (3) in each pluton, MME plot along Rb–Sr isochrons determined using granitoid samples only and do not display significant differences in either ε_{Nd} and ε_{Sr} values with host granitoids. MME can be thus called autoliths (Dodge and Kistler, 1990). In this model, differentiation (or unmixing) of a magma produced mafic, intermediate, and felsic magmas within a zoned magma chamber.

The model above explains the identical isotopic features of MME and host granitoids, but it does not explain why the isotopic values are intermediate between those of crustal and upper mantle materials

(Kistler and Peterman, 1973; DePaolo, 1981; Domenick et al., 1983; Kistler et al., 1986; Hill et al., 1988; DePaolo et al., 1992). As it is quite impossible to produce magmas with such isotopic features by melting crustal rocks only, addition of upper mantle magma is necessary. Furthermore, the upper mantle provides not only material but also the heat necessary to melt the crustal rocks (e.g., Didier and Lameyre, 1969; Brown and Fyfe, 1970; Reid et al., 1983; Huppert and Sparks, 1988; Annen and Sparks, 2002). Even if the MME of the Sierra Nevada batholith do not display a distinctive mantle signature, they are mixtures of crustal melt and mantle-derived magma like their calc-alkaline granitoid hosts (Kistler and Peterman, 1973; DePaolo, 1981; Domenick et al., 1983; Kistler et al., 1986; Hill et al., 1988; DePaolo et al., 1992). Some investigators have gone further to propose that calc-alkaline granitoids and MME were produced by the differentiation of upper mantle magmas without addition of crustal melt (e.g., Brown, 1977). Such models may be supported by the absence of aluminous mineral-bearing restites that imply involvement crustal rocks. Isotopic data, however, generally indicate incorporation of a crustal component even if the mantle-derived component is dominant (e.g., DePaolo, 1981; Kistler et al., 1986).

Considering that mafic dikes in the Sierra Nevada batholith may represent the mafic magma from which the MME were derived, isotopic systematics indicate that, through hybridization processes, isotopic equilibration for Rb–Sr may occur between the most hybridized blobs of mafic rocks and the host granitoid in the composite dikes; these then plot on the same isochron, whereas some darker and non-hybridized blobs plot off of it (Barbarin et al., 1989; Dodge and Kistler, 1990). After mingling, during ascent and emplacement, and until the pluton cooled off, diffusion processes were active between blobs of mafic magma and the enclosed granitoid magma and induced partial chemical and complete isotopic equilibration. Experimental studies indicate that diffusion processes induce chemical and isotope equilibration between silicates of contrasted compositions, and that isotopic equilibration is generally more easily achieved than chemical equilibration, because isotopic exchanges proceed more quickly than chemical exchanges (Lesher, 1990). Isotopic

equilibration between MME and host granitoids is a feature of mingling and mixing in the plutonic environment. In the volcanic environment, as eruption and cooling occur immediately after mixing and mingling, there is no time for isotopic equilibration: enclaves and host lavas may display distinct signatures.

Parallel variations in the chemistry of MME and granitoids are related to different contents of minerals having similar compositions (Barbarin, 1986). In each MME–granitoid pair, constant differences for FeOt and MgO contents result from the occurrence of hornblende and biotite with identical chemical compositions but in different proportions. Furthermore, mechanical mineral transfer helps to explain some chemical relationships between MME and hosts. Correlations between the abundance and fractionation of REE in MME and host granitoids probably result from mechanical transfer of accessory minerals. Exchanges of apatite crystals between MME and granitoids have been demonstrated in several plutons worldwide (e.g., Didier and Barbarin, 1991). In the Sierra Nevada batholith, granitoids rich in titanite enclose MME rich in titanite (e.g., Mount Givens granodiorite), whereas granitoids poor in titanite enclose MME in which titanite is rare. This correlation between titanite content in MME and host granitoids may either result from mechanical transfer of crystals or chemical transfer of the elements necessary for crystallization of titanite.

Identical compositions of minerals such as biotite and hornblende in MME and contiguous granitoid (Fig. 9) might be explained by crystallization under the same physical conditions (Barbarin, 1986). Even if mafic minerals are commonly more elongated and sometimes acicular in the MME, the differences in size or shape do not imply variations in composition. Most phenocrysts of mafic minerals and plagioclase were probably transferred either from the mafic to the felsic magmas or reversely (Barbarin, 1990). Unlike isotopic equilibration that occurred throughout the pluton, chemical compositions that are identical for the same minerals of MME and contiguous granitoids may change from one locality to another in the same pluton. Local variations in mineral compositions reflect local evolution of the host granitoids. Uniform mineral compositions and similar chemical affinities in the MME and hosts may also indicate that, concomitantly with mixing and mingling, diffusive processes affected the MME and host granitoids (e.g., Lesher, 1990; Baker, 1990).

Autoliths certainly exist but are relatively rare in the plutons of the Sierra Nevada batholith. Rare MME with large crystals and cumulate textures are autoliths and come from early disruption of cumulate layers in the plutons. Autoliths do not show signs of recrystallization even where they are associated with common MME in some swarms (Dorais et al., 1990; Barbarin, 1991). The rarity of these enclaves and their preserved cumulate texture exclude the model where MME with fine-grained texture are supposed to result from the recrystallization of coarse-grained cumulate rocks. Some unusual MME also have large grains and are identical with the rock that forms the schlieren layerings and mafic aggregates so common in the margins of the plutons (e.g., Courtright Reservoir: Bateman et al., 1984b). These MME come from the disruption of schlieren layerings or mafic aggregates, and differ from the usual fine-grained MME. When mechanical processes concentrate large crystals of biotite, hornblende, titanite, and other accessory minerals to schlieren layerings and mafic aggregates, MME were already present in the magma: they are invariably in sharp contact with the coarse-grained mafic mineral-rich matrix (Barbarin and Bateman, 1986). Some MME are commonly included in the schlieren layerings or mafic aggregates, as are large crystals and xenoliths that floated in the granitoid magma. Many polygenic swarms consist of MME enclosed in mafic aggregates (e.g., Courtright Reservoir: Bateman et al., 1984b). Disruption of these swarms by magma movement has locally produced isolated MME surrounded by granitoid rich in large crystals of mafic minerals (e.g., May Lake in the Tuolumne Intrusive Suite). Furthermore, in some polygenic swarms, contacts between MME may look more like contacts between pebbles than contacts between blobs of magma. These features suggest that when mechanical processes produced mafic aggregates and schlieren layerings, most MME were already formed and thus there is no genetic relationships between the common mafic aggregates or schlieren layerings and MME.

Rare MME display a cumulate texture and are autoliths (Dorais et al., 1990; Barbarin, 1991). One

problem is whether the more common and fine-grained MME enclosed in the plutons of the Sierra Nevada batholith are also autoliths. No petrographic, mineralogical, or chemical data clearly indicate whether the mafic magma that crystallized MME came from the upper mantle or from the same magma as the granitoid host. The above review of the different arguments suggests that the two origins are both possible. Actually, MME and host granitoids can be considered cogenetic, because they are both hybrid rocks resulting from the mixing of two magmatic components. Isotopic systematics indicate that the Sierra Nevada batholith granitoids consist of up to 30 vol.% of preexisting crustal material (DePaolo, 1981). MME probably contain from 10 vol.% to a few volume percent of crustal materials. The relatively limited difference in composition between the end members and the chemical equilibration that occurred during and after emplacement explain why so many similarities exist between the MME and host granitoids in the Sierra Nevada batholith. The mafic magma is presumed to have come from the upper mantle (e.g., Didier et al., 1982; Cantagrel et al., 1984) and to have been andesite (Reid et al., 1983) or high-alumina, calc-alkaline, or tholeiitic basalt in composition (Dorais et al., 1990).

7. General model for the formation of granitoids and associated mafic rocks

Considering the petrographic, mineralogical, geochemical, and isotopic data, a general model can be proposed for the origin and evolution of the Sierra Nevada granitoids and the associated mafic rocks including MME and dikes (Fig. 11). This model uses a triangular diagram in which granitoids, mafic aggregates, MME, and mafic dikes are plotted according to their content of mafic minerals (plus accessories), plagioclase, and quartz and K-feldspar (Fig. 3): (1) a mantle-derived mafic component M (basalt composition) thoroughly mixes with a felsic component F (graywacke composition) to create hybrid magmas H, which then fractionate to produce the different granitoids; (2) when contrasting physical conditions inhibit thorough mixing between the mafic and felsic components (Fig. 10), mingling occurs and

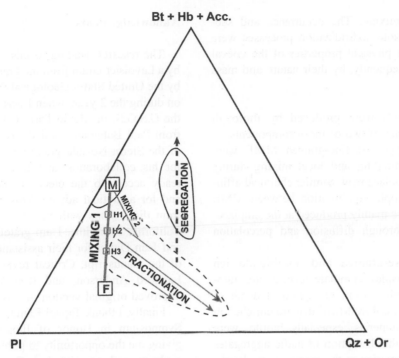

Fig. 11. Complex model proposed for the formation of the granitoids and various associated mafic rocks of the central Sierra Nevada batholith. This model is based on the modes of the different rocks (cf. Fig. 3) and is constrained by field, geologic, mineralogical, chemical, and isotopic data. (F) and (M) represent the supposed parent felsic and mafic end members and (H) the various homogeneous hybrid magmas that differentiated to produce the various granitoids. See text for detailed explanation.

leads to the formation of MME (local mixing between each MME and host results in enrichment of the MME in K-feldspar and quartz); (3) late surges of mafic magma mix with evolved granitoid magmas to produce the hybrid magmas of the composite dikes or monogenic swarms; (4) coarse-grained mafic aggregates represent mechanical concentrations of mafic minerals and accessory phases from evolved granitoid magmas. MME and xenoliths commonly are included in these segregations.

This model differs from the models proposed for plutons elsewhere (review in Didier and Barbarin, 1991), because it involves complex and multiple mixing and mingling processes instead of a single one-stage process. The sequential occurrence of the hybridization processes depends on the relative proportions of the two components and on their physical properties (Furman and Spera, 1985; Sparks and Marshall, 1986; Frost and Mahood, 1987; Barbarin, 1988; Barbarin and Didier, 1992). The complex model is consistent with the large diversity of MME in the same pluton, the existence of

composite enclaves that consist of a MME enclosed in another MME, and explains why the MME do not invariably plot on the granitoid trends in binary evolution diagrams (Fig. 6). In the Sierra Nevada batholith, hybridization processes cannot be either demonstrated or modeled using simple mixing tests, because they were of various types, commonly affected local areas and limited volumes of magmas, and occurred at different stages of the evolutionary history of the plutons.

8. Conclusions

A complex combination of hybridization, differentiation, and segregation processes governed the genesis and evolution of the granitoids and mafic rocks, including MME, in the plutons of the central Sierra Nevada batholith. Commonly, mixing and mingling of two contrasting end member components were followed by differentiation and segregation processes that affected the hybrid magmas produced during mixing. The occurrence and succession of the various hybridization processes were constrained by the physical properties of the coeval magmas and, consequently, by their nature and mass fractions:

(1) Host granitoids were produced by thorough mixing at depth of two or more components.
(2) The many types of fine-grained MME were formed by mingling and local mixing during ascent and emplacement. Similar chemical affinities and isotopic equilibration between MME and hosts were mainly produced at the emplacement level through diffusion and percolation processes.
(3) Scarce coarse-grained and hornblende-rich MME that display cumulate textures are autoliths and resulted from disruption at depth of early cumulates. Rare MME that are enriched in large mafic minerals, especially biotite, were produced by the disruption of mafic aggregates or schlieren layering at the emplacement level.
(4) In the evolved granitoid magmas, mafic aggregates and schlieren layering resulted from mechanical segregation and concentration of mafic minerals and other solid particles such as xenoliths or MME.
(5) Mafic dikes represent injection of mafic magma into early fractures in the cooling and crystallizing plutons. If some granitoid magma was still present, the mafic magma fragmented into blobs and formed composite dikes. The nature and extent of the fragmentation and exchanges depended on the rheology contrasts and relative abundance of the two components in the mixing systems.

Detailed field observations and petrographic, mineralogical, chemical, and isotope studies of the granitoids, MME, and other mafic rocks in some plutons of the central Sierra Nevada batholith provide information of their genesis and evolution. The model involving multiple hybridization processes can probably be applied to most calc-alkaline plutons worldwide (Barbarin, 1988; Didier and Barbarin, 1991), in particular Cordilleran batholiths that contain abundant associated mafic rocks and MME.

Aknowledgements

The research leading to this report was supported by a Lavoisier Grant from the French Government and by the United States Geological Survey. It was carried on during the 2 years when I was a visiting scientist at the U.S.G.S. in Menlo Park, CA. I largely benefited from Paul Bateman's and Frank Dodge's knowledge of the Sierra Nevada granitoids. I thank Ron Kistler for his collaboration and Lew Calk for providing ready access to the electron microprobe laboratory and for analytical advice. This paper also benefited from discussions with many geologists working on California batholiths. I am grateful to Paul Bateman and Jean Didier for their assistance on early versions of the manuscript. Critical reviews by John Hogan, Lawford Anderson, and Ron Vernon also largely improved original version of this report.

Finally, I thank Tapani Rämö for inviting me to the Symposium in Honor of Ilmari Haapala, and for giving me the opportunity to publish this paper on the enclaves and associated mafic rocks of the Sierra Nevada batholith.

References

Annen, C., Sparks, R.S.J., 2002. Effects of repetitive emplacement of basaltic intrusions on thermal evolution and melt generation in the crust. Earth and Planetary Science Letters 203, 937–955.

Bacon, C.R., 1986. Magmatic inclusions in silicic and intermediate volcanic rocks. Journal of Geophysical Research 91, 6091–6112.

Baker, D.R., 1990. Chemical interdiffusion of dacite and rhyolite: anhydrous measurements at 1 atm and 10 Kbar, application of transition state theory, and diffusion in zoned magma chambers. Contributions to Mineralogy and Petrology 104, 407–423.

Barbarin, B., 1986. Comparison of mineralogy of mafic magmatic enclaves and host granitoids, central Sierra Nevada, California. Abstracts with Programs-Geological Society of America 18, 83.

Barbarin, B., 1988. Field evidence for successive mixing and mingling between the Piolard Diorite and the Saint Julien-la-Vêtre Monzogranite (Nord Forez, Massif Central, France). Canadian Journal of Earth Sciences 25, 49–59.

Barbarin, B., 1989. Mélange de magmas et origine de la zonation normale des plutons granitiques crétacés du batholite de la Sierra Nevada, Californie. Comptes rendus de l'Académie des Sciences Paris 309, 1563–1569.

Barbarin, B., 1990. Plagioclase xenocrysts and mafic magmatic enclaves in some granitoids of the Sierra Nevada Batholith, California. Journal of Geophysical Research 95, 17747–17756.

Barbarin, B., 1991. Enclaves of the Mesozoic calc-alkaline granitoids of the Sierra Nevada batholith, California. In: Didier, J., Barbarin, B. (Eds.), Enclaves and Granite Petrology, Developments in Petrology, vol. 13. Elsevier, Amsterdam, pp. 135–153.

Barbarin, B., 1999. A review of the relationships between granitoid types, their origins and their geodynamic environments. Lithos 46, 605–626.

Barbarin, B., Bateman, P.C., 1986. Origin and evolution of mafic magmatic enclaves and mafic rocks associated with some granitoids of the central Sierra Nevada. 14th International Mineralogical Association Meeting, Stanford, California, 1986, Abstracts with Programs, p. 50.

Barbarin, B., Didier, J., 1992. Genesis and evolution of mafic microgranular enclaves through various types of interaction between coexisting felsic and mafic magmas. Transactions of the Royal Society of Edinburgh: Earth Sciences 83, 145–153.

Barbarin, B., Dodge, F.C.W., Kistler, R.W., 1985. REE contents and Rb–Sr systematics of mafic enclaves and other associated mafic rocks, central Sierra Nevada, California. EOS Transactions of the American Geophysical Union 66, 1150.

Barbarin, B., Dodge, F.C.W., Kistler, R.W., Bateman, P.C., 1989. Mafic inclusions and associated aggregates and dikes in granitoid rocks, central Sierra Nevada Batholith. Analytic Data, U.S. Geological Survey Bulletin, 1899.

Barnes, C.G., 1983. Petrology and upward zonation of the Wooley Creek Batholith, Klamath Mountains, California. Journal of Petrology 24, 495–537.

Bateman, P.C., 1992. Plutonism in the Central Part of the Sierra Nevada Batholith, California. U.S. Geological Survey Professional Paper, 1483.

Bateman, P.C., Chappell, B.W., 1979. Crystallization, fractionation, and solidification of the Tuolumne Intrusive Series, Yosemite National Park, California. Geological Society of America Bulletin 90, 465–482.

Bateman, P.C., Nokleberg, W.J., 1978. Solidification of the Mount Givens granodiorite, Sierra Nevada, California. Journal of Geology 86, 563–579.

Bateman, P.C., Dodge, F.C.W., Bruggman, P.E., 1984a. Major oxide analyses, CIPW norms, modes, and bulk specific gravities of plutonic rocks from the Mariposa 1°×2° sheet, central Sierra Nevada, California. U.S. Geological Survey Open-File, 84–162.

Bateman, P.C., Kistler, R.W., DeGraff, J.V., 1984b. Courtright intrusive zone: Sierra National Forest, Fresno County, California. California Geology 37, 91–98.

Best, M., 1963. Petrology of the Guadalupe igneous complex, southwestern Sierra Nevada foothills, California. Journal of Petrology 4, 223–259.

Brown, G.C., 1977. Mantle origin of Cordilleran granites. Nature 265, 21–24.

Brown, G.C., Fyfe, W.S., 1970. The production of granitic melts during ultrametamorphism. Contributions to Mineralogy and Petrology 28, 310–318.

Cantagrel, J.M., Didier, J., Gourgaud, A., 1984. Magma mixing: origin of intermediate rocks and "enclaves" from volcanism to plutonism. Physics of the Earth and Planetary Interiors 35, 63–76.

Chen, J.H., Moore, J.G., 1982. Uranium–lead isotopic ages from the Sierra Nevada batholith, California. Journal of Geophysical Research 87, 4761–4784.

Cloos, E., 1936. Der Sierra Nevada pluton in Californien, Neues Jahrbuch fur Mineralogie. Geologie und Palaontologie 76, 355–450.

Cobbing, E.J., Pitcher, W.S., 1972. The Coastal Batholith of central Peru. Journal of the Geological Society [London] 128, 421–460.

Coleman, D.S., Glazner, A.F., Miller, J.S., Bradford, K.J., Frost, T.P., Joyce, J.L., Bachl, C.A., 1995. Exposure of a Late Cretaceous layered mafic–felsic magma system in the central Sierra Nevada batholith, California. Contributions to Mineralogy and Petrology 120, 129–136.

Cruden, A.R., Tobisch, O.T., Launeau, P., 1999. Dinkey Creek pluton, central Sierra Nevada, California: magnetic evidence for conduit-fed emplacement of a tabular granite. Journal of Geophysical Research 104, 10511–10530.

DePaolo, D.J., 1981. A neodymium and strontium isotopic study of the Mesozoic calc-alkaline granitic batholiths of the Sierra Nevada and Peninsular Ranges, California. Journal of Geophysical Research 86, 10470–10488.

DePaolo, D.J., Perry, F.V., Baldridge, W.S., 1992. Crustal versus mantle sources of granitic magmas: a two-parameter model based on Nd isotopic studies. Transactions of the Royal Society of Edinburgh. Earth Sciences 83, 439–446.

Didier, J., 1973. Granites and their enclaves. The Bearing of Enclaves on the Origin of Granites, Development in Petrology, vol. 3. Elsevier, Amsterdam.

Didier, J., Barbarin, B. (Eds.), Enclaves and Granite Petrology, Developments in Petrology, vol. 13. Elsevier, Amsterdam.

Didier, J., Lameyre, J., 1969. Les granites du Massif Central Français. Etude comparée des leucogranites et granodiorites. Contributions to Mineralogy and Petrology 24, 219–238.

Didier, J., Duthou, J.L., Lameyre, J., 1982. Mantle and crustal granites, genetic classification of orogenic granites and the nature of their enclaves. Journal of Volcanology and Geothermal Research 14, 125–132.

Dodge, F.C.W., Kistler, R.W., 1990. Some additional observations on inclusions in the granitic rocks of the Sierra Nevada. Journal of Geophysical Research 95, 17841–17848.

Dodge, F.C.W., Papike, J.J., Mays, R.E., 1968. Hornblendes from granitic rocks of the central Sierra Nevada Batholith, California. Journal of Petrology 9, 378–410.

Dodge, F.C.W., Smith, C., Mays, R.E., 1969. Biotites from granitic rocks of the central Sierra Nevada Batholith, California. Journal of Petrology 10, 250–271.

Domenick, M.A., Kistler, R.W., Dodge, F.C.W., Tatsumoto, M., 1983. Nd and Sr study of crustal and mafic inclusions from the Sierra Nevada and implications for batholith petrogenesis. Geological Society of America Bulletin 94, 713–719.

Dorais, M.J., Whitney, J.A., Roden, M.F., 1990. The origin of mafic enclaves from the Dinkey Creek pluton, central Sierra Nevada Batholith. Journal of Petrology 31, 853–881.

Fourcade, S., Allègre, C.J., 1981. Trace elements behavior in granite genesis: a case study. The calc-alkaline plutonic association from the Quérigut complex (Pyrénées, France). Contributions to Mineralogy and Petrology 76, 177–195.

Frost, T.P., Mahood, G.A., 1987. Field, chemical, and physical constraints on mafic–felsic magma interaction in the Lamarck Granodiorite, Sierra Nevada, California. Geological Society of America Bulletin 99, 272–291.

Furman, T., Spera, T., 1985. Co-mingling of acid and basic magma with implications for the origin of mafic I-type xenoliths, field and petrochemical relations of an usual dike complex at Eagle Peak Lake, Sequoia National Park, California, U.S.A. Journal of Volcanology and Geothermal Research 24, 151–178.

Gilbert, G.K., 1906. Gravitational assemblages in granite. Geological Society of America Bulletin 17, 321–328.

Gilder, S.A., McNulty, B.A., 1999. Tectonic exhumation and tilting of the Mount Givens pluton, central Sierra Nevada, California. Geology 27, 919–922.

Hibbard, M.J., 1981. The magma mixing origin of mantled feldspars. Contributions to Mineralogy and Petrology 76, 158–170.

Hibbard, M.J., 1991. Textural anatomy of twelve magma-mixed granitoid systems. In: Didier, J., Barbarin, B. (Eds.), Enclaves and Granite Petrology, Developments in Petrology, vol. 13. Elsevier, Amsterdam, pp. 431–444.

Hibbard, M.J., Watters, R.J., 1985. Fracturing and diking in uncompletely crystallized granitic plutons. Lithos 18, 1–12.

Hill, M., O'Neil, J.R., Noyes, H., Frey, F.A., Wones, D.R., 1988. Sr, Nd, and O isotopic variations in compositionally zoned and unzoned plutons in the central Sierra Nevada batholith. American Journal of Science 288, 213–241.

Holden, P., Halliday, A.N., Stephens, W.E., 1987. Neodymium and strontium isotope content of microdiorite enclaves points to mantle input into granitoid production. Nature 330, 53–56.

Holmes, A., 1928. The Nomenclature of Petrology. Murby, London.

Huppert, H.E., Sparks, R.S.J., 1988. The generation of granitic magmas by intrusion of basalt into the crust. Journal of Petrology 29, 599–624.

Huppert, H.E., Sparks, R.S.J., 1989. Chilled margins in igneous rocks. Earth and Planetary Science Letters 92, 397–405.

Kistler, R.W., Peterman, Z.E., 1973. Variations in Sr, Rb, K, Na, and initial 87Sr/86Sr in Mesozoic granitic rocks and intruded wall rocks in central California. Geological Society of America Bulletin 84, 3489–3512.

Kistler, R.W., Chappell, B.W., Peck, D.L., Bateman, P.C., 1986. Isotopic variation in the Tuolumne Intrusive Suite, central Sierra Nevada, California. Contributions to Mineralogy and Petrology 94, 205–220.

Lacroix, A., 1893. Les enclaves des roches volcaniques. Protat, Mâcon.

Lesher, C.E., 1990. Decoupling of chemical and isotopic exchange during magma mixing. Nature 344, 235–237.

Link, A.J., 1969. Inclusions in the Half Dome quartz monzonite, Yosemite National Park, California. PhD Thesis, Northwestern University, Illinois, U.S.A.

Marsh, B.D., 1982. On the mechanics of igneous diapirism, stoping, and zone melting. American Journal of Science 282, 808–855.

McNulty, B.A., Tong, W., Tobisch, O.T., 1996. Assembly of a dike-fed magma chamber: the Jackass lakes pluton, central Sierra Nevada, California. Geological Society of America Bulletin 108, 926–940.

McNulty, B.A., Tobisch, O.T., Cruden, A.R., 2000. Multistage emplacement of the Mount Givens pluton, central Sierra Nevada batholith, California. Geological Society of America Bulletin 112, 119–135.

Noyes, H.J., Wones, D.R., Frey, F.A., 1983a. A tale of two plutons: geochemical evidence bearing on the origin and differentiation of the Red Lake and Eagle Peak plutons, central Sierra Nevada, California. Journal of Geology 91, 487–509.

Noyes, H.J., Wones, D.R., Frey, F.A., 1983b. A tale of two plutons: petrographic and mineralogic constraints on the petrogenesis of the Red Lake and Eagle Peak plutons, central Sierra Nevada, California. Journal of Geology 91, 353–379.

Pabst, A., 1928. Observations on inclusions in the granitic rocks of the Sierra Nevada. University of California Publications 17, 325–386.

Palivcová, M., 1978. Ocellar quartz leucogabbro (Central bohemian pluton) and genetic problems of ocellar rocks. Geologisky Zbornick Geologica Carpathica 29, 19–41.

Phillips, J.A., 1880. On concretionary patches and fragments of other rocks contained in granites. Quaternarly Journal of the Geological Society of London 36, 1–21.

Pin, C., Binon, M., Belin, J.M., Barbarin, B., Clemens, J.D., 1990. Origin of microgranular enclaves in granitoids: equivocal Sr–Nd evidence from Hercynian rocks in the Massif Central (France). Journal of Geophysical Research 95, 17821–17828.

Pitcher, W.S., 1983. Granite Type and Tectonic Environment. In: Hsu, K. (Ed.), Mountain Building Processes. Academic Press, London, pp. 19–40.

Pitcher, W.S., 1991. Synplutonic dykes and mafic enclaves. In: Didier, J., Barbarin, B. (Eds.), Enclaves and Granite Petrology, Developments in Petrology, vol. 13. Elsevier, Amsterdam, pp. 383–391.

Pitcher, W.S., Atherton, M.P., Cobbing, E.J., Beckinsale, R.D. (Eds.), Magmatism at a plate edge: the Peruvian Andes. Blackie and Son, Glasgow.

Piwinskii, A.J., 1968. Studies of batholithic feldspars: Sierra Nevada, California. Contributions to Mineralogy and Petrology 17, 204–223.

Piwinskii, A.J., 1973. Experimental studies of igneous rock series, central Sierra Nevada Batholith, California Part II. Neues Jahrbuch fur Mineralogie Monatshefte 5, 193–215.

Read, H.H., 1957. The Granite Controversy. Murby, London.

Reid, J.B., Hamilton, M.A., 1987. Origin of Sierra Nevadan granite: evidence from small scale composite dikes. Contributions to Mineralogy and Petrology 96, 441–454.

Reid, J.B., Evans, O.C., Fates, D.G., 1983. Magma mixing in granitic rocks of the central Sierra Nevada, California. Earth and Planetary Science Letters 66, 243–261.

Ryerson, F.J., Hess, P.C., 1978. Implications of liquid–liquid distribution coefficients to mineral–liquid partitioning. Geochimica et Cosmochimica Acta 42, 921–932.

Sawka, W.N., Chappell, B.W., Kistler, R.W., 1990. Granitoid compositional zoning by side-wall boundary layer differentiation: evidence from the Palisade Crest intrusive suite, central Sierra Nevada, California. Journal of Petrology 31, 519–553.

Sparks, R.S.J., Marshall, L.A., 1986. Thermal and mechanical constraints on mixing between mafic and silicic magmas. Journal of Volcanology and Geothermal Research 29, 99–124.

Stern, T.W., Bateman, P.C., Morgan, B.A., Newall, M.F., Peck, D.L., 1981. Isotopic U–Pb ages of zircon from granitoids of the central Sierra Nevada, California. U.S. Geological Survey Professional Paper, 1185.

Tindle, A.G., Pearce, J.A., 1983. Assimilation and partial melting of continental crust: evidence from the mineralogy and geochemistry of autoliths and xenoliths. Lithos 16, 185–202.

Tobisch, O.T., McNulty, B.A., Vernon, R.H., 1997. Microgranitoid enclave swarms in granitic plutons, central Sierra Nevada, California. Lithos 40, 321–339.

Vernon, R.H., 1983. Restite, xenoliths and microgranitoid enclaves in granites. Journal and Proceedings of the Royal Society of New South Wales 116, 77–103.

Vernon, R.H., 1984. Microgranitoid enclaves in granites–globules of hybrid magma quenched in a plutonic environment. Nature 309, 438–439.

Vernon, R.H., 1991. Interpretation of microstructures of microgranitoid enclaves. In: Didier, J., Barbarin, B. (Eds.), Enclaves and Granite Petrology, Developments in Petrology, vol. 13. Elsevier, Amsterdam, pp. 277–291.

Watson, E.B., 1976. Two-liquid partition coefficients: experimental data and geochemical implications. Contributions to Mineralogy and Petrology 56, 119–134.

Available online at www.sciencedirect.com

Lithos 80 (2005) 179–199

www.elsevier.com/locate/lithos

Pervasive assimilation of carbonate and silicate rocks in the Hortavær igneous complex, north-central Norway

Calvin G. Barnes[a,*], Tore Prestvik[b], Bjørn Sundvoll[c], Denny Surratt[a]

[a]*Department of Geosciences, Texas Tech University, Lubbock, TX 79409-1053, USA*
[b]*Department of Geology and Mineral Resources Engineering, Norwegian University of Science and Technology, N-7491 Trondheim, Norway*
[c]*Geological Survey of Norway, Mineralogical-Geological Museum, 0562 Oslo, Norway*

Received 26 February 2003; accepted 9 September 2004
Available online 11 November 2004

Abstract

The intrusive complex at Hortavær represents a magma transfer zone in which multiple pulses of gabbroic and dioritic magmas evolved along Fe- and alkali-enrichment trends. Extreme alkali enrichment resulted in nepheline-normative and sparse nepheline-bearing monzodioritic and monzonitic rocks. More evolved monzonitic and syenitic rocks are silica saturated and, in some cases, quartz bearing. Previous and current research recognized an abundance of clinopyroxene and other Ca-rich phases, such as scapolite, grossular-rich garnet, and igneous-textured calcite among the mafic and intermediate rocks. Even the most pyroxene-rich samples contain low Sc concentrations, which suggests early, intense fractionation of clinopyroxene. These features and the alkali enrichment are consistent with assimilation of carbonate-rich host rocks. Carbon isotope ratios of the igneous-textured calcite indicate an origin of the carbon from host rocks rich in calcite, consistent with assimilation. However, low ε_{Nd} values (−3.4 to −10.2) and moderate initial $^{87}Sr/^{86}Sr$ values (0.7052 to 0.7099) indicate the need for assimilation of quartzofeldspathic rocks as well. Models of combined assimilation and fractional crystallization indicate that assimilation of simple end members, either carbonate or silicate, cannot explain the entire data set. Instead, variable proportions of carbonate and silicate materials were assimilated, with the most pronounced assimilation effects in the mafic rocks. The reasons for variable degrees of assimilation are, as yet, uncertain. It is possible that assimilation of calc-silicate rocks with variable carbonate/silicate proportions resulted in the range of observed compositions. However, the importance of carbonate assimilation in mafic rocks compared to felsic ones suggests that assimilation of carbonates was predominant at high temperature and/or mafic magma compositions and assimilation of silicates was predominant at lower temperature and/or felsic magma compositions. We suggest that the ability of the mafic magma to dissolve higher proportions of carbonate contaminants is the result of the magma's ability to form clinopyroxene as a product of assimilation. In any case, extensive carbonate assimilation was possible because CO_2 escaped from the system.
© 2004 Elsevier B.V. All rights reserved.

Keywords: Assimilation; Carbonate; Alkaline; Norway

---

* Corresponding author. Tel.: +1 806 742 3106; fax: +1 806 742 0100.
  E-mail address: Cal.Barnes@ttu.edu (C.G. Barnes).

0024-4937/$ - see front matter © 2004 Elsevier B.V. All rights reserved.
doi:10.1016/j.lithos.2003.11.002

1. Introduction

Nearly a century ago, Daly (1910) proposed that alkaline intrusive rocks resulted from assimilation of carbonate rocks. Many petrologists described examples of this relationship, in which shallowly emplaced basaltic to granitic magma had interacted with carbonate rocks (e.g., Daly, 1910; Shand, 1930; Tilley, 1949, 1952; Sabine, 1975; Baker and Black, 1980). Most commonly, such interactions resulted in an increase in the alkalinity of the affected magma and in the development of skarn assemblages: endoskarns that were internal to the pluton and exoskarns at pluton and xenolith contacts. Alkali enrichment commonly accompanied desilication of the host magma (e.g., Daly, 1910, 1918; Tilley, 1952; Sabine, 1975) to form feldspathoid-bearing (silica-undersaturated) assemblages. Most published examples of magma–carbonate interaction zones are narrow (less than a few meters wide). In spite of this limited spatial extent, the relationships so impressed some workers, that they proposed an origin for alkaline magmas by assimilation of carbonate-rich metasedimentary rocks (Daly, 1910; Shand, 1930). This latter idea lacked widespread acceptance, particularly for cases in which silica-undersaturated alkaline rocks were thought to arise by interaction with granitic magmas (Tilley, 1949, 1952). Furthermore, isotopic, and trace element studies (summarized in Wilson, 1989) have shown that, in general, alkaline magmatism is unrelated to carbonate assimilation.

The hypothesis of carbonate assimilation as the origin of alkaline magmas was essentially rejected by experimental studies (Watkinson and Wyllie, 1969; Spera and Bergman, 1980). These studies showed that no more than about 0.5 wt.% CO_2 dissolves in subalkaline and mildly alkaline magmas at moderate to low pressure. Typical assimilation reactions should be like:

$$Mg_2SiO_4 + 3SiO_2 + 2CaCO_3 = 2CaMgSi_2O_6 + 2CO_2$$

Alkali enrichment (relative to silica) follows from this reaction due to the increased stability of clinopyroxene instead of olivine (e.g., Meen, 1990). However, it is clear that saturation of CO_2 will limit reaction progress, and therefore the ability of carbonate assimilation to modify magmatic compositions. Even instances in which stable isotope evidence strongly supported carbonate assimilation (e.g., Worley et al., 1995), carbonate assimilation was discounted because of the problem of low solubility of CO_2.

In this paper, we report on the 456 ± 8 Ma (U–Pb, titanite; Barnes et al., 2003) Hortavær igneous complex, a suite of gabbroic to syenitic and granitic rocks in the Caledonides of north-central Norway. In his original description of the complex, Vogt (1916) identified a group of clinopyroxene-rich, calcite- and idocrase-bearing dioritic rocks whose origin he ascribed to assimilation of host-rock marbles. Gustavson and Prestvik (1979) presented major element analyses and concurred with Vogt's conclusions concerning petrogenesis. What makes the Hortavær complex distinctive is that mineral assemblages characteristic of carbonate assimilation occur throughout the >6-km-wide body (Barnes et al., 2003). Furthermore, stable and radiogenic isotope ratios are consistent with assimilation of metasedimentary rocks, including carbonates. Thus, the Hortavær complex is one of the best known locations to study large-scale carbonate assimilation.

2. Geologic setting

The Hortavær igneous complex is part of the Bindal Batholith, a sequence of late Ordovician to Silurian plutons emplaced into the Helgeland Nappe Complex (HNC) of the Uppermost Allochthon of the Norwegian Caledonides (Nordgulen, 1993; Nordgulen and Sundvoll, 1992). The HNC is the structurally highest nappe complex in the province; its general geologic features were reviewed by Thorsnes and Løseth (1991) and Yoshinobu et al. (2002). The HNC and enclosed Bindal Batholith probably formed near the margin of Laurentia (Roberts et al., 2001; Yoshinobu et al., 2002), after which they were incorporated into the Caledonian nappe stack during the late Silurian to early Devonian Scandian stage of the orogeny (e.g., Roberts and Gee, 1985).

The Bindal Batholith ranges from gabbro to leucogranite, with calcic to alkali-calcic affinities. The Hortavær complex is one of a few truly alkalic intrusions in the Batholith (Nordgulen, 1993), and the only one to our knowledge that contains feldspathoids.

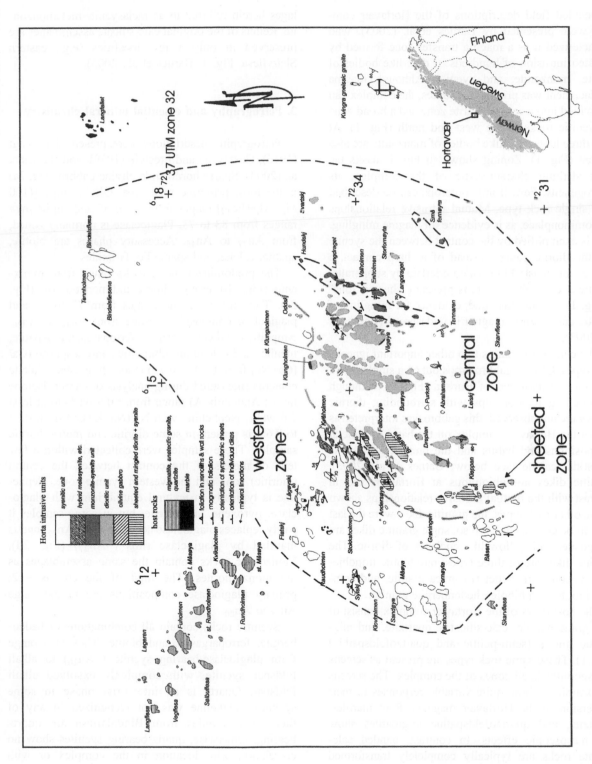

Fig. 1. Simplified geologic map of the Hortavær complex (after Barnes et al., 2003). The Kvingra gneissic granite was interpreted to represent a separate intrusive complex by Barnes et al. (2003). UTM grid marks are 3 km apart.

Detailed field descriptions of the Hortavær complex were presented by Barnes et al. (2003), who characterized it as a magma transfer zone formed by repeated intrusion of hundreds of dike-like bodies of diorite, monzonite, and syenite. Although magma emplacement was primarily as dikes, the complex can be divided into a central diorite zone and a broad zone of syenitic rocks to the west and north (Fig. 1). At least three large dike-like bodies of monzonite are also present (Fig. 1). Zoning shown in Fig. 1 masks the most striking characteristic of the complex, its heterogeneity. Few, if any, outcrops can be described by a single rock type. Mutual intrusive relationships are commonplace, as is evidence for magma mingling. This is exemplified by the contact between the syenite and the diorite zones. Instead of a sharp contact, a zone as much as 500 m wide underlain by subparallel dioritic and syenitic sheets is present ("sheeted zone" in Fig. 1). Within this zone, evidence for mingling of dioritic and syenitic magma is widespread (Barnes et al., 2003).

The complex contains two other important igneous rock types. Olivine gabbro crops out on a single island near the northern end of Burøya (Fig. 1). Although cross-cutting relationships with surrounding dioritic rocks were not observed, this gabbro was interpreted as part of the Hortavær complex on the basis of mineral compositions and habits and the presence of sparse interstitial calcite (see below; Barnes et al., 2003). Granitic dikes are ubiquitous at Hortavær, and in contrast with the mutual intrusive relationships shown by the other rock types, the granitic dikes were the last units to be emplaced. Even so, some granitic dikes are composite, with pillow-like enclaves of diorite. The granitic dikes are of three types: rare biotite trondhjemite, uncommon garnet two-mica granite, and common biotite+amphibole±hedenbergite granite.

The host rocks of the Hortavær complex consist of high-grade marble, calc-silicates, quartzite, and migmatitic gneiss (semi-pelitic and quartzofeldspathic) (Fig. 1). These same rock types are present as screens and xenoliths in all zones of the complex. The screens and xenoliths show quite variable responses to their immersion in the Hortavær magmas. Pure marbles, quartzite, and quartzofeldspathic migmatites show few mesoscopic effects. In contrast, banded calc-silicate rocks are typically completely transformed into garnet-bearing syenitic or monzonitic assemb-

lages herein referred to as melasyenite/melamonzonite. Relicts of the original calc-silicate assemblages are preserved in only a few localities (e.g., eastern Skarvflesa, Fig. 1; Barnes et al., 2003).

3. Petrography and essential mineral chemistry

Petrographic descriptions were presented in Vogt (1916), Gustavson and Prestvik (1979), and Barnes et al. (2003). Sparse hornblende olivine gabbro (Fig. 2a) is the most primitive rock type. Olivine mg# $[(100 \, Mg/(Mg+Fe_T)]$ ranges from 69 to 55 and augite mg# ranges from 83 to 73. Plagioclase is normally zoned, from An_{77} to An_{45}. Accessory phases are biotite, apatite, calcite, and sparse Fe–Ti oxides.

The predominant mafic rocks range from pyroxenite (Fig. 2b) through diorite and monzodiorite (Fig. 2c). The mg# of augite ranges from 77 to 37 and plagioclase compositions range from An_{48} to An_{20}. Amphibole, where present, is ferropargasitic. Apatite, titanite, and calcite are ubiquitous and scapolite (Ca/(Ca+Na) from 0.52 to 0.75) is locally present. Calcite contents (measured during analysis of carbon isotope ratios; Appendix A) range from <0.5 wt.% to at least 7.6 wt.%. Nepheline, with Na/(Na+K+Ca) from 0.82 to 0.65, is present in some dioritic and monzodioritic samples. These samples were collected within a few tens of meters of the contact between the central (dioritic) zone and the western syenite zone. Nepheline is typically in textural equilibrium with plagioclase, pyroxene, and calcite, but in some samples it partly replaces plagioclase cores (An_{42} to An_{26}) and is mantled by plagioclase rims ($\sim An_{27}$) (Fig. 2d). Monzonitic rocks contain the same assemblages as the monzodiorites. The mg# of the cpx is ~12; primary plagioclase compositions are in the range An_{25} to An_{20}.

Syenitic rocks contain all combinations of hedenbergite, ferropargasite, and biotite. They also range from plagioclase-bearing syenite ($\sim An_{20}$) to alkali feldspar syenite, with complexly exsolved alkali feldspar. Quartz is an interstitial phase in some syenites; nepheline was not recognized in any of them. All samples from Bindalsflesa are quartz bearing. Otherwise, quartz-bearing syenites show no correlation with location in the complex or with sequence of emplacement: we commonly sampled

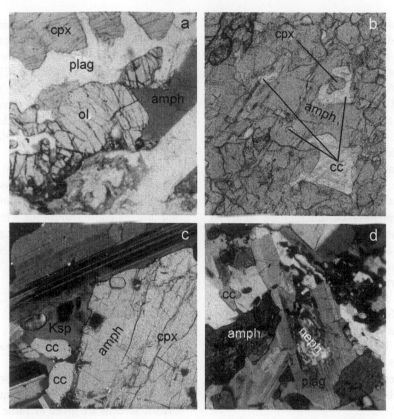

Fig. 2. Photomicrographs. Vertical field of view is 1 mm in all photos. Clinopyroxene (cpx), calcic amphibole (amph), olivine (ol), plagioclase (plag), K-feldspar (Ksp), nepheline (neph), calcite (cc). (a) Amphibole olivine gabbro. (b) Amphibole clinopyroxenite with magmatic-textured calcite. Amphibole has partly replaced clinopyroxene. (c) Amphibole monzodiorite with amphibole rimming clinopyroxene and calcite inclusions in K-feldspar. (d) Nepheline-bearing amphibole monzodiorite. Plagioclase cores are partly replaced by nepheline; this assemblage is mantled by more plagioclase.

quartz-bearing and quartz-free syenites from the same island. Accessory minerals in the syenites are calcite, titanite, apatite, zircon, fluorite, and rare allanite. A few syenitic samples contain grossular–andradite garnet.

The common type of granitic dike contains blue-green amphibole±biotite±hedenbergitic pyroxene. Accessory minerals are titanite, calcite, Fe–Ti oxides, tourmaline, fluorite, epidote, and zircon. These dikes are metaluminous to very slightly peraluminous, with A/CNK <1.02.

Melasyenite/melamonzonite screens and xenoliths are heterogeneous at all scales, consistent with the observation that they had calc-silicate protoliths. Clinopyroxene (mg# 44–5) and ferropargasitic amphibole (mg# 25–10) are typical mafic phases, along with grossular–andradite garnet. Plagioclase

ranges from An_{37} to An_{17}. Accessory minerals are interstitial calcite (as much as 6% of the rock), titanite, scapolite, apatite, and epidote, and some samples contain wollastonite, idocrase (Vogt, 1916), Fe-sulfides, magnetite, and zircon.

Carbonate-rich host rocks range from pure calcite marble through dolomitic marble (locally with forsterite and humite) to quartz- and alkali-feldspar-rich calc-silicates. Quartzo-feldspathic host rocks range from quartzite to semi-pelitic, two-mica migmatitic gneiss, to pelitic migmatitic gneiss with biotite+muscovite+quartz±sillimanite±cordierite±garnet±hercynite. Sillimanite is found as millimeter-scale poikilitic crystals, as needles, and as fibrolitic mats. These latter rocks are commonly diatexitic: migmatites that have lost their internal structure due to magma flowage and deformation (e.g., Sawyer, 1998). Similar rocks crop out as sheared screens or dikes (?) in the Hortavær complex. In the Velfjord area, such diatexitic dikes were shown to be disrupted pelitic migmatite from the nearby host rocks that had moved as a magma (Barnes and Prestvik, 2000; Barnes et al., 2002).

As demonstrated by Vogt (1916), interstitial calcite in Hortavær samples is in textural equilibrium with surrounding silicates. In some instances (Fig. 2b and c), calcite is included in silicate minerals (especially clinopyroxene and nepheline) and silicate minerals are included in calcite (especially clinopyroxene), suggesting coprecipitation of calcite with the silicate assemblage. Reconnaissance microprobe analysis (Table 1) shows the calcite to be nearly pure, with less than 0.28 wt.% FeO, 0.16 wt.% MnO, and 1.3 wt.% SrO. MgO and Na_2O were not detected.

Assemblages permitting estimates of emplacement pressure were not found. Some constraints can be applied, however. Sparse granitic dikes in the complex contain the assemblage sillimanite+cordierite+hercynitic spinel. In the absence of garnet and orthopyroxene, this assemblage is consistent with crystallization at pressures between 200 and 500 MPa and temperatures above about 750 °C (Spear et al., 1999).

Some melasyenitic screens contain the assemblage garnet+scapolite+calcite+plagioclase. If pressure and the activity of SiO_2 are fixed, then this assemblage is invariant. At a pressure of 400 MPa and unit activities, invariant conditions are at ~900 °C and $X(CO_2)$=0.63. Correction for activities of grossular in garnet and anorthite in plagioclase should have the effect of raising the equilibrium temperature and lowering $X(CO_2)$ slightly.

In addition, the assemblage plagioclase+calcite+scapolite is present in some monzodioritic rocks. This assemblage corresponds to a fluid-independent equilibrium: calcite + 3 anorthite = meionite. Preliminary data on scapolite compositions (primarily Me 90 to Me 63) suggest equilibrium temperatures between ~900 and ~1050 °C.

4. Geochemistry

4.1. Major and trace element variation

Compositions of granitic dikes and gneissic host rocks are presented in Table 2 and partial analyses of marble screens are given in Table 3. The overall major

Table 1
Compositions of interstitial calcite

	93.02H ave.	S.D.	93.06H ave.	S.D.	93.44H ave.	S.D.	93.46H ave.	S.D.
Oxide weight percent								
FeO	0.09	0.04	0.20	0.05	0.12	0.06	0.14	0.15
MnO	0.13	0.04	0.07	0.05	0.04	0.04	0.05	0.05
MgO	0.00	0.01	0.02	0.01	0.02	0.01	0.01	0.01
CaO	54.58	1.47	55.77	1.90	57.70	4.07	55.78	1.26
Na_2O	0.02	0.04	0.00	0.00	0.00	0.01	0.01	0.02
SrO	0.68	0.33	0.27	0.21	0.42	0.31	0.38	0.36
Total	55.50		56.33		58.30		56.37	
Cations per three oxygens								
Fe	0.001	0.001	0.003	0.001	0.002	0.001	0.002	0.001
Mn	0.002	0.001	0.001	0.001	0.001	0.001	0.001	0.001
Mg	0.000	0.000	0.000	0.000	0.000	0.000	0.000	0.000
Ca	0.979	0.023	0.997	0.029	1.026	0.064	0.997	0.019
Na	0.001	0.001	0.000	0.000	0.000	0.000	0.000	0.001
Sr	0.007	0.003	0.003	0.002	0.004	0.003	0.004	0.003
Total	0.990		1.003		1.032		1.004	
Sr (ppm)	9419	4626	5690	1617	6763	3949	7044	4564
N	10		9		15		8	

Key to rock types: amphibole biotite monzonite (93.02H), garnet hedenbergite melamonzonite (93.06H), amphibole augite nepheline diorite (93.44H), appinitic augite amphibole diorite (93.46H).

Table 2
Major and trace element analyses of quartzofeldspathic host rocks and granitic dikes

	93.84 H	93.88 H	93.94 H	93.98 H	73.23 H	73.24 H	93.82 H
Weight percent oxides, normalized volatile-free							
SiO$_2$	69.57	68.01	78.12	66.70	50.86	52.66	54.07
TiO$_2$	0.81	0.34	0.05	0.84	1.35	0.84	1.24
Al$_2$O$_3$	14.81	18.08	13.84	16.18	17.37	23.10	25.39
Fe$_2$O$_3$	0.77	0.32	0.71	0.69	6.36	7.65	10.39
FeO	4.01	1.57	n.a.	4.17	n.a.	n.a.	n.a.
MnO	0.07	0.02	0.01	0.07	.01	0.01	0.19
MgO	1.24	0.84	0.21	1.45	21.22	14.16	3.12
CaO	2.01	2.63	0.23	2.43	0.05	0.36	0.74
Na$_2$O	3.50	6.76	4.34	3.65	0.62	0.69	1.30
K$_2$O	3.09	1.20	2.34	3.59	2.10	0.30	3.48
P$_2$O$_5$	0.11	0.22	0.14	0.22	0.06	0.24	0.09
Total	100.00	100.00	100.00	100.00	100.00	100.00	100.00
LOI	0.85	0.54	0.97	0.88	4.63	4.34	0.89
Parts per million							
Rb	134	44	84	177	67	12	131
Sr	120	159	43	108	7	9	200
Zr	407	25	3	279	701	879	175
Y	67.0	106.4	13.0	37.7	57.0	70.0	35.3
Nb	21	14	4	23	n.a.	n.a.	20
Ba	404	175	228	374	173	68	943
Sc	10.7	5.8	1.3	10.2	n.a.	n.a.	22.0
V	49	26	3	55	78	75	137
Cr	28	14	b.d.	18	125	35	178
Ni	35	18	b.d.	4	45	39	78
Cu	1	0	b.d.	3	3	3	20
Zn	62	23	17	80	22	19	188
La	47.74	21.3	n.a.	40.3	n.a.	n.a.	n.a.
Ce	100.5	53.3	n.a.	83.2	n.a.	n.a.	n.a.
Nd	49.5	25.9	n.a.	38.4	n.a.	n.a.	n.a.
Sm	10.67	6.77	n.a.	8.29	n.a.	n.a.	n.a.
Eu	1.76	1.93	n.a.	1.39	n.a.	n.a.	n.a.
Tb	1.54	1.87	n.a.	1.03	n.a.	n.a.	n.a.
Yb	5.53	8.6	n.a.	3.2	n.a.	n.a.	n.a.
Lu	0.85	1.1	n.a.	0.44	n.a.	n.a.	n.a.
Co	10	5	n.a.	9	15	22	n.a.
Cs	4.38	1.34	n.a.	9.07	n.a.	n.a.	n.a.
Hf	10.4	0.43	n.a.	7.67	n.a.	n.a.	n.a.
U	n.a.	3.9	n.a.	3.6	n.a.	n.a.	n.a.
Th	n.a.	6.5	n.a.	17	26	13	n.a.
Ta	1.295	1.18	n.a.	1.80	n.a.	n.a.	n.a.

Key to rock types: undeformed two-mica garnet granite (93.84H), intrudes host rocks on Langflesa; undeformed biotite quartz diorite dike on Sølbuflesa (93.88H); muscovite alkali feldspar granite from lille Måsøya (93.94H); sillimanite-bearing biotite granite from Kviksholmen (93.98H); cordierite sillimanite two-mica gneiss from Legeren (73.23H); cordierite sillimanite two-mica gneiss from Legeren (73.24H); sillimanite biotite cordierite melagranite (diatexite) dike; cuts syenite north of Kvåholmen (93.82H).

and trace element compositional variations were described by Barnes et al. (2003). They recognized a possible magmatic trend that extends from olivine gabbro to syenite, with considerable scatter. Some of the compositional scatter is due to accumulation of

ferromagnesian minerals±feldspars, either by simple crystal accumulation or by reaction with, and assimilation of host rocks. Furthermore, Barnes et al. (2003) showed that, as a magma transfer zone, the Hortavær complex represents many pulses of magma in what

Table 3
Partial chemical analyses of marble from the Hortavær complex

	MgO (wt.%)	FeO (wt.%)	MnO (wt.%)	Na$_2$O (wt.%)	K$_2$O (wt.%)	Ba (ppm)	Sr (ppm)
02.25H	18.82	0.83	0.037	0.017	0.241	139	47
02.26H	19.17	0.24	0.012	0.043	0.295	325	49
02.31H	12.27	0.18	0.031	0.023	0.002	7	149
02.47H	8.79	0.46	0.067	0.022	0.002	6	78
02.52H	12.48	0.27	0.097	0.027	0.006	2	93
02.56H	1.13	0.06	0.011	0.031	0.062	9	1061
73.07H	0.10	0.08	0.002	0.016	0.027	4	2088
93.07H	0.05	0.13	0.003	0.034	0.014	n.d.	1531
93.47H	16.01	0.03	0.002	0.012	0.003	1	366
93.74H	5.97	0.04	0.005	0.010	0.002	n.d.	102

Analysis by ICP-AES at Texas Tech University using commercial multielement standards.

was probably a vertically extensive conduit system. Thus, bulk rock compositions reflect evolution of hundreds of individual magma batches, each of which evolved along similar but distinct liquid lines of descent because of variable amounts of assimilation and variable compositions of assimilated material. Such heterogeneity in emplacement history could explain the bulk compositional scatter combined with the similar mineral assemblages and compositions within specific rock types. Assuming a general magma evolution trend exists, it is characterized by enrichment of Fe and total alkalis, such that evolved diorites, monzonites, and syenites plot in the ferroan and alkaline fields of Frost et al. (2001; see Barnes et al., 2003). The trend is curved, which Barnes et al. (2003) took to indicate some control by crystal–liquid separation processes.

Fig. 3a shows total alkali contents plotted against SiO$_2$. The variation from olivine gabbro to syenite is characterized by a pronounced increase in total alkalis (<3 to >13 wt.%) with a change in SiO$_2$ content from 44 to 65 wt.%. Among the syenites, two groups can be identified. One extends to the highest total alkali contents (~13.5%) and lacks quartz syenite. The other contains quartz and has total alkali contents less than about 12%. Two groups of granitic rocks can also be distinguished. The first, the common granitic dikes in the complex, consists of predominantly metaluminous rocks that lack white mica. The second group consists of strongly peraluminous two-mica granites and diatexites from host rocks to the northwest of the complex (Fig. 1). One strongly peraluminous sample is from a sheared garnet cordierite two-mica screen or dike within the

complex (labeled with a "d" in Fig. 3). This sample has lower SiO$_2$ and total alkalis than other analyzed peraluminous granite. The peraluminous samples contain lower alkali contents than the metaluminous granitic dikes in the complex. Fig. 3a also shows the wide range of alkali contents of the melasyenite/melamonzonite screens and xenoliths.

Fig. 3b shows the variation of Zr content as a function of SiO$_2$. In general, Zr increases from olivine gabbro to monzonite. Among the syenites, two groups are apparent. One group contains <200 ppm Zr and consists predominantly of syenites that lack quartz. The other group contains >200 ppm Zr and extends to concentrations of ~800 ppm. These samples are predominantly quartz syenite. Among the quartz syenites, there is no correlation between Zr and SiO$_2$ contents, nor is there any relationship between sample location and Zr content. Metaluminous granitic dikes within the Hortavær complex range in Zr contents from ~300 to ~50 ppm. Peraluminous granites show an even wider range of Zr (~420 to ~10 ppm; Table 2).

Variation of Sr as a function of CaO is shown in Fig. 3c. From the olivine gabbro to the dioritic rocks, Sr concentrations show an overall increase, then decrease from the monzonites to a group of low-Sr syenites. However, another group of syenites and one monzonite plot at higher Sr contents (Fig. 3c); none of these samples contains quartz. There is no correlation between Sr content of the syenites and their geographic location.

A group of dioritic samples and one of the melasyenites plot at high CaO (>19 wt.%) and moderate to low Sr contents (Fig. 3c). These Ca-rich

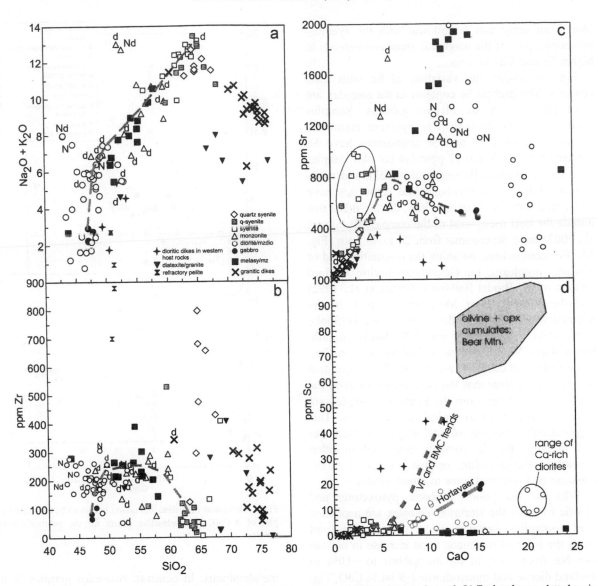

Fig. 3. (a) Total alkalis versus SiO$_2$ (in wt.%). The dashed line indicates one possible inferred magmatic trend. (b) Zr abundances plotted against wt.% SiO$_2$. Diamonds are quartz-bearing syenite; gray squares are syenite with normative quartz >1 wt.%. N indicates nepheline-bearing samples; d indicates a dike. The dashed line indicates one possible inferred magmatic trend. (c) Sr versus wt.% CaO. The dashed line indicates one possible inferred magmatic trend. The group of Sr-rich, quartz-free syenites is enclosed in an oval. (d) Sc abundances plotted against wt.% CaO. The short-dashed line is the compositional trend for the dioritic Velfjord plutons of the Bindal batholith and the gabbro/diorite plutons of the Jurassic Bear Mountain intrusive complex, northern California. The shaded field corresponds to compositions of augite±olivine-rich cumulates from the Bear Mountain complex.

rocks correspond to pyroxene-rich diorites and a few could be classified as pyroxenite; all have Al$_2$O$_3$ contents <15%. A second group of samples (diorite, monzodiorite, and two monzonite) plots at Sr contents higher than the inferred magmatic trend (>850 ppm)

and has Al$_2$O$_3$ contents >16%. In a chemical sense, the former group is equivalent to clinopyroxene cumulates, the latter to clinopyroxene+plagioclase cumulates. Except for the single melasyenite mentioned above, the melasyenite/melamonzonite samples plot in an array that is collinear with the syenite–monzonite part of the magmatic trend and extends to higher Sr and Ca contents.

Fig. 3d shows the variation of Sc with CaO contents. The highest Sc contents in the complex are circa 20 ppm in the olivine gabbros. Samples interpreted to be equivalent to pyroxene cumulates on the basis of CaO and Sr abundances have Sc contents between 8 and 17 ppm (all but two samples have Cr <250 ppm; Barnes et al., 2003), and rocks equivalent to pyroxene+plagioclase cumulates have less than 5 ppm Sc. In comparison, dioritic dikes that intrude the host rocks west of the complex (Barnes et al., 2003) have Sc contents from 26 to 45 ppm (Fig. 3d). For comparison, we show the magmatic trend of two arc complexes, the Caledonian dioritic Velfjord plutons of the Bindal Batholith (Barnes et al., 1992) and the Jurassic Bear Mountain complex in the Klamath Mountains of northwestern California (Snoke et al., 1981; C.G. Barnes and A.W. Snoke, unpublished data). We also show a field for Sc and CaO compositions of cumulate rocks in the Bear Mountain complex. It is clear that the Sc contents of even the most mafic samples from the Hortavær complex are quite low compared to typical arc intrusions. This suggests either that the parental magmas at Hortavær had initially low Sc contents compared to other diorites in the batholith, or that substantial differentiation had occurred prior to emplacement.

With two exceptions, gabbroic, pyroxenitic, and dioritic rocks in the Hortavær complex are nepheline (Ne) normative. Samples whose compositions plot along the inferred magmatic trend increase in normative Ne from ~4% in olivine gabbro to ~10% in evolved diorite and monzodiorite (~9 wt.% CaO; Fig. 4). At lower CaO concentrations, Ne values decrease, such that monzonitic and syenitic samples straddle the boundary between normative Ne and normative quartz (Q). Most monzonites are Ne normative whereas many of the syenites are Q normative or lie near the plane of critical undersaturation. In general, nepheline-bearing diorites and monzonites have the highest normative Ne values. Among the Q-normative, non-cumulate syenites there is a correlation between normative Q and aluminum saturation index (A/CNK), with A/CNK of ~0.85 at 0% Q and ~1.0 at 9% Q (Fig. 4 inset). Among the most evolved rocks, syenites and most granitic dikes in the complex are

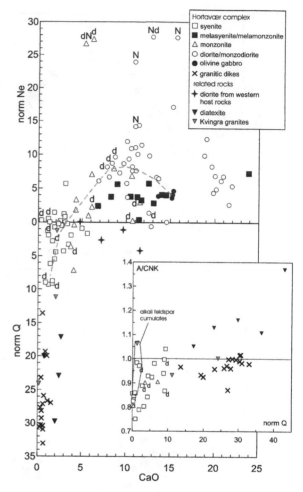

Fig. 4. Normative nepheline and quartz versus wt.% CaO. Inset is a plot of A/CNK and normative quartz for the quartz-normative samples.

metaluminous. In contrast, two-mica granites in the host rocks and garnet–two-mica granite dikes in the complex are strongly peraluminous.

4.2. Radiogenic isotopes

Bulk-rock Nd and Sr isotope ratios are presented in Table 4; analytical methods are given in Appendix A. The olivine gabbro has ε_{Nd} (at 456 Ma) of -3.4, whereas other samples from the complex range from -7.5 to -10.2 (Fig. 5a). Metaluminous granitic dikes in the complex show the same range of ε_{Nd}, as does a peraluminous granite collected from the host migmatitic rocks west of the complex. However, a peralu-

Table 4
Nd and Sr isotope analyses

	Sm (ppm)	Nd (ppm)	Sm/Nd	^{147}Sm/^{144}Nd	^{143}Nd/^{144}Nd	ε_{Nd} (456 Ma)	Rb (ppm)	Sr (ppm)	Rb/Sr	^{87}Rb/^{86}Sr 456 Ma	^{87}Sr/^{86}Sr	^{87}Sr/^{86}Sr	Rock types
Gabbro and diorite													
70-27	4.35	17.96	0.242	0.14746	0.512113	−7.51	41.1	513.0	0.080	0.23183	0.709330	0.7078	mzdiorite
89.05H	7.52	38.37	0.196	0.11924	0.512009	−7.86	194.2	1253.1	0.155	0.44831	0.708280	0.70532	augite amph mzdiorite
91.49H	3.64	13.457	0.271	0.16479	0.512375	−3.42	13.09	481.8	0.027	0.07860	0.707370	0.70685	amph olivine gabbro
93.44H	7.28	36.77	0.198	0.12057	0.512034	−7.45	102.4	1007.14	0.102	0.29422	0.709540	0.7076	amph augite neph mzdiorite
Monzonite													
89.20H	6.61	33.13	0.200	0.12151	0.511944	−9.27	264.2	437.7	0.604	1.74873	0.721040	0.70948	biotite monzonite
Melasyenite melamonzonite													
73-29	3.46	19.19	0.180	0.10977	0.511999	−7.50	166.9	1525.7	0.109	0.22173	0.739970	0.7085	hedenbergite melasy
89.06H	9.85	44.04	0.224	0.13622	0.512062	−7.84	330.3	809.7	0.408	1.18088	0.712970	0.70516	garnet hedenbergite melasy
Syenitic rocks													
89.09H	4.59	34.23	0.134	0.08165	0.511888	−7.99	231.5	208.3	1.111	3.22137	0.726750	0.70545	hedenbergite syenite
89.14H	0.765	4.107	0.186	0.11340	0.511873	−10.20	334.2	194.9	1.715	4.978557	0.752760	0.70985	biotite leucosyenite
89.16H	3.05	21.72	0.141	0.08559	0.511894	−8.11	337.9	41.99	8.047	23.64040	0.864790	0.70851	amphibole quartz
Biotite quartz-bearing leucosyenite/monzonite													
91.34H(L)	3.00	11.68	0.257	0.15643	0.512139	−7.53	253.11	357.53	0.708	1.92086	0.721350	0.70865	metaluminous granitic dikes
89.03H	3.80	25.39	0.150	0.09109	0.511903	−8.26	1.48	193.5	0.008	0.02206	0.705450	0.7053	plagiogranite
89.15H	0.847	5.720	0.148	0.09015	0.511908	−8.11	440.2	19.99	22.017	66.43697	1.145940	0.70775	biotite granite
89.18H	0.497	1.796	0.277	0.16848	0.512140	−8.23	667.2	5.84	114.307	405.08630	3.405560	0.72872	biotite granite
89.19H	1.75	8.31	0.211	0.12798	0.512026	−8.05	572.1	94.1	6.077	17.79209	0.830000	0.71238	biotite granite
Peraluminous granites													
93.82H	11.29	64.696	0.175	0.10625	0.511567	−15.72	124.92	176.99	0.706	2.04802	0.735690	0.72415	diatexitic sillimanite
Cordierite biotite granite													
93.84H(L)	7.52	37.76	0.199	0.12116	0.512040	−7.37	130.76	119.48	1.094	3.17802	0.745400	0.72439	garnet two-mica
Kvingra													
73-18	4.63	24.50	0.189	0.11506	0.512001	−7.77	530.0	300.7	1.763	5.11602	0.742330	0.70911	syenite
89.22H	3.73	15.82	0.236	0.14352	0.512468	−0.34	784.64	6.95	112.9	398.76	2.964300	0.32878	biotite granite
89.22H(L)	2.132	10.328	0.206	0.12569	0.512350	−1.59	738.77	3.27	225.92	n.d.	6.539300	–	biotite granite
93.31H(L)	1.439	6.991	0.206	0.12532	0.511977	−8.85	307.67	193.76	1.588	4.60942	0.741820	0.71135	chloritized biotite quartz mz

Analytical methods and data precision described in Appendix A. Samples with an L in the sample name were leached prior to analysis (see Appendix A). Key to abbreviations: monzonite (mz), syenite (sy), calcic amphibole (amph).

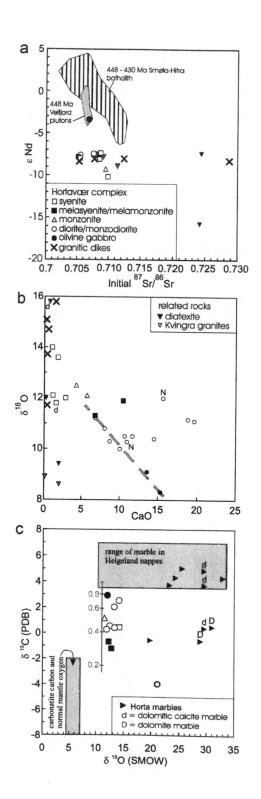

minous garnet two-mica granitic screen or dike (sample 93.82H) has a value of −15.7 (Fig. 5a).

Initial $^{87}Sr/^{86}Sr$ (at 456 Ma) ranges from 0.7052 to 0.7287. Within this range, there is no correlation between initial $^{87}Sr/^{86}Sr$ and bulk composition except that only granitic rocks yield initial $^{87}Sr/^{86}Sr$ >0.710 (Fig. 5a). In spite of their different ε_{Nd} values, the two peraluminous granites are virtually identical in their initial $^{87}Sr/^{86}Sr$ of ~0.724.

4.3. Stable isotopes

Bulk-rock $\delta^{18}O$ values increase as a function of decreasing CaO content (Fig. 5b) from values of +8.3‰ to +9.1‰ in olivine gabbro to values >+11.8‰ among the syenites. Two of the three syenite samples with normative quartz have $\delta^{18}O$ >13.6‰, whereas the other three syenites have $\delta^{18}O$ almost two per mil lower. Melasyenites and one of the nepheline-bearing monzodiorites also have $\delta^{18}O$ in the +12‰ range (Fig. 5b). All of the metaluminous granitic dikes in the complex have $\delta^{18}O$ greater than +11.7‰ and most are >+13.7‰. The peraluminous granite screen or dike (93.82H) has $\delta^{18}O$ of +15.8‰, yet the range of host rock migmatites and associated peraluminous granites is from +9.7‰ to +12.4‰.

The $\delta^{13}C$ and $\delta^{18}O$ values of calcite with igneous textures in the Hortavær complex were measured using methods presented in Appendix A. This calcite shows a narrow range of $\delta^{18}O$ from +11.5‰ to +14.2‰ (VSMOW; Table 5, Fig. 5c); there is no correlation with bulk composition. In contrast, $\delta^{13}C$ varies from +2.9‰ to −1.2‰ (PDB), with the highest values among gabbro and diorite and the lowest among syenitic and melasyenitic rocks.

Samples from marble screens in the complex have $\delta^{18}O$ (VSMOW) from +19.8‰ to +33.0‰ and $\delta^{13}C$ (PDB) from −0.4‰ to +5.1‰ (Table 5, Fig. 5c). Most of the calcite marbles have $\delta^{13}C$ >4‰, whereas the dolomite-bearing marbles show a wider range of $\delta^{13}C$, from +5.1‰ to −0.4‰. Fig. 5c also shows

Fig. 5. (a) Initial Nd and Sr isotopic compositions calculated at 456 Ma. (b) Bulk-rock oxygen isotopes versus wt.% CaO. The dashed line indicates an inferred magmatic trend. (c) Carbonate $\delta^{18}O$ versus $\delta^{13}C$. The vertical line illustrates the Rayleigh effect on calcite $\delta^{13}C$ during assimilation, assuming α(calcite-fluid)=1.04 and initial $\delta^{13}C$ (calcite) of 4‰. Tic marks indicate proportions of calcite remaining.

Table 5
Stable isotope analyses

Carbon and oxygen isotope compositions of igneous carbonate

	$\delta^{18}O$ (‰VSMOW)	$\delta^{13}C$ (‰PDB)	%carb.		Rock type
89.03H	21.15	−3.97	0.5	(1)	plagiogranite dike
89.06H	12.28	−0.67	0.6	(1)	garnet hedenbergite melasyenite
91.20H	14.20	0.49	7.3	(1)	wollastonite garnet hedenbergite melamonzonite
91.48H	12.46	0.56	1.5	(1)	amphibole augite monzodiorite
91.50H	13.32	0.42	1.6	(1)	amphibole augite gabbro
93.28H	11.51	1.17	1.7	(1)	biotite amphibole quartz monzonite
93.44H	14.12	2.53	7.6	(1)	amphibole augite nepheline diorite
93.46H	13.27	2.04	<0.5	(1)	appinitic amphibole augite monzodiorite
93.49H	12.02	2.94	0.4	(1)	amphibole olivine gabbro
93.52H	11.87	0.29	<0.5	(1)	biotite amphibole augite nepheline diorite
93.65H	12.76	−1.21	1.6	(1)	hedenbergite monzonite

Carbon and oxygen isotope compositions of calcite and dolomitic marble

73.07H	25.53	5.03	n.d.	(1)	marble
93.07H	24.30	4.31	n.d.	(1)	calcite marble
93.47H	23.24	3.75	n.d.	(1)	calcite marble
93.74H	19.84	−0.55	n.d.	(1)	dolomitic forsterite marble
02.25H	28.88	−0.43	89.9	(1)	dolomitic marble
02.26H	30.96	0.69	98.1	(1)	dolomitic marble
02.31H	29.59	0.546	72.9	(1)	dolomitic marble
02.47H	29.54	5.09	95.7	(1)	dolomitic marble
02.52H	29.62	3.94	95.1	(1)	dolomitic marble
02.56H	32.97	4.27	91.9	(1)	calcite marble

Bulk-rock oxygen isotopic compositions (silicate rocks). Gabbro, diorite, and monzodiorite

89.05H	10.5			(2)	augite amphibole monzodiorite
91.49H	8.3			(1)	amphibole olivine gabbro
93.44H	12.0			(1)	amphibole augite nepheline diorite

Monzonite

89.20H	12.5				biotite monzonite

Melasyenite

89.06H	12.0			(1)	garnet hedenbergite melasyenite

Syenite

89.09H	13.6			(2)	hedenbergite syenite
89.14H	14.0			(2)	biotite leucosyenite
89.16H	12.1			(2)	amphibole quartz syenite/monzonite
91.34H	12.0			(1)	biotite tourmaline quartz syenite

Granitic dikes

89.03H	14.8			(1)	plagiogranite dike
89.15H	13.7			(2)	biotite granite dike
89.18H	15.1			(2)	biotite granite dike
89.19H	15.8			(2)	biotite granite dike

Peraluminous granites

93.82H	15.8			(1)	diatexitic sillimanite cordierite biotite granite
93.84H	9.4			(1)	garnet two-mica granite

Kvingra granites

89.22H	8.9			(1)	granite

(1) Analyzed at Texas Tech University; (2) analyzed at the University of South Carolina.

that most calcite marbles from the Hortavær screens plot in the field of isotopic compositions of calcite marbles from the Helgeland Nappe Complex (Trønnes, 1994).

5. Discussion

Previous workers assumed that the range of rock types at Hortavær were genetically related. However, recognition of both nepheline- and quartz-bearing rock units in the complex raises the possibility that two or more parental magma types may have been present. Although this possibility cannot be entirely ruled out, we suggest it is not the case and that compositional variation at Hortavær reflects open-system behaviour, specifically host-rock assimilation, in numerous magma batches of mafic magma that evolved to syenitic compositions. This is supported by the continuous variation in clinopyroxene and plagioclase compositions and, more significantly, by the remarkably low ε_{Nd} and high $\delta^{18}O$ values displayed by all rocks in the complex except olivine gabbro. It is difficult to imagine how unrelated magmas acquired these isotopic compositions without being related by some specific differentiation process, in this case by carbonate assimilation.

All previous studies concluded that carbonate assimilation caused the Ca-rich nature of the complex, as manifested in the presence of igneous-textured calcite, scapolite, grossular-rich garnet, idocrase, and abundant clinopyroxene±Ca-amphibole (Vogt, 1916; Gustavson and Prestvik, 1979). Carbonate assimilation can also explain the compositional variation from mildly undersaturated olivine gabbro to alkaline, critically silica-undersaturated rocks like the nepheline-bearing monzodiorites and monzonites. However, as mentioned in the introduction, carbonate assimilation in a closed system is severely limited by the solubility of CO_2. Barnes et al. (2003) showed that the Hortavær complex was not a closed system with regard to influx of magma, in fact it was a magma transfer zone, probably emplaced in an extensional setting. They suggested that this environment allowed for loss of CO_2-rich fluids as soon as they evolved during the assimilation process. The consequence of such fluid loss was the ability of carbonate assimilation reactions to proceed when heat for assimilation

was available. The mechanism of CO_2-rich fluid loss is unclear. We suggest that either:

(1) CO_2 was removed (upward) from the complex in a mixed H_2O–CO_2 fluid entrained in, or streaming through magmas, or
(2) CO_2-rich fluids caused formation of an immiscible carbonate melt, which was removed by upward migration.

Volcanic rocks provide possible examples of both cases: the intra-Apennine volcanic rocks for case (1) (Peccerillo, 1998) and immiscible trachyte-carbonate ash flow tuff from Mount Suswa, Kenya for case (2) (Macdonald et al., 1993). In either case, we suggest that CO_2-producing reactions operated primarily at high-T conditions, because evidence for late-stage migration of carbonate-rich fluids (e.g., calcite veins or carbonatite dikes) is lacking.

The carbon isotope data support the conclusion that carbonate assimilation was pervasive in the Hortavær complex. However, the Sr and particularly Nd isotope data complicate a simple picture of carbonate assimilation. We discuss these complications below, and begin by recognizing that any explanation of the mineral assemblages and compositional variation in the Hortavær complex must account for the following constraints. First, field relations show that dioritic, monzonitic, and syenitic magmas were coeval in the complex. A consistent mafic-to-felsic sequence of emplacement was not recognized, except that granitic dikes were the last to be emplaced. Second, magma emplacement consisted of repeated injection of hundreds of dike-like bodies (Barnes et al., 2003). Third, although a general magmatic trend can be inferred from the geochemical data, the increase in normative Ne from olivine gabbro to evolved diorite, followed by a decrease in normative Ne and the appearance of normative and modal quartz in the syenites, cannot be explained by closed-system differentiation. The fact that the Hortavær complex was a magma transfer zone that consisted of numerous magma batches means that any differentiation process would not result in a single liquid line of descent, especially when the influence of assimilation of heterogeneous host rocks is considered. We will first review evidence for the influence of carbonate assimilation in the complex, then consider the broader implications of the isotope data. This will be followed by simple assimilation models and then by some of the implications of the results.

5.1. Evidence for carbonate assimilation

The evidence for substantial assimilation of carbonate material at Hortavær can be divided into three groups. The first is the abundance of Ca-rich minerals, particularly clinopyroxene, in the diorite zone. Unlike the olivine gabbro, none of the dioritic rocks contain olivine, nor do they show petrographic evidence of high-T olivine stability. However, numerical models of closed-system fractionation using MELTS (Ghiorso and Sack, 1995) and PELE (Boudreau, 1999) indicate that magmas with the compositions of dioritic and monzonitic rocks along the inferred magmatic trend should crystallize olivine through much of their cooling history. As indicated by Tilley (1952), carbonate assimilation reduces olivine stability in favor of clinopyroxene. Unfortunately, neither program can model carbonate assimilation because thermodynamic data for carbonates have not yet been incorporated into their databases.

The second line of evidence is the evolution of the complex to alkaline compositions, and in some instances to critically silica-undersaturated ones. This also fits the classic observations of carbonate assimilation (Daly, 1910; Shand, 1930), in which alkali enrichment is a function of Ca depletion of the melt, largely due to excess clinopyroxene fractionation. Even the most mafic non-cumulate Hortavær samples are characterized by low Sc and Cr contents (Barnes et al., 2003) which is consistent with early clinopyroxene fractionation.

The third line of evidence involves the presence of interstitial calcite in apparent textural equilibrium with igneous silicates (Fig. 2), as inclusions in igneous phases, and locally enclosing igneous clinopyroxene and other silicates. The isotopic composition of carbon in this calcite is unlike mantle-derived carbon (Fig. 5c); instead, it is similar to the carbon isotopic composition of marble screens within the complex. Furthermore, $\delta^{13}C$ decreases with differentiation in the complex. This variation is not compatible with contamination (mixing) of mantle-derived carbon with metasedimentary carbon, because in that case the mafic rocks should have lower $\delta^{13}C$ and the evolved rocks should have higher $\delta^{13}C$. Instead, variation of $\delta^{13}C$ is consistent with Rayleigh fractionation and loss of CO_2 during carbonate assimilation. Modeling of this process is hampered by the temperature-dependent nature of ^{13}C fractionation; however, a simple Rayleigh model with the fractionation factor α(calcite-fluid)=1.04 and an initial $\delta^{13}C$ (calcite) of 4‰ suggests that the amount of carbonate processed was in the range of 20% to 60% for the igneous rocks and as much as 70% among the melasyenite/melamonzonite screens (Fig. 5c).

5.2. Variation of Sr, Nd, and oxygen isotopes

The range of Sr isotopic compositions of calcite marbles from the HNC is quite small, from 0.7067 to 0.7076 (Trønnes, 1994); it is clear that assimilation of marble in this compositional range cannot explain the wider range of values seen in the Hortavær complex. Furthermore, the vanishingly small abundances of REE in calcite marbles precludes their having significant influence on Nd isotopic variation. Therefore, the low ε_{Nd} of the Hortavær complex, which is lower than ε_{Nd} in nearly all other plutons in the Bindal Batholith (Birkeland et al., 1993), must result from assimilation of material with significant Nd content and low ε_{Nd}. The most straightforward choice for such material is the local quartzofeldspathic metasedimentary rocks (commonly migmatitic gneiss).

The need for assimilation of quartzofeldspathic material is also emphasized by the evolution from nepheline-normative monzodioritic and monzonitic compositions to quartz-normative and quartz-bearing syenitic and monzonitic compositions. In a closed system, the well-known thermal divide between silica over- and undersaturated magmas prohibits this from happening (e.g., Bowen, 1937). However, assimilation of quartzofeldspathic metasedimentary rocks may allow such a transition (e.g., Fowler, 1988; Foland et al., 1993; Landoll et al., 1994; Landoll and Foland, 1996). Finally, it is noteworthy that the quartz-bearing and quartz normative syenites are typically enriched in Zr (Fig. 3b) as well as Th, Hf, and U (not shown) relative to the rest of the syenites. These enrichments are too large to result entirely from fractional crystallization and are, in fact, not correlated with most indices of differentiation. Furthermore, neither carbonate assimilation, nor bulk assimilation of known quartzofeldspathic samples can explain the highest Zr contents observed in the quartz syenites.

5.3. Assimilation models

The effects of assimilation on trace element and isotope variations can be modeled using two distinct formulations. One involves mass balance calculations (e.g., DePaolo, 1981), whereas the other involves energy balance (Spera and Bohrson, 2001). The former makes no assumptions concerning the nature of the assimilation reaction(s); it requires end member compositions, partition coefficients, and the ratio of assimilation to crystallization rates ($=r$). The latter assumes that assimilation cannot occur until the assimilated material is partly molten. Energy from the magma is assumed to heat and melt the assimilated material prior to assimilation. Both methods assume two end members: the parental magma and the assimilated material. Algorithms for assimilation that involves three or more end members have not been developed.

In the Hortavær complex, at least five end members must be considered: a parental mafic magma, metacarbonate rocks (marbles), carbonate-rich calc-silicates, metapelitic rocks with low ε_{Nd} and high $^{87}Sr/^{86}Sr$ (e.g., −16 and 0.725, respectively), and rocks or magmas similar to the evolved metaluminous granitic dikes with $\varepsilon_{Nd} \sim -8$ and $^{87}Sr/^{86}Sr$ as high as 0.729. In addition, the changes in bulk partition coefficients and r during magma evolution must be accounted for. At present, we have too few data to constrain all of the variables of assimilation.

The lowered ε_{Nd} of all samples in the complex relative to other mafic plutons in the Norwegian Caledonides requires early input of Nd-bearing, low-ε_{Nd} material. Calcite or dolomite marble cannot be responsible for lowered ε_{Nd} because such rocks lack appreciable Nd. However, calcite-rich calc-silicate rocks with enough quartzofeldspathic material could lower the ε_{Nd} of the contaminated magmas.

Fig. 6 shows AFC models that use the mass-balance approach of DePaolo (1981); input data are in Table 6. These models are meant to illustrate general processes by which many hundreds of magma batches may have evolved. For the first three models, the parental isotopic composition is similar to the most primitive

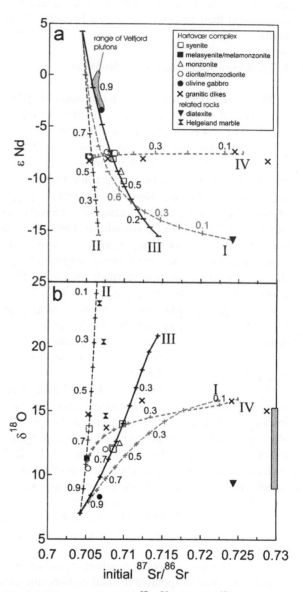

Fig. 6. Initial ε_{Nd} versus initial $^{87}Sr/^{86}Sr$ (a) and $\delta^{18}O$ versus initial $^{87}Sr/^{86}Sr$ (b) diagrams illustrating assimilation-fractional crystallization (AFC) models. AFC calculated according to DePaolo (1981). Input parameters are given in Table 6. Model I assumes a metapelitic contaminant. It fails to fit the oxygen isotope data. Model II assumes a predominantly carbonate component with minor quartzofeldspathic material and Model III assumes an approximately equal mixture of carbonate and metapelitic contaminant(s). Model IV shows the effect of initial carbonate-dominated contamination followed by silicate-dominated contamination. See text for further explanation. In panel b, the shaded bar shows the range of $\delta^{18}O$ in migmatitic gneisses from the western host rocks.

Table 6
Input data for assimilation-fractional crystallization models

Model	r^a	ε_{Nd}	Nd	$D(Nd)^b$	$^{87}Sr/^{86}Sr$	Sr	$D(Sr)^b$	$\delta^{18}O$
Parent								
Models I–III	–	4	10	–	0.7043	500	–	7
Model IV	–	−8	15	–	0.7052	700	–	11
Contaminant								
Model I	0.5	−16	50	0.8	0.7250	250	0.9	15
Model II	0.5	−16	25	0.8	0.7066	1000	0.7	25
Model III	0.5	−16	35	0.8	0.7150	700	0.7	20
Model IV	0.5	−7.5	20	1.5	0.725	100	1.5	15

Nd and Sr abundances in ppm. $\delta^{18}O$ in ‰ (VSMOW).

[a] r is the ratio of rate of assimilation to rate of crystallization.

[b] D is the bulk partition coefficient of the element indicated.

rocks the Caledonian Smøla-Hitra batholith west of Trondheim (Lindstrøm, 1995), which is consistent with isotope data for other mafic plutons in the Bindal Batholith (Nordgulen and Sundvoll, 1992; Nordgulen et al., 1993; Barnes et al., 2002, 2003). This composition was chosen instead of that of the olivine gabbro because some Hortavær samples have initial $^{87}Sr/^{86}Sr$ values lower than the olivine gabbro.

The first type of model (type I) uses only quartzofeldspathic contaminants. It results in strongly curved AFC paths in the Sr and Nd isotope space (Fig. 6a) so that a specific model may intersect some part of the Hortavær sample set, but cannot relate one part to another. The oxygen isotope data are not consistent with such models (Fig. 6b), even when the assumed contaminant has $\delta^{18}O$ values at the high end of the measured range, and such models cannot explain evolution toward nepheline-bearing monzodioritic and monzonitic compositions.

The second model (type II) uses a contaminant with a large carbonate component and a relatively small quartzofeldspathic one. This contaminant should have the necessary carbonate for calcite assimilation as well as sufficient Nd (in the silicate component) to lower ε_{Nd}. Input values for Nd abundance and ε_{Nd} values are poorly constrained and can range from <5 to >35 ppm and −8 to −16, respectively. The effect of varying these values is on the value of F (fraction of melt remaining), not on the AFC path. Fig. 6 shows an example in which the contaminant ε_{Nd} is −16. This model explains the Nd, Sr, and oxygen isotope values of the cluster of Hortavær samples with the lowest $^{87}Sr/^{86}Sr$.

The third model (type III) uses another mixed carbonate/silicate contaminant with a larger quartzofeldspathic component and $^{87}Sr/^{86}Sr$ intermediate between the marbles and the most evolved granites. Such models explain the isotopic composition of the olivine gabbro and many other Hortavær samples (Fig. 6).

These models show that neither assimilation of quartzofeldspathic rocks lacking carbonate, nor assimilation of pure carbonate rocks can explain the isotopic data. A mixture of carbonate and quartzofeldspathic material is required. This raises a number of interesting questions. Were two distinct rock types assimilated? If so, were they marble and pelitic migmatite? Alternatively, was the assimilated material already mixed (i.e., calc-silicate rocks with a range of carbonate-to-silicate proportions)? Calcite-rich calcsilicates correspond to model II; silicate-rich calcsilicates correspond to model III. A calc-silicate contaminant is consistent with a characteristic feature of the Hortavær complex: screens of marble, quartzite, and quartz-rich gneiss show virtually no interaction with the host magmas, whereas banded calc-silicate rocks are completely "made over" into melasyenite/melamonzonite.

The significant lowering of ε_{Nd} and increase in $\delta^{18}O$ from gabbro to nepheline-bearing monzonite requires assimilation of both carbonate and quartzofeldspathic rocks. Continued compositional variation from nepheline-bearing monzonite to quartz-bearing syenite requires a greater influence of a silicate contaminant. This can be modeled by assimilation along a type II path followed by assimilation of a silicate component similar to the metaluminous granites (model IV; Fig. 6).

Models of type IV are consistent with observations made by Tilley (1949, 1952) and Tilley and Harwood (1931), that the CaO component of carbonates is more readily assimilated into mafic magmas than into granitic ones. This is in part due to the higher temperature of mafic magmas. We suggest that it is also due to the higher Mg and Fe contents of mafic magmas, which allows them to process assimilated CaO by making calcic pyroxene and amphibole. The low Mg and Fe in felsic magmas restrict significant assimilation of CaO, but permit assimilation of quartzofeldspathic rocks and consequent increase in silica activity.

Was the assimilated material solid or liquid? A calc-silicate contaminant would probably have contained hydrous silicates, so it is possible that calc-silicates screens produced hydrous, carbonate-rich partial melts (e.g., Wyllie, 1965). Such melts would have low viscosity and could percolate as intergranular melts through a crystallizing, mushy, mafic magma. This could explain why all mafic rocks in the Hortavær complex show evidence of carbonate assimilation. Assimilation of solid carbonate would result in piecemeal contamination, whereas assimilation of low-viscosity, carbonate-rich melts would be pervasive.

As noted above, some syenites have Zr contents too high to be explained by assimilation of any analyzed granitic rocks (Table 2). We suggest that the high Zr contents result from local percolation of granitic melts through syenitic crystal mushes. This problem may be resolved by investigating possible inheritance in zircons from the quartz syenites (in progress).

6. Conclusions and consequences

Chemical evolution of magmas in the Hortavær complex involved sequential emplacement of mafic magma batches into which both carbonate and silicate material was assimilated. Assimilation of carbonate material was possible because the magmas were emplaced in an extensional environment; thus, CO_2 that was evolved during assimilation could escape upward, out of the system (Barnes et al., 2003). Evolution of the mafic magmas was toward greater alkalinity (monzonite and syenite). However, as evolution proceeded, the effects of carbonate assimilation reached a maximum among monzodioritic magmas (nepheline stability) and then diminished as the ability of magmas to assimilate carbonates decreased and assimilation of quartzofeldspathic material became predominant (quartz stability). Dioritic magmas were able to process Ca by making clinopyroxene and Ca-amphibole, whereas the syenitic magmas had too little Fe and Mg to process appreciable quantities of carbonate. The principal contaminants were probably calc-silicate rocks because they carried all of the isotopic components necessary to explain compositions in the Hortavær

system. The silicate component necessary to produce quartz stability may also have been introduced by late-stage granitic dikes (Barnes et al., 2003). It is additionally possible that the effects of carbonate assimilation are so pervasive because of intergranular migration of carbonate melts, rather than because of piecemeal stoping of solid carbonate.

A cursory view of the radiogenic isotope data does not suggest a need for carbonate assimilation in the Hortavær complex. Only the carbon isotopic composition of igneous-textured calcite provides strong evidence for the presence of assimilated carbonates. If some of the Hortavær magmas had erupted and lost their dissolved CO_2, then no obvious isotopic, elemental, or physical evidence for carbonate assimilation would be preserved. Only the low Sc contents and the unusually low initial $^{87}Sr/^{86}Sr$ coupled with low ε_{Nd} would hint at such assimilation. In rare instances where carbonate-rich lavas are erupted, it is possible to infer a sedimentary origin for the carbonates (e.g., Peccerillo, 1998). Otherwise, identification of cryptic carbonate assimilation in volcanic rocks requires the use of isotopic and trace element data that permit multiple interpretations. For example, Yogodzinski et al. (1996) used Sr and Nd isotopic data to suggest that carbonate assimilation had affected basaltic magmas in the Great Basin region of Nevada.

We do not espouse the idea of carbonate assimilation as a universal, or perhaps even as a widespread process. However, we conclude that the Hortavær complex provides an interesting example of carbonate assimilation on a pluton scale, quite different from the contact effects commonly cited in the literature. Other plutonic sequences show nearly identical effects, for example the Dismal Nepheline Syenite of Antarctica (Worley and Cooper, 1995; Worley et al., 1995). We therefore conclude that given the appropriate thermal, petrologic, and tectonic conditions, carbonate assimilation can occur and can influence significant volumes of magma.

Acknowledgments

It is a pleasure to contribute to this volume honoring Professor Ilmari Haapala. He has provided an excellent example for all field petrologists by illustrating the benefits of thorough, thoughtful integration of geology, mineralogy, geochemistry, and geophysics. We wish him the best in his retirement.

We thank Jostein Hiller, Reidar Berg, and Arne Lysø for their able seamanship. Melanie Barnes, J. G. deHaas, Ingrid Vokes, and James Browning provided expert assistance in the laboratory. Ø. Nordgulen, the late K. Skjerlie, and O.T. Rämö provided reviews, which greatly improved the manuscript. This research was initially funded by a grant from Nansenfondet to Prestvik and received partial support from National Science Foundation grant EAR-9814280 to Barnes.

Appendix A. Methods

Analysis of silicate bulk-rock oxygen isotope ratios in the Texas Tech University Stable Isotope Laboratory followed the method of Clayton and Mayeda (1963). Oxygen was liberated from powdered samples by fluorination with bromide pentafluorine (BrF_5) in nickel tubes at 500–600 °C. The liberated O_2 was reacted with heated graphite rods to form CO_2, which was analyzed on a VG SIRA 12 gas-source mass spectrometer. The measured gas yields were compared to theoretically expected yields determined on the basis of bulk chemical analysis. With one exception all measured yields were within ±3% of expected yields.

Analysis of marble samples followed the standard methods of McCrea (1950). Powdered samples of host-rock marble were reacted with H_3PO_4 at 25 °C for 1–2 days to release CO_2 gas, which was then analyzed for carbon and oxygen isotope ratios.

Analysis of the calcite fraction of the igneous rock samples was similar to that for the marbles except that larger masses of rock powder and H_3PO_4 were used to obtain sufficient CO_2 gas for analysis. To test for the ability of the method of quantitatively extract calcite from the rock powder, synthetic mixtures of a calcite-free basaltic lab standard were mixed with pure calcite. Gas yields were 91±1.7% over a range of calcite mass fractions of 0.5% to 10.0%. For these tests, $\delta^{13}C$ of the extracted calcite fraction was $-6.94\pm0.01\%o$(PDB) and $\delta^{18}O$ was $25.78\pm0.06\%o$(VSMOW), compared to expected values of -6.94 and 25.75, respectively. Details of the methodology are available from the first author.

The Nd and Sr isotope ratios were determined by isotope dilution methods at the Mineralogical-Geological Museum, University of Oslo and using a VG Isotope VG354 TIMS instrument. The analytical procedure for Nd was identical to that reported by Mearns (1986) and for Sr identical to that reported by Sundvoll et al. (1992). During the analytical work the JM-reference standard gave a value of: $^{143}Nd/^{144}Nd=$ 0.51111 ± 0.00005 and the NBS987 Sr standard yielded a value of $^{87}Sr/^{86}Sr=0.71023\pm0.00003$. Errors quoted are 2 sigma standard error.

Four samples were leached prior to analysis. In the case of low-Sr sample 89.22 H, leaching was done to test for effects of marine carbonate contamination on the isotope ratios. Leaching consisted of immersion of rock chips in 6 N distilled HCl for 2 h at 40 °C. The sample was then washed three times in deionized water and dried. In the case of sample 89.22 H, a biotite granite from Kvingra, leaching removed enough Sr to make the $^{87}Rb/^{86}Sr$ ratio meaningless. Leaching also shifted the Nd isotope ratios slightly. In all other cases, the rocks are Sr-rich and leaching did not shift the analyzed values from those expected on the basis of analyses from the rest of the complex.

References

Baker, C.K., Black, P.M., 1980. Assimilation and metamorphism at a basalt–limestone contact, Tokatoka, New Zealand. Mineralogical Magazine 43, 797–807.

Barnes, C.G., Prestvik, T., 2000. Conditions of pluton emplacement and anatexis in the Caledonian Bindal Batholith, north-central Norway. Norsk Geologisk Tidsskrift 80, 259–274.

Barnes, C.G., Prestvik, T., Nordgulen, Ø., Barnes, M.A., 1992. Geology of three dioritic plutons in Velfjord, Nordland. Bulletin-Norges Geologiske Undersøkelse 423, 41–54.

Barnes, C.G., Yoshinobu, A.S., Prestvik, T., Nordgulen, Ø., Karlsson, H.R., Sundvoll, B., 2002. Mafic magma intraplating: anatexis and hybridization in arc crust, Bindal Batholith, Norway. Journal of Petrology 43, 2171–2190.

Barnes, C.G., Prestvik, T., Barnes, M.A.W., Anthony, E.Y., 2003. Geology of a magma transfer zone: the Hortavær Igneous Complex, north-central Norway. Norsk Geologisk Tidsskrift 83, 187–208.

Birkeland, A., Nordgulen, Ø., Cumming, G.L., Bjørlykke, A., 1993. Pb–Nd–Sr isotopic constraints on the origin of the Caledonian Bindal Batholith, central Norway. Lithos 29, 257–271.

Boudreau, A.E., 1999. PELE—a version of the MELTS software program for the PC platform. Computers & Geosciences 25, 201–203.

Bowen, N.L., 1937. Recent high temperature research on silicates and its significance in igneous petrology. American Journal of Science 33, 1–21.

Clayton, R.N., Mayeda, T.K., 1963. The use of bromine pentafluoride in the extraction of oxygen from oxides and silicates for isotopic analysis. Geochimica et Cosmochimica Acta 27, 43–52.

Daly, R.A., 1910. Origin of the alkaline rocks. Geological Society of America Bulletin 21, 87–118.

Daly, R.A., 1918. Genesis of the alkaline rocks. Journal of Geology 26, 7–134.

DePaolo, D.J., 1981. Trace element and isotopic effects of combined wallrock assimilation and fractional crystallization. Earth and Planetary Science Letters 53, 189–202.

Foland, K.A., Landoll, J.D., Henderson, C.M.B., Jiangfeng, C., 1993. Formation of cogenetic quartz and nepheline syenites. Geochimica et Cosmochimica Acta 57, 697–704.

Fowler, M.B., 1988. Elemental evidence for crustal contamination of mantle-derived Caledonian syenite by metasediment anatexis and magma mixing. Chemical Geology 69, 1–16.

Frost, B.R., Barnes, C.G., Collins, W.J., Arculus, R.J., Ellis, W.J., Frost, D.J., 2001. A geochemical classification for granitic rocks. Journal of Petrology 42, 2033–2048.

Ghiorso, M.S., Sack, R.O., 1995. Chemical mass transfer in magmatic processes: IV. A revised and internally consistent thermodynamic model for the interpolation and extrapolation of liquid–solid equilibria in magmatic systems at elevated temperatures and pressures. Contributions to Mineralogy and Petrology 119, 197–212.

Gustavson, M., Prestvik, T., 1979. The igneous complex of Hortavær, Nord-Trøndelag, central Norway. Bulletin-Norges Geologiske Undersøkelse 348, 73–92.

Landoll, J.D., Foland, K.A., 1996. The formation of quartz syenite by crustal contamination at mont Shefford and other Monteregian complexes, Quebec. Canadian Mineralogist 34, 301–324.

Landoll, J.D., Foland, K.A., Henderson, C.M.B., 1994. Nd isotopes demonstrate the role of contamination in the formation of coexisting quartz and nepheline syenites at the Abu Khruq Complex, Egypt. Contributions to Mineralogy and Petrology 117, 305–329.

Lindstrøm, M., 1995. Petrology of the Caledonian Hitra plutonic complex, central Norway. PhD Thesis, University of Trondheim, Trondheim, Norway.

Macdonald, R., Kjarsgaard, B.A., Skilling, I.P., Davies, G.R., Hamilton, D.L., Black, S., 1993. Liquid immiscibility between trachyte and carbonate in ash flow tuffs from Kenya. Contributions to Mineralogy and Petrology 114, 276–287.

McCrea, J.M., 1950. The isotopic chemistry of carbonates and a paleotemperature scale. Journal of Chemical Physics 18, 849–857.

Mearns, E.W., 1986. Sm–Nd ages from Norwegian garnet peridotite. Lithos 19, 269–278.

Meen, J.K., 1990. Elevation of potassium content of basaltic magma by fractional crystallization: the effect of pressure. Contributions to Mineralogy and Petrology 104, 309–331.

Nordgulen, Ø., 1993. A summary of the petrography and geochemistry of the Bindal Batholith. Geological Survey of Norway, Report 92.111.

Nordgulen, Ø., Sundvoll, B., 1992. Strontium isotope composition of the Bindal Batholith, Central Norwegian Caledonides. Bulletin-Norges Geologiske Undersøkelse 423, 19–39.

Nordgulen, Ø., Bickford, M.E., Nissen, A.L., Wortman, G.L., 1993. U–Pb zircon ages from the Bindal Batholith, and the tectonic history of the Helgeland Nappe Complex, Scandinavian Caledonides. Journal of the Geological Society (London) 150, 771–783.

Peccerillo, A., 1998. Relationships between ultrapotassic and carbonate-rich volcanic rocks in central Italy: petrogenetic and geodynamic implications. Lithos 43, 267–279.

Roberts, D., Gee, D.G., 1985. An introduction to the structure of the Scandinavian Caledonides. In: Gee, D.G., Sturt, B.A. (Eds.), The Caledonide Orogen-Scandinavia and related areas. John Wiley and Sons, Chichester, pp. 55–68.

Roberts, D., Heldal, T., Melezhik, V.A., 2001. Tectonic structural features of the Fauske conglomerates in the Løvgavlen quarry, Nordland, Norwegian Caledonides, and regional implications. Norwegian Journal of Geology 81, 245–256.

Sabine, P.A., 1975. Metamorphic processes at high temperature and low pressure: the petrogenesis of the metasomatized and assimilated rocks of Carneal, Antrim. Philosophical Transactions of the Royal Society of London 280, 225–269.

Sawyer, E.W., 1998. Formation and evolution of granitic magmas during crustal reworking: the significance of diatexites. Journal of Petrology 39, 1147–1167.

Shand, S.J., 1930. Limestone and the origin of felspathoidal rocks: an aftermath of the Geological Congress. Geological Magazine 67, 415–427.

Snoke, A.W., Quick, J.E., Bowman, H.R., 1981. Bear Mountain igneous complex, Klamath Mountains, California: an ultrabasic to silicic calc-alkaline suite. Journal of Petrology 22, 501–552.

Spear, F.S., Kohn, M.J., Cheney, J.T., 1999. P–T paths from anatectic pelites. Contributions to Mineralogy and Petrology 134, 17–32.

Spera, F.J., Bergman, S.C., 1980. Carbon dioxide in igneous petrogenesis: I. Aspects of the dissolution of CO2 in silicate liquids. Contributions to Mineralogy and Petrology 74, 55–66.

Spera, F.J., Bohrson, W.A., 2001. Energy-constrained open-system magmatic processes I: General model and energy-constrained assimilation and fractional crystallization (EC-AFC) formulation. Journal of Petrology 42, 999–1018.

Sundvoll, B., Larsen, B.T., Wandås, B., 1992. Early magmatic phase in the Oslo Rift and its related stress regime. Tectonophysics 208, 37–54.

Thorsnes, T., Løseth, H., 1991. Tectonostratigraphy in the Velfjord-Tosen region, southwestern part of the Helgeland Nappe Complex, Central Norwegian Caledonides. Bulletin-Norges Geologiske Undersøkelse 421, 1–18.

Tilley, C.E., 1949. An alkali facies of granite at granite–dolomite contacts in Skye. Geological Magazine 86, 81–93.

Tilley, C.E., 1952. Some trends of basaltic magma in limestone syntexis. American Journal of Science, Bowen 250A, 529–545.

Tilley, C.E., Harwood, H.F., 1931. The dolerite–chalk contact of Scawt Hill, Antrim. The production of basic alkali-rocks by the assimilation of limestone by basaltic magma. Mineralogical Magazine 22, 439–468.

Trønnes, R.G., 1994. Marmorforekomster i Midt-Norge: Geologi, isotopgeokjemi og industrimineralpotensiale. Norges Geologiske Undersøkelse-Rapport 94.042 (in Norwegian), 21 pp.

Vogt, T., 1916. Petrographisch-chemische studien an einigen assimilations-gesteinen der nord-Norwegishen gebirgskette. Videnskaps Selskapets Kristiania, Skrifter 1 (8), 1–33.

Watkinson, D.H., Wyllie, P.J., 1969. Phase equilibrium studies bearing on the limestone-assimilation hypotheses. Geological Society of America Bulletin 80, 1565–1576.

Wilson, M., 1989. Igneous Petrogenesis. Unwin Hyman, London.

Worley, B.A., Cooper, A.F., 1995. Mineralogy of the dismal nepheline syenite, southern Victoria Land, Antarctica. Lithos 35, 109–128.

Worley, B.A., Cooper, A.F., Hall, C.E., 1995. Petrogenesis of carbonate-bearing nepheline syenites and carbonatites from Southern Victoria Land, Antarctica: origin of carbon and the effects of calcite-graphite equilibrium. Lithos 35, 183–199.

Wyllie, P.J., 1965. Melting relationships in the system CaO–MgO–CO_2–H_2O, with petrological applications. Journal of Petrology 6, 101–123.

Yogodzinski, G.M., Naumann, T.R., Smith, E.I., Bradshaw, T.K., Walker, J.D., 1996. Evolution of a mafic volcanic field in the central Great Basin, south central Nevada. Journal of Geophysical Research 101, 17425–17445.

Yoshinobu, A.S., Barnes, C.G., Nordgulen, Ø., Prestvik, T., Fanning, M., Pedersen, R.-B., 2002. Ordovician magmatism, deformation, and exhumation in the Caledonides of central Norway: an orphan of the Taconic orogeny? Geology 30, 883–886.

Available online at www.sciencedirect.com

LITHOS

Lithos 80 (2005) 201–227

www.elsevier.com/locate/lithos

ELSEVIER

Quartz and feldspar zoning in the eastern Erzgebirge volcano-plutonic complex (Germany, Czech Republic): evidence of multiple magma mixing

Axel Müller[a,*,1], Karel Breiter[b], Reimar Seltmann[a], Zoltán Pécskay[c]

[a]*Natural History Museum, Department Mineralogy, Cromwell Road, London SW7 5BD, UK*
[b]*Czech Geological Survey, Geologicka 6, CZ-15200 Praha 2, Czech Republic*
[c]*Institute of Nuclear Research of the Hungarian Academy of Sciences, Debrecen, Hungary*

Received 14 April 2003; accepted 9 September 2004
Available online 5 November 2004

Abstract

Zoned quartz and feldspar phenocrysts of the Upper Carboniferous eastern Erzgebirge volcano-plutonic complex were studied by cathodoluminescence and minor and trace element profiling. The results verify the suitability of quartz and feldspar phenocrysts as recorders of differentiation trends, magma mixing and recharge events, and suggest that much heterogeneity in plutonic systems may be overlooked on a whole-rock scale. Multiple resorption surfaces and zones, element concentration steps in zoned quartz (Ti) and feldspar phenocrysts (anorthite content, Ba, Sr), and plagioclase-mantled K-feldspars etc. indicate mixing of silicic magma with a more mafic magma for several magmatic phases of the eastern Erzgebirge volcano-plutonic complex. Generally, feldspar appears to be sensitive to the physicochemical changes of the melt, whereas quartz phenocrysts are more stable and can survive a longer period of evolution and final effusion of silicic magmas. The regional distribution of mixing-compatible textures suggests that magma mingling and mixing was a major process in the evolution of these late-Variscan granites and associated volcanic rocks.

Quartz phenocrysts from 14 magmatic phases of the eastern Erzgebirge volcano-plutonic complex provide information on the relative timing of different mixing processes, storage and recharge, allowing a model for the distribution of magma reservoirs in space and time. At least two levels of magma storage are envisioned: deep reservoirs between 24 and 17 km (the crystallisation level of quartz phenocrysts) and subvolcanic reservoirs between 13 and 6 km. Deflation of the shallow reservoirs during the extrusion of the Teplice rhyolites triggered the formation of the Altenberg-Teplice caldera above the eastern Erzgebirge volcano-plutonic complex. The deep magma reservoir of the Teplice rhyolite also has a genetic relationship to the younger mineralised A-type granites, as indicated by quartz phenocryst populations. The pre-caldera biotite granites and the rhyodacitic Schönfeld volcanic rocks represent temporally and spatially separate magma sources. However, the deep magma reservoir of both is assumed to have been at a depth of 24–17 km. The drastic chemical contrast between the pre-caldera

---

* Corresponding author. Tel.: +47 7390 4216; fax: +47 7392 1620.
E-mail address: Axel.Muller@ngu.no (A. Müller).
[1] Present address: Norges geologiske undersøkelse, N-7491 Trondheim, Norway.

Schönfeld (Westfalian B–C) and the syn-caldera Teplice (Westfalian C–D) volcanic rocks is related to the change from late-orogenic geotectonic environment to post-orogenic faulting, and is considered an important chronostratigraphic marker.

Keywords: Quartz; Feldspar; Trace elements; Magma mixing; Resorption textures; Erzgebirge

---

1. Introduction

Granitic magmas may be affected by mingling and mixing with mantle-derived magma, and there is manifold evidence for magma mingling and mixing to occur in the plutonic environment, at least on a local scale (e.g., Didier and Barbarin, 1991; Hibbard, 1991; Wilcox, 1999; Baxter and Feely, 2002). Often, it is whole-rock chemistry and isotopic composition that provide much of the evidence for magma mixing (hybridisation). If the percentage of the mafic melt input is very low; however, it is hard to prove mixing events chemically, and whole-rock geochemical data can lead to an incomplete picture of true chemical variability. Fundamental to such interpretations is the recognition of petrographic textures that may form during magma mixing and mingling (cf. Hibbard, 1981, 1991; Andersson and Eklund, 1994). Whole-rock chemical composition of granites can be interpreted as reflecting a mixture of crystals, but comprehensive understanding of magma evolution should be based on microchemical and microtextural investigations of phenocrysts and groundmass minerals.

Mingling and mixing are widespread in late-Variscan granitic batholiths of the Variscan orogenic belt in central Europe (Siebel et al., 1995; Janoušek et al., 2002; Müller and Seltmann, 2002; Słaby et al., 2002). Here we show that the integration of textures and chemical zoning of phenocrysts in the igneous rocks of the Eastern Erzgebirge granite pluton, which is a part of the Erzgebirge/Krušne Hory granite batholith astride the border of Germany and Czech Republic (Fig. 1), can be used to track multiple mingling and mixing and recharge events. The study is focused on the characterisation of quartz phenocrysts by cathodoluminescence (CL) and by trace element (Al, Ti, K, Fe) composition and by major and trace element profiling of coexisting feldspars. We evaluate the extent to which variations in mineral chemistry reflect either magma mixing or changes in intensive parameters during the evolution of the felsic systems.

2. Geologic setting and chemical evolution

Eastern Erzgebirge is the easternmost part of the Fichtelgebirge–Erzgebirge anticline where late- to post-orogenic Variscan uplift and exhumation processes accompanied intense felsic (rhyolitic and granitic) magmatism that was controlled by brittle-fracture tectonics (Seltmann et al., 1996) and formed the eastern Erzgebirge volcano-plutonic complex (Štemprok et al., 2003). The orogenic collapse of the Saxothuringian zone as part of the Variscan orogenic belt contributed to the formation of volcano-tectonic intramontane depressions (Olbernhau and Schönfeld depressions) and strike-slip related pull-apart basins (Erzgebirge and Döhlen basins) in the marginal and central parts of Erzgebirge. Caldera complexes developed in the Altenberg block (Altenberg-Teplice caldera) and in Tharandter Wald (Fig. 1). The 650-km^2 Altenberg-Teplice caldera forms the main part of the eastern Erzgebirge volcano-plutonic complex. The caldera collapse was triggered by the extrusion of 70–120 km^3 Teplice volcanic rocks (Benek, 1991) whose chemical characteristics resemble those of the Bishop Tuff, CA (Hildreth, 1979). An idealised profile through the volcanic pile of the Altenberg-Teplice caldera has been compiled from the borehole Mi-4 in the southwestern part of the caldera (Jiránek et al., 1987; Breiter, 1997; Breiter et al., 2001) (Fig. 2).

Our study includes 14 igneous phases of the eastern Erzgebirge volcano-plutonic complex: the Niederbobritzsch granites (NB1–NB4), the Schönfeld basal rhyolite (SBR) and the Schönfeld rhyodacite (SDC), the Teplice rhyolites (TR1–TR3), the Altenberg-Frauenstein microgranites (GP1 and GP2), and the Schellerhau granite complex (SG1–SG3). Geochemical composition of representative samples (Fig. 3)

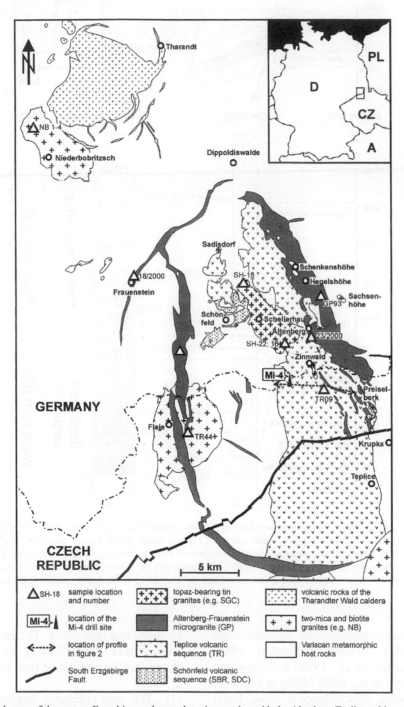

Fig. 1. Geological sketch map of the eastern Erzgebirge volcano-plutonic complex with the Altenberg-Teplice caldera, Mesozoic and Cenozoic uncovered. The caldera is framed by the Altenberg-Frauenstein microgranite (dark grey). Inset shows area relative to east-central Europe [Germany (D), Poland (PL), Czech Republic (CZ), Austria (A)]. Modified from Hoth et al. (1995) and Mlčoch (1994).

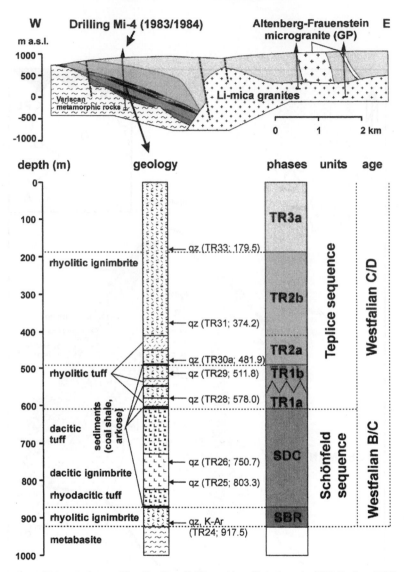

Fig. 2. Geological cross section of the central part of the Altenberg-Teplice caldera (Jiránek et al., 1987; Breiter, 1997) and geological section of Mi-4 borehole and its division into volcanic phases and subphases (events) based on whole-rock chemistry. The stratigraphic position of samples from the Schönfeld-Teplice volcanic rocks is indicated by the arrows along the drill log. Sample number and sample depth (in metres) is in brackets. Whole-rock chemistry and phenocrysts were studied on samples with the abbreviation "qz". Dating was carried out on the sample with the abbreviation "K–Ar". Key to abbreviations: Schönfeld basal rhyolite (SBR), Schönfeld rhyodacite (SDC), subphases of the Teplice rhyolite (TR1a–TR3a).

confirms the general petrogenetic characteristics known from earlier studies (Tischendorf, 1989; Breiter, 1995; Breiter et al., 1999; Seltmann and Breiter, 1995; Förster et al., 1999; Müller et al., 2000; Müller and Seltmann, 2002; Štemprok et al., 2003) and delineates magmatic evolutionary trends that are well documented by changes in some major and trace

elements and their ratios. The investigated magmatic phases have the following characteristics (order determined by age, oldest first; cf. Fig. 3):

(1) The post-kinematic Niederbobritzsch granites (NB) belong to a group of late-collisional low-F biotite monzogranites and form a multiphase

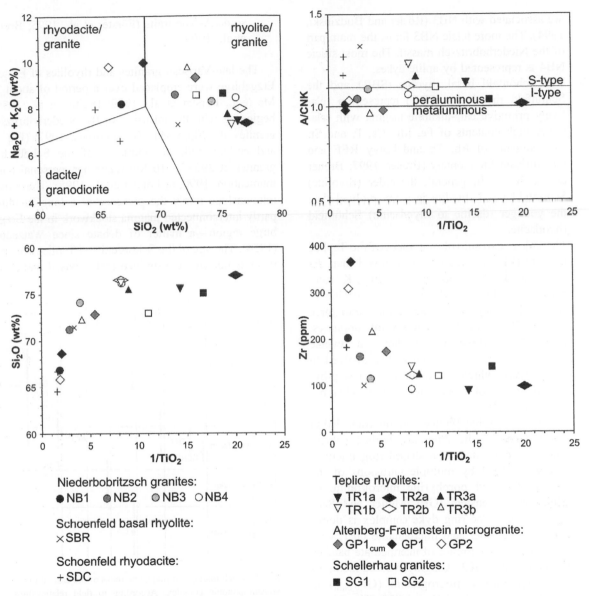

Fig. 3. Variation diagrams for selected major and trace elements of the the late Variscan volcano-plutonic associations in Eastern Erzgebirge. In most of the plots, the Niederbobritzsch granites and the Schönfeld volcanic rocks exhibit similar characteristics. The intrusive phases (Niederbobritzsch granites and Schellerhau granites) show internal fractionation trends towards more evolved end members, whereas the volcanic rocks (Schönfeld basal rhyolite, Schönfeld rhyodacite, Teplice rhyolite) and subvolcanic microgranites (Altenberg-Frauenstein microgranite) exhibit, for each eruption cycle, a trend from evolved to less evolved phases (indicated by increase in Zr and decrease in SiO₂). The initial eruption products of the Teplice rhyolites (TR1a and TR2a) show similar compositional traits as the strongly peraluminous Schellerhau granite complex (SG1 and SG2), suggesting a link between the two. Date from Förster et al. (1999), Müller et al. (2000), Müller and Seltmann (2002), and K. Breiter (unpublished data).

intrusion that consists of four intrusive phases (NB1–NB4) with a transitional I- to S-type character (Förster et al., 1999). NB2 is more heterogeneous than NB1 and is interpreted as a

hybrid granite, resulting from mingling and mixing with more mafic magma (Rösler and Bothe, 1990). Texturally, mafic enclaves in NB2 correspond texturally to lamprophyre dykes that are associated with NB3 (Rösler and Budzinski, 1994). The more felsic NB3 forms the main part of the Niederbobritzsch massif. The most silicic NB4 is represented by aplite dykes.

(2) The Schönfeld basal rhyolite (SBR) and the Schönfeld rhyodacite (SDC) represent a relatively primitive calc-alkaline magma with relatively high contents of Fe, Mg, Ca, P, and Sr, low contents of Rb, Th and heavy REE and insignificant Eu anomaly (Breiter, 1997; Breiter et al., 2001). In general, the older (rhyolitic) Schönfeld basal rhyolite is more evolved than the younger (dacitic to rhyodacitic) Schönfeld rhyodacite.

(3) The volcanic to subvolcanic rocks of the Teplice rhyolite (TR) are much more evolved than the older Schönfeld rocks and are high-K calc-alkaline in character (Breiter, 1997; Breiter et al., 2001). They can be subdivided into three phases (TR1, TR2, TR3) and several subphases. Each of the three phases shows a reverse geochemical evolution from highly evolved to less evolved which is explained by a step-by-step deflation of a stratified magma chamber (Breiter, 1997).

(4) The eruption of the Teplice volcanic rocks led to the collapse of the Altenberg-Teplice caldera with N–S elongated, pear-shaped ring fractures that were filled by multiple intrusions of the rapakivi-textured porphyritic microgranite of Altenberg-Frauenstein (GP) (Müller and Seltmann, 2002). The ring dyke complex is dominated by the older acid microgranite GP1, with less common occurrence of intermediate batches of microgranite GP2. A highly evolved medium- to coarse-grained microgranite (GP1$_{cum}$) is found as enclaves in GP1 and GP2 (Fig. 3).

(5) The topaz-bearing rare metal granites (post-caldera stage) are represented by the Schellerhau granite complex (SGC) and several small intrusions (Sadisdorf, Altenberg, Schenkenshöhe, Hegelshöhe, Zinnwald-Cínovec, Loupený, Preiselberk, and Knötel). The Schellerhau granite complex is characterised by porphyritic (SG1) to weakly porphyritic (SG2) biotite syeno- to monzogranites and seriate albite granites (SG3) (Müller et al., 2000). The SG1, SG2 and SG3 rocks are P-poor and Li–F-enriched and show

some A-type traits (Förster et al., 1995; Breiter et al., 1999).

The late-Variscan granites and rhyolites of Eastern Erzgebirge were emplaced over a period of about 25 Ma (e.g., Förster et al., 1999; Breiter et al., 1999), beginning with the intrusion of the Niederbobritzsch granites at 320±6 Ma (Tichomirowa, 1997) (Fig. 4) and ending with the intrusion of the Schellerhau granites at 295+7/−10 Ma (Hammer, personal communication, 1990, in Förster et al., 1998). These time relationships imply the presence of a long-standing, partly interconnected magma stockwork in the Erzgebirge region—a matter of debate since Watznauer (1954). The age of the Schönfeld and Teplice volcanic rocks is not well constrained and is based mainly on

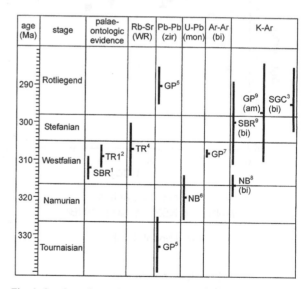

Fig. 4. Geochronology of magmatic rocks of the eastern Erzgebirge volcano-plutonic complex. According to field relationships, the relative timing of magmatic events is Niederbobritzsch granite (NB, oldest); Schönfeld basal rhyolite (SBR); Schönfeld rhyodacite (SDC); Teplice rhyolite (TR); Altenberg-Frauenstein microgranite (GP); Schellerhau granite complex (SGC, youngest). GP[5] sample (Kempe et al., 1999) yielded single zircon ^{207}Pb/^{206}Pb evaporation ages of 290±5 and 333±4 Ma. The older age is in contrast to the field relationships and to the other age determinations and is probably hampered by inheritance. Key to abbreviations: amphibole (am), biotite (bi), whole rock (WR), zircon (zir). Data sources: (1) Lobin (1983), (2) Šimùnek in Jiránek et al. (1987), (3) Hammer (personal communication, 1990) in Förster et al. (1998), (4) R. Seltmann (unpublished data), (5) Kempe et al. (1999), (6) Tichomirowa (1997), (7) Seltmann and Schilka (1995), (8) Werner and Lippolt (1998), (9) this study.

palaeontological evidence (Fig. 4). We have acquired the first K–Ar age of unaltered biotite from the Schönfeld basal rhyolite (borehole Mi-4, 917.5 m), 300 ± 11 Ma. Because the K–Ar systematics of biotite can be affected by later thermal events, this age should be considered as a minimum. However, it fits the field relationships well. A K–Ar dating of amphibole of the Altenberg-Frauenstein microgranite yielded an age of 297 ± 13 Ma.

3. Analytical techniques

3.1. Scanning electron microscope cathodoluminescence (SEM-CL)

About 320 quartz phenocrysts from each magmatic phase and subphase (8–38 from each) were SEM-CL imaged using a JEOL electron microprobe system at the Geowissenschaftliches Zentrum Göttingen. A CL detector with an extended wavelength range from 200 to 900 nm and slow beam scan rates of 20 s per image at processing resolution of 1024×860 pixels and 256 grey levels were used. The electron beam voltage and current were 20 kV and 200 nA, respectively. CL imaging before and after electron probe microanalysis allowed positioning of the analytical points in relation to CL structures.

3.2. Electron probe microanalysis (EPMA)

Trace element abundances of Al, K, Ti, and Fe in quartz were performed with a JEOL 8900 RL electron microprobe with five wavelength dispersive detectors at the Geowissenschaftliches Zentrum Göttingen equipped. Synthetic Al_2O_3 (52.9 wt.% Al), orthoclase from Lucerne, Switzerland (12.2 wt.% K), synthetic TiO_2 (59.9 wt.% Ti), and haematite from Rio Marina, Elba (69.9 wt.% Fe) were used for standards. A beam current of 80 nA, an accelerating voltage of 20 kV, diameter of 5 μm, and counting times of 15 s for Si, and of 300 s for Al, Ti, K, and Fe were used. Raw analyses were converted into concentrations using the phi–rho–Z matrix correction method of Armstrong (1995). Analytical errors were calculated from the counting statistics of peak and background signals, following the Gauss law of error propagation. At low element concentrations, the background forms the main part of the total signal. On the other hand, the background signal is nearly constant for a given quartz matrix and the absolute error based on counting statistic is thus nearly constant. The precision of the analyses (at 3σ) was 12 ppm absolute standard deviation for Al, 12 ppm for K, 18 ppm for Ti, and 21 ppm for Fe. Limits of detection (7 ppm for Al, 11 ppm for Ti, 7 ppm for K, and 12 ppm for F) were calculated with a confidence level of 95% on the basis of the Student's t-distribution (Plesch, 1982) from 107 background measurements.

Quantitative point analyses for Na, Al, Si, P, K, Ca, Ti, Fe, Sr and Ba in feldspar were carried out with a Cameca SX 50 microprobe at the Natural History Museum London, using an accelerating voltage of 20 kV and a beam current of 50 nA. Alkalies were analysed first, which lead to a slight relative enrichment of Si during electron exposure due to the migration of alkalies. However, the loss of Na and K was not significant over the applied counting period of 10 s. P, Sr, and Ba were analysed over a 30-s counting time on peak, Ti over 20 s, and Al, Si, Ca, and Fe over 15 s. Representative detection limits (3σ) for minor and trace elements were 50 ppm for P, 61 ppm for Ti, 129 ppm for Fe, 176 ppm for Sr, and 130 ppm for Ba.

4. Results

4.1. Growth patterns in quartz phenocrysts

SEM-CL of quartz reveals intra-granular textures that cannot be observed using optical microscopy or backscattered electrons. These textures yield insight into the relative timing of crystal growth, dissolution and precipitation, crack formation and healing, and diffusion. Quartz phenocrysts are present in almost all magmatic rocks of the pre-caldera to post-caldera stages. Phenocrysts exhibit growth patterns including oscillatory zoning, resorption surfaces, and growth embayments, which are successfully imaged by CL. For instance, resorption surfaces can be easily identified as they truncate older growth zones, whereas the growth zones adapt to the shape of lobate and drop-like embayments (e.g., Müller et al., 2003). The classification of quartz phenocrysts into different types is interpretative, but phenocrysts of the same stratigraphic phase show generally similar growth patterns, which allows their classification. Fig. 5 shows an example of neighbouring quartz phenocrysts of the same population in the Schellerhau granite SG1. The similarity (i.e., correlative zoning patterns) of quartz phenocrysts from the same phase is consistent with previous studies (Müller et al., 2000; Peppard et al., 2001). In Fig. 6, quartz phenocrysts of the Niederbobritzsch granites, the Schönfeld-Teplice volcanic rocks, the Altenberg-Frauenstein microgranites, and the Schellerhau granites are characterised based on their growth patterns.

4.1.1. Niederbobritzsch granites (NB)

Quartz phenocrysts (type 1) with an average diameter of 4.3 mm are abundant in Niederbobritzsch granite NB2 (hybrid granite). The phenocrysts exhibit weak growth zoning, which does not allow detailed characterisation of the zoning pattern (Fig. 6). A dense network of healed cracks and neocrystallised domains of non-luminescent secondary quartz makes the detection of the zoning more difficult. However, resorption surfaces are detected in the core zone and at the margin, which is overgrown by unzoned groundmass quartz. Similar phenocrysts also occur sporadically in the younger Niederbobritzsch granite NB3.

4.1.2. Schönfeld volcanic rocks (SBR, SDC)

Quartz phenocrysts of the ignimbritic Schönfeld basal rhyolite are mostly fragmented. The high degree of phenocryst fragmentation points to a powerful explosive eruption. The thickness of the unit in the drill core Mi-4 is about 55 m. Two maxima in the

crystal size distribution are observed, one at 0.2 mm (small fragments) and an other at 0.9 mm (almost complete, unbroken phenocrysts). The type 2 crystals (Fig. 6) have a weak red-brown (dark grey in SEM-CL) luminescent core. The luminescence changes from red-brown to blue (bright grey in SEM-CL) towards the crystal margin. Unbroken phenocrysts show wave-shaped crystal surfaces (Fig. 7a), which developed prior to effusion. The wavy surface truncates older growth zones, implying solution.

Phenocrysts from the rhyodacite (type 3) are much stronger resorbed than the phenocrysts in the basal rhyolite (Fig. 7b). The rounded phenocrysts have an average diameter of 1.0 mm and are overgrown by a thin layer of bright blue luminescent (bright grey in SEM-CL) quartz, similar to the groundmass quartz. The luminescence contrast between growth zones is low, as in the phenocrysts from the Niederbobritzsch granites (type 1). Growth zone boundaries in most of the crystals are smudged, which may be the result of secondary redistribution and healing of defect centres responsible for visible zoning. Crystals with strong smudging show a higher intensity of blue CL (brighter grey in SEM-CL). Owing to the smudging of the growth pattern it is not clear if all crystals represent one population. However, they do not exhibit a red-brown luminescent core (dark grey in SEM-CL) like the types 2 and 5 (see below). The crystals exhibit healed cracks (nonluminescent) formed by contraction during eruption (Figs. 6, 7b).

4.1.3. Teplice rhyolites (TR)

The type *TR1a* Teplice rhyolite contains the smallest phenocrysts (type 4) with an average size

Fig. 5. Scanning electron microscope cathodoluminescence (SEM-CL) image of quartz phenocrysts (type 5 in Fig. 6) from the Schellerhau granite SG1. The growth patterns of the phenocrysts with a dark grey core and bright grey oscillatory overgrowth are comparable, indicating a similar origin and crystallisation history.

of 0.4 mm. The euhedral crystals have planar growth zones and crystal faces and show a blue luminescent core (bright grey in SEM-CL), which is overgrown by bluish violet luminescent zones (grey in SEM-CL). The area around the core often contains melt inclusions (indicated by arrows in Fig. 6). Only a few crystals are fragmented and type 4 phenocrysts can also be found sporadically in the TR2a and TR3a subphases. The TR1a, TR2a and TR3a rhyolites represent the first and chemically most evolved subphases of an eruptive sequence and thus show also the highest degree of crystal fragmentation within the Teplice sequence. From the *TR1b* subphase on, the phenocrysts of type 5 are present in each subphase. Thus, different phenocryst populations are found in TR1a and TR1b despite the small chemical variation observed (e.g., Breiter et al., 2001). Type 5 has a weak red-brown luminescent core (dark grey in SEM-CL) with common small zircon and apatite inclusions causing bright radiation halos (indicated by arrows in the TR2b, TR3a and TR3c panels in Fig. 6). After resorption, the core was overgrown by zones of violet to blue luminescent quartz (bright grey in SEM-CL). The average crystal size increases from 0.8 mm in the TR1b subphase to 1.4 mm in the TR3c subphase. The average crystal size in the TR3a is smaller than in the older TR2b subphase because of a higher degree of fragmentation in the former. Moreover, the TR3a crystals underwent resorption shortly before deposition. Small-scale, tooth-like crystal surfaces are developed in type 5 quartz grains in TR2b (Figs. 6, 7c).

The type 5 phenocrysts show numerous lobate growth embayments that may have been caused by foreign phases, such as small crystals or melt droplets, which stack to the crystal surface; morphological instability due to large growth rates may also have been a contributing factor (e.g., Lowenstern, 1995; Jamtveit, 1999). Fig. 8 shows an inclusion of groundmass in the margin of a type 5 phenocryst from the Teplice rhyolite TR3c. A small foreign quartz crystal with bright blue luminescence (white in SEM-CL) attached to the growing phenocryst and disturbed the planar growth of the crystal face. The phenocryst grew around the foreign crystal, but in the shadow of the nearly enclosed crystal a cavity was left, later filled by magma.

Like the type 1 phenocrysts (Niederbobritzsch granites), type 5 phenocrysts in the Teplice rhyolite TR3c show a dense network of neo-crystallised quartz, which is not developed in the phenocrysts of the Schönfeld basal rhyolite, Schönfeld rhyodacite and Teplice rhyolites. However, similar structures were found in all the studied subvolcanic and plutonic rocks (Niederbobritzsch granite, Altenberg-Frauenstein microgranite, Schellerhau granite complex). BSE images show that the porosity of a subvolcanic phenocryst can be up to 2 vol.% (Fig. 9a). The quartz in the Teplice rhyolite TR3c is exceptionally porous, particularly in non-luminescent domains (Fig. 9b). In contrast, phenocrysts of the effusive rocks have a low porosity (0.2 vol.%). The porosity in the effusive phenocrysts is mainly the result of incompletely healed microcracks, which were formed during effusion (Fig. 9c).

The large TR3c crystal shown in Fig. 6 has a more complex growth pattern than most of the other TR3c crystals of the same type. However, the most significant textures, such as resorption surfaces and repeated step zones, are found in all crystals. Thus, some crystals grew faster and some were resorbed more strongly than others.

4.1.4. Altenberg-Frauenstein microgranite (GP)

Type 5 phenocrysts with resorbed margins and plagioclase-mantled K-feldspar phenocrysts are the most characteristic of these microgranites (Figs. 6, 7d). A high degree of resorption is generally found in the low-silica varieties of the Altenberg-Frauenstein microgranite (GP2, $GP2_{hbl}$; see Müller and Seltmann 2002), except for the silica-rich cumulate variety ($GP1_{cum}$). The phenocrysts of GP2 are partly mantled by microcrystalline plagioclase, biotite and titano-magnetite. The average crystal size is 2.2 mm and thus much larger than that of the type 5 crystals in Teplice rhyolite TR3c.

The marginal resorption surface is overgrown by bright blue luminescent (bright grey in SEM-CL) microcrystalline quartz in the GP1 and by graphic quartz intergrown with feldspar in the GP2 (Fig. 6). Moreover, the growth zoning of strongly resorbed crystals is slightly smudged. The strong marginal resorption, the overgrowth of bright blue luminescent groundmass quartz and the smudging of growth

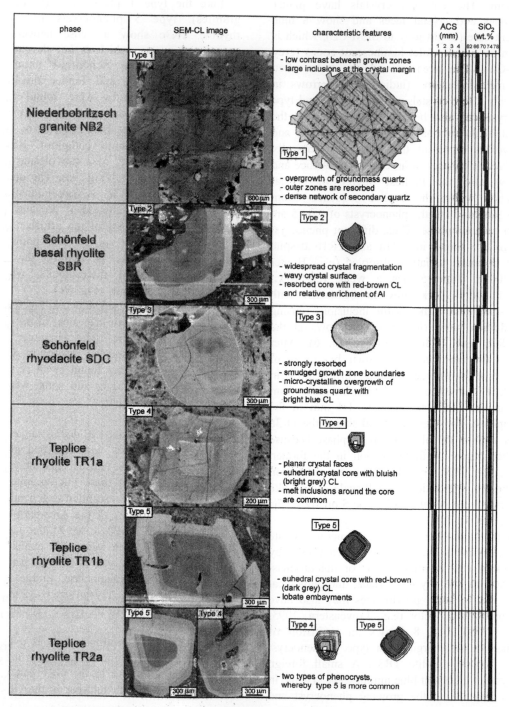

Fig. 6. Textures of quartz phenocrysts from 13 igneous phases and subphases of the eastern Erzgebirge volcano-plutonic complex. In the left column, a CL image of a phenocryst from each phase is shown. The sketches in the middle summarise the characteristic features of the phenocrysts. The average crystal (phenocryst) size (ACS) and the whole-rock silica content are plotted in the right columns. Phenocrysts in the Niederbobritzsch granite NB3 and Teplice rhyolite TR3b (not shown) are similar to the phenocrysts in NB2 and TR3a (type 5), respectively.

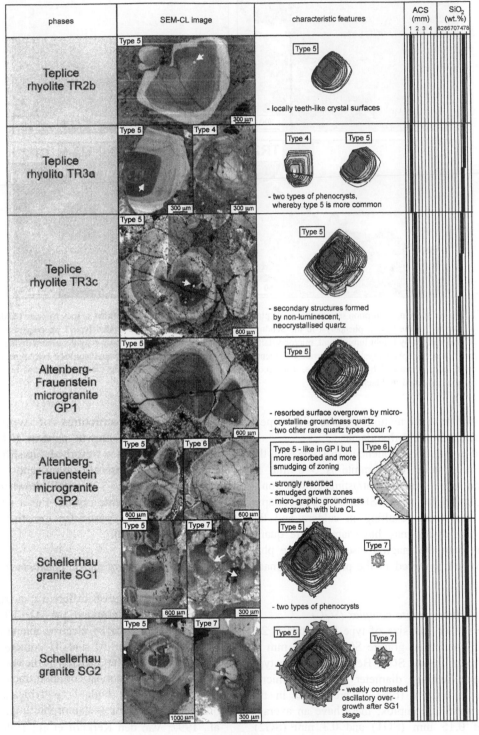

Fig. 6 (*continued*).

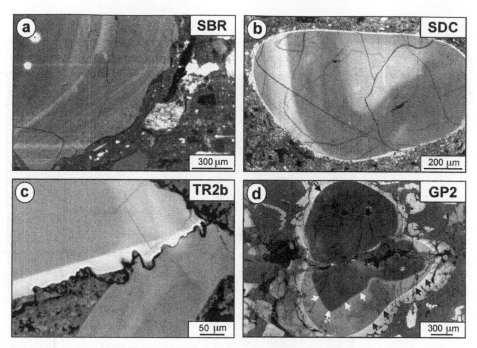

Fig. 7. SEM-CL images showing resorption surfaces of quartz phenocrysts. These can be easily identified as they truncate older growth zones. (a) Wavy crystal surface of the type 2 phenocryst in the Schönfeld basal rhyolite (SBR). (b) Rounded type 3 phenocryst in the Schönfeld rhyodacite (SDC) overgrown by bright luminescent, microcrystalline groundmass quartz. (c) Undulatory crystal surfaces in the Teplice rhyolite (TR2b). (d) Strongly resorbed phenocryst in the Altenberg-Frauenstein microgranite (GP2). The first main resorption event resulting in rounding of the dark crystal core is marked by white arrows. The second main resorption event predating the overgrowth of bright luminescent groundmass quartz is indicated by black arrows.

zoning are similar to the textural features of the type 3 crystals in the Schönfeld rhyodacite.

A second population of phenocrysts (type 6) in the Altenberg-Frauenstein microgranite is 5–8 mm in diameter and have weakly contrasting growth zones with blue CL (bright grey in SEM-CL). The collapse of magma reservoirs resulting in caldera and the associated ring dykes may have tapped rare quartz phenocrysts from the magma reservoirs; these phenocrysts are not observed in the previous magmatic stages.

4.1.5. Schellerhau granite complex (SGC)

The average crystal size of type 5 crystals is 2.3 mm in the Schellerhau granite SG1 and 2.6 mm in the Schellerhau granite SG2. Besides the large type 5 phenocrysts (average diameter ~2.5 mm), microphenocrysts of quartz (type 7) are found in the Schellerhau granite complex; these have an average diameter of 0.15 mm (SG1) and 0.3 mm (SG2). Similarities in the outer growth zones of type 5

and the growth structures of type 7 justify correlation of their growth patterns (Müller et al., 2000). Type 7 has a resorbed bluish luminescent core and inclusions of zircon and apatite. For example, the type 7 crystal shown in Fig. 6 nucleated on an apatite crystal. For a more detailed description of this particular growth pattern, see Müller et al. (2000).

4.2. Trace elements of quartz phenocrysts

To further distinguish different quartz phenocryst generations, we determined the Al, Ti, K and Fe distributions in quartz by electron microprobe (Table 1). In Fig. 10a, the Al and Ti contents of quartz phenocrysts from the different volcanic subphases are plotted. Al is the most common trace elements in quartz, and relative high Ti is typical of igneous quartz and high-grade metamorphic quartz (Perny et al., 1992; Van den Kerkhof et al., 1996; Watt et al., 1997; Mullis and Ramseyer, 1999; Müller et al.,

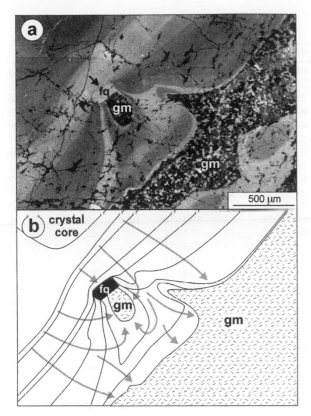

Fig. 8. (a) SEM-CL image of a quartz phenocryst margin (type 5, Teplice rhyolite TR3c) showing an inclusion of groundmass (gm) in a growth embayment. The formation of the cavity was promoted by a foreign quartz crystal (fq; arrow) that was stuck at a previously formed crystal surface. (b) Sketch of the growth zoning of the quartz phenocryst margin shown in panel a. The grey arrows indicate the relative growth direction. The indentation on the older crystal surface existed already before the foreign quartz crystal was attached to the phenocryst.

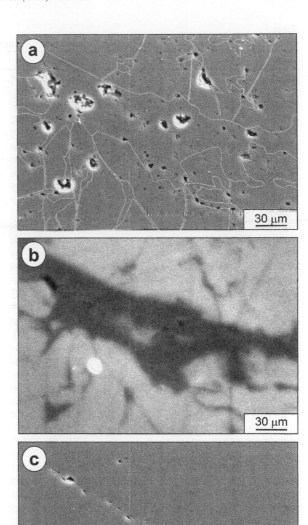

Fig. 9. Images showing porosity of quartz phenocrysts. (a) Back-scattered electron (BSE) image of a highly porous (up to 2 vol.%) type 5 phenocryst from the Teplice rhyolite TR3c. Such highly porous quartz has been found only in Teplice rhyolite TR3c. The dotted lines indicate areas of low-luminocity quartz representing neocrystallised domains-most of the porosity is found in these domains. (b) CL image of the area shown in panel a. Neocrystallised quartz appears dark grey. (c) Thin section surface of the type 4 phenocryst showing low porosity (<0.2 vol.%). The images reveal a partially healed crack that caused the porosity. Such low porosity is typical of phenocrysts in the effusive rocks.

2000, 2003). About 75% of the data yield concentrations between 70 and 140 ppm for Al and between 35 and 90 ppm for Ti, independent of phenocryst type. The red-brown luminescent crystal core of the type 2 and 5 phenocrysts is depleted in Ti (<25 ppm). The core of type 2 quartz crystals has higher average Al than the core of the type 5, implying a different origin of the crystal cores. Relatively high Al content of the cores is more typical of late-magmatic quartz (Müller et al., 2000). The bluish luminescent core of type 4 is slightly enriched in Ti (40 to 60 ppm) in comparison to the rim (30–40 ppm). In type 2 and 5 phenocrysts, the opposite is observed. Ti increases tendentiously from the core to

Table 1
Electron microprobe analyses of trace element concentrations (in ppm) in quartz phenocrysts from the Schönfeld-Teplice volcanic sequence

Sample	Al	Ti	K	Fe	Sample	Al	Ti	K	Fe	Sample	Al	Ti	K	Fe
SBR-01	171	<11	72	86	SDC-18	107	44	10	38	TR2b-18	73	22	7	17
SBR-02	279	14	67	51										
SBR-03	122	<11	14	36	TR1a-01	101	55	16	15	TR3c-01	105	18	7	36
SBR-04	165	16	21	26	TR1a-02	102	44	12	22	TR3c-02	103	22	<7	20
SBR-05	127	20	7	17	TR1a-03	95	47	9	25	TR3c-03	120	<11	11	24
SBR-06	124	<11	11	20	TR1a-04	106	52	23	17	TR3c-04	98	<11	10	21
SBR-07	133	<11	7	14	TR1a-05	99	53	18	16	TR3c-05	39	<11	10	21
SBR-08	310	<11	105	33	TR1a-06	96	56	18	<12	TR3c-06	106	<11	19	15
SBR-09	391	19	154	32	TR1a-07	98	61	15	19	TR3c-07	128	14	19	17
SBR-10	196	<11	48	17	TR1a-08	83	49	15	23	TR3c-08	103	13	15	19
SBR-11	156	<11	13	14	TR1a-09	82	41	16	16	TR3c-09	109	<11	7	<12
SBR-12	147	21	<7	26	TR1a-10	94	63	21	<12	TR3c-10	186	19	35	38
SBR-13	130	29	8	30	TR1a-11	88	52	12	27	TR3c-11	112	52	23	31
SBR-14	141	14	17	26	TR1a-12	94	58	12	24	TR3c-12	92	63	12	15
SBR-15	145	23	16	12	TR1a-13	79	55	14	27	TR3c-13	112	84	7	15
SBR-16	173	24	25	24	TR1a-14	97	43	17	31	TR3c-14	112	95	13	15
SBR-17	102	51	7	25	TR1a-15	101	49	19	34	TR3c-15	106	86	9	25
SBR-18	91	43	15	13	TR1a-16	79	29	17	44	TR3c-16	118	83	22	26
SBR-19	129	38	39	53	TR1a-17	86	26	27	77	TR3c-17	562	79	245	129
										TR3c-18	114	63	15	18
SDC-01	148	59	26	89	TR2b-01	101	74	43	65	TR3c-19	110	71	7	<12
SDC-02	127	63	14	47	TR2b-02	104	76	12	32	TR3c-20	45	17	16	22
SDC-03	127	71	14	38	TR2b-03	86	76	21	19	TR3c-21	90	58	15	17
SDC-04	107	25	18	17	TR2b-04	89	77	10	33	TR3c-22	94	50	14	24
SDC-05	113	59	<7	16	TR2b-05	83	73	<7	25	TR3c-23	84	50	18	35
SDC-06	113	54	12	<12	TR2b-06	84	60	25	30	TR3c-24	95	44	8	21
SDC-07	119	52	16	13	TR2b-07	89	47	17	28	TR3c-25	115	85	14	26
SDC-08	132	37	7	<12	TR2b-08	87	64	12	26	TR3c-26	108	70	17	17
SDC-09	120	40	15	28	TR2b-09	88	62	20	19	TR3c-27	92	67	25	28
SDC-10	128	41	8	17	TR2b-10	87	79	17	27	TR3c-28	103	59	10	27
SDC-11	118	56	18	15	TR2b-11	82	61	<7	<12	TR3c-29	100	49	7	14
SDC-12	117	49	12	20	TR2b-12	84	58	16	22	TR3c-30	95	56	12	22
SDC-13	119	53	<7	15	TR2b-13	85	55	7	<12	TR3c-31	111	86	20	18
SDC-14	110	66	18	<12	TR2b-14	73	64	16	25	TR3c-32	127	116	15	18
SDC-15	118	39	18	<12	TR2b-15	84	53	14	82	TR3c-33	122	123	20	21
SDC-16	119	44	15	<12	TR2b-16	80	61	9	68	TR3c-34	119	122	8	39
SDC-17	116	44	9	19	TR2b-17	82	22	7	35	TR3c-35	104	104	7	83

the rim. Highest Ti (up to 120 ppm) is observed in the outer growth zones of the type 5 phenocrysts in the Teplice rhyolite TR3c representing the last crystallisation stage before eruption. The average Ti concentration also increases from type 2 to type 5 phenocrysts. The heterogeneous distribution of Ti is related to growth zoning imaged by CL; thus, the visible growth zoning reflects the Ti distribution in quartz phenocrysts. Al concentrations above 150 ppm, which are found only in the red-brown luminescent cores of type 2 and 5 phenocrysts, correlate positively with K (Fig. 10b).

4.3. Composition of feldspar phenocrysts

Growth patterns and chemical composition of feldspar phenocrysts of the Schönfeld basal rhyolite, Schönfeld rhyodacite, Teplice rhyolites and Altenberg-Frauenstein microgranite were studied to complement the data acquired from the quartz phenocrysts. Major element concentrations in feldspar phenocrysts from ignimbrites and tuffs of the Schönfeld-Teplice sequence show evidence of syn- to post-extrusive re-equilibration. However, well preserved, zoned feldspars were found in the

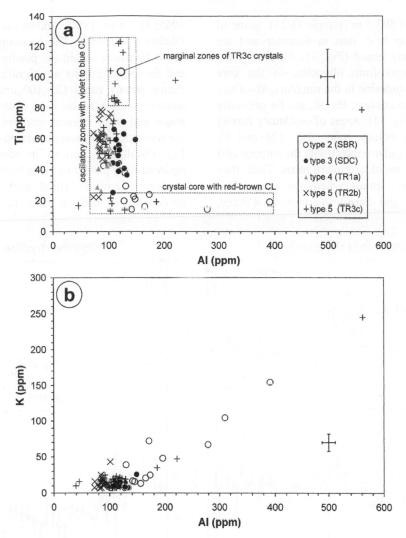

Fig. 10. Ti versus Al (a) and K versus Al (b) diagrams showing the composition of quartz phenocrysts of the Schönfeld-Teplice volcanic rocks. Error bars (at 3σ) of measurements are shown. (a) Oscillatory-zoned parts of phenocrysts are characterised by wide variation in Ti, which is probably responsible for the visible CL zoning. The red-brown luminescent cores of the type 2 and type 5 phenocrysts have low Ti and relative high Al in comparison to the oscillatory overgrowths. In panel b, correlation of Al and K at concentrations >200 and >50 ppm, respectively, corresponds to the general assumption that K^+ behaves as charge compensator of Al^{3+} defects in the quartz lattice. The analysed areas are free of visible (>0.5 µm) inclusions.

Schönfeld rhyodacite (sample TR25), which probably represents a magma flow, and in the subvolcanic Teplice rhyolite TR3c. The presence of zoned feldspar phenocrysts with a pre-eruptive element signature in both subphases implies that the homogeneous feldspar phenocrysts of ignimbrites and tuffs were re-equilibrated during or after eruption.

The re-equilibrated K-feldspars of the Schönfeld basal rhyolite unit have a high average Ba concen-

tration (1168 ppm). As diffusion of Ba is low in feldspar (e.g., Afonina and Shmakin, 1970; Cherniak, 1999), the high Ba content should be a relic of pre-eruptive crystallisation.

Plagioclase is the most abundant phenocryst population in the Schönfeld rhyodacite samples and its composition at different depths of the Mi-4 drill core varies between An_{15} and An_{40} (K. Breiter, unpublished data). Well-preserved plagioclase is found at a depth of 803.3 m (sample TR25). Some of the crystals are up to 5 mm in diameter and are chemically inversely zoned (Fig. 11). Generally, the composition changes from oligoclase in the core ($An_{27}Ab_{67}Or_6$) to andesine in the rim ($An_{38}Ab_{58}Or_4$). With increasing An content, Ba, Sr, and Fe generally increase as well (Fig. 11). Areas of oscillatory zoning are interrupted by patchy zones (Fig. 12a and b), which have an irregular contact with the interior and penetrate and truncate older growth zones. Thus, they represent resorption zones where the plagioclase became thermally and chemically unstable (Vance,

1965; Barbarin, 1990; Andersson and Eklund, 1994; Ginibre et al., 2002). The An component increases by about 10 mol% within the patchy zones and also Ba and Sr concentrations are significantly higher. Concentric patchy zones (20–100 μm) are developed in some crystals, but there is no concentration step in the major and trace element concentration between the pre-resorption and post-resorption growth zones (Fig. 11). Oscillatory zoning in some phenocrysts is replaced by a patchy or sieve texture (Fig. 12c), both containing cavities filled with groundmass. The absence of K-feldspar may be explained by the

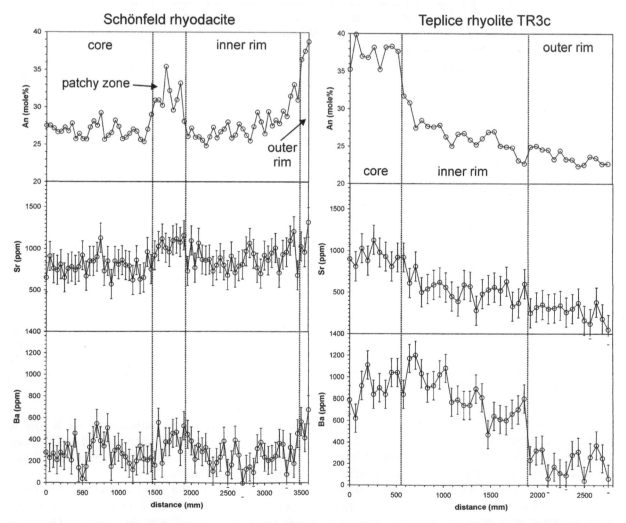

Fig. 11. Element profiles of plagioclase phenocrysts in the Schönfeld rhyodacite SDC and the Teplice rhyolite TR3c. The distinction between the 'outer rim' and the 'inner rim' is based on the concentration step of Ba. Profile location of the plagioclase from the Schönfeld rhyodacite is shown in Fig. 12a.

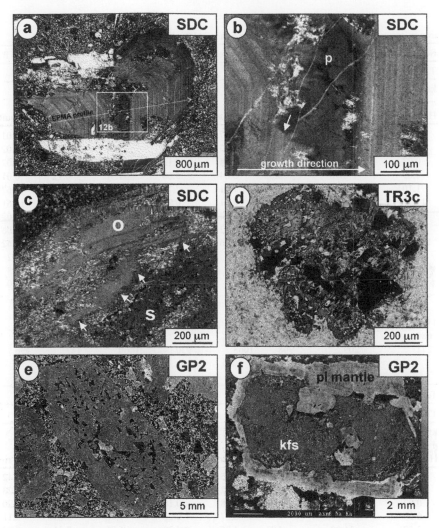

Fig. 12. Microtextures of feldspar phenocrysts. (a) Euhedral plagioclase with a patchy zone in the crystal centre (white rectangle) in the Schönfeld rhyodacite. The EPMA profile corresponds to the profile in Fig. 11. (b) Enlargement of the patchy zone (p) shown in panel a. The inner irregular boundary of the patchy structures penetrates and replaces older growth zones (arrow) whereas the outer boundary is flat. (c) Detail of a plagioclase phenocryst showing a sieve-textured core (s). The sieve-textured zone replaces older oscillatory growth zones (o) and thus implies feldspar dissolution. The arrows indicate the replacement front. (d) Microgranular clot of mafic minerals (biotite, epidote, apatite, titanite, titanomagnetite, and ilmenite (surrounded by fine-grained quartz-feldspar corona). (e) Sieve-textured plagioclase with biotite, epidote, titanomagnetite, and quartz inclusions (Altenberg-Frauenstein microgranite GP2). The spongy inner part of the crystal results from dissolution. After re-equilibration, a normally zoned outermost rim developed forming a boxy cellular texture. (f) Distribution of Na in plagioclase-mantled K-feldspar (rapakivi texture) from the Altenberg-Frauenstein microgranite. Oligoclase mantle (pl mantle) encloses the K-feldspar phenocryst (kfs).

relatively high crystallisation temperature (>800 °C) and the low water content (≤2%) of the melt, because low water content favours plagioclase crystallisation (e.g., Long and Luth, 1986).

K-feldspars from the eruptive Teplice rhyolite subphases TR1a to TR3b have a relative homoge-neous composition of Or_{94-98} with low Sr (532–680 ppm), Fe (136–271 ppm), and Ba (166–199 ppm) (Table 2). Non-equilibrated parts of K-feldspar (e.g., in the TR1a rocks) have a much lower Or (Or_{76}), but similar Ba and Sr contents as the re-equilibrated feldspar. Similar Ba and Sr contents in re-equilibrated

Table 2
Representative electron microprobe analyses of feldspar from the Schönfeld-Teplice volcanic sequence

Units and phases	Schönfeld basal rhyolite			Schönfeld rhyodacite			Teplice rhyolite TR1a		
Sample	TR24	TR24	TR25	TR25	TR25	TR25	TR28	TR28	TR28
Mineral	kfs	ab	pl	pl	pl	pl	kfs	kfs	ab
	n=29	n=25	n=30	n=8	n=32	n=3	n=19	n=28	n=16
	Homogeneous	Homogeneous	Core	Patchy zone	Inner rim	Outer rim	Primary?	Re-equilibr.	Homogeneous
SiO_2	65.76	69.95	61.16	59.37	60.81	58.07	66.40	65.01	67.90
Al_2O_3	18.28	19.48	24.22	25.23	24.38	25.78	18.33	18.02	20.58
FeO	0.03	0.03	0.12	0.14	0.12	0.19	0.06	0.02	0.03
CaO	0.09	0.18	5.65	6.64	5.64	7.96	0.12	0.01	1.53
Na_2O	1.10	11.78	7.84	7.31	7.77	6.82	2.72	0.67	11.13
K_2O	15.77	0.34	0.98	0.91	0.98	0.75	13.56	16.61	0.31
BaO	0.13	0.02	0.03	0.04	0.03	0.06	0.02	0.02	0.02
SrO	0.08	0.05	0.09	0.12	0.11	0.13	0.06	0.06	0.09
Sum	101.26	101.84	100.09	99.76	99.84	99.76	101.25	100.42	101.59
Or	90.0	1.9	5.6	5.2	5.6	4.2	76.3	94.2	1.7
Ab	9.5	97.3	67.5	63.1	67.4	58.2	23.2	5.8	91.4
An	0.4	0.8	26.9	31.7	27.0	37.6	0.5	0.0	6.9
Ti	<61	<61	<61	84	70	210	82	<61	<61
P	68	<50	<50	59	<50	<50	<50	<50	<50
Ba	1168	152	273	360	267	557	169	166	185
Sr	680	455	788	1043	900	1103	483	545	728
Fe	271	209	908	1063	930	1460	447	136	247

Units and phases	Teplice rhyolite TR2a				Teplice rhyolite TR3c				
Sample	TR30a	TR30a	TR09	TR09	TR09	TR09	TR09	TR09	TR09
Mineral	kfs	ab	kfs	kfs	kfs	pl	pl	pl	pl
	n=16	n=24	n=22	n=6	n=4	n=9	n=21	n=9	n=5
	Homogeneous	Homogeneous	Core	Inner rim	Outer rim	Core	Inner rim	Inner rim II	Outer rim
SiO_2	65.18	69.73	67.77	67.65	67.50	59.73	63.55	64.42	64.69
Al_2O_3	18.10	19.87	18.59	18.52	18.56	25.52	23.71	23.23	23.17
FeO	0.02	0.03	0.09	0.08	0.10	0.20	0.22	0.20	0.20
CaO	0.00	0.26	0.30	0.27	0.22	7.98	5.65	5.11	4.95
Na_2O	0.57	11.89	4.41	4.35	4.23	6.95	7.91	8.20	8.43
K_2O	16.78	0.18	10.73	10.85	11.14	0.66	1.11	1.19	1.14
BaO	0.02	0.02	0.02	0.02	0.02	0.10	0.09	0.02	0.02
SrO	0.06	0.02	0.06	0.05	0.06	0.11	0.06	0.04	0.02
Sum	100.73	101.98	101.98	101.81	101.83	101.25	102.31	102.40	102.62
Or	95.1	1.0	60.7	61.3	62.7	3.7	6.2	6.6	6.3
Ab	4.9	97.9	37.9	37.4	36.2	58.9	67.3	69.4	70.8
An	0.0	1.2	1.4	1.3	1.1	37.4	26.5	23.9	23.0
Ti	<61	<61	78	88	<61	186	169	91	84
P	<50	<50	<50	<50	90	81	<50	<50	<50
Ba	199	147	193	188	203	900	828	211	196
Sr	532	<176	479	428	525	932	545	311	178
Fe	151	234	717	638	775	1591	1708	1562	1554

Key to abbreviations: albite (ab), K-feldspar (kfs), plagioclase (pl).

and non-equilibrated K-feldspar indicates that Ba and Sr were little affected by alteration associated with eruption.

Plagioclase in the subvolcanic Teplice rhyolite *TR3c* exhibits normal chemical zoning (Fig. 11) with an andesine core ($An_{37}Ab_{59}Or_4$) and distinctly higher average Sr (932 ppm) than in the younger overgrowths. A core and an inner and the outer rim can be distinguished. The boundary of the rims is indicated by a significant drop in Ba content (Fig. 11). The composition of the inner rim is $An_{27}Ab_{67}Or_6$ and that of the outer rim $An_{23}Ab_{71}Or_6$. The average Sr content drops from 545 ppm (inner rim) to 311 ppm (outer rim) and the Ti concentration halves. Thus, plagioclase phenocrysts from the Schönfeld rhyodacite and Teplice rhyolite TR3c are rather different from each other. K-feldspar phenocrysts from Teplice rhyolite TR3c are more homogeneous in composition, ranging from $An_1Ab_{38}Or_{61}$ in the core to $An_1Ab_{36}Or_{63}$ at the rim.

The composition of feldspar in the Altenberg-Frauenstein microgranites was described in detail by Müller and Seltmann (2002). Plagioclase phenocrysts (up to 2 cm in diameter) consist of an andesine core ($An_{36}Ab_{58}Or_6$) overgrown by an oligoclase rim ($An_{27}Ab_{64}Or_9$), and are thus identical with the plagioclase of the Teplice rhyolite TR3c. Another type of phenocrysts (An_{26-30}) forms ovoid to euhedral sieve-textured crystals with biotite, epidote, titanomagnetite, and quartz inclusions (Fig. 12e). The plagioclase-mantled K-feldspar phenocrysts (rapakivi texture; Fig. 12f) have a core composition of $An_0Ab_2Or_{98}$ and $An_2Ab_{40}Or_{58}$ at the margin, and the content of total FeO increases and BaO and SrO decrease. Some K-feldspar phenocrysts contain subhedral to euhedral inclusions of oligoclase (An_{26-29}).

5. Discussion

5.1. Disequilibrium textures in phenocrysts-indicators for magma mixing?

We use the term "disequilibrium textures" for obvious resorption surfaces truncating preexisting growth zones of quartz phenocrysts, plagioclase-mantled K-feldspar, patchy zones and sieve-textured zones in feldspar phenocrysts. Three kinds of resorption surfaces have been observed in quartz phenocrysts: tooth-like, wavy, and rounded (Fig. 7). The tooth-like surfaces were formed after crystal fragmentation, i.e., after effusive deposition. However, the narrow dovetailing of two the phenocrysts shown in

Fig. 7c indicates that the rocks were still at a viscoplastic stage when silica-undersaturated fluids circulated in the hot volcanic beds and dissolved some of the quartz crystal surfaces. The wavy surfaces (Fig. 7a) may have been caused by the same process that was responsible for the tooth-like surfaces, but at a higher degree of resorption. However, they could also result from rapid pressure release during magma ascent and explosive eruption. We favour the latter, because the wavy surfaces are only observed in rocks with a high degree of crystal fragmentation.

Strongest resorption of crystals was observed in the core of the type 2 and type 5 phenocrysts and at the margins of type 3 from Schönfeld rhyodacite and type 5 from Altenberg-Frauenstein microgranite (Fig. 7b, d). The weak red-brown luminescent core (dark grey in SEM-CL) of the type 2 and 5 quartz phenocrysts may have derived from the country rock or magma educt and thus be xenocrysts. However, the cores show a weak contrasted zoning with a pattern similar to the younger bluish luminescent overgrowths and they contain common inclusions of magmatic zircon and apatite. The resorbed cores and the marginal resorption surfaces are overgrown by more bluish luminescent quartz (indicating a higher Ti content; Müller et al., 2003). Based on the assumption that the concentration of trace elements in the solid phase is mainly controlled by the concentration of the elements in the melt, the high Ti in quartz implies that, after resorption, the crystal was incorporated into a more Ti-rich melt. This may reflect movement of the crystal into a deeper part of the magma reservoir (settling) or mixing with a more mafic magma. The smudging of growth zoning observed in phenocrysts of the Schönfeld rhyodacite and Altenberg-Frauenstein microgranite is associated with strong rounding of crystals. The resorption surface itself is sharp indicating that the smudging of zoning predates the resorption event. The smudging of zoning and the high-Ti quartz overgrowths in the Schönfeld rhyodacite and Altenberg-Frauenstein microgranite are likely to result from redistribution and healing of defect centres in the quartz lattice, as the crystals moved from a relatively Ti-poor (silica-rich) magma to a relatively Ti-rich (silica-poor) and hotter magma. Crystal settling is an unlikely explanation for the resorption, because of the small density contrast between quartz crystals and felsic magma and the buoyancy of quartz crystals in a compositionally zoned magma reservoir.

The major-element composition of *feldspar* is generally controlled by the T–P–H$_2$O conditions of the magma, whereas the concentration of minor and trace elements reflects the concentration of these elements in the melt and the partition coefficients between crystal and melt; a contributing factor may be the order of feldspar precipitation (e.g., Long and Luth, 1986). The wide variation of plagioclase chemistry of the Schönfeld rhyodacite suggests that the different parts of the reservoir were mingled. A number of plagioclase phenocrysts in the Schönfeld rhyodacite have a sieved-textured core and up to two patchy zones indicating a drastic change of P–T–H$_2$O conditions, which is also reflected in the crystal chemistry. The similar major element and trace element concentration of the pre-resorption and post-resorption growth zones evidence that the crystals moved temporarily into a hotter part of the magma reservoir. After a relative short period the crystals moved back to the preresorption environment. Moreover, the steep increase of the An content at the margin of some Schönfeld rhyodacite plagioclases probably formed when the more sodic plagioclase from the felsic melt was introduced into the more mafic melt (cf. Wiebe, 1968). Such reverse core to rim trend with increasing An, Fe, Ti, and Sr, points to mixing with a less differentiated magma (e.g., Barbarin, 1990; Ginibre et al., 2002). Thus, the plagioclase phenocrysts probably originated from a magma, which was more silicic than the Schönfeld rhyodacite magma itself.

The general decrease of An, Sr, and Ba in the Teplice rhyolite TR3c phenocrysts from core to rim can be related to differentiation of the magma. However, the abrupt change in An content between the core (An$_{37}$) and inner rim (An$_{27}$), and the Ba concentration step between the inner rim and the outer rim are not consistent with continuous differentiation. The most feasible process to produce such rapid changes in the growth environment is movement of the crystal in a melt with a strong gradient of water content and/or temperature. Moreover, mafic microgranular enclaves (MME) containing biotite, epidote, apatite, titanite, titanomagnetite, and ilmenite surrounded by fine-crystalline quartz-feldspar corona (Fig. 12d), support the idea of magma mixing and

mingling as probable processes for the Teplice rhyolite TR3c. On the other hand, evidence for magma mixing does not exist in the Teplice rhyolite TR1 to TR3b.

Further textures indicating magma mixing and mingling (plagioclase-mantled K-feldspar, sieve-textured plagioclase, MME) are dominant in the Altenberg-Frauenstein microgranite (Müller and Seltmann, 2002). The formation of plagioclase mantles around K-feldspar phenocrysts is related to changes in physicochemical conditions (P, T, composition) of the magma, which allowed plagioclase to nucleate on K-feldspar crystals. Experimental studies and petrographic observations indicate that two of the most realistic mechanisms to form plagioclase-mantled K-feldspar crystals are (1) mixing of two magmas of different composition (Hibbard, 1981; Wark and Stimac, 1992), and (2) crystallisation of granite melt under conditions involving a marked decrease of pressure combined with a small change in temperature (Nekvasil, 1991; Eklund and Shebanov, 1999). The resorbed round quartz crystals mantled by microcrystalline plagioclase, biotite, and titanomagnetite indicate textural adjustment of quartz phenocrysts that were suddenly placed in an environment in which they were out of equilibrium (e.g., Hibbard, 1991; Schreiber et al., 1999; Baxter and Feely, 2002). According to Andersson and Eklund (1994) sieve-textured plagioclase crystals (such as those in the Altenberg-Frauenstein microgranite; Fig. 12e) can be interpreted as xenocrysts, which developed a spongy cellular texture in response to interaction with a hotter magma. Inclusions of mafic minerals in the cellular texture imply penetration of the xenocrysts by mafic magma and, after re-equilibration, a normally zoned outermost rim developed and formed a boxy cellular texture. The common occurrence of these textures in the Altenberg-Frauenstein microgranites is explained as a result of interaction of felsic and mafic magmas (cf. Hibbard, 1981, 1991; Andersson and Eklund, 1994).

5.2. Implications for magma chamber structure

We have found seven different quartz phenocryst types in the 14 studied magmatic phases of the eastern Erzgebirge volcano-plutonic complex. Correlation of growth patterns reveals that each type represents a quartz population grown in the same magma reservoir in similar physicochemical conditions. On the other hand, growth patterns of different types are dissimilar. These may originate from different levels of a stratified magma reservoir or from different subreservoirs, or may represent older quartz populations that had dissolved completely before new populations nucleated. The last hypothesis is unable explain the presence of the same quartz type in different phases and several quartz types within one subphase.

In Fig. 13, cross sections showing evolutionary stages of the eastern Erzgebirge volcano-plutonic complex are illustrated. The Niederbobritzsch granites, Schönfeld basal rhyolite and Schönfeld rhyodacite contain a unique quartz type. An obvious chemical relationship does not exist between the older biotite granites (Niederbobritzsch granites, Flaje granite; Fig. 13a) and the effusive Schönfeld basal rhyolite and Schönfeld rhyodacite (Fig. 13b). Therefore, the magma reservoirs of both were spatially and temporally separated. During the early stage of the evolution of the eastern Erzgebirge volcano-plutonic complex, magma mixing and mingling textures are documented in the Niederbobritzsch granite NB2. The Niederbobritzsch granites were intruded at a depth between 5 and 8 km (Benek, 1991) and emplaced into a transpressional tectonic environment with compression in a NNE–SSW direction (Wetzel, 1985; Müller et al., 2001). At the time of emplacement, the thickness of the internal part of the Variscan orogen is assumed to have been at least 60 km (e.g., Gayer, 1994; Behr et al., 1994; Kröner and Willner, 1998) (Fig. 13a).

The Schönfeld basal rhyolite represents the upper part and the rhyodacite the bottom part of the chamber that was erupted first. In the rhyodacite, rounded quartz phenocrysts with smudged growth zoning and chemically reverse zoned plagioclases point to multiple recharge of more mafic magma into the Schönfeld reservoir.

The base of the Teplice rhyolite TR1 marks an important change in the composition of the volcanic rocks and is associated with a change from a late orogenic to a post-orogenic regime. Transtensional faulting caused the ascent and extrusion of large volumes of magma from the middle to lower crust (e.g., Benek, 1995) resulting in the extrusion of the Teplice volcanic rocks at the first stage. A similar compositional change in the volcanic rocks was reported from the late-Variscan volcanoes in the Polish Sudetes (Awdankiewicz, 1999). This change is also reflected in the quartz phenocryst populations. The quartz phenocryst type 5 is present in three different phases (10 subphases, see Fig. 6), which indicates that these three units (Teplice rhyolite, Altenberg-Frauenstein microgranite and Schellerhau granite complex) come from the same magma reservoir (Fig. 13c through e). The reservoir containing type 5 phenocrysts is interpreted as the main reservoir (Teplice) of the eastern Erzgebirge volcano plutonic complex. Types 4, 6 and 7 occur together with type 5. Type 4 occurs in the initial effusive Teplice rhyolite subphases (TR1a, TR2a, TR3a), which represent the most silica-rich rocks of Teplice rhyolite—the "silica cap" of the Teplice reservoir. The subsequent volcanic subphases (TR1b, TR2b, TR3b, and TR3c) are chemically less evolved and contain only type 5 phenocrysts. Thus, type 5 originates from the lower parts of the reservoir. These observations indicate a strong stratification of quartz phenocryst populations within the reservoir and show that exchange of crystals between the different levels was restricted.

The Altenberg-Frauenstein microgranite seems to be the chemical continuation of the Teplice rhyolite TR3b and TR3c (Fiala, 1959; Bolduan et al., 1967). Both magmas were geochemically identical, but the Teplice rhyolite TR3b and TR3c have no plagioclase-mantled K-feldspars like the microgranites of Altenberg-Frauenstein do. We interpret the magma of the Altenberg-Frauenstein microgranite to represent "frozen" residual Teplice rhyolite magma remobilized due to the caldera collapse (Fig. 13e). The shape and size of the collapsed caldera may be used as a rough gauge of the dimensions of the underlying Teplice reservoir. The shallow Teplice reservoir that finally caused the collapse of the Altenberg-Teplice trapdoor caldera is assumed to have been at 6–10 km depth (Benek, 1991). This is consistent with the depth of other silicic magma chambers related to calderas of similar size (e.g., Lipman, 2000). Magma mixing produced again a hybrid magma represented by Altenberg-Frauenstein microgranite GP2. Mixing was restricted to the lower part of the magma chamber, resulting in amplification of the preexisting magma stratification (Campbell and Turner, 1987).

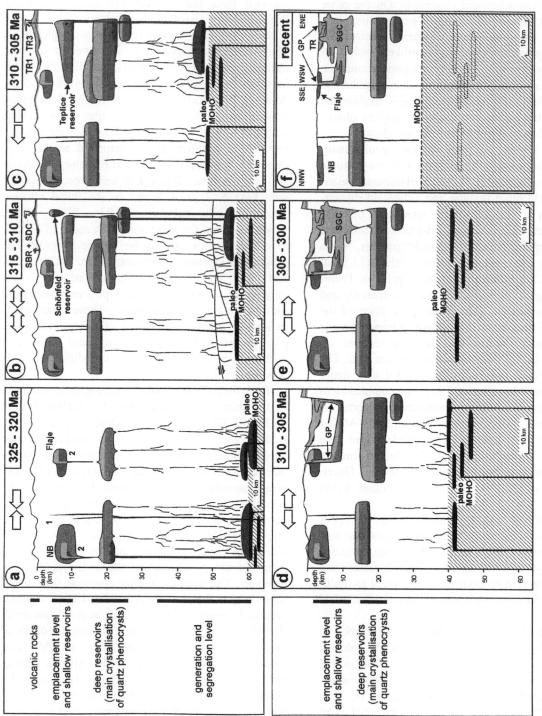

Fig. 13. Schematic cross sections showing the magmatectonic evolution of the eastern Erzgebirge volcano-plutonic complex. (a) Namurian stage (325–320 Ma): (1) intrusion of lamprophyres and (2) biotite granites (Niederbobritzsch granite NB and Flaje granite). (b) Westfalian A–B–C (315–310 Ma) pre-caldera stage: extrusion of the Schönfeld sequence (SBR and SDC). (c) Westfalian C–D (310–305 Ma) caldera stage: extrusion of the Teplice rhyolites TR1–TR3. (d) Westfalian C–D (310–305 Ma) caldera collapse: formation of the trapdoor caldera and intrusion of the Altenberg-Frauenstein microgranites GP1 and GP2 into the ring dyke faults. (e) Westfalian D-Stefanian (305–300 Ma) post-caldera stage: intrusion of the Li-mica granites (e.g., Schellerhau granite complex, SGC). (f) Current situation.

Save for the mixing textures in Teplice rhyolite TR3c and Altenberg-Frauenstein microgranite, no temporally and spatially associated mafic rocks are exposed. However, a large number of mafic dykes (lamprophyres) with Ar–Ar biotite ages between 330 and 315 Ma have been found in the Erzgebirge region (e.g., Werner and Lippolt, 1998). These dykes may have acted both as a mafic end member and a heat source.

The Schellerhau granite complex is totally decoupled from the proposed mingling and mixing process because of the high fractionation degree; the feldspar in it shows an almost perfect re-equilibration (microclinisation). On the other hand, the Schellerhau granite complex contains the type 5 phenocrysts, which indicates a genetic relationship between the Altenberg-Frauenstein microgranite and Teplice rhyolite (Fig. 13f). After extrusion of the Teplice rhyolite, the reservoir collapsed and the collapse structures were intruded by the Altenberg-Frauenstein microgranite representing the bottom (restite) of the Teplice reservoir. Only one plausible explanation remains: a deeper reservoir beneath the Altenberg-Teplice caldera (at a level where the type 5 phenocrysts were crystallized) that was not affected by the caldera collapse. Comparable deep magma source that connects several shallow magma chambers of different magma cycles is known from the Yellowstone Plateau volcanic field and the Timber Mountain-Oasis Valley caldera (Christiansen, 1979).

Thomas (1992) calculated the maximum depth crystallisation for the quartz phenocryst in the older biotite granites (e.g., Niederbobritzsch granites), the younger Li-mica granites (e.g., Schellerhau granite complex), and the Teplice rhyolites to be between 17 and 24 km (Fig. 14). He used the relation of the water content of melt inclusions in quartz and their homogenisation temperature to calculate the water pressure and, therefore, the minimum depth of melt inclusion entrapment. Thus, the supposed deeper magma reservoirs of the Niederbobritzsch granites, Schellerhau granite complex and Teplice rhyolite (possibly also of the Schönfeld basal rhyolite and Schönfeld rhyodacite) should have been situated at this crustal level. This is postulated for the main crystallisation of quartz phenocrysts of all the magmatic stages that built the eastern Erzgebirge volcano-plutonic complex (Fig. 13).

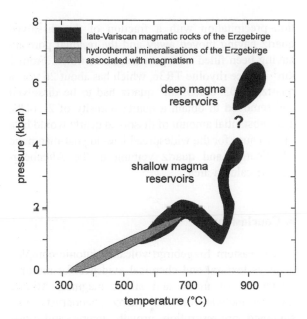

Fig. 14. Scheme of melt crystallization pathways based upon geothermobarometric investigations of melt inclusions predominantly in quartz from late-Variscan magmatic rocks of the Erzgebirge (modified after Thomas 1994). No data were obtained for depth between 17 and 13 km (question mark) which may indicate that no magma was stored at this depth.

5.3. Secondary quartz mobilisation

Quartz in the subvolcanic units of the eastern Erzgebirge volcano-plutonic complex (Niederbobritzsch granite, Teplice rhyolite TR3c, Altenberg-Frauenstein microgranite, Schellerhau granite complex) exhibits a dense network of healed cracks connecting domains (usually smaller than 10 µm) of non-luminescent secondary (post-magmatic) quartz. Previous studies show that these domains are nearly trace element free (Van den Kerkhof et al., 2001; Müller et al., 2002). In our samples these domains contain fluid inclusions showing homogenisation temperatures between 280 and 350 °C and high NaCl equiv. contents of up to 20 wt.%. A large number of these inclusions can result in a porosity of up to 2 vol.% and has been observed in the Teplice rhyolite TR3c. The formation of pure, secondary quartz may have been twofold. On one hand, defect structures in quartz immediately surrounding fluid inclusions may have lost trace elements by diffusion and have lower CL intensity. On the other hand, highly reactive fluids preserved in the inclusions may have led to dissolution–precipitation with a fraction of the dissolved quartz having moved away and the residual porosity having been filled with post-magmatic fluids. From 1 km³ Teplice rhyolite TR3c, which has about 26 vol.% quartz, 0.003–0.005 km³ quartz had to be dissolved and removed to obtain a quartz porosity of 2 vol.%. The substantial amount of dissolved quartz would be a likely cause for the widespread late- to post-magmatic silicification and quartz veining in the Altenberg-Teplice caldera area.

6. Conclusions

The eastern Erzgebirge volcano-plutonic complex displays textural and chemical evidence for multiple interaction of mafic and silicic magmas. Mixing textures include rounded quartz phenocrysts with smudged pre-resorption growth zoning and post-resorption quartz overgrowth with high Ti, plagioclase-mantled K-feldspar (rapakivi texture), sieve-textured plagioclase and patchy zones in plagioclase. In addition, concentration steps of trace and minor elements in quartz (Ti) and feldspar (anorthite content, Ba, Sr) support magma mixing. The heterogeneous nature of mixing is indicated by the varying distribution of textures between and within the magmatic stages. The observed plagioclase textures are consistent with models involving magma mixing proposed for the origin and evolution of other granitoids (Hibbard, 1981; Barbarin, 1990; Andersson and Eklund, 1994). Currently, a number of studies indicate that such textures, especially rapakivi, are present on a regional scale within the granite belts of the Variscan orogen.

Mainly based on the classification of quartz phenocrysts using their growth patters, we present a model for the evolution of the eastern Erzgebirge volcano-plutonic complex. Considering thermobarometric melt inclusion studies of quartz (Thomas, 1992), two levels of magma storage may be distinguished: deep reservoirs between 17 and 24 km and shallow reservoirs between 13 and 6 km. A main deep magma reservoir of the eastern Erzgebirge volcano-plutonic complex served as a source for the Teplice rhyolite, Altenberg-Frauenstein microgranite and the Schellerhau granites, which are associated with the Altenberg-Teplice caldera. The reservoir was strongly stratified, as the phases show a wide chemical variation but contain the same quartz phenocryst population. The older Niederbobritzsch granites and Schönfeld volcanic rocks represent temporally and spatially separate magma sources. However, the deep magma reservoir of both is assumed to have been at a depth between 17 and 24 km.

The compositional gap between the older primitive calc-alkaline Schönfeld volcanic rocks and the evolved high-K calc-alkaline Teplice rhyolites marks an important change not only in chemistry and magma sources, but also in terms of a change from a late-orogenic to a post-orogenic geotectonic environment. Similar changes in the chemistry of volcanic rocks reported from other parts of the Variscan belt (Awdankiewicz, 1999) imply an important chronostratigraphic marker between Westfalian B and D.

Acknowledgements

This study was supported by the Deutsche Forschungsgemeinschaft (MU 1717/2-1), the Bundesministerium für Bildung und Forschung (CZE 00/011) and the Natural History Museum of London. Technical support was provided by J. Spratt and T. Williams from the Natural History Museum of London. We appreciate the assistance of A. Kronz from the Geowissenschaftliches Zentrum Göttingen to obtain microprobe analyses of trace elements in quartz. We thank R. Thomas, C. Stanley and H. Kämpf for comments on previous versions of the manuscript. The reviews of B. Barbarin, O. Eklund and T. Rämö improved the manuscript significantly. This paper is dedicated to Professor Ilmari Haapala on the occasion of his retirement and in acknowledgement of his contribution to granite petrology and metallogeny, as well many inspiring discussions and fruitful cooperation with some of the authors.

References

Afonina, G.G., Shmakin, B.M., 1970. Inhibition of lattice ordering of potassic feldspar by barium ions. Doklady Akademii Nauk Sojus Sowjezkich Sozialistitscheskich Respublik, Moscow 195, 133–135. (in Russian).

Andersson, U.B., Eklund, O., 1994. Cellular plagioclase inter-growths as a result of crystal-magma mixing in the Proterozoic Åland rapakivi batholith, SW Finland. Contributions to Mineralogy and Petrology 117, 124–136.

Armstrong, J.T., 1995. CITZAF: a package of correction programs for the quantitative electron microbeam X-ray analysis of thick polished materials, thin films, and particles. Microbeam Analysis 4, 177–200.

Awdankiewicz, M., 1999. Volcanism in a late Variscan intramontane through: the petrology and geochemistry of the Carboniferous and Permian volcanic rocks of the Intra-Sudetic basin, SW Poland. Geologia Sudetica 32/2, 83–111.

Barbarin, B., 1990. Plagioclase xenocrysts and mafic magmatic enclaves in some granitoids of the Sierra Nevada batholith, California. Journal of Geophysical Research 95, 17,747–17,756.

Baxter, S., Feely, M., 2002. Magma mixing and mingling textures in granitoids: examples from the Galway Granite, Connemara, Ireland. Mineralogy and Petrology 76, 63–74.

Behr, H.-J., DEKORP Research Group B, 1994. Crustal structure of the Saxothuringian Zone: results of the deep seismic profile MVE-90 (East). Zeitschrift für Geologische Wissenschaften 22, 647–770.

Benek, R., 1991. Aspects of volume calculation of paleovolcanic eruptive products—the example of the Teplice rhyolite (east Germany). Zeitschrift für Geologische Wissenschaften 19, 379–389.

Benek, R., 1995. Late Variscan calderas/volcanotectonic depressions in eastern Germany. Terra Nostra 7/95, 16–19.

Bolduan, H., Lächelt, A., Malasek, K., 1967. Zur Geologie und Mineralisation der Lagerstätte Zinnwald (Cinovec). Freiberger Forschungshefte C218, 35–52.

Breiter, K., 1995. Geology and geochemistry of the Bohemian part of the Teplice rhyolite and adjacent post-rhyolite granites. Terra Nostra 7/95, 20–24.

Breiter, K., 1997. The Teplice rhyolite (Krušné Hory Mts., Czech Republic)—chemical evidence of a multiply exhausted stratified magma chamber. Věstník Českého geologického ústavu 72, 205–213.

Breiter, K., Förster, H.-J., Seltmann, R., 1999. Variscan silicic magmatism and related tin–tungsten mineralization in the Erzgebirge-Slavkovsky les metallogenic province. Mineralium Deposita 34, 505–531.

Breiter, K., Novák, J.K., Chulpáčová, M., 2001. Chemical evolution of volcanic rocks in the Altenberg-Teplice Caldera (Eastern Krušné Hory Mts. Czech Republic, Germany). Geolines 13, 17–22.

Campbell, I.H., Turner, J.S., 1987. A laboratory investigation of assimilation at the top of a basaltic magma chamber. Journal of Geology 95, 155–172.

Cherniak, D.J., 1999. Ba diffusion in feldspars. AGU Fall Meeting 1999, EOS Transactions of American Geophysical Union 80 suppl. 46, pp. F1078.

Christiansen, R.L., 1979. Cooling units and composite sheets in relation to caldera structure. Geological Society of America Special Paper 180, 29–41.

Didier, J., Barbarin, B., 1991. Enclaves and Granite Petrology. Developments in Petrology, vol. 13. Elsevier, Amsterdam.

Eklund, O., Shebanov, A.D., 1999. Origin of the rapakivi texture by sub-isothermal decompression. Precambrian Research 95, 129–146.

Fiala, F., 1959. The Teplice rhyolite between Krupka, Cinovec, Dubi and Mikulov and its surrounding rocks. Ustredni Ustav Geologicky Praha, 445–494. (in Czech).

Förster, H.-J., Seltmann, R., Tischendorf, G., 1995. High-fluorine, low phosphorus A-type (post-collision) silicic magmatism in the Erzgebirge. 2nd Symp. Permocarboniferous Igneous Rocks, vol. 7. Terra Nostra, Bonn, pp. 32–35.

Förster, H.-J., Tischendorf, G., Seltmann, R., Gottesmann, B., 1998. Die variszischen Granite des Erzgebirges: neue Aspekte aus stofflicher Sicht. Zeitschrift für Geologische Wissenschaften 26, 31–60.

Förster, H.-J., Tischendorf, G., Trumbull, R.B., Gottesmann, B., 1999. Late-collisional granites in the Variscan Erzgebirge, Germany. Journal of Petrology 40, 1613–1645.

Gayer, R., 1994. Evolution of coal-bearing Variscan foreland basins. Nachrichten der Deutschen Geologischen Gesellschaft 52, 83.

Ginibre, C., Wörner, G., Kronz, A., 2002. Minor- and trace-element zoning in plagioclase: implications for magma chamber processes at Parinacota volcano, northern Chile. Contributions to Mineralogy and Petrology 143, 300–315.

Hibbard, M.J., 1981. The magma mixing origin of mantled feldspar. Contributions to Mineralogy and Petrology 76, 158–170.

Hibbard, M.J., 1991. Textural anatomy of twelve magma-mixed granitoid systems. In: Didier, J., Barbarin, B. (Eds.), Enclaves and Granite Petrology, Developments in Petrology, vol. 13. Elsevier, Amsterdam, pp. 431–444.

Hildreth, W., 1979. The Bishop Tuff: evidence for the origin of compositional zonation in silicic magma chambers. Geological Society of America Special Paper 180, 43–75.

Hoth, K., Berger, H.-J., Breiter, K., Mlčoch, B., Schovánek, P., 1995. Geologische Karte Erzgebirge/Vogtland 1:100 000. Sächsisches Landesamt für Umwelt und Geologie Freiberg.

Jamtveit, B., 1999. Crystal growth and intracrystalline zonation patterns in hydrothermal environments. In: Jamtveit, B., Meakin, P. (Eds.), Growth, Dissolution and Pattern Formation in Geosystems. Kluwer Academic Publishers, Dordrecht-Boston-London, pp. 65–84.

Janoušek, V., Vrana, S., Erban, V., 2002. Petrology, geochemical character and petrogenesis of a Variscen post-orogenic granite; case study from the Sevetin Massif, Moldanubian Batholite, southern Bohemia. Journal of Czech Geological Society 47, 1–22.

Jiránek, J., Køíbek, B., Mlčoch, B., Procházka, J., Schovánek, P., Schovánková, D., Schulmann, K., Šebesta, J., Šimůnek, Z., Štemprok, M., 1987. The Teplice rhyolite. MS Czech Geological Survey Praha. (in Czech).

Kempe, U., Wolf, D., Ebermann, U., Bombach, K., 1999. 330 Ma Pb/Pb single zircon evaporation ages for the Altenberg Granite Porphyry, Eastern Erzgebirge (Germany): implications for Hercynian granite magmatism and tin mineralisation. Zeitschrift für geologische Wissenschaften 27, 385–400.

Kröner, A., Willner, A.P., 1998. Time of formation and peak of Variscan HP-HT metamorphism of quartz–feldspar rocks in the central Erzgebirge, Saxony, Germany. Contributions to Mineralogy and Petrology 132, 1–20.

Lipman, P.W., 2000. Calderas. In: Sigurdsson, H. (Ed.), Encyclopedia of Volcanoes. Academic Press, San Diego, pp. 643–662.

Lobin, M., 1983. Pflanzenfunde aus den Tuffiten des Teplicer Quarzporphyrs. Exkursionsführer "Fortschritte der Paläontologie und Stratigraphie des Rotliegenden". Gesellschaft Geologischer Wissenschaften der DDR, Berlin, p. 31.

Long, P.E., Luth, W.C., 1986. Origin of K-feldspar megacrysts in granitic rocks: implication of a partitioning model for barium. American Mineralogist 71, 367–375.

Lowenstern, J.B., 1995. Applications of silicate-melt inclusions to the study of magmatic volatiles. In: Thompson, J.F.H. (Ed.), Magmas, Fluids and Ore Deposits, Mineralogical Association of Canada Short Course Series vol. 23, pp. 71–99.

Mlčoch, B., 1994. The geological structure of the crystalline basement below the North Bohemian brown coal basin. In: Hirschmann, G., Harms, U. (Eds.), KTB Report 93/4. Niedersächsisches Landesamt für Bodenforschung, Hannover, pp. 39–46.

Müller, A., Seltmann, R., 2002. Plagioclase-mantled K-feldspar in the Carboniferous porphyritic microgranite of Altenberg-Frauenstein, Eastern Erzgebirge/Krušné Hory. Bulletin of the Geological Society of Finland 74, 53–79.

Müller, A., Seltmann, R., Behr, H.-J., 2000. Application of cathodoluminescence to magmatic quartz in a tin granite—case study from the Schellerhau Granite Complex, Eastern Erzgebirge, Germany. Mineralium Deposita 35, 169–189.

Müller, A., Müller, B., Behr, H.-J., 2001. Structural contrasts in granitic rocks of the Lusatian Granodiorite Complex and the Erzgebirge, Germany—in commemoration of Hans Cloos. Zeitschrift für Geologische Wissenschaften 29, 521–544.

Müller, A., Lennox, P., Trzebski, R., 2002. Cathodoluminescence and micro-structural evidence for crystallisation and deformation processes of granites in the Eastern Lachlan Fold Belt (SE Australia). Contributions to Mineralogy and Petrology 143, 510–524.

Müller, A., René, M., Behr, H.-J., Kronz, A., 2003. Trace elements and cathodoluminescence of igneous quartz in topaz granites from the Hub Stock (Slavkovský Les Mts., Czech Republic). Mineralogy and Petrology 79, 167–191.

Mullis, J., Ramseyer, K., 1999. Growth related Al-uptake in fissure quartz, Central Alps, Switzerland. Terra Nostra 99/6, 209.

Nekvasil, H., 1991. Ascent of felsic magmas and formation of rapakivi. American Mineralogist 76, 1279–1290.

Peppard, B.T., Steele, I.M., Davis, A.M., Wallace, P.J., Anderson, A.T., 2001. Zoned quartz phenocrysts from the rhyolitic Bishop Tuff. American Mineralogist 86, 1034–1052.

Perny, B., Eberhardt, P., Ramseyer, K., Mullis, J., Pankrath, R., 1992. Microdistribution of aluminium, lithium and sodium in a quartz: possible causes and correlation with short lived cathodoluminescence. American Mineralogist 77, 534–544.

Plesch, R., 1982. Auswerten und Prüfen in der Röntgenspektrometrie. G-I-T Verlag Ernst Giebeler, Darmstadt.

Rösler, H.J., Bothe, M., 1990. Bemerkungen zur Petrologie des Granits von Niederbobritzsch bei Freiberg und zur Bildung der Allanite. Abhandlungen des Staatlichen Museum für Mineralogie und Geologie Dresden 37, 73–101.

Rösler, H.J., Budzinski, H., 1994. Das Bauprinzip des Granits von Niederbobritzsch bei Freiberg (Sa.) auf Grund seiner geochemischen Analyse. Zeitschrift der geologischen Wissenschaften 22, 307–324.

Schreiber, U., Anders, D., Koppen, J., 1999. Mixing and chemical interdiffusion of trachytic and latitic magma in a subvolcanic complex of the Tertiary Westerwald (Germany). Lithos 46, 695–714.

Seltmann, R., Breiter, K., 1995. Late-Variscan crustal evolution and related tin-tungsten mineralization in the Altenberg-Teplice caldera. In: Breiter, K., Seltmann, R. (Eds.), Ore mineralizations of the Krušné Hory Mts. (Erzgebirge). Third Biennial SGA Meeting Praha, Excursion Guide, pp. 65–76.

Seltmann, R., Schilka, W., 1995. Late-Variscan crustal evolution in the Altenberg-Teplice caldera. Evidence from new geochemical and geochronological data. Terra Nostra 7/95, 120–124.

Seltmann, R., Breiter, K., Schilka, W., Benek, R., 1996. The Altenberg-Teplice caldera: reversed zonation of a stratified magma chamber. V.M. Goldschmidt Conference Heidelberg 1996. Journal of Conference Abstracts 1 (1), 556.

Siebel, W., Höhndorf, A., Wendt, I., 1995. Origin of late Variscan granitoids from NE Bavaria, Germany, exemplified by REE and Nd isotope systematics. Chemical Geology 125, 249–270.

Słaby, E., Galbarczyk-Gąsiorowska, L., Baszkiewicz, A., 2002. Mantled alkali-feldspar megacrysts from the marginal part of the Karkonosze granitoid massif (SW Poland). Acta Geologica Polonica 52, 501–519.

Štemprok, M., Holub, F.V., Novák, J.K., 2003. Multiple magmatic pulses of the Eastern Erzgebirge Volcano-Plutonic Complex, Krušne hory/Erzgebirge batholith, and their phosphorus contents. Bulletin of Geosciences 78, 277–296.

Thomas, R., 1992. Results of investigations on melt inclusions in various magmatic rocks from the northern border of the Bohemian Massif. In: Kukal, Z. (Ed.), Proceedings of the 1st International Conference about the Bohemian Massif, Prague 1988. Czech Geological Survey, Prague, pp. 298–306.

Thomas, R., 1994. Fluid evolution in relation to the emplacement of the Variscan granites in the Erzgebirge region: a review of the melt and fluid inclusion evidence. In: Seltmann, R., Möller, P., Kämpf, H. (Eds.), Metallogeny of Collisional Orogens. Czech Geological Survey, Prague, pp. 70–81.

Tichomirowa, M., 1997. [207]Pb/[206]Pb-Einzelzirkondatierungen zur Bestimmung des Intrusionsalters des Niederbobritzscher Granites. Terra Nostra 8, 183–184.

Tischendorf, G. (Comp.), 1989. Silicic magmatism and metallogenesis of the Erzgebirge. Veröffentlichungen Zentralinstitut für Physik der Erde 107.

Vance, J.A., 1965. Zoning in igneous plagioclase: patchy zoning. Journal of Geology 73, 636–651.

Van den Kerkhof, A.M., Scherer, T., Riganti, A., 1996. Cathodoluminescence and EPR analysis of Archean quarzites from the Nondweni Greenstone Belt, South Africa. Conference Abstracts SLMS International Conference on Cathodoluminescence. Society for Luminescence Microscopy and Spectroscopy, Nancy, p. 75.

Van den Kerkhof, A.M., Kronz, A., Simon, K., 2001. Trace element redistribution in metamorphic quartz and fluid inclusion modification: observations by cathodoluminescence. XVI ECROFI, Porto 2001, Abstracts, Memória vol. 7. Faculdade de Ciências do Porto, Departamento de Geologia, pp. 447–450.

Wark, D.A., Stimac, J.A., 1992. Origin of mantled (rapakivi) feldspars: experimental evidence of a dissolution- and diffusion-controlled mechanism. Contributions to Mineralogy and Petrology 111, 345–361.

Watt, G.R., Wright, P., Galloway, S., McLean, C., 1997. Cathodoluminescence and trace element zoning in quartz phenocrysts and xenocrysts. Geochimica et Cosmochimica Acta 61, 4337–4348.

Watznauer, A., 1954. Die erzgebirgischen Granitintrusionen. Geologie 3, 688–706.

Werner, O., Lippolt, H.J., 1998. Datierung von postkinematischen magmatischen Intrusionssubphasen des Erzgebirges: thermische und hydrothermale Überprägung der Nebengesteine. Terra Nostra 98/2, 160–163.

Wetzel, H.-U., 1985. Interpretation von Bruchzonen im Osterzgebirge und ihre Fortsetzung in Deckgebirgseinheiten der Lausitz. Zeitschrift für Geologische Wissenschaften 13, 111–121.

Wiebe, R.A., 1968. Plagioclase stratigraphy: a record of magmatic conditions and events in a granite stock. American Journal of Science 266, 690–703.

Wilcox, R.E., 1999. The idea of magma mixing: history of a struggle for acceptance. Journal of Geology 107, 421–432.

Available online at www.sciencedirect.com

Lithos 80 (2005) 229–247

www.elsevier.com/locate/lithos

Prolonged postcollisional shoshonitic magmatism in the southern Svecofennian domain – a case study of the Åva granite–lamprophyre ring complex

Olav Eklund[a,*], Alexey Shebanov[b]

[a]*Department of Geology, University of Turku, FI-20014, Turku University, Finland*
[b]*Department of Earth Sciences, Uppsala University, Villavägen 16, Uppsala SE-75236 Sweden*

Received 23 April 2003; accepted 9 September 2004
Available online 13 November 2004

Abstract

The Åva ring complex is one of four Paleoproterozoic postcollisional shoshonitic ring complexes in southwestern Finland. It is composed of ring dykes of K-feldspar megacryst-bearing granite, mingled in places with a shoshonitic monzonite, and lamprophyre dykes crosscutting all the rocks in a radial pattern. A survey was undertaken to trace the magma chamber beneath the ring complex to date it and measure some intensive parameters to clarify the crystallisation conditions at depth before the granite was emplaced in the upper crust. Mineral separates were extracted from the core zones of K-feldspar megacrysts in the granite, heavy mineral fractions (including zircons) from these separates were used for P-T assessment and age determinations, and the results were compared to data obtained from bulk rock samples. It appears that magma differentiation took place in a midcrustal magma chamber (at 4 to 7 kbar) possibly ~30 Ma before the emplacement of the ring complex in the upper crust (deep assemblage ~1790 Ma, shallow assemblage ~1760 Ma). Relatively high activity of the alkalies and a low oxygen fugacity characterised the midcrustal chamber. The juvenile Svecofennian crust was invaded by shoshonitic magmas from an enriched lithospheric mantle over a long period of time. Some of these magmas were stored and differentiated in the middle crust before transportation to the upper crust. The results also show that coarse-grained granites may provide evidence for several magmatic evolutionary episodes, e.g., differentiation and crystallisation in different environments prior to final emplacement.

Keywords: Ring complex; Postcollisional; Shoshonite; Thermobarometry; U-Pb geochronology; Megacryst

1. Introduction

Granites with resorbed megacrysts in a fine-grained matrix, emplaced as thin dykes in the uppermost crust, carry information on several stages of

---

* Corresponding author. Tel.: +358 2 333 5484; fax: +358 2 333 6580.
E-mail address: olav.eklund@utu.fi (O. Eklund).

0024-4937/$ - see front matter © 2004 Elsevier B.V. All rights reserved.
doi:10.1016/j.lithos.2004.06.012

magmatic history. The megacrysts presumably grew in a deep magma chamber and were transported into the upper crust, interacting with the melt and becoming partially dissolved or abraded by other crystals or country rock fragments. The megacrysts, thus, record processes that took place both in a deep-seated magma chamber and during magma ascent. Magmatic history comprising several stages of crystal growth, especially in coarse-grained phenocrystic rocks, means that minerals subjected to thermobarometric work have to be selected under strict textural control. The same approach must be taken concerning age determinations, i.e., zircons in different textural positions may yield different ages (cf. Shebanov et al., 2000).

This paper presents a thermobarometric and geochronological study carried out under rigorous textural control on a postcollisional K-feldspar-porphyritic, high Ba and Sr (HiBaSr) granite ring complex (Åva) in southwestern Finland. We present a model for the magmatic evolution of the ring complex and discuss the regional implications of postcollisional magmatism in the Svecofennian domain of southern Finland.

2. Geological background

Southwestern Finland belongs to the accretionary arc complex of southern Finland in the Svecofennian domain (Korsman et al., 1999). The magmatic history of this arc complex begins with the precollisional volcanic arc magmatism dated at 1.90 Ga. Volcanic rocks with more mature arc signatures formed at 1.88 Ga. The collision of the arc complex towards the accretionary arc complex of western Finland took place at 1.88 to 1.86 Ga. These ages were obtained from the syn-collisional calc-alkaline granitoids in the area (Väisänen and Mänttäri, 2002). The syn-orogenic granitoids intruded simultaneously with the regional D_2. During postcollisional convergence (D_3) at 1.84 to 1.82 Ga, regional high-T and low-P metamorphism was associated with extensive migmatisation and crustal melting, mainly of S-type (Väisänen and Hölttä, 1999). Roughly simultaneously with the metamorphic culmination, the Svecofennian crust of southern Finland and Russian Karelia was intruded by at least 14 small, P-, F-, Ba-, Sr- and LREE-enriched

bimodal shoshonitic intrusions (Eklund et al., 1998; Väisänen et al., 2000). These small intrusions are found in a 600-km long belt extending from Lake Ladoga in Russian Karelia to the Åland archipelago in southwestern Finland (Fig. 1a). Based on geochemistry, Eklund et al. (1998) concluded that the shoshonitic magmas stemmed from an enriched lithospheric mantle. Age determinations indicate that this magmatism occurred between 1815 and 1770 Ma, most ages center around 1800 Ma (Eklund et al., 1998; Väisänen et al., 2000). The oldest shoshonitic rocks were emplaced at midcrustal levels (~4.5 kbar; Väisänen et al., 2000) and the younger in the upper crust, ~2 kbar and less (Eklund et al., 1998; Niiranen, 2000).

Some of the granites in these postcollisional intrusions are coarse porphyritic with K-feldspar megacrysts up to 3 cm in size. It appears that they crystallized in deeper magma chambers prior to emplacement into the upper crust.

3. The Åva ring complex

3.1. Geochemistry, petrography, and age

The Åva ring complex is one of three bimodal shoshonitic ring complexes (Åva, Seglinge, and Mosshaga) situated along a northeast trending shear zone in soutwestern Finland (Fig. 1a). The Åva ring complex is approximately 7 km across (Fig. 1b). The complex comprises hundreds of ring dykes of coarse porphyritic HiBaSr granite, monzodiorite, monzonite, quartz-monzonite, and granodiorite pillows having a shoshonitic geochemical affinity (Fig. 2). These more mafic rocks are collectively referred to as "monzonite". The ring complex is cut by shoshonitic lamprophyres in a radial pattern. By using the geochemical method developed by Liégeois et al. (1998), it is possible to separate rocks with alkaline affinity from those with shoshonitic affinity. Compared to a reference rock series (Fig. 3), the former is relatively enriched in Zr, Ce, Sm, Y, and Yb, the latter in Rb, Th, U, and Ta. The postcollisional ring complexes in southwestern Finland plot within the field of shoshonitic rocks and may, according to Liégeois et al. (1998), have stemmed from a previously enriched phlogopite- and amphibole-bearing source in the lithospheric mantle.

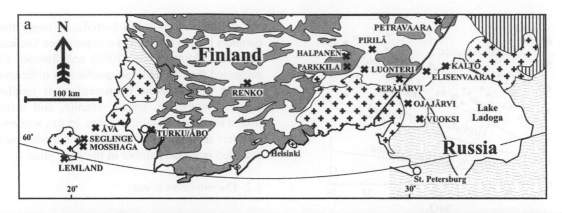

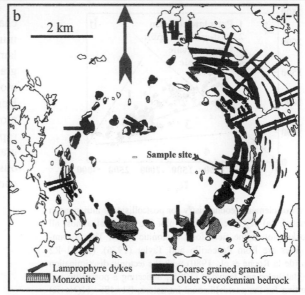

Fig. 1. (a) Geological sketch map and distribution of postcollisional shoshonitic intrusions in southern Finland and Russian Karelia. Modified after Eklund et al. (1998). (b) The Åva ring complex in the Åland archipelago exposed on islands and skerries. The sample site on Notholm is marked by an arrow. The map is modified after Kaitaro (1953), Ehlers and Bergman (1984), Branigan (1989), and Eklund et al. (1998).

The investigated granites carry K-feldspar mega-crysts up to 3 cm in size. The megacrysts are rounded, corroded, and sometimes, subhedral (tabular) in shape (Fig. 4). They contain small fairly rare inclusions of variable amphibole and mica, coexisting plagioclase, titanite, and accessory apatite and zircon. Plagioclase occurs as small phenocrysts or forms a mantle around the K-feldspar megacrysts. The matrix consists of quartz (often bluish) and biotite, minor plagioclase, K-feldspar, and Fe-Ti oxides (mainly titanomagnetite). Occasionally, amphibole is present as inclusions in mica within the matrix of the granite. Both K-feldspar megacrysts

and groundmass contain accessory apatite, pyrite, titanite, zircon, and fluorite.

In places, HiBaSr granite and, sometimes, also S-type granite form radial dykes (Skyttä, 2002). The HiBaSr granite was emplaced almost simultaneously with crustal melts, forming a mingled zone between the two granites in the central part of the intrusion. Ehlers and Bergman (1984) and Bergman (1986) suggested that the Åva intrusion centers on a gneissic ring structure formed by diapiric emplacement of a late orogenic S-type microcline granite pluton that deformed the surrounding gneisses into a steep inward dipping ring.

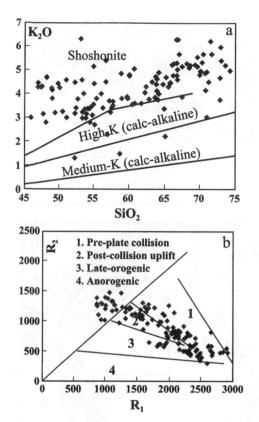

Fig. 2. Major element geochemistry for postcollisional intrusions in SW Finland. (a) K_2O versus SiO_2 plot showing that almost all these rocks plot in the field for shoshonitic rocks. The discrimination lines are from Peccerillo and Taylor (1976). (b) R1–R2 diagram (de la Roche et al., 1980) with the tectonic discrimination grid by Batchelor and Bowden (1985), indicating a postcollisional uplift setting for the postcollisional intrusions. R1=4Si - 11(Na+K) - 2(Fe+Ti), R2=6Ca+2Mg+Al. Geochemical data from Rutanen et al. (1997) and Eklund et al. (1998).

The monzonite in Åva was dated at 1799 ± 13 Ma and the granite at 1803 ± 10 Ma (U-Pb from zircon; Suominen, 1991). Magma mingling structures indicate that the monzonite and granite magmas intruded contemporaneously. Suominen (1991) also reported younger titanite $^{207}Pb/^{206}Pb$ ages between 1754 and 1789 Ma for the monzonites and between 1759 and 1782 Ma for the granites.

In the Åva area, boulders of a rock consisting of plagioclase, orthopyroxene, biotite, amphibole, apatite, and minor clinopyroxene have been found. A SIMS age determination shows that rims of zircons have the same age as the monzonite (Table 1). Mineralogical correlations considering substitution

mechanisms in mica and hornblende imply that the boulders belong to the Åva complex (see Sections 6.1 and 6.2). Eklund et al. (1998) and Rutanen (2001) suggested that the HiBaSr granite is a differentiation product after crystal fractionation of amphibole, biotite, plagioclase, apatite, titanite, and magnetite from a mafic shoshonitic magma. In this scenario, the boulders represent the high-density cumulus part of a stratified magma chamber in the deep crust.

3.2. The intrusion event

In the investigated area, the multiple ring system is characterised by coarse porphyritic granite, sometimes mingled with monzonite, and crosscutting radial lamprophyre dykes that are sometimes mingled with granite. These observations have led to at least two models about the intrusive event. (1) Ehlers and Bergman (1984) presented a model where the monzonite intruded before the granite and the lamprophyres. (2) Branigan (1989a) concluded that the granite and the monzonite intruded contemporaneously as evidenced by the net-veined structures, composite dykes, and in-situ mingling/mixing fea-

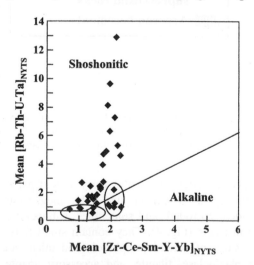

Fig. 3. Trace element diagram separating rocks with shoshonitic affinity from rocks with alkaline affinity. The values of the trace elements are normalized to the Yenchichi-Telabit rock series in the Tuareg shield (NYTS normalization) after Liégeois et al. (1998). The postcollisional rocks follow a well-developed shoshonitic trend with most of the mafic rocks having a prominent shoshonitic affinity, while some of the granites (encircled) are marginally alkaline. Geochemical data from Rutanen et al. (1997) and Eklund et al. (1998).

Fig. 4. A slice of coarse porphyritic Åva granite with drill holes in the cores of K-feldspar megacrysts. Heavy fractions for P-T and age determinations were separated from about 100 drilled cores. The white dot is 13 mm across.

tures. The latter model has become generally accepted (cf. Eklund et al., 1998; and references therein).

Using textural and geochemical data, Hubbard and Branigan (1987) presented a model in which the Åva ring complex is connected to a more extensive magma chamber in the middle crust. This chamber was later reactivated by replenishment with fresh magma from the mantle. Branigan (1989b) constructed velocity profiles of ring dykes based on their fabric and concluded that the granite emplacement in the upper crust was rapid and forceful. After finding a mixed population of K-feldspars showing both plastic and brittle deformation features, Branigan (1989b) concluded that the granite magma was derived through disaggregation of an already crystalline granite at depth.

3.3. Relative timing

Magma mingling and mixing between the coarse porphyritic HiBaSr granite and monzonite is recognized in the ring dykes and in the central part of the intrusion (Branigan, 1989a; Eklund et al., 1998). Monzonite pillows contain partially resorbed K-feldspar and quartz xenocrysts from the granite. The quartz xenocrysts have reacted with the monzonite melt, and amphibole rims have formed around them (quartz ocelli). The granite is peppered with small, centimeter-sized mafic shoshonitic enclaves distributed over extensive areas. These textures are strong evidence that the granitic end member of the mixing system was crystal-saturated at the time of the vigorous mixing/mingling event. In the mingling relation between the monzonite and the granite, we consider the monzonite to represent the invasive magma. Thus, the crystallisation age of the monzonite is supposed to represent the age of the emplacement in the upper crust. In contrast, the cores of K-feldspar megacrysts in the granite represent an earlier crystallisation event during the evolution of the magma. The contacts between the radial lamprophyre dykes and the granite vary from lobate to angular. This indicates that the radial lamprophyre dykes intruded during and after the consolidation of the granite and represent the youngest intrusive phase of the complex.

4. Sample preparation and experimental methods

About 15 kg of the Åva granite was sampled from a shoreline outcrop on the island of Notholm (Fig. 1b). The sample was taken from an area where the porphyritic granite shows no change in texture and chemistry over extensive areas. The same granite variety is found all around the ring complex, and thus, the sample can be considered representative of the whole ring complex.

Only a few mineral inclusions can be seen in the core zone of the K-feldspar megacrysts. To receive a representative assemblage of core zone minerals for

Table 1
U-Pb SIMS data for zircons from Åva ring complex

Sample/ Spot#	Domain	Derived ages, Ma						Corrected ratios						r	Disc. [%]	Concentrations			Th/U
		$^{207}Pb/^{206}Pb$	±s	$^{207}Pb/^{235}U$	±s	$^{206}Pb/^{238}U$	±s	$^{207}Pb/^{206}Pb$	±s, %	$^{207}Pb/^{235}U$	±s, %	$^{206}Pb/^{238}U$	±s, %			(U) ppm	(Th) ppm	(Pb) ppm	
Monzonite																			
n966-08a	Rim	1769.5	17.1	1790.5	26.1	1808.7	46.0	0.10821	0.94	4.8323	3.06	0.32389	2.91	0.95	2.5	65	39	27	0.600
n966-10a	Rim	1778.4	16.6	1779.2	26.0	1780.0	45.6	0.10874	0.91	4.7678	3.06	0.31800	2.92	0.95	0.1	113	131	51	1.160
n966-10b	Rim	1775.9	23.9	1796.2	27.3	1813.7	46.1	0.10859	1.32	4.8650	3.19	0.32493	2.91	0.91	2.4	119	139	54	1.175
n966-12a	Rim	1750.2	19.0	1787.0	26.6	1818.7	46.8	0.10707	1.05	4.8119	3.12	0.32594	2.94	0.94	4.5	49	42	21	0.872
n966-12b	Midzone	1783.2	16.6	1806.1	26.5	1826.1	47.2	0.10902	0.91	4.9225	3.10	0.32746	2.96	0.96	2.8	77	90	35	1.177
n966-10b	Midzone	1760.3	26.5	1751.4	18.5	1744.0	25.2	0.10767	1.46	4.6120	2.20	0.31067	1.65	0.75	-1.1	72	75	31	1.039
n967-10c	Midzone	1770.0	24.7	1770.4	18.2	1770.7	25.9	0.10824	1.36	4.7177	2.15	0.31610	1.67	0.77	0.0	73	79	32	1.080
Granite																			
n966-02a	Rim	1756.7	18.8	1765.5	26.2	1773.0	45.2	0.10745	1.03	4.6903	3.09	0.31658	2.91	0.94	1.1	114	56	45	0.490
n966-03a	Rim	1752.9	13.3	1799.7	25.6	1840.4	46.7	0.10723	0.73	4.8851	3.00	0.33041	2.91	0.97	5.7	172	15	65	0.089
n966-04a	Rim	1750.9	21.0	1793.9	26.7	1831.2	46.6	0.10711	1.16	4.8519	3.13	0.32852	2.91	0.93	5.3	108	91	47	0.846
n966-15b	Rim	1750.2	11.9	1770.4	25.4	1787.7	45.8	0.10707	0.66	4.7180	3.00	0.31958	2.93	0.98	2.5	97	52	39	0.535
n967-09a	Rim	1779.0	14.2	1768.2	15.8	1759.1	26.2	0.10877	0.78	4.7054	1.87	0.31374	1.70	0.91	-1.3	112	83	46	0.739
n967-08c	Core	1794.1	6.3	1790.6	14.2	1787.7	25.7	0.10968	0.35	4.8328	1.68	0.31958	1.65	0.98	-0.4	596	538	262	0.902
n967-08d	Core	1781.4	4.5	1806.1	14.2	1827.5	26.4	0.10892	0.25	4.9221	1.67	0.32776	1.66	0.99	3.0	894	907	412	1.014
n967-08e	Core	1794.7	5.2	1799.9	14.2	1804.4	25.9	0.10972	0.28	4.8863	1.67	0.32301	1.64	0.99	0.6	970	1016	441	1.047
n967-13a	Core	1791.4	5.4	1801.9	30.8	1811.0	56.8	0.10952	0.30	4.8982	3.60	0.32437	3.58	1.00	1.3	765	886	359	1.158
Lamprophyre																			
n967-01a	Rim	1690.3	16.2	1719.5	15.6	1743.5	25.2	0.10364	0.88	4.4380	1.87	0.31057	1.64	0.88	3.6	269	139	104	0.516
n967-02a	Rim	1735.2	14.4	1686.5	15.1	1647.6	24.0	0.10620	0.79	4.2641	1.83	0.29121	1.65	0.90	-5.7	879	383	313	0.436
n967-03a	Rim	1734.0	9.4	1644.2	14.1	1574.8	23.0	0.10613	0.52	4.0493	1.73	0.27671	1.65	0.95	-10.3	561	299	197	0.534
n967-04a	Rim	1771.5	10.7	1741.1	15.4	1715.8	26.3	0.10833	0.59	4.5549	1.84	0.30495	1.74	0.95	-3.6	289	163	111	0.562
n967-05a	Rim	1771.5	8.1	1679.3	14.1	1606.5	23.5	0.10833	0.44	4.2271	1.71	0.28300	1.65	0.97	-10.5	643	206	218	0.321
n967-06a	Rim	1763.9	10.0	1746.0	14.7	1731.2	25.3	0.10788	0.55	4.5822	1.75	0.30807	1.66	0.95	-2.1	742	412	284	0.556
n967-01b	Core	1821.0	18.4	1809.7	16.5	1799.8	26.0	0.11132	1.02	4.9431	1.94	0.32206	1.65	0.85	-1.3	340	182	139	0.533
n967-02b	Core	1800.3	5.9	1788.0	14.2	1777.4	25.6	0.11005	0.32	4.8175	1.68	0.31748	1.65	0.98	-1.5	998	72	357	0.072
n967-06b	Core	1805.6	8.1	1786.4	14.4	1770.1	25.5	0.11037	0.44	4.8087	1.70	0.31598	1.64	0.97	-2.2	542	423	226	0.780
Mafic cumulate																			
n966-06a	Homogenous	1730.6	15.1	1747.3	25.5	1761.2	45.0	0.10594	0.83	4.5890	3.02	0.31418	2.91	0.96	2.0	123	161	56	1.309

All spots made in the more metamict parts of the crystals yielded low $^{206/204}Pb$ ratios. Due to the high content and unknown age of the common lead, all data with $^{206}Pb/^{204}Pb$ less than 3000 were omited from further considerations.
All omited data had a very high degree of uncertainty in measured ratios. The whole data set is available upon request from the authors.

thermobarometry and geochronology, a lengthy separation process was undertaken. The sample was cut into about 5- to 7-mm thick slices. From these slices, cores from more than 100 K-feldspar megacryst were carefully drilled out (cf. Fig. 4) and crushed. After separation with heavy liquids, the recovered heavy mineral fraction was prepared for electron microprobe and ion microprobe studies. The recovered assemblage of several thousand grains contained both single and intergrown mineral grains. Magnetite, biotite, titanite, amphibole, apatite, and plagioclase were obtained from the core zones of the K-feldspar megacrysts.

The mineral chemical data were acquired using the Cameca Camebax SX50 microprobe at the Department of Earth Sciences, Uppsala University (mainly 3 to 5 μm spot-mode, ZAF correction followed the Cameca PAP procedure). U-Th-Pb isotope analyses were performed on a high-mass resolution, high-sensivity Cameca IMS 1270 ion microprobe (NORDSIM facility) at the Museum of Natural History, Stockholm. The analysed spots were 25 to 30 μm across. Details of the analytical procedures and data reduction are given in Whitehouse et al. (1997) and Zeck and Whitehouse (1999). The calibration of the raw data were based on the analyses of the Geostandards zircons 91500 with an accepted age of 1065 Ma (Wiedenbeck et al., 1995). Ages were calculated from overlapping concordant data according to the "concordia age" method of Ludwig (1998), using the Isoplot/Ex software (Ludwig, 1999).

5. U-Pb SIMS geochronology

Lamprophyres often contain mineral assemblages that were formed during the early stages of magmatic evolution, making it notoriously difficult to date their intrusive event (Rock, 1990). Coarse-grained or porphyritic granites emplaced in the upper crust are often formed as a result of multistage magmatic evolution (Eklund and Shebanov, 1999). However, if lamprophyric magmatism accompanies the granitoid suite coevally or slightly later as in this case, U-Pb zircon geochronology on the late-formed phases of granite may provide an important constraint on the crystallisation time of the lamprophyric dykes. For dating source components or multiple magmatic events, the SIMS technique is preferred. In the present study, all spots made in the more metamict parts of the crystals yielded low $^{206}Pb/^{204}Pb$ ratios. Because of unknown origin and age of the common lead, all data with $^{206}Pb/^{204}Pb$ less than 3000 (in most cases below 1000) were omitted from further consideration. All the omitted data had a very high degree of uncertainty in the measured ratios. After careful evaluation, 26 analytical spots out of 46 were considered suitable for further elaboration (Table 1). The whole data set is available upon request from the authors.

We consider the age of the monzonite mingled with the granite to represent the crystallization age of the Åva intrusion. Eight analytical spots from zircon rims yielded a concordia age of 1766±13 Ma for the monzonite (Fig. 5a). An overlapping age was obtained from rims of zircons from the granite (1762±13 Ma; Fig. 5b). Four spots from zircon cores from the granite yielded an age of 1790.6±7.5 Ma (Fig. 5c). Overlapping ages were obtained from zircon cores in the lamprophyre (1801±10 Ma; grey ellipses in Fig. 5d), while the rims yielded an age of 1769±11 Ma (open ellipses in Fig. 5d). The reference line on Fig. 5d comprises the only three statistically equal data points (open ellipses). Dashed ellipses in Fig. 5d represent analytical data obtained from the lamprophyre zircons that scatter towards younger ages on the concordia diagram. If all the data for the rims were included in the age calculation, they would yield an upper intercept age of 1753±28 Ma (MSWD=6.1). We, however, consider the three statistically most coherent points (n967-04a, n967-05a, n967-06a) to yield a relatively reliable age for the rims. The origin of the scatter for the others remains unclear.

Ion beam spots in core and rim positions for zircons from the granites, monzonites, lamprophyres, and cumulates are shown in Fig. 6. The results indicate two distinct age groups: (1) 1801±10 to 1790±7.5 Ma from zircon cores that represent crystallisation of the early granite and the earliest crystallisation stage of the lamprophyre material; and (2) 1769±11 to 1762±13 Ma from the rims. The second age group corresponds to the time of emplacement for the granite, monzonite, and lamprophyre in the upper crust. The same age groups can be distinguished from core rim analyses in single zircon crystals. The broadest variations are exhibited by zircons from lamprophyres (Fig. 5d).

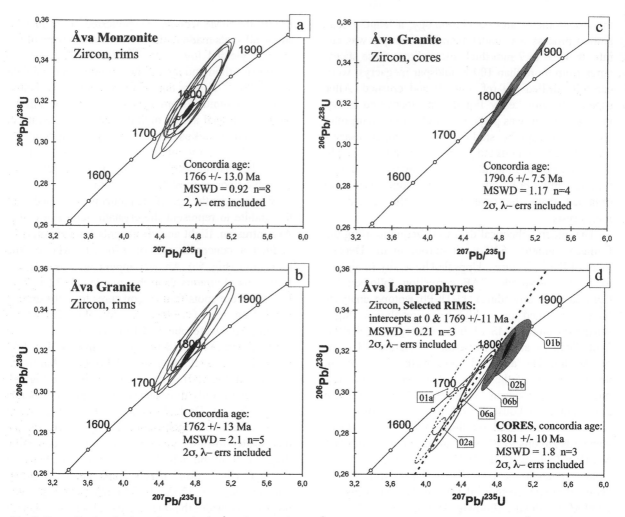

Fig. 5. Concordia plots for the rock types in the Åva ring-complex. (a) Åva monzonite, analyses from zircon rims; (b) Åva granite, analyses from zircon rims; (c) Åva granite, analyses from zircon cores; (d) Åva lamprophyres, analyses from zircon cores—grey ellipses; analyses from rims with statistically equal data points—open ellipses. Analyses that scatter towards younger ages—dashed ellipses. See text for further explanation.

Zircons from the lamprophyre have a considerably higher content of U, Th, and Pb compared with zircons from the monzonite and granite. Exceptions are the cores from zircons in the granite. These exhibit high contents of U (590 to 970 ppm), Th (530 to 1020 ppm), and Pb (260 to 440 ppm; Table 1), while their rims have significantly lower concentrations: 90 to 170, 15 to 90, and 45 to 65 ppm, respectively. These data show that the inherited cores formed in a U- and Th-rich environment, while the rims crystallised from melts depleted in U and Th or together with other U-Th-rich minerals, such as allanite. This trend is opposite to

what would be expected for zoned zircons formed in a closed granitic system, where enrichment in U and decrease in the Th/U ratio commonly occur in the course of crystallisation (Krasnobaev, 1986). As the partition coefficients of U in rock-forming minerals (K-feldspar, biotite, plagioclase, quartz) precipitating from felsic melt are relatively small (1.2 and less; Dostal and Capedri, 1975; Nash and Crecraft, 1985), simultaneously precipitated zircons are expected to concentrate U and Th (predominantly U) in their rims.

Our data show a drastic drop in U and Th content and a decrease in the Th/U ratio from cores to rims in

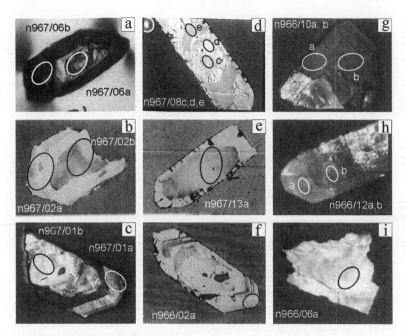

Fig. 6. Photographs (reflected light) of selected analysed zircons. Ion microprobe spots are shown as ellipses (30 μm in length). (a), (b), (c) Zircons from lamprophyres; (d), (e), (f) zircons from granite; (g), (h) zircons from monzonite; (i) zircon from the cumulate. Analysed spots are numbered as in Table 1.

the granite. One process responsible for this might be a contamination of the initial, enriched shoshonitic magmas by depleted crustal material, i.e., dilution. Another process might be precipitation of allanite prior to or simultaneously with crystallisation of the zircon rims. Allanite, which is a common accessory mineral in Åva monzonite and granite, is an efficient concentrator of U and Th. After the formation of zircon cores in a midcrustal reservoir, the onset of allanite crystallisation would decrease the amount of U and, particularly, that of Th in the evolved melts. Allanite has a high Th/U ratio (Amelin et al., 1997), and precipitation of allanite prior to or simultaneously with the second generation of zircon (rims), thus, might explain the low Th/U ratio of the rims (0.09 to 0.84), compared with early formed cores (0.90 to 1.16). These observations seem to be applicable to the felsic rocks of the Åva complex. However, the U-Th-Pb contents in zircons from the lamprophyres remain high regardless of the nature of accompanying minerals.

Suominen (1991) reported conventional TIMS ages for the Åva granite and monzonite. All the fractions in Suominen (1991) have U and Pb contents that vary between the two limits registered here for the

rims and cores of zircons from Åva. Many of the fractions had very high U (220 to 880 ppm) and Pb (90 to 160 ppm) close to the concentrations typical of the cores in the present study. As all the investigated rims have lower concentrations of U and Pb, it is possible that high proportions of the older component were present in the fractions used in the conventional analyses. In our study, the rims were relatively poor in both U and Pb. Hence, even a moderate portion of the older high U and high Pb core material may have caused the rather high conventional upper intercept ages (centering around 1800 Ma).

6. Mineral chemical traits and thermobarometry

Mafic silicates in different textural positions were investigated from the Åva granite and lamprophyres (representative compositions and recalculated formulas are given in Tables 2 and 3; the whole data set is available upon request; mineral chemical data for the lamprophyres are quoted from Rutanen et al., 1997). These minerals show broad chemical variations that imply important constraints to the magmatic evolution of the Åva complex.

Table 2
Representative analyses of amphibole from the Àva ring complex

	Àva granite											Mafic cumulate		
Rock														
Sample	Àva-C	Àva-C	Àva-C	Àva-C	Ava2-SC	Ava2-SC	Ava-LC	Ava-LC	Ava-LC	Ava-LC	Ava-SC	Àva-BMC	Àva-BMC	Àva-BMC
Anlysis	R3p.12	R3p.4	R3p.5	R3p.6	OEp.42	OEp.42a	OEp.12	OEp.14	OEp.19	OEp.50	p.21	OEp.63	OEp.64	OEp.64b
Location[a]	CZ	CZ	CZ	CZ	MC	MC	MC	MC	MC	MC	MC	BR-OPX	BR-OPX	BR-OPX
SiO_2	41.852	41.645	37.836	36.764	41.583	40.403	38.998	39.673	41.599	43.188	41.074	42.375	42.344	43.033
TiO_2	0.332	0.398	0.587	0.452	1.211	0.685	0.328	0.26	0.62	1.035	0.708	1.949	1.617	1.581
Al_2O_3	9.406	9.176	11.388	11.206	8.705	10.485	10.473	10.205	8.13	9.273	8.949	9.626	9.714	9.576
FeO_{tot}	19.947	20.226	29.945	28.949	18.058	20.428	21.987	21.776	19.583	16.169	21.012	12.052	11.923	11.441
MnO	0.955	0.922	0.496	0.452	0.564	0.531	0.959	0.997	0.778	0.294	0.927	0.072	0.106	0.056
MgO	8.897	8.941	2.267	1.988	10.465	9.053	7.977	7.61	9.474	11.414	8.868	14.055	14.421	14.376
CaO	11.591	11.744	10.955	11.07	11.597	11.682	11.355	11.455	11.346	11.937	11.446	11.856	11.924	11.826
BaO	0.071	0.053	–	–	0.013	–	–	0.0005	0.018	0.001	–	0.079	0.084	0.066
Na_2O	1.555	1.44	1.219	1.988	1.45	1.433	1.665	1.629	1.636	1.05	1.443	1.296	1.505	1.44
K_2O	1.59	1.568	1.991	2.062	1.564	1.591	1.75	1.668	1.509	0.94	1.497	1.348	1.255	1.218
Cl	–	–	–	0.956	0.209	0.209	0.471	0.471	0.193	0.02	0.176	0.215	0.216	0.216
F	–	–	–	0.889	1.637	1.637	1.327	1.327	1.687	0.001	1.835	1.3	1.307	1.307
Total	96.196	96.113	96.684	96.776	97.056	98.137	97.29	97.072	96.573	95.322	97.935	96.223	96.416	96.136
a.f.u., based on 23 (O, OH, F, Cl)														
Si	6.516	6.483	6.151	6.147	6.474	6.26	6.168	6.293	6.571	6.579	6.419	6.422	6.394	6.506
T-site $Al_{(IV)}$	1.484	1.517	1.849	1.853	1.526	1.74	1.832	1.707	1.429	1.421	1.581	1.578	1.606	1.494
$Al_{(VI)}$	0.24	0.165	0.331	0.353	0.07	0.173	0.119	0.2	0.084	0.243	0.066	0.14	0.122	0.211
Ti	0.039	0.047	0.072	0.057	0.142	0.08	0.039	0.031	0.074	0.119	0.083	0.222	0.184	0.18
Fe^{3+}	0.382	0.513	0.577	0.302	0.425	0.662	0.771	0.606	0.392	0.448	0.614	0.352	0.435	0.267
C-site Fe^{2+}	2.215	2.12	3.47	3.745	1.926	1.985	2.137	2.282	2.195	1.599	2.133	1.11	1.014	1.103
Mg	2.065	2.075	0.549	0.495	2.429	2.091	1.881	1.8	2.231	2.592	2.066	3.176	3.246	3.24
Mn	0.059	0.08	–	0.047	0.009	0.009	0.053	0.081	0.024	–	0.039	–	–	–
Ca	–	–	–	–	–	–	–	–	–	–	–	–	–	–
Mn	0.067	0.041	0.068	0.017	0.066	0.061	0.076	0.053	0.08	0.038	0.084	0.009	0.014	0.007
Fe^{2+}	–	–	0.024	–	–	–	–	–	–	0.014	–	0.066	0.057	0.077
B-site Ca	1.933	1.959	1.908	1.983	1.934	1.939	1.924	1.947	1.92	1.948	1.916	1.925	1.929	1.916
Na	–	–	–	–	–	–	–	–	–	–	–	–	–	–
Ba	–	–	–	–	–	–	–	–	–	–	–	–	–	–
A-site Na	0.469	0.435	0.384	0.587	0.438	0.431	0.511	0.501	0.501	0.31	0.437	0.381	0.441	0.422
K	0.316	0.316	0.311	0.413	0.311	0.314	0.353	0.338	0.304	0.183	0.298	0.261	0.242	0.235
Total	15.785	15.751	15.694	15.999	15.75	15.745	15.864	15.839	15.805	15.494	15.736	15.642	15.684	15.658
(Fe+Mn)/(Fe+Mn+Mg)	57.581	57.991	88.289	90.173	50.062	56.616	62.425	63.742	54.941	44.745	58.607	32.612	31.893	30.976

[a] CZ—core zones of the K-feldspar megacrysts from drilled cores. MC—K-feldspar megacrysts, analysed in thin sections. BR-OPX—intergrowths with orthopyroxene phenocrysts.

Table 3
Representative analyses of mica from the Åva ring complex

Rock	Åva granite											Mafic cumulate		
Sample	Åva-C	Åva-C	Åva-C	ÅVA-C	ÅVA-C-5	ÅVA-C-5	Ava-LC	Åva-SC	Ava2-SC	Ava2-SC	Ava-InterC	Åva-BMC	Åva-BMC	Åva-BMC
Analysis/spot#	R3p.43	R3p.45b	R3p.47b	p.13b	p.22	p.42	OEp.20	p.1	prof.p.40	prof.p.9	OEp.77b	OEp.55	OEp.68b	OEp.71b-2
Location[a]	CZ	CZ	CZ	CZ	CZ	CZ	MC	MC	MC	GM	GM	BR	BR	BR
SiO_2	34.776	35.243	35.059	38.427	37.599	38.403	37.104	38.623	39.062	36.595	37.839	38.803	38.058	36.213
TiO_2	3.167	3.014	2.964	1.298	0.865	0.737	1.329	1.054	1.006	1.581	0.823	3.897	4.354	4.402
Al_2O_3	13.803	13.554	14.039	12.337	12.874	12.338	12.489	12.432	12.548	13.162	11.739	13.384	12.437	13.048
Fe_2O_3	–	–	–	–	–	–	–	–	–	–	–	–	–	–
FeO_{tot}	26.071	26.071	26.595	16.870	16.562	14.086	15.851	15.181	15.404	16.073	15.334	8.333	9.333	10.432
MnO	0.322	0.357	0.357	0.392	0.558	0.741	0.492	0.519	0.438	0.536	0.650	0	0.034	0.074
MgO	7.277	7.057	6.782	14.88	15.771	16.765	15.359	15.416	16.511	16.191	16.415	19.576	19.382	18.42
CaO	0.018	bdl	bdl	bdl	0.016	0.007	bdl	bdl	0.015	0.004	0.465	bdl	bdl	bdl
BaO	0.617	0.755	0.520	0.290	0.171	0.262	0.291	0.524	0.180	0.279	bdl	1.095	0.997	2.423
SrO	0.234	0.086	bdl	bdl	bdl	bdl	bdl	bdl	bdl	bdl	0.125	bdl	bdl	bdl
Na_2O	0.075	0.016	0.075	0.056	0.025	0.086	0.057	0.032	bdl	bdl	0.123	0.084	0.151	0.003
K_2O	9.243	8.896	9.366	9.944	8.889	9.283	9.079	10.132	9.302	8.213	9.350	9.166	9.176	8.691
Cl	0.642	0.734	–	–	0.138	0.119	0.371	0.113	–	0.167	0.123	0.175	0.217	0.224
F	1.504	1.766	–	–	2.767	2.960	3.867	3.315	–	1.912	3.395	1.557	1.428	1.452
Total	97.749	97.549	95.237	77.334	96.235	92.827	96.289	97.341	94.286	94.713	86.236	96.07	95.567	95.382
O-F-Cl	0.778	0.909	–	–	1.196	1.273	1.712	1.421	–	0.843	1.453	0.695	0.650	0.662
a.f.u., based on 24 (O, OH, F, Cl)														
Si	5.182	5.283	5.112	5.390	5.464	5.577	5.545	5.614	5.406	5.297	5.579	5.314	5.273	5.122
$Al_{(IV)}$	2.422	2.393	2.411	2.038	2.203	2.110	2.198	2.128	2.045	2.243	2.033	2.159	2.029	2.173
Ap.Def.[b]	0.396	0.324	0.477	0.572	0.333	0.313	0.257	0.258	0.549	0.460	0.383	0.527	0.698	0.705
$Al_{(VI)}$	0	0	0	0	0	0	0	0	0	0	0	0	0	0
Ti	0.355	0.34	0.325	0.137	0.095	0.081	0.149	0.115	0.105	0.172	0.092	0.401	0.454	0.468
Fe^{3+}	–	–	–	–	–	–	–	–	–	–	–	–	–	–
Fe^{2+}	3.248	3.269	3.243	1.979	2.013	1.711	1.981	1.845	1.783	1.946	1.891	0.954	1.082	1.234
Mg	1.616	1.577	1.474	3.112	3.417	3.629	3.422	3.340	3.406	3.494	3.603	3.997	4.004	3.884
Mn	0.041	0.045	0.044	0.046	0.069	0.091	0.062	0.064	0.051	0.016	0.081	0	0.04	0.009
Ca	0.003	0	0	0	0.003	0.001	0	0	0.002	0.001	0.027	0	0	0
Ba	0.036	0.044	0.030	0.016	0.010	0.015	0.017	0.030	0.010	0.016	0.035	0.059	0.054	0.134
Na	0.022	0.005	0.021	0.015	0.007	0.024	0.017	0.009	0	0	0.035	0.022	0.041	0.001
K	1.757	1.701	1.742	1.780	1.648	1.720	1.731	1.879	1.642	1.517	1.759	1.601	1.622	1.684
Total	14,682	14,657	14,402	14,513	14,929	14,959	15,122	15,024	14,450	14,702	15,111	14,507	14,599	14,709
Cl^-	0.162	0.187	–	–	0.034	0.029	0.094	0.030	–	0.041	0.032	0.041	0.051	0.054
F^-	0.709	0.837	–	–	1.272	1.359	1.827	1.606	–	0.875	1.583	0.674	0.626	0.65
(Fe+Mn)/(Fe+Mn+Mg)	67.050	67.755	69.039	39.429	37.858	33.1755	37.389	36.368	34.999	36.536	35.339	19.276	21.330	24.2411

[a] CZ—core zones of the KFsp-megacrysts, recovered from the drilled cores. GM—groundmass. MC—K-feldspar megacrysts, analyzed in thin sections. BR—bulk rock, no specified position in the texture.

[b] Apparent deficiency, 8−(Si+Al).

6.1. Amphibole

At least two different generations of amphibole, both coexisting with titanite, biotite, quartz, and sometimes clinopyroxene, are found as inclusions in the K-feldspar megacrysts. Amphibole in the matrix is present as rare inclusions in mica. Compositionally, the generations differ significantly from each other (Fig. 7a and b). The first (amphibole-I) is found as small (20 to 40 μm) bluish green subhedral grains in the core zones of K-feldspar megacrysts and is characterised by a very high Fe index [Fe#; molar $Fe/(Fe+Mg) \times 100$] of ~90 and high Al content of about 2.2 a.f.u. (Fig. 7). In terms of overall chemistry, this generation is ferrohastingsite. The second (amphibole-II) comprises green euhedral grains up to 0.1 to 0.4 mm in diameter, varying between magnesiohastingsitic and tschermakitic hornblende.

Amphibole-II has lower Fe# (44 to 65) and Al-content (1.5 to 2.0 a.f.u.) than amphibole-I.

Amphibole-II from K-feldspar megacrysts in the granite is compositionally similar to the most Fe- and Al-rich amphibole in the lamprophyres (see Fig. 7a,b; Table 2). Although the composition of the early formed hornblende in lamprophyres and amphibole-II in K-feldspar megacrysts overlap, hornblende from lamprophyres exhibits substantially broader variation, ranging from tschermakitic hornblende (~1.5 a.f.u. Al_{IV}, Fe# 50) through magnesio-hornblende to more silicic, Mg-rich, late magmatic, tremolitic hornblende (0.3 to 1.1 a.f.u. Al_{tot}).

The major substitution mechanisms seem to be common for the two groups of amphibole from the Åva granites, as well as for the hornblende in the lamprophyres and the mafic cumulates. Chemical variations are mainly controlled by the Al- (and

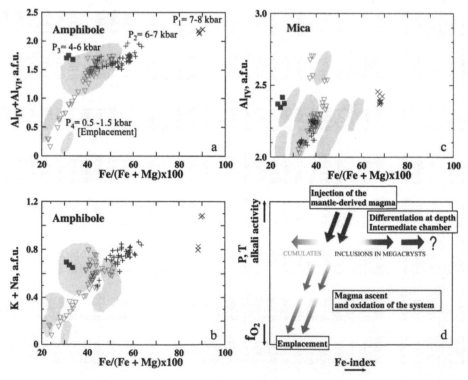

Fig. 7. (a–b) Amphibole and (c) mica mineral chemistry from rock types in the Åva ring complex. Key to symbols: solid blue squares—mafic cumulates; inclined red crosses—granites, separates from core zones of K-feldspar; red crosses—granites, K-feldspar megacryst core zones and the matrix; green inverted triangles—lamprophyres (data from Rutanen et al., 1997). Grey areas represent analytical results from several other shoshonitic postcollisional intrusions in southern Finland and Russian Karelia (Rutanen et al., 1997). (d) A sketch illustrating the magmatic evolution of the Åva ring complex based on mineral chemistry. See text for further explanation.

possibly in part Fe^{3+}) Tschermak substitution (1) combined with incorporation of alkalies via edenite-hastingsite component (and/or its K-analogue) $Na_A Al_T$ (2):

$$R_{VI}^{2+} + Si \Leftrightarrow Al_{VI} + Al_{IV} \qquad (1)$$

$$R^+ + Al_{VI} \Leftrightarrow Si + \square \qquad (2)$$

where $\square$ denotes vacancy in A-site; R^+ are Na and K. As the latter substitution has been calibrated as a thermobarometer (Blundy and Holland, 1990, 1992), we utilized it to assess the intensive parameters of formation of the amphibole-bearing assemblages (Fig. 8).

Amphibole-I is intergrown with a fairly calcic plagioclase ($An_{20-41}Or_{0-3}$) and forms an assemblage that yields pressures of 7 to 8 kbar and temperatures of 750 to 840 °C. The same pressures were obtained using Al-in-Hbl calibration by Schmidt (1992). Amphibole-II from the Åva granite coexists with a wide range of plagioclase compositions ($An_{08-42}Or_{0-2}$, average $An_{30-34}Or_1$). More Fe-rich varieties (Fe# 62 to 65) were formed under higher pressure (6 to 7 kbar) and temperatures ranging from 840 to 740 °C, while Fe-poor varieties (Fe# 45 to 50) yield lower pressures (4 to 5 kbar) and temperatures between 740 and 760 °C (Fig. 7). The same pressures were obtained for early formed Fe-rich varieties of amphibole in the lamprophyres and for Mg-rich hornblende from the

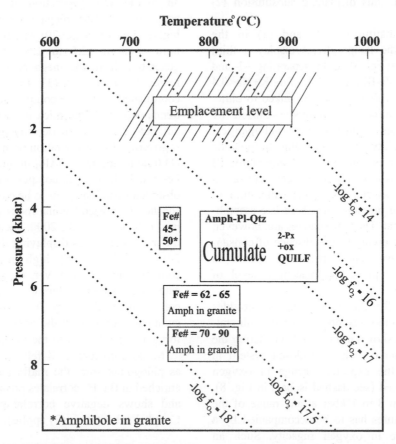

Fig. 8. Pressure versus temperature diagram summarising our thermobarometrical assessment for the Åva ring complex. Boxes correspond to approximate temperatures and pressures obtained for different amphibole-bearing assemblages in the Åva granite and mafic cumulates; dashed area shows parameters obtained for late-formed assemblages in the lamprophyres. Fe index values are for amphibole. Oxygen fugacities for the average amphibole-biotite-bearing granite are shown as dotted lines (according to calculations by Perchuk and Fed'kin (1976). Note that the rise of the magma is accompanied by significant oxidation of the system. Abbreviations: Amph—amphibole; Pl—plagioclase; Qtz—quartz; 2-Px—clinopyroxene and orthopyroxene; ox—Fe-Ti oxides.

mafic cumulate. This rather high-pressure, high-temperature regime is in accordance with elevated Na- and K-content in the amphibole from the megacrysts and earlier Fe-rich hornblende in the lamprophyres (Krylova et al., 1991; Deer et al., 1997). It is worth noting that the plots shown in Fig. 7a and b are fairly similar to each other, although the former reflects hornblende compositions adjusted to pressure and temperature change [mainly pressure, as it follows from the total Al-content (Blundy and Holland, 1990, 1992; Anderson, 1996)], while the latter depicts variation in alkali activity. The high chemical potential of alkali elements enhances the incorporation of alkalis (K in particular) in hornblende at elevated P-T conditions (Perchuk and Fed'kin, 1976; Helz, 1979), thus displacing substitution (2) towards $R^+ + Al_{VI}$.

Both major substitutions (1) and (2) in the amphiboles from the lamprophyres display a third compositional gap at Fe# 35 to 38, separating Al- and alkali-poor amphibole-III ($Al_{tot} < 1.1$ a.f.u., $K+Na < 0.4$ a.f.u.). The nature of this gap can be linked to either break in crystallisation of amphibole caused by volatile or ascent rate fluctuation or crystal chemical factors (e.g., miscibility gap for tremolite–hornblende series). If the latter is true, then the amphibole-III series represents frozen stages of recrystallisation during the magma ascent. Using the calibration of Schmidt (1992), the pressures for these late assemblages are 0.5 to 2 kbar. The temperatures are, however, difficult to estimate owing to the absence of consanguineous plagioclase. These pressures appear to correspond to the supposed emplacement level of the complex.

The P-T path for the Åva complex provides insights into the factors responsible for the mineral-chemical variations shown in Fig. 7a. Considering the evolution of the initial melt as a closed system in respect to water, the expected change in oxygen fugacity can be traced (see dashed isobars in Fig. 8). Magma ascent from 8 to 1 kbar at any range of the calculated temperatures has to be accompanied by a significant increase in oxygen fugacity. Such an increase appears to facilitate precipitation of Fe-Ti oxides and results in a relative enrichment of Mg in the residual melt. This, in turn, leads to a decrease in Fe-content in the crystallising hydrous silicates (cf. Fig. 7).

Considering substitution mechanism (1), it is important to notice that the content of Fe^{2+} correlates positively with Al_{tot}, while Mg shows a strong negative correlation. The same is also true for all micas. This confirms that the variations in composition of the late stage amphiboles described above were caused by oxidation during ascent rather than by crystallisation at a more shallow level at constant f_{O2} (Ague, 1989).

6.2. Mica

Representative analyses of micas in the Åva complex are given in Table 3. Mica in Åva shows compositional trends similar to those displayed by amphibole. One generation of mica from the core zones of K-feldspar megacrysts in the granite has the highest Fe# (average~70) and relatively high aluminum contents (2.5 to 2.6 a.f.u.). Another mica generation from the megacrysts and the matrix are lower in their Fe# (33 to 40) and tetrahedral-coordinated Al. These compositions overlap with the trends shown by the mica in the lamprophyres (Fig. 7c). Fe# decreases with decreasing total Al content and is governed by the operation of modifying substitution (1) (due to absence of Al_{VI} in micas, incorporation of Fe- and Ti-Tschermak components is presumed). The observed variation seems to be common for mica in all shoshonitic postcollisional rocks in southwestern Finland (Shebanov et al., 2000; see Fig. 7c). Mica from the mafic cumulates displaying the lowest Fe# (19 to 24) are still relatively high in Al_{IV} content. Their compositions are, however, rather uniform and indicate that oxidation during magma ascent did not control their chemistry. This implies that mica in cumulates, along with the coexisting amphibole, was formed at depth prior to the magma ascent.

Micas from cores of K-feldspar megacrysts, as well as phlogopite from the mafic cumulate, are slightly enriched in Ba. Ba correlates positively with Al and Ti and shows negative correlation with K and Si. Coupled substitution (3) applies,

$$Ba_{XII}^{2+} + R_{IV}^{3+} \Leftrightarrow K_{XII}^+ + Si_{IV}^{4+} \qquad (3)$$

where R_{IV}^{3+} has traditionally been considered to be Al (Wendlandt, 1977; Mansker et al., 1979). This scheme is the dominating substitution mechanism for Ba-rich micas, which commonly occurs in almost all Ba- and Sr-rich postcollisional rocks in southwestern Finland (Shebanov et al., 2000). The relative timing of feldspar and mica precipitation possibly dictates the degree of Ba enrichment in the mica. In many cases, mica remains relatively Ba poor despite high bulk rock Ba content. Ba is preferably hosted by K-feldspar even in the case if it is simultaneously precipitated with mica. It is worth noting that most of the micas, both from the K-feldspar megacrysts and in the matrix of the Åva granite, are Ba poor (Table 3), while the K-feldspar itself contains up to 3% BaO. Another feature more regularly in common for micas in the Åva rocks and other Ba-Sr-rich postcollisional rocks in the region (Shebanov et al., 2000) is the "apparent" deficiency in T-site occupancy (Table 3).

Ti contents of the mica are rather high, particularly for early formed deep-seated biotite, (Table 3) indicating that they equilibrated at relatively high temperatures (Krylova et al., 1991; Tronnes et al., 1985). Crystallisation temperatures of the mica were calculated using their Ti-saturation levels at the corresponding Fc# (cf. Henry and Guidotti, 2002). Estimated temperatures for mica in the cumulate (Fe# 0.2 to 0.25, Ti 0.42 to 0.52 a.f.u.) are at the upper limit of possible determinations (i.e., 800 °C). Fe-Ti-rich biotite inclusions in the K-feldspar megacrysts and early formed Ti-rich mica in the lamprophyres yield estimates ranging from 680 to 720 °C. Late-formed mica coexisting with Fe-Ti oxides in K-feldspar megacrysts and in the matrix of the granite exhibit lower Ti contents (between 0.08 up to 2.0 a.f.u.). Correspondingly, the estimated temperatures drop to 550 °C and seem to be unreasonably low. These results must be treated with caution owing to strongly Al-undersaturated character of the micas.

Considering all the mineral series together, we can summarise that variations for the deep-seated early formed mafic hydrous silicates can be described as displaying isovalent

$$Fe^{2+} \Leftrightarrow Mg \text{ isomorphism} \tag{4}$$

(as shown by the horizontal arrows in Fig. 7d). This variation can be explained by differentiation at depth during cooling of the parental magma chamber. Mineral chemical variations for the later formed assemblages take place along the exchange vectors (1) and (2). These changes are mainly attributed to a decrease in pressure accompanied by a decrease in alkali activity and increase of oxygen fugacity during magma ascent (green inclined arrows in Fig. 7d).

6.3. Pyroxene-bearing assemblages in mafic cumulates

Subhedral crystals of clinopyroxene and orthopyroxene coexist with hydrous mafic silicates. Clinopyroxene displays some variation in Ca, Na, and Al contents; however, the proportion of the jadeite component does not exceed 4%. Compositional fluctuations within the pyroxene quadrilateral are also fairly limited: $En_{39.2-40.4}Fs_{11.0-16.1}Wo_{42.2-45.4}$. Orthopyroxene displays an even more narrow compositional variation at around $En_{61.6-64.2}Fs_{33.5-35.5}Wo_{2.9-3.1}$. Binary two-pyroxene thermometry (Lindsley, 1983) and estimates using QUILF (Lindsley and Frost, 1992; Andersen et al., 1993) yield the highest equilibrium temperatures centering at around 940 ± 60 °C (shown as shaded area for the cumulate in Fig. 8). Some of the pairs are not at equilibrium in terms of the QUILF system. Using monzonite bulk composition to model the liquidus orthopyroxene-clinopyroxene-plagioclase (all from the cumulate) equilibrium, temperature is estimated to have been at 1070–1080 °C by the combined calibrations of Drake (1976) and Nielsen and Drake (1979).

In some cases, orthopyroxene contains up to 2.6 wt.% Al_2O_3, which corresponds to ~5 to 6 kbar according to the experiments of Longhi et al. (1993). Although the composition of the parental melt for the Åva cumulates was undoubtedly different from the high Al basalts in those experiments, the estimates obtained for orthopyroxene corroborate with pressures obtained from hornblende-plagioclase-quartz thermobarometry.

7. Discussion

In their study on late Svecofennian magmatism and tectonism in southwestern Finland, Hubbard and Branigan (1987) concluded that the postcollisional ring complexes were emplaced after the cessation of the Svecofennian orogeny during a prolonged period of continuing mantle–crust interaction and repeated crustal adjustment by shearing, which ended in the rapakivi granite event. They suggested a scenario where magma, heat, and volatiles from the mantle initiated anatexis of the lower crust. This magma ascended and spread out laterally along preexisting zones of weakness at midcrustal level. This lateral layer was later activated by impounding of alkali-basaltic - lamprophyric magma that hybridised and fluidised the layer and triggered a forceful intrusive event in the upper crust (Hubbard and Branigan, 1987).

Our results support this model and add geochronologic and thermobarometric constrains to it. The lateral layer was probably formed at pressures between 5 and 7 kbar. However, it was not solely a granitic layer but rather a bimodal, presumably layered magma chamber including also mafic cumulates. The ages of the core zones of zircons from lamprophyres and granitic megacrysts overlap (1801 to 1790 Ma; Fig. 5c, d). Correspondingly, assemblages from the K-feldspars megacrysts and Fe- and Al-enriched assemblages in lamprophyres yield overlapping pressure estimates (5 to 7 kbar; Figs. 7a and 8), which are also close to the estimated depth for the mafic cumulate. Cumulation–fractionation processes resulted in the formation of a rather heterogeneous intermediate magmatic reservoir, as illustrated by the widespread variation in the Fe index of the early formed mafic silicates at similar level of saturation in Al (as depicted by horizontal arrows on Fig. 7d). Compositional overlapping of early formed mica and amphibole from the granite and lamprophyres emphasises a similarity in crystallisation conditions for these deep-seated assemblages, presumably reflecting the P-T-X conditions of a parental magma (greyish arrows in Fig. 7d).

This midcrustal layer was later reactivated by new pulses of shoshonitic magma, which fluidised the layer and caused its forceful migration to the upper crust. Field evidence and stress theory indicate that the ring intrusions were formed prior to the radial dykes. Initially, the granitic part of the midcrustal chamber intruded the ring system, while the remobilised lamprophyric material subsequently filled the radial dykes. The rise of the magmas from the midcrustal layer caused a dramatic increase in the oxygen fugacity in the system, resulting in precipitation of Fe-Ti oxides, which led to a decrease in the Fe# of the mafic silicates (greenish arrows in Fig. 7d).

The 1766±13 Ma age of the Åva monzonite is considered to be the best age estimate of its emplace-

ment in the upper crust. This age overlaps with the $^{207}Pb/^{206}Pb$ titanite ages reported by Suominen (1991): 1754 to 1789 Ma for the monzonites and 1759 to 1782 Ma for the granites. The age of the monzonite also overlaps with the conventional age (1770±2 Ma; Suominen, 1991) of the neighbouring 150 km^2 Lemland intrusion (Fig. 1; Lindberg and Eklund, 1988). In addition to shoshonitic mafic magmatic enclaves, the Lemland intrusion carries mafic autoliths with mineralogical composition similar to the mafic boulders in Åva (Rutanen et al., 1997; Eklund, 2000). Data from Rutanen et al. (1997) imply that the crystallisation pressure for the autoliths was 5 to 7 kbar, the same as that for the deep-seated assemblages in the Åva complex.

Väisänen et al. (2000) reported areas of bimodal magmatism near Turku, east of our study area (Fig. 1), where shoshonitic rocks have mingled and mixed with palingenetic S-type granites at 1815 Ma and register an emplacement pressure of 4.5 kbar. This may give indication of how the midcrustal bimodal magma chamber was initially formed: shoshonitic magma invaded midcrustal levels, initiated melting of the country rock, and developed a zoned magma chamber. The formation of zoned magma chambers by invasion of mafic magmas into the crust with subsequent partial melting has been described by Huppert and Sparks (1988).

These issues raise further questions: how extensive and how long was the invasion of shoshonitic magmas during the Svecofennian postcollisional period? Can it be traced in other areas and in later magmatic events? Here are some aspects on these questions.

- The Obbnäs granite in southern Finland (Kosunen, 1999) has a composition transitional between post- and anorogenic granites. Associated dykes south of the granite intrusion have shoshonitic affinities (unpublished data by the authors). The U-Pb zircon age from Obbnäs is 1645±5 Ma (Vaasjoki, 1977). However, Heeremans (1997) reported two ~1780 Ma Ar-Ar ages from K-feldspar megacrysts.
- The mafic rocks of the 1635±7 Ma Abja intrusion in Estonia are shoshonitic in composition (Kirs and Petersell, 1994).
- Mineralogically, inherited minerals from the core zones of K-feldspar megacrysts from the Järppilä

quartz-feldspar porphyry dyke in the Vehmaa rapakivi batholith in southwestern Finland can be correlated to a shoshonitic event. The inherited assemblage contains barian Ti-rich micas with fairly low iron index (0.58) and has low $Sr_i=0.7018\pm0.0004$ (mineral isochron) and unusually high ϵ_{Nd} (at 1.58 Ga) values of 0 to +2.0 (Shebanov et al., 2000). U-Pb zircon ages of the inherited component in the K-feldspar ovoid from Järppilä is 1616 ± 7 Ma while the intrusion age of the anorogenic pluton is 1579 ± 6 Ma. (Shebanov et al., 2000).

These examples suggest that shoshonitic material from an enriched lithospheric mantle invaded the Svecofennian crust over a long period, before the mafic magmatism switched character to intraplate tholeiitic magmatism ~1600 Ma ago.

8. Conclusions

U-Pb geochronology of rim zones in zircons from several rock types of the Åva ring complex shows that the complex was emplaced in the upper crust at 1770 to 1760 Ma. The overlapping ages of the different rock facies are supported by field evidence for mingling and mixing of the crystal-saturated granitic magma and shoshonitic monzonitic magma, in places also lamprophyre. Before their emplacement in the upper crust, the rocks resided in a midcrustal magma chamber at depths of 18 to 25 km (5 to 7 kbar). This is indicated by thermobarometric estimates from amphibole-bearing assemblages within K-feldspar megacrysts of the granite. Also, data from the most Al-rich amphiboles from the lamprophyres and mafic cumulate indicate emplacement at midcrustal levels. These early formed assemblages presumably grew in the deep-seated, partly mixed magma chamber at 1800 to 1790 Ma. The most Al-poor and Mg-rich amphibole of the lamprophyres records an emplacement level of 1 to 2 kbar.

Acknowledgements

Mineral separations were made by Rafael Jimenez Garrido and Paqui Garcia Garcia (Granada) during an EU employee exchange in Turku. SIMS work was performed as part of the NORDSIM project "The origin of rapakivi granites of Fennoscanian shield" (Alexey Shebanov et al.). Dr. Karin Högdahl is acknowledged for her supervision at the SIMS laboratory at the Museum of Natural History, Stockholm. Hans Harysson assisted us during electron-microprobe sessions at the Department of Earth Sciences, Uppsala University. Paula Kosunen (University of Helsinki) provided us with a sample of a dyke from Obbnäs. Jeremy Woodard is acknowledged for correcting the language. This work was partially financed by the NorFA network "Transition from Orogenic to Anorogenic Magmatism in the Fennoscandian Shield" and was also supported by the NorFA mobility scholarship #020138 (Alexey Shebanov). We gladly acknowledge Dr. Mikko Nironen and Dr. Brent Elliott for their reviews. We appreciate Sören Fröjdö and Dr. Ulf B. Andersson for comments on the first draft of the manuscript. The Lithos Guest Editor Prof. Tapani Rämö is acknowledged for his constructive scientific and editorial comments on the manuscript.

References

Ague, J.J., 1989. The distribution of Fe and Mg between biotite and amphibole in granitic rocks: effects of temperature, pressure and amphibole composition. Geochemical Journal 23, 279–293.

Amelin, Yu.V., Larin, A.M., Robert, D.T., 1997. Chronology of multiphase emplacement of the Salmi rapakivi granite-anorthosite complex, baltic shield: implications for magmatic evolution. Contributions to Mineralogy and Petrology 127, 353–368.

Andersen, D.J., Lindsley, D.H., Davidson, P.M., 1993. QUILF: a Pascal program to assess equilibria among Fe-Mg-Mn-Ti oxides, pyroxenes, olivine, and quartz. Computers & Geosciences 19, 1333–1350.

Anderson, J.L., 1996. Status of thermobarometry in granitic batholiths. Transactions of the Royal Society in Edinburgh. Earth Sciences 87, 125–138.

Batchelor, R.A., Bowden, P., 1985. Petrogenetic interpretation of granitoid rock series using multicationic parameters. Chemical Geology 48, 43–55.

Bergman, L., 1986. Structure and mechanism of intrusion of postorogenic granites in the archipelago of southwestern Finland. Acta Academiae Aboensis. Ser. B Mathematica et Physica 46 (5), 1–74.

Blundy, J.T., Holland, J.B., 1990. Calcic amphibole equilibria and a new amphibole-plagioclase geothermometer. Contributions to Mineralogy and Petrology 104, 208–224.

Blundy, J.T., Holland, J.B., 1992. "Calcic amphibole equilibria and a new amphibole-plagioclase geothermometer": reply to the comments of Hammarstrom and Zen, and Rutherford and Johnson. Contributions to Mineralogy and Petrology 111, 269–272.

Branigan, N.P., 1989a. Internal deformation, flow profiles and emplacement velocities of granitic dykes, southwestern Finland. Lithos 22, 199–211.

Branigan, N.P., 1989b. Hybridisation in the middle Proetrozoic high-level ring complexes, Åland SW Finland. Precambrian Research 45, 83–95.

Deer, W.A., Howie, R.A., Zussman, J., 1997. Rock-forming minerals. Double-Chain Silicates, vol. 2 B. Geological Society.

de la Roche, H., Leterrier, J., Grand Claude, P., Marchal, M., 1980. A classification of volcanic and plutonic rocks using R1-R2 diagrams and major element analyses—its relationship with current nomenclature. Chemical Geology 29, 183–210.

Dostal, J., Capedri, S., 1975. Partition coefficients of uranium for some rock-forming minerals. Chemical Geology 15, 285–294.

Drake, M.J., 1976. Plagiclase-melt equilibria. Geochimica et Cosmochimica Acta 40, 457–465.

Ehlers, C., Bergman, L., 1984. Structure and mechanism of two postorogenic granitic massifs, southwestern Finland. In: Kröner, A., Greiling, R. (Eds.), Precambrian Tectonics Illustrated. Schwitzerbart, Stuttgart, pp. 173–190.

Eklund, O., 2000. A NorFA sponsored Nordic Network on the transition from orogenic to anorogenic magmatism in the Fennoscandian shield. The Archipelago of SW Finland and the Turku Granulite Area, Excursion Guide, Geocenter Tiedottaa—Geocenter Informerar Report, vol. 17. Turku University-Åbo Akademi University.

Eklund, O., Shebanov, A.D., 1999. Origin of the rapakivi texture by subisothermal decompression. Precambrian Research 95, 129–146.

Eklund, O., Konopelko, D., Rutanen, H., Fröjdö, S., Shebanov, A.D., 1998. 1.8 Ga Svecofennian post-collisional shoshonitic magmatism in the Fennoscandian Shield. Lithos 45, 87–108.

Heeremans, M., 1997. Silicic magmatism and continental lithospheric thinning. Netherlands Research School of Sedimentary Geology (NGS) publication no. 970122.

Helz, R.T., 1979. Alkali exchange between hornblende and melt: a temperature-sensitive reaction. American Mineralogist 64, 953–965.

Henry, D.J., Guidotti, C.V., 2002. Titanium in biotite from metapelitic rocks: temperature effects, crystal-chemical controls, and petrologic applications. American Mineralogist 87, 375–382.

Hubbard, F., Branigan, N., 1987. Late Svecofennian magmatism and tectonism, Åland, Southwest Finland. Precambrian Research 35, 241–256.

Huppert, H.E., Sparks, S.J., 1988. The generation of granitic magmas by intrusion of basalt into continental crust. Journal of Petrology 29, 599–624.

Kaitaro, S., 1953. Geologic structure of the late pre-Cambrian intrusives in the Åva area, Åland Islands. Bulletin de la Commision Géologique de Finlande 62, 1–69.

Kirs, J., Petersell, V., 1994. Age and geochemical character of plagiomicrocline granite veins in the Abja gabbro-dioritic massif. Acta et Commentationes Universitatis Tartuensis 956, 3–14.

Korsman, K., Korja, T., Pajunen, M., Virransalo, P., 1999. The GGT/SVEKA Transect: structure and evolution of the continental crustin the Paleoproterozoic Svecofennian orogen in Finland. International Geological Review 41, 287–333.

Kosunen, P., 1999. The rapakivi granite plutons of Bodom and Obbnäs, southern Finland: petrography and geochemistry. Bulletin of the Geological Society of Finland 71, 275–304.

Krasnobaev, A.A., 1986. Zircon as Indicator of Geological Processes. Nauka, Moscow. in Russian.

Krylova, M.D., Galibin, V.A., Krylov, D.P., 1991. Major Mafic Minerals of the High-Metamorphism Complexes. Nedra, Leningrad. in Russian.

Liégeois, J.-P., Navez, J., Hertogen, J., Black, R., 1998. Contrasting origin of post-collisional high-K calc-alkaline and shoshonitic versus alkaline and peralkaline granitoids. The use of sliding normalization. Lithos 45, 1–28.

Lindberg, B., Eklund, O., 1988. Interaction between basaltic and granitic magmas in a Svecofennian postorogenic granitoid intrusion, Åland, southwest Finland. Lithos 22, 13–23.

Lindsley, D.H., 1983. Pyroxene thermometry. American Mineralogist 68, 477–493.

Lindsley, D.H., Frost, B.R., 1992. Equilibria among Fe-Ti oxides, pyroxenes, olivine and quartz: Part I. Theory. American Mineralogist 77, 987–1003.

Longhi, D., Fram, M.S., Vander Auwera, J., Montieth, J.N., 1993. Pressure effects, kinetics, and rheology of anorthositic and related magmas. American Mineralogist 78, 1016–1030.

Ludwig, K.R., 1998. Isoplot/Ex Version 1.00. Berkley Geochronological Centre, Special Publication, vol. 1.

Ludwig, K.R., 1999. Isoplot/Ex version 2.00. A geochronological Toolkit for Microsoft Excel. Berkeley Geochronological Center, Special, vol. 2.

Mansker, W.L., Ewing, R.C., Keil, K., 1979. Barian-titanian biotites in nephelinites from Oahu, Hawaii. American Mineralogist 64, 156–159.

Nash, W.P., Crecraft, H.R., 1985. Partitioning coefficients for trace elements in silicic magmas. Geochimica et Cosmochimica Acta 49, 2309–2322.

Nielsen, R.L., Drake, M.J., 1979. Pyroxene-melt equilibria. Geochimica et Cosmochimica Acta 43, 1259–1279.

Niiranen, T., 2000. Svekofennisen orogenian jälkeinen ekshumaatio ja isostaattinen tasapainottuminen Kaakkois-Suomessa. M.Sc. Thesis, Department of Geology, University of Turku, Finland (in Finnish).

Peccerillo, A., Taylor, S.R., 1976. Geochemistry of Eocene calc-alkaline volcanic rocks from the Kastomonon area, northern Turkey. Contributions to Mineralogy and Petrology 58, 63–81.

Perchuk, L.L., Fed'kin, V.V., 1976. Thermal and gas regime of the formation of granitoids. In: Krats, K.O., Glebovitsky, V.A. (Eds.), Thermodynamic Regime of Metamorphism. Nedra, Leningrad, pp. 97–105. in Russian.

Rock, N.M.S., 1990. Lamprophyres. Van Nostrand Reinhold, New York.

Rutanen, H., 2001. Geochemistry and petrogenesis of the 1.8 Ga old post-collisional intrusions in southern Finland and Russian Karelia. Lic.Phil. Thesis, Department of Geology and Mineralogy, Åbo Akademi University, Finland.

Rutanen, H., Eklund, O., Konopelko, D., 1997. Rock and mineral analyses of Svecofennian postorogenic 1.8 Ga intrusions in southern Finland and Russian Karelia. Geocenter Raportti—Geocenter Rapport, vol. 15. Turku University-Åbo Akademi University.

Schmidt, M.W., 1992. Amphibole composition in tonalite as a function of pressure: an experimental calibration of the Al-in-hornblende barometer. Contributions to Mineralogy and Petrology 110, 304–310.

Shebanov, A., Savatenkov, V., Eklund, O., Andersson, U.B., Annersten, H., Claesson, S., 2000. Regional mineralogical correlation linking post- and anorogenic magmatic events from unusual barian biotites in the Järppilä rapakivi porphyries, Vehmaa batolith (S. Finland). Meeting on Advances on Micas, Rome 2-3.11 2000. Università degli Studi Roma Tre MURST—Progetto Fillosilicati: aspetti cristallochimici strutturali e petrologici, pp. 183–186.

Skyttä, P., 2002. Emplacement of the Åva ring intrusion: some structural and magmatic aspects. M.Sc. Thesis, Department of Geology, University of Turku, Finland.

Suominen, V., 1991. The chronostratigraphy of southwestern Finland with special reference to Postjotnian and Subjotnian diabases. Geological Survey of Finland, 356.

Tronnes, R.G., Edgar, A.D., Arima, M., 1985. A high pressure–high temperature study of TiO2 solubility in Mg-rich phlogopite: implications to phlogopite chemistry. Geochimica et Cosmochimica Acta 49, 2323–2329.

Vaasjoki, M., 1977. Rapakivi granites and other postorogenic rocks in Finland: their age and the lead isotopic composition of certain associated galena mineralozations. Geological Survey of Finland, 294.

Väisänen, M., Hölttä, P., 1999. Structural and metamorphic evolution of the Turku migmatite complex, southwestern Finland. Bulletin of the Geological Society of Finland 71, 177–218.

Väisänen, M., Mänttäri, I., 2002. 1.90–1.88 Ga arc and back-arc basin in the Orijärvi area, SW Finland. Bulletin of the Geological Society of Finland 74, 185–214.

Väisänen, M., Mänttäri, I., Kriegsman, L.M., Hölttä, P., 2000. Tectonic setting of post-collisional magmatism in the Paleoproterozoic Svecofennian orogen, SW Finland. Lithos 54, 63–81.

Wendlandt, R.F., 1977. Barium-phlogopite from haystack butte, Highwood Mountains, Montana. Year Book-Carnegie Institution of Washington 76, 534–539.

Whitehouse, M.J., Claesson, S., Sunde, T., Vestin, J., 1997. Ion microprobe U-Pb zircon geochronology and correlation of Archaean gneisses from the lewisian complex of gruinard bay, northwestern Scotland. Geochimica et Cosmochimica Acta 61, 4429–4438.

Wiedenbeck, M., Alle, P., Corfu, F., Griffin, W., Meier, M., Oberli, F., von Quadt, A., Roddick, J.C., Speigel, W., 1995. Three natural zircon standards for U-Th-Pb, Lu-Hf, trace element and REE analysis. Geostandard Newsletter 19, 1–23.

Zeck, H.P., Whitehouse, M.J., 1999. Hercynian, pan-African, Proterozoic and Archaean ion-microprobe zircon ages for a betic-rif core complex, alpine belt, W Mediterranean—consequences for its P-T-t path. Contributions to Mineralogy and Petrology 134, 134–149.

Available online at www.sciencedirect.com

ELSEVIER

Lithos 80 (2005) 249–266

www.elsevier.com/locate/lithos

Origin of chemically distinct granites in a composite intrusion in east-central Sweden: geochemical and geothermal constraints

Anders Lindh*

GeoBiosphere Science Centre II, Department of Geology, Sölvegatan 12, SE-223 62 Lund, Sweden

Received 31 March 2003; accepted 9 September 2004
Available online 10 November 2004

Abstract

A few tens of millions of years after the intrusion of the Early Svecofennian (~1.87–1.85 Ma) granitoids in central Sweden, a renewed magmatic activity resulted in the emplacement of the Late Svecofennian granites, the tectonic setting of which remains obscure. S-type granites dominate this group, but both I-type and transitional granites are common. This study deals with one of these intrusions in east-central Sweden; a composite pluton that is insignificantly deformed and hosts both I- and S-type granites. One of the I-type granites shows a compositional trend from granodiorite to granite, which is uncommon among the Late Svecofennian granites. Major element and incompatible trace element compositions and ϵ_{Nd} data show that two different sources, one igneous and one sedimentary, were involved. An important conclusion is that nearly coeval granites derived from different sources are found in close connection. The granites are suggested to have formed by partial melting in a thickened continental crust that was formed in an early stage of the Svecofennian event. Thermal models suggest that the slightly older, high-temperature I-type granite (granodiorite) was formed deeper in the crust than the S-type granite. The coexistence of essentially pure I- and S-type granites, rather than transitional mixtures, reflects the relative depths of the proposed sources and the varying thermal parameters of the lithologic units in the Svecofennian crust.
© 2004 Elsevier B.V. All rights reserved.

Keywords: Granite; I-type; S-type; Geochemistry; Magma modelling; Thermal modelling

1. Introduction

The Palaeoproterozoic Svecofennian orogen makes up the central part of the Fennoscandian shield (cf. Fig. 1). In the west, slightly younger but still Palaeoproterozoic rocks of the Southwest Scandinavian domain are found, and in the east, Archaean rocks are exposed. The Southwest Scandinavian rocks were metamorphosed in the Sveconorwegian orogeny at ~1.1–0.95 Ga. There is no general consensus of how and when these crustal units accreted to the preexisting part of the Fennoscandian shield. This uncertainty, however, does not affect the conclusions of this study.

---

* Tel.: +46 46 2227876; fax: +46 46 2224419.
E-mail address: anders.lindh@geol.lu.se.

doi:10.1016/j.lithos.2004.06.013

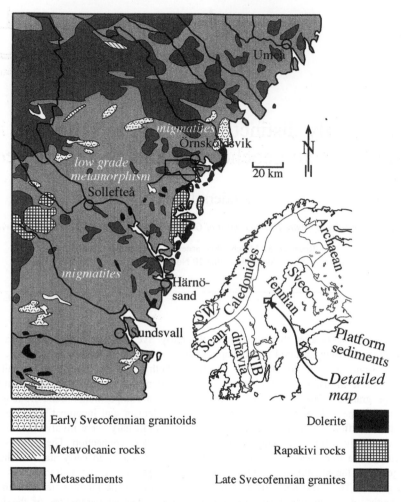

Fig. 1. Map of east-central Sweden showing the central part of the Bothnian Basin with the sampling area indicated at the boundary between low-grade metamorphism and migmatites. The rectangle in the principal map outlines Fig. 2. The word "migmatites" refers only to the metasediments; the younger granites are not metamorphosed. Simplified after Lundqvist et al. (1990). The inset shows the position of three domains of the Fennoscandian shield: Svecofennian, ~1.89–1.82 Ga; Transscandinavian Igneous Belt (TIB), ~1.81–1.66 Ga; SW Scandinavia: major periods of granitoid intrusions at ~1.7, ~1.6, ~1.4, ~1.3 and ~0.9 Ga, finally metamorphosed in the Sveconorwegian orogeny (~1 Ga).

The granites discussed here are located in the Bothnian Basin, which constitutes the central part of the Svecofennian orogen in Sweden. It is dominated by metaturbidites with depositional ages between >1.95 and 1.87 Ga (Lundqvist et al., 1990; Claesson et al., 1993; Wasström, 1993; Welin et al., 1993). In the central part of the basin, Svecofennian granitic rocks intrude metaturbidites and have been traditionally split into an early (synkinematic) and a late (postkinematic) group (cf. Lundqvist et al., 1990). In the east-central part of the orogen, the early granites have been dated at 1.89–1.85 Ga, the late

granites at 1.82–1.80 Ga. They are followed by post-orogenic (1.80–1.77 Ga) granites and, finally, by the anorogenic, rapakivi-type granites (1.58–1.50 Ga). The Early Svecofennian granitoids consist of calc-alkaline I-type granitoids comprising dominant granodiorite and more sparse granite, quartz diorite and tonalite. Granite almost exclusively makes up the late group, which is here referred to as the Late Svecofennian granites; small amounts of granodiorite are also found. This late group includes small rounded plutons of typical S-type granite that has intruded the metasediments.

Claesson and Lundqvist (1995) discussed the source rocks of the Late Svecofennian granites of the Bothnian Basin, including rocks from the present study area. According to them, the Bothnian Basin metasediments were an important, but not the only, source of the Late Svecofennian granites; also the Early Svecofennian granitoids may have contributed. Claesson and Lundqvist studied rocks from a relatively wide area. Many of these granites show a faint foliation, which may locally be even stronger than that found in the Early Svecofennian granitoids. A faint foliation is also observed in the restricted area of this study.

Sampling for the present study was limited to the area south and southwest of Örnsköldsvik (Figs. 1–3). In this area, Lundqvist et al. (1990) distinguished a reddish and greyish, sometimes muscovite-bearing granite (both even-grained and porphyritic varieties) from a granitoid with rectangular plagioclase and some microcline megacrysts 2–3 cm in diameter. Lundqvist et al. (1990) distinguished a granodioritic variety of the latter granite (cf. Fig. 2). In this contribution, this

granite is referred to as the Själevad granite. Part of the area was remapped in 2003 using criteria slightly different from those of Lundqvist et al. (1990) and the resultant map is shown in Fig. 3. Besides the Själevad granite, two additional granites, Härnö and Bergom were distinguished. Claesson and Lundqvist (1995) used Härnö granite for all Late Svecofennian granites. Deliberately, they excluded the Själevad granite from their discussion, as it appeared too different from the others (Thomas Lundqvist, oral communication, 2003).

Lundqvist et al. (1990) considered the Själevad granite as an early member of the Late Svecofennian group. They based their conclusion on the essentially undeformed character of the granite. In this area, the Early Svecofennian granitoids are strongly deformed. This study presents geochemical criteria that help to resolve the identity of the granites and granodiorites in the Bothnian Basin, focusing on a relatively small area to study relations of different granite types in detail. In particular, the present study addresses the question of why chemically different and spatially

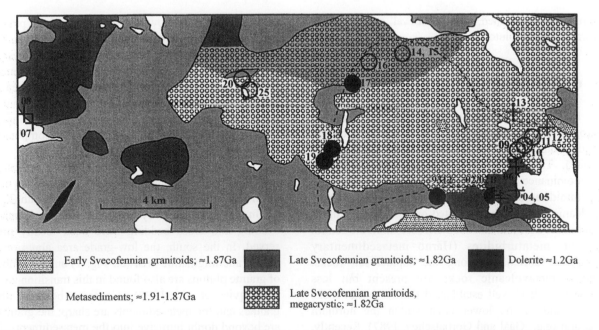

| | Early Svecofennian granitoids; ≈1.87Ga | | Late Svecofennian granitoids; ≈1.82Ga | | Dolerite ≈1.2Ga |
| | Metasediments; ≈1.91-1.87Ga | | Late Svecofennian granitoids, megacrystic; ≈1.82Ga | | |

Fig. 2. Map showing the local geology and sampling points for this study: Själevad granite (○), Bergom granite (●), Härnö granite (+). The shaded part of the megacrystic Late Svecofennian granites has a granodioritic composition (cf. Lundqvist et al., 1990). The dashed line displays the area mapped in 2003 (see Fig. 3). Modified after Lundqvist et al. (1990) who did not discriminate between the Själevad, Bergom and Härnö granites as in this paper.

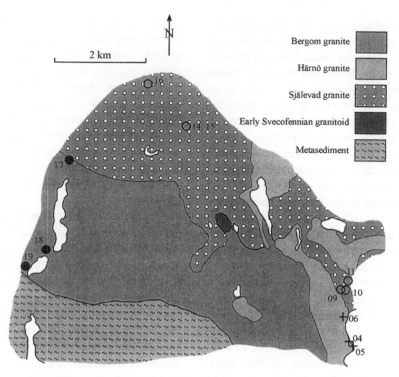

Fig. 3. Map of the eastern part of the study area showing the distribution of the three granites and some sampling localities. Själevad granite (○), Bergom granite (●), Härnö granite (+). The map forms part of an undergraduate thesis by Paula Lindgren (Lund University).

intimately associated, almost coeval granites were formed in one pluton.

2. Geologic setting

Based on supracrustal rocks, the Svecofennian orogen may be subdivided into a northern, central and southern section (cf. Gaál and Gorbatschev, 1987). This subdivision also coincides with the predominance of either Early or Late Svecofennian granitoids among the associated plutonic rocks. Felsic volcanic rocks with associated ore bodies are important in the northern and southern sectors. In the central sector, metaturbidites (Härnö metasedimentary gneisses) dominate among the supracrustal rocks. Mafic metavolcanic rocks are present but less important. It is well established that Archaean rocks are found in the lower crust within the northern section (e.g., Gaál and Gorbatschev, 1987). Recently, Andersson (1997), Claesson et al. (1997), Persson (1997), Lindh (2001), Lindh et al. (2001) and Andersson et al. (2002) argued that Archaean rocks

are also found in the lowermost crust in at least part of the central section, where they constitute one source component for rapakivi rocks. Lindh (2001) suggested that the Archaean component decreases in importance from the west to the east. The present study area is close to the Nordingrå rapakivi granite in the southeast (Fig. 1). Consequently, Archaean rocks cannot a priori be eliminated as a possible source component for the granites under discussion.

The sampled granite plutons belong to a series of intrusions that separate migmatised gneisses in the north from rather low-grade gneisses in the south (Fig. 1). In the latter, primary structures such as graded bedding and cross bedding have been locally preserved. In the south, the low-grade area abuts on a high-grade migmatitic area (cf. Fig. 1) and a number of granite plutons are also found in this transition zone (Lundqvist et al., 1990). Contacts between the granites and the metasediments are sharp; the granites are beyond doubt intrusive into the metasediments. In addition to metasediments, granitoids belonging to the Early Svecofennian group occur sparsely in the area (cf. Figs. 1–3). Besides the 1.58 Ga Nordingrå rapakivi complex, rocks younger than the Late Svecofennian granite group include a dolerite lopolith (~1.26 Ga; Fig. 1) and occasional dolerite dykes.

3. Petrologic and field description

The three studied granites are associated with abundant pegmatite, the Bergom granite in particular. The presence of pegmatite is also a clear difference between the Själevad granite and the Early Svecofennian granites, giving support to the suggestion of Lundqvist et al. (1990) that the Själevad granite is not part of the Early Svecofennian suite.

3.1. Härnö granite

The Härnö granite is a slightly reddish, medium-grained rock with a few percent of muscovite. It is either unfoliated or shows a faint orientation of minerals, especially of ~1-cm-long perthitic microcline megacrysts. The megacrysts are normally composed of a few crystals. Compared to microcline, plagioclase is subordinate and smaller in grain size. Reddish-brown biotite is found in small amounts with no discernable orientation and typical are also rather large crystals of bended or twisted muscovite. Accessory minerals include apatite, zircon and opaque minerals. From other localities, Lundqvist et al. (1990) and Claesson and Lundqvist (1995) reported a number of additional Al-rich accessory phases such as sillimanite, almandine garnet and cordierite. A few enclaves of metaturbidite are found in the Härnö granite.

3.2. Själevad granite

The Själevad granite has a medium-grained grey matrix and rectangular white megacrysts of feldspar. The plagioclase/microcline ratio is higher than in the Härnö granite. Plagioclase almost always occurs as megacrysts, often together with megacrysts of microcline. Biotite is the dominant mafic mineral, while muscovite is absent except for secondary, accessory crystals in altered plagioclase. In the granodioritic varieties, green hornblende is prominent. The most important accessories are zircon, apatite and, to a lesser extent, titanite. Dykes of Härnö granite fre-

quently occur in the Själevad granite giving a clear indication of their relative ages. The Själevad granite also has abundant enclaves of the Härnö gneiss, especially in the contact zones. These vary from angular or rounded enclaves to almost totally absorbed, ghost-like, greyish shadows in the normal rock.

3.3. Bergom granite

The Bergom granite is a pegmatite-associated granite, red to grey, coarse-grained, approximately equigranular and leucocratic. Major minerals are quartz, microcline and, in smaller amounts, plagioclase. Sparse biotite is the only major mafic mineral, zircon and apatite are typical accessories. Age relations are far from clear but the Bergom granite appears younger than the Härnö granite.

4. Geochemistry

4.1. Analytical methods

The major element analyses were performed at the Geochemical Laboratory, Institute of Geology, Lund University (except for samples 8802B and 9312). After crushing, homogenization and melting to glass beads, SiO_2, TiO_2, Al_2O_3, total iron as Fe_2O_3, CaO and K_2O were determined with X-ray fluorescence spectrometry (XRF). MgO, MnO and Na_2O were analysed by atomic absorption spectrophotometry, FeO with potentiometric titration, and P_2O_5 with optical spectrophotometry. Major elements of samples 8802B and 9312 were analysed with XRF at the University of Hamburg. The results from the major-element analyses are presented in Table 1. Trace elements (Table 2) were determined with the ICP-MS method at the ACME laboratories in Vancouver, Canada.

The Nd isotope determinations were made using a Finnigan MAT 261 TIMS at the Museum of Natural Science, Stockholm using total spiking with a mixed $^{147}Sm/^{150}Nd$ spike. Concentrations and ratios were reduced assuming exponential fractionation. Neodymium was run in multidynamic mode and normalised to $^{146}Nd/^{144}Nd=0.7219$. Sm was analysed in static mode and normalised to

Table 1
Major element composition of the Härnö, Bergom and Själevad granites (in wt.%)

Sample	Härnö									Bergom												
	8802	8803	8804	8805	8806	8807	8808	8812	8813	8802B	8817	8818	8819	9312	8809	8810	8811	8814	8815	8816	8820	8825
SiO$_2$	72.93	72.91	72.72	72.79	72.25	72.74	68.86	71.05	74.13	66.77	69.42	74.35	74.26	76.30	68.28	67.75	69.80	65.18	59.33	62.54	64.00	65.84
TiO$_2$	0.29	0.28	0.30	0.27	0.23	0.21	0.34	0.26	0.17	0.20	0.33	0.20	0.17	0.08	0.56	0.59	0.48	0.66	1.09	0.87	0.82	0.66
Al$_2$O$_3$	13.83	13.99	14.04	14.06	14.50	14.50	16.26	15.04	13.96	14.50	14.88	12.87	13.00	12.10	14.54	14.54	14.20	15.66	16.80	15.87	15.51	15.04
Fe$_2$O$_3$	0.70	0.45	0.50	0.45	0.36	0.39	1.03	0.45	0.26	1.95[a]	0.17	0.38	0.09	2.05[a]	0.60	0.47	0.39	0.36	0.62	0.76	0.62	0.44
FeO	1.15	1.26	1.33	1.27	1.11	0.98	0.94	1.13	0.90		2.53	1.70	1.81		3.53	3.99	3.33	4.58	6.80	5.40	5.14	4.44
MnO	0.02	0.02	0.02	0.01	0.02	0.01	0.01	0.03	0.01	0.06	0.04	0.03	0.03	0.07	0.06	0.07	0.06	0.07	0.11	0.09	0.08	0.07
MgO	0.50	0.44	0.44	0.44	0.56	0.30	0.35	0.42	0.24	0.58	0.62	0.23	0.24	0.30	1.33	1.41	1.18	1.26	1.73	1.34	1.44	1.46
CaO	0.53	0.62	0.68	0.60	0.50	0.52	1.04	0.74	0.69	1.74	1.63	0.44	0.85	0.80	2.64	2.73	1.98	2.40	3.75	3.02	2.95	2.83
Na$_2$O	2.94	2.90	2.95	2.86	3.66	3.23	3.19	3.13	3.04	3.16	2.86	2.70	2.55	2.83	2.67	2.69	2.78	2.96	3.33	3.17	3.36	3.17
K$_2$O	5.35	5.40	5.30	5.61	5.43	5.43	6.20	6.18	5.22	7.11	6.08	5.79	5.71	3.93	4.04	3.95	3.81	5.01	4.20	4.58	3.69	4.15
P$_2$O$_5$	0.20	0.22	0.21	0.20	0.22	0.33	0.32	0.20	0.10	0.31	0.11	0.05	0.04	0.05	0.14	0.14	0.11	0.17	0.34	0.26	0.23	0.17
LOI	1.06	1.05	1.04	1.02	1.10	0.95	0.95	0.90	0.83	1.59	0.84	0.90	0.95	1.09	1.18	1.21	1.44	1.23	1.42	1.59	1.72	1.27
Total	99.50	99.54	99.53	99.58	99.94	99.59	99.49	99.53	99.55	97.97	99.51	99.64	99.70	99.60	99.57	99.54	99.56	99.54	99.52	99.49	99.56	99.54

[a] Total Fe as Fe$_2$O$_3$.

Table 2
Trace element compositions of the Härnö, Bergom and Själevad granites (in ppm)

Sample	Ba	Cs	Ga	Hf	Nb	Rb	Sr	Ta	Th	U	W	Zr	Y	La	Ce	Pr	Nd	Sm	Eu	Gd	Tb	Dy	Ho	Er	Tm	Yb	Lu
Härnö																											
8802	224	6.6	23	4.7	13	336	58	1.5	41	11	326	134	8.6	37	92	12	43	9.8	0.33	5.0	0.51	2.0	0.27	0.65	0.09	0.55	0.08
8803	169	16.1	24	4.5	14	414	47	2.0	36	6	319	121	8.4	34	83	10	39	8.9	0.28	4.4	0.48	2.0	0.28	0.69	0.09	0.56	0.09
8804	167	16.6	24	4.6	15	409	47	1.6	41	12	292	131	9.0	36	89	11	40	9.4	0.32	4.6	0.50	2.2	0.33	0.81	0.10	0.58	0.08
8805	183	12.8	24	4.1	14	414	49	1.5	37	7	296	122	8.2	31	76	10	36	7.9	0.32	4.3	0.44	1.8	0.28	0.72	0.08	0.52	0.07
8806	280	7.0	23	5.4	14	347	120	1.6	48	5	267	153	9.1	43	105	13	48	10.5	0.36	5.9	0.58	2.3	0.32	0.75	0.09	0.61	0.08
8807	126	21.0	24	3.4	17	465	42	2.2	16	11	314	84	7.6	18	44	5	21	5.8	0.20	4.0	0.47	2.1	0.27	0.54	0.07	0.28	0.05
8808	127	22.3	26	3.5	17	474	38	2.2	19	10	333	91	6.9	19	48	6	22	6.7	0.24	3.9	0.44	1.8	0.23	0.49	0.06	0.33	0.04
8812	342	9.2	24	4.5	12	360	86	2.0	39	3	304	137	9.5	38	94	12	44	8.9	0.48	4.7	0.47	2.1	0.34	0.83	0.13	0.90	0.13
8813	175	7.5	23	3.3	12	344	52	1.5	24	7	477	82	10.2	32	72	8	30	5.9	0.32	3.2	0.42	2.2	0.37	0.84	0.11	0.65	0.09
Bergom																											
8802B	686	9.7	21	11.6	16	286	82	2.7	24	12	161	367	25.6	161	335	36	124	18	0.98	9.8	1.40	6.0	0.83	1.99	0.25	1.59	0.21
8817	1774	4.0	18	8.3	10	120	171	0.8	9	3	363	295	12.8	54	103	11	41	6.2	1.37	3.6	0.47	2.3	0.46	1.36	0.19	1.16	0.20
8818	666	2.3	15	8.5	12	134	57	0.8	17	2	438	277	10.1	94	194	21	74	10.6	0.80	5.4	0.64	2.5	0.38	0.90	0.11	0.72	0.12
9312	1110	3.2	17	7.7	4.4	122	121	0.5	24	3	273	257	69	102	231	24	84	13	1.43	9.6	1.81	12.2	2.5	6.91	0.99	6.17	0.98
Själevad																											
8809	976	7.1	19	8.5	12	127	179	0.8	13	2	252	297	19.2	65	134	15	51	8.5	1.11	5.8	0.73	4.1	0.72	1.96	0.27	1.61	0.21
8810	890	5.8	19	8.0	14	127	161	1.1	11	3	287	289	23.6	51	108	12	44	8.0	1.06	5.8	0.81	4.5	0.85	2.57	0.34	2.30	0.34
8811	832	5.5	19	8.2	13	134	150	1.1	10	3	278	271	21.2	70	143	16	55	8.8	1.07	5.4	0.72	3.9	0.71	2.09	0.33	2.18	0.34
8814	1889	5.2	20	15.9	17	117	192	1.2	14	2	215	609	22.9	65	135	15	56	9.2	1.67	5.9	0.81	4.3	0.83	2.42	0.33	2.13	0.35
8815	2820	4.3	21	25.6	25	111	247	1.4	5	2	152	1046	30.0	37	8	10	41	8.2	2.14	6.2	0.89	5.1	1.04	3.28	0.47	3.22	0.50
8816	2207	5.2	21	16.1	19	112	215	1.3	11	2	204	664	26.3	63	122	13	48	8.5	1.95	6.2	0.86	5.0	0.95	2.86	0.41	2.61	0.43
8819	688	7.8	18	7.2	11	161	64	0.6	15		428	240	11.0	103	212	23	83	11.9	0.76	6.2	0.74	3.2	0.47	1.00	0.11	0.68	0.10
8820	1317	8.3	20	14.3	20	112	184	1.3	13	2	193	554	24.2	56	116	13	49	8.2	1.52	5.8	0.76	4.4	0.88	2.53	0.35	2.25	0.39
8825	1196	7.4	19	10.7	16	114	186	1.2	13	2	251	392	21.6	67	133	15	55	9.1	1.41	5.9	0.72	4.2	0.78	2.20	0.32	2.08	0.33

$^{149}Sm/^{152}Sm=0.51686$. The external precision for $^{143}Nd/^{144}Nd$ as judged from values for the La Jolla standard is 15 ppm (±0.3 ε-units). The measured $^{143}Nd/^{144}Nd$ ratios were normalised to $^{143}Nd/^{144}Nd$ =0.511854 of the La Jolla standard. Initial ε-values were calculated applying present day $^{147}Sm/^{144}Nd=0.1967$ and $^{143}Nd/^{144}Nd=0.512638$ as the chondritic values and an age of 1.82 Ga (cf. Claesson and Lundqvist, 1995).

4.2. Major elements

In the normative diagram shown in Fig. 4, the Härnö and Bergom granites plot in the granite field close to the Ab–Or side, while the Själevad granite is higher in An, approaching the granodiorite field. This is also evident from the differentiation index (DI): the Själevad granite varies between 63 and 77, while the Härnö and Bergom granites vary less and have higher values (88–91 and 84–90, respectively).

In Harker diagrams (Fig. 5), at the same SiO_2 content, Al_2O_3 is higher in the Härnö than in the Själevad and Bergom granites. $Fe_2O_3^{tot}$ decreases rapidly with increasing SiO_2 in the Själevad but not in the Härnö and Bergom granites. K_2O is much lower in the Själevad than in the Härnö and Bergom granites (Fig. 5). The differences in Al_2O_3 contents also show up in the molar $Al_2O_3/(Na_2O+K_2O+CaO)$ ratio: the Härnö granite averages at 1.17$\pm$0.02 (1σ), typical of S-type granites (cf. Chappell and White, 1974), whereas the Själevad (1.05$\pm$0.05; 1σ) and Bergom granites (1.06$\pm$0.10; 1σ) are only marginally peraluminous.

Fig. 6 displays the major element chemistry of the three granites expressed as Debon and Le Fort (1982)

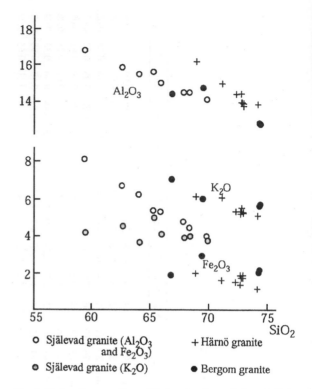

Fig. 5. Harker diagrams (in wt.%) showing the variation of Al_2O_3, total Fe as Fe_2O_3 and K_2O in the studied granites.

parameters. In these diagrams, the B parameter measures the amount of mafic minerals; the Härnö and Bergom granites have low and almost constant values, whereas the Själevad granite has values with a negative correlation with A. There is also an overlap in the B parameter for the two granites. A similar relation is found in the BQF diagram: the Härnö and Bergom granites follow the QF join, whereas in the Själevad granite decreasing Q accompanies increasing B. The latter behaviour is similar to what is usually found in orogenic granitoid suites (e.g., the Early Svecofennian granitoids; cf. Fig. 6). In the PQ diagram, the Härnö and Bergom granites are granitic (alkali granitic) and the Själevad suite includes adamellitic (monzogranitic) and quartz monzonitic compositions.

The chemical composition of the Early Svecofennian granites of the study area is different from that of the Late Svecofennian granites; for example, they are lower in SiO_2. In the BA diagram (Fig. 6), the Själevad granites could be considered a continuation of the Early Svecofennian granitoid trend towards more evolved compositions. However, in both the PQ

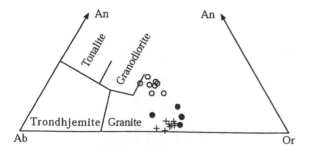

Fig. 4. Rock classification based on normative feldspar compositions (Barker, 1979). Själevad granite (O), Bergom granite (●), Härnö granite (+).

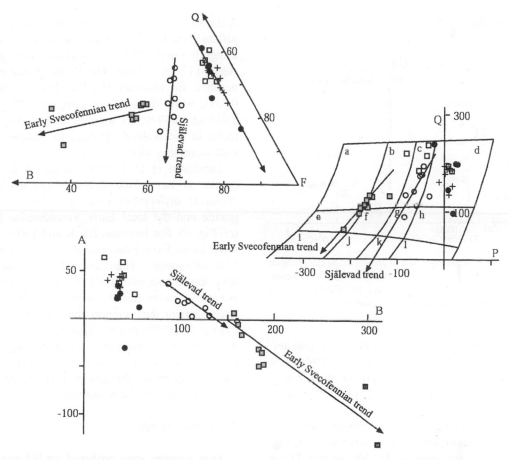

Fig. 6. Composition of the studied granites plotted in the classification diagrams of Debon and Le Fort (1982). Additional data from the granites of the same region are also shown. Själevad granite (O), Bergom granite (●), Härnö granite (+), muscovite-bearing Late Svecofennian granites from Claesson and Lundqvist (1995) (□), Early Svecofennian granitoids from Lindh (in press) (■). Fields in the PQ-diagram: tonalite (a), granodiorite (b), adamellite (c), granite (d), quartz diorite, gabbro (e), quartz monzodiorite (f), quartz monzonite (g), quartz syenite (h), gabbro, diorite (i), monzogabbro (j), monzonite (k), syenite (l).

and BQF diagrams, the trend of the Själevad granite is significantly offset from that of the Early Svecofennian granitoids. At a corresponding quartz content (Q, Fig. 6), the Själevad granite is more potassic and less mafic than the Early Svecofennian granitoids.

4.3. Trace elements

According to the division by Whalen et al. (1987), the three granites straddle the line separating the A-type granite field from the fields of other granites (Fig. 7). In Fig. 7a, the Själevad and Bergom granites, and, in Fig. 7b, the Härnö granite have an A-type character. Even if these granites may show certain A-type characteristics, they are clearly different from the

nearby Nordingrå rapakivi granites that are A-type by definition (cf. Lindh and Johansson, 1996, Fig. 9; see also Haapala and Rämö, 1992).

The trace elements Rb, Ba and Sr behave differently in the three granites (Fig. 8). The Själevad and Bergom granites cluster in the field of normal granites (cf. El Bouseily and El Sokkary, 1975), whereas the Härnö granite is 'evolved'. The more plagioclase-rich Själevad granite has higher Sr/Ba than the Bergom granite. There is a distinct compositional gap between the Härnö granite and the other two granites.

Thiéblemont and Tegyey (1994) used Nb and Zr to study the influence of continental crust on magmatic rocks, noting that Zr/Nb is high in continental rocks. As can be seen from Table 2, Zr/Nb is much lower in

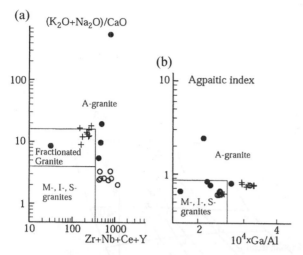

Fig. 7. Comparison of the three studied granites in the classification diagrams of Whalen et al. (1987). (a) $(K_2O+Na_2O)/CaO$ versus $Zr+Nb+Ce+Y$; (b) Agpaitic index versus $10^4 \times Ga/Al$. Själevad granite (O), Bergom granite (●), Härnö granite (+).

the Härnö granite than in the other two granites. Following Thiéblemont and Tegyey (1994), this would mean that the Härnö granite had not a prominent source component in an evolved continental crust.

Differences among the three granites are also obvious in the multielement diagrams shown in Fig. 9. As in almost any granite, Ta, Nb, Sr and Ti are low; the Härnö and Bergom granites are particularly low in Sr and Ti. Ba is much lower and Rb much higher in Härnö than in the two other granites. U and Th are higher in the Härnö than in the Själevad granite, while the Bergom granite shows larger scatter. Hf and Zr are highest in the Själevad granite and lowest in the Härnö granite, the Bergom granite is intermediate between the two, in spite of the fact that it is the most silica rich of the three. These inconsistencies between SiO_2 and Hf and Zr among the three granites cannot be explained in terms of crystal fractionation or different degrees of partial melting, rather they reflect different source materials.

The REE patterns are basically different for the three granites (Fig. 10). The Bergom granite has the highest total contents of REE. Together with the Själevad granite, it has a steeper LREE pattern [(La/Sm)$_N$ for Bergom=127±7 (1σ), for Själevad 103±16)] than the Härnö granite (58±11). The

Själevad granite has an almost horizontal HREE pattern [(Gd/La)$_N$ 2.1±0.6], the Bergom granite has a slightly steeper (4.5±2.7) and the Härnö granite the steepest (7.6±2.5) one. The Härnö granite has a deep negative Eu anomaly (Eu/Eu* between 0.030 and 0.051), while those of the Bergom (0.20–0.82) and Själevad (0.44–0.88) granites are smaller. Furthermore, in the Själevad granite, Eu anomaly deepens with increasing silica content.

Lundqvist et al. (1990) considered the Själevad granite to be a Late Svecofennian granite. Clear chemical differences exist between the Själevad granite and the local Early Svecofennian granitoids (cf. Fig. 6). For instance, Ba, K and LREE are much lower in the Early Svecofennian granitoids than in the Själevad granite. The Early Svecofennian granites have lower contents of Hf, Zr, Ta and Nb than the younger granites (Lindh, in press). Thus, major and trace element data support the suggestion of Lundqvist et al. (1990) that the Själevad granite belongs to the Late Svecofennian granites. Following Lundqvist et al., the discussion below is based on the suggestion that the Själevad, Bergom and Härnö granites are roughly, if not precisely, coeval.

4.4. Isotope results

Four samples were analyzed for Nd isotopes; two of these are from the Härnö and one from the Bergom and Själevad granites each (Table 3). In general, the initial ε$_{Nd}$ values of the samples are slightly negative

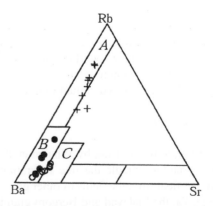

Fig. 8. Rb–Ba–Sr variation in the three studied granites. The fields [(A) Evolved granites, (B) Normal granites, (C) Anomalous granites] according to El Bouseily and El Sokkary (1975). Själevad granite (O), Bergom granite (●), (+) Härnö granite.

Sample/primordial mantle

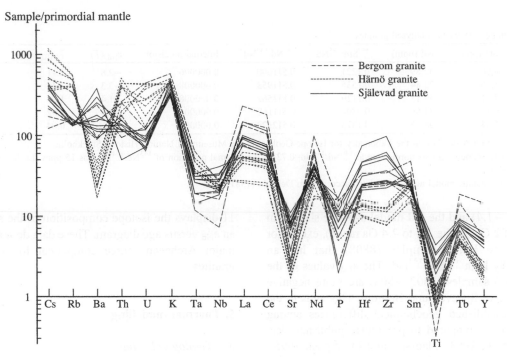

Fig. 9. Multielement diagrams for the Härnö and Själevad granites. The data are normalised to the primordial values of Wood (1979).

Sample/chondrite

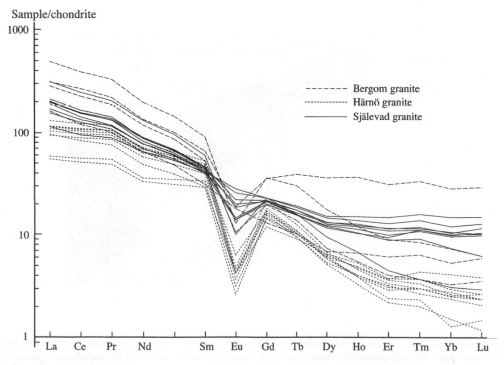

Fig. 10. Chondrite-normalised REE diagram for the studied granites. The data are normalised to the chondrite of Nakamura (1974).

Table 3
Sm–Nd isotope data for the analysed granites

Sample	Sm (ppm)	Nd (ppm)	$^{147}Sm/^{144}Nd$	$^{143}Nd/^{144}Nd$[a]	Internal precision	ε_{Nd} (T)[b]	T_{CHUR} (Ga)	T_{DM} (Ga)[c]
8802	8.66	42.89	0.1221	0.511600	0.000006	−2.8	2.11	2.41
8808	5.70	22.44	0.1535	0.511953	0.000008	−3.3	2.41	2.80
8815	7.31	39.13	0.1129	0.511562	0.000008	−1.4	1.95	2.25
8818	9.52	71.59	0.0804	0.511168	0.000006	−1.5	1.92	2.15
BCR-1	6.54	28.75	0.1375	0.512623	0.000008		0.04	0.85

BCR-1 is an internal standard at the Laboratory for Isotope Geology at the Museum of Natural History, Stockholm.
[a] The Nd isotope ratio was normalised to $^{146}Nd/^{144}Nd=0.7219$; the external precision of $^{143}Nd/^{144}Nd$ was 15 ppm (La Jolla standard).
[b] $T=1.82$ Ga.
[c] Depleted mantle model age is calculated according to DePaolo (1981).

(−3.3 to −1.4) and the Nd model ages are slightly in excess of 2 Ga (in the 2.2 to 2.4 Ga range), except for one of the Härnö samples (8808) that has an abnormally high $^{147}Sm/^{144}Nd$. The ε_{Nd} values of the two Härnö samples (8802, 8808) are more negative than those of Själevad and Bergom. This adds to the already established geochemical differences among the granites. Compared to previously published data (Welin et al., 1993; Claesson and Lundqvist, 1995), the Härnö granite has similar or slightly more negative ε_{Nd} values, whereas the Själevad data are within the normal range of the Late Svecofennian granites. Fig.

11 displays the isotope composition of the samples in an ε_{Nd} versus age diagram. These data do not require a major Archaean source component for any of the granites.

5. Thermal modelling

5.1. Geological setup

In the following, some preliminary melting scenarios are presented. The present Svecofennian crust is at

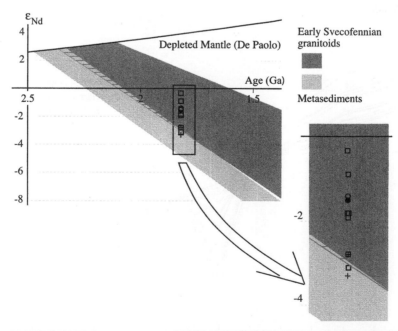

Fig. 11. Time integrated evolution of ε_{Nd} for the Härnö metasediments and the Early Svecofennian granitoids. The area within the rectangle is magnified in the lower right. Själevad granite (O), Härnö granite (+), Bergom granite (●), data for Late Svecofennian granites from Claesson and Lundqvist (1995) (□).

many places thicker (see compilation by Puura and Flodén, 2000) than the 'standard continental crust' normally used for modelling (35 km). The model devised assumes an over-thickened crust resulting from collision of island arcs and thrusting at a continental margin (cf. Nironen, 1997). The Early Svecofennian granitoids were formed at this stage. Thickening of the lithosphere due to collision may occur by (1) thrusting of one crustal slice on top of the other (England and Thompson, 1984; Patiño Douce et al., 1990) or by homogeneous thickening of the crust, (2) either leaving the lithospheric mantle unaffected or (3) affecting the whole lithosphere (England and Thompson, 1984). These three models lead to different heat distribution patterns and thus different melting scenarios. The modelling presented here intends to show whether the discussed granites could have formed in a thickened crust, and to give insights into the possible source materials. The model assumes homogeneous thickening of the lithosphere.

5.2. Model parameters

Parameters that control changes of lithospheric temperature with time include mantle heat flux, amount and distribution of radiogenic heat producing elements, uplift of the crust subsequent to erosion and thermal conductivity and diffusivity. The ratio between the latter two parameters is the heat content (heat capacity). As assessing present-day data is difficult, data for Proterozoic rocks certainly pose a greater challenge still.

Radioactive material responsible for heat production in the crust may be distributed in different ways and, in general, becomes concentrated in the upper parts of the crust. An exception to this is a newly formed continental crust, which has not undergone high-grade metamorphism or anatexis. One, extreme, way to model the heat distribution is to assume that heat production is entirely confined to a specific part of the crust (e.g., England and Thompson, 1984; Thompson, 1999). However, the model tested in this work assumes heat production to occur within a newly formed continental crust, which has not been chemically differentiated; thus, the heat producing elements are supposed to be evenly distributed at all levels of the crust. Thermal conductivity (K) and diffusivity (κ) are constrained to give realistic values of surface heat

flux (55 mW m^{-2}) and heat content (heat capacity $\sim 2.5 \times 10^6$ J K^{-1} m^{-3}) of the rocks (e.g., Thompson, 1999); this estimate of surface heat flux is rather conservative. Finally, the rate of uplift, u, is determined by assuming that the crust after the collision had a thickness of 70 km (e.g., England and Thompson, 1984). It is further supposed that the rapakivi granites (1.58 Ga) intruded into a crust with a thickness of ~50 km (today's value 48 km, Puura and Flodén, 2000) implying a linear uplift and erosion rate of merely 20 km in 200 Ma.

5.3. Mathematical model

The basic equation for modelling of temperature–depth–time relations is (cf. England and Thompson, 1984):

$$\frac{\partial T(z,t)}{\partial t} = \kappa \frac{\partial^2 T(z,t)}{\partial z^2} - u \frac{\partial T(z,t)}{\partial z} + \frac{\kappa}{K} A \frac{fC - ut}{fC}, \quad (1)$$

where $T(z,t)$ describes the temperature (T) as a function of depth (z) and time (t). A is the radioactive heat production, C the initial thickness before thickening (in this case 35 km) and u the uplift rate due to erosion. This equation assumes that heat transport is vertical, i.e., that horizontal dissipation of heat is negligible compared to the vertical transport. Applied boundary conditions are (1) temperature distribution after rapid thickening:

$$T(z,0) = \frac{A\left(\frac{z}{f}\right)\left(C - \frac{z}{2f}\right)^2}{K} + \frac{Q\frac{z}{f}}{K} + 5 \quad (2)$$

where Q is the mantle heat flow (here 25 mW m^{-2}; England and Thompson, 1984), f the thickening factor (here 2), and the number 5 refers to the surface temperature (degrees centigrade); (2) the Dirichlet condition of a constant surface temperature:

$$T(0,t) = 5 \quad (3)$$

and (3) a Neuman condition of a constant temperature gradient in the lithospheric mantle:

$$\frac{\partial T(2C,t)}{\partial z} = \frac{q}{K} \quad (4)$$

Time is given in Ma and vertical distances in km. Heat transport by convection in the asthenospheric mantle gives the time-integrated, approximately constant value of the temperature gradient $[=(Q/K)]$ in the lower lithospheric mantle. This model does not recognize that the thickness of the lithosphere changes with time. However, the error introduced from this simplification is probably negligible. The equation is solved with the help of the program Mathcad applying the procedure PDE to solve partial differential equations. The solution is obtained for a grid with spatial (vertical) resolution of 500 m and temporal resolution of 0.025 Ma, i.e., a grid of 7.2×10^5 points.

5.4. Assessment of parameters

Thermal parameters of the crust have a profound influence on the temperature distribution and are difficult to estimate. England and Thompson (1984) and Thompson (1999) used values between 1.5 and 3.0 W m^{-1} K^{-1} for the thermal conductivity. More recent data (Popov et al., 2003) suggest values around 2–2.5 W m^{-1} K^{-1} for granites and gneisses. The conductivity decreases at higher temperatures (Seipold and Huenges, 1998), approaching 2 W m^{-1} K^{-1} at 800 °C. However, κ decreases faster with increasing temperature than K does (Seipold and Huenges, 1998); κ reaches a value of 0.5×10^{-6} m^2 s^{-1} at 800 °C. This means that the ratio (K/κ) (heat capacity) increases with increasing temperature. Keeping K and κ constant leads to an error in temperature. Using a K of 2.0 W m^{-1} K^{-1} and κ of 0.5×10^{-6} m^2 s^{-1} gives a heat capacity of 4×10^6 J K^{-1} m^{-3} at 800 °C. After 30 Ma, a very conservative estimate of the internal crustal heat production (0.85 µW m^{-3}) results in a temperature of 840 °C at 60 km depth and 965 °C at 70 km depth. The applied thermal parameters are listed in Table 4 and the results of the calculations are shown in Fig. 12.

5.5. Influence of parameter values

With the above values of K and κ, which are typical for high temperatures, too high model temperatures are obtained using a high rate of crustal heat production. Using a value of 2 µW m^{-3} (cf. Patiño Douce et al., 1990) should give a temperature of 1500 °C at 70 km after 30 Ma. The

Table 4
Parameters used for thermal modelling

Parameter	
A (crustal heat production)	0.85×10^{-6} W m^{-3}
C (initial crustal thickness)	35×10^3 m
K (crustal thermal conductivity)	2.0 W m^{-1} K^{-1}
κ crustal thermal diffusivity)	0.5×10^{-6} m^2 s^{-1}
f (crustal thickening factor)	2
Q (mantle heat flow)	25×10^{-3} W m^{-2}
u (uplift rate)	100 m Ma^{-1}

reason for using a rather low value for crustal heat production is that the crust is assumed to be juvenile, not segregated into upper and lower crust, and to have a homogeneous distribution of radioactive elements. It is unrealistic to assume that a heat production of 2 µW m^{-3} could apply to the total thickness of the crust. Using a faster exhumation rate lowers the obtained model temperatures. The slow exhumation rate used is constrained by the present-day thickness of the local crust. The values used in the model result in a conservative crustal heat flow of 60 mW m^{-2}, when the mantle heat flow is 25 mW m^{-2}.

Assuming a thicker initial crust (45 km) and an increased exhumation rate (40 km in 200 Ma) to obtain a crustal thickness of 50 km at the time of rapakivi intrusion results in even higher temperatures. At the bottom of the then 84-km-thick crust, temperatures should rise to 1150 °C in 30 million years. These very high temperatures are unrealistic, as melting would have started long before they had been reached. This model assuming two over-thickened crust slices are totally unrealistic from a thermal point of view; thus, crustal thinning due to erosion must be very slow if the reported crustal thicknesses (Puura and Flodén, 2000) are correct.

5.6. Result

In a metapelite, a temperature of ~800 °C is needed to produce 30% melt at a pressure of 1–1.5 GPa by dehydration melting (Clemens and Vielzeuf, 1987). The model with an initial crustal thickness of 35 km gives a temperature of ~50 °C in excess of the melting temperature; for an extremely thick (2×45; 90 km) crust the excess temperature is 350 °C. Latent heat of melting for silicate rocks is on the order of 200–300 times their heat capacity (e.g.,

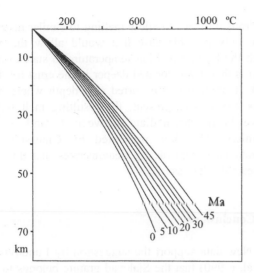

Fig. 12. Temperature versus depth diagram showing the results from modelling with K=2.0 W m^{-1} K^{-1}, κ=0.5×10^{-6} m^2 s^{-1}, u=0.1 km Ma^{-1}, A=0.85 µW m^{-3} (evenly distributed within the crust), initial crustal thickness of 35 km and rapid doubling of crustal thickness. Erosion starts immediately after thickening of the crust. The curves corresponds to 0, 5, 10, (15), 20, (25), 30, (35) and 45 Ma after doubling of the crust; curves within brackets are displayed but not labelled in the figure. During the 45 Ma of erosion, the crust is thinned to 66.5 km.

Robie et al., 1978). Thus, a thermal model producing a temperature of ~60–80 °C in excess of the melting temperature is necessary for obtaining 30% melting. It is important that a model of the type presented here, even with conservative estimates of the relevant parameters, suggests that extensive melting can be induced by crustal thickening. This is in accordance with the results obtained by Thompson (1982), Patiño Douce et al. (1990) and Thompson (1999), but contradicts with those of Clemens (2003), who based his discussion on the early England and Thompson (1984) model for crustal thickening.

6. Discussion

6.1. Isotopic and chemical evidence for different sources

The isotope and chemical evidence points to at least two different sources that were involved in the generation of the three granites. Claesson and

Lundqvist (1995) suggested the mantle below the Bothnian Basin to be depleted to approximately two thirds of that given by the DePaolo (1981) model. No data exist suggesting significant enrichment or strong depletion of the mantle below the Bothnian Basin after the Svecofennian orogeny.

Claesson (1987), Welin et al. (1993), Patchett et al. (1987) and Andersson et al. (2002) presented Nd isotope data on the Bothnian sediments and the Early Svecofennian granitoids. Generally, initial ε_{Nd} values are more negative for the Bothnian sediments than for the Early Svecofennian granitoids (cf. Fig. 11). The Härnö granites plot within the evolution path of the metasediments, whereas the Själevad and Bergom granites fall above it, within the granitoid path. The chemical and isotope data suggest that the three granites were formed from at least two different sources, one of which resembles the metasediments of the Bothnian Basin (Härnö granite) and the other one the Early Svecofennian granitoids (Själevad and Bergom granites). This conclusion is similar to that of Claesson and Lundqvist (1995). In view of the data presented by Claesson and Lundqvist, it seems viable that, in a regional context, such two overall sources may have contributed in any proportions. The granites discussed here are probably rather close to being formed from the end members.

Trace element data on the possible source rocks for the Härnö and Själevad granites are scarce. Claesson and Lundqvist (1995) obtained, but did not publish, trace element data from the Härnö metasediments and the Early Svecofennian granitoids from the study area. These rocks or their equivalents could represent sources for the discussed granites. Expected relative differences for granites derived from these two potential sources can be obtained by using Claesson and Lundqvist's unpublished data and data from Lindh (in press). A rough estimate is obtained by assuming similar degrees of partial melting and distribution coefficients for both sources. This means that the mineralogical differences between the metasediments and Early Svecofennian granitoids are disregarded. The overall chemical composition of greywacke and granodiorite is rather similar, save for the Al content. This means that at the high P and T conditions of the lowermost crust, the mineral composition should be roughly similar in both sources, except for a larger garnet content in the metasediments. This would imply a decreased amount of HREE in magmas formed from the metasedimentary source compared to those from the igneous source. LREE contents should mimic the relative abundances of these elements in the two sources. In view of these considerations, the Härnö granite should show higher LREE and lower HREE than the other two granites. The REE patterns (Fig. 10) are consistent with this. The Eu anomaly is probably more influenced by restite mineralogy than by the content of Eu in the source. It seems that the amount of restitic feldspar in the Härnö source was larger than in the source of the other two granites (Fig. 10).

Disregarding (rather improbable) differences in the degree of melting, differences in other trace elements should, to a large extent, mirror differences in the source rocks. From Claesson's and Lundqvist's (1995) unpublished data, higher Sr should be expected for rocks derived from the igneous than from the sedimentary source. The composition of the Härnö and Själevad granites agrees with this expectation. The REE patterns (Fig. 10) are also consistent with the Härnö granite having a source resembling the Härnö metasediments and the Själevad granite having a source similar to the Early Svecofennian granitoids; the essential source rocks of the Själevad granite were probably the felsic members of these granitoids.

6.2. Constraints on melting from the thermal modelling

When a sufficient amount of melt has formed, it assembles, leaves its site of generation and moves heat and heat-producing elements (K, Th, U) from the source area. This means that the remaining crust becomes simultaneously impoverished in both fertile material and heat-producing elements, making this part of the crust more unlikely for further granite melt generation. The loss of thermal energy also effectively stops the melting.

The I-type Själevad granite is the oldest and the most mafic of the three granites. Its magma probably came from a more mafic igneous source that required a higher temperature for melting than the metasediments, which were the probable source for the Härnö granite. No substantial temperature difference can be inferred for the Bergom and Härnö granites, provided that their sources both had a sufficient amount of

hydrous minerals to allow melting to similar degrees. The only possible case that would allow the rock with the highest melting temperature to start to melt first is that it was located deeper in the crust (cf. Fig. 12). If melting had occurred at a depth where both rock types were present, the resulting melt would have been intermediate between I- and S-type granites. The data presented by Claesson and Lundqvist (1995) clearly demonstrates that this was repeatedly the case.

7. Conclusions

- New data support the suggestion by Lundqvist et al. (1990) that the Själevad granite belongs to the Late Svecofennian granites.
- The chemical and isotopic data suggest different source rocks for the Härnö and Själevad granites, the Härnö granite having a predominantly metasedimentary and the Själevad granite a predominantly metaigneous source. The Bergom granite probably had a rather felsic yet otherwise similar source as the Själevad granite. This shows that granites of contrasting origin intruded closely together in space and time without forming intermediate rocks.
- The hiatus between the Early and Late Svecofennian granites, combined with contrasting chemical compositions, suggests that the two granite suites intruded in different tectonic settings. Modelling of the thermal evolution suggests that the Late Svecofennian magmas could have formed in a continental crust that had been thickened after collision.
- Thermal modelling implies that the source of the Själevad granite was deeper than that of the two other granites.

Acknowledgements

This investigation was financed by a grant from the former Natural Science Research Council of Sweden (NFR). I thank Tapani Rämö for inviting me to the Symposium Granitic Systems – State of the Art and Future Avenues where parts of this contribution were presented. Fig. 3 is used here by the courtesy of Paula

Lindgren. Zoltan Solyom and Ingrid Johansson performed the major element analyses at the Geolaboratory, Lund University. Hans Schöberg made the isotope analyses at the Laboratory for Isotope Geology, Stockholm. I acknowledge Stefan Claesson and Thomas Lundqvist for letting me use their unpublished chemical data. I also thank U.-B. Andersson, M. Vaasjoki and an anonymous referee for comments on the manuscript.

References

Andersson, U.B., 1997. An overview of the Fennoscandian rapakivi complexes with emphasis on the Swedish occurrences. In: Ahl, M., Andersson, U.B., Lundqvist, T., Sundblad, K. (Eds.), Rapakivi Granites and Related Rocks in Central Sweden, Sveriges Geologiska Undersökning Ca. Geological Survey of Sweden, vol. 87, pp. 33–49.

Andersson, U.B., Neymark, L.A., Billström, K., 2002. Petrogenesis of Mesoproterozoic (Subjotnian) rapakivi complexes of central Sweden: implications from U–Pb zircon ages, Nd, Sr, and Pb isotopes. Transactions of the Royal Society of Edinburgh. Earth sciences 92, 201–228.

Barker, F., 1979. Trondhjemite: definition, environment and hypotheses of origin. In: Barker, F. (Ed.), Trondhjemites, Dacites and Related Rocks. Elsevier, New York, pp. 1–11.

Chappell, B.W., White, A.J.R., 1974. Two contrasting granite types. Pacific Geology 8, 173–174.

Claesson, S., 1987. Nd isotope data on 1.9–1.2 Ga basic rocks and metasediments from the Bothnian Basin, central Sweden. Precambrian Research 35, 115–126.

Claesson, S., Lundqvist, T., 1995. Origins and ages of Proterozoic granitoids in the Bothnian Basin, central Sweden: isotopic and geochemical constraints. Lithos 36, 115–140.

Claesson, S., Huhma, H., Kinny, P.D., Williams, I., 1993. Svecofennian detrital zircon ages—implications for the Precambrian evolution of the Baltic Shield. Precambrian Research 64, 109–130.

Claesson, S., Andersson, U.B., Schumacher, M., Sunde, T., Whitehouse, M., Vestin, J., 1997. Inherited Archaean components in a Mesoproterozoic rapakivi complex from Central Sweden: implications from SIMS U–Pb imaging and spot analyses of a zircon. Terra Nova, Abstract Supplement 9, 356.

Clemens, J.D., 2003. S-type granitic magmas—petrogenetic issues, models and evidence. Earth-Science Reviews 61, 1–18.

Clemens, J.D., Vielzeuf, D., 1987. Constraints on melting and magma production in the crust. Earth and Planetary Science Letters 86, 287–306.

Debon, F., Le Fort, P., 1982. A chemical–mineralogical classification of common plutonic rocks and associations. Transactions of the Royal Society of Edinburgh. Earth sciences 73, 135–149.

DePaolo, D.J., 1981. Neodymium isotopes in the Colorado Front Range and crust–mantle evolution in the Proterozoic. Nature 291, 193–196.

El Bouseily, A.M., El Sokkary, A.A., 1975. The relation between Rb, Ba, and Sr in granitic rocks. Chemical Geology 16, 207–219.

England, P.C., Thompson, A.B., 1984. Pressure–temperature–time paths of regional metamorphism: I. Heat transfer during the evolution of regions of thickened continental crust. Journal of Petrology 25, 928–984.

Gaál, G., Gorbatschev, R., 1987. An outline of the Precambrian evolution of the Baltic Shield. Precambrian Research 35, 15–52.

Haapala, I., Rämö, O.T., 1992. Tectonic setting and origin of the Proterozoic rapakivi granites of the southeastern Fennoscandia. Transactions of the Royal Society of Edinburgh. Earth sciences 83, 165–171.

Lindh, A., 2001. Microgranular enclaves in the Nordingrå rapakivi granite, east central Sweden—the early rapakivi development. Neues Jahrbuch fur Mineralogie. Abhandlungen 176, 299–322.

Lindh, A., in press. Die Granitoide des Bottnischen Beckens, Zentralschweden. Zeitschrift für geologische Wissenschaften.

Lindh, A., Johansson, I., 1996. Rapakivi granites of the Baltic Shield: the Nordingrå granite, its chemical variation and Sm–Nd isotope variation. Neues Jahrbuch fur Mineralogie. Abhandlungen 170, 291–312.

Lindh, A., Andersson, U.B., Lundqvist, T., Claesson, S., 2001. Evidence of crustal contamination of mafic rocks associated with rapakivi rocks, an example from the Nordingrå complex, Central Sweden. Geological Magazine 138, 371–386.

Lundqvist, T., Gee, D.G., Kumpulainen, R., Karis, L., 1990. Beskrivning till berggrundskartan över Västernorrlands län. Sveriges Geologiska Undersökning. Geological Survey of Sweden Ba 31. 429 pp.

Nakamura, N., 1974. Determination of REE, Ba, Fe, Mg, Na, and K in carbonaceous and ordinary chondrites. Geochimica et Cosmochimica Acta 38, 757–775.

Nironen, M., 1997. The Svecofennian orogen: a tectonic model. Precambrian Research 86, 21–44.

Patchett, J., Todt, W., Gorbatschev, R., 1987. Origin of continental crust of 1.9–1.7 Ga age; Nd isotopes in the Svecofennian orogen terrains of Sweden. Precambrian Research 35, 145–160.

Patiño Douce, A., Humphreys, E.D., Johnston, A.D., 1990. Anatexis and metamorphism in tectonically thickened continental crust exemplified by the Sevier hinterland, western North America. Earth and Planetary Science Letters 97, 290–315.

Persson, A., 1997. The Ragunda rapakivi complex. In: Ahl, M., Andersson, U.B., Lundqvist, T., Sundblad, K. (Eds.), Rapakivi granites and related rocks in central Sweden, Sveriges Geologiska Undersökning Ca. Geological Survey of Sweden, vol. 87, pp. 49–58.

Popov, Y., Tertychnyi, V., Romushkevich, R., Korobkov, D., Pohl, J., 2003. Interrelations between thermal conductivity and other physical properties of rocks: experimental data. Pure and Applied Geophysics 160, 1137–1161.

Puura, V., Flodén, T., 2000. Rapakivi-related basement structures in the Baltic Sea area; a regional approach. GFF 122, 257–272.

Robie, R.A., Hemmingway, B.S., Fisher, J., 1978. Thermodynamic properties of minerals and related substances at 298.15 K and 1 bar (10^5 Pascals) pressure and at higher temperatures. US Geological Survey Bulletin 1452. 456 pp.

Seipold, U., Huenges, E., 1998. Thermal properties of gneisses and amphibolites—high pressure and high temperature investigations of KTB-rock samples. Tectonophysics 291, 173–178.

Thiéblemont, D., Tegyey, M., 1994. Une discrimination géochimique des roches différenciées témoin de la diversité d'origine et de situation tectonique des magmas calco-alkalins. Compte Rendu des Academies des Sciences, Paris, Science, Série II 319, 87–94.

Thompson, A.B., 1982. Dehydration melting of pelitic rocks and the generation of H$_2$O-undersaturated granitic liquids. American Journal of Science 282, 1567–1595.

Thompson, A.B., 1999. Some time–space relationships for crustal melting and granitic intrusion at various depths. In: Castro, A., Fernández, C., Vigneresse, J. (Eds.), Understanding Granites: Integrating New and Classical Techniques, Special Publication-Geological Society of London, vol. 168, pp. 7–25.

Wasström, A., 1993. The Knaften granitoids of Västerbotten County, northern Sweden. In: Lundqvist, T. (Ed.), Radiometric

Dating Results, Sveriges Geologiska Undersökning. Geological Survey of Sweden C, vol. 823, pp. 60–64.

Welin, E., Christiansson, K., Kähr, A.-M., 1993. Isotopic investigation of metasedimentary and igneous rocks in the Palaeoproterozoic Bothnian Basin, central Sweden. Geologiska Föreningens i Stockholm Förhandlingar 115, 285–296.

Whalen, J.B., Currie, K.L., Chappell, B.W., 1987. A-type granites: geochemical characteristics, discrimination and petrogenesis. Contributions to Mineralogy and Petrology 95, 407–419.

Wood, D.A., 1979. A variably veined suboceanic upper mantlēgenetic significance for mid-ocean ridge basalts from geochemical evidence. Geology 7, 499–503.

Whalen, J.B., Currie, K.L., Chappell, B.W., 1987. A-type granites: geochemical characteristics, discrimination and petrogenesis. Contributions to Mineralogy and Petrology 95, 407–419.

Wood, D.A., 1979. A variably veined suboceanic upper mantle – genetic significance for mid-ocean ridge basalts from geochemical evidence. Geology 7, 499–503.

Available online at www.sciencedirect.com

ELSEVIER

Lithos 80 (2005) 267–280

www.elsevier.com/locate/lithos

The effect of water on accessory phase solubility in subaluminous and peralkaline granitic melts

Robert L. Linnen[*]

Department of Earth Sciences, University of Waterloo, Waterloo, Ontario, N2L 3G1, Canada

Received 14 April 2003; accepted 9 September 2004
Available online 5 November 2004

Abstract

The solubilities of columbite, tantalite, wolframite, rutile, zircon and hafnon were determined as a function of the water contents in peralkaline and subaluminous granite melts. All experiments were conducted at 1035 °C and 2 kbar and the water contents of the melts ranged from nominally dry to approximately 6 wt.% H_2O. Accessory phase solubilities are not affected by the water content of the peralkaline melt. By contrast, solubilities are affected by the water content of the subaluminous melt, where the solubilities of all the accessory phases examined increase with the water content of the melt, up to ~2 wt.% H_2O. At higher water contents, solubilities are nearly constant. It can be concluded that water is not an important control of accessory phase solubility, although the water content will affect diffusivities of components in the melt, thus whether or not accessory phases will be present as restite material. The solubility behaviour in the subaluminous and peralkaline melts supports previous spectroscopic studies, which have observed differences in the coordination of high field strength elements in dry vs. wet subaluminous granitic glasses, but not for peralkaline granitic glasses. Lastly, the fact that wolframite solubility increases with increasing water content in the subaluminous melt suggests that tungsten dissolved as a hexavalent species.
© 2004 Elsevier B.V. All rights reserved.

Keywords: Solubility; Granite; Melt; Water; Nb; Ta; W; Ti; Zr; Hf; Columbite; Tantalite; Wolframite; Rutile; Zircon; Hafnon

1. Introduction

It has long been known that water has a profound affect on the physical and chemical properties of magma. In granitic systems, water lowers melt viscosity several orders of magnitude (e.g., Shaw,

---

* Tel.: +1 519 888 4567x6929; fax: +1 519 746 7484.
E-mail address: rlinnen@uwaterloo.ca.

0024-4937/$ - see front matter © 2004 Elsevier B.V. All rights reserved.
doi:10.1016/j.lithos.2004.04.060

1963). Water also acts as a flux, lowering eutectic, solidus and liquidus temperatures (e.g., Tuttle and Bowen, 1958) as well as being an important factor controlling phase equilibria (e.g., Whitney, 1975). One aspect that remains poorly understood is whether the water content of a melt affects the solubilities of accessory phases. This is important for three reasons. Firstly, the solubility of the accessory phase itself is important. For example, phases such as zircon or apatite may dominate the rare earth element content of a whole rock analysis, thus, it is important to understand the controls on the solubilities of accessory minerals. Secondly, some accessory phases may be ore minerals, e.g., columbite–tantalite. Knowledge of the solubilities of ore minerals in silicate melts is essential to generating and/or evaluating ore deposit models in magmatic rocks such as granites and pegmatites. Thirdly, the solubility of an accessory phase is related to the activity coefficient of a trace element component in a melt, e.g., ZrO_2 activity in the case of zircon solubility (e.g., Linnen and Keppler, 2002). Because mineral–melt partition coefficients change with changes in the activity coefficients of elements in either minerals or melts, the solubilities of accessory phases can be used to indicate whether the activity coefficient of an element in a melt, and hence its partitioning behaviour, changes.

There is very little experimental data on the effect of water on the partitioning of elements between solids and melts. For the experiments that have been conducted, water does not have a uniform affect on the partitioning of all elements. Righter and Drake (1999) determined partition coefficients for siderophile elements (Ni, Co, Mo, W and P) between an iron-rich metal alloy and hydrous basaltic melt at 1300 °C and 10 kbar. The partition coefficients for Ni, Co, Mo and W were not affected by the water content of the melt over the range of 0.5–3.4 wt.% H_2O. However, the partitioning of P did increase with increasing H_2O content of the melt. Wood and Blundy (2002) used an indirect approach to distinguish between temperature and H_2O effects and concluded that water does affect partitioning of REE between basaltic melt and pyroxene, but does not affect REE partitioning in garnet.

There have been very few studies on the effect of water on accessory phase solubility and trace element partitioning in silicate melts. Harrison and Watson (1983) examined the effect of water on zircon solubility and zirconium diffusion in granitic melts at 1020 to 1500 °C and 8 kbar. Their results clearly showed increasing zirconium diffusivity with increasing H_2O content of the melt, however, the dependence of zircon solubility on the H_2O content of the melt was much less well defined. More recently Baker et al. (2002) determined zircon solubility as a function of the H_2O contents of granitic melt at 1050 to 1400 °C and 10 kbar. In one series of experiments in particular

(a halogen-free melt composition at 1200 °C), zircon solubility increases with the water content of the melt up to a water content of ~2 wt.% H_2O. Harrison and Watson (1984) examined apatite solubility as a function of the H_2O content of a granitic melt at 1100 to 1400 °C and 8 kbar and did not observe a significant variation of apatite solubility at these temperatures for melts with 0.1 to 10.0 wt.% H_2O. By contrast, Rapp and Watson (1986) and Montel (1993) did observe some dependence of monazite solubility on the water content of the melt for granitic melts at 1000 to 1400 °C and 8 kbar for the former study and 800 °C and 2 kbar for the latter study. Lastly, Linnen and Keppler (1997) determined columbite and tantalite solubility at 1035 °C and 2 kbar in subaluminous, peraluminous and peralkaline granitic melts. They concluded that columbite and tantalite solubilities were lower in the subaluminous and peraluminous melt with H_2O contents lower than ~3 wt.%, but that H_2O did not affect solubility in the peralkaline melt. However, that study involved a limited amount of data and additional points are needed to confirm these trends.

It is apparent that the effect of water on accessory phase solubility and the activities of trace elements in melts remain poorly understood. This study examines the effect of water on behaviour of high field strength elements, Ti, Zr, Hf, Nb, Ta and W, in a subaluminous and a peralkaline granitic melt. This is accomplished by determining the effect of water on the solubilities of rutile, zircon, hafnon, columbite, tantalite and wolframite in the two melt compositions.

2. Experimental

2.1. Starting materials

The Mn end-members of columbite ($MnNb_2O_6$) and tantalite ($MnTa_2O_6$) were synthesized hydrothermally by sealing a stoichiometric oxide mixture plus a 5% HF solution in Au capsules (length 30, i.d. 4.8, o.d. 5.0 mm), then placing the capsules in cold-seal autoclaves at 850 °C and 2 kbar for 2 weeks. Hübnerite ($MnWO_4$) was synthesized by sealing a stoichiometric mixture of tungstic acid (H_2WO_4) and MnO, together with H_2O in an Au capsule. The capsule was placed in an air-quench cold-seal autoclave at 800 °C and 2 kbar for 1 week. Hafnon ($HfSiO_4$) was synthesized in a piston cylinder apparatus in a Pt–Rh capsule. A stoichiometric oxide mixture plus a 5 wt.% HF solution was placed at 1400 °C and 15 kbar for 24 h. Rutile was synthesized by melting reagent-grade TiO_2 in an Ir crucible in a high-temperature 1 atmosphere furnace, then crystallizing the melt by cooling. The zircon used is from Miask, Ural (cf. Keppler, 1993). The intended phases were the only phases identified in X-ray powder diffraction patterns of all of the synthesis experiments. The XRD patterns specifically for the case of manganocolumbite and manganotantalite indicate that the crystals are fully ordered (cf. Linnen and Keppler, 1997).

Two granitic glass compositions were used. One is subaluminous, with an alumina saturation index of 1.0 [ASI is the molar $Al_2O_3/(Na_2O+K_2O)$ ratio]. This glass has an anhydrous composition of 4.41% Na_2O, 4.21% K_2O, 12.01% Al_2O_3 and 79.21% SiO_2. The second glass is peralkaline (ASI=0.6) with an anhydrous composition of 5.41% Na_2O, 5.21% K_2O, 9.31% Al_2O_3 and 80.91% SiO_2. These compositions were originally prepared from gels (Linnen and Keppler, 1997), and then were melted at 1200 °C and 1 bar in a covered Pt crucible for 2 h to produce nominally anhydrous glasses. Glasses with 6 wt.% H_2O were produced by sealing the appropriate proportions of anhydrous glass and distilled H_2O in Au capsules. The capsules were placed in a drying oven at 120 °C to evenly distribute the H_2O as well as to check for leaks. A homogeneous hydrous glass was produced by placing the capsules in cold-seal auto-claves, and running experiments at 800 °C and 4.5 kbar (H_2O-undersaturated conditions) for 8 days. Glasses with 4 wt.% H_2O were produced by a similar method, except that experiments were run at 850 °C and 2 kbar for 6 days. Glasses with 3 and 2 wt.% H_2O were produced by mixing equal proportions of nominally anhydrous glasses with 6 and 4 wt.% H_2O glasses, respectively. Finally, glasses with 1 wt.% H_2O were made by mixing equal proportions of nominally anhydrous glasses and 2 wt.% H_2O glasses.

The intended water contents of the H_2O under-saturated experiments in this study were confirmed by Fourier transform infrared spectroscopy (Tables 1–6) using a Bruker IFS 120 spectrometer equipped with a Bruker IR microscope, tungsten source, CaF_2 beamsplitter and a MCT detector. Glass densities

Table 1
Manganocolumbite solubility

Experiment	ASI	H_2O (wt.%)	MnO (wt.%)	Nb_2O_5 (wt.%)	$K_{sp} \times 10^{-4}$ (mol²/kg²)
Nb06-0[a]	0.64	0.1	1.73(0.21)	7.34(1.51)	675(193)
Nb06gel[a]	0.64	0.3	1.51(0.29)	5.53(2.47)	443(289)
Nb06-1	0.64	1.2	1.37(0.04)	4.98(0.15)	361(19)
Nb06-2	0.64	2.0	1.45(0.05)	5.28(0.13)	405(18)
Nb06-3[a]	0.64	3.1	0.65(0.05)	2.34(0.08)	80.7(7.2)
Nb06-4	0.64	3.8	1.43(0.06)	5.15(0.12)	391(16)
Nb06-6[a]	0.64	5.0	1.68(0.08)	6.94(0.09)	618(34)
Nb10-0[a]	1.02	0.1	0.08(0.05)	0.01(0.04)	0.04(0.2)
Nb10-1	1.02	1.3	0.37(0.05)	1.04(0.04)	20.5(3.0)
Nb10-2	1.02	2.5	0.42(0.03)	1.24(0.05)	27.5(2.4)
Nb10-3[a]	1.02	3.1	0.34(0.03)	1.15(0.05)	20.5(1.7)
Nb10-4	1.02	4.6	0.48(0.03)	1.54(0.06)	39.5(3.1)
Nb10gel[a]	1.02	5.8	0.35(0.01)	1.37(0.09)	25.5(4.4)
Nb10-6[a]	1.02	5.9	0.36(0.04)	1.45(0.07)	27.7(3.3)

All experiments were conducted at 1035 °C and 2 kbar (Ar pressure). ASI is the molar Al/(Na+K) ratio, K_{sp} is the molar solubility product [MnO]·[Nb_2O_5], numbers in parentheses are 2σ standard deviation.

[a] From Linnen and Keppler (1997).

were calculated with the equation $\rho_{(g/L)}=2347-12.6*C_{(wt.\% \ H2O)}$ (Holtz et al., 1995). Extinction coefficients of 1.68, 1.48 and 67 l·mol⁻¹·cm⁻¹ and a flexicurve baseline correction were used to calculate concentrations of OH (4500 cm⁻¹), H_2O (5200 cm⁻¹) and total water (3550 cm⁻¹), respectively, from peak heights (Stolper, 1982; Withers and Behrens, 1999; Ohlhorst et al., 2001). The use of these densities and extinction coefficients will introduce some error in the calculated water content of the peralkaline glasses. In addition, water contents were only determined for one representative sample of each hydrous glass composition. However, the water contents determined by infrared spectroscopy are close to those estimated by weighing, and for the purpose of this study approximate water contents are adequate.

2.2. High-temperature–pressure equipment and experimental conditions

Experiments were conducted in rapid quench TZM (molybdenum alloy) bombs with Ar as the pressure medium at 1035 °C and 2 kbar. Temperatures were measured with NiCr–Ni thermocouples in the sheath that surrounds the TZM bomb, calibrated against an internal thermocouple. Temperature measurements are

Table 2
Manganotantalite solubility

Experiment	ASI	H_2O (wt.%)	MnO (wt.%)	Ta_2O_5 (wt.%)	$K_{sp} \times 10^{-4}$ (mol^2/kg^2)
Ta06-0[a]	0.64	0.1	3.00(0.79)	5.90(4.08)	565(482)
Ta06gel[a]	0.64	0.3	2.16(0.15)	10.19(2.75)	703(228)
Ta06-1	0.64	1.2	1.49(0.09)	9.91(1.67)	471(98)
Ta06-2	0.64	2.0	1.53(0.07)	9.66(0.48)	472(24)
Ta06-3[a]	0.64	3.1	1.01(0.04)	6.10(0.10)	195(22)
Ta06-4	0.64	3.8	1.52(0.06)	9.39(0.32)	454(29)
Ta06-6[a]	0.64	5.0	1.53(0.22)	9.90(1.37)	483(132)
Ta10-0[a]	1.02	0.1	0.07(0.05)	0.30(0.26)	0.68(4.5)
Ta10-1	1.02	1.3	0.48(0.03)	2.72(0.18)	41.7(4.6)
Ta10-2	1.02	2.5	0.57(0.03)	3.26(0.15)	58.8(5.2)
Ta10-3[a]	1.02	3.1	0.44(0.01)	2.75(0.07)	39.0(3.3)
Ta10-4	1.02	4.6	0.62(0.03)	3.61(0.22)	71.8(7.6)
Ta10gel[a]	1.02	5.8	0.54(0.01)	3.45(0.10)	59.5(4.7)
Ta10-6[a]	1.02	5.9	0.42(0.02)	2.62(0.19)	35.2(4.2)

All experiments were conducted at 1035 °C and 2 kbar (Ar pressure). ASI is the molar Al/(Na+K) ratio, K_{sp} is the molar solubility product [MnO]·[Ta_2O_5], numbers in parentheses are 2σ standard deviation.

[a] From Linnen and Keppler (1997).

accurate to within ±15 °C, including the effects of temperature gradients and intrinsic EMF errors of the thermocouples. Pressures were measured using pressure transducers calibrated against a Heise gauge and are accurate to within ±50 bars. Experiments were quenched by dropping the capsule into the cold end of the autoclave. Quenching was close to isobaric and occurred in 1 to 2 s. All capsules were weighed before and after experiments to ensure that no leaks occurred during the experiments. The oxygen fugacity of the experiments was not controlled, but the intrinsic oxygen fugacity of the autoclaves is believed to be

Table 3
Hübnerite solubility

Experiment	ASI	H_2O (wt.%)	MnO (wt.%)	WO_3 (wt.%)	$K_{sp} \times 10^{-4}$ (mol^2/kg^2)
W10-0	1.02	0.3	0.14(0.02)	0.16(0.11)	1.4(0.5)
W10-1	1.02	1.3	0.49(0.02)	1.52(0.09)	45.0(2.0)
W10-2	1.02	2.5	0.55(0.03)	1.72(0.07)	57.7(2.3)
W10-3	1.02	3.0	0.53(0.03)	1.71(0.13)	54.8(3.5)
W10-4	1.02	4.6	0.87(0.03)	2.76(0.08)	146(3)
W10-6[a]	1.02	5.9	0.85(0.03)	2.68(0.17)	139(6)

All experiments were conducted at 1035 °C and 2 kbar (Ar pressure). ASI is the molar Al/(Na+K) ratio, K_{sp} is the molar solubility product [MnO]·[WO_3], numbers in parentheses are 2σ standard deviation.

[a] From Linnen (1998).

Table 4
Rutile solubility

Experiment	ASI	H_2O (wt.%)	TiO_2 (wt.%)	TiO_2 (mol/kg)
Ti10-0	1.02	0.3	0.40(0.30)	0.05
Ti10-1	1.02	1.3	1.96(0.59)	0.25
Ti10-2	1.02	2.5	1.76(0.18)	0.22
Ti10-3	1.02	3.0	1.64(0.17)	0.21
Ti10-4	1.02	4.6	1.58(0.30)	0.20
Ti10-6	1.02	5.9	2.16(0.15)	0.27

All experiments were conducted at 1035 °C and 2 kbar (Ar pressure). ASI is the molar Al/(Na+K) ratio, numbers in parentheses are 2σ standard deviation.

constant because Ar of the same purity was used in all of the experiments and the P–T conditions were constant. Although the intrinsic oxygen fugacity was not determined, the use of Ar as the pressure medium means that the experiments were likely conducted at moderately oxidized conditions (cf. Popp et al., 1984; Matthews et al., 2003). However, if different amounts of water were absorbed on the walls inside the autoclave of each experiment then slightly different intrinsic oxygen fugacities of each experiment would result (Matthews et al., 2003).

Approximately 30 mg of glass and 8 mg of crystals were gently mixed, then loaded into Pt capsules (length 10, i.d. 2.8, o.d. 3.0 mm). The experiment duration was typically 2 weeks, although run times ranged from 10 to 19 days. After the experiment, a portion of the run products (glass+crystals) was analysed by electron microprobe and a portion was analysed by X-ray diffraction, to ensure that no unexpected reactions had

Table 5
Zircon solubility

Experiment	ASI	H_2O (wt.%)	ZrO_2 (wt.%)	ZrO_2 (mol/kg)
Zr06-1	0.64	1.2	3.08(0.3)	0.25
Zr06-2	0.64	2.0	2.87(0.10)	0.23
Zr06-4	0.64	3.8	2.51(0.09)	0.20
Zr06-6[a]	0.64	5.0	3.79(0.12)	0.31
Zr10-1	1.02	1.3	0.24(0.08)	0.02
Zr10-2	1.02	2.5	0.25(0.11)	0.02
Zr10-4	1.02	4.6	0.25(0.02)	0.02
Zr10-6[1]	1.02	5.9	0.27(0.06)	0.02

All experiments were conducted at 1035 °C and 2 kbar (Ar pressure). ASI is the molar Al/(Na+K) ratio, numbers in parentheses are 2σ standard.

[a] From Linnen and Keppler (2002).

Table 6
Hafnon solubility

Experiment	ASI	H$_2$O (wt.%)	HfO$_2$ (wt.%)	HfO$_2$ (mol/kg)
Hf06-1	0.64	1.2	5.92(0.35)	0.28
Hf06-2	0.64	2.0	5.72(0.71)	0.27
Hf06-4	0.64	3.8	5.27(0.15)	0.25
Hf06-6[a]	0.64	5.0	6.84(0.23)	0.33
Hf10-1	1.02	1.3	0.87(0.15)	0.04
Hf10-2	1.02	2.5	0.84(0.10)	0.04
Hf10-4	1.02	4.6	0.92(0.24)	0.04
Hf10-6[a]	1.02	5.9	0.88(0.22)	0.04

All experiments were conducted at 1035 °C and 2 kbar (Ar pressure). ASI is the molar Al/(Na+K) ratio, numbers in parentheses are 2σ standard deviation.

[a] From Linnen and Keppler (2002).

occurred. None of the glasses had any apparent colour, although because the amount of crystals is fairly high, the sample colour is the same as that of the crystals, e.g., pink for manganotantalite and gold–brown for manganocolumbite and hübnerite. For most of the experiments, the duration was long enough to obtain equilibrium saturation values. However, for some of the experiments in nominally anhydrous melts, the diffusion of the high field strength elements was significantly slower, and solubilities were determined by inverting diffusion profiles (cf. Harrison and Watson, 1983; Watson, 1994).

2.3. Analytical methods

The concentrations of MnO, Nb$_2$O$_5$, Ta$_2$O$_5$, WO$_3$, ZrO$_2$, HfO$_2$ and TiO$_2$ in the experimental glasses were determined with a JEOL JXA 8600 (University of Western Ontario, London, Canada) or a Cameca SX-50 electron microprobe (Bayerisches Geoinstitut, Bayreuth, Germany) using a 20-kV accelerating voltage, 60- to 100-nA beam current, and a 1- to 10-μm beam diameter. The respective standards were MnTiO$_3$, Nb, Ta and W metal and synthetic ZrSiO$_4$ and HfSiO$_4$, and counting times ranged from 60 to 300 s. Na$_2$O, K$_2$O, Al$_2$O$_3$ and SiO$_2$ (albite and orthoclase standards, 10 s counting times each) were analysed at the same time for correction purposes. At least 15 analyses for each experiment were obtained. Analyses were typically at a distance of ~10 to 20 μm from crystals of the accessory phase, and where possible analyses were obtained from glass in the

centre between two or more crystals. This distance was selected because at a closer distance, roughly 5 μm, fluorescence can be a problem resulting in a spurious analysis, and at larger distances the metals may not have had enough time to diffuse the required length. However, for the majority of the experiment, diffusion rates were sufficiently fast enough for the experimental run times to produce saturation metal values in the melt.

Diffusion profiles away from crystals were measured to ensure that the analyses at a distance of ~10 to 20 μm from crystals of the accessory phase were representative of saturation values (flat profiles). In a few experiments, slow diffusivities, or coupled alkali diffusion did cause problems, and these cases are discussed in the text below. In order to quantify the alkali diffusion, additional profiles were measured using electron microprobe conditions of 15 kV, 8 nA and a 5-μm beam diameter. The counting times were 10 s for Na and K, 20 s for Al and Si and 30 s for Ti for these measurements and secondary glass standards were periodically analysed and a correction factor was used to correct for alkali loss and/or migration during the analysis. The ASI ratio is accurate to within approximate ±0.05 (cf. Linnen et al., 1996).

Glasses doped with Nb$_2$O$_5$, Ta$_2$O$_5$, ZrO$_2$ and HfO$_2$ were also synthesized and independently analysed by ICP-AES in order to verify the accuracy of the electron microprobe analyses (cf. Linnen and Keppler, 1997, 2002). Powder diffraction patterns (10–80°2θ) of all run products were measured on a Stoe STADIP powder diffractometer at the Bayerisches Geoinstitute, Germany, using monochromatised Co Kα1 radiation.

3. Results

3.1. Columbite and tantalite

The solubility of manganocolumbite in a granitic melt depends on the activity of MnO, as well as Nb$_2$O$_5$. The dissolution reaction can be written as:

$$MnNb_2O_6^{Xl} = MnO^{melt} + Nb_2O_5^{melt} \qquad (1)$$

where the superscript 'Xl' represents the crystalline phase and 'melt', the metal oxide component dissolved in the silicate liquid. The reaction constant for Eq. (1) is:

$$K^{Nb} = \frac{\gamma_{MnO}^{melt} \cdot \gamma_{Nb_2O_5}^{melt} \cdot X_{MnO}^{melt} \cdot X_{Nb_2O_5}^{melt}}{a_{MnNb_2O_6}^{Xl}} \qquad (2)$$

where γ represents activity coefficients, X mole fractions and a activities. The solubility product of columbite K_{sp}^{col} is the product of the oxide components in the melt, which are determined by measuring the molar oxide concentrations in quenched melt (glass) by electron microprobe, $[MnO] \cdot [Nb_2O_5]$. In this study, the crystalline phase is pure $MnNb_2O_6$ and thus it has an activity of one at the pressure and temperature of the experiments. The reaction constant will not change for the experiments in this study because they were all conducted at the same temperature and pressure. Thus, for a given anhydrous melt composition, the solubility product will only change if one or both of the activity coefficients of MnO and Nb_2O_5 in the melt change as a function of the water content of the melt.

Some of the manganocolumbite and manganotantalite solubilities were previously determined by Linnen and Keppler (1997). These results, together with the solubilities from melts from this study with 1, 2 and 4 wt.% H_2O are shown in Tables 1 and 2 and Figs. 1 and 2. Fully ordered manganocolumbite was the only crystalline phase identified by X-ray dif-

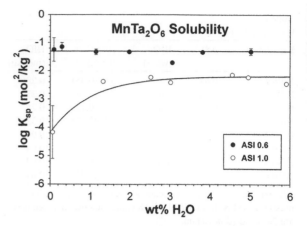

Fig. 2. Solubility of manganotantalite ($MnTa_2O_6$) in peralkaline (solid circles) and subaluminous (open circles) haplogranitic melts at 1035 °C and 2 kbar. The solubility is expressed as a solubility product, molar $[MnO]^*[Ta_2O_5]$. The data for the subaluminous composition was fitted to an equation of the form $y = a + b \cdot e^{-x}$. Error bars represent 2σ values.

fraction and electron microprobe analysis in all of the columbite dissolution experiments. Fully ordered manganotantalite was the only crystalline phase identified by X-ray diffraction and electron microprobe analysis of the tantalite dissolution experiments in subaluminous melts. In the peralkaline melts, however, fully ordered manganotantalite coexists (in textural equilibrium) with a hydrous Ta–silicate phase or $NaTaO_3$ (cf. Linnen and Keppler, 1997).

Figs. 1 and 2 show that H_2O has no apparent effect on the solubilities of manganocolumbite and manganotantalite in the peralkaline melt. The averages of these data indicate solubility products of 425×10^{-4} and 478×10^{-4} mol^2/kg^2, respectively. The Mn/Nb and Mn/Ta ratios of the glasses with >1 wt.% H_2O were 0.45 to 0.52 and 0.48 to 0.51, respectively, i.e., very close to columbite/tantalite stoichiometry (i.e., Mn/Nb of 1:2). For glasses with <1 wt.% H_2O, these values are similar for manganocolumbite, 0.44 to 0.51, but slightly higher for manganotantalite, 0.66 to 1.58. Although these experiments were not reversed, the consistency of the solubility products, even at low H_2O contents, indicates that equilibrium was probably attained. Excess alkalis and water both strongly decrease melt viscosity (Hess et al., 1995; Dingwell et al., 1998), which will increase the diffusivities of Mn, Nb and Ta in the melt. Because the solubilities are the same, even in the nominally dry samples, it can be concluded that diffusivities were sufficiently

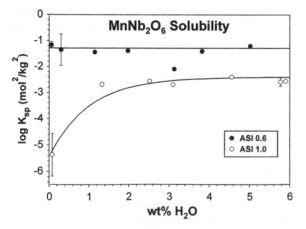

Fig. 1. Solubility of manganocolumbite ($MnNb_2O_6$) in peralkaline (solid circles) and subaluminous (open circles) haplogranitic melts at 1035 °C and 2 kbar. The solubility is expressed as a solubility product, molar $[MnO]^*[Nb_2O_5]$. The data for the subaluminous composition was fitted to an equation of the form $y = a + b \cdot e^{-x}$. Error bars represent 2σ values.

fast such that the melt was saturated with respect to columbite/tantalite at a distance of 10 to 20 μm from the nearest crystal. This could not be confirmed by measuring diffusion profiles because the sizes of the synthetic manganocolumbite and manganotantalite (and hübnerite) crystals are too small.

The solubilities of manganocolumbite and manganotantalite in the subaluminous melt appear to be constant for H_2O concentrations in the melt of >1.0 wt.% (Figs. 1 and 2), with average solubility products of 26.9×10^{-4} and 51.0×10^{-4} mol^2/kg^2, for manganocolumbite and manganotantalite, respectively. The Mn/Nb and Mn/Ta ratios of the glasses in these experiments are 0.47 to 0.67 and 0.49 to 0.55, respectively, supporting the interpretation that these represent saturation values. However, the solubilities in the nominally dry compositions are apparently significantly lower, 0.04×10^{-4} and 0.68×10^{-4} mol^2/kg^2, respectively. A potential problem with these two data points is that the glass at a distance of 10 to 20 μm from the nearest columbite/tantalite crystal, the values of tantalum and niobium may not represent saturation values, i.e., the diffusion of Ta and Nb were too slow (cf. Mungall et al., 1999). The Mn/Nb and Mn/Ta ratios of the glasses in the nominally dry experiments are 14.9 for manganocolumbite and 0.74 for manganotantalite. Based on these stoichiometries, the nominally dry experiment for manganocolumbite may not represent a saturation value, but by contrast, the value for the manganotantalite experiment is likely close to being a saturation value. Unfortunately, experiments at higher temperatures could not be run with the equipment in this study, so reversals were not possible.

3.2. Wolframite

Wolframite, like columbite and tantalite, is a double oxide, thus, its solubility is reported as a solubility product $[MnO] \cdot [WO_3]$. There is nearly a threefold increase in hübnerite solubility at 1.3 wt.% H_2O to water-saturated conditions , from 45×10^{-4} to 139×10^{-4} mol^2/kg^2, respectively (Table 3 and Fig. 3). For all of these experiments, the molar Mn/W ratio ranges from 1.0 to 1.05 (i.e., stoichiometric hübnerite), suggesting that these values represent saturation concentrations. The concentration in the nominally dry melt composition contains significantly

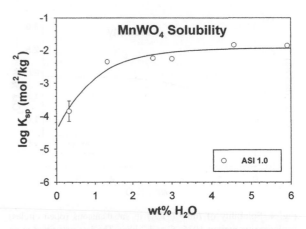

Fig. 3. Solubility of hübnerite ($MnWO_4$) in subaluminous (open circles) haplogranitic melt at 1035 °C and 2 kbar. The solubility is expressed as a solubility product, molar $[MnO]*[WO_3]$. The data was fitted to an equation of the form $y=a+b \cdot e^{-x}$. Error bars represent 2σ values.

less MnO and WO_3, 1.4×10^{-4} mol^2/kg^2, and also has a molar Mn/W ratio of 2.9. This value may underestimate the hübnerite saturation value, similar to manganocolumbite.

Experiments with the peralkaline composition had a very different behaviour. For all of the water undersaturated experiments (0%, 1%, 2%, 3% and 4% H_2O) wolframite was not stable and tungsten–manganese-rich spheres were present in the quench products. This is interpreted as tungstate and silicate liquids having been immiscible during the course of the experiments. The water saturated experiment, at ~6 wt.% H_2O, did contain hübnerite crystals instead of W-rich spheres. However, because hübnerite was not stable in the other experiments, this glass was not analysed and the effect of water on hübnerite solubility in peralkaline melt at 1035 °C could not be established.

3.3. Rutile

The results of the rutile solubility experiments are shown in Table 4 and Fig. 4. The experiments for the peralkaline composition were not successful because rutile solubility in each experiment was >4 wt.% TiO_2, and owing to the high titanium contents, coupled alkali diffusion occurred in all experiments. The ASI composition near the rutile glass boundary is different in each experiment and consequently the apparent solubility of rutile varies. However, it is

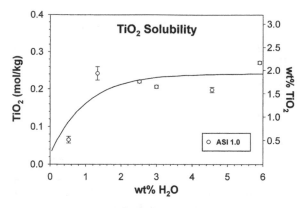

Fig. 4. Solubility of rutile (TiO_2) in subaluminous (open circles) haplogranitic melt at 1035 °C and 2 kbar. The data was fitted to an equation of the form $y = a + b \cdot e^{-x}$. Error bars represent 2σ values.

interesting that the highest solubility, roughly 8 wt.% TiO_2 (determined by diffusion profiles) was observed for the nominally dry composition. This suggests that it is likely that rutile solubility is independent of water content in the peralkaline melt, similar to columbite and tantalite.

Coupled diffusion of alkalis was not a problem in the experiments with the subaluminous composition. The ASI ratio was constant at 1.0 ± 0.05 along traverses away from rutile crystals. However, for the 'dry' and 1% H_2O experiments of this series, titanium contents decreased systematically with distance from rutile crystals and solubilities had to be determined by the diffusion profile method (cf. Harrison and Watson, 1983; Watson, 1994). For the 'dry' experiments solubilities determined from six diffusion profiles ranged from 0.24 to 0.80 wt.%, with an average of 0.52 wt.% TiO_2. For the 1 wt.% H_2O experiments solubilities determined from five diffusion profiles ranged from 1.52 to 2.41 wt.%, with an average of 1.93 wt.% TiO_2. Additional profiles were measured for the other experiments in this series, but it was established that profiles were not necessary to determine rutile solubilities in these samples. For the 2 to 6 wt.% H_2O experiments, rutile solubility is nearly independent of the water content of the melt, and has a value of approximately 1.8 wt.% TiO_2. This is in surprisingly good agreement with a value of 2.1% TiO_2, predicted from the equation of Ryerson and Watson (1987), who conducted experiments at much higher pressure, and on melts with compositions different from the present study.

It is very significant that rutile is less soluble in the 'dry' melt. There was a question as to whether the diffusivities of Nb, Ta and W were fast enough for solubility measurements in the columbite, tantalite and wolframite solubility experiments in dry melts, respectively. The lower solubility of rutile in the 'dry' melt implies that the lower solubilities of columbite, tantalite and wolframite in similar melts are also reasonable.

3.4. Zircon and hafnon

The results of the zircon and hafnon solubility experiments are shown in Tables 5 and 6 and Figs. 5 and 6, respectively. Zircon or hafnon were the only crystalline phases identified by XRD in the run products, although electron microprobe analyses indicate that some of the hafnon dissolution experiments contain a minor amount of unreacted HfO_2. For the experiments with the peralkaline composition, the experiments of zircon solubility in 'dry' melts were unsuccessful, and consequently, similar experiments for hafnon were not attempted. There are two reasons why the zircon dissolution experiments were unsuccessful. First, coupled diffusion accompanied zircon dissolution, similar to that which occurred with rutile dissolution. Secondly, in contrast to all other phases investigated, zirconium concentrations drop to the detection limit (approximately 100 ppm) at a distance of only 15 to 20 μm away from crystal faces and it was not possible to collect enough data points to fit a diffusion profile and calculate solubility. The solubilities of zircon and hafnon in peralkaline melts with 1 to 6 wt.% H_2O are both nearly constant, an average of 3.1 wt.% ZrO_2 and 5.9 wt.% HfO_2, respectively.

The diffusivities of zirconium and hafnium in dry subaluminous melts are considerably slower compared to dry peralkaline melt, so experiments for the dry subaluminous composition were not conducted. Zircon solubility in subaluminous melt with 1 to 6 wt.% H_2O appears to be constant, at 0.25 ± 0.02 wt.% ZrO_2 (Table 5 and Fig. 5). The solubility for the 1 wt.% H_2O composition was determined by the diffusion profile method, whereas the other experiments were from analyses roughly 20 μm from zircon crystals. Alkali diffusion was not observed around zircon crystals in any of the experiments in this series and the similarity of the solubilities determined by

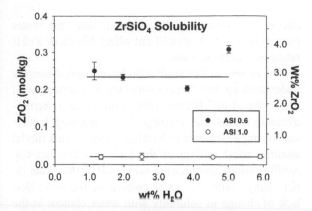

Fig. 5. Solubility of zircon ($ZrSiO_4$) in peralkaline (solid circles) and subaluminous (open circles) haplogranitic melts at 1035 °C and 2 kbar. Error bars represent 2σ values.

two different methods indicates that these are equilibrium values. Hafnon solubility in subaluminous melts with 1 to 6 wt.% H_2O also appears to be constant, at 0.88 ± 0.04 wt.% HfO_2.

Zircon is the only phase investigated in this study for which there is previous experimental data on the effect of the water content of the melt on its solubility. Fig. 7 compares the data from the 1200 °C experiments of Baker et al. (2002) to this study. There is a moderate amount of scatter in the former data, but this is largely explained by the fact that the composition of the melt was not constant. In particular, the iron contents are variable, and iron lowers zircon solubility Baker et al. (2002). The data from this study was supplemented by two points. The first point is an estimate of zircon solubility for the ASI 1.0 melt with

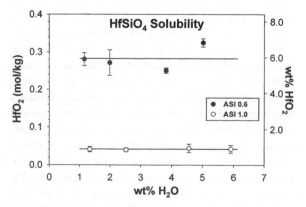

Fig. 6. Solubility of hafnon ($HfSiO_4$) in peralkaline (solid circles) and subaluminous (open circles) haplogranitic melts at 1035 °C and 2 kbar. Error bars represent 2σ values.

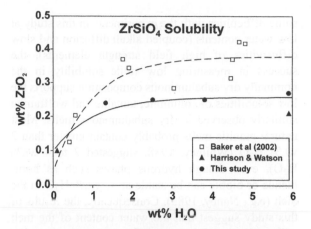

Fig. 7. Comparison of the results from Baker et al. (2002) at 1400 °C and 10 kbar (open squares) with this study, 1035 °C and 2 kbar (solid circles). The work from this study is supplemented by two points. The point at 0.1 wt.% H_2O (solid triangle) is a value extrapolated to 1035 °C from the data of Harrison and Watson (1983). The second point (also a solid triangle) is estimated using the equation of Watson and Harrison (1983). The data was fitted to an equation of the form $y=a+b\cdot e^{-x}$.

0.1 wt.% H_2O. The estimate was made by extrapolating the 1200 to 1500 °C data for 0.1 wt.% H_2O from Harrison and Watson (1983). This was done using a linear fit of their data plotted as log solubility vs. $1/T$ (K). The second data point is for comparison. It is the calculated zircon solubility in the ASI 1.0 melt using the equation of Watson and Harrison (1983). The calculated solubility, 0.21 wt.% ZrO_2, is in good agreement with the measurement of this study, 0.27 wt.% ZrO_2. The general shape of both data sets suggests that water has little effect on zircon solubility for melts with greater than ~2 wt.% H_2O, but for subaluminous melt compositions with less than ~2 wt.% H_2O, zircon solubility increases with increasing water content.

4. Discussion

The results from this study indicate that the water content of the melt has no effect on accessory phase solubility in the peralkaline granite composition, at any concentration of H_2O in the melt. Water has very little effect on accessory phase solubilities in subaluminous melts that contain >2 wt.% H_2O, but, at water contents less than 2 wt.% accessory phase solubilities decrease markedly. Although there are some difficulties with the experiments in this study at low water contents (coupled alkali diffusion and slow diffusivities of high field strength elements), the success in measuring low rutile solubility in the nominally dry subaluminous composition supports the low solubilities of columbite, tantalite and wolframite similarly observed in dry subaluminous melts. Most natural granitic melts probably contain greater than 2 wt.% H_2O (Whitney, 1988, suggested 2 to 4 wt.% H_2O), even though hydrous phases such as hornblende or biotite can be stable at <2 wt.% H_2O in the melt (e.g., Naney, 1983). Consequently, the results of this study suggest that the water content of the melt will not have an important influence on the saturation of accessory phases in natural granitic melts. Nevertheless caution should be exercised in using the solubilities of accessory phases such as zircon to estimate temperature (e.g., Barrie, 1995) because there are a variety of factors controlling solubility including the abundances of fluorine (Keppler, 1993), lithium (Linnen, 1998) and iron (Baker et al., 2002). This is particularly true for the peralkaline granite composition because most natural peralkaline granites contain several wt.% iron. It is also apparent that, whereas water has little effect on accessory phase solubility, it has a very large effect on diffusivity and consequently restite crystals are more likely to be present in relatively dry melt (e.g., Harrison and Watson, 1983; Baker et al., 2002), or if melts at the source are extracted quickly, they could be undersaturated with respect to accessory phases.

High field strength element ratios are also used in geochemical modelling, e.g., Weyer et al. (2003). Changes in the relative solubilities of columbite–tantalite and zircon–hafnon can be used to infer changes in Nb/Ta and Zr/Hf activity coefficient ratios in melts (Linnen and Keppler, 1997, 2002). In the strongly peralkaline melt composition, columbite and tantalite, as well as zircon and hafnon have very similar molar solubilities (compare Figs. 1 and 2, and Figs. 5 and 6, respectively). Because water does not affect accessory mineral solubility in peralkaline melt, it can be concluded that the water content of the melt, similarly, will not affect the Nb/Ta or Zr/Hf ratios of minerals that crystallize from the melt. Water does affect accessory phase solubility in the subaluminous melt composition, but within the accuracy of this study the solubilities of columbite–tantalite and

zircon–hafnon are equally affected, thus, the water content of the melt should not affect Nb/Ta or Zr/Hf ratios of a granitic suite.

The results of this study also have implications for understanding the thermodynamics and structure of melts. Wood and Blundy (2002) proposed a thermodynamic model that predicts the partitioning behaviour of trace elements in basaltic systems. This model assumes ideal mixing in the melt and that activity coefficients of components such as REE-diopside do not change with the water content of the melt. The lack of change in solubility with water content in the peralkaline (depolymerized) melt in this study suggests that the activity coefficients of the high field strength oxide components in the melt do not change with the water content of the melt, in agreement with the Wood and Blundy (2002) model, and imply that water does not have a role in complexing these oxide components in peralkaline melts. Farges and Rossano (2000) interpreted that XAFS spectra showed that water had no effect on the coordination of zirconium in sodium trisilicate glass. The XAFS study of Piilonen et al. (2003) investigated peralkaline and subaluminous glasses of the same composition as this study. They did not find any affect of water on the coordination of Nb in these glasses. The lack of change of solubilities in this study thus supports this interpretation.

By contrast, the lower accessory phase solubilities observed for low water contents in the subaluminous composition imply that activity coefficients of the high field strength oxide components in the melt do change with the water content of the subaluminous melt. In a Raman spectroscopic study, Mysen et al. (1997) interpreted that the bonding of phosphorous in metaluminous to peraluminous glasses changes between anhydrous and hydrous samples. Farges and Rossano (2000) observed a similar phenomenon for zirconium coordination in albite and sodium trisilicate (peralkaline) glasses. Zirconium was six-fold coordinated with oxygen for both albite and sodium trisilicate hydrous glasses. However, for the anhydrous composition, a change was only observed for the albite glass, where zirconium is seven- to eight-fold coordinated. The change in the activity coefficients of high field strength elements in melts with low water contents is thus consistent with the change in coordination observed by Farges and Rossano (2000). There is an additional complication when considering the solubilities of columbite, tantalite and wolframite because these minerals are double-oxides and water may also affect the activity of MnO in the melt. The coordination of some of the transition metals has been shown to be dependent on the water content of the melt, e.g., Ni (Nowak and Keppler, 1998; Farges et al., 2001). Kohn et al. (1990) did observe an increase in the coordination number of Mn with increasing water content and suggested that one possible explanation was an increase in octahedrally coordinated Mn with H_2O. However, this increase is apparently independent of the non-bridging oxygen to tetrahedral cation ratio (NBO/T), and thus ASI composition of the melt. This last observation does not fit well with the data from this study. Columbite and tantalite solubilities in the peralkaline composition did not change with increasing water content. Thus, the change in solubility in the subaluminous composition may be more related to changes in Nb_2O_5 and Ta_2O_5 activity coefficients rather than MnO activity coefficients, although Piilonen et al. (2003) did not observe any change of Nb coordination in dry vs. hydrous subaluminous granitic glasses. In natural melts, the solubilities of columbite, tantalite and wolframite will also depend on the activity coefficient of FeO in the melt. The activity of iron will vary with oxygen fugacity, in contrast to manganese, which is dominantly divalent, except at very oxidized conditions (Kohn et al., 1990; see also below).

Previous studies have shown that the solubilities of accessory phases are related to the amount of excess alkalis in the melt, and thus may be related to the amount of NBO in the melt (Watson, 1979; Hess, 1991; Linnen and Keppler, 1997, 2002). A potential explanation for the data in this study is that for the peralkaline composition, there is a sufficient excess of NBO in the anhydrous melt such that the creation of more NBO caused by the addition of water has no affect on the accessory phase solubility. The lack of an effect of water on trace element solubility has been observed elsewhere for melt compositions with high NBO contents. Blaine et al. (2003) observed that the water content of haplobasaltic melt (high NBO content) has no affect on platinum solubility, although varying the amount of water can change the oxygen fugacity of the melt, which indirectly changes platinum solubility.

By contrast, the anhydrous subaluminous melt does not have any NBO (or at least has very little). The creation of NBO or modification of the melt structure caused by the addition of water in this case does increase solubility. However, once a finite number of non-bridging oxygens have been created, the further addition of water will no longer increase accessory phase solubility. Water apparently only affects the solubilities of accessory phases in melts with low NBO contents. It is interesting to note that in the preliminary data of Linnen and Keppler (1997), the effect of water on columbite–tantalite solubility in a peraluminous composition (also low NBO content) was similar to that of the subaluminous composition. There is some spectroscopic evidence to support this idea. Farges and Rossano (2000) interpret a slight structural relaxation of the tetrahedral framework around Zr for albitic glasses with >1wt.% H_2O, suggesting that once a melt has relaxed to a certain degree, accessory phase solubility will no longer increase with increasing relaxation (i.e., water content).

The variation of oxygen fugacity with the water content at H_2O-undersaturated conditions was also considered by Harrison and Watson (1984), although they did not observe a variation of apatite solubility as a function of fO_2 in that study. All of the experiments in this study were conducted in TZM autoclaves at the same temperature and pressure, and it is probable that the fugacity of hydrogen for these experiments was nearly constant. However, any variation in the amount of absorbed H_2O in the autoclave prior to an experiment will affect the intrinsic fO_2 and fH_2 of the autoclave (Matthews et al., 2003), as discussed previously. If it is assumed that the fH_2 of all the experiments was the same, the oxygen fugacities of the experiments will be progressively higher with progressively more water. Using the equations of Zhang (1999) it can be estimated that the oxygen fugacity of the experiments in this study increased by approximately 5 orders of magnitude for water contents increasing from 0.1 to 5.7 wt.% H_2O (saturation). It can therefore be concluded that oxygen fugacity has no affect on the solubilities of the accessory phases examined in this study, at least for the peralkaline melt composition. The effect of oxygen fugacity on the solubilities of the accessory phases in the subaluminous composition is more difficult to ascertain, because at low water contents, it is not possible to separate the potential effects of non-bridging oxygens from those of oxygen fugacity. However, it can be stated that oxygen fugacity has no apparent affect on the solubilities of the accessory phases examined in this study at greater than ~2 wt.% H_2O.

The only elements examined in this study that could have changed their valences are Mn and W. The fact that the solubility products of manganotantalite and manganocolumbite remained constant (at >2 wt.% H_2O), particularly in the peralkaline composition which will favour higher redox states of cations (e.g., Dickenson and Hess, 1981), indicates that Mn was divalent in all of the experiments in this study. Tungsten in the melt likely has a charge of +4, +5 or +6 and hübnerite contains W^{+6}, thus if tungsten remains hexavalent during dissolution into the melt, hübnerite solubility will be independent of fO_2:

$$MnWO_4^{Xl} = MnO^{melt} + WO_3^{melt} \qquad (3)$$

By contrast, if tungsten enters the melt as either a pentavalent or quadrivalent species, then hübnerite solubility will be dependent on fO_2:

$$MnWO_4^{Xl} = MnO^{melt} + 0.5W_2O_5^{melt} + 0.25O_2 \qquad (4)$$

$$MnWO_4^{Xl} = MnO^{melt} + WO_2^{melt} + 0.5O_2 \qquad (5)$$

In the case of reactions (4) and (5), hübnerite solubility will be greater at lower fO_2, i.e., greater with decreasing water contents of the melt. The opposite trend is observed in this study, hübnerite solubility increases with increasing water in the melt. This indicates that water has a direct effect on hübnerite solubility. It is likely that tungsten is hexavalent for the conditions of the experiments, although it is possible that the increase in solubility with increasing water content is masking a tungsten redox reaction. Support for the interpretation that hexavalent tungsten is the dominant species in the melt comes from the fact that W is expected to behave similar to Mo, and Siewert et al. (1997) interpret that molybdenum in hydrous albite glasses is hexavalent. An additional consideration is that it is only assumed that the fH_2 of experiments was constant, and this assumption may be erroneous. Future experiments, with systematic redox control, are required to conclusively establish the redox state of tungsten in granitic silicate melts.

5. Conclusions

Water has different effects on the solubilities of accessory minerals in peralkaline and subaluminous granitic melts. For the peralkaline composition, water has no effect on solubility, whereas in the subaluminous composition, solubility appears to be independent or weakly dependent on water content at concentrations greater than ~2 wt.% H_2O, and decrease dramatically at less than ~2 wt.% H_2O. Since natural granitic melts typically contain greater than ~2 wt.% H_2O, it can be concluded that water has little effect on the saturation of accessory phases. Independent of solubility, water will still strongly influence whether accessory phases are metastable as restite phases because of the control water has on diffusivity.

Acknowledgements

A portion of this work was conducted while the author was employed at the Bayerisches Geoinstitut, who funded the experiments and analytical costs. A portion of the analytical costs were also supported by an NSERC grant. I am grateful to Philippe Courtial for providing the synthetic rutile. I also thank Hans Keppler and Bruno Scaillet for their insightful comments as well as discussions with Harald Behrens.

References

Baker, D.R., Conte, A.M., Freda, C., Ottolini, L., 2002. The effect of halogens on Zr diffusion and zircon dissolution in hydrous metaluminous granitic melts. Contributions to Mineralogy and Petrology 142, 666–678.

Barrie, C.T., 1995. Zircon thermometry of high-temperature rhyolites near volcanic-associated massive sulfide deposits, Abitibi subprovince, Canada. Geology 23, 169–172.

Blaine, F., Linnen, R.L., Holtz, F., Bruegmann, G.E., 2003. The effect of water on platinum solubility in a haplobasaltic melt at 1250 °C and 0.2 GPa, European Union of Geosciences–American Geophysical Union–European Union of Geophysics Joint Assembly, Nice, France [CD-ROM].

Dickenson, M.P., Hess, P.C., 1981. Redox equilibria and the structural role of iron in alumino-silicate melts. Contributions to Mineralogy and Petrology 78, 352–357.

Dingwell, D.B., Hess, K.-U., Romano, C., 1998. Extremely fluid behaviour of hydrous peralkaline rhyolites. American Mineralogist 83, 31–38.

Farges, F., Rossano, S., 2000. Water in Zr-bearing synthetic and natural glasses. European Journal of Mineralogy 12, 1093–1107.

Farges, F., Brown Jr., G.E., Petit, P.-E., Munoz, M., 2001. Transition elements in water-bearing silicate glasses/melts: Part I. A high-resolution and anharmonic analysis of Ni coordination environments in crystals, glasses, and melts. Geochimica et Cosmochimica Acta 65, 1665–1678.

Harrison, T.M., Watson, E.B., 1983. Kinetics of zircon dissolution and zirconium diffusion in granitic melts of variable water content. Contributions to Mineralogy and Petrology 84, 66–72.

Harrison, T.M., Watson, E.B., 1984. The behavior of apatite during crustal anatexis: equilibrium and kinetic considerations. Geochimica et Cosmochimica Acta 48, 1467–1477.

Hess, K.-U., Dingwell, D.B., Webb, S.L., 1995. The influence of excess alkalis on the viscosity of a haplogranitic melt. American Mineralogist 80, 297–304.

Hess, P., 1991. The role of high field strength cations in silicate melts. In: Perchuk, L.L., Kushiro, I. (Eds.), Physical Chemistry of Magmas, Advances in Physical Geochemistry, vol. 9, pp. 152–191.

Holtz, F., Behrens, H., Dingwell, D.B., Johannes, W., 1995. H_2O solubility in haplogranitic melts: compositional, pressure, and temperature dependence. American Mineralogist 80, 94–108.

Keppler, H., 1993. Influence of fluorine on the enrichment of high field strength trace elements in granitic rocks. Contributions to Mineralogy and Petrology 114, 479–488.

Kohn, S.C., Charnock, J.M., Henderson, C.M.B., Greaves, G.N., 1990. The structural environments of trace elements in dry and hydrous silicate glasses; a manganese and strontium K-edge X-ray adsorption spectroscopic study. Contributions to Mineralogy and Petrology 105, 359–368.

Linnen, R.L., 1998. The solubility of Nb–Ta–Zr–Hf–W in granitic melts with Li and Li+F: constraints for mineralization in rare-metal granites and pegmatites. Economic Geology 93, 1013–1025.

Linnen, R.L., Keppler, H., 1997. Columbite solubility in granitic melts: consequences for the enrichment and fractionation of Nb and Ta in the Earth's crust. Contributions to Mineralogy and Petrology 128, 213–227.

Linnen, R.L., Keppler, H., 2002. Melt composition control of Zr/Hf fractionation in magmatic processes. Geochimica et Cosmochimica Acta 66, 3293–3301.

Linnen, R.L., Pichavant, M., Holtz, F., 1996. The combined effects of fO_2 and melt composition on SnO_2 solubility and tin diffusivity in haplogranitic melts. Geochimica et Cosmochimica Acta 60, 4965–4976.

Matthews, W., Linnen, R.L., Guo, Q., 2003. A filler-rod technique for controlling redox conditions in cold-seal pressure vessels. American Mineralogist 88, 701–707.

Montel, J.M., 1993. A model for monazite/melt equilibrium and application to the generation of granitic magmas. Chemical Geology 110, 127–146.

Mungall, J.E., Dingwell, D.B., Chaussidon, M., 1999. Chemical diffusivities of 18 trace elements in granitoid melts. Geochimica et Cosmochimica Acta 63, 2599–2610.

Mysen, B.O., Holtz, F., Pichavant, M., Beny, J.-M., Montel, J.-M., 1997. Solution mechanisms of phosphorous in quenched hydrous and anhydrous granitic glass as a function of peraluminosity. Geochimica et Cosmochimica Acta 61, 3913–3926.

Naney, M.T., 1983. Phase equilibria of rock-forming ferromagnesian silicates in granitic systems. American Journal of Science 283, 993–1033.

Nowak, M., Keppler, H., 1998. The influence of water on the environment of transition metals in silicate glasses. American Mineralogist 83, 43–50.

Ohlhorst, S., Behrens, H., Holtz, F., 2001. Compositional dependence of molar absorptivities of near-infrared OH- and H_2O bands in rhyolitic to basaltic glasses. Chemical Geology 174, 5–20.

Piilonen, P., Farges, F., Linnen, R., Brown Jr., G.E., 2003. Sn and Nb in dry and fluid-rich (H_2O, F) silicate glasses. XAFS-12 Conference, Malmö, Sweden. Physica Scripta. 3 pp.

Popp, R.K., Nagy, K.L., Hajash Jr., A., 1984. Semiquantitative control of hydrogen fugacity in rapid-quench hydrothermal vessels. American Mineralogist 69, 557–562.

Rapp, R.P, Watson, E.B., 1986. Monazite solubility and dissolution kinetics: implications for the thorium and light rare earth chemistry of felsic magmas. Contributions to Mineralogy and Petrology 94, 304–316.

Righter, K., Drake, M.J., 1999. Effect of water on metal–silicate partitioning of siderophile elements: a high pressure and temperature terrestrial magma ocean and core formation. Earth and Planetary Science Letters 171, 383–399.

Ryerson, F.J., Watson, E.B., 1987. Rutile saturation in magmas: implications for Ti–Nb–Ta depletion in island-arc basalts. Earth and Planetary Science Letters 86, 225–239.

Shaw, H.R., 1963. Obsidian-H_2O viscosities at 1000 and 2000 bars in the temperature range 700 ° to 900 °C. Journal of Geophysical Research 68, 6337–6343.

Siewert, R., Farges, F., Behrens, H., Buettner, H., Brown Jr., G.E., 1997. The effect of water on the coordination of Mo in albite glass. A high-resolution XANES spectroscopic study. Eos 78, F768.

Stolper, E., 1982. Water in silicate glasses: an infrared spectroscopic study. Contributions to Mineralogy and Petrology 81, 1–17.

Tuttle, O.F., Bowen, N.L., 1958. Origin of granite in light of experimental studies in the system $NaAlSi_3O_8$–$KAlSi_3O_8$–SiO_2–H_2O. Memoir-Geological Society of America 74. 153 pp.

Watson, E.B., 1979. Zircon saturation in felsic liquids: experimental results and applications to trace element geochemistry. Contributions to Mineralogy and Petrology 70, 407–419.

Watson, E.B., 1994. Diffusion in volatile-bearing magmas. In: Carroll, M.R., Holloway, J.R. (Eds.), Volatiles in Magmas, Reviews in Mineralogy vol. 30. Mineralogical Society of America, pp. 371–411.

Watson, E.B., Harrison, T.M., 1983. Zircon saturation revisited: temperature and composition effects in a variety of crustal magma types. Earth and Planetary Science Letters 64, 295–304.

Weyer, S., Munker, C., Mezger, K., 2003. Nb/Ta, Zr/Hf and REE in the depleted mantle: implications for the differentiation history of the crust–mantle system. Earth and Planetary Science Letters 205, 309–324.

Whitney, J.A., 1975. The effects of pressure, temperature, and XH$_2$O on phase assemblage in four synthetic compositions. Journal of Geology 83, 1–31.

Whitney, J.A., 1988. The origin of granite: the role and source of water in the evolution of granitic magmas. Geological Society of America Bulletin 100, 1886–1897.

Withers, A.C., Behrens, H., 1999. Temperature-induced changes in the NIR spectra of hydrous albitic and rhyolitic glasses between 300 and 100 K. Physics and Chemistry of Minerals 27, 119–132.

Wood, B.J., Blundy, J.D., 2002. The effect of H$_2$O on crystal-melt partitioning of trace elements. Geochimica et Cosmochimica Acta 66, 3547–3656.

Zhang, Y., 1999. H$_2$O in rhyolitic glasses and melts: measurement, speciation, solubility, and diffusion. Reviews of Geophysics 37, 493–516.

Available online at www.sciencedirect.com

Lithos 80 (2005) 281–303

LITHOS

www.elsevier.com/locate/lithos

Granitic pegmatites: an assessment of current concepts and directions for the future

David London*

School of Geology and Geophysics, University of Oklahoma, Norman, OK 73019, USA

Received 18 March 2003; accepted 9 September 2004
Available online 23 December 2004

Abstract

Although many explanations have been proposed for the internal zonation of granitic pegmatites, the most widely accepted model is attributed to R.H. Jahns. Jahns and Burnham [Jahns, R.H., Burnham, C.W., 1969. Experimental studies of pegmatite genesis: I. A model for the derivation and crystallization of granitic pegmatites. Econ. Geol. 64, 843–864] said that pegmatites owe their distinctive textural and zonal characteristics to the buoyant separation of aqueous vapor from silicate melt, giving rise to K-rich pegmatitic upper portions and Na-rich aplitic lower zones of individual pegmatites. Jahns and Tuttle [Janhs, R.H., Tuttle, O.F., 1963. Layered pegmatite–aplite intrusives. Spec. Pap.-Miner. Soc. Am. 1, 78–92] cited experiments as confirmation of this effect, but several experimental studies contradict the partitioning behavior that was the premise of Jahns' model. More recent work indicates that pegmatite-forming melts should cool quickly, or in any case, more quickly than crystallization can keep pace with. The distinctive textural and zonal features of pegmatites have been replicated in experiments that employ constitutional zone refining of melts that are substantially undercooled before crystallization commences. Melt boundary layers formed by this process would represent the last silicate liquids to crystallize in pegmatites, which explains the tendency in pegmatites for abrupt transitions from simple to evolved mineral and rock compositions. The sources of pegmatite-forming melts and of the causes of regional zonation within pegmatite groups represent important directions for future research.
© 2004 Elsevier B.V. All rights reserved.

Keywords: Granite; Pegmatite; Fractional crystallization; Experimental petrology

1. Introduction

This paper on granitic pegmatites honors Ilmari Haapala's contributions to the study of rapakivi and orbicular granites. Haapala's investigations of the metallogeny, geochemistry, and petrology of the rapakivi granites (e.g., Haapala, 1977; Haapala and Rämö, 1990, 1992, 1999) are an extension of his early interests in granitic pegmatites (e.g., Haapala, 1966). Whether discussing pegmatites or rapakivi and orbicular granites, the problem is the same: how to explain the origins of complex fabrics, zonation, and geochemical and mineralogical peculiarities that distinguish these rocks from ordinary granites.

* Tel.: +1 405 325 7626; fax: +1 405 325 3140.
E-mail address: dlondon@ou.edu.

0024-4937/$ - see front matter © 2004 Elsevier B.V. All rights reserved.
doi:10.1016/j.lithos.2004.02.009

This analysis of past, present, and future directions for the study of granitic pegmatites evaluates the proposed origins of fabrics and zonation that distinguish pegmatites from granites (Fig. 1). Recent petrologic experiments as analogues to natural pegmatites are presented, and advances in understanding particular features of pegmatites, such as layered aplites, are included. The manuscript is a synthesis with interpretation, which distinguishes it from a simple review. It is modeled closely after the manuscript by Jahns (1982), which was written in much the same style and served the same purpose—to show the steps and sources of data that led the

author to his particular view of the origins of pegmatites. This manuscript includes some personal communications from Jahns to me that he never published, and which I believe he intended for me to bring forward at some appropriate time in the future. These experiments figured prominently in his thinking about pegmatites and his own model for their internal evolution.

2. The past

Most geologists accept that granitic pegmatites are derived from larger masses of granite. The medium that forms pegmatite is thought to represent a late residual fraction of silicic melt that is squeezed out of a crystal-laden mush to form small pools within the source granite or dikes extruded into the surrounding rocks. Virtually every conceivable mechanism to explain pegmatites had been considered by the end of the 19th century (reviewed by Jahns, 1955). The most widely held view, which can be traced to Brögger (1890), is that the distinctive features of pegmatites arise from the interplay of coexisting silicate melt and water vapor. For nearly half a century, however, geologists have ascribed that model to Richard H. Jahns (Jahns, 1955, 1982; Jahns and Tuttle, 1963; Jahns and Burnham, 1969) because Jahns and Tuttle (1963, p. 90) cited experimental laboratory studies as confirmation of the essential role of water vapor in the formation of pegmatite:

"...their genesis can be explained in terms of the model proposed by Jahns and Burnham (1963) for the crystallization of granitic pegmatites. According to this model, segregation of major alkalies can occur in significant degree if a pegmatitic magma becomes saturated with volatile constituents, i.e., if both silicate melt and vapor are present in the system. Experimental evidence indicates that potassium is extracted from the liquid by the vapor in preference to sodium, and that potassium and other constituents can travel rapidly through the vapor in response to a temperature gradient. If the composition of the magma were at or near the thermal minimum for the confining pressure imposed upon it, preferential loss of the potassium feldspar component would promote crystallization of an albite-rich rock from the melt, probably in the form

Fig. 1. A pegmatitic segregation within granite (A), Reformatory Granite, Granite, OK. Many of the common features of pegmatites are evident in this sample, including a fine-grained zone (B) that is finer than the surrounding granite, a coarse-grained granophyric grading to graphic intergrowth of alkali feldspar and quartz (C), a pure-quartz zone (D), a clay and quartz-filled miarolitic cavity (E), and a region of hydrothermal alteration in the surrounding granite (F), where late-stage hydrothermal fluids exited the pegmatite from the miarolitic cavity.

Fractionated Granites and Pegmatites

of an aplite. Potash feldspar could crystallize from the vapor, either in the immediate vicinity or elsewhere in the system." (Jahns and Tuttle, 1963, p. 90)

The citation to Jahns and Burnham (1963) is given in the references as:

"Jahns RH, Burnham CW (1963). Experimental studies of pegmatite genesis. I. A model for the crystallization of granitic pegmatites (in preparation)."

Only one experiment from this period was presented (Wyllie, 1963), and that photo shows a texturally and spatially zoned experimental result of glass plus crystals. Jahns (1982) offered an interpretation of that experiment that was consistent with the Jahns and Burnham (1969) model for melt-vapor fractionation in a closed system. Wyllie (1963), however, attributed the heterogeneity of the same experiment to a thermal gradient across the experiment. In 1978, Jahns (personal communication, Geological Society of America National Meeting, San Diego, CA) told me:

"All I ever learned about pegmatites from experiments was from experiments that failed."

Jahns explained that "failed" referred only to the fact that these runs were intended as typical isothermal–isobaric experiments in a closed system (the precious metal capsule), but they leaked during the experiments and the silicate contents were redistributed along the length of the vessel and filler rod. The redistribution of Na, K, and Si in these experiments appeared to confirm the validity of his model for mass transfer through aqueous vapor, which emerged as he began to elucidate his own views of pegmatite genesis in his review article on pegmatites (Jahns, 1955). To my knowledge, the quotation above from Jahns and Tuttle (1963) referred to these open-system experiments as providing confirmation of the role of aqueous vapor in the formation of paired pegmatite–aplite dikes. Jahns and Tuttle (1963) recognized that the cause of the element redistribution was the thermal gradient that existed in the vessels (e.g., see Orville, 1963). The redistribution of components also occurs when thermal gradients exist in the melt, albeit at much slower rates because of the high viscosity of granitic silicate liquids (e.g., Lesher and Walker,

1991; Cygan and Carrigan, 1992). One important feature of Jahns' model that changed over time, however, was the role of a thermal gradient in promoting the segregation of alkalis. Jahns gradually came to accept that fractionation of alkalis between melt and vapor is an intrinsic feature of closed granitic melt–vapor systems at equilibrium (Jahns, 1982). The first experimental test of this hypothesis (Kilinc, 1969, as cited by Burnham and Nekvasil, 1986) showed congruent dissolution of alkalis into vapor in equilibrium with metaluminous granitic melt. More recent work shows a slight enrichment of Na over K in vapor, similar to aqueous fluids equilibrated with feldspars (Anovitz and Blencoe, 1999; Schäfer et al., 1999; Student and Bodnar, 1999). Jahns was a pioneer in crystallization experiments that started far from equilibrium because he recognized their value and application to pegmatites. It is very unfortunate that more of Jahns' experimental work was not published. What became the fundamental precept of the Jahns and Burnham (1969) model, however, the incongruent fractionation of K over Na into an aqueous fluid phase in a closed granitic system close to equilibrium, has not been corroborated by several experimental programs employing various methods with pure and saline aqueous fluids. London (1992, 1996) raised and discussed other problematic features of the Jahns model, which include numerous and significant discrepancies in the vertical zoning sequence and the distribution of Na- and K-rich domains within pegmatite dikes (e.g., London, 1985).

3. The present

3.1. Pressure and temperature of crystallization

Many geologists infer that pegmatites crystallize at the temperatures of the water-rich granite solidus, near 650–700 °C. Disparities commonly appear, however, when $P–T$ estimates derived from mineral assemblages, isotope systematics, or fluid inclusions are compared with hypothetical magmatic conditions. From field observations, Jahns (1955) suggested that pegmatites crystallize below 550 °C, although he later revised that upward to the range of the hydrous granite liquidus to be consistent with the magmatic equilibrium model. London (1986a) used the liquidus melting temperatures of crystal-rich inclusions to conclude that lithium-rich portions of the giant Tanco pegmatite, Manitoba, crystallized at approximately 450 °C. London (1986a), however, delineated an isobaric cooling path to ∼350–375 °C, followed by cooling on a geotherm. Chakoumakos and Lumpkin (1990) proposed an identical cooling path for the Harding pegmatite, New Mexico, although at higher pressure and entirely in the stability field of spodumene. Morgan and London (1999) found that feldspars from the Little Three pegmatite, California, record temperatures of 425 °C at the pegmatite margins to 375 °C near the pegmatite pocket zone. Such low temperatures determined by feldspar geothermometry or by fractionation of stable isotopes have been discounted by some on the presumption that they have been reset by reequilibration (e.g., Taylor and Friedrichsen, 1983), but the K-feldspar samples studied by Morgan and London (1999) are nonperthitic and the lack any other textural evidence of exsolution (e.g., albite rims). Sirbescu and Nabelek (2003) proposed that the Tin Mountain pegmatite, South Dakota, crystallized at ∼340 °C, and they noted that stable isotope systematics of quartz reflect crystallization near 350 °C as well (Nabelek et al., 1992). Thomas et al. (1988) advocated that silicate melt persisted down to 262 °C in the Tanco pegmatite, Manitoba. What these studies reveal is that the temperatures recorded by pegmatite mineral assemblages, fluid inclusions, and calibrated solid solutions commonly point to temperatures of crystallization well below the solidus temperatures of hydrous granitic melts.

The lithium aluminosilicate phase diagram (London and Burt, 1982; London, 1984) has been used as a petrogenetic grid for lithium-rich pegmatites. The primary lithium aluminosilicates, spodumene and petalite, are stable with quartz up to about 700 °C, which places an upper limit on the crystallization temperatures for Li-rich pegmatites. London (1984) confirmed that eucryptite (a feldspathoid) is stable with quartz at geologically feasible subsolidus conditions. London (1986b) observed that miarolitic gem-producing pegmatites from Afghanistan and California contain spodumene+quartz, which constrains these pegmatites to have solidified within the stability field of this assemblage. Assuming that the pegmatites crystallized close to the solidus temperatures of 550–

600 °C for a fluxed granitic melt, London (1986b) proposed crystallization at pressures near 275–350 MPa (depths near 7 to 10 km), the same depths as other, chemically similar but otherwise nonmiarolitic LCT pegmatites (Černý, 1991a; London, 1986a). These conditions are deeper than those normally inferred for the miarolitic class, which presented Černý (1991b, 2000) with a conundrum in genetic classification. The problem exists, however, only if spodumene-bearing miarolitic pegmatites crystallize near a feasible magmatic solidus. If, instead, such pegmatite melts experience appreciable cooling before crystallization commences at, say, 400 °C, then primary spodumene+quartz assemblages could form at pressures as low as ∼200 MPa, which is in line with estimates from other data sources (e.g., Taylor et al., 1979).

3.2. The importance of fluxes

Over most of the 20th century, geologists have reconciled the igneous composition of pegmatites with their apparent low temperatures of formation by suggesting that exotic components play a key role in lowering the liquidus temperatures of pegmatite-forming melts. These components are cited as "volatiles", or "hyperfusible" components (Jahns, 1953), but a more accurate and familiar term for them is "fluxing" components. The commonly cited fluxing components in pegmatite magmas are H_2O, B, F, and P (London, 1997). As fluxes, they lower the melting and crystallization temperatures (e.g., London et al., 1989, 1993; London, 1997), and they enhance miscibility among otherwise less soluble constituents (e.g., London, 1986a; London et al., 1987; Keppler, 1993; Thomas et al., 2000; Sowerby and Keppler, 2002). To the limited extent that they have been studied, increasing concentrations of these fluxing components appear to suppress the nucleation of quartz and especially feldspars (e.g., Fenn, 1977; London et al., 1989; Swanson and Fenn, 1992). London (1987) proposed that this effect stemmed from complex-forming (i.e., speciation) reactions between the fluxes and the major components of the melt. For example, F and P exhibit a strong tendency to associate with Na and Al in the melt (Manning, 1981; London et al., 1993; Wolf and London, 1994). Except for its strongly hygroscopic nature in melts, the behavior of B is less clear (Pichavant, 1981; Morgan et al., 1990), but the impact on crystallization is similar.

The vast majority of pegmatites, however, consists almost exclusively of quartz and feldspars and lacks exotic minerals, miarolitic cavities, and hydrothermal alteration envelopes. Their compositions plot close to the haplogranite minimum at elevated H_2O content (Jahns and Tuttle, 1963; Norton, 1966). Rare-element pegmatites that are manifestly rich in these or other exotic components constitute $< \sim 2\%$ of any particular pegmatite group (e.g., Černý, 1991c). Even the evolved rare-element pegmatites, including the well-known Tanco pegmatite, which may be the most fractionated igneous body on Earth, contains <2 wt.% total of the components B_2O_3, P_2O_5, and F (Černý, 1991c). In the most recent and precise estimate by Stilling (1998), the Tanco pegmatite whole rock contains 0.94 wt.% P_2O_5, 0.05 wt.% B_2O_3, and 0.18 wt.% F. A successful paradigm for pegmatites, therefore, must reconcile the apparent need for highly fluxed melts to produce crystals of pegmatitic size when only minor quantities of these fluxes are present in the bulk composition of all but a negligible few pegmatite bodies. Two current lines of inquiry have revised our understanding of pegmatites and provide a means to this end. One pertains to cooling histories, and the other follows from the effects of cooling on crystallization.

3.3. Cooling history

Recent efforts to model the cooling histories of some chemically evolved pegmatite dikes indicate that they cooled quickly. Based on inferred liquidus and host-rock temperatures, the large Harding pegmatite (20 m thick), New Mexico, would have cooled completely to its solidus in approximately 3 to 5 months after emplacement (Chakoumakos and Lumpkin, 1990). The gem-bearing Little Three dike (2 m thick), Ramona, CA, appears to have cooled to its solidus in about 25 days (Morgan and London, 1999), and the famous gem-producing Himalaya-San Diego pegmatites (~ 30 cm thick), Mesa Grande, CA, may have cooled to their solidi in just over 1 week (Webber et al., 1999). The principal source of uncertainty in the modeled cooling curves is the temperature of the host rocks, which is usually set at ~ 150–350 °C (for the

pegmatites cited above), based on the depth of emplacement, estimated geotherms, fluid inclusion analysis, and retrograde mineral assemblages in pegmatite host rocks (e.g., Morgan and London, 1987).

These calculated cooling rates are nothing short of revolutionary because most textbooks and dictionaries define pegmatites as the products of very slow cooling. Geologists have presumed that a positive correlation exists between the size of a crystal and the time frame in which it grows; therefore, in comparison with the much finer grain size of slowly cooled granite plutons, crystals of pegmatitic dimensions should require geologic eons to form. What the synthetic crystal industry has taught geologists, however, is that giant crystals of normally insoluble silicate and oxide crystals can be grown in weeks to months, given a properly fluxed growth medium and compositional or thermal gradients (e.g., O'Donoghue, 1987; Hughes, 1990).

3.4. Crystallization response

Early in the history of magmatic experimentation, Bowen (1913) demonstrated that crystallization studies of anhydrous silicic liquids would be impractical because the liquids tended to persist metastably below their liquidus temperatures, or else transformed to glass upon cooling. Few petrologists since Bowen (and Jahns, as noted above) have actually tried to study the crystallization of hydrous silicic melts at geologically relevant pressures. Those who have (e.g., Fenn, 1977; Swanson and Fenn, 1986; London et al., 1989; MacLellan and Trembath, 1991; Baker and Freda, 2001; Evensen et al., 2001) report two important observations: (1) there is a lag time, also referred to as the nucleation delay or incubation time, between cooling and the onset of visible crystallization, which is measurable in days to months for these liquid compositions (e.g., Dowty, 1980; Lofgren, 1980), and (2) as the magnitude of liquidus undercooling increases, crystal habits evolve progressively from euhedral to skeletal to radial spherulitic (e.g., Lofgren, 1974). Among these studies, the lag time is not well constrained, although all workers cite lag times of days to months. Previously, experimental cooling histories measured in days to months would have been considered irrelevant to natural pegmatites, which were thought to crystallize (and hence, cool) very slowly. With the new understanding that pegmatite dikes cool rapidly, the central questions surrounding the textural development of pegmatite are when crystallization commences and how it keeps up with cooling.

The undercooling of a hydrous silicic melt by more than ∼75 to 100 °C below the liquidus temperature results in prominent changes in crystal nucleation patterns and growth morphologies. For example, London et al. (1989) observed that progressively larger liquidus undercoolings promoted a gradual transition from crystal nucleation within the melt to sidewall-nucleation only, from crystals distributed through interstitial melt to solid crystallization fronts, and from random crystal orientations at low undercooling to increasingly anisotropic, oriented (comb-structure) fabrics with increased liquidus undercooling. In short, the greater the supersaturation of the melt obtained via rapid liquidus undercooling, the more pegmatite-like the fabric and texture. With increasing liquidus undercooling and delayed response of nucleation, all of the authors cited above have noted the gradual transition from euhedral crystal forms to increasingly skeletal morphologies, graphic intergrowths, and radial crystal growth patterns. These crystal habits are so prevalent in pegmatites that they may be taken as diagnostic.

3.5. Temperature gradients

The concepts presented above reintroduce thermal gradients into the pegmatite paradigm. Thermal gradients provided the driving force for mass transport and chemical fractionation in the experiments described by Jahns (personal communication cited above). In the cooling models described above, thermal gradients of several hundred degrees will develop within pegmatites shortly after emplacement.

The conduction of heat from pegmatite to host rock should produce axially symmetric thermal gradients and, hence, might be expected to produce pegmatites that are concentrically zoned or bilaterally symmetrical to the axial plane of the dike. A very large majority of granitic pegmatites manifests this zoning pattern (Jahns, 1982). Some pegmatites, however, do not, and in most instances, the coarse pegmatitic portions are displaced toward the top of a pegmatite body. Using

the corresponding compositions of garnets and tourmaline, Morgan and London (1999) determined that the lower two thirds of the Little Three dike (mostly aplite with some included microcline–quartz graphic intergrowths) had solidified before crystallization commenced on the pegmatitic hanging wall. The dike "pocket line", which is displaced above the axial centerline of the pegmatite, represents not only the last portion to crystallize, but also the last to cool. As heat loss drives crystallization, either the flow of heat from this body was not symmetrical or the upper portions of the pegmatite were or became hotter than the footwall parts. The existence and causes of a vertical thermal asymmetry are speculative at this point. One possible mechanism might be convection, which Burnham and Nekvasil (1986) deemed necessary to promote the chemical zonation within pegmatites. Given their mostly ordinary granitic compositions and low temperatures of formation, however, magmas of thin pegmatite dikes would be far too viscous to convect (Jahns, 1982). The reaction of hydrous melt to crystals+H_2O vapor is exothermic, but the heat of vaporization of H_2O from the silicate melt is "negligible" (Burnham and Davis, 1974; Burnham, 1979). Buoyant ascent of aqueous vapor could transfer heat from the crystallizing footwall, but that does not explain why crystallization should commence only along the footwall in the first place.

3.6. Sequential crystallization and zonation of feldspars

In isothermal experiments that entailed liquidus undercooling before crystallization commenced, London et al. (1989) observed that the sequence of crystallization of feldspars followed, more or less, the line of descent upon the liquidus surface, although the system began crystallization far from that surface. The tendency, which we have seen repeatedly (e.g., Morgan and London, 2003), is that the melt system responds to rapid undercooling (and consequent supersaturation of melt) by the precipitation of the phase or solid solution composition that is most metastable (farthest from equilibrium), not the most stable, at the conditions of growth. A possible explanation for this tendency is illustrated schematically in Fig. 2, which depicts a hypothetical T–X phase diagram for the system AB with no solid

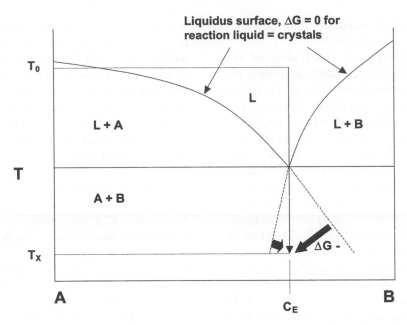

Fig. 2. Schematic illustration of crystallization response in an undercooled silicate liquid of eutectic composition in the system AB. Heavy solid arrows depict the magnitude of ΔG $(-)$ for the reactions A(liquid)=A(crystals) and B(liquid)=B (crystals) along the metastable extensions of their respective liquidus surfaces. See the text for further explanation.

solution between A and B. For the reaction liquid= crystals, the liquidus surface represents the locus of T–X points for which ΔG=0. Consider a melt of eutectic composition C_E that is undercooled to temperature T_X before crystallization commences. Although a eutectic assemblage might be expected, the undercooled melt lies farther from the metastable extension of the liquidus surface for A than for B. Assuming that ΔG increases similarly for phases A and B away from their respective liquidus surfaces, then ΔG $(-)$ for the reaction A(liquid)=A(crystal) is greater than for B. Ignoring kinetic effects, the driving force to form A is greater, and A should precipitate first. As the crystallization of A shifts the bulk composition toward B, then the relative magnitudes of ΔG $(-)$ for A and B reverse, B will crystallize, and sequential, possibly oscillating crystal assemblages are produced from the initially eutectic composition.

In our experiments, an assemblage containing ternary sodic feldspar (oligoclase) is the first formed because that feldspar composition is farther from equilibrium than the solvus pair of albite and orthoclase, which are stable at the growth temperature. Fig. 3 shows this evolutionary trend from ternary sodic feldspar as the first-formed phase to albite–

orthoclase pairs, whose compositions match exactly the strain-free solvus compositions at those conditions. Because the crystallization is sequential, a sharp

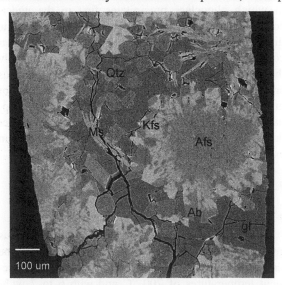

Fig. 3. Backscattered electron image of experiment MAC-149, Macusani rhyolite obsidian+3 wt.% added H_2O, preconditioned at 750 °C, then dropped in a single isobaric step to 575 °C (liquidus temperature ~ 725 °C). Phases include ternary alkali feldspar (Afs), albite (Ab), K-feldspar (Kfs), quartz (Qtz), and muscovite (Ms).

mineralogical zonation is manifested from the margins to the center of the solidified melt.

Recently, Acosta-Vigil et al. (2002) observed rapid diffusion and fractionation of alkalis in granitic melt when a chemical gradient was induced by reaction coupling at a crystal–melt interface. Although these experiments were conducted at 800 °C, Na diffused through the entire melt column up to 5 mm in length to achieve a steady-state profile in as little as 5 days. Rates of diffusion for most components should fall substantially with increasing viscosity (e.g., Watson and Baker, 1991, and references therein). The diffusion of alkalis, however, is reportedly independent of melt viscosity (Mungall, 2002). Therefore, the experiments of Acosta-Vigil et al. (2002) opened the possibility that the fractionation of alkalis in pegmatites might be promoted by the same mechanism and might be rapid even at the low temperatures and high viscosities inferred for the bulk pegmatite melt. In the case of pegmatites, the driving force for diffusion would be the chemical gradient in Na caused by the crystallization of plagioclase (usually oligoclase) at the pegmatite margins. Once Na was removed to the locus of plagioclase crystallization, the remaining pegmatite would become relatively enriched in K and produce an abundance of K-feldspar. By this process, in which oligoclase precipitates first in the under-cooled regions, the hotter portions of the pegmatite dike may become enriched in potassic feldspar components (London, 1992). This trend is predictably the opposite of that achieved with the simultaneous crystallization of alkali feldspars from aqueous vapor in a thermal gradient, wherein sodic feldspar precipitates in hotter regions and potassic feldspar in cooler domains (Orville, 1963, and in Jahns' open-system experiments described above).

3.7. Viscosity of pegmatite-forming melts

3.7.1. Hydrous granitic liquid

Most pegmatite-forming melts are simple hydrous granitic liquids, and their melt viscosities will be similar to those of granitic melt. At likely emplacement temperatures of 700 °C and with 6 to 7 wt.% H_2O, corresponding to H_2O-saturated melts at ~ 200 MPa, the viscosity of such melt will be on the order of 10^5 to 10^6 Pa·s (e.g., Baker, 1996; Dingwell et al., 1996). If the same melts cool to ~ 450 °C, which

would be close to their glass transition temperatures (e.g., Seifert et al., 1981; Dingwell et al., 1996), without loss of H_2O, then the viscosity climbs to $\sim 10^8$ Pa·s. These would be exceedingly viscous liquids through which mass transfer of most components, in particular, the network-forming components Al and Si, via diffusion or by the upward migration of water bubbles should be exceedingly slow on the time frame of continued cooling.

3.7.2. Fluxed granitic melts

In contrast, the common petrologic fluxes lower the viscosity of granitic melts, but only in proportion to their mole fractions (e.g., Bagdassarov et al., 1993; Dingwell et al., 1992; Thomas, 1994; Thomas and Klemm, 1997; Baker and Vaillancourt, 1995). On an equimolar basis at constant P and T, the fluxing components reduce the viscosity of haplogranite melt in the order H>F>B>P, with viscosity differences between consecutive components of about 10–100 Pa·s (Dingwell et al., 1996). Thomas and Webster (2000) calculated the viscosity of a very highly fluxed melt containing $H_2O+F=17.3$ wt.% to be 4.7 Pa·s at 600 °C, which they equated to the viscosity of castor oil at 20 °C. London (1986a) estimated the viscosity of highly fluxed melt in the system $Li_2B_4O_7$–$NaAlSi_3O_8$–SiO_2–H_2O at 500–540 °C and 200 MPa from the settling of quartz crystals in capsules. The solution of the Stokes equation yielded a maximum viscosity of 10 Pa·s.

3.8. Comparative assessment of models for internal evolution

3.8.1. Vapor exsolution and mass transfer

The basic hypothesis of the Jahns model (Jahns and Tuttle, 1963; Jahns and Burnham, 1969) was that the incongruent partitioning of alkalis between the melt and an upwardly buoyant vapor phase generated the chemical and textural segregation that are the hallmark of granitic pegmatites. The partitioning of alkalis, which was the foundation of this model, is not supported by any of several experimental tests, and numerous other conflicts of zonation, crystal morphology and rock fabric, and mineralogical zonation remain unanswered (London, 1992, 1996). Jahns (1982) later characterized the vapor phase as a film at the crystal–melt boundary, and as such, he advocated a boundary-layer effect for selective partitioning and crystal growth. What Jahns may not have anticipated is the extent to which the build-up of B, P, and F could enhance the silicate–H_2O miscibility and change the chemical and physical properties of the boundary layer medium.

3.8.2. Constitutional zone refining

Athough highly fluxed silicate melts seem the most likely materials to produce pegmatite, the problem remains that most pegmatites are essentially simple granites in composition. Although highly fluxed silicate inclusions representing former liquids have been documented in pegmatites (e.g., London, 1986a; Thomas and Klemm, 1997; Thomas et al., 2000), these tend to be far from the bulk compositions of the pegmatites that contain them. A scenario that couples a delayed response of crystallization to the relatively rapid cooling of silicic melts at elevated pressures (London, 1990, 1992, 1996) provides the means to develop an initially small volume of flux-rich melt from a far less evolved bulk composition. When delayed crystallization commences, it usually does so as a solid crystal front along the margins of an undercooled melt body, where phase boundaries may facilitate crystal nucleation (Fig. 3). The driving force to grow crystals may be very large (cf. Durant and Fowler, 2002), but if the melt is viscous, the diffusion of essential constituents through that medium becomes the initial rate-limiting step in crystal growth. When crystal growth commences, incompatible components, including the fluxes, are rejected at the growing crystal interface of quartz and feldspar, and they concentrate along the margins of the growing crystal front to form a fluxed boundary layer of melt. Even if the bulk concentration of fluxes is low, such a boundary layer can form if the diffusion of excluded components through the bulk melt is slow in relation to crystal growth. Because of its increasingly flux-rich composition, the boundary layer liquid will acquire a low solidus temperature and enhanced silicate–H_2O miscibility. The scenario described above applies the metallurgical process of constitutional zone refining to igneous solidification (also see McBirney, 1987; McBirney and Russell, 1987). The flux-rich boundary layer advances into a solid or semisolid body ahead of the crystal growth front. In constitutional zone refining, normally, incompatible elements are concen-

trated in the melt boundary layer, where they act as fluxes to lower the solidus of the melt. The crystalline assemblage possesses a rather ordinary composition as the boundary layer liquid becomes more evolved and grows in volumetric proportion. The process provides an excellent explanation for zonation, especially in thin miarolitic pegmatites. Approaching the end stages of crystal growth, the boundary layers would merge and become the principal fluid medium left in the pegmatite. Because the boundary layer composition is substantially different from that of the bulk melt, the crystallization of the boundary layer itself will produce an abrupt change in mineral chemistry (Fig. 4), from rather ordinary mineral assemblages to highly evolved and unusual minerals. This is the case in many gem-producing miarolitic pegmatites around the world (e.g., California: Morgan and London, 1999; Elba, Italy: Pezzotta, 2000).

Evidence for boundary layers in natural igneous rocks has been more theoretical than empirical, and boundary layers have not been documented in relevant crystal growth experiments until recently (Fig. 5). The B-rich boundary layer shown in Fig. 5 is interesting for three reasons: The glass composition is exceedingly alkaline, it is uncommonly H_2O rich, and the alkali ratio is very sodic compared with the bulk melt (Table 1). Note also that, despite containing $\sim$ 13.42 wt.% H_2O and 18.30 wt.% B_2O_3, equivalent to a combined 48.77 mol% of these two components on a simple oxide basis, the boundary layer liquid quenched to a water-insoluble glass that takes a polish and does not evidently dehydrate (usually manifested by the development of cracks) under vacuum in the electron microprobe. In all respects, the composition of this experimentally generated boundary layer is similar to the bulk composition of trapped melt inclusions from the Tanco pegmatite, Manitoba (London, 1986a). This important correlation explains why the exotic crystal-rich inclusions studied by London (1986a) are prevalent in primary spodumene and petalite throughout the Tanco pegmatite, although the estimated compositions of those inclusions are far from any plausible bulk melt composition (Černý, 1991c). The inclusions represent trapped boundary layer melts.

Included crystals that were produced only by local saturation at the crystal–melt interface may also reveal the former presence of fluxed boundary layers. For

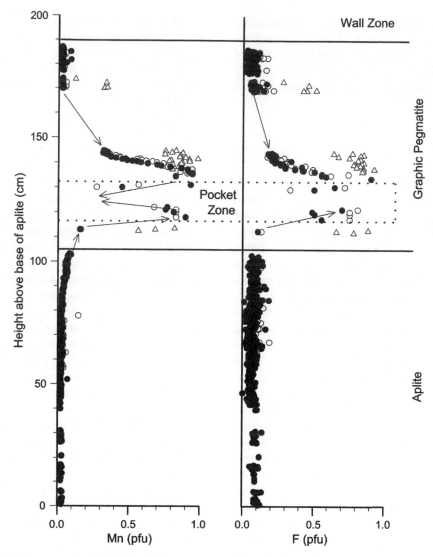

Fig. 4. Chemical variations in tourmaline from the Little Three pegmatite, Ramona district, San Diego County, CA. The flat patterns for Mn and F, and their abrupt increases approaching the pocket zone, cannot be accounted for by Rayleigh fractionation or simple zone refining. From Morgan and London (1999).

example, Fig. 6 shows cryolite inclusions in quartz that were generated by the crystallization of a haplogranitic melt initially containing <2 wt.% F. Cryolite was deposited along the quartz growth surfaces at the inception of crystallization, although the F content of the bulk melt was well below cryolite saturation. This is attributable to the creation of an alkaline sodic boundary layer sufficiently enriched in F to have achieved cryolite saturation at the quartz growth surfaces only. In such cases, it may be possible

to identify the former presence of boundary layers, either by the unusual bulk compositions of crystal-rich inclusions or simply by the prevalence of an unusual mineral (saturated only in the boundary layer melt) as a commonly included phase.

3.8.3. Internal zonation: an equilibrium or disequilibrium process?

A fundamental difference in the Jahns–Burnham model (Jahns and Burnham, 1969; Burnham and

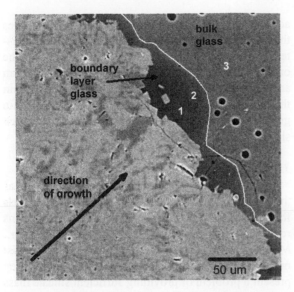

Fig. 5. Backscattered electron image of a boundary layer of melt (quenched to glass) at a crystal growth front. The experiment (PEG-25) comprised of a mixture of quartz, albite, and orthoclase (200 MPa H_2O minimum composition)+3 wt.% B_2O_3 (no added H_2O), was preconditioned to 850 °C for 25 h, then cooled in a single isobaric step to 450 °C for 670 h. The composition of the melts (1–3) in and away from the boundary layer shows marked changes in Na/K, B_2O_3, and H_2O contents (see Table 1). The outer edge of the boundary layer is delimited by a solid white line.

Nekvasil, 1986) and that advocated by London (e.g., London et al., 1989; London, 1990, 1992, 1996) concerns the tendency of pegmatites to represent systems near equilibrium on the liquidus (Jahns and Burnham, 1969; Burnham and Nekvasil, 1986) or not (London, 1990, 1992, 1996). Jahns (1982) suggested that pegmatites generally are not vapor saturated upon emplacement, but become so with crystallization (also termed second boiling). In this case, the liquid line of descent for a typical complex pegmatite will be represented by the experimental liquidus diagram for the bulk composition of the Harding pegmatite, New Mexico (Burnham and Nekvasil, 1986). For equilibrium crystallization on or near the liquidus, the crystalline assemblage will evolve from saturation in a single crystalline phase to multiphase crystalline assemblages in the sequence, with decreasing temperature, of:

quartz
quartz+muscovite
quartz+muscovite+K-feldspar+plagioclase
quartz+muscovite+K-feldspar+plagioclase+
  spodumene

The tendency of zonation in granitic pegmatites, however, is to evolve from multiphase mineral assemblages at the onset of crystallization (border zones) to singly saturated units at the end (e.g., block microcline zones and quartz cores). This is evident in the zoning sequence for the Black Hills pegmatite district, reported by Cameron et al. (1949):

quartz I muscovite+K-feldspar+plagioclase±
  tourmaline, garnet, biotite
K-feldspar+quartz
plagioclase+quartz+spodumene
quartz+spodumene
quartz+K-feldspar
quartz

The tendency to form monomineralic segregations of alkali feldspar or quartz is even more prevalent in compositionally simpler pegmatite districts (Cameron et al., 1949). The general sequence of crystallization

Table 1
Compositions[a] of glasses in experiment PEG-25[b]

Point		wt.%	mol%
3	SiO_2	68.94	62.10
	Al_2O_3	12.01	6.38
	Na_2O	5.10	4.45
	K_2O	3.37	1.94
	B_2O_3	2.93	2.28
	H_2O	7.61	22.86
2	SiO_2	57.04	46.79
	Al_2O_3	9.69	4.68
	Na_2O	5.77	4.59
	K_2O	2.87	1.50
	B_2O_3	12.22	8.65
	H_2O	12.35	33.78
1	SiO_2	50.88	40.87
	Al_2O_3	7.79	3.69
	Na_2O	6.66	5.19
	K_2O	2.90	1.49
	B_2O_3	18.30	12.69
	H_2O	13.42	36.08

[a] Electron microprobe analyses, Cameca SX-50 microprobe, University of Oklahoma, with beam conditions as in Morgan and London (1996); H_2O calculated by difference.
[b] See numbered points in Fig. 5 for locations.

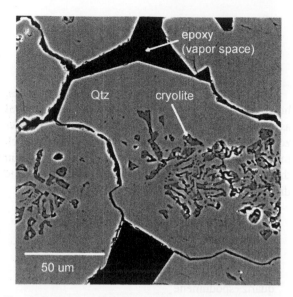

Fig. 6. Backscattered electron image of experiment PEG 07, composed of a mixture of quartz, albite, and orthoclase (200 MPa H_2O minimum composition)+2 wt.% F+11 wt.% H_2O, preconditioned to 850 °C, single-step isobaric undercooling to 550 °C, H_2O-saturated on quench. Note that cryolite was included at the centers of quartz (Qtz) crystals and presumably grew at the earliest stages of quartz crystallization, when the melt contained <2 wt.% F.

observed in our experiments that have been appreciably undercooled, such that crystallization began far from the equilibrium of the liquidus, can be generalized by:

alkali feldspar+quartz±biotite
plagioclase+quartz±muscovite
K-feldspar+quartz±muscovite
K-feldspar
quartz±petalite, if Li-rich
quartz

This sequence more accurately mirrors the zoning sequence found in natural rocks.

The essential feature of the pegmatite-forming model advocated by London (1990, 1992, 1996) pertains to low rates of crystal nucleation from a fluxed boundary layer, which, in turn, is generated from viscous granitic melt via crystallization at conditions away from the equilibrium liquidus boundary. Although rapid liquidus undercooling is indicated by the cooling models for pegmatite dikes, all that would be necessary is for crystallization to lag behind cooling, whatever that rate, such that flux-enriched residual melts can be generated in

advance of the crystal growth front. The fluxes themselves appear to increase the metastable persistence of melt and hence increase the lag time to the onset of crystal nucleation and growth (e.g., Morgan and London, 2003). This effect alone might explain the occurrence of pegmatite segregations within granite masses, wherein cooling presumably is much slower.

3.8.4. Gels

Merritt (1924a,b) advocated viscous silicate gels as the pegmatite-forming medium, and gel-based models for pegmatites and other cavity fillings in igneous rocks have been proposed again recently (e.g., Merino, 1999; Taylor et al., 2002). Gels consist mostly of water (e.g., Henisch, 1988), and they present a problem of providing sufficient silicate mass and, at the same time, only minor volume reduction upon crystallization. Aside from the question of their origins, the emplacement of pegmatite dikes as rigid gels is problematic. When the gel-forming medium is itself a component of the growing crystals, as it would be in the case of granitic pegmatites, then the gel decomposes, rapid diffusion is reestablished, and the result is a fine-grained crystalline product (e.g., summarized in Eitel, 1954). This is especially true for amorphous silica colloids at elevated P and T (Betterman and Liebau, 1975; Oehler, 1976). The concentrations of H_2O or other fluxing components needed to form gels, as they are presently known, would most likely occur in a boundary layer of melt adjacent to a crystal surface (e.g., see Williamson et al., 1997) or wherever the boundary layer liquids accumulate. Now, however, a disparity is introduced—whether highly fluxed silicate liquids at low temperatures (~ 400 °C) might be present as viscous gels, or rather as exceedingly low-viscosity fluids. From the physical evidence of the quenched boundary layer in Fig. 5, it might be construed that the low-viscosity fluxed melts solidify or gel rapidly over a narrow temperature interval.

Some features of pegmatites suggest gel-like properties of the original fluid media. London (1987) noted textures in the Tanco pegmatite, Manitoba, that resembled plastic or viscous liquid–liquid contacts between pure quartz bodies and layered aplites. Teardrop-shaped layers of aplite appeared to have been suspended in a viscous medium that is now pure quartz, and pure quartz filled what appeared to be delaminations or shrinkage cracks in layered aplites (London, 1992). For the most part, these are restricted to the latest stages of pegmatite consolidation, when boundary layer fluids may comprise a large proportion of the remaining amorphous silicate material.

3.8.5. Liquid miscibility and immiscibility

An accumulation of B, F, and P to high concentrations in silicate melt can enhance H_2O solubility to the point of complete miscibility (e.g., Fig. 5), leading to a supercritical transition from silicate melt to aqueous fluid at P–T conditions appropriate for pegmatites (e.g., London, 1986a; Thomas et al., 2000; Sowerby and Keppler, 2002). Most of this evidence is from melting experiments on natural glass- or crystal-rich inclusions from pegmatites (e.g., London, 1986a; Thomas and Klemm, 1997; Thomas et al., 2000), and hence, supercritical behavior from silicate to aqueous fluid is a real possibility where flux-rich melts accumulate. In highly fluxed melts, the components of B, P, and F dominate the behavior and miscibility of H_2O and silicate components and forestall aqueous vapor saturation of melts (London, 1986a,b).

The Jahns and Burnham (1969) model hinges on the experimental observation of immiscibility between silicate melt and aqueous vapor in the haplogranite–H_2O system at low to moderate pressures. Simple Rayleigh fractionation predicts that a granitic melt starting with 2 to 3 wt.% H_2O should achieve the H_2O saturation of the melt at ~75% crystallization and moderate crustal pressures (e.g., Jahns and Burnham, 1969), and the addition of less melt-soluble components, such as CO_2 and NaCl (e.g., Blank et al., 1993; Webster and De Vivo, 2002), that are miscible with H_2O signify that the immiscible melt-vapor phenomena should be present at lower percentages of crystal fractionation.

Despite a model that predicts the aqueous vapor saturation of silicate melt at the pegmatite-forming stage of granitic plutons, evidence that pegmatites possess an aqueous vapor from their inception, in addition to silicate melt, is conspicuously missing. Indeed, Jahns, who noted that most pegmatites possess a granitic border unit, argued that granitic texture signified and originated from magmatic crystallization in the absence of an aqueous vapor and, hence, that pegmatites themselves were vapor undersaturated upon emplacement (Jahns and Burnham, 1969; Jahns, 1982). London (1992, 1996) has assessed the attributes of pegmatites and their enclosing wallrocks in search for evidence of an aqueous vapor in the early stages of pegmatite consolidation, and the evidence is mostly absent or contradictory to the concept. This issue could be regarded as one of the last major questions surrounding the internal differentiation of pegmatites, and it should be considered as part of future work on the origins of pegmatites, as discussed in a following section.

3.9. Layered aplites

Although pegmatites are thought of mostly as exceedingly coarse-grained rocks, they commonly contain very fine-grained quartzofeldspathic units, which are termed aplites. These may or may not be layered (Fig. 7), but the layered rocks have attracted the most geologic scrutiny. Origins by replacement of preexisting rock (see Jahns, 1955), mineral segregation during viscous flow, crystal settling (e.g., Kleck and Foord, 1999), and formation from vapor-saturated melt (e.g., Jahns and Tuttle, 1963; Rockhold et al., 1987) have been deemed viable mechanisms. Although one argument or another might prevail in a particular setting, the most widely accepted explanation has been that of Jahns (Jahns, 1982; Jahns and Tuttle, 1963). The quotation from Jahns and Tuttle (1963) cited above pertains specifically to the formation of aplite. In that model, the volumetrically important aplites in paired aplite–pegmatite intrusions form because the loss of potassium to a buoyant aqueous phase shifts the granitic melt away from the (thermal) minimum composition toward the albite–quartz join. As a result of "compositional quench", the crystals nucleate rapidly because of pronounced liquidus undercooling of the melt. Jahns (personal communication, 1978) explained layering in aplite in terms of migration of the quartz–alkali feldspar cotectic in response to oscillations of internal H_2O vapor pressure in pegmatites (Fig. 8). Episodic loss of vapor overpressure through rupturing of the dike was Jahns' mechanism for oscillation of internal H_2O pressure. Crystal settling is either plausible (e.g., Kleck and Foord, 1999) or implausible (e.g., Webber et al., 1997), depending on the estimates of melt viscosities, for which temper-

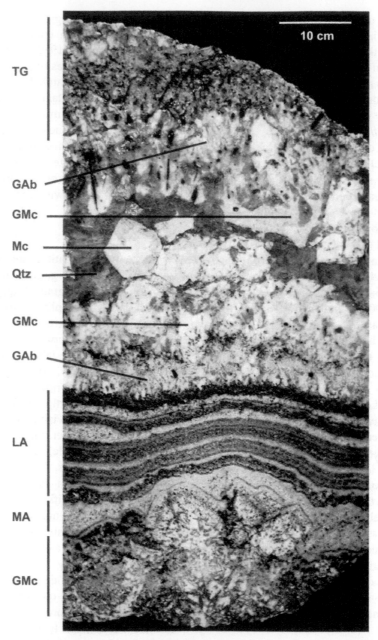

Fig. 7. Complete section through a subhorizontal layered aplite–pegmatite dike, Little Three mine property, Ramona, CA. The lower margin (dike footwall) begins abruptly with graphic and highly skeletal microcline–quartz intergrowths (GMc) that trap tourmaline within the intergrowths and along a sharply defined massive aplitic (MA) boundary above the terminus of the crystals. Some layers of tourmaline (the same as those that define layering in the aplite) are also evident within the graphic microcline–quartz intergrowths. The massive aplite is succeeded inwardly by layered aplite (LA), defined by alternations of black tourmaline with fine-grained granitic layers. These are followed by graphic microcline (GMc) crystals and graphic albite–quartz intergrowths (GAb) that flair upward and fill the lower portions of the pegmatite dike. A quartz-rich zone (Qtz), with a cluster of perthitic microcline crystals (Mc) occupy the center of the dike. Down from the hanging wall, the border zone is tourmaline granite (TG) that evolves downward into a comb-structured intergrowth of quartz, albite, and tourmaline (GAb) with minor muscovite. Note that the graphic albite–quartz intergrowths (GAb) appear to have nucleated around thin black tourmaline crystals that project downward. Although a few large, blocky crystals of graphic and perthitic microcline–quartz crystals (GMc) grew down from the hangingwall, these are volumetrically subordinate to similar crystals that radiate up from the footwall.

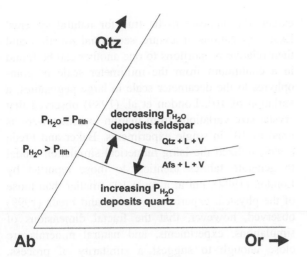

Fig. 8. Schematic portion of the quaternary liquidus diagram for Ab–Or–Qtz–H_2O, showing the displacement of the cotectic as a function of variations in H_2O pressure (based on work by Tuttle and Bowen, 1958).

atures and, especially, flux contents are poorly known but important variables.

London et al. (1989) were consistently able to generate solidification fronts of alkali feldspar, followed by a solid overgrowth of pure quartz, but they did not produce repetitions of this pattern, i.e., banding similar to layered aplites. The sequential development of these monominerallic feldspar–quartz layers, however, led London (1992) to propose that pronounced liquidus undercooling might promote layering in aplites, i.e., an oscillatory feedback between the crystallization front and its corresponding boundary layer of melt (e.g., as portrayed in Fig. 2). London likened the process to that proposed by Fenn (1986) for the localized saturation of quartz and microcline, which creates graphic granite, but noted that it might include any phase that reached local saturation in the boundary layer of the melt that advanced before the crystallization front. Webber et al. (1997) reached a similar conclusion, suggesting that the liquidus undercooling that generated layered aplites within the George Ashley Block, San Diego County, CA, might have been triggered by the devolatilization of melt accompanying a pressure drop. In the classic subhorizontal pegmatites from San Diego County, CA, layered aplites also typically contain abundant graphic quartz–microcline intergrowths. The simultaneity of graphic granite and layered aplite is sometimes evident

within individual microcline crystals when layering that is conspicuous in adjacent aplites runs right through the microcline crystals (Fig. 7).

Fine-grained alternations of quartz and alkali feldspar resembling layered aplites (Fig. 9a) have now been produced in experiments involving hydrous granitic melt plus added boron (London, 1999). These experiments began with a mixture of quartz, albite, and orthoclase made to the 200 MPa hydrous granite minimum plus 3 wt.% B_2O_3 added as glass. The compositions were taken to 850 °C (above the liquidus) for 96 h, then dropped in a single isobaric cooling step to 450 °C for 650 h. Elsewhere in the same experiment, two interesting types of chemical zonation are manifested (Fig. 9b). The layered aplite grew toward a zone dominated by potassic feldspar, and this was succeeded, in turn, with the growth of a texturally similar graphic zone in which the proportions of feldspar to quartz remain the same but the normative Or and Ab contents are precisely reversed. The graphic feldspar–quartz zone was succeeded inwardly by a zone of pure alkali feldspar (corresponding to "block feldspar" zones of some pegmatite classification schemes) and, finally, centrally located pure quartz masses (the cores).

In experiments of this sort, which go to completion, it is always difficult to uniquely interpret cause and effect (cf. Jahns, 1982, p. 302). Fluctuations in pressure, however, played no part in generating the layered aplite. No water was added to the experiment (only that adsorbed to the borate glass powder), and the charge may not have been water saturated throughout most of its history of crystallization (the fully crystalline end product, however, did contain free water upon opening the capsule). The product formed by crystallization far from the equilibrium of the liquidus, and diffusion-limited saturation in alternating phases is a likely explanation of the process.

3.10. Physical scaling of experiments and natural pegmatites

The experimental results shown in Fig. 9 replicate most aspects of internal fabrics and zoning in a single, typical granitic pegmatite to a remarkable extent. This and similar experiments, however, also must be reconciled with physical scale: the experimental

a

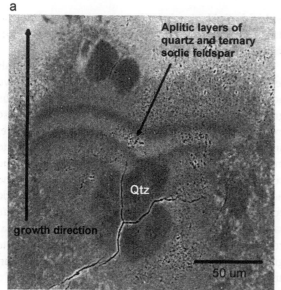

b

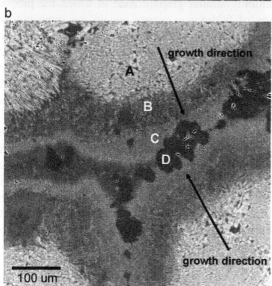

Fig. 9. (a) Backscattered electron image of alternating layers of quartz and ternary sodic feldspar. The experiment (PEG-16) was composed of a mixture of quartz, albite, and orthoclase (200 MPa H_2O minimum composition)+3 wt.% B_2O_3 (no added H_2O), was preconditioned to 850 °C for 72 h, then cooled in a single isobaric step to 450 °C for 650 h. (b) Zoned crystallization elsewhere in PEG-16. Four sequential zones, Parts A–D can be recognized. Their normative compositions are $Ab_{13}Or_{62}Qtz_{25}$ (A), $Ab_{62}Or_{12}Qtz_{26}$ (B), $Ab_{61}Or_{35}Qtz_4$ (C), and $Ab_{00}Or_{00}Qtz_{100}$ (D).

products constitute very fine-grained rocks, and their direct application to pegmatites requires that the same process operates at widely different scales. To a large

extent, this appears to be true for natural systems. Exact replications of texture, sequential zonation, and their relative proportions to one another can be found in a continuum from the millimeter scale in granophyres to the decameter scale in large pegmatites, a variation of 10^4. London et al. (1989) observed that crystal size variations from margin to center were as great as 10^3 in single experiments. Baker and Freda (1999) utilized the Ising numerical simulation model to generate fabrics identical to those reported by London (1992), but at a scale 10^2 smaller than those of the physical experiments. Baker and Freda (1999) observed, however, that the fractal dimensions of simulations, experiments, and natural minerals are close enough to suggest a similarity of process. Although the fractal dimension of natural crystals from pegmatites was not well established, Baker and Freda (1999) suggested that the observed similarities augur well for the applications of simulations and experiments to natural pegmatites.

3.11. Why are most pegmatites granitic in composition?

A model based on high viscosity of the pegmatite-forming medium (supercooled silicate liquid, gel, or glass) explains why granitic pegmatites are far more common than those of basic or alkaline composition. The higher viscosity of granitic liquids inhibits the diffusion necessary to nucleate crystals, as evidenced by long incubation times before the onset of crystallization in silicic melts. The higher viscosity of granitic liquids also impedes the diffusion of excluded components back into the melt and hence makes boundary layer formation more likely. When crystallization finally commences, granitic melts tend to be farther below their liquidus temperatures, and pegmatitic fabrics are the result. Fluxes are important to the process, and the more these can be concentrated in boundary layers, as opposed to dispersed or lost from the pegmatite body, the more effective they become.

4. The future

My view of the future for the study of granitic pegmatites is influenced by my belief that constitutional zone refining in an undercooled melt body accounts for most of the essential features of pegmatites, with or without the involvement of an aqueous vapor phase. Cameron et al. (1949) came to this same conclusion, and Jahns (1953) initially embraced it as well. Much remains to be understood about how this process works, but it provides a theoretical and now an experimental framework that relates the diverse textures of pegmatites along with their chemical fractionation and zonation trends. Virtually all of the features of pegmatites can be reconciled with the constitutional zone refining model, and all of the essential and diagnostic features (texture, fabric, crystal morphology, and spatial zonation) have been generated by experiments that employ this process of melt fractionation. Evidence for boundary layers can be found in natural silicate inclusions, whose fluxed and unusual compositions lie far from any feasible pegmatite bulk composition, trapped in pegmatite minerals. The presence of an aqueous or other immiscible fluid phase in addition to melt requires more investigation, but I believe it plays a secondary role in any case. Crystallization from an aqueous vapor in the presence of melt does not promote the graphic or skeletal habits, the unidirectional fabrics, or the fractionation trends seen in pegmatites (London et al., 1989). For these reasons, I believe that the major thrust of pegmatite research should lie outside of individual pegmatite bodies and should aim to relate pegmatites to their sources and surroundings.

The current paradigm for pegmatites entails the emplacement of crystal-free melts along a dike system that emanates from a pluton, such that those dikes cool substantially below their (near liquidus) emplacement temperatures before crystallization commences. Regardless of cooling rate, the farther the melts cool below their liquidi before crystallization commences, the more pegmatitic are their crystalline fabrics and zonation. If pegmatites are derived only after extended fractional crystallization of their source pluton, then heat loss from the pluton itself will facilitate the migration of melts along fractures away from the main body (e.g., Baker, 1998). Baker (1998), however, concluded that the spatial relations of zoned pegmatites to source plutons are consistent with emplacement at normal magmatic temperatures and viscosities. In the parameterization model used by Baker (1998), the source

overpressure is probably the least constrained variable, and it is crucial to the rate of dike propagation in determining how far a melt may migrate before an increase in viscosity caused by heat loss to host rocks will impede its flow.

In general, however, the most distal (and chemically evolved) pegmatites appear to be injected into rocks hundreds of degrees below the temperatures of the pegmatite-forming melts at emplacement. Some of the important questions to solve lie embedded within this scenario.

4.1. How are pegmatite-forming melts generated?

Two competing hypotheses have explained pegmatites either as the direct products of partial melting (anatexis), or as the products of extended fractional crystallization of granitic magmas. Proponents of the former argument cite three factors in support of their hypothesis. First is the difficulty of relating highly evolved magma compositions to the comparatively primitive chemistry of the likely plutonic sources (e.g., Stewart, 1978; Norton and Redden, 1990). Second is a tendency for the compositions of some pegmatite groups to mirror those of their host rocks with respect to major elements (e.g., Stugard, 1958; Novak et al., 1999). Third is the common isolation of thin pegmatite dikes from any known plutonic source (e.g., Simmons et al., 1995). Proponents of large-scale magmatic differentiation note the strong correlation of trace element enrichment trends between pegmatites and source granites, where exposed (e.g., O'Connor et al., 1991). There are field occurrences where pegmatites, even highly fractionated dikes, can be continuously traced back to a granitic source (e.g., London, 1985). Where studied, pegmatite dikes show pronounced isotopic disequilibrium with their host rocks; Sr isotope data, for instance, preclude the derivation of pegmatites from their immediate hosts (e.g., Brookins, 1986; Taylor and Friedrichsen, 1983). Trace-element fractionation models and other isotope systems can provide additional constraints. An experimentally based fractionation model for beryllium (e.g., Evensen and London, 2002; London and Evensen, 2003) shows little possibility that beryl-saturated pegmatites, which are common, can be derived by direct anatexis of any likely sources. Initial Be contents of undifferentiated partial melts are only a few ppmw, and London and Evensen (2003) proposed a three-stage process of crystallization and melt extraction to bring pegmatitic magmas to beryl saturation after ∼95% crystallization of the starting granitic melt. The same is true of Cs and its eventual saturation as pollucite, which requires an increase in Cs of ∼10^4 from the likely concentrations at source (London et al., 1998). Trace-element fractionation models that attempt to relate such chemically evolved pegmatites to their source granite might be expected to fail because the available partition coefficients are not calibrated for the boundary layer compositions and conditions inherent in the constitutional zone refining model. Thus, Shearer et al. (1992) found that chemically simple pegmatites could be related by Rb/Ba fractionation trends to the main facies of the Harney Peak Granite, South Dakota, but not the Li-, Cs-, and Ta-rich complex pegmatite bodies. Shearer et al. (1992) proposed a more direct anatectic source for these evolved pegmatites at low degrees of partial melting. Radiogenic isotope systems provide a means to date and associate genetically related magma bodies, and the few such studies to date (e.g., Krogstad and Walker, 1996; Tomascak et al., 1998) show promise for establishing genetic links between pegmatites and their sources.

Pegmatites of the LCT class (peraluminous, mostly S-type and evolved I-type sources: Černý, 1991a) tend to emanate from the roof zones of their plutons, whereas those of the NYF class (mostly A-types and chemically similar rocks) appear to originate from within the centers of their source plutons (e.g., Simmons et al., 1987; Černý, 1991a,b). These observations might reflect a fundamental difference in how these plutons crystallize, e.g., from the bottom up for the LCT types and from the margins to center in the NYF sources, although there is no immediate explanation for why they should be different. Zoning of the sort seen in the LCT granite types is reported, for example, by Mahood et al. (1996) and Bea et al. (1994). Bea et al. (1994) demonstrated how sharply bounded chemical fractionation occurred upward within a zoned granite sheet at Pedrobernardo, Spain. The model combined, in sequence, crystallization during convection, crystallization of a stagnant mush, and either crystal settling or compaction to extract residual melt upward in the sheet-like pluton. The last and most chemically evolved melt phase, however,

formed a fine-grained leucogranite, not a pegmatite dike system.

4.2. How are crystal-free, pegmatite-forming melts extracted?

If most pegmatites originate by extended crystal fractionation of large magma bodies, why do pegmatites lack evidence of entrained phenocrysts? Porphyritic texture, as it is typically manifested in other igneous rocks, is unknown among pegmatites, as are restite crystal populations. The implication is that the extraction of pegmatite-forming melt from a mostly crystalline magma body is so efficient that no crystals are entrained in the dikes. To frame the problem, consider that the compositions of most pegmatites are not far from the H_2O-saturated minimum of the granite system and differ from their source granite melts only by an increase in their H_2O content. Next, evaluate the viscosities of two liquids: a melt representing the bulk composition of H_2O-undersaturated granite (X_{H2O}=0.3) at 1073 K (800 °C) and 200 MPa, and a pegmatite-forming boundary layer of H_2O-saturated melt (X_{H2O}=0.5) at 973 K (700 °C) and the same pressure that forms as crystallization proceeds along the margins of the pluton. The viscosities calculated from Baker (1996) are $10^{4.7}$ Pa·s for the bulk melt and $10^{5.2}$ Pa·s for the boundary layer melt. Because of cooling, the pegmatite-forming boundary layer melt may be even more viscous than the bulk melt.

One new possibility is that pegmatite dikes actually do contain inherited phenocrysts upon emplacement, but that these are obliterated by the zone refining process. As a fluxed boundary layer progresses ahead of an advancing crystallization front, it might dissolve alkali aluminosilicate crystals as effectively as alkali aluminosilicate melt or glass.

4.3. What is the origin of regional zonation?

I believe that the most intriguing question yet to be solved pertains to the causes of regional zonation in pegmatite groups. The fractionation seen in a group of pegmatite dikes is thought to originate from the fractionation of their source pluton, such that continuous or episodic melt extraction, coupled with continuous crystallization on dike margins, produces the regional zonation toward more fractionated, distal pegmatites (e.g., Trueman and Černý, 1982; Černý, 1991c). The problem of open conduits and connectivity of pegmatite dikes, however, has been and continues to be a problematic issue. The majority of field evidence indicates that genetically related pegmatite dikes are emplaced in a single pulse, that they progressively neck down to increasingly local scales before crystallization commences (especially in schistose hosts), and that their fractionation occurs largely within closed melt bodies (Jahns, 1953, 1955, 1982). If chemical evolution of pegmatites originates from the progressive fractionation of their granitic source, then the dike system must remain open and connected over a period of time comparable to that of the solidification of the granite itself (possibly 10^3 to 10^4 years). This is implausible in light of the current understanding of pegmatite cooling upon dike emplacement. These observations raise the possibility that pegmatites inherit a vertical chemical zonation from a zoned magmatic source (e.g., Simmons et al., 1987; Poli and Tommasini, 1990; London, 1990, 1992; Bea et al., 1994; Mahood et al., 1996) that is now well documented by rhyolite eruptions from large silicic magma chambers (e.g., Hildreth, 1979). This chemical zonation, which is inverted upon eruption of rhyolites, may be preserved in its original orientation within the pegmatite group. Whereas the variations of rhyolite chemistry are defined partly by accumulated crystal content, those of pegmatites may reflect chemical stratification in the liquid phase only.

Acknowledgments

Funding for this work has been provided by the National Science Foundation, in particular, grants EAR-8516753, EAR-8720498, EAR-8821950, EHR-9108771, EAR-9618867, EAR-990165, and EAR-0124179. The Electron Microprobe Laboratory, which figures prominently in the analysis of experimental products, was created by a grant from the U.S. Department of Energy (DE-FG22-87FE1146), with upgrades from the National Science Foundation (EAR-9404658) and annual operational support from the Vice President of Research, University of Oklahoma. Thanks to Don Baker, Rainer Thomas, and Karen Webber for valuable review comments on the original manuscript draft, and to Bob Linnen and George Morgan for additional suggestions.

References

Acosta-Vigil, A., London, D., Dewers, T.A., Morgan VI, G.B., 2002. Dissolution of corundum and andalusite in H_2O-saturated haplogranitic melts at 800 °C and 200 MPa: constraints on diffusivities and the generation of peraluminous melts. J. Petrol. 43, 1885–1908.

Anovitz, L.M., Blencoe, J.G., 1999. Experimental determination of the compositions of aqueous fluids coexisting with silicate melts. Abstr. Programs-Geol. Soc. Am. 31, 354.

Bagdassarov, N.S., Dingwell, D.B., Webb, S.L., 1993. Effect of boron, phosphorus, and fluorine on shear stress relaxation in haplogranite melts. Eur. J. Mineral. 5, 409–425.

Baker, D.R., 1996. Granitic melt viscosities: empirical and configurational entropy models for their calculation. Am. Mineral. 81, 126–134.

Baker, D.R., 1998. The escape of pegmatite dikes from granitic plutons: constraints from new models of viscosity and dike propagation. Can. Mineral. 36, 255–263.

Baker, D.R., Freda, C., 1999. Ising models of binary undercooled system crystallization: comparison with experimental and pegmatite textures. Am. Mineral. 84, 725–732.

Baker, D.R., Freda, C., 2001. Eutectic crystallization in the undercooled orthoclase–quartz–H_2O system: experiments and simulations. Eur. J. Mineral. 13, 453–466.

Baker, D.R., Vaillancourt, J., 1995. The low viscosities of F+H_2O-bearing granitic melts and implications for melt extraction and transport. Earth Planet. Sci. Lett. 132, 199–211.

Bea, F., Pereira, M.D., Corretge, L.G., Fershtater, G.B., 1994. Differentiation of strongly perphosphorous granites: the Pedrohernardo pluton, central Spain. Geochim. Cosmochim. Acta 58, 2609–2627.

Betterman, P., Liebau, F., 1975. The transformation of amorphous silica to crystalline silica under hydrothermal conditions. Contrib. Mineral. Petrol. 53, 25–36.

Blank, J.G., Stolper, E.M., Carroll, M.R., 1993. Solubilities of carbon dioxide and water in rhyolite melt at 850 degrees C and 750 bars. Earth Planet. Sci. Lett. 119, 27–36.

Bowen, N.L., 1913. The melting phenomena of the plagioclase feldspars. Am. J. Sci. 35, 577–599.

Brögger, W.C., 1890. Die mineralien der syenitpegmatitgänge der Südnorwegischen augit und nephelinsyenite. Zeitschrift für Kristallographie und Mineralogie 16, 1–63.

Brookins, D.G., 1986. Rubidium–strontium geochronologic studies of large granitic pegmatites. Neues Jahrb. Mineral., Abh. 156, 81–97.

Burnham, C.W., 1979. Magmas and hydrothermal fluids. In: Barnes, H.L. (Ed.), Geochemistry of Hydrothermal Ore Deposits. John Wiley and Sons, New York, pp. 71–136.

Burnham, C.W., Davis, N.F., 1974. The roles of H_2O in silicate melts: II. Thermodynamic and phase relations in the system

NaalSi₃O₈–H₂O to 10 kilobars, 700 to 1100 °C. Am. J. Sci. 274, 902–940.

Burnham, C.W., Nekvasil, H., 1986. Equilibrium properties of granite pegmatite magmas. Am. Mineral. 71, 239–263.

Cameron, E.N., Jahns, R.H., McNair, A.H., Page, L.R., 1949. Internal structure of granitic pegmatites. Econ. Geol. Monogr. 2. 115 pp.

Černý, P., 1991a. Fertile granites of Precambrian rare-element pegmatite fields: is geochemistry controlled by tectonic setting or source lithologies? Precambrian Res. 51, 429–468.

Černý, P., 1991b. Rare-element granite pegmatites: Part I. Anatomy and internal evolution of pegmatite deposits. Geosci. Can. 18, 49–67.

Černý, P., 1991c. Rare-element granite pegmatites: Part II. Regional to global environments and petrogenesis. Geosci. Can. 18, 68–81.

Černý, P., 2000. Constitution, petrology, affiliations and categories of miarolitic pegmatites. In: Pezzotta, F. (Ed.), Mineralogy and Petrology of Shallow Depth Pegmatites; Papers from the First International Workshop. Memorie della Società di Scienze Naturali e del Museo Civico di Storia Naturale di Milano, vol. 30, pp. 5–12.

Chakoumakos, B.C., Lumpkin, G.R., 1990. Pressure–temperature constraints on the crystallization of the Harding pegmatite, Taos County, New Mexico. Can. Mineral. 28, 287–298.

Cygan, R.T., Carrigan, C.R., 1992. Time-dependent Soret transport: applications to brine and magma. Chem. Geol. 95, 201–212.

Dingwell, D.B., Knoche, R., Webb, S.L., Pichavant, M., 1992. The effect of B₂O₃ on the viscosity of haplogranitic liquids. Am. Mineral. 77, 457–461.

Dingwell, D.B., Hess, K.-U., Knoche, R., 1996. Granite and granitic pegmatite melts: volumes and viscosities. Trans. Royal Soc. Edinb.: Earth Sci. 87, 65–72.

Dowty, E., 1980. Crystal growth and nucleation theory and the numerical simulation of igneous crystallization. In: Hargraves, R.B. (Ed.), Physics of Magmatic Processes. Princeton University Press, Princeton, New Jersey, pp. 419–485.

Durant, D.G., Fowler, A.D., 2002. Origin of reverse zoning in branching orthopyroxene and acicular plagioclase in orbicular diorite, Fisher Lake, California. Mineral. Mag. 66, 1003–1019.

Eitel, W., 1954. The Physical Chemistry of the Silicates. University of Chicago Press, Chicago.

Evensen, J.M., London, D., 2002. Experimental silicate mineral/melt partition coefficients for beryllium, and the beryllium cycle from migmatite to pegmatite. Geochim. Cosmochim. Acta 66, 2239–2265.

Evensen, J.M., London, D., Dewers, T.A., 2001. Effects of starting state and superliquidus–subliquidus pathways on crystal growth from silicic melts. 11th Annual Goldschmidt Conference Abstract 3729. Lunar Planetary Institute Contribution 1088, Lunar Planetary Institute, Houston [CD-ROM].

Fenn, P.M., 1977. The nucleation and growth of alkali feldspars from hydrous melts. Can. Mineral. 15, 135–161.

Fenn, P.M., 1986. On the origin of graphic granite. Am. Mineral. 71, 325–330.

Haapala, I., 1966. On the granitic pegmatites in the Peräseinäjoki-Alavus area, south Pohjanmaa, Finland. Bull. Comm. Géol. Finl. 224. 98 pp.

Haapala, I., 1977. Petrography and geochemistry of the Eurajoki stock, a rapakivi–granite complex with greisen-type mineralization in southwestern Finland. Geol. Surv. Finl. 286. 128 pp.

Haapala, I., Rämö, O.T., 1990. Petrogenesis of the Proterozoic rapakivi granites of Finland. In: Stein, H.J., Hannah, J.L. (Eds.), Ore-bearing Granite Systems; Petrogenesis and Mineralizing Processes. Geological Society of America Special Paper, vol. 246, pp. 275–286.

Haapala, I., Rämö, O.T., 1992. Tectonic setting and origin of the Proterozoic rapakivi granites of southeastern Fennoscandia. In: Brown, P.E., Chappell, B.W. (Eds.), The Second Hutton Symposium on the Origin of Granites and Related Rocks. Geological Society of America Special Paper, vol. 272, pp. 165–171.

Haapala, I., Rämö, O.T., 1999. Rapakivi granites and related rocks; an introduction. In: Haapala, I., Rämö, O.T. (Eds.), Rapakivi Granites and Related Rocks: An Introduction. Precambrian Research, vol. 95, pp. 1–7.

Henisch, H.K., 1988. Crystals in Gels and Liesegang Rings. Cambridge University Press, New York.

Hildreth, W., 1979. The Bishop Tuff: evidence for the origin of compositional zonation in silicic magma chambers. In: Chapin, C.E., Elston, W.E. (Eds.), Ash-Flow Tuffs. Geological Society of America Special Paper, 180, pp. 43–75.

Hughes, R.W., 1990. Corundum. Butterworths, Great Britain, p. 314.

Jahns, R.H., 1953. The genesis of pegmatites: I. Occurrence and origin of giant crystals. Am. Mineral. 38, 563–598.

Jahns, R.H., 1955. The study of pegmatites. Econ. Geol., 1025–1130. (50th Anniversary Volume).

Jahns, R.H., 1982. Internal evolution of pegmatite bodies. In: Černý, P. (Ed.), Granitic Pegmatites in Science and Industry. Mineralogical Association of Canada Short Course Handbook, vol. 8, pp. 293–327.

Jahns, R.H., Tuttle, O.F., 1963. Layered pegmatite–aplite intrusives. Spec. Pap.-Miner. Soc. Am. 1, 78–92.

Jahns, R.H., Burnham, C.W., 1969. Experimental studies of pegmatite genesis: I. A model for the derivation and crystallization of granitic pegmatites. Econ. Geol. 64, 843–864.

Keppler, H., 1993. Influence of fluorine on the enrichment of high field strength trace elements in granitic rocks. Contrib. Mineral. Petrol. 114, 479–488.

Kilinc, I.A., 1969. Experimental metamorphism and anatexis of shales and graywackes. PhD thesis, Pennsylvania State University, University Park, PA, USA.

Kleck, W.D., Foord, E.E., 1999. The chemistry, mineralogy, and petrology of the George Ashley Block pegmatite body. Am. Mineral. 84, 695–707.

Krogstad, E.J., Walker, R.J., 1996. Heterogeneous sources for the Harney Peak granite, South Dakota: Nd isotopic evidence. Trans. Royal Soc. Edinb.: Earth Sci. 87, 331–337.

Lesher, C.E., Walker, D., 1991. Thermal diffusion in petrology. In: Ganguly, J. (Ed.), Diffusion, Atomic Ordering, and Mass Transport; Selected Topics in Geochemistry. Advances in Physical Geochemistry, vol. 8, pp. 396–451.

Lofgren, G.E., 1974. An experimental study of plagioclase crystal morphology: isothermal crystallization. Am. J. Sci. 274, 243–273.

Lofgren, G.E., 1980. Experimental studies on the dynamic crystallization of silicate melts. In: Hargraves, R.B. (Ed.), Physics of Magmatic Processes. Princeton University Press, Princeton, New Jersey, pp. 487–551.

London, D., 1984. Experimental phase equilibria in the system LiAlSiO$_4$–SiO$_2$–H$_2$O: a petrogenetic grid for lithium-rich pegmatites. Am. Mineral. 69, 995–1004.

London, D., 1985. Pegmatites of the Middletown district, Connecticut. 77th Annual Meeting, New England Intercollegiate Geological Conference, Yale University. Connecticut Geological and Natural History Survey Guidebook, vol. 6, pp. 509–533.

London, D., 1986a. The magmatic–hydrothermal transition in the Tanco rare-element pegmatite: evidence from fluid inclusions and phase equilibrium experiments. Am. Mineral. 71, 376–395.

London, D., 1986b. Formation of tourmaline-rich gem pockets in miarolitic pegmatites. Am. Mineral. 71, 396–405.

London, D., 1987. Internal differentiation of rare-element pegmatites: effects of boron, phosphorus, and fluorine. Geochim. Cosmochim. Acta 51, 403–420.

London, D., 1990. Internal differentiation of rare-element pegmatites: a synthesis of recent research. In: Stein, H.J., Hannah, J.L. (Eds.), Ore-Bearing Granite Systems; Petrogenesis and Mineralizing Processes. Geological Society of America Special Paper, vol. 246, pp. 35–50.

London, D., 1992. The application of experimental petrology to the genesis and crystallization of granitic pegmatites. Can. Mineral. 30, 499–540.

London, D., 1996. Granitic pegmatites. Trans. Royal Soc. Edinb.: Earth Sci. 87, 305–319.

London, D., 1997. Estimating abundances of volatile and other mobile components in evolved silicic melts through mineral-melt equilibria. J. Petrol. 38, 1691–1706.

London, D., 1999. Melt boundary layers and the growth of pegmatitic textures. Can. Mineral. 37, 826–827.

London, D., Burt, D.M., 1982. Lithium aluminosilicate occurrences in pegmatites and the lithium aluminosilicate phase diagram. Am. Mineral. 67, 483–493.

London, D., Evensen, J.M., 2003. Beryllium in silicic magmas and the origin of beryl-bearing pegmatites. In: Grew, E.S. (Ed.), Beryllium: Mineralogy, Petrology, and Geochemistry. Mineralogical Society of America Reviews in Mineralogy and Geochemistry, vol. 50, pp. 445–486.

London, D., Zolensky, M.E., Roedder, E., 1987. Diomignite: natural Li$_2$B$_4$O$_7$ from the Tanco pegmatite, Bernic Lake, Manitoba. Can. Mineral. 25, 173–180.

London, D., Morgan VI, G.B., Hervig, R.L., 1989. Vapor-undersaturated experiments in the system macusanite–H$_2$O at 200 MPa, and the internal differentiation of granitic pegmatites. Contrib. Mineral. Petrol. 102, 1–17.

London, D., Morgan VI, G.B., Babb, H.A., Loomis, J.L., 1993. Behavior and effects of phosphorus in the system Na$_2$O–K$_2$O–Al$_2$O$_3$–SiO$_2$–P$_2$O$_5$–H$_2$O at 200 MPa (H$_2$O). Contrib. Mineral. Petrol. 113, 450–465.

London, D., Morgan VI, G.B., Icenhower, J., 1998. Stability and solubility of pollucite in granitic systems at 200 MPa H$_2$O. Can. Mineral. 36, 497–510.

MacLellan, H.E., Trembath, L.L., 1991. The role of quartz crystallization in the development and preservation of igneous texture in granitic rocks: experimental evidence at 1 kbar. Am. Mineral. 76, 1291–1305.

Mahood, G.A., Nibler, G.E., Halliday, A.N., 1996. Zoning patterns and petrologic processes in peraluminous magma chambers: Hall Canyon pluton, Panamint Mountains, California. Geol. Soc. Amer. Bull. 108, 437–453.

Manning, D.A.C., 1981. The effect of fluorine on liquidus phase relationships in the system Qz–Ab–Or with excess water at 1 kb. Contrib. Mineral. Petrol. 76, 206–215.

McBirney, A.R., 1987. Constitutional zone refining of layered mafic intrusions. In: Parsons, I. (Ed.), Origins of Igneous Layering. D. Reidel Publishing, Holland, pp. 437–452.

McBirney, A.R., Russell, W.J., 1987. Constitutional zone refining of magmatic intrusions. In: Loper, D.E. (Ed.), Structure and Dynamics of Partially Solidified Systems, NATO ASI Series E, Applied Sciences, vol. 125, pp. 349–365.

Merino, E., 1999. Origin of agates and other supposed vesicle fillings; an overlooked reaction to produce silica gels in basalt flows. EOS Trans. Am. Geophys. Union 80, 1122.

Merritt, C.A., 1924a. The function of gels in the formation of pegmatites and of quartz and carbonate veins. MSc thesis, University of Manitoba, Winnipeg, Manitoba, Canada.

Merritt, C.A., 1924b. The function of colloids in pegmatitic growths. Proc. Trans. R. Soc. Can. 17, 61–68.

Morgan VI, G.B., London, D., 1987. Alteration of amphibolitic wallrocks around the Tanco rare-element pegmatite, Bernic Lake, Manitoba. Am. Mineral. 72, 1097–1121.

Morgan VI, G.B., London, D., 1996. Optimizing the electron microprobe analysis of hydrous alkali aluminosilicate glasses. Am. Mineral. 81, 1176–1185.

Morgan VI, G.B., London, D., 1999. Crystallization of the Little Three layered pegmatite-aplite dike, Ramona District, California. Contrib. Mineral. Petrol. 136, 310–330.

Morgan VI, G.B., London, D., 2003. Trace element partitioning at conditions far from equilibrium: Ba and Cs distributions between alkali feldspar and undercooled hydrous granitic liquid at 200 MPa. Contrib. Mineral. Petrol. 144, 722–738.

Morgan VI, G.B., London, D., Kirkpatrick, R.J., 1990. Reconnaissance spectroscopic study of hydrous sodium aluminum borosilicate glasses. Abstr. Programs-Geol. Soc. Am. 22, A167.

Mungall, J.E., 2002. Empirical models relating viscosity and tracer diffusion in magmatic silicate melts. Geochim. Cosmochim. Acta 66, 125–143.

Nabelek, P., Russ-Nabelek, C., Denison, J.R., 1992. The generation and crystallization conditions of the Proeterozoic Harney Peak Leucogranite, Black Hills, South Dakota, USA: petrological and geochemical constraints. Contrib. Mineral. Petrol. 110, 173–191.

Norton, J.J., 1966. Ternary diagrams of the quartz–feldspar content of pegmatites in Colorado. U.S. Geol. Surv. Bull. 1241, D1–D16.

Norton, J.J., Redden, J.A., 1990. Relations of zoned pegmatites to other pegmatites, granite, and metamorphic rocks, in the southern Black Hills, South Dakota. Am. Mineral. 75, 631–655.

Novak, M., Selway, J.B., Černý, P., Hawthorne, F.C., Ottolini, L., 1999. Tourmaline of the elbaite–dravite series from an elbaite-subtype pegmatite at Bližná, southern Bohemia, Czech Republic. Eur. J. Mineral. 11, 557–568.

O'Connor, P.J., Gallagher, V., Kennan, P.S., 1991. Genesis of lithium pegmatites from the Leinster Granite Margin, southeastern Ireland: goechemical constraints. Geol. J. 26, 295–305.

O'Donoghue, M., 1987. Quartz. Butterworths, Great Britain, p. 110.

Oehler, J.H., 1976. Hydrothermal crystallisation of silica gel. Geol. Soc. Amer. Bull. 87, 1143–1152.

Orville, P.M., 1963. Alkali ion exchange between vapor and feldspar phases. Am. J. Sci. 261, 201–237.

Pezzotta, F., 2000. Internal structures, parageneses and classification of the miarolitic Li-bearing complex pegmatites of Elba Island (Italy). Memorie della Società di Scienze Naturali e del Museo Civico di Storia Naturale di Milano 30, 29–43.

Pichavant, M., 1981. An experimental study of the effect of boron on a water saturated haplogranite at 1 kbar vapour pressure, geological applications. Contrib. Mineral. Petrol. 76, 430–439.

Poli, G., Tommasini, S., 1990. A geochemical approach to the evolution of granitic plutons: a case study, the acid intrusions of Punta Falcone (northern Sardinia, Italy). Chem. Geol. 92, 87–105.

Rockhold, J.R., Nabelek, P.I., Glascock, M.D., 1987. Origin of rhythmic layering in the Calamity Peak satellite pluton of the Harney Peak Granite, South Dakota: the role of boron. Geochim. Cosmochim. Acta 51, 487–496.

Schäfer, B., Frischknecht, R., Günther, D., Dingwell, D.B., 1999. Determination of trace-element partitioning between fluid and melt using LA-ICP-MS analysis of synthetic fluid inclusions in glass. Eur. J. Mineral. 11, 415–426.

Seifert, F.A., Mysen, B.O., Virgo, D., 1981. Structural similarity of glasses and melts relevant to petrological processes. Geochim. Cosmochim. Acta 45, 1879–1884.

Shearer, C.K., Papike, J.J., Jolliff, B.L., 1992. Petrogenetic links among granites and pegmatites in the Harney Peak rare-element granite–pegmatite system, Black Hills, South Dakota. Can. Mineral. 30, 785–809.

Simmons, W.B., Lee, M.T., Brewster, R.H., 1987. Geochemistry and evolution of the South Platte granite–pegmatite system, Jefferson County, Colorado. Geochim. Cosmochim. Acta 51, 455–472.

Simmons, W.B., Foord, E.E., Falster, A.U., King, V.T., 1995. Evidence for an anatectic origin of granitic pegmatites, western Maine, USA. Abstr. Programs-Geol. Soc. Am. 27, 411.

Sirbescu, M.-L., Nabelek, P., 2003. Crystallization conditions and evolution of magmatic fluids in the Harney Peak Granites and associated pegmatites, Black Hills, South Dakota—evidence from fluid inclusions. Geochim. Cosmochim. Acta 67, 2443–2465.

Sowerby, J.R., Keppler, H., 2002. The effect of fluorine, boron, and excess sodium on the critical curve in the albite–H_2O system. Contrib. Mineral. Petrol. 143, 32–37.

Stewart, D.B., 1978. Petrogenesis of lithium-rich pegmatites. Am. Mineral. 63, 970–980.

Stilling, A., 1998. Bulk composition of the Tanco pegmatite at Bernic Lake, Manitoba, Canada. MSc thesis, University of Manitoba, Winnipeg, Manitoba, Canada.

Student, J.J., Bodnar, R.J., 1999. Synthetic fluid inclusions: XIV; Coexisting silicate melt and aqueous fluid inclusions in the haplogranite–H_2O–NaCl–KCl system. J. Petrol. 40, 1509–1525.

Stugard Jr., F., 1958. Pegmatites of the Middletown area, Connecticut. U.S. Geol. Surv. Bull., B 1042-Q, 613–683.

Swanson, S.E., Fenn, P.M., 1986. Quartz crystallization in igneous rocks. Am. Mineral. 71, 331–342.

Swanson, S.E., Fenn, P.M., 1992. The effect of F and Cl on albite crystallization: a model for granitic pegmatites? Can. Mineral. 30, 549–559.

Taylor, B.E., Friedrichsen, H., 1983. Light stable isotope systematics of granitic pegmatites from North America and Norway. Isot. Geosci. 1, 127–167.

Taylor, B.E., Foord, E.E., Friedrichsen, H., 1979. Stable isotope and fluid inclusion studies of gem-bearing granitic pegmatite-aplite dikes, San Diego Co., California. Contrib. Mineral. Petrol. 68, 187–205.

Taylor, M.C., Sheppard, J.B., Walker, J.N., Kleck, W.D., Wise, M.A., 2002. Petrogenesis of rare-element granitic aplite–pegmatites: a new approach. International Mineralogical Association General Meeting Programme with Abstracts, vol. 18, pp. 260.

Thomas, R., 1994. Estimation of the viscosity and the water content of silicate melts from melt inclusion data. Eur. J. Mineral. 6, 511–535.

Thomas, R., Klemm, W., 1997. Microthermometric study of silicate melt inclusions in Variscan granites from SE Germany: volatile contents and entrapment conditions. J. Petrol. 38, 1753–1765.

Thomas, R., Webster, J.D., 2000. Strong tin enrichment in a pegmatite-forming melt. Miner. Depos. 35, 570–582.

Thomas, A.V., Bray, C.J., Spooner, E.T.C., 1988. A discussion of the Jahns–Burnham proposal for the formation of zoned granitic pegmatites using solid–liquid–vapour inclusions from the Tanco pegmatite, S.E. Manitoba, Canada. Trans. Royal Soc. Edinb.: Earth Sci. 79, 299–315.

Thomas, R., Webster, J.D., Heinrich, W., 2000. Melt inclusions in pegmatitic quartz: complete miscibility between silicate melts and hydrous fluids at low pressure. Contrib. Mineral. Petrol. 139, 394–401.

Tomascak, P.B., Krogstad, E.J., Walker, R.J., 1998. Sm–Nd isotope systematics and the derivations of granitic pegmatites in southwestern Maine. Can. Mineral. 36, 327–337.

Trueman, D.L., Černý, P., 1982. Exploration for rare-element granitic pegmatites. In: Černý, P. (Ed.), Granitic Pegmatites in Science and Industry. Mineralogical Association of Canada Short Course Handbook, vol. 8, pp. 463–494.

Tuttle, O.F., Bowen, N.L., 1958. Origin of granite in the light of experimental studies in the system $NaAlSi_3O_8$–$KAlSi_3O_8$–SiO_2–H_2O. Geol. Soc. Am. Memoir 74. 153 pp.

Watson, E.B., Baker, D.R., 1991. Chemical diffusion in magmas: an overview of experimental results and geochemical applications. In: Perchuk, L.L., Kushiro, I. (Eds.), Physical Chemistry of Magmas. Springer-Verlag, New York, pp. 120–151.

Webber, K.L., Falster, A.U., Simmons, W.B., Foord, E.E., 1997. The role of diffusion-controlled oscillatory nucleation in the formation of line rock in pegmatite–aplite dikes. J. Petrol. 38, 1777–1791.

Webber, K.L., Simmons, W.B., Falster, A.U., Foord, E.E., 1999. Cooling rates and crystallization dynamics of shallow level pegmatite–aplite dikes, San Diego County, California. Am. Mineral. 84, 708–717.

Webster, J.D., De Vivo, B., 2002. Experimental and modeled solubilities of chlorine in aluminosilicate melts, consequences for magma evolution, and implications for exsolution of hydrous chlorine melt at Mt. Somma-Vesuvius. Am. Mineral. 87, 1046–1061.

Williamson, B.J., Stanley, C.J., Wilkinson, J.J., 1997. Implications from inclusions in topaz for greisenization and mineralisation in the Hensbarrow topaz granite, Cornwall, England. Contrib. Mineral. Petrol. 127, 119–128.

Wolf, M.B., London, D., 1994. Apatite dissolution into peraluminous haplogranitic melts: an experimental study of solubilities and mechanisms. Geochim. Cosmochim. Acta 58, 4127–4145.

Wyllie, P.J., 1963. Applications of high pressure studies to the earth sciences. In: Bradley, R.S. (Eds.), High Pressure Physics and Chemistry, vol. 2. Academic Press, New York, pp. 1–89.

Available online at www.sciencedirect.com

LITHOS

Lithos 80 (2005) 305–321

www.elsevier.com/locate/lithos

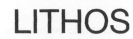

The Greer Lake leucogranite, Manitoba, and the origin of lepidolite-subtype granitic pegmatites

Petr Černý[a,*], Morgan Masau[a,b], Bruce E. Goad[a,c], Karen Ferreira[a]

[a]*Geological Sciences, University of Manitoba, Winnipeg, MB, Canada R3T 2N2*
[b]*Geology and Geophysics, University of New Orleans, New Orleans, LA 70148, USA*
[c]*Inukshuk Exploration Inc., 21861 44A Avenue, Langley, BC, Canada V3A 8E1*

Received 31 March 2003; accepted 9 September 2004
Available online 14 November 2004

Abstract

The Archean Greer Lake leucogranite intruded metabasalts of the Bird River Greenstone Belt in the southwestern part of the Superior Province of southeastern Manitoba. The considerably evolved, multiphase, peraluminous, B-, P-, and S-poor leucogranite (K/Rb 132 to 24) was probably generated by fault-friction-assisted anatexis of dominantly metatonalitic rocks and subsequent differentiation. The leucogranite produced interior, transitional, non-crosscutting pods of barren, beryl-columbite- and lepidolite-subtype pegmatites that solidified from local segregations of highly fractionated residual melt. Steep fractionation gradients characterize the granite-to-pegmatite transition, most conspicuously so in the case of the most evolved, Li, Rb, Cs, Be, Mn, Sn, Nb-Ta, F-rich, lepidolite-subtype pod AC #3 (with K/Rb $\geq$ 16 and Cs 330 ppmwt in accessory K-feldspar, $\geq$2.5 and $\leq$11,200 ppmwt, respectively, in lepidolite, Cs $\leq$28,000 ppmwt in beryl, and Ta/(Ta+Nb) at. $\leq$ 0.95 in manganotantalite). The Greer Lake example documents beyond any doubt the igneous derivation of lepidolite-subtype pegmatites from a plutonic parent. Most cases of generally very scarce lepidolite-subtype pegmatites obscure this relationship, as the volatile-rich, highly fluid melts stable to relatively low temperatures commonly migrate to great distances from their plutonic sources.
© 2004 Elsevier B.V. All rights reserved.

Keywords: Leucogranite; Granitic pegmatite; Lepidolite; Petrology; Geochemistry; Petrogenesis

1. Introduction

Complex granitic pegmatites of the lepidolite subtype are characterized by lepidolite as the main (up to the only) Li-bearing rock-forming mineral. They are prominently enriched in Li, F, Rb, Cs, commonly also in B and P, and they are prone to carry significant concentrations of minerals of Sn, Nb, Ta, and Be. They belong to the rare-element class of granitic pegmatites, generally located in the environment of lower-amphibolite (to upper-greenschist) facies of metamorphism (e.g., Ginsburg et al., 1979; Černý, 1991a).

* Corresponding author. Tel.: +1 204 474 8765; fax: +1 204 474 7623.
E-mail address: p_cerny@umanitoba.ca (P. Černý).

0024-4937/$ - see front matter © 2004 Elsevier B.V. All rights reserved.
doi:10.1016/j.lithos.2003.11.003

Granitic pegmatites of the rare-element class are generally accepted today as products of magmatic differentiation from pluton-sized granite bodies (as reviewed by Černý, 1991b, 1992a). However, speculations about a direct anatectic origin still persist in some cases, notably concerning lepidolite-subtype pegmatites. Charoy and Noronha (1996) discussed the problem at some length and agreed that the albite+quartz+lepidolite+amblygonite assemblage, late as it is in the evolution of the examined microgranite, is magmatic and produced by igneous fractionation. However, Zasedatelev (1974, 1977) denied origin by magmatic differentiation to lithium-bearing pegmatites in general, favoring instead speculative models of direct anatexis. Stewart (1978) expressed similar views, based on, i.e., bimodal distribution of lithium abundances in granitic pegmatites. Recently, Gordiyenko et al. (1996) developed a complex postmagmatic scheme specifically for lepidolite pegmatites: they are supposed to consolidate from an undefined medium that is otherwise alleged to generate rare-element mineralization superimposed on preexisting barren pegmatite bodies.

We present here the Greer Lake leucogranite, Manitoba, Canada, and its interior pegmatites as a case of derivation of lepidolite-subtype pegmatites from an igneous granitic parent. We also review other cases and arguments supporting this mode of origin, and the reasons why the generally rare lepidolite-subtype pegmatites have been misinterpreted in the past.

2. Regional geology

The Greer Lake leucogranite is located 135 km east–northeast of Winnipeg, Manitoba, at latitude 50°20′39″N, longitude 95°19′W (Fig. 1). The leucogranite and its pegmatite aureole are part of the Cat Lake–Winnipeg River pegmatite field situated in the volcano-plutonic Bird River subprovince of the Superior Province of the Canadian Shield. The Bird River Greenstone Belt, host of the pegmatite field, consists of six formations of metavolcanic and proximally derived metasedimentary rocks intercalated with, or intruded by, syn- and subvolcanic intrusions (Černý et al., 1981). The metasedimentary–metavolcanic edifice forms a broad and complex synclinorium. Two major episodes of folding affected the greenstone belt; the second correlates with the diapiric intrusion of the Maskwa Lake and Marijane Lake batholiths and with the peak of regional metamorphism. Metamorphism attained the greenschist-facies level over most of the area of Fig. 1, but

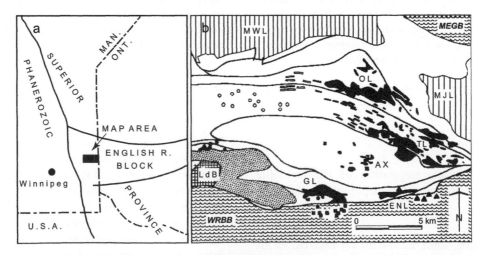

Fig. 1. (a) Location of the Winnipeg River pegmatite district at the Manitoba–Ontario boundary in central Canada. (b) Geology of the district: open, greenstone belt; WRBB, Winnipeg River batholithic belt; MEGB, Manigotagan–Ear Falls Gneiss Belt; MWL, Maskwa Lake; MJL, Marijane Lake tonalitic diapirs; LdB, complex eastern end of Lac du Bonnet batholith; most pegmatite groups (each marked by a different spot symbol) are associated with leucogranites (in black; TL, Tin Lake; ENL, Eagle Nest Lake; AX, Axial; OL, Osis Lake; and GL, Greer Lake, which is also shown in Fig. 2). Modified from Černý et al. (1986).

amphibolite to low granulite in parts of its eastern to northeastern areas and to the south of the greenstone belt. The greenstone belt is dissected by numerous E–W-striking subvertical faults and locally offset by NNW–SSE-striking cross-faults.

The dominant tonalite and trodhjemite components of the Maskwa Lake (2779±32 Ma) and Marijane Lake diapirs are rimmed and penetrated by largely undeformed biotite granites. The Lac du Bonnet batholith consists predominantly of an analogous late biotite granite (2665±20 Ma). The field relationships of the Greer Lake leucogranite and related intrusions (dated at 2640±7 Ma via derived pegmatites) show that they postdate the emplacement of the biotite granites of the batholiths (see Černý et al., 1998, for a review of ages and references).

3. The Greer Lake leucogranite

3.1. Intrusive relationships

The Greer Lake leucogranite is the southwestern one among the five leucogranites of the Winnipeg River district that were emplaced along subvertical faults and adjacent décollement structures late in the history of regional events (Fig. 1; Goad and Černý, 1981; Černý et al., 1981). The EW-elongated intrusion, about 4×0.8 km in size, is hosted largely by metabasalts. At the eastern and western extremities of the intrusion and its southwestern offshoot, the contacts trend northward at a high angle to the layering and foliation of the metabasalts (Fig. 2). These contacts are sharp, truncating both the layering and foliation with no evident deformation. Inclusions of host rocks are rare in the leucogranite, except for angular blocks of metabasalt in the southwestern offshoot, indicative of stoping. All contacts of the leucogranite with metabasalt are marked by a thin zone in which mafic components of the wallrock are replaced by biotite.

A WNW-striking fault separates the leucogranite into two segments and displays evidence of subvertical movement (Fig. 2). The fault sliced off the southwestern corner of the intrusion and also affected the southeastern margin, but the full extent and course of the fault could not be traced there. In the southwestern part, the fault is running parallel with, and in part within, a subvertical screen of iron formation, incorporated into the leucogranite from the largely

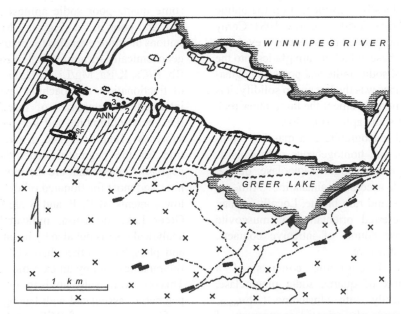

Fig. 2. The Greer Lake leucogranite (open, heavy outline; see GL in Fig. 1 for regional location) in metabasalt (diagonal ruling) N of gneissic belt (x's). The Annie Claim #2 and #3 pegmatites are marked 2 and 3 at ANN. The Greer Lake group of exterior pegmatites (heavy bars) intruded mostly the gneissic belt, except the Silverleaf pegmatite marked SF. Modified from Černý et al. (1986).

displaced metabasalt. The leucogranite shows extensive shearing adjacent to the fault, but the effects fade out over a distance of ~10 m. The geochemical signature of the leucogranite and pegmatites within a wedge marked by this fault and the other one along the metabasalt-gneiss contact (Fig. 2) is distinctly more fractionated than in the main body of the leucogranite to the north. This strongly suggests a graben structure with the wedge below its original elevation: general geochemical, mineralogical, and textural evidence from fertile granites generating rare-element pegmatites shows strong vertical gradients with fractionation increasing upwards (e.g., Černý and Meintzer, 1988; Černý, 1992b).

3.2. Internal constitution

The Greer Lake leucogranite is compositionally and texturally heterogeneous: it consists of four phases that outcrop rather randomly across the current erosional exposure: (i) minor fine-grained leucogranite encountered mainly along the northern margin of the intrusion; (ii) dominant pegmatitic leucogranite (Fig. 3a); (iii) layered sodic aplite facies, slightly more abundant in the southern part of the intrusion (Fig. 3b and c); and (iv) potassic pegmatite facies, commonly associated with the aplite and locally consolidated on the aplitic substrate (Fig. 3b and c) (Goad and Černý, 1981; Černý et al., 1981). The fine-grained phase is early, evolving into the pegmatitic phase, which in turn grades into the layered sequences of sodic aplite and potassic pegmatite. This potassic facies was the last one to solidify: it is observed to cross-cut and cement the three other rock types, including the sodic aplite (Fig. 3d).

The fine-grained leucogranite is a massive, equigranular (1–3 mm), and homogeneous rock consisting of microcline-perthite, sodic plagioclase, quartz, and muscovite, with accessory almandine (Alm_{86-78} Sps_{14-22}) and zircon, and rare biotite. Locally, it may show indistinct preferred orientation of muscovite parallel to the gross course of intrusive contacts, resembling a flow structure.

The pegmatitic leucogranite consists of a coarse-grained matrix of quartz, sodic plagioclase, and muscovite, locally with almandine (Alm_{86-84} Sps_{14-16}), and of randomly oriented megacrysts of microcline-perthite (5–100 cm in size) intergrown with graphic quartz (Fig. 3a).

The sodic aplite typically consists of albite, quartz, muscovite, and garnet (Alm_{85-70} Sps_{15-30}) with variable but generally minor microcline-perthite and accessory fluorapatite, gahnite, monazite-(Ce), zircon, and rare opaque minerals (i.a., magnetite and ferrocolumbite). The aplite is commonly layered, with garnet conspicuously enriched in layers alternating with those of the felsic components; quartz is locally enriched in the garnetiferous layers and albite+muscovite in the intervening layers (Fig. 3b, c, and d).

The potassic pegmatite component consists of equant to club-shaped microcline-perthite and quartz, locally with minor muscovite. The proportion of albite is highly variable: it is mostly minor but locally it may attain the abundance of microcline-perthite. The pegmatite is commonly paired with aplite, and it shows a more or less distinct comb texture, indicative of growth from the substrate of aplite layers outwards (Fig. 3b and c). However, the potassic pegmatite phase also forms diffuse masses that penetrate all other rock types (Fig. 3d).

Petrochemistry of the four leucogranite phases closely reflects their strongly felsic character in general, and the modal and mesonormative variations (Fig. 4a) in the rock-forming minerals in particular (Table 1). All four rock types are highly silicic (except some quartz-poor sodic aplites), peraluminous (Fig. 4b), and poor in Ca (Fig. 4a), Mg, and Fe. Concentrations of minor elements (particularly Rb, Cs, Ga) and geochemically meaningful element ratios (such as K/Rb, K/Cs, K/Ba, Mg/Li, Al/Ga) indicate a high degree of fractionation, approaching levels characteristic of intermediate categories of rare-element pegmatites (i.e., beryl-columbite and beryl-columbite-phosphate subtypes) (Table 1; Fig. 4c, d, and e). This is especially true for the pegmatitic-leucogranite and potassic-pegmatite phases that are the most strongly peraluminous and most fractionated of all four rock types. Very low contents of B, P, and S are characteristic of the Greer Lake intrusion, indicated not only by the analytical results but also by the absence of tourmaline and phosphates (other than very sparse microscopic fluorapatite), and by an extreme scarcity of sulphides. This observation also holds for the interior and exterior pegmatites associated with leucogranite.

Concentrations of REE are low and do not rise above 25× chondritic. The patterns are quite flat with very prominent negative Eu anomalies in virtually all

Fig. 3. Textural features of selected leucogranite phases: (a) pegmatitic leucogranite with subrectangular white megacrysts of K-feldspar (intergrown with graphic quartz) in coarse-grained matrix; the central megacryst is 30×40 cm in size; (b) layers of garnetiferous sodic aplite alternating with poorly comb-structured potassic pegmatite; (c) layered garnetiferous sodic aplite forms substrate for upwards-growing potassic pegmatite; 10 cm across; (d) contorted and disrupted layers of garnetiferous sodic aplite, cemented by potassic pegmatite; dark rounded patches at the bottom and top are lichens.

rock types (Fig. 4f). The high content of garnet, which is known for its strong preference for heavy REE, does not shift the patterns of sodic aplites from those of other rock facies. On the contrary, potassic pegmatites differ from the other three rocks as they are the last rock type to solidify and virtually devoid

of accessory minerals: consequently, the potassic pegmatites are the most strongly depleted of REE. Oxygen isotope data range between +8.1‰ and +8.7‰ $\delta^{18}O$ (SMOW) for all four rock phases (Longstaffe et al., 1981). These low values are not only within the range of +6.7‰ to +8.9‰ obtained for

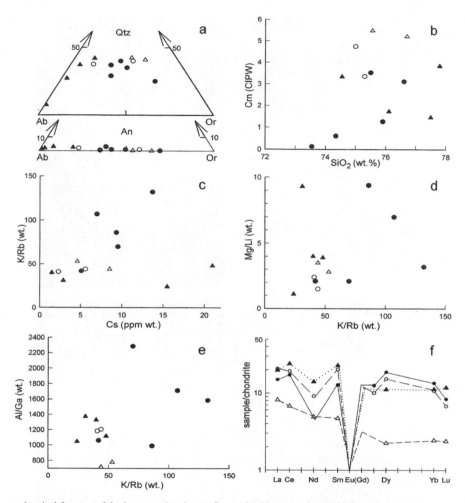

Fig. 4. Selected petrochemical features of the leucogranite phases: fine-grained leucogranite (solid dots), pegmatitic leucogranite (open circles), sodic aplite facies (solid triangles), potassic pegmatite facies (open triangles): (a) mesonormative ratios Ab-Or-Qtz and Ab-Or-An (arrows mark coincidence of pegmatitic leucogranite plots with those of fine-grained leucogranite); (b) CIPW corundum versus silica; (c) K/Rb versus Cs; (d) Mg/Li versus K/Rb; (e) Al/Ga versus K/Rb; (f) chondrite-normalized REE contents of single representatives of the four rock phases, Gd is approximated by interpolation.

the host metabasalts, but also close to the +7.0‰ to +7.5‰ data obtained for basement tonalitic gneisses. In contrast, $\delta^{18}O$ values of metapelites, metagreywackes, and metarhyolites of the adjacent greenstone belt formations range from +9.3‰ to +15.8‰, as do those of the derived B-, P-, and S-enriched leucogranites synchronous with the Greer Lake leucogranite.

3.3. Interior pegmatites

Besides the potassic pegmatite facies of the leucogranite, four categories of discrete pegmatite

bodies reside inside the Greer Lake leucogranite, distinguished by mineral assemblages and geochemical signatures: (i) barren, (ii) beryl±columbite bearing, (iii) lithium-enriched Annie Claim #2 (AC#2), and (iv) the lepidolite-subtype Annie Claim #3 (AC#3). The first two types grade from (i) to (ii) by the local appearance of the rare-element accessory minerals and are treated together as a single pegmatite category in the following text. In contrast, the AC#2 and AC#3 bodies show substantial differences; as will be shown below, the AC#2 pegmatite (iii) represents an intermediate step in the evolution of interior pegmatites from types (i) and (ii) toward the lepidolite-subtype AC#3 (iv). Thus, the AC#2 and AC#3 bodies are largely dealt with separately. Table 2 lists mineral assemblages and relative abundances of individual minerals in all four pegmatite categories.

Barren and beryl-columbite-subtype pegmatites form ellipsoidal pods, a few decameters to ~1 m across, that are randomly dispersed throughout the leucogranite body (Fig. 5a). Pegmatites gradually evolve from the pegmatitic leucogranite by an increase in size and abundance of microcline-perthite, disappearance of graphic quartz (commonly along a boundary within a single feldspar crystal), and transition into euhedral K-feldspar against a central core of bluish-grey quartz. Sheafs of cleavelandite and books of muscovite adorn locally the K-feldspar–quartz interface. A minority of these pegmatite pods also contain altered beryllian cordierite (Goad and Černý, 1981; Jobin-Bevans and Černý, 1998), green columnar beryl, and lath-shaped crystals of ferrocolumbite; however, these pegmatites do not differ from the barren ones in any other field-observable features.

The AC#2 pegmatite is located at the southern margin of the leucogranite intrusion, about 40 m from the exposed contact of the leucogranite with the host metabasalts (Fig. 2). The pegmatite forms a flat ellipsoidal body 6×10 m in plan and about 2–3 m thick. The pegmatite is surrounded by a fine- to medium-grained granitic zone, with gradual reduction of grain size from the surrounding pegmatitic leucogranite on the outside, and coarsening of a somewhat irregular comb structure of K-feldspar and quartz into the interior (Fig. 5b). Blocky microcline-perthite (up to 1.2 m across) and quartz compose most of the AC#2 body, with only small local pods of quartz resembling a segmented quartz core. In the internal parts, cleavelandite and three generations of micas are found mainly along the K-feldspar–quartz interface and in interstices and fractures in the K-feldspar. Silvery flat tabular muscovite is followed by curvilamellar brownish Li-bearing mica and by late local nests of fine-flaked greenish muscovite. Stumpy crystals of yellowish beryl are associated with the early muscovite, whereas brick-shaped manganocolumbite is found with the second type of mica. Rare wodginite, occasionally intergrown on microscopic scale with manganotantalite, is associated with microcline-perthite, quartz, and cleavelandite (Černý et al., 1986).

The lepidolite-subtype pegmatite AC#3 is located 75 m WSW of the AC#2 body, separated by about 3 m of pegmatitic leucogranite from the contact of this phase with host metabasalts. The AC#3 pegmatite was extensively excavated in the past and is in part covered by very coarse blasting rubble. Dimensions and internal structure can be reconstructed from exposed outcrops and broken-up material, although a true-to-scale plan of pegmatite would be difficult to draft. The roughly ellipsoidal body of AC#3 was at least 7×10 m in plan, and at least 3 m thick. A zone of fine-grained aplitic character separates the surrounding pegmatitic granite from the pegmatite proper, with gradual transitions of the same types as observed around AC#2.

The AC#3 pegmatite proper consists of five concentric zones, from the outside inwards (Masau, 1999): (i) border zone, (ii) wall zone, (iii) outer intermediate zone, (iv) inner intermediate zone, and (v) lepidolite+quartz core. Complete mineral assemblages of zones (ii) to (v) are listed in Table 2.

The border zone (i) consists of an aplitic layer, 20–30 cm thick, fine to medium grained, and steeply transitional from the surrounding pegmatitic leucogranite. Albite, muscovite, and quartz are associated with accessory microcline-perthite and garnet. Albite- and almandine-rich layers parallel to the outline of the pegmatite alternate locally at a millimeter scale. Rapid increase of grain size, platy silvery muscovite, and irregular comb structure of quartz and microcline-perthite mark the transition into the barren wall zone (ii), 30–40 cm thick. It grades into the voluminous outer intermediate zone (iii) with substantial increase in grain size and in the content of muscovite. Cleavelandite, quartz, and twisted plus globular varieties of green muscovite are characteristic. Minerals of Nb, Ta, and Sn (Masau et al., 2000a), typical of this zone, distinctly increase in variety and volume in the inner intermediate zone (iv), which is also substantial in volume. The first appearances of beryl, spessartine (Sps_{54-74} Alm_{46-26}), and silvery curvilamellar lithian muscovite are typical of this zone, as are minerals of U and rare phosphates of REE (Masau et al., 2000b, 2002). Lepidolite core (v) carries mainly purple, fine-flaky, and globular lepidolite with smoky quartz, spessartine (Sps_{77-91} Alm_{23-9}), and minor minerals of Nb, Ta, Zr, and U; small pods of pure quartz are observed locally. Late clusters of green Fe-

Table 1
Petrochemistry of the main phases of the Greer Lake leucogranite

	Fine-grained Leucogranite		Pegmatitic leucogranite		Sodic aplite		Potassic pegmatite	
	$x(5)^a$	Range	$x(2)$	Range	$x(4)$	Range	$x(2)$	Range
wt.%								
SiO_2	75.18	73.55–76.60	75.15	75.00–75.30	76.49	74.55–77.80	76.13	75.55–76.70
TiO_2	0.02	0.01–0.06	0.04	0.03–0.04	0.02	0.01–0.06	0.03	0.01–0.04
Al_2O_3	13.89	13.48–14.24	14.55	14.30–14.80	13.91	12.79–14.54	14.75	14.48–15.01
Fe_2O_3	0.33	0.04–0.46	0.54	0.21–0.86	0.48	0.11–0.76	0.58	0.56–0.60
FeO	0.41	0.16–0.94	0.92	0.92–0.92	0.76	0.48–1.16	0.30	0.20–0.40
MnO	0.07	0.01–0.26	0.09	0.08–0.09	0.17	0.02–0.33	0.04	0.01–0.06
MgO	0.05	0.01–0.09	0.03	0.03–0.03	0.04	0.03–0.06	0.06	0.05–0.07
CaO	0.22	0.04–0.40	0.14	0.06–0.22	0.28	0.22–0.40	0.24	0.18–0.30
Na_2O	3.85	2.62–4.97	3.98	2.75–5.20	5.70	4.55–6.70	3.43	2.90–3.95
K_2O	5.23	3.50–8.81	3.76	2.19–5.32	1.57	0.66–3.25	3.52	2.38–4.65
P_2O_5	0.06	0.04–0.08	0.05	0.03–0.06	0.06	0.03–0.08	0.02	0.02–0.02
CO_2	0.04	0.02–0.07	0.09	0.08–0.09	0.04	0.01–0.10	0.16	0.10–0.22
F_2	n.d.	n.d.–0.01	0.05	0.03–0.06	0.02	0.01–0.03	0.07	0.06–0.07
H_2O	0.43	0.27–0.71	0.63	0.55–0.71	0.72	0.28–1.56	1.10	1.04–1.15
-O=F	0.00	0.00–0.00	0.02	0.01–0.02	0.00	0.00–0.01	0.03	0.03–0.03
Total	99.78	–	100.04	–	100.26	–	100.46	–
ppm								
Li	58	29–86	98	76–119	117	19–324	114	107–120
Rb	562	340–1050	722	442–1001	355	138–559	594	453–735
Cs	8.9	5.1–13.7	4.0	2.3–5.6	10.2	1.5–21.0	6.6	4.6–8.5
Be	3.8	1.2–10.0	1.8	1.7–1.8	2.4	1.6–4.0	2.1	1.7–2.5
Sr^b	18	11–28	28	25–31	13	5–21	10	6–13
Ba	112	12–150	5	1–9	74	2–113	n.d.	n.d.
Pb	23	5–33	8	5–11	7	4–10	5	4–5
Ga	53	32–73	65	64–65	62	51–71	105	102–107
U	15	8–30	9	8–9	7	2–13	5	5
Th	7	n.d.–25	4	2–5	10	7–12	5	3–6
Zr	44	11–58	8	n.d.–16	50	14–93	2	1–2
Hf	0.24	n.d.–1.20	0.63	n.d.–1.25	0.5	n.d.–1.8	0.2	n.d.–0.4
Sn	14	5–25	28	23–33	16	8–21	23	20–25
K/Rb	87	42–132	43	41–44	36	24–48	48	44–53
K/(Csx100)	51	33–77	79	79	22	7–37	54	23–84
K/Ba	767	215–2425	11544	4907–18,180	3417	–13,490	–	–
Ba/Rb	0.27	0.02–0.42	0.01	0.00–0.01	0.16	–0.35	–	–
Mg/Li	4.7	1.7–9.4	2.0	1.5–2.4	3.8	1.1–8.3	3.2	2.8–3.5
Zr/Sn	4.98	0.55–11.00	0.35	–0.7	3.8	1.1–8.3	0.1	0.0–0.1
Zr/Hf	48.3	–	–	–	10.8	–43.3	150	–300
Al/Ga	1526	990–2284	1194	1182–1205	1215	1046–1373	747	716–779
ppm								
La	5.50	–	7.50	–	7.35	–	3	–
Ce	16.5	–	18.5	–	23	–	6.5	–
Nd	5	–	6.5	–	10	–	3.5	–
Sm	2.950	–	4.685	–	5.3	–	1.09	–
Eu	0.08	–	0.05	–	0.10	–	0.15	–
Tb	0.730	–	0.585	–	–	–	–	–
Dy	7.15	–	5.85	–	4.25	–	0.85	–
Yb	3.35	–	2.70	–	2.75	–	0.6	–
Lu	0.320	–	0.260	–	0.45	–	0.09	–
Y	30	17–45	16	n.d.–32	41	14–82	7	n.d.–14

Table 2
Mineral assemblages in the interior pegmatites of the Greer Lake leucogranite

	Beryl–columbite	AC#2	AC#3 per zone				
			(ii) wall	(iii) outer intermediate	(iv) inner intermediate	(v) lepidolite core	Late assemblages[a]
K-feldspar	●	●	●	–	✚	–	–
Albite	▲	▲	✚	●	●	✚	–
Quartz	●	●	–	▲	●	●	●
Muscovite	✚	▲	●	●	–	–	●
Li-muscovite	–	▲	–	–	●	▲	–
Lepidolite	–	–	–	–	✚	●	–
Cordierite	✚	–	–	–	–	–	–
Garnet	•	•	–	–	✚	•	–
Beryl	•	✚	–	–	✚	✚	–
Zircon	–	–	–	–	–	•	–
Coffinite[b]	–	–	–	–	–	•	–
Apatite	•	•	–	–	•	•	–
Monazite-(Sm)	–	–	–	–	•	–	–
Xenotime-(Y)	–	–	–	–	•	–	–
Florencite-(Ce)	–	–	–	–	•	–	–
Ferrocolumbite	•	•	–	–	✚	–	✚
Manganocolumbite	–	✚	–	✚	✚	–	✚
Ferrotantalite	–	–	–	–	–	✚	–
Manganotantalite	–	–	–	✚	✚	–	✚
Ferrowodginite	–	–	–	•[c]	–	–	–
Wodginite	–	✚	–	✚[c]	✚[c]	✚[c]	–
Ferrotapiolite	–	✚	–	–	•	•	–
Cassiterite	–	–	–	•	✚	✚	–
Uranpyrochlore	–	–	–	–	•	–	–
Microlite	–	–	–	–	–	–	•
Uranmicrolite	–	✚	–	–	–	–	–
Formanite(?)	–	–	–	–	•	–	–
Uraninite[d]	–	–	–	–	•	–	–

●, rock forming; ▲, subordinate; ✚, accessory; •, rare; –, absent.
[a] Late assemblages include green ferrous muscovite, zinnwalditic mica, and fracture-filling quartz.
[b] Exsolution product in zircon.
[c] Exsolution product in cassiterite.
[d] Breakdown product in zircon.

enriched muscovite, zinnwalditic mica, and fracture-filling quartz, all with minor columbite-group minerals, are found in the outer and, much less so, inner intermediate zones.

It should be briefly noted here that all of the above categories of interior pegmatites have their textural and paragenetic analogs in the aureole of exterior pegmatites, spread to the south of the leucogranite (Fig. 2).

3.4. Geochemical signature of the interior pegmatites

Geochemical features of some of the rock-forming and accessory minerals of the interior pegmatites are shown in Fig. 6. In consequence of the high degree of compositional evolution attained already by the rock phases of the leucogranite (Table 1, Fig. 4), the alkali fractionation in K-feldspar and micas is very advanced in the barren and beryl-columbite pegma-

Notes to Table 1:
REEs are based on a single analysis of each rock type; n.d.=not detected.
[a] x(N)=arithmetic mean (number of samples averaged).
[b] Largely to entirely radiogenic (cf. Clark and Černý, 1987).

Fig. 5. Transitions from pegmatitic leucogranite into interior pegmatite pods. (a) Pegmatitic leucogranite at the top (coarse-grained matrix with two megacrysts of K-feldspar dotted by graphic quartz) grades through a feldspar+muscovite layer into a barren potassic pegmatite pod with white to grey cleavable K-feldspar and darker-grey quartz. (b) Pegmatitic leucogranite (lower right corner) evolves into the aplitic border zone and indistinctly comb-textured wall zone of the AC#2 pegmatite pod; grey quartz and white K-feldspar at the level of the hammer handle mark the beginning of the predominant blocky zone.

tites, despite being the most primitive members of the pegmatite suite (Fig. 6a and b). Alkali fractionation culminates at remarkable levels in minerals of Li-bearing pegmatites and especially in AC#3.

Parallel to the alkali fractionation, the Mn content of garnet is remarkably increasing (Fig. 6d). The general trend of enrichment in Mn is also indicated in minerals of the columbite group throughout the three pegmatite categories (Fig. 6c). Within each of the AC#2 and AC#3 pegmatites, the columbite-group minerals show Mn increasing to virtual exclusion of Fe, and a broad range of high Ta/(Ta+Nb) values. Similar tendencies are shown by the Ta-dominant

wodginite-group minerals, primary ones, as well as those exsolved from the AC#3 cassiterite, which itself trends to distinctly Mn-dominant compositions (Masau et al., 2000a).

The general behavior of Nb and Ta is not so easy to track, as there is a strong trend of Ta enrichment within each individual pegmatite category (Fig. 6c). However, taking into account the relative abundances of columbite- and wodginite-group minerals of different compositions in individual pegmatite categories, and the composition of subordinate (Ta>Nb)-bearing phases in AC#3, a gradual increase in Ta/(Ta+Nb) from the beryl-columbite pegmatites to AC#3 becomes evident.

It should be mentioned that the geochemical characteristics of the interior pegmatites are closely reflected in their exterior paragenetic counterparts, from barren through beryl-columbite, Li-mica-bearing and lepidolite-petalite dikes crosscutting the metabasalts and gneisses (Černý et al., 1981).

4. Discussion

4.1. The leucogranite

The Greer Lake leucogranite is one of numerous similar, internally diversified, highly fractionated, and in part pegmatitic bodies parental to rare-element granitic pegmatites, identified in most continents in medium-grade terrains of diverse ages. Besides southeastern Manitoba, classic Archean examples are known from northwestern Ontario, Northwest Territories, Quebec, and western Australia. Proterozoic complexes are reported from South Dakota, Colorado, Finland, Sweden, and southern Greenland; Grenvillian edifices occur in Sweden, Damaran in Namibia. Fertile multifacies leucogranites in the New England states, Spain, Italy, Czech Republic, and Transbaikalia are of Hercynian age, whereas the youngest bodies of Laramide and Alpine ages are known from California and the Himalayas (reported and in part reviewed by Haapala, 1966; Blockley, 1980; Černý and Meintzer, 1988; Lahti, 1989; Černý, 1989b; Wise and Francis, 1992; Mulja et al., 1995; Smeds et al., 1998; London et al., 1999; Wilke et al., 2002; Visonà and Lombardo, 2002; Breaks et al., 2003). Specific mechanisms of origin of these

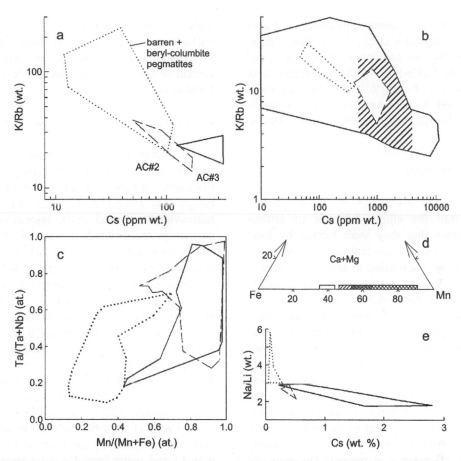

Fig. 6. Geochemical features of selected minerals from the barren plus beryl-columbite-subtype (dotted line), AC#2 (dashed line), and AC#3 (solid line) interior pegmatites; numbers of data points are given in parentheses for the three respective categories of pegmatites: (a) K/Rb versus Cs in the bulk composition of blocky K-feldspar (11, 8, 3); (b) K/Rb versus Cs in book muscovite, lithian muscovite, and lepidolite (a few extremely Cs-poor compositions in AC#3 correspond to muscovite from the outermost zone, and to Cs-depleted muscovite coexisting with Cs-rich lepidolite; most AC#3 micas plot in the diagonally ruled area) (11, 9, 58); (c) compositions of columbite-group minerals in the columbite quadrilateral (12, 21, 41); (d) composition of garnet in the Fe-Mn-(Ca+Mg) system (Ca and Mg are negligible and the thickness of the bars exaggerates their combined contents; open, beryl-columbite pegmatites; diagonal ruling, AC#2; cross-hatched, AC#3) (4, 3, 13); (e) Na/Li versus Cs in beryl (7, 3, 7).

leucogranites, locally known to cap plutonic intrusions of more primitive facies, are much debated in individual cases (e.g., the Himalayan leucogranites), but the principal process of partial melting of metasedimentary or meta-igneous protolith followed by extensive differentiation of the ensuing granitic melt is beyond reasonable doubt (e.g., Černý and Meintzer, 1988).

In the case of the Greer Lake leucogranite, it was initially interpreted, along with its close analogs in the Winnipeg River district, as a product of partial melting of the greenstone belt metasediments and

subsequent differentiation. Anatexis took place much below the level of emplacement and current exposure, and was probably aided by friction heat along subvertical faults that channeled the magma intrusions (Goad and Černý, 1981; Anderson et al., 1998). In view of the presence of cordierite and petalite in related pegmatites, the leucogranite was probably emplaced and consolidated at ~2.8 kbar P(H₂O), within a T span of ~700–600 °C (extended to ~450 °C in AC #3; Anderson et al., 1998; cf. London, 1992). The relatively low values of $\delta^{18}O$ were interpreted as resulting from extensive exchange with the host metabasalts during and after emplacement (Longstaffe et al., 1981). However, in view of extremely low contents of B, P, and S, and proximity to the K/Rb-low, biotite-rich metatonalite basement, anatexis of these $\delta^{18}O$-low juvenile rocks may have yielded the parent magma (F.J. Longstaffe, personal communication, 1989; Černý, 1989a; 1991b).

4.2. Interior pegmatites

Gradual textural transition of the pegmatitic leucogranite into the ellipsoidal pods of interior pegmatites shows that they were formed by local segregation of residual, low-temperature pegmatite-generating melt and solidified in situ. The coarse to giant grain size of the barren and beryl-columbite pegmatites indicates local increase in H_2O content of the residual melt, an increase which was undoubtedly distinctly higher in the Li- and F-rich melts that consolidated into the AC#2 and AC#3 pegmatites. Dramatic increase in solubility of H_2O, due to substantial concentration of Li and F, and consequent depression of liquidus and solidus were proven experimentally for such melts (London et al., 1989; London, 1992). The F content of the melt and μKF (dominant over μHF) must have been particularly high in the AC #3 magma, as the resulting pegmatite is very poor in K-feldspar, does not carry topaz or anhydrous Li-Al-silicates, and widespread micas are the only K-, Li-, and F-bearing phases (cf. London, 1982).

Determination of bulk composition of the AC#3 pegmatite or of its individual zones is not possible in the present disturbed state of the exposure. However, the overall mineral assemblage and its abrupt change from that of the surrounding granite and border zone invite speculation about possible F-induced immiscibility between the pegmatitic-leucogranite and lepidolite-pegmatite melts (Melentyev and Delitsyn, 1969; Melentyev et al., 1971). Experimental evidence indicates that F contents of ≥4 wt.% are required to induce immiscibility, at least in haplogranitic melts (Bailey, 1977; Kovalenko, 1977; Manning and Henderson, 1981). However, the effect of F on more complex melts enriched in Li, and also Rb, Cs, or Be has not been examined experimentally. Also, neither topaz nor amblygonite-montebrasite are

found in AC#3, and micas moderate but do not buffer F in melt; thus the presence and composition of lepidolite cannot lead to an estimate of the F content of the AC#3-generating melt (London, 1997; London et al., 1999).

4.3. Geochemical signature of the interior pegmatites

The barren and beryl-columbite-subtype pegmatites exceed the level of fractionation attained by the pegmatitic phases of leucogranite by a significant margin. The extreme K/Rb values for K-feldspar and muscovite (≥20 and ≥10, respectively) of these pegmatites are distinctly lower than the whole-rock values for leucogranites (132–24), in which K/Rb is controlled by these two minerals and consequently closely comparable to the mineral values.

Concerning the geochemistry of minor and trace elements in the AC#2 and particularly AC#3 pegmatites, the extreme values of K/Rb and Cs in the K-feldspar (≥14 and ≤330 ppm) and micas (≥2.5 and ≤11,200 ppm, respectively) compare favorably with those in some of the most fractionated pegmatites worldwide (e.g., Tanco, Manitoba; Varuträsk, Sweden; Himalaya, California; see Foord, 1976; Wise, 1995; Černý et al., 1998, 2004). The same applies to beryl that is modestly enriched in Li in AC#2, but both Li and particularly Cs dramatically increase in beryl from the AC #3 pegmatite (Fig. 6e).

The compositional evolution of the columbite-group minerals (Masau, 1999) does not follow a sequential path of increasing Mn and Ta typical of the lepidolite-subtype pegmatites, as generalized by Černý (1989b) and exemplified by, e.g., Foord (1976), Raimbault (1998), and Novák and Černý (1998). However, the cumulative effect of the overall columbite-group population corresponds to the initial Fe depletion and subsequent Ta enrichment of the typical trend, except a somewhat lower than extreme Mn enrichment (Fig. 6c). The elevated Fe content in the Ta-rich segment of the trend is probably due to contamination from the very closely adjacent metabasalt, which is also reflected in local aggregates of Fe-enriched muscovite and zinnwalditic mica. The contamination is closely analogous to that shown by columbite-group minerals in lepidolite-subtype pegmatites of the Jihlava district, Czech Republic, that reacted with enclosed xenoliths of mafic syenitic wallrocks (Černý and Němec, 1995).

4.4. The petrogenetic question

The AC#3 lepidolite-subtype pegmatite was derived from a leucogranitic parent by internal in situ differentiation, as were the less typical but Li-bearing AC#2 and the more primitive barren and beryl-columbite-subtype pegmatite pods. What are the reasons that lepidolite-subtype pegmatites have been denied this mode of origin, although magmatic differentiation is generally accepted for other categories of rare-element pegmatites? The reasons are, in principle, twofold: general scarcity of the lepidolite subtype, and long distance of emplacement from potential fertile granites.

In the overall spectrum of rare-element pegmatites, empirical data show that the most fractionated and rare-element-enriched complex pegmatites constitute only about 2% of their regional populations (Ginsburg et al., 1979). No numbers are available for the relative abundance of the lepidolite-subtype pegmatites, but they undoubtedly constitute only a small fraction of the above ~2%. Lepidolite-subtype pegmatites may constitute, in rare cases, a substantial to dominant component of local pegmatite groups (Quartz Creek group in Colorado, Staatz and Trites, 1955; Černý, 1982; Jihlava district in Czech Republic, Staněk, 1962; Němec, 1990; Mongolia, Vladykin et al., 1974), but they are missing in most cogenetic pegmatite populations or they are represented by only a few dikes compared to the multitude of pegmatites of other categories. The general paucity of lepidolite-subtype pegmatites greatly reduces the chance of observing them at erosional levels that would reveal their relationship to granitic parents (Černý, 1992b).

This fact is compounded by considerable distances separating these pegmatites from their proven, or potential, granitic parents. Besides the Greer Lake occurrences, Li-bearing minerals are known from inside pegmatite-generating leucogranites only as rare accessories from the Wekusko Lake area (lithian muscovite: Černý et al., 1981) and from Separation Lake (Breaks and Tindle, 1997). Lithium-rich melts solidifying as lepidolite-subtype pegmatites are invariably strongly enriched not only in Li but also in Rb, Cs, and in most cases B and P, locally more so than melts generating any other categories of granitic pegmatites. All of the above components increase solubility of H_2O (e.g., London, 1992), and the lepidolite-subtype-generating melts are probably the most volatile-rich and consequently the most fluid pegmatite-generating melts of them all. The extreme reduction of viscosity, combined with the depressed liquidus–solidus range due to the same spectrum of volatile components (London et al., 1989; London, 1992; Dingwell et al., 1992), translates into high mobility and thermal stability of these melts that facilitate intrusion far away from plutonic parents (Černý, 1992b).

Thus, it is not surprising that lepidolite-subtype pegmatites are found at outcrop distances from identified or potential granitic parents closely corresponding to the 2 km empirical maximum of Beus (1960): these are the cases of the Quartz Creek group (Černý, 1982: 2 km), Chèdeville (Raimbault, 1998: 2 km), Phangnga (Garson et al., 1969: 3 km), and Phuket (Suwimonprecha et al., 1995: 3 km). Baker (1998) developed a quantitative model for escape of pegmatite melts from granitic parents based on viscosity of the melts and propagation mechanisms (Rubin, 1995), and concluded that distances of up to 10 km are possible. Thus, even the 7-km distance separating the lepidolite-subtype Dobrá Voda pegmatite from its nearest outcropping potential parent seems to be realistic, populated as the intervening region is by a roughly zoned pegmatite sequence (Novák and Staněk, 1999).

All of the above show that field-based proof of plutonic derivation of lepidolite-subtype pegmatites is very difficult to obtain. The case of the Annie Claim #3 is therefore extremely valuable in this respect: the highly mobile melts with low liquidus temperature that generate lepidolite-subtype pegmatites can be expected to easily escape from parental plutons given a chance by either buoyant ascent in a plastic medium or by tectonic disturbance of solidified granitic envelope. The interior retention of AC#2 and AC#3 is therefore truly exceptional. The only other case we are aware of, which shows a direct physical link between the parent and product, is the lepidolite-subtype pegmatite from the Aksu–Pushtiru pegmatite field in Turkestan, illustrated by Beus (1948). This mountainous terrain with excellent vertical exposure reveals a mapped granite-to-pegmatite link (Fig. 7), which could hardly be surpassed by any schematic

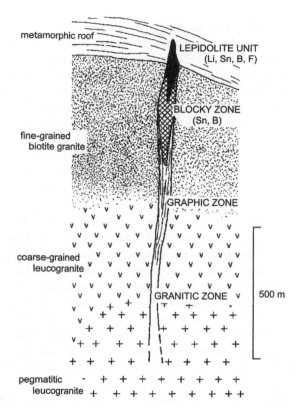

Fig. 7. Telescoped vertical zoning of a lepidolite-subtype pegmatite dike in the Aksu–Pushtiru pegmatite field of Turkestan. The zoning was generated by differentiation of residual volatile-rich melt inside consolidating granitic pluton; the dike penetrates the outer shells of more primitive types of granite into the metamorphic roof. Modified from Beus (1948).

textbook idealization, but which is virtually unknown to most pegmatite investigators.

5. Summary

The multiphase Greer Lake leucogranite provides a rare example of interior lepidolite-subtype pegmatites, with steep fractionation gradients from the host granite into the Li-, Rb-, Cs-, Be-, Mn-, Sn-, Nb-Ta-, and F-rich pegmatite pods. The textural, paragenetic, and geochemical features of the lepidolite-subtype bodies, combined with evidence from less fractionated barren and beryl-columbite pegmatite pods, leave no doubt about in situ derivation of pegmatites from local pods of residual, volatile-rich, and rare-element-enriched melt trapped in the leucogranite. General

rarity of lepidolite-subtype pegmatites, low liquidus and high fluidity of the parent melts, and their consequent ease of long-range migration from plutonic sources have been the main obstacles to general recognition of these pegmatites as products of plutonic differentiation.

Acknowledgments

The field work and laboratory research were supported by NSERC of Canada Operating, Research, Equipment, and Major Installation grants to P.Č. in 1971–2003, by Equipment and Infrastructure grants to F.C. Hawthorne, by the Canada-Manitoba Subsidiary Agreement on Mineral Exploration and Development 1973–1978, and by the Canada Department of Energy, Mines and Resources Research Agreements 1971–1975. Logistic support in the field provided by the Tantalum Mining Corporation of Canada Ltd. is also much appreciated. R. Chapman, K. Ramlal, W. Blonski, and G. Morden provided the wet chemical analyses, and R. Chapman supervised and in part performed the electron-microprobe work. S. Mejia processed photographs and some of the graphics. The constructive reviews by S.I. Lahti and an anonymous reviewer, and the editorial effort of T. Rämö were much appreciated.

This paper is dedicated to Professor Dr. Ilmari Haapala: his classic 1966 paper introduced the first author to the world of internally diversified fertile leucogranites that proved later to be fundamental components of uncountable pegmatite fields on global scale in general, and in Fennoscandia and Canadian Shield in particular.

Appendix A. Analytical methods

Chemical whole-rock and mineral compositions were collected in part during the Canada-Manitoba MDA project 1973–1978 and remained largely unpublished, except a few examples and summaries (Černý et al., 1981; Goad and Černý, 1981). Most of the data on the AC#3 minerals come from the BSc Hon's thesis of Masau (1999). Experimental conditions for atomic absorption spectrophotometry, X-ray fluorescence spectrometry, neutron activation, and other techniques applied to whole-rock analysis, and for the atomic absorption spectrophotometry and electron-microprobe analysis used for minerals are quoted in the two works cited above, and are available on request from the first author.

References

Anderson, S.D., Černý, P., Halden, N.M., Chapman, R., Uher, P., 1998. The YITT-B pegmatite swarm at Bernic Lake, southeastern Manitoba: a geochemical and paragenetic anomaly. Canadian Mineralogist 36, 283–301.

Bailey, J.C., 1977. Fluorine in granitic rock and melts. Chemical Geology 19, 1–42.

Baker, D.R., 1998. The escape of pegmatite dikes from granitic plutons: constraints from new models of viscosity and dike propagation. Canadian Mineralogist 36, 255–264.

Beus, A.A., 1948. Vertical zoning of pegmatites on the example of the pegmatite field Aksu Pushtiru (Turkestan Range). Doklady of the Academy of Sciences of the USSR 60, 1235–1238 (in Russian).

Beus, A.A., 1960. Geochemistry of Beryllium and the Genetic Types of Beryllium Deposits. Academy of Sciences of the USSR, Moscow. (English translation, Freeman, San Francisco, California 1966).

Blockley, J.G., 1980. The tin deposits of Western Australia, with special reference to the associated granites. Australia Geological Survey. Mineral Resources 12.

Breaks, F.W., Tindle, A.G., 1997. Rare-element exploration potential of the Separation Lake area: an emerging target for Bikita-type mineralization in the Superior Province of northwest Ontario. Summary of Field Work and Other Activities 1997, Ontario Geological Survey Miscellaneous Paper, vol. 168, pp. 72–88.

Breaks, F.W., Selway, J.B., Tindle, A.G., 2003. Fertile peraluminous granites and related rare-element mineralization in pegmatites, Superior Province, northwest and northeast Ontario: operation treasure hunt. Ontario Geological Survey, Open File Report 6099.

Černý, P., 1982. Petrogenesis of granitic pegmatites. In: Černý, P. (Ed.), Granitic Pegmatites in Science and Industry, Mineralogical Association of Canada Short Course Handbook, vol. 8, pp. 405–461.

Černý, P., 1989a. Contrasting geochemistry of two pegmatite fields in Manitoba: products of juvenile Aphebian crust and polycyclic Archean evolution. Precambrian Research 45, 215–234.

Černý, P., 1989b. Characteristics of pegmatite deposits of tantalum. In: Möller, P., Černý, P., Saupé, F. (Eds.), Lanthanides, Tantalum and Niobium. Springer-Verlag, Berlin-Heidelberg, pp. 195–239.

Černý, P., 1991a. Rare-element granitic pegmatites. I. Anatomy and internal evolution of pegmatite deposits. Geoscience Canada 18, 49–67.

Černý, P., 1991b. Fertile granites of Precambrian rare-element pegmatite fields: is geochemistry controlled by tectonic setting or source lithologies? Precambrian Research 51, 429–468.

Černý, P., 1992a. Geochemical and petrogenetic features of mineralization in rare-element granitic pegmatites in the light of current research. Applied Geochemistry 7, 393–416.

Černý, P., 1992b. Regional zoning of pegmatite populations and its interpretation. Mitteilungen der Österreichischen Mineralogischen Gesellschaft 137, 99–107.

Černý, P., Meintzer, R.E., 1988. Fertile granites in the Archean and Proterozoic fields of rare-element pegmatites: crustal environment, geochemistry and petrogenetic relationships. In: Taylor, R.P., Strong, D.F. (Eds.), Recent Advances in the Geology of Granite-Related Mineral Deposits, Canadian Institute of Mining and Metallurgy Special, vol. 39, pp. 170–207.

Černý, P., Němec, D., 1995. Pristine vs. contaminated trends in Nb,Ta-oxide minerals of the Jihlava pegmatite district, Czech Republic. Mineralogy and Petrology 55, 117–130.

Černý, P., Trueman, D.L., Ziehlke, D.V., Goad, B.E., Paul, B.J., 1981. The Cat Lake-Winnipeg River and the Wekusko Lake pegmatite fields, Manitoba. Manitoba Department of Energy and Mines, Mineral Resources Division, Economic Geology Report ER80-1.

Černý, P., Goad, B.E., Hawthorne, F.C., Chapman, R., 1986. Fractionation trends of the Nb- and Ta-bearing oxide minerals in the Greer Lake pegmatitic granite and its pegmatite aureole, southeastern Manitoba. American Mineralogist 71, 501–517.

Černý, P., Ercit, T.S., Vanstone, P.J., 1998. Mineralogy and petrology of the Tanco rare-element pegmatite deposit, southeastern Manitoba. International Mineralogical Association, 17th General Meeting Toronto, Field Trip Guidebook B6.

Černý, P., Chapman, R., Ferreira, K., Smeds, S.A., 2004. Geochemistry of oxide minerals of Nb, Ta, Sn and Sb in the Varuträsk granitic pegmatite, Sweden: the case of an anomalous columbite-tantalite trend. American Mineralogist 89, 505–518.

Charoy, B., Noronha, F., 1996. A multistage model for growth of a rare-element, volatile-rich microgranite: the example of Argemela (Portugal). Journal of Petrology 37, 73–94.

Clark, G.S., Černý, P., 1987. Radiogenic ^{87}Sr, its mobility, and the interpretation of Rb–Sr fractionation trends in rare-element granitic pegmatites. Geochimica et Cosmochimica Acta 51, 1011–1018.

Dingwell, D.B., Knoche, R., Webb, S.L., 1992. Effects of boron, fluorine and phosphorus on the viscosity and density of haplogranitic melts. Abstracts, Geological Association of Canada-Mineralogical Association of Canada Annual Meeting Wolfville 17, A-27.

Foord, E.E., 1976. Mineralogy and petrogenesis of layered pegmatite-aplite dikes in the Mesa Grande district, San Diego County, California. PhD thesis, Stanford University, California, USA.

Garson, M.S., Bradshaw, N., Rattawong, S., 1969. Lepidolite pegmatites in the Phangnga area of peninsular Thailand. In: Fox, W. (Ed.), 2nd Technical Conference on Tin, International Tin Council, Bangkok, pp. 327–339.

Ginsburg, A.I., Timofeyev, I.N., Feldman, L.G., 1979. Principles of Geology of Granitic Pegmatites. Nedra, Moscow (in Russian).

Goad, B.E., Černý, P., 1981. Peraluminous pegmatitic granites and their pegmatite aureoles in the Winnipeg River district, Southeastern Manitoba. Canadian Mineralogist 19, 177–194.

Gordiyenko, V.V., Ilyina, A.N., Timochina, L.A., Badanina, E.B., Staněk, J., 1996. Geochemical model of evolution of a pegmatite-forming ore-magmatic system of western Moravia. Proceedings of the Russian Mineralogical Society 125, 38–48. (in Russian).

Haapala, I., 1966. On the granitic pegmatites in the Peräseinäjoki–Alavus area, South Pohjanmaa, Finland. Bulletin de la Commission géologique de Finlande 224.

Jobin-Bevans, S., Černý, P., 1998. The beryllian cordierite+beryl+spessartine assemblage, and secondary beryl in altered cordierite, Greer Lake granitic pegmatites, southeastern Manitoba. Canadian Mineralogist 36, 447–462.

Kovalenko, N.I., 1977. The reactions between granite and aqueous hydrofluoric acid in relation to the origin of fluorine-bearing granites. Geochemistry International 1977, 503–515.

Lahti, S.I. (Ed.), Symposium Precambrian granitoids. Petrogenesis, geochemistry and metallogeny, August 14–17, 1989, Helsinki, Finland. Excursion C 1: lateorogenic and synorogenic Svecofennian granitoids and associated pegmatites of southern Finland, Geological Survey of Finland Guide, vol. 26.

London, D., 1982. Stability of spodumene in acidic and saline fluorine-rich environments. Annual Report, Carnegie Institution Geophysical Laboratory 81, 331–334.

London, D., 1992. The application of experimental petrology to the genesis and crystallization of granitic pegmatites. Canadian Mineralogist 30, 499–540.

London, D., 1997. Estimating abundances of volatile and other mobile components in evolved silicic melts through mineral-melt equilibria. Journal of Petrology 38, 1691–1706.

London, D., Morgan VI, G.B., Hervig, R.L., 1989. Vapor-undersaturated experiments with Macusani glass+ H$_2$O at 200 MPa, and the internal differentiation of granitic pegmatites. Contributions to Mineralogy and Petrology 102, 1–17.

London, L., Wolf, M.B., Morgan VI, G.B., Garrido, M.G., 1999. Experimental silicate-phosphate equilibria in peraluminous granitic magmas, with a case study of the Albuquerque batholith at Tres Arroyos, Badajoz, Spain. Journal of Petrology 40, 215–240.

Longstaffe, F.J., Černý, P., Muehlenbachs, K., 1981. Oxygen isotope geochemistry of the granitoids in the Winnipeg River pegmatite district, southeastern Manitoba. Canadian Mineralogist 19, 195–204.

Manning, D.A.C., Henderson, C.M.B., 1981. The effect of the addition of fluorine on liquidus phase relationships in the system Qz-Ab-Or with excess water at 1 kb. Progress in Experimental Petrology 1978–1980, 5th Report, NERC Publication Series D, No. 18, pp. 16–23.

Masau, M., 1999. Mineralogy and geochemistry of the Annie Claim No. 3 pegmatite pod at Greer lake, southeastern Manitoba. B.Sc. Hon's thesis, University of Manitoba, Winnipeg, Canada.

Masau, M., Černý, P., Chapman, R., 2000a. Exsolution of zirconian-hafnian wodginite from manganoan-tantalian cassiterite, Annie Claim #3 granitic pegmatite, southeastern Manitoba, Canada. Canadian Mineralogist 38, 685–694.

Masau, M., Černý, P., Chapman, R., 2000b. Dysprosian xenotime-(Y) from the Annie Claim #3 granitic pegmatite, southeastern Manitoba, Canada: evidence of the tetrad effect? Canadian Mineralogist 38, 899–905.

Masau, M., Černý, P., Cooper, M.A., Chapman, R., 2002. Monazite-(Sm), a new member of the monazite group from the Annie Claim #3 granitic pegmatite, southeastern Manitoba. Canadian Mineralogist 40, 1649–1655.

Melentyev, B.N., Delitsyn, L.M., 1969. Problems of liquation in magma. Doklady of the Academy of Sciences of the USSR Earth Sciences Series, vol. 186. American Geological Institute, New York, pp. 215–217.

Melentyev, B.N., Martyanov, N.N., Alekseyeva, Y.A., 1971. New data on the pegmatite fields of Soviet Central Asia and the concentration of tantalum, cesium and transparent tourmaline in them. Doklady of the Academy of Sciences of the USSR Earth Sciences Series 200, 1437–1440 (in Russian).

Mulja, T., Williams-Jones, A.E., Woods, S.A., Boily, M., 1995. The rare-element-enriched monzonite–pegmatite–quartz vein systems in the Preissac–Lacorne batholith, Quebec. II. Geochemistry and petrogenesis. Canadian Mineralogist 33, 817–834.

Němec, D., 1990. Fertile leucogranites and aplites of the Jihlava pegmatite field, western Moravia, Czechoslovakia. Acta Universitatis Carolinae. Geologica 1990, 25–34.

Novák, M., Černý, P., 1998. Niobium-tantalum oxide minerals from complex granitic pegmatites in the Moldanubicum, Czech Republic: primary versus secondary compositional trends. Canadian Mineralogist 36, 659–672.

Novák, M., Staněk, J., 1999. Lepidolite pegmatite from Dobrá Voda near Velké Meziříčí, western Moravia. Acta Musei Moraviae. Scientiae Geologicae 84, 3–44.

Raimbault, L., 1998. Composition of complex lepidolite-type granitic pegmatites and of constituent columbite-tantalite, Chèdeville, Massif Central, France. Canadian Mineralogist 36, 563–584.

Rubin, A.M., 1995. Getting granite dikes out of the source region. Journal of Geophysical Research 100, 5911–5929.

Smeds, S.-A., Uher, P., Černý, P., Wise, M.A., Gustafsson, L., Penner, P., 1998. Graftonite-beusite in Sweden: primary phases, products of exsolution, and distribution in zoned populations of granitic pegmatites. Canadian Mineralogist 36, 377–394.

Staatz, M.H., Trites, A.F., 1955. Geology of the Quartz Creek pegmatite district, Gunnison County, Colorado. US Geological Survey Professional Paper 265.

Staněk, J., 1962. Mineralogically significant pegmatites of the Jihlava region. Folia Facultatis Scientiarum Natturalium, Universitas J.E. Purkyně Brno, vol. 3, pp. 79–105.

Stewart, D.B., 1978. Petrogenesis of lithium-rich pegmatites. American Mineralogist 63, 970–980.

Suwimonprecha, P., Černý, P., Friedrich, G., 1995. Rare metal mineralization related to granites and pegmatites, Phuket, Thailand. Economic Geology 90, 603–615.

Visonà, D., Lombardo, B., 2002. Two-mica and tourmaline leucogranites from the Everest–Makalu region (Nepal–Tibet), Himalayan leucogranite genesis by isobaric heating? Lithos 62, 125–150.

Vladykin, N.V., Dorfman, M.D., Kovalenko, V.I., 1974. Mineralogy, geochemistry and genesis of rare-element topaz–lepidolite–albite pegmatites of the Mongolian People's Republic. Trudy of the Mineralogical Museum of the Academy of Sciences of the USSR 23, 6–49 (in Russian).

Wilke, M., Nabelek, P.I., Glascock, M.D., 2002. B and Li in Proterozoic metapelites from the Black Hills, USA: implications for the origin of leucogranitic magmas. American Mineralogist 87, 491–500.

Wise, M.A., 1995. Trace element chemistry of lithium-rich micas from rare-element granitic pegmatites. Mineralogy and Petrology 55, 203–214.

Wise, M.A., Francis, C., 1992. Distribution, classification and geological setting of granitic pegmatites in Maine. Northeastern Geology 14, 82–93.

Zasedatelev, A.M., 1974. Possible accumulation of lithium in host rocks of lithium pegmatite veins during old sedimentation processes. Doklady of the Academy of Sciences of the USSR Earth Sciences Series 218, 196–198 (in Russian).

Zasedatelev, A.M., 1977. Quantitative model of metamorphic generation of rare-metal pegmatites with lithium mineralization. Doklady of the Academy of Sciences of the USSR, Earth Sciences Series 236, 219–221 (in Russian).

Available online at www.sciencedirect.com

Lithos 80 (2005) 323–345

www.elsevier.com/locate/lithos

Textural and chemical evolution of a fractionated granitic system: the Podlesí stock, Czech Republic

Karel Breiter[a],*, Axel Müller[b], Jaromír Leichmann[c], Ananda Gabašová[a]

[a]*Czech Geological Survey, Geologická 6, CZ-15200 Praha 5, Czech Republic*
[b]*Natural History Museum, Dept. Mineralogy, Cromwell Road, London SW75BD, United Kingdom*
[c]*Department of Geology and Paleontology, Masaryk University, Kotlářská 2, CZ-61137 Brno, Czech Republic*

Received 17 March 2003; accepted 9 September 2004
Available online 18 November 2004

Abstract

The Podlesí granite stock (Czech Republic) is a fractionated, peraluminous, F-, Li- and P-rich, and Sn, W, Nb, Ta-bearing rare-metal granite system. Its magmatic evolution involved processes typical of intrusions related to porphyry type deposits (explosive breccia, comb layers), rare-metal granites (stockscheider), and rare metal pegmatites (extreme F–P–Li enrichment, Nb–Ta–Sn minerals, layering). Geological, textural and mineralogical data suggest that the Podlesí granites evolved from fractionated granitic melt progressively enriched in H_2O, F, P, Li, etc. Quartz, K-feldspar, Fe–Li mica and topaz bear evidence of multistage crystallization that alternated with episodes of resorption. Changes in chemical composition between individual crystal zones and/or populations provide evidence of chemical evolution of the melt. Variations in rock textures mirror changes in the pressure and temperature conditions of crystallization. Equilibrium crystallization was interrupted several times by opening of the system and the consequent adiabatic decrease of pressure and temperature resulted in episodes of nonequilibrium crystallization. The Podlesí granites demonstrate that adiabatic fluctuation of pressure ("swinging eutectic") and boundary-layer crystallization of undercooled melt can explain magmatic layering and unidirectional solidification textures (USTs) in highly fractionated granites.
© 2004 Elsevier B.V. All rights reserved.

Keywords: Granite; Breccia; UST; Magmatic layering; Feldspars; Quartz

1. Introduction

Granitic pegmatites, tin granites and subvolcanic porphyry deposit-related intrusions represent three different results of fractionation of silicic magmas. For porphyry-related intrusions, evolution in an open system was established a long time ago, but granites and pegmatites were, until the 1980s, considered to be products of slow closed-system crystallization. However, many structural as well as compositional links among these systems have been found recently. Explosive brecciation in a subvolcanic tin–granite

---

* Corresponding author. Tel.: +420 251085508; fax: +420 251818748.
E-mail address: breiter@cgu.cz (K. Breiter).

was recognised in Krušné Hory/Erzgebirge and Corn-wall (Schust and Wasternack, 1972; Allman-Ward et al., 1982). Magmatic layering with crystals oriented perpendicular to the individual layers (unidirectional solidification texture, UST[1]; Shannon et al., 1982) was found to be typical of subvolcanic granitic stocks with associated Sn, W and Mo mineralisation. Such layering is typically parallel to the contact of the intrusions and is composed of layers of euhedral quartz crystals alternating with layers of fine-grained aplite. A special type of UST are comb quartz layers (Kirkham and Sinclair, 1988). Simple magmatic layering with orientation of minerals parallel to the layers is common in aplite–pegmatite bodies (e.g., Duke et al., 1992; Breaks and Moore, 1992; Morgan and London, 1999), but in true granites it is much less abundant (Zaraisky et al., 1997). Calculations based on conductive cooling of some flat aplite–pegmatite dykes intruded into relatively cool country rocks suggest a rapid rate of crystallization (5 to 340 days for 1- to 8-m-thick dykes; Webber et al., 1999).

All these phenomena suggest crystallization of granitic magma in an open system involving rapid changes in p–T–X conditions. Fractionated granitic and pegmatitic systems have been intensively studied during the last two decades and a considerable amount of new natural and experimental data are now available (cf. London, 1990, 1992, 1995, 1996). Magmatic layering has been explained by several models (Brigham, 1983; London, 1992, 1999; Bala-shov et al., 2000; Fedkin et al., 2002) with strong undercooling and/or a sudden decrease of pressure having a crucial role. Nevertheless, there are few well-exposed natural examples that verify these experi-mental predictions.

---

[1] In this paper, we use the term "layering" or "layered rock" for rocks with layers differing in grain size and/or mineral composition regardless of the orientation of individual crystals. The term "UST" (unidirectional solidification texture) refers to a layer composed of crystals oriented perpendicularly to the plain of layering and the term "comb mineral" is an oriented crystal within a UST layer. "Snow-ball quartz" (late quartz phenocryst with abundant albite laths arranged concentrically along its growth planes) is considered to be an igneous texture. "Snow-ball fabric" applies here also to comb orthoclase. "Stockscheider" is an old German mining term for K-feldspar-dominated pegmatite at the upper contact of some tin-bearing granites (in the geological literature since 1865; Jarchovský, 1962, and references therein).

One of these is the Podlesí granite stock (Czech Republic)—an extremely fractionated, peraluminous, F-, Li- and P-rich, and Sn, W, Nb, Ta-bearing rare-metal granite with an early intrusive/explosive brec-cia, marginal pegmatite, feldspar- and zinnwaldite-dominated USTs, and late intra-dyke brecciation. Major minerals bear evidence for multistage crystal-lization that alternated with episodes of resorption. Changes in chemical composition between individual crystal zones and/or populations provide evidence for the chemical evolution of the melt, while variation in rock textures mirrors changes in the pressure and temperature conditions of crystallization.

2. Geologic setting and shape of the intrusion

The Podlesí granite stock (0.1 km^2) is situated in the western part of the Krušné Hory Mts., Czech Republic. It is the youngest intrusion (313 to 310 Ma) of the multistage late-Variscan tin-specialised Eibenstock-Nejdek pluton that intruded Ordovician phyllites and a Variscan biotite granite. Several boreholes drilled by the Czech Geological Survey penetrate the intrusion in a >400-m-long vertical section (Breiter, 2002).

The overall shape of the intrusion is tongue-like (Fig. 1). The stock consists primarily of an albite–protolithionite–topaz granite (stock granite), which can be divided into two subfacies. The "upper facies" forms the uppermost 30- to 40-m-thick carapace of the stock and is fine-grained and porphyritic. The "lower facies" comprising the main part of the stock is medium-grained and equigranular. The uppermost part of the cupola is bordered by a 50-cm-thick marginal pegmatite (stockscheider).

In the uppermost 100 m, the stock granite has been intruded by several subhorizontal dykes of albite–zinnwaldite–topaz granite (dyke granite). Upper and lower contacts of the dykes are sharp and slightly uneven. Thin (mm- to cm-scale) steeply dipping apophyses are found around the larger dykes. The biggest dyke (hereafter referred to as "major dyke") is 7 m thick. Its western part is well exposed and shows prominent magmatic layering with USTs. In the eastern part, a late-magmatic intra-dyke breccia is also found. The dyke generally dips 5° to 10° to the southwest, with the eastern part of the dyke represent-ing the uppermost apical part. The western and eastern

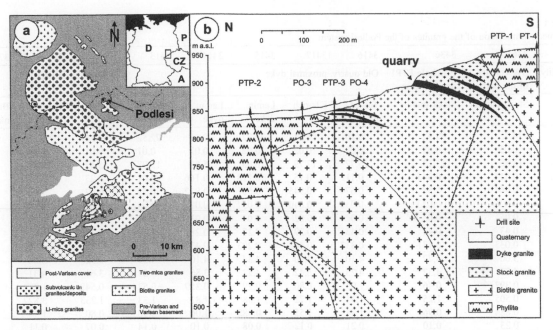

Fig. 1. (a) Simplified geological map of the Western Krušné Hory/Erzgebirge Mountains and location of the Podlesí granite system. Inset shows area relative to eastern Europe; A—Austria, CZ—Czech Republic, D—Germany, P—Poland. (b) Geological cross section through the Podlesí granite stock. The "major dyke" crops out in the quarry.

parts of the dyke are here termed the proximal and distal portions of the dyke, respectively.

3. Data

We present scanning electron microscope and optical cathodoluminescence as well as mineral chemical and whole-rock geochemical data on the various rock types and mineral assemblages of the Podlesí stock. The pertinent analytical methods are described in Appendix A.

3.1. Whole-rock chemistry

Both facies of the stock granite are strongly peraluminous (A/CNK 1.15–1.25; including Li, Rb and Cs as charge balancing cations in the denominator, the peraluminity index will decrease to 1.1–1.2) and, in comparison with common Ca-poor granites, enriched in incompatible elements such as Li, Rb, Cs, Sn, Nb, U, W, and poor in Mg, Ca, Sr, Ba, Fe, Sc, Zr, Pb, and V (Breiter, 2002). They are also rich in P (0.4–0.8 wt.% P_2O_5) and F (0.6–1.8 wt.%). A high degree of magmatic fractionation is demonstrated by

low K/Rb (22–35) and Zr/Hf (12–20) and high U/Th (4–7). The dyke granite is even more enriched in Al (A/CNK 1.2–1.4; 1.1–1.3 taking Li, Rb and Cs into account), P (0.6–1.5 wt.% P_2O_5), F (1.4–2.4 wt.%), Na, Rb, Li, Nb, Ta, and depleted in Si, Zr, Sn, W and REE. The K/Rb (14–20) and Zr/Hf (9–13) ratios are lower than in the stock granite (Table 1).

All the Podlesí granite types are rich in P, which shows a positive correlation with F, Al, Li, Rb, Nb, Ta and peraluminosity, and a negative correlation with Si, Zr and Sn. They are also relatively B-poor (20–60 ppm). However, tourmalinisation evolved in phyllite in a broader contact aureole of the granite stock suggests that B content in the melt may have been substantially higher.

A high degree of fractionation is documented also by high concentrations of rare metals. Nb and Ta are more abundant in the dyke granite (50–95 ppm Nb and 30–55 ppm Ta) than in the stock granite (25–50 and 10–25 ppm, respectively). On the other hand, contents of Sn and W are distinctly higher in the stock granite (10–50 ppm Sn, 20–80 ppm W) than in the dyke granite (5–20 and 35–60 ppm, respectively). The REE contents of the stock granite are low and chondrite-normalised patterns relatively flat [(Ce/

Table 1
Chemical composition of the granites of the Podlesí system

No.	3361	3436	3416	3417	3415	3414	3413	3739	3742	3741
Locality	SW contact of the body	Borehole PTP1 depth 200 m	Old quarry, proximal dyke					Outcrop, distal dyke		
Rock	Stockscheider	Stock granite	Top of the dyke	Kfs-rich UST	Lamin. dyke	Lamin. dyke	Base of the dyke	Homogen. dyke	Breccia fragm.	Breccia matrix
SiO_2	74.78	72.52	70.10	66.54	70.72	70.66	72.52	72.10	62.53	67.90
TiO_2	0.05	0.02	0.02	0.02	0.02	0.02	0.01	0.01	0.04	0.03
Al_2O_3	13.80	15.13	15.77	17.79	16.03	15.89	15.98	15.42	18.65	17.31
Fe_2O_3	0.414	0.345	0.111	0.15	0.118	0.166	<0.01	0.23	0.37	0.22
FeO	0.464	0.58	0.674	0.945	0.551	0.822	0.55	0.42	1.84	0.74
MnO	0.026	0.023	0.036	0.046	0.037	0.08	0.041	0.054	0.122	0.053
MgO	0.05	0.05	0.02	0.02	0.01	0.03	0.04	0.02	0.01	0.02
CaO	0.43	0.44	0.43	0.39	0.27	0.28	0.17	0.18	0.1	0.06
Li_2O	0.033	0.181	0.332	0.442	0.293	0.432	0.29	0.348	1.502	0.608
Na_2O	3.60	3.77	4.10	2.90	4.34	4.10	4.89	4.92	3.84	5.46
K_2O	4.23	4.32	4.25	6.42	4.22	3.80	3.51	3.61	4.31	3.90
P_2O_5	0.397	0.46	1.03	0.80	0.79	0.859	0.524	0.549	1.272	0.996
F	0.505	1.25	1.31	1.72	1.38	1.82	1.33	1.233	3.822	1.529
H_2O+	0.884	0.966	1.34	1.55	0.882	1.06	0.90	0.80	1.34	1.15
H_2O-	0.23	0.10	0.21	0.12	0.08	0.10	0.14	0.07	0.11	0.05
Total	99.72	99.63	99.22	99.16	99.18	99.38	100.36	99.96	99.85	100.03
Ba	23	4	13	12	5	24	9	1	1	5
Bi	3	3	8	49	57	24	8	9	22	20
Cs	56	82	143	198	128	170	159	150	343	172
Ga	31	39	60	56	50	51	52	59	69	74
Hf	2.0	1.9	4.6	2.2	3.5	3.1	1.8	2.1	4.2	3.5
Nb	56	35	99	89	90	83	55	59	153	104
Pb	10	4	<2	8	<2	2	<2	<2	3	<2
Rb	551	1260	2162	3025	2084	2175	1970	2370	4380	2870
Sn	18	69	29	31	41	27	34	24	78	27
Sr	16	27	182	99	86	102	60	11	15	14
Ta	19	15	68	54	53	54	20	20	65	42
Th	7.5	5.3	5.4	12.1	6	5.8	6.3	7.3	11.4	6.9
Tl	2.5	2.9	4.3	6.2	4.3	4	3.5	1.7	6.3	3.4
U	43	35	47	44	37	35	20	20	47	32
W	24	38	49	55	49	63	23	30	93	48
Zn	47	43	79	97	60	69	73	69	224	100
Zr	22	25	34	13	29	24	18	14	37	34

Major elements (in wt.%) determined at the Czech Geological Survey by wet chemistry. Nb, Pb, Rb, Sn, Sr, Zn and Zr (in ppm) by XRF at the Czech Geological Survey, the other trace elements (in ppm) by ICP-MS in ACMElab in Vancouver, Canada.

Yb)$_N$=4–5] with distinct negative Eu anomalies. The dyke granite is even more depleted in REE and shows the lanthanide tetrad effect (Breiter et al., 1997).

3.2. Textural evidence of crystallization conditions

3.2.1. Stockscheider

At Podlesí, a 30- to 50-cm-thick stockscheider is found along the southern and southwestern margin of the body. It is composed of large prismatic microcline

crystals oriented perpendicular to the contact plane of the granite with phyllite. The space between individual microcline crystals is filled with albite-rich fine-grained matrix and, in places, with large grains of milky quartz. The stockscheider changes downwards into fine-grained stock granite. In a transition zone several metres thick, the granite contains K-feldspar crystals about 5 cm long and 1 to 2 cm thick and local layers with comb quartz. Phenocrysts of quartz and microcline within the stockscheider are markedly altered. In one sample, fragments of large microcline crystals are cemented by fine-grained granitic matrix containing fragmented topaz crystals (Fig. 2A). This suggests that crystallization of at least some parts of the stockscheider predated the first episode of brecciation (see below).

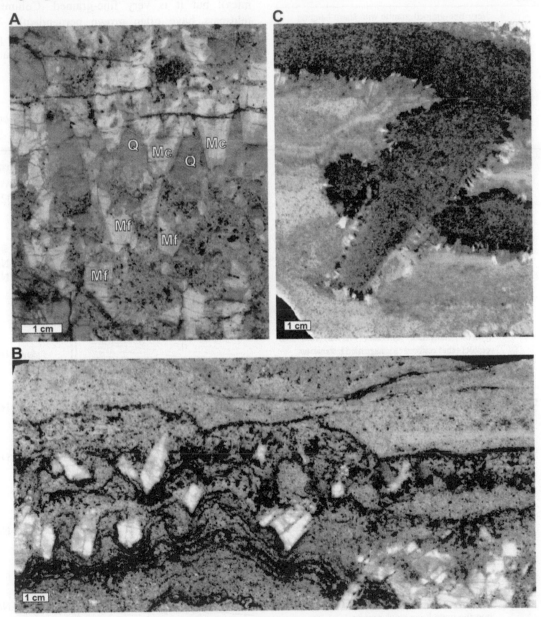

Fig. 2. Macrotextures in the Podlesí stock. (A) Stockscheider, individual crystals of microcline (Mc) in the upper part of the sample have grown downwards, space between the crystals is filled by milky quartz (Q). In the lower part, fragments of microcline crystals (Mf) have been cemented with fine-grained granitic matrix. (B) K-feldspar (orthoclase) dominated UST layer near the upper contact of the major dyke. Light laminae are composed of quartz, albite, K-feldspar and topaz, dark laminae are pure zinnwaldite. (C) Intra-dyke breccia: horizontal mica-rich facies in the upper part of the sample is rimmed with quartz–zinnwaldite UST at its lower margin. Fragments of the mica-rich facies in the central part of the sample are rimmed by zinnwaldite–orthoclase UST. Space between the fragments is filled with albite–quartz–topaz matrix. In all these cases, direction of crystallization is downwards.

Table 2
Petrography of the major dyke in Podlesí (cf. Fig. 3)

No. of layers	Thickness (cm)	Description
1	1	Very fine-grained granite with sharp hanging wall contact.
2	2–3	K-feldspar-dominated UST layer, columnar orthoclase crystals overgrown by fine-grained granitic matrix.
3	30	Thin laminae consisting of quartz, albite and P-rich K-feldspar in different proportions. Zinnwaldite is partly disseminated, partly concentrated in thin nearly monomineralic laminae. Contacts of the laminae are generally irregular and/or folded. Small-scale protrusions of not fully crystallized material in the hanging wall of some laminae are present.
4	10	Pegmatite-like UST layer consisting of large (up to 7×1 cm) oriented subhedral columns of twinned K-feldspar crystals rimmed by fan-like aggregates of zinnwaldite (Fig. 2B). The fine-grained granitic matrix is composed of anhedral quartz, K-feldspar, albite and zinnwaldite. Topaz is common. This layer is the most enriched in Nb–Ta minerals.
5	15	Laminated fine-grained layer with scarce individual oriented quartz crystals. This layer is enriched in phosphates. Brown apatite and hydrous phosphates form thin (1–2 mm) late veinlets that conform to magmatic lamination. Intergrowths of zinnwaldite with brown apatite and small nests of pegmatite with tourmaline are also typical.
6	500–700	Homogeneous, fine-grained (0.1–0.5 mm) granite. Quartz, K-feldspar and albite grains are mostly anhedral; both feldspars also occur as short subhedral prisms. Grains of quartz, K-feldspar, mica and topaz are markedly zoned. Rims of large feldspar grains have been leached in many places and replaced by late quartz, albite and topaz. Zinnwaldite forms subhedral flakes. Topaz (up to 0.5 mm) is subhedral to euhedral, but small grains (about 0.1 mm) are interstitial. Phosphates—green apatite, amblygonite and childrenite—are the most common accessory minerals.
7	10	Laminated fine-grained layer of granite (similar to layer 1) is present at the base of the dyke with thin veinlets that have intruded into the underlying stock granite.

3.2.2. Early breccia

Early breccia in an isolated block at the south-western contact of the stock consists of fragments of phyllite (diameter <5 cm) cemented with fine-grained

granitic matrix. Some of the fragments are rounded; others are not. Composition of the matrix is similar to that of the stock granite (protolithionite as the dark mica), but it is very fine-grained. Columnar K-feldspars have often grown perpendicular to the surface of the phyllite fragments. Topaz, often present as fragments of older zoned crystals, and apatite are the accessory minerals.

3.2.3. Unidirectional solidification textures

USTs have been developed both in the stock and the dyke granites. Within the stock granite, a UST is found several metres above the major dyke. It is composed of a fine-grained, 1.5-cm-thick quartz layer accompanied by a 3-cm-thick layer of downward-oriented columns of orthoclase (Breiter, 2002). The

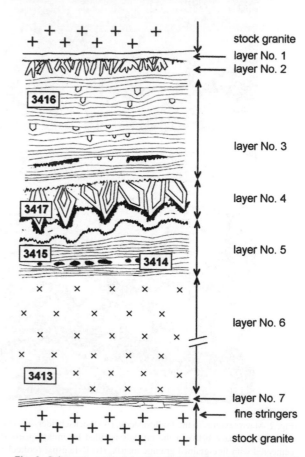

Fig. 3. Schematic vertical section through the western (proximal) part of the major dyke within the quarry (individual layers not to scale). Chemical composition in Table 1, petrographic description in Table 2.

stock granite immediately above and below the UST layer is medium-grained and chemically identical to the stock granite.

USTs occur prominently in the upper part of the major dyke. The western, proximal part of the dyke is exposed across an area of 25 m by 7 to 9 m and consists of seven different layers (Table 2; Figs. 3–5). Mineralogical and chemical composition and grain size of individual laminae in the thinly laminated layers of the dyke (Layers No. 3 and No. 5 in Table 2) differ significantly (Table 3). Boundaries between laminae are commonly sharp (Fig. 6A), but in places diffuse. The newly crystallized fine-grained layers were not entirely solid when the interstitial melt was still present. As a consequence, individual layers were ptygmatically deformed (Fig. 5). Layers of fine-grained zinnwaldite are often overgrown on large comb orthoclase crystals (Fig. 2B). Centimetre-scale intrusions of residual melt into the older layers have also occurred (Fig. 6B).

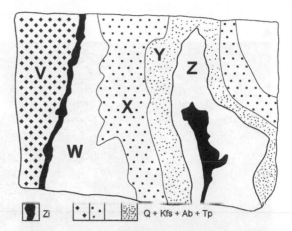

Fig. 5. Detail of the laminated part of the major dyke from Fig. 4. V through Z: location of analysed domains (Table 3). The size of the area described is 2.5×3.5 cm.

3.2.4. Intra-dyke brecciation

The eastern, distal part of the major dyke is not well exposed and a vertical section cannot be

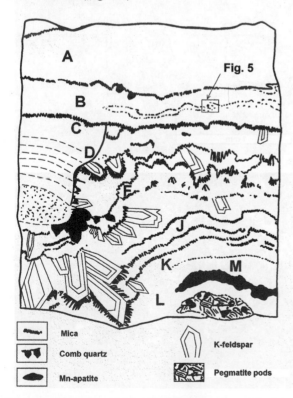

Fig. 4. Detailed structure of the UST in the upper part of the major dyke (natural size 15×13 cm). A through M: location of analysed domains (cf. Table 3). The upper contact of the dyke is 15 cm upwards from the upper margin of the picture.

Table 3
Contents of selected elements in the UST layer and laminated domains and brecciated domains of the major dyke at Podlesí

Domain	Laminated dyke in the quarry (Layer No. 3)								
	A	B	C	D	F	J	K	L	M
SiO_2	68.4	71.0	71.2	68.2	68.7	65.7	67.5	71.4	75.8
Al_2O_3	15.2	13.9	15.7	18.4	15.6	16.5	16.7	11.7	10.4
FeO	0.6	0.5	0.3	0.0	1.1	4.1	1.1	0.8	1.0
CaO	<0.1	0.1	0.1	<0.1	0.2	<0.1	1.5	3.0	0.3
Na_2O	5.1	4.8	5.5	7.2	6.5	3.2	1.0	0.6	3.1
K_2O	4.5	4.1	1.7	0.7	2.3	5.1	5.3	4.4	3.4
P_2O_5	1.0	0.7	0.5	0.5	0.6	0.3	1.8	3.2	0.9

Domain	Laminated dyke (Layer No. 3), detail					Intra-dyke breccia			
	V	W	X	Y	Z	B1	B2	B3	B4
SiO_2	74.2	67.5	67.8	65.5	66.1	66.7	69.0	68.7	67.7
Al_2O_3	12.4	16.9	16.5	15.8	17.5	18.5	15.9	17.2	16.5
FeO	0.4	0.1	0.6	0.2	0.2	2.2	1.6	0.4	0.4
CaO	<0.1	0.1	<0.1	<0.1	<0.1	<0.1	<0.1	0.1	<0.1
Na_2O	4.9	8.0	6.6	1.6	7.3	3.4	3.8	4.9	6.4
K_2O	2.9	1.7	2.9	10.8	3.0	3.4	4.6	2.9	3.4
P_2O_5	0.4	0.7	0.6	1.1	1.0	0.8	0.2	0.8	0.6

Contents (in wt.%) determined at the Czech Geological Survey using defocused electron microprobe analysis (energy-dispersive mode). All analyses are normalised to 95% total oxides (5% reserved for F, Li, Rb and other elements not determined). For location of samples A through M, see Fig. 4; for V through Z, Fig. 5 and for samples B1 through B4, Fig. 7.

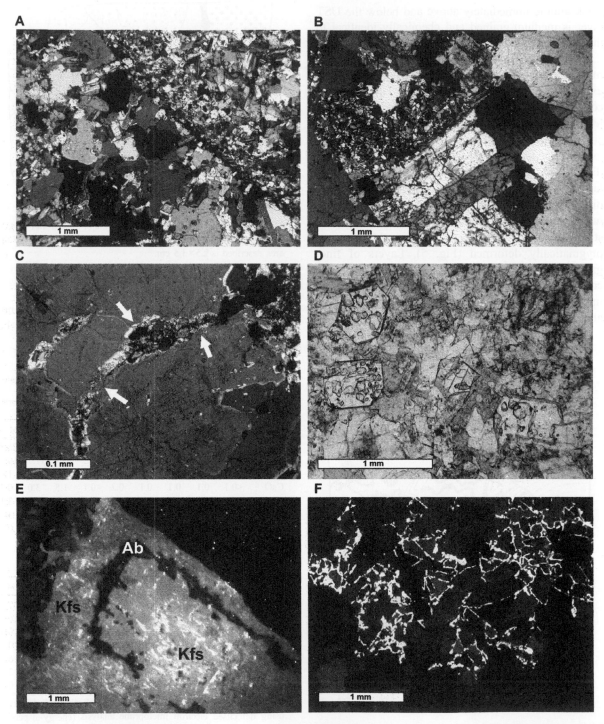

Fig. 6. Photomicrographs of textures in the Podlesí stock. (A) Contact between fine- and very fine-grained laminae in the dyke granite. (B) Invasion of dyke melt into stock granite. (C) Veinlet of late melt in a fractured quartz phenocryst. (D) Topaz from the stock granite with inclusions of quartz. (E) Zone of albite (Ab) in K-feldspar crystal (Kfs) from UST in the brecciated part of the major dyke. (F) CL image of late amblygonite upon feldspar grains in the brecciated dyke.

reconstructed with certainty. The lower part of the dyke consists of layers similar to those described above. The upper part of the dyke consists of 1-mm to 10-cm-thick layers of mica-rich facies (quartz+ zinnwaldite+albite+topaz) alternating with layers of mica-poor facies (quartz+albite+K-feldspar>topaz> zinnwaldite). Some of the mica-rich layers are rimmed with a zinnwaldite-dominated UST layer on their footwall. Below this laminated sequence, a 0.5-m-thick layer of magmatic breccia is present (Figs. 2C and 7). Clasts of this breccia are composed of mica-rich facies (fragments of mica-rich layers) and are all around rimmed by zinnwaldite-rich UST (Fig. 8A). The space between the fragments is filled with at least two types of mica-poor matrix, differing in grain size and quartz–albite ratio.

3.2.5. Late cracks in quartz crystals

Quartz and K-feldspar phenocrysts near the upper contact in the proximal part of the major dyke are cracked and the cracks have been filled with fine veinlets of residual melt (Fig. 6C). At least two different types of veinlets are recognised, one Na+F-rich (Q+Ab+Tp) and the other K+P-rich (Kfs+Ap). This fits well the two types of melt inclusions in quartz in the dyke granite (Breiter et al., 1997).

3.3. Mineralogical evidence for magmatic evolution

3.3.1. Alkali feldspars

Alkali feldspars are represented by nearly pure end members—albite (<3% An) and K-feldspar (<5% Ab)—and are rich in P (Frýda and Breiter, 1995; Breiter et al., 2002). The K-feldspars are more significant from a genetic point of view and will be described in detail. The following textural types can be distinguished:

Kfs1 is represented by phenocrysts with perthitic cores and homogeneous rims and contains 0.15% to 0.20% Rb and about 0.4% P_2O_5. It is common in the stock granite. Scarce phenocrysts from the dyke granite reflect a more complicated history: perthitic cores are covered by zones rich in entrapped albite crystals and later rimmed by zones of pure P-rich K-feldspar (Fig. 8B). Such phenocrysts exhibit very bright blue to white colour in cathodoluminescence (CL).

Kfs2, non-perthitic groundmass K-feldspar, is the most abundant type. Larger grains are subhedral, late small grains are anhedral and interstitial. Kfs2 from the stock granite is homogeneous (0.3–0.7% P_2O_5, 0.15–0.25% Rb) and exhibits very bright blue to white CL, whereas Kfs2 from the dyke granite is zoned with P-enriched rims (0.6–1.75% P_2O_5, 0.35–0.5% Rb). In CL, the core is white to bright yellow and the rim

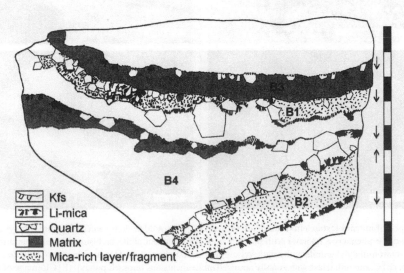

Fig. 7. Detailed structure of the eastern, brecciated part of the major dyke. B1 through B4: location of analysed samples (Table 3). Direction of UST-growth is marked by arrows. Kfs denotes K-feldspar. Scale bar is 9 cm long.

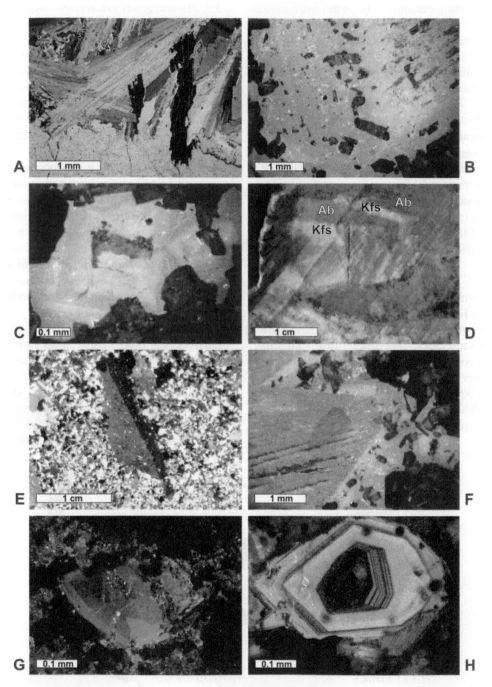

Fig. 8. Photomicrographs of mineral textures from the Podlesí stock. (A) Zinnwaldite-dominated UST from the dyke breccia (crossed polars). (B) CL image of K-feldspar phenocryst (yellow) with perthitic core and abundant albite inclusions (blue) around the core. (C) CL image of zoned matrix K-feldspar from the dyke granite, the dark yellow zones are richest in P. (D) Growth zones of albite in microcline crystal (Kfs) from the stockscheider; (E) Comb orthoclase with zonally arranged albite inclusions (crossed polars). (F) CL image of a comb K-feldspar crystal from the brecciated dyke—the core is perthitic (dark yellow) and the P-rich rim (light yellow) contains euhedral albite inclusions (blue). (G) CL image of fragmented and/or altered topaz crystal from the explosive breccia. (H) CL image of late topaz from the dyke granite.

yellow to brownish (Fig. 8C). In the laminated part of the major dyke, the content of Rb in Kfs2 reaches 0.7%.

Kfs3 forms large crystals that have grown perpendicular to the contact in the stockscheider. The inner parts of the crystals are moderately enriched in P (0.6–0.8% P_2O_5); the outer zones have been partially leached and depleted in P (0.1–0.3% P_2O_5) and contain small inclusions of apatite visible in the CL images. Of all the K-feldspar types, only Kfs3 is triclinic. It exhibits a very bright blue to white colour in CL. A remarkable feature of some feldspar megacrysts is the appearance of albitic growth zones inside the microcline crystals (Fig. 8D). Microclines in the brecciated stockscheider are altered, P-free and do not exhibit any CL.

Kfs4 is represented by large crystals from the UST layer of the dyke granite. These crystals typically have macroscopically visible zones with dominant pale cores upon which thin (up to 2 mm) colourless rims have crystallized. Microscopically, the cores contain numerous inclusions of small albite crystals oriented parallel to the growth zones of K-feldspar, the rims contain only scarce albite inclusions (Fig. 8E). No apatite occurs within the Kfs4 crystals. Distribution of P (0.6–0.9% P_2O_5) and Rb (0.3–0.4% Rb) is homogeneous, sometimes with slight enrichment in the rims. The content of P in the albite inclusions is also very homogeneous and only slightly lower than in K-feldspar (Fig. 9). Comb K-feldspar in the

brecciated part of the dyke is less abundant and forms up to 5-mm-long prismatic crystals, slightly zoned in CL. Some crystals have a perthitic core and clusters of Ab-inclusions in the outer zone (Fig. 8F), others have a homogeneous core and a rim divided by an albite zone (Fig. 6D). Both of these subtypes of Kfs4 are rich in P.

Albite is found in the granite types only in the groundmass. The P content in albite is generally lower and more uniform than in K-feldspar, typically between 0.05 and 0.7 wt.% P_2O_5 in the stock granite. Albite from the dyke granite contains 0.2–1.0 wt.% P_2O_5. This albite is zoned but the zoning is weaker than in the K-feldspar. In perthite, the admixed albite is usually depleted in P relative to the surrounding K-feldspar. In CL, albite exhibits a poorly developed zoning and medium to dark blue colours.

3.3.2. Quartz

Quartz was studied in detail by Müller et al. (2002). Based on CL and textural studies five magmatic quartz populations can be distinguished. The stock granite contains rare euhedral quartz phenocrysts (Q1, 1–5 mm) that show complex growth zoning. A red-brown luminescent core is overgrown by blue luminescent oscillatory growth zones. Marginally, the phenocrysts are strongly embayed. The zoning parallels the shape of lobate embayments (Fig. 10a). The phenocrysts are overgrown by anhedral fine-grained groundmass quartz (Q2) which

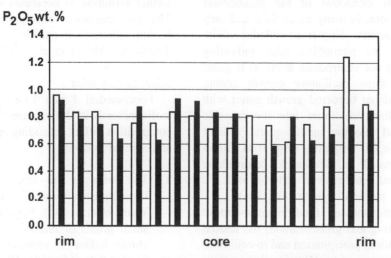

Fig. 9. Distribution of P in comb orthoclase (white columns) with albite inclusions (black columns) from the proximal dyke (cf. Fig. 8E). At each point, albite inclusion and adjacent orthoclase were analysed. Distance between the individual points is about 0.1 mm.

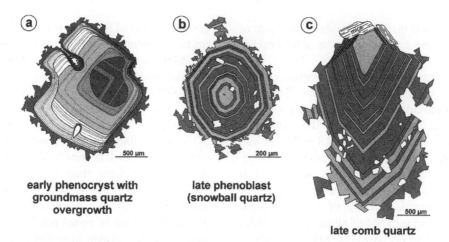

early phenocryst with
groundmass quartz
overgrowth

late phenoblast
(snowball quartz)

late comb quartz

Fig. 10. Growth patterns in quartz: (a) phenocryst (Q1) with groundmass quartz overgrowth (Q2); (b) snowball quartz (Q3); (c) comb quartz (Q5). (a) and (b) modified from Müller et al. (2002).

is free of growth zoning and shows weak red-brown CL. Secondary structures are characterized by bright halos around radioactive micro-inclusions and non-luminescent, neocrystallized domains. Like groundmass quartz, stockscheider quartz (Q3) is free of growth zoning, shows red-brown luminescence, and contains a number of secondary structures.

In the dyke, the quartz forms snowball-textured phenoblasts (Q4) 0.2 to 1.5 mm in size. Generally, the snowball quartz has a red- to red-brown CL and shows continuous growth into the matrix quartz, recognisable by the ramified, amoebic grain boundaries, and penetrates the matrix. The crystals commonly contain inclusions of the groundmass minerals. Furthermore, in many cases fluid and melt inclusions are present. Numerous tabular albite crystals envelope the phenoblast edge indicating that the quartz did not incorporate albite as it grew. The phenoblasts show oscillatory growth zoning characterized by planar bordered growth zones with α-quartz habit, which continues into the amoebic crystal margin and into the matrix quartz without any changes in the CL properties (Fig. 10b). Comb quartz (Q5) nucleated in the fan-like zinnwaldite layers of UST exhibits growth patterns and CL colours similar to snowball quartz (Fig. 10c).

A late- to post-magmatic fluid-driven overprint (e.g., micro-fracturing and greisenisation) has caused small-scale dissolution, precipitation and re-equilibration of pre-existing quartz (Q1–5) along grain boundaries, intra-granular micro-cracks, and around

fluid inclusions. This neocrystallized quartz (Q6) may be of different generations not distinguishable by the methods used in this study.

In order to better understand the crystallization conditions of the different quartz types, contents of the trace elements Ti, Al, K and Fe were determined by Müller et al. (2002). The plot of Al and Ti contents in the magmatic quartz populations (1 through 5) yields a trend reflecting the evolution of the magma (Fig. 11). The early crystallization stage represented by the quartz phenocrysts is characterized by high Ti and low Al concentrations in the quartz lattice. During further evolution Ti decreases whereas Al increases. The post-magmatic neocrystallized quartz of the healing structures is outside the magmatic trend (for details, see Müller et al., 2002).

3.3.3. Li-rich mica

Trioctahedral F-rich Li–Fe micas are the only mafic minerals in all the granite types. Their chemical composition differs depending on the host granite type and textural position:

1. The mica in the stock granite is generally homogeneous protolithionite (in the sense of Weiss et al., 1993). In many places, it contains small grains of radioactive minerals with pleochroic haloes. F contents reach 5 to 7 wt.%, Li_2O 2.0 to 2.5 wt.%, Rb 0.8 to 1.0 wt.% and Cs 0.2 wt.%.

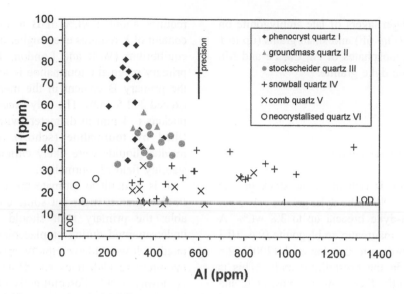

Fig. 11. Plot of Ti versus Al concentrations in the different quartz populations (I through VI—see text for more details). LOD indicates limit of detection. Modified from Müller et al. (2002).

2. The dyke granite contains a distinctly zoned zinnwaldite in the groundmass with cores relatively enriched in Fe, Mg and Ti, and rims higher in Si and Li. There is no zoning in the F, Rb and Cs contents. Compared with protolithionite, zinnwaldite is enriched in F (6–8 wt.%), Li_2O (3.5–4.7 wt.%) and Rb (1.0–1.1 wt.%), but depleted in Cs (about 0.1 wt.%).

3. Fan-like UST aggregates of zinnwaldite (Fig. 8A) occur within the brecciated parts of the dyke. Zinnwaldite is the oldest mineral within the zinnwaldite-rich sections of UST and the crystals often extend into older quartz–feldspar laminae. In K-feldspar rich sections, zinnwaldite borders comb feldspar. The UST zinnwaldite contains 7.5 to 7.7 wt.% F, 4.5 to 4.6 wt.% Li_2O, 1.2 to 1.3 wt.% Rb and about 0.1wt.% Cs. Zoning of the UST-zinnwaldite in the dyke breccia is the opposite compared to the zinnwaldite in the groundmass: the roots are Fe-poor, the terminal parts of aggregates are relatively Fe- and locally also Mg- and Ti-enriched.

Hydrothermal F-rich Li-biotite (<1 wt.% Li_2O) is common in greisen stringers. Muscovite was found only as a rare product of hydrothermal alteration in the stockscheider and near the greisen stringers.

3.3.4. Topaz

Topaz is present in two types. Euhedral to subhedral crystals with intensive oscillatory CL-zoning (Tp1) are found in all the rock types. This type contains numerous irregular inclusions of quartz (Fig. 6E). Fragments of crushed and altered zoned topaz crystals are present in the matrix of the early breccia (Fig. 8G). Tp1 forms, together with the QI, the earliest crystallized minerals. Late, interstitial topaz crystals (Tp2) occur only in the dyke granite and exhibit intensive blue CL-zoning (Fig. 8H). Both types of topaz are rich in F (90–97% of theoretical F-saturation), the late topaz is also rich in phosphorus (up to 1 wt.% of P_2O_5; Breiter and Kronz, 2003).

3.3.5. Phosphates

Two generations of fluorapatite are present: early Mn-poor euhedral green crystals and later Mn-rich interstitial flakes. Brown variety of late Mn-rich apatite forms small nests and aggregates with zinnwaldite in a laminated zone immediately below the UST layer. All types are poor in Cl (mostly below 0.1 wt.% Cl) and exhibit intensive yellow CL. Small prismatic crystals of childrenite–eosphorite, amblygonite and zwiesselite, and irregular grains of triphylite are found in the dyke granite (Breiter, 2002). Their texture suggests a relatively late, yet magmatic origin. Thin layers of late amblygonite on the feldspar grains (Fig. 6F) and small grains (up to 1 mm^2) of hydrated phosphates of Al, Ca, Fe and Mn are also found in the dyke granite.

4. Discussion

4.1. F, P, B and H_2O saturation

The Podlesí stock is rich in F: the stock granite contains 0.6 to 1.8 wt.% F, the dyke granite 1.4 to 2.4 wt.% and the intra-dyke breccia up to 3.8 wt.%. A high content of F in magmatic amblygonite (9.4–10.3 wt.%) reflects, according to London et al. (2001), 2.5 to 3.0 wt.% of F in the crystallized melts. This is consistent also with the F contents in zinnwaldite and topaz, which are greater than 90% of the theoretical maxima.

P was already enriched in the parental stock granite melt (0.5% P_2O_5). High peraluminosity coupled with low Ca content suppressed nucleation of apatite and promoted further enrichment of P (about 1 wt.% P_2O_5) in the dyke granite melt. London et al. (1993) calibrated the ratio between alkali feldspar and coexisting melt as $D_P^{Afs/melt}$= 2.05ASI−1.75 and $D_P^{Or/Ab}$=1.2. Also, according to the P content in the rims of K-feldspars in the dyke groundmass, the maximum P content of the crystallized melt reached about 2 wt.% in restricted domains. Although the substitution of P in the feldspars is coupled with Al, any other component that decreases the activity of Al, F for instance, will also decrease the partition coefficient. Thus in the fragments of the intra-dyke breccia, the most F-rich and P-rich domains of the dyke (3.8 wt.% F, 1.3 wt.% P_2O_5), the P content in K-feldspars is relatively low (~1 wt.%).

The B content in the granites is negligible (20–60 ppm), but an extensive outer contact aureole of tourmalinisation in the phyllite several hundreds of metres thick indicates that a large amount of B emanated from the crystallized magma. Tourmaline is found along numerous fractures in phyllite, locally along the foliation in the whole rock, ultimately changing the chlorite–sericite phyllite into quartz–tourmaline rock with up to 0.5 wt.% B. Crystallization of tourmaline from granitic melt in equilibrium with Fe-mica in relevant conditions (500–600 °C, 1 kbar)

requires about 2 wt.% of B in the melt and a high content of F requires even higher contents of B at this equilibrium (Wolf and London, 1997). Because no primary magmatic tourmaline is found in the granite, the primary B content in the melt probably did not exceed 2–2.5 wt.%. The only exceptions are the small pockets (<4 mm in diameter) within the UST layers filled by tourmaline (schorl). Here, droplets of residual liquids were likely enriched in B up to the stability field of tourmaline.

It is difficult to estimate the water content in the Podlesí granite melt, but some constraints are available: the primary melt should have been water-undersaturated during emplacement as a water-saturated melt would have rapidly crystallized at decreasing pressure (Johannes and Holz, 1996; Sykes and Holloway, 1987). Forceful early brecciation followed by extensive fluid escape shows that the melt became water-saturated shortly after emplacement. As deduced from the composition of mica, feldspars and topaz, the dyke granite melt was relatively rich in F, P, Li and Al—all these elements enhance the water solubility and lower the solidus (Johannes and Holz, 1996). Thus, the dyke granite magma was probably also undersaturated. Crystallization of the assemblage quartz–K-feldspar–albite–zinnwaldite–topaz enhanced the water content of melt (as the OH/F sites in mica and topaz were saturated with F). Breiter et al. (1997) determined 7 eq.-wt.% water from melt inclusions in the dyke granite quartz, but it is not clear whether this melt was water-saturated. From the melt inclusions in the stockscheider in Ehrenfriedersdorf (comparable to Podlesí), Thomas et al. (2003) deduced a full transition from B-rich silicate melt to B-rich water-based vapour. Thus, no water-saturation limit seems to exist in such a specialised melt. In the Podlesí dyke granite, the presence of late brecciation suggests that water saturation was reached.

4.2. Magmatic brecciation

Most ore-related breccias described from granitic bodies are related to late magmatic-hydrothermal stages of porphyry-type intrusion (Sillitoe, 1985). These breccias (their roots), which typically occur inside the plutons, were formed by aqueous fluid and cemented by hydrothermal quartz, chlorite, etc. Less abundant are early magmatic breccias in which the fragments are cemented by the rapidly cooled granitic melt. Small intrusions of tin-bearing granites in Krušné Hory/Erzgebirge contain related pipes of early magmatic breccias as a typical constituent (Oelsner, 1952; Schust and Wasternack, 1972; Seltmann and Schilka, 1991; Jarchovský and Pavlů, 1991). This type of breccia typically occurs at the top of the granite intrusions and is composed of country rock fragments cemented by rapidly cooled granitic melt and greisenised. In some cases, several episodes of brecciation may be superimposed in the same pipe (Gottesberg: Gottesmann et al., 1994; Sadisdorf: Seltmann, 1994).

In Podlesí, the top of the intrusion with supposed breccia pipe has already been removed. Extensive fracturing of the phyllite envelope, presence of an isolated block of phyllite breccia and brecciated stockscheider, and an overall similarity with explosive breccias described by Seltmann and Schilka (1991) support our interpretation of the early breccia from Podlesí as a forceful explosive event. Microclinisation of K-feldspars, loss of Li from mica and alteration of topaz in stockscheider indicates intensive reaction with fluid. Thus, the probable timing of the early brecciation was during the crystallization of the stockscheider. Escape of hydrous fluids during the early brecciation was probably also responsible for B loss from the magma, which led to tourmalinisation of the outer contact aureole. The later intra-dyke breccia from Podlesí, although situated within the granite body, should be also assigned to the class of "early breccias" according to Sillitoe (1985), as it is cemented by granite matrix without visible influence of hydrothermal fluids.

Another possible mechanism for brecciation and opening of the Podlesí system could have been an input of external water into the crystallizing magma (Bardintzeff, 1999). Oxygen isotope data, however, do not favour influence of meteoritic water (Žák et al., 2001).

4.3. Zoned crystals

The different populations and zoning of quartz, K-feldspar, micas and topaz reflect several stages of crystallization history of the rocks. Sharp boundaries between the zones and zones of partial resorption indicate rapid and substantial changes in the p–T–X

conditions of the magma. The zoning of feldspars and mica grains in the dyke granite imply two major stages in the crystallization history of this rock, whereas only one major crystallization stage can be deduced for the stock granite. The unzoned, mostly perthitic K-feldspar and unzoned mica crystals of the stock granite are nearly identical to the K-feldspar and mica cores in the dyke granite and represent an older crystallization stage from the melt. In contrast, the P-enriched rims of K-feldspar and Li-enriched rims of mica in the dyke granite document a distinctly more evolved environment enriched in P and F. The inherited cores of K-feldspar and mica crystals in the dyke granite indicate that the magma intruded into flat joints was a mixture of melt and entrapped crystals.

Another style of K-feldspar zoning is developed in the UST layer. The comb orthoclase is rich in albite inclusions and has inclusion-poor rims with sharp borders between the two domains. This points to abrupt changes in crystallization conditions. The early growth of K-feldspar crystals promoted local saturation of incompatible constituents in the boundary layer. This resulted in the crystallization of zonally arranged albite crystals in the inner parts of comb K-feldspar (Fig. 8B) and snowball quartz (Fig. 10b). Later, the rims crystallized in conditions that allowed Na to diffuse away and contribute to the growth albite crystals in the matrix.

The contents of P_2O_5 in albite inclusions and adjacent K-feldspar (Fig. 9) fit with experimental observation (London et al., 1993) that implies a $D_P^{Or/Ab}=1.2$. This supports magmatic origin for the comb feldspars via crystallization from silicate melt without later re-equilibration. Lamellae of albite inside the microcline crystals within the stockscheider (Fig. 8B) are similar to crystals obtained by Petersen and Lofgren (1986) during eutectic experiments. According to Petersen and Lofgren (1986), these intergrowths may be explained as a product of boundary layer crystallization from an undercooled melt.

Because the Podlesí system is rich in P, it might be expected that P would have entered the lattice of quartz according to the berlinite substitution $P^{5+}+Al^{3+}=Si^{4+}+Si^4$ (e.g., Maschmeyer and Lehmann, 1983). Larsen et al. (2002; personal communication, 2002) reported P and Al contents of about 40 and 175 ppm, respectively, in pegmatitic quartz in south Norway. Quartz from Podlesí, although Al-rich, contains less than 60 ppm P (the detection limit of the electron microprobe).

4.4. Stockscheider and K-feldspar-dominated UST layers in the dyke

Both the stockscheider and the K-feldspar-dominated UST layer are similar in that they contain large oriented K-feldspar crystals together with quartz and a fine-grained granitic matrix. However, the whole-rock chemical composition, internal fabric and composition of quartz and K-feldspar of these rock types differ significantly. The Ti content in the stockscheider quartz is twice as high as that in the comb quartz, whereas Al is much more enriched in the comb quartz (Fig. 11). This illustrates the course of magmatic evolution from the early crystallization of stockscheider at the top of the intrusion to the late crystallization of UST layers within the dyke.

The bulk composition of the stockscheider (3361 in Table 1) is similar to the bulk composition of the stock granite. This means that crystallization of the large K-feldspar crystals in the stockscheider was complemented by crystallization of the quartz–albite matrix with no substantial addition of K. The bulk composition of the K-feldspar-dominated UST layer (3417 in Table 1) is markedly different from the bulk composition of the dyke granite—it is strongly enriched in K and Al and depleted in Si and Na. The bulk composition of the UST layer can be modelled as a mixture of the dyke-granite melt with 25% K-feldspar added. According to London (1999), the growth of comb K-feldspar crystals may be explained by crystallization in the boundary layer, which effectively removed K from the adjacent melt. Such preferential consumption of K would be expected to deplete K in the melt from which the adjacent fine-grained layer was crystallized. This is true for the stockscheider but not for the UST.

4.5. Magmatic layering and UST

Simple magmatic layering with non-oriented crystals or with orientation of minerals parallel to the layers is common in aplite–pegmatite bodies (Breaks and Moore, 1992; Morgan and London, 1999). A well-known example of magmatic layering is the B-rich Calamity Peak granite–pegmatite complex in South Dakota, USA, which includes a 400-m-thick series of alternating pegmatite and layered aplite, in 0.1- to 2-m-thick layers (Rockhold et al., 1987). Magmatic layering in B-poor granites is much less frequently reported. When present, it has crystallized inward from the outer contacts (Baluj, 1995; Zaraisky et al., 1997; Frindt, 2002). Magmatic layering with crystals oriented perpendicular to the individual layers (UST) is typically found in subvolcanic granitic stocks associated with Mo and/or W±Sn mineralisation. Layering is typically parallel to the contact and is expressed by alternating layers of euhedral comb quartz crystals and fine-grained aplite (Kormilicyn and Manujlova, 1957; Shannon et al., 1982; Kirkham and Sinclair, 1988). K-feldspar UST and also rare tourmaline UST have been described from aplite–pegmatites (Duke et al., 1992; London, 1992).

In layered granites, the bulk composition of layered domains is similar to the bulk composition of the whole body (Zaraisky et al., 1997). The layered aplite–pegmatites are commonly differentiated into a slightly Na-enriched lower aplite and K-enriched upper pegmatite (Duke et al., 1992; Morgan and London, 1999). The oldest model for layering and UST in aplite–pegmatites involved preferential partitioning of Na into lower aplite and K into aqueous fluid that precipitated as the upper pegmatite (Jahns and Tuttle, 1963; Jahns and Burnham, 1969). This model was later discredited and replaced by models based on the undercooling of H_2O-rich melt (see overview in London, 1992, 1996). Fenn (1977) stated that the nucleation density of feldspar fell sharply with increasing H_2O, which resulted in the growth of large crystals from the H_2O-saturated melt. At the same degree of undercooling, the nucleation density of alkali feldspars is suppressed much more than that of quartz (London et al., 1989), so the K-feldspar grains grow larger than the associated quartz grains. Also Webber et al. (1997) stressed the significance of undercooling for heterogeneous nucleation and oscillatory crystal growth. London (1999), after experiments with B, F and P-doped granitic melt, preferred the boundary-layer effect in undercooled melt (>100 °C below the liquidus) as the cause for the layered textures.

Part of the problem of all the models based on undercooling is how to explain repetitive UST layers. For this, repetitive episodes of "undercooling" (=crystallization of UST layer) and "warming" minimally up to eutectic temperature (=crystallization of aplitic or granitic layers) will be necessary. Such fluctuation inside a relatively thin dyke cannot be explained via cooling at a constant pressure. Alternatively, a sudden adiabatic drop of pressure during opening of the system causing "undercooling" could have been followed by a pressure increase, restoring the conditions of "standard" granitic crystallization (Fig. 12). This model fits well the observed intra-dyke brecciation.

Oscillation of fluid pressure due to episodic degassing is also the basis for the "swinging eutectic" model (term coined by D. Jahns in 1982). Changes in pressure resulted in expansion of either the quartz field at high pressure or the albite field at low pressure and further crystallization of albite–quartz line rock (Balashov et al., 2000). Fedkin et al. (2002) made experiments with P- and F-doped granite from Podlesí and obtained a glass specimen with thin bands differing in the contents of Al, Si, F, P and alkalis.

Cox et al. (1996) showed that the rate of nucleation and size of K-feldspar crystals depend on the water content of a granitic melt. Depletion in water lowered the nucleation rate and promoted the formation of large K-feldspar crystals. Thus, the presence of large K-feldspar crystals in the UST layers in Podlesí may indicate water deficiency.

The UST domains in porphyry-type systems may contain >60% of modal quartz and thus these layers are not consistent with the overall granitic composition of the parental porphyry melt. The strong silica enrichment within the comb layers has been explained by addition of quartz from an aqueous fluid (Kirkham and Sinclair, 1988; Lowenstern and Sinclair, 1996). Recently, Taylor et al. (2002) introduced a model for crystallization of giant quartz crystals via alumosilicic hydrogel supported by an internal aqueous fluid, while D. London (personal communication, 2003) preferred, also in this case, crystallization from the boundary layer of fluxes-enriched silicate melt.

At Podlesí, UST layers are defined by K-feldspar or zinnwaldite. This mineralogical feature, together with strong enrichment of large-ion lithophile elements (LILE), makes the Podlesí system more comparable to aplite–pegmatite bodies with UST layers than to porphyry type intrusions. Nevertheless, with the exception of the 10-cm-thick K-feldspar-dominated UST, there is no K-enrichment and Na-depletion in the upper part of the dyke. Also the lack of B is an important difference between the Podlesí granites and other layered aplite–pegmatites.

The trigonal habit of the zoning in the snowball quartz, the frequently enclosed groundmass minerals and the occurrence of melt inclusions indicate growth of the snowball quartz in a nearly non-convecting and fluid-saturated crystal mush at <600 °C (<1 kbar). The extrapolated solidus temperature (Breiter et al., 1997) is 610±26 °C for the melt of the dyke granite, which is in agreement with the trigonal habit of the quartz crystals. Although the dyke granite magma was rapidly cooled (undercooled as indicated by the USTs), no classic mineralogical and textural markers of rapid cooling, such as needle-like apatite, glassy spherules or very fine-grained marginal facies (cf. Best and Christiansen, 2001) are present. This reflects the substantial differences in behaviour during rapid cooling between an ordinary granitic magma and a peraluminous magma strongly enriched in water and fluxes.

Individual laminae show a large scatter of normative compositions (Fig. 13), another feature supporting non-eutectic crystallization in an open system. Mica-rich fragments of breccia are shifted towards the Ab-Q join, which may be attributed to high contents of Li and dissolved water in melt (Johannes and Holtz,

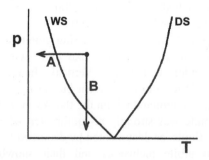

Fig. 12. Diagram indicating possible ways to produce repeated UST layers. A—cooling path at constant pressure; this process is irreversible without an additional heat source. B—adiabatic cooling path of water-rich melt after opening of the system. Later, when the joints are filled with crystallizing melt, the pressure increases again. This process is almost reversible and may produce repeated UST layers. WS—solidus of wet melt, DS—solidus of dry melt.

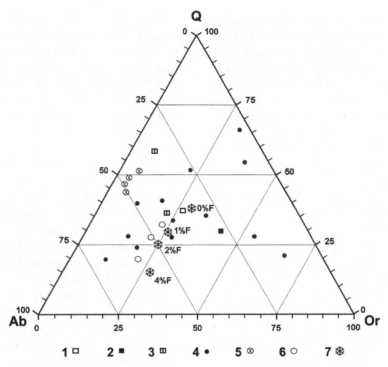

Fig. 13. Normative Ab–Or–Q triangle showing the composition of individual samples in the major dyke of Podlesí. Key to the symbols: 1—laminated rock above the UST layer; 2—UST; 3—laminated rock under the UST layer; 4—individual laminae in laminated rock; 5—fragments in the brecciated dyke; 6—individual laminae in matrix in brecciated dyke; 7—position of the minima for the water-saturated system without F and with 1, 2 and 4 wt.% F added (p_{H_2O}=1 kbar) (according to Manning, 1981). Modified from Breiter (2002).

1996). The breccia matrix lies near the expected evolutionary path of granitic melt with 2% to 4% F (Manning, 1981). The actual F content in the breccia matrix is lower but, at the time of crystallization, it was probably higher, i.e., more than 3 wt.%. Because of a high melt/water partition coefficient, F did not escape with the fluids during opening of the system. Thus, the matrix, although mica-poor, crystallized topaz-rich.

4.6. Comparison with other P-rich fractionated granites

Rare-metal granites enriched in P are typical for the European Variscan belt. Only a few, however, are comparable to the Podlesí system in the degree of P and F enrichment. In the Beauvoir granite (France), metre-wide zones of layers composed of lepidolite and albite with marked grain size variation are associated mainly with the transitional zones between individual intrusive units. One of these layers is brecciated and cemented by later injection of magma. The layering was interpreted as the product of magmatic flow (Jacquot, 1987; Cuney et al., 1992). No stockscheider has been found, but the so-called "fringe" zone near its upper contact contains thin layers enriched in lepidolite, albite, columbite and cassiterite (Raimbault et al., 1995).

The P-rich Mesozoic Ta-bearing granite at Yichun (China) consists of a sheet-like body with a huge stockscheider at the top. Vertical changes in K-feldspar/albite ratio are evident but no real stratification was encountered within the flat, 43-m-thick body. The granite was strongly re-equilibrated with hydrothermal fluids, as documented by disturbed P contents between albite inclusions and their snowball-type K-feldspar host, but no indications of opening have been found (Huang et al., 2002).

The mica-rich layer of the distal dyke in Podlesí contains 24% quartz, 33% albite, 34% zinnwaldite, 5% topaz and 4% K-feldspar, which may be transitional between F-rich granites and a group of F-rich rocks termed "ongonite" (Kovalenko and Kovalenko, 1976), "topazite" (Eadington and Nashar, 1978; Johnston and Chappell, 1992) or "magmatic greisens" (Xiong et al., 1999). The very high F content is manifested by high modal contents of topaz and zinnwaldite and the absence of K-feldspar, because all K is bound in the mica. The F content in the distal dyke is higher than in ongonites and the presence of magmatic albite in Podlesí fragments distinguishes them from topazite. Experiments by Xiong et al. (1999) using albite granite with up to 6% of F produced quartz, topaz and mica together with an alkali-rich fluid at <600 °C. Albite appeared in the crystallized assemblage when F in the system was below 4 wt.% (Xiong et al., 1999). The mica-rich facies from Podlesí substantiates these experiments and stability of albite near solidus up to 4 wt.% of F. Moreover, incorporation of nearly all K into micas inhibited K-feldspar crystallization, which indicates a possible way to produce magmatic mica-rich "greisen". Thus, the Podlesí system seems to be quite unique in its complexity, including crystallization of stockscheider, early and late brecciation, magmatic layering and UST.

5. Genetic scenario and conclusions

The Podlesí granites were crystallized from fractionated granitic melt progressively enriched in F, P, Li and H_2O. Crystallization conditions changed abruptly as documented by complicated zonation of quartz, feldspars and mica crystals. Stockscheider, magmatic layering and UST give evidence for repeated episodes of nonequilibrium crystallization probably from undercooled melt. Two episodes of magmatic brecciation and extensive escape of B-bearing fluids resulting in tourmalinisation of surrounding phyllite indicate that the system was opened at least twice and exposed to an abrupt decrease of pressure. Mineralogical evidence of post-magmatic, aqueous fluid-related processes is restricted, so we believe that the system has preserved its magmatic condition. Although brecciation has occurred, the Podlesí rocks are more comparable to layered aplite–pegmatites than to porphyry-type intrusions with comb layers. Among the evolutionary models for layering and UST, a combination of undercooling

(Webber et al., 1997; London, 1999) and pressure changes ("swinging eutectic"; Balashov et al., 2000) seems to be the most plausible.

We envisage the following four evolutionary stages for the Podlesí system:

1. Emplacement of the stock granite melt at a shallow level, crystallization of the stockscheider, enrichment of water and fluxes at the top of the magma body, and brecciation of the stockscheider and overlying phyllites, caused by the escape of the exsolved water (first opening of the system) and followed by escape of fluid and tourmalinisation of the phyllites.

2. Fine-grained granitic matrix of the breccia (and the granite) crystallized from a volatile-poor melt and is composed of Li-biotite, P-poor feldspars and homogeneous quartz, all without zoning. Topaz is only an accessory, often in fragments. Beneath this rapidly crystallized "cork", water and fluxes became again slowly enriched. The stock is generally formed of albite–protolithionite–topaz granite (A/CNK=1.15–1.25, 0.4–0.8 wt.% P_2O_5, 0.6–1.8 wt.% F, 0.15–0.20 wt.% Li). Subsequent crystallization of parental melt produced a small amount of a more F, P, Li- and water-rich residual magma. When the upper part of the stock had crystallized and cooled sufficiently to allow brittle fracturing, this residual magma (A/CNK=1.2–1.4, 0.6–1.5 wt.% P_2O_5, 1.4–2.4 wt.% F, 0.2–0.3 wt.% Li) penetrated upward, forming a set of flat dykes.

3. Crystallization of the major dyke proceeded from the contacts inwards. Crystallization from the bottom proceeded more rapidly and crystallization from the upper contact produced layered rock. When a substantial part of the dyke had crystallized (probably more than 80 vol.%), the rest of the magma become water-oversaturated, the liberated vapour escaped (the second opening) and the pressure dropped. Decrease of pressure caused undercooling and crystallization of the UST. Repetition of UST layers is explained by fluctuation of pressure in nearly adiabatic conditions. Escape of fluids also promoted local intra-dyke brecciation.

4. Cementation of dyke fragments with late magma. Late amblygonite crystallized from P-rich fluid.

During the final consolidation of the deeper parts of the system, F-rich and Li-poor aqueous fluids were released. These fluids, also enriched in Sn and W, ascended along steep joints causing small-scale greisenisation.

In contrast to layered aplite–pegmatites, layering and UST in Podlesí dyke granite developed near the upper contact and crystallized downwards. Sudden opening of the system followed by rapid decrease of pressure led to undercooling and crystallization of UST. Differences in the nucleation lag time of the major minerals promoted evolution of orthoclase- or zinnwaldite-dominated UST layers. Changes in chemical composition of the boundary layer produced the oscillating mica- and feldspar-rich laminae. The Podlesí granites demonstrate that combination of models of adiabatic fluctuation of pressure ("swinging eutectic"; Balashov et al., 2000) and boundary-layer crystallization of undercooled melt (Webber et al., 1997; London, 1999) can explain the formation of magmatic layering and UST in highly fractionated granitic rocks. However, the accumulation of K in the orthoclase-dominated UST layer, similar to the K accumulation in the upper parts of layered aplite–pegmatites, remains to be resolved.

Acknowledgements

This work has been supported by the state scientific program "Complex geochemical research of interaction and migration of organic and inorganic chemicals in rock environment" supported by the Ministry of the Environment of the Czech Republic by Grant Project No. VaV/3/630/00. We thank many colleagues who supported the investigation of the Podlesí granite system through years, namely J. Frýda and I. Vavřín for technical assistance at the microprobe, and the staff of the chemical laboratory of CGS for the whole-rock analyses. Financial support for A.M. was provided by the Deutsche Forschungsgemeinschaft (MU 1717/2-1) and for J.L. by the Ministry of Education, Youth and Sports of the Czech Republic (MSM J/07-981431000004). D. London and W.D. Sinclair are thanked for detailed revision of the manuscript, which helped us to improve the article significantly. We are grateful for the improvement of the English to C. Stanley. We are also grateful to O. Tapani Rämö for invitation to take part in the Prof. Haapala celebration meeting and for many inspiring comments and editorial improvement of the manuscript.

Appendix A. Analytical methods

SEM cathodoluminiscence and electron microprobe analysis of trace elements in quartz

SEM-CL images were collected from a JEOL 8900 RL electron microprobe at the University of Göttingen with a CL detector (CLD40 R712) using slow beam scan rates of 20 s at processing resolution of 1024 × 860 pixels and 256 grey levels. The electron beam voltage and current was 30 keV and 200 nA, respectively. Trace element abundances of Al, Ti, K, and Fe in quartz were determined using the same JEOL system which is fitted with five wavelength-dispersive spectrometers. Synthetic Al_2O_3 and TiO_2, orthoclase, Lucerne, Switzerland and haematite, Rio Marina, Elba, were used as standards. Matrix corrections ware made using the phi-rho-z method of Armstrong (1991). Accelerating voltage of 20 kV, beam current of 80 nA, beam diameter of 5 μm, and counting times of 15 s for Si and of 300 s for Ti, K, and Fe were chosen. Limits of detection are 15 ppm for Ti, 11 ppm for K, 15 ppm for Fe, 51 ppm for Na, and 10 ppm for Al. For more explanations, see "Appendix A" in Müller et al. (2002).

Optical cathodoluminiscence

The samples were analysed using cathodoluminescence equipment with hot cathode HC₂-LM, Simon Neuser, Bochum, accelerating voltage 14 kV, beam density 10 mA/mm² in the laboratory of Masaryk University Brno and in the laboratory of University of Göttingen.

Whole-rock chemical analyses

Major elements were analysed using standard methods of wet chemistry at the Laboratory of the Czech Geological Survey, Prague. Pb, Rb, Sn, Zn, and

Zr were analysed by XRF at the Laboratory of the Czech Geological Survey; other trace elements were analysed using ICP-MS in ACME Laboratory, Vancouver.

Estimation of Li-content in Li–Fe micas

For the estimation of Li-content in micas analysed by micropobe, we used the equation Li_2O (wt.%) = $0.335 \times$ (wt.% of SiO_2)-12.5, which is based on statistical evaluation of 42 chemically analysed monomineralic mica concentrates from the studied locality. This equation fit in the area of Li-rich micas (>1.5 wt.% Li_2O) better than the equations published by Stone et al. (1988), Tindle and Webb (1990) and Tischendorf et al. (1997), all fitted for micas with much broader interval of Li-contents.

References

Allman-Ward, P., Halls, C., Rankin, A.H., Bristov, C.M., 1982. An intrusive hydrothermal breccia body at Wheal Remfry in the western part of the St. Austel granite pluton, Cornwall, England. In: Evans, A.M. (Ed.), Metallization Associated with Acid Magmatism. John Wiley, London, pp. 1–28.

Armstrong, J.T., 1991. Quantitative elemental analysis of individual microparticles with electron beam instruments. In: Heinrich, K.F.J., Newbury, D.E. (Eds.), Electron Probe Quantification. Plenum, New York, London, pp. 261–315.

Balashov, V.N., Zaraisky, G.P., Seltmann, R., 2000. Fluid–magma interaction and oscillatory phenomena during crystallization of granitic melt by accumulation and escape of water and fluorine. Petrology 8, 505–524.

Baluj, G.A., 1995. Unique example of layered granitic melts in intrusions in the Coast Zone of the Primorie region. Doklady Akademii Nauk 341, 83–88. (in Russian).

Bardintzeff, J.M., 1999. Vulkonologie. Enke, Stuttgart.

Best, M.G., Christiansen, E.H., 2001. Igneous Petrology. Blackwell Science, Malden.

Breaks, F.W., Moore Jr., J.M., 1992. The Ghost lake batholith, Superior province of northwestern Ontario: a fertile, S-type, peraluminous granite-rare element pegmatite system. Canadian Mineralogist 30, 835–875.

Breiter, K., 2002. From explosive breccia to unidirectional solidification textures: magmatic evolution of a phosphorus- and fluorine-rich granite system (Podlesí, Krušné Hory Mts., Czech Republic). Bulletin of the Czech Geological Survey 77, 67–92.

Breiter, K., Kronz, A., 2003. Chemistry of phosphorus-rich topaz. In: Cempírek, J. (Ed.), International Symposium on Light Elements in Rock-Forming Minerals, Abstracts. Masaryk University Brno, pp. 9–10.

Breiter, K., Frýda, J., Seltmann, R., Thomas, R., 1997. Mineralogical evidence for two magmatic stages in the evolution of an extremely fractionated P-rich rare-metal granite: the Podlesí stock. Journal of Petrology 38, 1723–1739.

Breiter, K., Frýda, J., Leichmann, J., 2002. Phosphorus and rubidium in alkali feldspars: case studies and possible genetic interpretation. Bulletin of the Czech Geological Survey 77, 93–104.

Brigham, R.H., 1983. A fluid dynamic appraisal of a model for the origin of comb layering and orbicular structure. Journal of Geology 91, 720–724.

Cox, R.A., Dempster, T.J., Bell, B.R., Rogers, G., 1996. Crystallization of the Shap granite: evidence from zoned K-feldspar megacrysts. Journal of the Geological Society 153, 625–635.

Cuney, M., Marignac, C., Weisbrod, A., 1992. The Beauvoir topaz-lepidolite albite granite (Massif central, France): the disseminated magmatic Sn–Li–Ta–Nb–Be mineralization. Economic Geology 87, 1766–1794.

Duke, E.F., Papike, J.J., Laul, J.C., 1992. Geochemistry of a boron-rich peraluminous granite pluton: the Calamity Peak layered granite–pegmatite complex, Black Hills, South Dakota. Canadian Mineralogist 30, 811–833.

Eadington, P.J., Nashar, B., 1978. Evidence for the magmatic origin quartz–topaz rocks from the New England batholith, Australia. Contributions to Mineralogy Petrology 67, 433–438.

Fedkin, A., Bezmen, A., Seltmann, R., Zaraisky, G., 2002. Experimental simulation of a layered texture in the F–P-rich granite system. Bulletin of the Czech Geological Survey 77, 113–125.

Fenn, P.M., 1977. The nucleation and growth of alkali feldspars from hydrous melt. Canadian Mineralogist 15, 135–161.

Frindt, S., 2002. Petrology of the Cretaceous anorogenic Gross Spizkoppe granite stock, Namibia. PhD thesis, University of Helsinki, Finland.

Frýda, J., Breiter, K., 1995. Alkali feldspars as a main phosphorus reservoirs in rare-metal granites: three examples from the Bohemian Massif (Czech Republic). Terra Nova 7, 315–320.

Gottesmann, B., Wasternack, J., Märtens, S., 1994. The Gottesberg tin deposit (Saxony): geological and metallogenic characteristic. In: Seltmann, R., Kämpf, H., Möller, P. (Eds.), Metallogeny of Collisional Orogens, Czech Geological Survey. Prague, pp. 110–115.

Huang, X.L., Wang, R.C., Chen, X.M., Hu, H., Liu, C.S., 2002. Vertical variations in the mineralogy of the Yichun topaz-lepidolite granite Jiangxi province, southern China. Canadian Mineralogist 40, 1047–1068.

Jahns, R.H., Tuttle, O.F., 1963. Layered pegmatite–aplite intrusives. Mineralogical Society of America Special Paper 1, 78–92.

Jahns, R.H., Burnham, C.W., 1969. Experimental studies of pegmatite genesis: I. A model for the derivation and crystallization of granitic pegmatites. Economic Geology 64, 843–864.

Jacquot, Th., 1987. Dynamique de Íorganisation séquentielle du magma de Beauvoir. Apport de la pétrologie structurale. In: Cuney, M., Autran, A.Géologie profonde de la France forage scientifique dÉchassiéres, vol. 2–3. Géologie de la France, pp. 209–222.

Jarchovský, T., 1962. Entstehung der Feldspatsäume (Stock-scheider) an den Kontakten von Granit-und Greisen-Stöcken im Erzgebirge. Krystalinikum 1, 71–80.

Jarchovský, T., Pavlů, D., 1991. Albite–topaz microgranite from Horní Slavkov (Slavkovský les Mts.) NW Bohemia. Věstnik Ústředního Ústavu Geologického 66, 13–22.

Johannes, W., Holz, F., 1996. Petrogenesis and experimental petrology of granitic rocks. Minerals and Rocks vol. 22. Springer-Verlag, Berlin.

Johnston, C., Chappell, B.W., 1992. Topaz-bearing rocks from Mount Gibson, North Queensland, Australia. American Mineralogist 77, 303–313.

Kirkham, R.V., Sinclair, W.D., 1988. Comb quartz layers in felsic intrusions and their relationship to porphyry deposits. In: Taylor, R.P., Strong, D.F. (Eds.), Recent Advances in the Geology of Granite-Related Mineral Deposits, Canadian Institute of Mining and Metallurgy, Special Volume, vol. 39, pp. 50–71.

Kormilicyn, V.S., Manujlova, M.M., 1957. Rhythmic banded quartz porphyry at Bugdai Peak, southeast Transbaykal region. Zapiski Vsesoyuznovo Mineralogiceskovo Obschestva 86, 355–364. (in Russian).

Kovalenko, V.I., Kovalenko, N.I., 1976. Ongonites (topaz-bearing quartz keratophyre)—subvocanic analogue of rare-metal Li–F granites. Joint Soviet-Mongolian scientific-research geological expedition. Transaction 15 (Nauka, Moscow (in Russian)).

Larsen, R.B., Henderson, I., Ihlen, P.M., Jacamon, F., Flem, B., 2002. Application of trace elements in granitic quartz to unravel petrogenetic links in complex igneous fields: examples from South Norway. In: Parsons, I. (Ed.), 18th General Meeting of the International Mineralogical Association, 1–6 September 2002, Edinburgh, Scotland. Programme with Abstracts, pp. 261.

London, D., 1990. Internal differentiation of rare-element pegmatites: a synthesis of recent research. In: Stein, H.J., Hannah, J.L. (Eds.), Ore-Bearing Granite Systems; Petrogenesis and Mineralizing Processes, Geological Society of America Special Paper, vol. 246, pp. 35–50.

London, D., 1992. The application of experimental petrology to the genesis and crystallization of granitic pegmatites. Canadian Mineralogist 30, 499–540.

London, D., 1995. Geochemical features of peraluminous granites, pegmatites, and rhyolites as sources of lithophile metal deposits. In: Thompson, J.F.H. (Ed.), Magmas, Fluids, and Ore Deposits vol. 23. Mineralogical Asociation of Canada Short Course, pp. 175–202.

London, D., 1996. Granitic pegmatites. Transaction of the Royal Society in Edinburgh: Earth Sciences 87, 305–319.

London, D., 1999. Melt boundary-layers and the growth of pegmatitic textures. Canadian Mineralogist 37, 826–827.

London, D., Morgan, G.B., Hervig, V.I., 1989. Vapor-undersaturated experiments with Macusani glass+H_2O at 200 MPa, and the internal differentiation of granitic pegmatites. Contributions to Mineralogy and Petrology 102, 1–17.

London, D., Morgan, G.B., Babb, V.I., Loomis, H.A., 1993. Behaviour and effect of phosphorus in system Na_2O–K_2O–Al_2O_3–SiO_2–P_2O_5–H_2O at 200 MPa (H_2O). Contributions to Mineralogy and Petrology 113, 175–202.

London, D., Morgan, G.B., Wolf, V.I., 2001. Amblygonite–montebrasite solid solutions as monitors of fluorine in evolved granitic and pegmatitic melts. American Mineralogist 86, 225–233.

Lowenstern, J.B., Sinclair, W.D., 1996. Exsolved magmatic fluid and its role in the formation of comb-layered quartz at the Cretaceous Logtung W–Mo deposit, Yukon Territory, Canada. Transactions of the Royal Society of Edinburgh. Earth Sciences 87, 291–303.

Manning, D.A.C., 1981. The effect of fluorine on liquidus phase relationships in the system Qz–Ab–Or with excess water at 1 kbar. Contributions to Mineralogy and Petrology 76, 206–215.

Maschmeyer, D., Lehmann, G., 1983. A trapped-hole center causing rose coloration in natural quartz. Zeitschrift für Kristallographie 163, 181–196.

Morgan, G.B., London, V.I., 1999. Crystallization of the little three layered pegmatite–aplite dike, Ramona District, California. Contributions to Mineralogy and Petrology 136, 310–330.

Müller, A., Kronz, A., Breiter, K., 2002. Trace elements and growth patterns in quartz: a fingerprint of the evolution of the subvolcanic Podlesí Granite System (Krušné Hory, Czech Republic). Bulletin of the Czech Geological Survey 77, 135–145.

Oelsner, O.W., 1952. Die pegmatitisch-pneumatolitischen Lagerstätten des Erzgebiges mit Ausnahme der Kontaktlagerstatten. Freibergische Forschungshefte C4. 80 pp.

Petersen, J.S., Lofgren, G.E., 1986. Lamellar and patchy intergrowths in feldspars; experimental crystallization of eutectic silicates. American Mineralogist 71, 343–355.

Raimbault, L., Cuney, M., Azencott, C., Duthou, J.L., Joron, J.L., 1995. Geochemical evidence for a multistage magmatic genesis of Ta–Sn–Li mineralization in the granite at Beauvoir. French Massif Central. Economic Geology 90, 548–596.

Rockhold, J.R., Nabelek, P.I., Glascock, M.D., 1987. Origin of rhythmic layering in the Calamity Peak satellite pluton of the Harney Peak granite, South Dakota: the role of boron. Geochimica et Cosmochimica Acta 51, 487–496.

Schust, F., Wasternack, J., 1972. Über das Auftreten von Schlotformigen Brekzienkorpern bei Gottesberg und Mühlleithen im Granitmassiv von Eibenstock, Erzgebirge. Zeitschrift für angewandte Geologie 18, 400–410.

Seltmann, R., 1994. Sub-volcanic minor intrusions in the Altenberg caldera and their metallogeny. In: Seltmann, R., Kämpf, H., Möller, P. (Eds.), Metallogeny of Collisional Orogens. Czech Geological Survey, Prague, pp. 198–206.

Seltmann, R., Schilka, W., 1991. Metallogenic aspects of breccia-related tin granites in the eastern Erzgebirge. Zeitschrift für geologische Wissenschaften 19, 485–490.

Shannon, J.R., Walker, B.M., Carten, R.B., Geraghty, E.P., 1982. Unidirectional solidification textures and their significance in determining relative ages of intrusions at the Hederson Mine, Colorado. Geology 10, 293–297.

Sillitoe, R.H., 1985. Ore-related breccias in volcanoplutonic arcs. Economic Geology 80, 1467–1514.

Stone, M., Exley, C.S., George, M.C., 1988. Compositions of trioctahedral micas in the Cornubian batholith. Mineralogical Magazine 52, 175–192.

Sykes, M.L., Holloway, J.R., 1987. Evolution of granitic magmas during ascent: a phase equilibrium model. In: Mysen, B.O. (Ed.), Magmatic Processes: Physicochemical Principles, Geochemical Society Special Publication, vol. 1, pp. 447–461.

Taylor, M.C., Sheppard, J.B., Walker, J.N., Kleck, W.D., Wise, M.A., 2002. Petrogenesis of rare-element granitic pegmatites: a new approach. In: Parsons, I. (Ed.), 18th General Meeting of the International Mineralogical Association, 1–6 September 2002, Edinburgh, Scotland. Programme with Abstracts, pp. 260.

Thomas, R., Förster, H.-J., Heinrich, W., 2003. The behaviour of boron in a peraluminous granite–pegmatite system and associated hydrothermal solutions: a melt and fluid-inclusion study. Contributions to Mineralogy and Petrology 144, 457–472.

Tindle, A.G., Webb, P.C., 1990. Estimation of lithium contents in trioctahedral micas using microprobe data: application to micas from granitic rocks. European Journal of Mineralogy 2, 595–610.

Tischendorf, G., Gottesmann, B., Förster, H.-J., Trumbull, R.B., 1997. On Li-bearing micas: estimation Li from electron microprobe analyses and an improved diagram for graphical representation. Mineralogical Magazine 61, 809–834.

Webber, K.L., Falster, A.U., Simmons, W.B., Foord, E.E., 1997. The role of diffusion-controlled oscillatory nucleation in the formation of line rock in pegmatite–aplite dikes. Journal of Petrology 38, 1777–1791.

Webber, K.L., Simmons, W.B., Falster, A.U., Ford, E.E., 1999. Cooling rates and crystallization dynamics of shallow level pegmatite–aplite dikes, San Diego County, California. American Mineralogist 84, 708–717.

Weiss, Z., Rieder, M., Smrčok, L., Petříček, V., Bailey, S.W., 1993. Refinement of the crystal structures of two "protolithionites". European Journal of Mineralogy 5, 493–502.

Wolf, M.B., London, D., 1997. Boron in granitic magmas: stability of tourmaline in equilibrium with biotite and cordierite. Contributions to Mineralogy and Petrology 130, 12–30.

Xiong, X.L., Zhao, Z.H., Zhu, J.C., Rao, B., 1999. Phase relations in albite granite–H_2O–HF system and their petrogenetic application. Geochemical Journal 33, 199–214.

Žák, K., Pudilová, M., Breiter, K., 2001. Oxygen isotope study of the highly fractionated Podlesí granite system—preliminary data. In: Breiter, K. (Ed.), International Workshop Phosphorus- and Fluorine-rich Fractionated Granites, Podlesí. Czech Geological Survey, Praha, pp. 34–35.

Zaraisky, G.P., Seltmann, R., Shatov, V.V., Aksyuk, A.M., Shapovalov., Yu.B., Chevychelov, V.Yu., 1997. Petrography and geochemistry of Li–F granites and pegmatite–aplite banded rocks from the Orlovka and Etyka tantalum deposits in Eastern Transbaikalia, Russia. In: Papunen, H. (Ed.), Mineral Deposits. Balkema, Rotterdam, pp. 695–698.

Available online at www.sciencedirect.com

Lithos 80 (2005) 347–362

www.elsevier.com/locate/lithos

Petrological and geochemical evolution of the Kymi stock, a topaz granite cupola within the Wiborg rapakivi batholith, Finland

Ilmari Haapala*, Sari Lukkari

Department of Geology, P.O. Box 64 (Kumpula, Physicum), FI-00014, University of Helsinki, Finland

Received 28 July 2003; accepted 9 September 2004
Available online 23 November 2004

Abstract

The 6×3 km Kymi monzogranite stock represents the apical part of an epizonal late-stage pluton that was emplaced within the 1.65 to 1.63 Ga Wiborg rapakivi batholith. The stock has a well-developed zonal structure, from the rim to the center: stockscheider pegmatite, equigranular topaz granite, porphyritic topaz granite. The contact between the two granites is usually gradational within a few centimeters, but local inclusions of the porphyritic granite in the equigranular granite indicate that the latter solidified later. Hydrothermal greisen and quartz veins, some of which contain genthelvite, beryl, wolframite, cassiterite, and sulfides, cut the granites of the stock and the surrounding country rocks. The equigranular granite contains 1 to 4 vol.% topaz, and its biotite is lithian siderophyllite; the porphyritic granite has 0 to 3 vol.% topaz, and the mica is siderophyllite. The equigranular granite is geochemically highly evolved with elevated Li, Rb, Ga, Ta, and F, and very low Ba, Sr, Ti, and Zr. The REE patterns show deep negative Eu anomalies and tetrad effects indicating extreme magmatic fractionation and aqueous fluid–rock interaction. The zonal structure of the stock is interpreted as a result of differentiation within the magma chamber. Internal convection in the crystallizing magma chamber and upward flow of residual melt as a boundary layer along sloping contacts resulted in accumulation of a layer of highly evolved, volatile-rich magma in the apical part of the chamber. Crystallization of this apical magma produced the stockscheider pegmatite and the equigranular granite; the underlying crystal mush solidified as the porphyritic granite. Much of the crystallization took place from volatile-saturated melt, and episodic voluminous degassing expelled fluids into opened fractures where they or their derivatives reacted with country rocks and caused alteration and mineralization.
© 2004 Elsevier B.V. All rights reserved.

Keywords: Rapakivi granite; Stockscheider pegmatite; Zonal intrusion; Topaz granite; Greisen; Genthelvite

1. Introduction

Several Proterozoic rapakivi granite complexes in Finland and Russian Karelia have topaz-bearing granites as late-stage intrusive phases (Laitakari, 1928; Haapala and Ojanperä, 1972; Haapala, 1977a;

---
* Corresponding author. Tel.: +358 9 590400; fax: +358 9 19150826.
  E-mail address: ilmari.haapala@helsinki.fi (I. Haapala).

0024-4937/$ - see front matter © 2004 Elsevier B.V. All rights reserved.
doi:10.1016/j.lithos.2004.05.012

Amelin et al., 1991; Edén, 1991; Rämö, 1991; Lukkari, 2002). These evolved granites deviate in texture, mineral composition, and geochemical characteristics from the more primitive "normal" rapakivi granites. They may be porphyritic or equigranular, and plagioclase-mantled alkali feldspar megacrysts are not present. Many of these granites show the mineralogical and geochemical characteristics of tin granites, and in several cases, greisen-type (Haapala and Ojanperä, 1972; Haapala, 1977b, 1995; Edén, 1991) and rare skarn-type (Amelin et al., 1991) Sn–Be–W–Zn mineralization is associated with them. Corresponding tin-mineralized topaz-bearing granites have been described from the Proterozoic rapakivi complexes of the Amazonian craton, Brazil (Horbe et al., 1991; Bettencourt et al., 1999; Costi et al., 2000; Lenharo et al., 2003).

The ~1.64 Ga Kymi topaz granite stock near Kotka is one of the youngest intrusive phases in the Wiborg rapakivi granite batholith (Vaasjoki, 1977; Rämö, 1991). The stock is unique in its well-developed zonal structure, imposing marginal pegmatite (stockscheider), and associated mineralogically diverse (genthelvite, beryl, cassiterite, wolframite, and sulfide minerals) greisen and quartz veins. Only a few short reports (e.g., Haapala and Ojanperä, 1972; Haapala, 1974, 1988) and two master's theses (Viita, 1988; Kaartamo, 1996) dealing with the main outlines of the stock and associated mineralization are available. The second author is currently studying the composition of topaz granite magmas by analysing solidified melt inclusions and crystallization of corresponding fluorine-rich magmas using the Kymi granites as research material. The aim of this paper is to present petrographic and geochemical data of the Kymi topaz granite stock and to discuss the origin of its zonal structure.

2. The rocks of the Kymi stock

The Kymi stock is a 6-km-long and 3-km-wide ovoid intrusive body (Fig. 1). The contacts generally dip 20° to 30° radially outward beneath the country rocks that comprise older rapakivi granites, mainly wiborgite (mantled alkali feldspar ovoids), with minor pyterlite (unmantled ovoids) and sparcely porphyritic or equigranular rapakivi varieties. The three-dimen-

sional form and subsurface dimensions of the Kymi stock are not known, but the gently dipping contacts and certain indirect hints suggest that the stock represents the apical parts of a laccolith or a thick, upward doming subhorizontal sheet. In the rapakivi complexes of Finland, we have observed numerous subhorizontal aplite and granite dikes, and the contacts between different granite types are in many cases subhorizontal. It is also known that greisen veins and zones are generally found in the endocontact and exocontact zones, close to roofs of evolved granite plutons (e.g., Lehmann, 1990). In the Kymi area, mineralized greisen veins extend at least 5 km outside the exposed part of the stock, which suggest that the roof of the stock continues at shallow depths several kilometers from the exposed stock. From the dip observations and surface dimensions, it can be estimated that the supposed laccolith is at least 0.5 to 1 km thick, but may be much thicker.

The Kymi stock consists of two types of granite—a porphyritic granite at the centre and an equigranular granite at the margin—and the whole stock is rimmed by a pegmatite zone, stockscheider ("stock separator" in English). Such topaz-bearing marginal pegmatites are typically found at the apical contacts of topaz granite cupolas (e.g., in the Ertzgebirge; Rosenbusch, 1896, p. 66), and their formation is attributed to the accumulation of residual melts and fluids below the roof of the magma chamber or into contraction fractures at the contact between the granite cupola and its country rocks (e.g., McCarthy and Fripp, 1980; Schmitz and Burt, 1990). No mafic or intermediate rocks, not even as enclaves, are found in the Kymi stock.

2.1. The porphyritic granite

The porphyritic granite is a monzogranite that contains 1- to 3-cm-long alkali feldspar megacrysts together with ~0.5 to 1 cm quartz crystals in a medium- to fine-grained matrix (Table 1). The only mafic mineral is biotite. It is usually present as reddish brown subhedral to anhedral grains, but also euhedral crystals and grain aggregates are found. According to a wet chemical analysis, the biotite is siderophyllite with 0.31 wt.% Li_2O and 1.29 wt.% F (Table 2). Biotite is commonly partly altered to chlorite and muscovite. Topaz is a minor constituent, typically subhedral, with

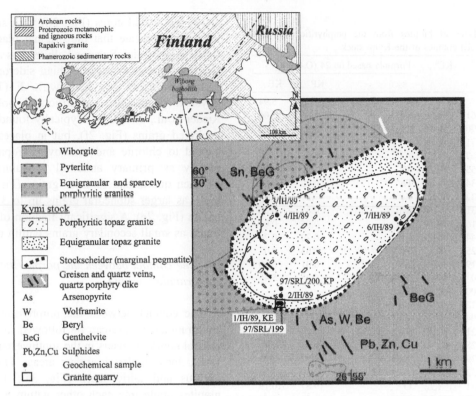

Fig. 1. Position relative to southern Finland and geological map of the Kymi stock. Sample sites are indicated. Modified from Haapala and Ojanperä (1972; Fig. 6) and Kaartamo (1996).

crystal faces against late quartz and alkali feldspar (Fig. 2a). In addition to primary (magmatic) topaz, there is a later, secondary topaz that occurs as small grains replacing plagioclase. Characteristic accessory miner-

Table 1
Modal compositions of the Kymi granites

	Porphyritic granite $n=11$	Equigranular granite $n=10$
Quartz	35.5 (21.9–50.0)	34 (40.7–31.2)
Microcline (perthite)	32.3 (29.8–42.2)	31.3 (25.9–44.2)
Plagioclase (+sericite)	21.7 (18.9–29.4)	27.6 (13.7–34.5)
Biotite	3.0 (0.9–8.9)	1.7 (0.8–4.3)
Chlorite	2.3 (0.5–7.0)	1.6 (0.4–5.1)
Muscovite	1.4 (0.3–2.3)	0.2
Topaz	1.05 (0.1–3.0)	2.95 (4.3–1.2)
Fluorite	0.7 (0.1–3.1)	0.35 (0.1–0.8)
Opac	1.9 (0.2–4.0)	0.1 (0.1–0.3)
Others	0.15	0.2
Total	100	100

Note: about 1200 points per sample calculated.

als include fluorite, zircon, ilmenite, anatase, columbite, monazite, thorite, and molybdenite.

Alkali feldspar is string or vein perthite and is found as subhedral, often tabular grains. Plagioclase is commonly anhedral; however, the turbid cores are idiomorphic. Well-preserved plagioclase is zoned with oligoclase (up to An_{25}) cores and albitic margins (Fig. 2b), whereas plagioclase grains that contain small secondary inclusions of sericite, topaz, fluorite, and quartz (induced by deanorthitization) are albite throughout. Swapped albite rims occur between two alkali feldspar grains (Fig. 2b, c), and plagioclase grains have water-clear albite rims against plagioclase (Fig. 2b). The rim albite was obviously produced by exsolution from alkali feldspar (see Haapala, 1977a, 1997).

2.2. The equigranular granite

The equigranular granite is a light grey, locally pink, medium-grained monzogranite. The main min-

Table 2

Chemical analyses of biotites from the porphyritic (KP) and equigranular (KE) granites of the Kymi stock

Sample	KP[a]	KE[a]	Formula based on 24 (O,OH,F)		
				KP	KE
			Si	5.19	5.94
SiO_2	32.61	38.57	Al^{IV}	2.81	2.06
Al_2O_3	18.90	18.90	T-site (Σ)	8.00	8.00
TiO_2	1.71	0.18			
Fe_2O_3	7.67	5.82	Al^{VI}	0.74	1.39
FeO	24.53	19.53	Ti	0.21	0.02
MnO	0.34	0.60	Fe^{3+}	0.92	0.67
MgO	0.22	0.02	Fe^{2+}	3.27	2.51
Li_2O	0.31	1.19	Mn	0.05	0.08
Na_2O	0.16	0.19	Mg	0.05	0.01
K_2O	8.44	9.42	Li	0.20	0.74
Rb_2O	0.25	0.77	Zn	0.03	0.03
CaO	0.19	0.13	Nb	0.01	0.00
BaO	0.0	0.00	Y-site (Σ)	5.48	5.45
SrO	0.00	0.00			
Nb_2O_5	0.12	0.06	Na	0.05	0.06
SnO_2	0.04	0.04	K	1.71	1.85
ZnO	0.21	0.29	Rb	0.03	0.08
H_2O+	3.28	1.65	Ca	0.03	0.02
F	1.29	4.36	X-site (Σ)	4.13	2.01
sum	100.52	101.72			
O=F2	0.51	1.72	O	19.87	20.19
Total	99.97	99.88	OH	3.48	1.69
			F	0.65	2.12
			$Fe^{2+}/(Fe^{2+}+Mg)$	0.98	1.00

[a] Wet chemical analyses by Pentti Ojanperä in 1972 (Haapala, 1988; Rieder et al., 1996).

erals are quartz, perthitic alkali feldspar, albitic plagioclase; biotite and topaz are ubiquitous minor constituents (Table 1). Accessory minerals include fluorite, monazite, bastnäsite, columbite, a pyrochlore-group mineral (X-ray identification, probably microlite), and apatite (Haapala, 1974). In terms of texture and mineral composition, the rock is homogeneous; only a few greisen veins and hydrothermally altered fracture zones, as well as some enclaves of the porphyritic granite, exposed on the walls of a granite quarry, disturb the monotonous character of the granite.

The equigranular granite is granular-hypidiomorphic. Serrate grain margins between the main minerals indicate recrystallization at the grain boundaries (Fig. 2d), and the rock appears to have suffered a stronger postmagmatic recrystallization than the porphyritic granite. Alkali feldspar is string, vein, or patch perthite. Plagioclase is compositionally homogeneous albite, but turbid cores (Fig. 2d) with small mineral inclusions indicate that the original plagioclase was richer in anorthite. Swapped albite rims are common. Biotite is greenish brown lithian siderophyllite with 1.19 wt.% Li_2O and 4.36 wt.% F (Table 2; for physical properties and structural polytypes, see Rieder et al., 1996). It is typically found as anhedral unaltered grains (Fig. 2f), but in places it is partly altered to chlorite and muscovite. Topaz is present mainly as primary anhedral to subhedral grains between quartz and feldspar grains (Fig. 2f), sometimes as larger subhedral grains up to 3 to 4 mm in length (Fig. 2e). A significant fraction of the topaz is found as small secondary grains within plagioclase.

2.3. The contact between the porphyritic granite and equigranular granite

The contact between the porphyritic granite and equigranular granite can be followed in outcrops for several hundred meters in the southwestern part of the stock, locally also in other areas. The contact is sinuous and fairly sharp (Fig. 3a). Typically, the granites grade into each other within a few cm, but locally the contact is more diffuse. During the geological mapping in 1970s, no unequivocal field evidence was found regarding the age relations of the two granites. However, during a visit to the Kymi stock in July 2003, we detected inclusions of the porphyritic granite in the equigranular granite on a newly blasted northwestern wall of the granite quarry (Fig. 3b). Some of the enclaves are rounded, others more irregular, and some are pierced by the equigranular granite. The contacts of the enclaves are generally sharp, but in places diffuse (Fig. 3b). These observations suggest that the equigranular granite solidified after, but only slightly after, the porphyritic granite, and that the two rocks or magmas were at least locally in contact with each other in a partially molten state allowing some mingling. However, any corroded quartz grains with biotite or amphibole rims—a characteristic feature for mingled rocks (e.g., Salonsaari, 1995)—were not detected.

The grain size of the granites does not change markedly when approaching the contact. Alkali feldspar megacrysts are more abundant in the porphyritic granite near the contact than in the central parts of the stock. Near the contact, the equigranular

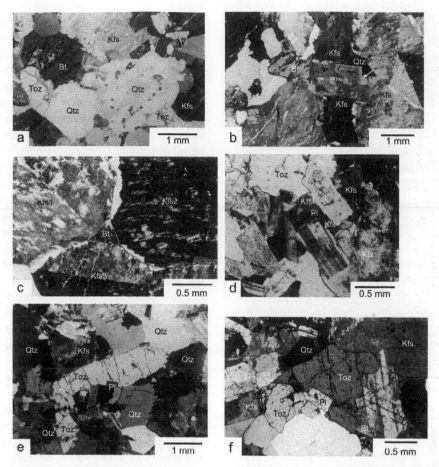

Fig. 2. Microtextures of the Kymi granites. (a) In the porphyritic granite, zonally arranged mineral inclusions mark a growth zone of a quartz crystal that was later granulated; sample 73/TV/70/AK; crossed nicols and 1/4 λ glimmer plate. (b) Plagioclase crystal in contact with alkali feldspar, quartz, and biotite. The plagioclase is mainly oligoclase (An$_{25}$), but, adjacent to alkali feldspar, it shows later growth of albite (arrow), obviously formed by exsolution from the alkali feldspar. Swapped albite rims are found between alkali feldspar grains; porphyritic granite, sample 185/TV/70/IH; crossed nicols. (c) Swapped albite rims between adjacent alkali feldspar grains (Kfs 1–3). Interstitial biotite grain is at the triple junction between the alkali feldspar grains. Porphyritic granite, same sample as in Fig. 2b; crossed nicols. (d) Albite grains with turbid cores in the equigranular granite; sample 97/SRL/199; crossed nicols. (e) Subhedral topaz grain in the equigranular granite, same sample as in (d). Note the euhedral form of plagioclase crystals against topaz; crossed nicols. (f) Anhedral to subhedral topaz grains in contact with plagioclase, alkali feldspar, quartz, and biotite. Equigranular granite, sample 321/TV/70/IH; crossed nicols. Key to abbreviations: Toz-topaz; Bt-biotite; Qt-quartz; Kfs-alkali feldspar; Pl-plagioclase.

granite contains single megacrysts of alkali feldspar (Fig. 3a). Some pegmatite-lined druses, ~20 to 40 cm in diameter, have also been found in both the granites at the border zone.

2.4. The stockscheider

In most places where the contact between the equigranular granite of the Kymi stock and the surrounding rapakivi granites is exposed, a pegmatite zone, stockscheider, is present. Rarely, the equigranular granite is in direct contact with the country rock. In terms of structure and texture, the stockscheider is heterogeneous, consisting of pegmatite granite (grain size ≤2 cm) that contains coarser pegmatite pockets and pegmatite-lined druses. The thickness of the stockscheider is usually 2 to 4 m. In the southwestern and northeastern margins of the stock, the stock-sheider dips 20° to 30° below the wiborgitic and pyterlitic country rocks (Kaartamo, 1996). The stock-

Fig. 3. Photographs showing (a) a relatively sharp contact between the porphyritic granite and equigranular granite. A single alkali feldspar megacryst in the equigranular granite is marked with arrow. (b) Inclusions of porphyritic granite in the equigranular granite on the northern wall of the granite quarry (cf., Fig. 1). Note that the contact between the inclusions and host rock varies from sharp to diffuse. (c) Dendritic biotite in the stockscheider pegmatite at the southeastern margin of the Kymi stock, ~300 m east of the granite quarry. The biotite dendrites show unidirectional growth from the contact (up, not visible in the picture) toward the interior of the stock.

scheider pegmatite can be followed 300 m in the southwestern margin of the stock, southeast of the granite quarry. In these outcrops, some aplite–pegma-

tite dikes are seen to cut the country rock wiborgite in the immediate vicinity of the contact and parallel to it.

The main minerals in the stockscheider are alkali feldspar, quartz, albite, biotite, and topaz. More rare minerals include muscovite, fluorite, tourmaline, monazite, bastnaesite, arsenopyrite, and molybdenite (Haapala, 1974; Haapala and Ojanperä, 1972), as well as beryl, columbite, cassiterite (Kaartamo, 1996), and plumbomicrolite (Hytönen, 1999). The stockscheider thus belongs to the NYF pegmatites of Černý, 1991. The topaz is, especially in the miarolitic cavities, bluish and sometimes of gem quality, and there have been some attempts to exploit the topaz pegmatite.

A characteristic textural feature of the stockscheider pegmatite is the dendritic growth of biotite perpendicularly from the country rock contact toward the interior of the stock (Fig. 3c). Such uniformly oriented dendritic growth indicates rapid growth from undercooled melt along the thermal gradient (e.g., Shannon et al., 1982) and is thus strong evidence for the magmatic origin of the pegmatite.

2.5. The hydrothermal greisen and quartz veins

Greisen and quartz veins cut the granites of the Kymi stock, as well as surrounding rapakivi granites up to 3 km from the stock contact (Fig. 1). The veins run mostly in a 130°–150° direction. Apparently, the prevailing stress field of the area determined the direction of fractures, which were then variably mineralized (Haapala, 1977b). The main minerals in the greisen veins are quartz, phengitic muscovite, and chlorite (chemical analyses in Haapala and Ojanperä, 1972), biotite, topaz, and fluorite. Ore or rare-element minerals are genthelvite (Zn, Mn, Fe)$_4$Be$_3$Si$_3$O$_{12}$S (analyses in Haapala and Ojanperä, 1972), wolframite, cassiterite, arsenopyrite, sphalerite, galena, and chalcopyrite. Most of the mineralized greisen veins are found in the country rocks; in the porphyritic granite, only quartz or quartz–fluorite fracture-filling veins have been detected.

3. Geochemistry

Chemical analyses of the porphyritic and equigranular granites are presented in Table 3 and Figs. 4–8. For comparison, chemical analyses of typical wiborgite (Summa, Hamina) and pyterlite (Pyterlahti, Vironlahti) from Haapala et al. (2005) are shown. The Kymi granites are evolved SiO_2-rich granites, and plot into the fields of A-type granites of Whalen et al. (1987) and within-plate granites of Pearce et al. (1984). According to Taylor's (1992) classification, they represent "low-P" subtype of topaz granites.

In Fig. 4, the granites are plotted in a haplogranitic albite–orthoclase–quartz diagram showing the effect of fluorine (0, 1, 2, and 4 wt.% F) on the ternary minima at 1 kb and water-saturated conditions (Manning, 1982). The analyses of the porphyritic granite (0.39–1.11 wt.% F) plot near the minimum for 0 wt.% F, but deviate from it to more K-rich compositions (possibly an effect of normative anorthite; see James and Hamilton, 1969). The equigranular granites (0.73–1.66 wt.% F) plot between 0 and 1 wt.% F. Although the applicability of the experimental diagrams is decreased by several variables (normative anorthite, dissolved CO_2 and B, activity of water, pressure, etc.), the diagram does suggest that the equigranular granite crystallized from a magma that contained more fluorine than the porphyritic granite magma. This is further supported by unpublished analyses of melt inclusions in quartz of the two granites [average 1.0 wt.% F in porphyritic granite, 1.9 wt.% F in equigranular granite (unpublished data by Sari Lukkari and Rainer Thomas)].

Both phases of the Kymi granite show chemical features typical of evolved or highly evolved fluorine-rich granites affected by postmagmatic reactions. However, there are also marked differences between the porphyritic and equigranular granites (Figs. 5–8). The porphyritic granite is metaluminous to slightly peraluminous (alumina saturation index A/CNK 0.79 to 1.06), whereas the equigranular granite rich in topaz is peraluminous (A/CNK 1.05 to 1.16) and has higher Na_2O/K_2O. In the Harker variation diagrams (Fig. 5a through e) and in the Rb–Sr–Ba diagram (Fig. 6), the porphyritic granite generally plots in the evolved field of "normal" (non-topaz-bearing) rapakivi granites, whereas the equigranular granite has extreme compositions resembling the topaz-bearing granite of the Eurajoki stock (cf., Haapala, 1977a, 1997). Regarding mobile elements, such as Rb, Sr, and Ba (Figs. 6 and 7), the "more evolved" character of the equigranular granite may be related in part to postmagmatic autometasomatic reactions, but it is

more difficult to assign the decreased contents of the immobile elements Ti and Zr to such metasomatism.

The REE patterns (Fig. 8) show deep negative Eu anomalies: Eu/Eu* is 0.2 to 0.1 in the porphyritic granite and 0.1 to 0.0 in the equigranular granite. This suggests extreme crystal fractionation or preferential partitioning of Eu^{2+} in an aqueous fluid phase coexisting with the melt (see Irber, 1999) and is compatible with petrographic observations. Local miarolitic cavities and larger pegmatite-lined druses and mineral alteration in the granites, as well as associated mineralized hydrothermal veins and altered fracture zones, indicate fluid saturation and loss during the late-magmatic and postmagmatic stages.

Except for the Eu anomaly, the REE patterns of the porphyritic granite are smooth and similar to other highly evolved F-rich granites. The REE patterns of the equigranular granite have elevated (Ce, Pr) and (Tb, Dy) values in relation to the neighboring REE (tetrad effect $TE_{1,3}$=1.32 to 1.52; see Irber, 1999). This effect is typical of transitional magmatic–hydothermal systems, and is often found in highly evolved volatile-rich granites. According to Irber (1999) and Jahn et al. (2001), it is unlikely that mineral fractionation alone could generate the tetrad effect in evolved granites, and they emphasized the increasing importance of aqueous fluid during the final stages of granite crystallization. This is apparently the case in the Kymi equigranular granite. The granite crystallized from highly evolved H_2O- and F-rich residual magma, and late-magmatic to postmagmatic aqueous fluid–rock interactions only modified the texture and composition of the already originally anomalous granite.

4. Origin of the zonal structure

4.1. Possible models

In the lack of knowledge of the three-dimensional form and subsurface dimensions of the magma chamber, quantitative or even semiquantitative modeling of the crystallization history of the Kymi stock is not possible. Thus, the origin of the zonal structure will be discussed here on a qualitative basis, considering five alternative models.

Table 3

Major and trace element compositions of the Kymi granites and the type samples of wiborgite (Summa, Hamina) and pyterlite (Pyterlahti, Virolahti)

Sample	Wiborgite 1A/IH/2001	Pyterlite 2A/IH/2001	Porphyritic granite 2/IH/89	4/IH/89	6/IH/89	7/IH/89	97/SRL/200	KP	Equigranular granite 1/IH/89	3/IH/89	97/SRL/199	KE
SiO_2 (wt.%)	69.93	75.69	74.40	75.40	72.50	74.00	74.30	75.08	73.50	74.30	73.70	73.40
TiO_2	0.48	0.22	0.11	0.04	0.14	0.08	0.13	0.09	0.02	0.02	0.04	0.00
Al_2O_3	13.41	11.70	12.60	13.00	12.90	12.90	12.40	11.52	15.10	14.70	14.70	14.20
Fe_2O_3	0.45	0.31	0.63	0.87	1.15	0.96	0.80	0.68	0.19	0.20	0.42	0.39
FeO^w	3.4	1.5	1.3	0.7	1.6	0.9	1.6	1.32	0.6	0.3	0.5	0.57
MnO	0.06	0.03	0.02^1	0.02^1	0.03^1	0.02^1	0.01	0.030	0.02^1	0.03^1	0.01	0.020
MgO	0.27	0.10	0.14	0.14	0.20	0.14	0.02	0.04	0.10	0.10	0.00	n.a.
CaO	2.07	1.08	1.24	0.70	1.19	1.12	1.61	1.17	0.76	0.53	0.90	0.73
Na_2O	2.97	2.49	2.79	3.27	2.42	3.00	2.73	2.81	4.05	4.65	4.01	4.15
K_2O	5.55	5.76	5.28	4.88	5.69	5.20	5.03	5.66	4.55	4.22	4.60	4.91
P_2O_5	0.13	0.03	0.01^1	0.01^1	0.02^1	0.01^1	0.01	0.02	0.01^1	0.01^1	0.01	0.02
F^w	0.24	0.43	0.86	0.68	0.67	0.93	1.11	0.39	>1.00	1.00	1.66	0.73
Cl^w (ppm)	800	300	310	190	230	190	218	n.a.	110	50	59	n.a.
$H2O+^w$	0.31	0.17	0.50	0.50	0.70	0.50	0.80	0.14	0.40	0.40	0.30	0.37
LOI	0.33	0.52	1.31	1.08	1.85	1.54	0.95	n.a.	1.08	1.08	0.40	n.a.
Total	99.35	99.70	100.10	100.30	100.10	100.10	100.00	99.61	100.20	100.20	99.40	99.63
Q	30.3	39.6	38.5	38.8	37.9	37.5	39.9	37.6	33.2	31.6	33.4	30.9
Ab	30.2	23.1	26.4	29.9	23.5	28.2	26.3	25.9	37.4	41.8	36.9	37.8
Or	39.5	37.3	35.0	31.3	38.6	34.2	33.9	36.5	29.4	26.6	29.7	31.3
A/CNK	0.91	0.95	1.00	1.09	1.05	1.02	0.96	0.89	1.16	1.12	1.11	1.05
NK\ALPHA	0.81	0.88	0.82	0.82	0.79	0.82	0.80	0.93	0.77	0.83	0.79	0.86
Ga/Al	3.8	3.9	6.6	7.4	6.8	7.0	7.9	7.9	7.7	9.9	8.5	7.9
Rb/Ba	0.2	0.6	2.1	4.6	0.9	2.7	2.8	2.8	6.5	9.2	49.5	2.8
Rb/Sr	1.7	4.6	12.7	28.0	9.4	17.6	14.2	14.2	44.5	51.5	90.0	14.2
Be^4	5^1	5^1	46	52	11	34	42.2^1	22	17	10	14^1	8
Li^1	n.a.	n.a.	114	117	59	103	118	110	194	351	181	170
Rb	271	349	610	700	533	688	641	550	978	1340	990	990
Cs^2	10	9	24	20	10	18	12	n.a.	24	22	12	n.a.
Sr^1	155	76	48	25	57	39	45	<40	22	26	11	<40
Ba	1144	541	286	153	591	257	232	200	151	146	<20	100
Ga^1	27^3	24^3	44	51	47	48	52^3	61	61	77	66^3	92
Y^1	63^3	75^3	348	221	312	329	300^3	n.a.	7	17	9^3	n.a.
Sc^1	8	3	6	8	7	7	4	n.a.	8	6	7	n.a.
Zr	460	300	255	98	358	182	262	190	12	25	29	30
Sn	8^5	9^5	52	34	25	44	29^5	27	15	<2	10^5	8
Nb	26.4^3	25.8^3	188	71	229	159	185^3	100	58	59	64^3	50
Ta^2	2	2	26	28	26	28	13	n.a.	45	62	21	n.a.
W^2	3	3	25	19	17	27	22	n.a.	23	18	14	n.a.
Th^2	25	39	83	83	95	81	49	n.a.	31	25	17	n.a.
U^2	8	15	35	18	33	43	20	n.a.	35	11	20	n.a.
Hf^2	13	10	23	14	27	21	10	n.a.	8	8	3	n.a.
Zn^1	125	86	95	90	137	71	90	n.a.	67	60	60	n.a.
Cr	11^2	10.6^2	29	23	28	28	$<2^2$	n.a.	9	5	$<2^2$	n.a.
V^4	8^1	$<5^1$	6	2	7	5	3^1	n.a.	4	2	$<2^1$	n.a.
Pb^1	38	55	122	90	85	71	76	n.a.	130	73	118	n.a.
As^2	25	7	11	15	13	9	6	n.a.	36	8	9	n.a.
La^3	95.9	99.0	166.0	43.0	274.0	136.0	211.0	n.a.	65.0	17.0	65.5	n.a.
Ce^3	182.0	182.0	367.0	95.0	544.0	301.0	433.0	n.a.	154.0	52.3	154.0	n.a.
Pr^3	21.5	21.2	39.8	10.1	55.7	33.1	48.4	n.a.	15.0	5.0	15.4	n.a.
Nd^3	80.6	76.2	151.0	39.4	208.0	128.0	182.0	n.a.	34.9	12.6	39.5	n.a.
Sm^3	14.6	13.7	32.3	11.5	40.2	30.6	36.7	n.a.	8.1	3.2	7.9	n.a.

Table 3 (*continued*)

Sample	Wiborgite 1A/IH/2001	Pyterlite 2A/IH/2001	Porphyritic granite 2/IH/89	4/IH/89	6/IH/89	7/IH/89	97/SRL/200	KP	Equigranular granite 1/IH/89	3/IH/89	97/SRL/199	KE
Eu[3]	2.6	1.3	1.9	0.4	2.7	1.4	2.1	n.a.	<0.05	<0.05	0.0	n.a.
Gd[3]	12.6	12.2	30.1	10.1	33.5	28.1	34.7	n.a.	3.2	2.2	4.2	n.a.
Tb[3]	2.1	2.2	5.8	2.6	6.2	5.5	6.2	n.a.	1.4	0.7	1.3	n.a.
Dy[3]	12.7	13.8	38.7	22.0	40.5	36.8	36.8	n.a.	9.7	4.9	8.7	n.a.
Ho[3]	2.7	3.0	8.5	5.2	8.1	8.0	8.5	n.a.	2.0	1.1	2.1	n.a.
Er[3]	7.3	8.5	28.9	22.7	28.3	28.1	29.8	n.a.	8.6	5.2	8.8	n.a.
Tm[3]	1.1	1.3	4.6	4.6	4.3	4.9	4.8	n.a.	2.5	1.2	2.4	n.a.
Yb[3]	7.3	8.6	36.3	39.8	33.5	38.6	31.6	n.a.	27.9	14.9	26.2	n.a.
Lu[3]	1.1	1.2	5.2	6.4	4.9	5.9	5.0	n.a.	4.4	2.4	4.2	n.a.
La_N/Yb_N	8.8	7.8	3.1	0.7	3.5	2.4	4.5	–	1.6	0.8	1.7	
Eu/Eu*	0.6	0.3	0.2	0.1	0.2	0.1	0.2	–	0.0	0.1	0.0	–
Ce/Ce*	0.96	0.96	1.09	1.09	1.04	1.08	1.03	–	1.39	1.63	1.33	–
Pr/Pr*	0.99	1.01	1.01	0.98	0.96	1.00	1.00	–	1.38	1.42	1.30	–
TE1	0.98	0.98	1.05	1.03	1.00	1.00	1.04	–	1.39	1.52	1.32	–

n.a.—Not analyzed. $Eu/Eu^* = Eu_N/Sm_N \cdot Gd_N$; $Ce/Ce^* = Ce_N/(La_N^{2/3} \cdot Nd_N^{1/3})$; $Pr/Pr^* = Pr_N/(La_N^{1/3} \cdot Nd_N^{2/3})$. Degree of the tetrad effect: $TE1 = (Ce/Ce^* \cdot Pr/Pr^*)^{1/2}$. A/CNK = molecular $Al_2O_3/(CaO+Na_2O+K_2O)$. Samples were analyzed at X-ray Assay Laboratories (Canada). Trace elements in ppm. Analytical methods: XRF, if not otherwise mentioned; [1]ICP, [2]NA, [3]ICP-MS, [4]DCP, [5]XRP, [w]wet chemical, n.a. not analyzed. Analyses KE and KP from Rieder et al. (1996).

4.1.1. Model 1

Crystallization started with solidification of the porphyritic granite against the country rock. After nearly complete crystallization of the apical part of the magma chamber, the contact of the cupola was fractured, and highly evolved residual melt was intruded from the deeper parts of the chamber into the fracture. The stockscheider and the equigranular granite crystallized from this melt. If it is assumed that the residual melt was intruded into solidified rocks, this model has difficulties in explaining the geometry of the stock and the gradational contacts between the granites. The situation would be different, however, if the evolved magma flowed between the country rock

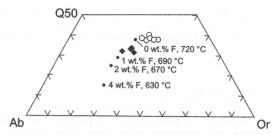

Fig. 4. Experimental albite–orthoclase–quartz system showing the effect of fluorine (0, 1, 2, 4 wt.% F) on the ternary minima at 1 kb pressure under water-saturated conditions (Manning, 1982).

and porphyritic granite that still contained a significant amount of melt (cf. model 2).

4.1.2. Model 2

This model presumes that the Kymi magma chamber was a convecting system, probably in the form of a laccolith (flat, subhorizontal igneous body with domal upper contact). Internal convection (see Worster et al., 1990) in the crystallizing magma chamber led to accumulation of evolved interstitial melts into the apical part of the chamber (Fig. 9a). Differentiation was enhanced by the flow of this low-density residual melt along the sloping contacts up to the cupola, where it replaced a less-evolved, denser magma (see Nilson et al., 1985; Huppert et al., 1986; Sawka et al., 1990; Mahood and Cornejo, 1992). Thus, a highly evolved, volatile-rich, possibly stratified magma zone was developed at the top of the crystallizing chamber. In addition, the low viscosity of the H_2O- and F-rich apical melt may have allowed settling of larger phenocrysts into the underlying, less fractionated phenocryst-bearing magma (see Audétat et al., 2000), which then crystallized as the porphyritic granite (Fig. 9b). The stockscheider and equigranular granite crystallized from this highly evolved apical melt. Solidification of the cupola margin may have started with the crystallization of the stockscheider pegmatite as a

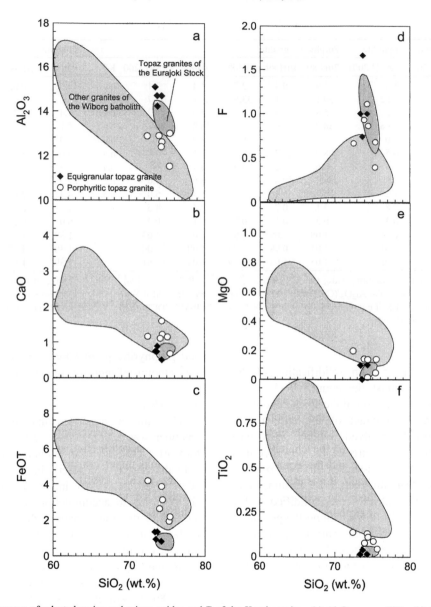

Fig. 5. Variation diagrams of selected major and minor oxides and F of the Kymi granites. (a) Al_2O_3 versus SiO_2, (b) CaO versus SiO_2, (c) FeO_{tot} versus SiO_2, (d) F versus SiO_2, (e) MgO versus SiO_2, and (f) TiO_2 versus SiO_2. The fields of "normal" (not topaz-bearing) granites (from data file of the Department of Geology, University of Helsinki) and the topaz granites of the Eurajoki stock (Haapala, 1997) are marked for comparison.

result of undercooling (related to rapid heat loss by conduction to the country rock and possible degassing), and nearly penecontemporaneous crystallization of the outer margin of the equigranular granite. The smaller grain size of the matrix of the porphyritic granite might be a result of pressure quenching related to opening of fractures and

vigorous escape of volatiles (see Model 3) or simply of voluminous crystallization related to a small decrease of temperature or pressure in fluid-saturated low-temperature magma (see Whitney, 1988). Pegmatite-lined druses in the contact zone between the porphyritic and equigranular granites indicate fluid saturation.

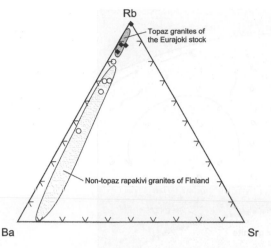

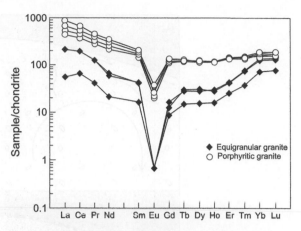

Fig. 8. Chondrite-normalized (Boynton, 1984) REE patterns of the equigranular and porphyritic granites of Kymi.

Fig. 6. Ba–Sr–Rb diagram showing the composition of the Kymi granites (labels as in Fig. 5) compared to the fields of the non-topaz-bearing rapakivi granites of Finland and the topaz granites of the Eurajoki stock. References as in caption of Fig. 5.

4.1.3. Model 3

Many epizonal felsic intrusions hosting por-phyry-type copper and molybdenum deposits have a zonal structure with an equigranular margin and a porphyritic core. The origin of the zonal structure is often explained (Phillips, 1973; Carten et al., 1988; Burnham, 1997) by the equigranular granitoid having crystallized from the margins inward, increasing the volatile content of the residual magma chamber and, when the internal volatile pressure exceeded the confining pressure and tensile

strength of the rock, causing vigorous degassing and rapid crystallization of the central part of the intrusion (porphyritic granitoid). Two features are against application of this model to the Kymi stock: (1) the new field observations indicate that the equigranular granite finished solidification after the porphyritic granite; (2) the geochemical character of the granites does not fit. For example, because fluorine concentrates into the melt phase in granite melt–vapor system (e.g., Webster and Holloway, 1990), the fluorine content should not be lower in the porphyritic granite than in the equigranular granite. In addition, care must be exercised in applying this model—it has been pointed out (Candela, 1997) that significant undercooling related to degassing can take place only in ascending magma chambers, not in isobarically crystallizing magma chambers.

4.1.4. Model 4

It can also be thought that the Kymi porphyritic granite represents a later intrusion phase, emplaced into cooler solid or semisolid equigranular granite (cf., Carten et al., 1988). However, the contact relations (no apophyses of the porphyritic granite in the equigranular granite, no xenoliths of the equigranular granite in the porphyritic granite) do not support this.

4.1.5. Model 5

Several evolved granite plutons (e.g., in Erzge-birge) have roughly similar zonal structures as the

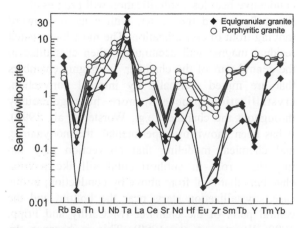

Fig. 7. Abundances of certain trace elements in the Kymi granites, normalized to the composition of wiborgite (Table 3).

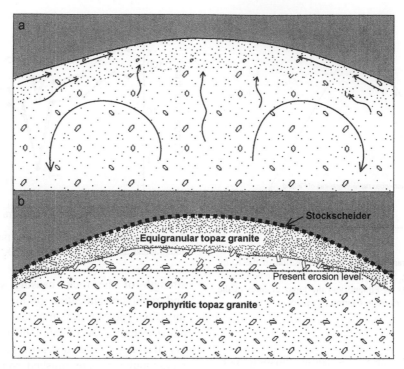

Fig. 9. Cross-sections showing the formation of the zonal structure of the Kymi stock (Model 2, see text). (a) Internal convection in the crystallizing magma chamber and upward flow of the residual boundary layer melt accumulate a highly evolved H_2O- and F-rich melt layer in the apical part of the chamber. Low viscosity of the melt allows megacrysts to sink into the underlying less-evolved, more viscose magma. (b) Crystallization of the highly evolved melt produces the stockscheider pegmatite and equigranular granite, solidification of the underlying crystal mush the porphyritic granite.

Kymi stock, with topaz and LiFe-mica granites and a stockscheider in the apical parts of the intrusions. In the 1960s through the 1980s, the origin of the topaz granites was often attributed to metasomatism along the cupola contacts (e.g., Beus and Zalachkova, 1964; Štemprok, 1971; Rundquist, 1980). Subsequently, detailed petrographic, experimental, and melt inclusion studies have provided evidence for the magmatic origin of the late-stage granites and their subvolcanic equivalents (e.g., Kovalenko et al., 1971; Haapala, 1977a; Manning, 1982; Raimbault et al., 1995; Breiter et al., 1997; Haapala and Thomas, 1999). We regard the equigranular granite of Kymi as an originally anomalous granite that crystallized from a highly evolved volatile-rich magma. Postmagmatic fluid–rock reactions only modified the texture and composition of the granite. The zonal structure of the stock is not a product of metasomatism, although autometasomatic processes have affected the late-stage granites more than earlier "dry" rapakivi granites.

5. Discussion

Theoretically, cooling and crystallization of magma chambers may be a stagnant or a convective process. As a stagnant homogenous magma chamber cools by conductive heat loss, solidification will progress from the walls toward the center, with a narrow inward moving front of crystallization. The most fractionated residual magma will accumulate, then crystallize in the central part of the chamber. In magma chambers that are mixed thermally by internal convection, crystallization proceeds more homogeneously throughout the chamber (e.g., Worster et al., 1990). It has been shown by experimental, thermodynamic, and chemical modeling that convection plays an important role in subhorizontal sill-like magma chambers that cool from above by conduction; in this case, crystallization starts near the bottom and the central parts of the chamber (McCarthy and Fripp, 1980; Worster et al., 1990). This is because the temperature along the adiabat for convecting magma decreases less with decreasing pressure than does the liquidus temperature (e.g., McCarthy and Fripp, 1980). With increasing crystallinity, the convection of the magma will slow down and finally stop, possibly at about 50 to 65 vol.% crystals (Shinohara and Hedenquist, 1997; Audétat et al., 2000), and crystallization will continue to completion in a relatively stagnant chamber. High volatile (H_2O, F) contents will decrease the viscosity and extend the convection to lower temperatures.

Of the models above, model 2 is the most compatible with the petrologic and geochemical observations, as well as theoretical considerations. In a thick, laccolith-shaped evolved magma chamber, internal convection and up-slope flow of the residual boundary layer melt will result in accumulation of low-density, highly evolved melt at the top of the magma chamber, above the crystallizing main magma body. In this model, the stockscheider and equigranular granite crystallized from the highly evolved F-rich melt and the porphyritic granite from the underlying crystal mush. Field observations indicate that the porphyritic granite solidified slightly before the adjacent equigranular granite. This conclusion is supported by geochemical data (e.g., the effect of F on the ternary minima in albite–orthoclase–quartz system; Fig. 4), which—although modified by subsolidus reactions—suggest that the porphyritic granite had a higher solidus temperature than the equigranular granite. The zircon saturation geothermometer (Watson and Harrison, 1983) yields magma temperatures of 753 to 864 °C for the porphyritic granite, and 608 to 661 °C for the equigranular granites. Corresponding temperatures of the wiborgite and pyterlite of Table 3 are 866 and 739 °C, respectively. Although the zircon saturation thermometer has serious limitations (e.g., Miller et al., 2003; Clemens, 2003), these numbers may be taken as indications of relative temperatures of the Kymi magmas. Obviously, when the crystal mush of the porphyritic granite solidified, the apical F-rich magma was still well above its solidus temperature.

Miarolitic cavities and pegmatite-lined druses in the stockscheider and in the contact zone between the porphyritic and equigranular granite suggest that much of the crystallization took place under fluid-saturated conditions, and the hydrothermal greisen and quartz veins and altered fracture zones further

suggest vigorous fluid loss. Because there are some quartz veins and quartz–fluorite veins in the porphyritic granite, it is possible that some degassing occurred during the crystallization of both granites. Strong dilation in connection with the fracturing and degassing and/or crystallization of fluid-saturated magma may have affected the texture of the granites, e.g., by producing the fine- to medium-grained matrix of the porphyritic granite (cf., Whitney, 1988; Burnham, 1997).

6. Conclusions

Field, petrologic, and geochemical data allow the following conclusions regarding the structure, composition, and magmatic–hydrothermal evolution of the Kymi stock:

1. The Kymi topaz granite stock probably represents the apical part of an evolved epizonal laccolith or thick (>0.5 km) subhorizontal sheet that has a domal upper contact. The stock intrudes the less evolved rapakivi granites of the Wiborg batholith.
2. The Kymi stock has a zonal structure, from the margin to the center: stockscheider pegmatite, equigranular topaz granite, porphyritic topaz granite. Hydrothermal greisen and quartz veins cut the country rocks around the stock, and some also cut the granites of the stock. Several of the veins, especially those in the country rocks, are mineralized (Be, W, Sn, Zn, Pb).
3. The equigranular granite is geochemically more evolved (e.g., lower Ti, Zr, Ba, and Sr; higher Li, Rb, Ga, and F; deeper negative Eu anomaly) than the porphyritic granite. The granites are interpreted to have crystallized from evolved volatile-rich magmas; postmagmatic fluid–rock reactions have modified the texture and composition of the rocks, especially those of the equigranular granite.
4. Cooling of the magma chamber took place mainly by conduction from above, but heat distribution, crystallization, and formation of the zonal structure were largely controlled by convection in the chamber. Convection in the crystallizing magma chamber and upward flow of residual melt along sloping contacts caused accumulation of low-

density, evolved melt in the apical parts of the magma chamber.

5. The stockscheider and equigranular granite crystallized from the highly evolved apical magma and the porphyritic granite from the underlying crystal mush. Contact observations and geochemical data indicate that the porphyritic granite solidified before the adjacent equigranular granite. The unidirectional solidification structures of the stockscheider suggest that it crystallized from undercooled melt (rapid loss of heat, possible fluid loss) by rapid growth from the rim toward the centre. Pegmatite-lined druses in the contact zone between the equigranular and porphyritic granites indicate fluid saturation. Possibly the solidification of the main granites of the stock was finished by the crystallization of the inner parts of the equigranular granite.

6. The hydrothermal fluids that contributed to the formation of the greisen and quartz veins were probably separated from the crystallizing fluid-saturated granite magmas and expanded into the opened fractures, where they caused hydrothermal alteration and mineralization. The evolution of the fluids has not been studied, but obviously their composition and temperature changed by mixing with ground water, fluid–rock interaction, and mineral crystallization.

Acknowledgments

The financial support from the University of Helsinki (Research Fund 2105005) and Foundation of Research of Natural Resources in Finland (Research Funds 1599/01 and 1628/02) is acknowledged with gratitude. The manuscript has been reviewed by Eric H. Christiansen, Hanna Nekvasil, Gail A. Mahood, and Tapani Rämö; their many valuable comments and suggestions are gratefully acknowledged.

References

Amelin, Yu., Beljaev, A., Larin, A., Neymark, L., Stepanov, K., 1991. Salmi batholith and Pitkäranta ore field in Soviet Karelia. In: Haapala, I., Rämö, O.T., Salonsaari, P.T. (Eds.), IGCP Project 315 Symposium Rapakivi Granites and Related Rocks, Excursion Guide Geological Survey of Finland, 33, 57 pp.

Audétat, A., Günther, D., Heinrich, C.A., 2000. Magmatic-hydrothermal evolution in a fractionating granite: a microchemical study of the Sn–W–F-mineralized Mole Granite (Australia). Geochimica et Cosmochimica Acta 64, 3373–3393.

Bettencourt, J.S., Tosdal, R.M., Leite Jr, W.B., Payolla, B.L., 1999. Mesoproterozoic rapakivi granites of the Rondonia Tin Province, southwest border of the Amazonian craton, Brazil—I. Reconnaissance U–Pb geochronology and regional implications. Precambrian Research 95, 41–67.

Beus, A.A., Zalachkova, N.E., 1964. Post-magmatic high temperature metasomatic processes in granitic rocks. International Geology Review 6, 668–681.

Boynton, W.V., 1984. Geochemistry of the rare earth elements: meteorite studies. In: Henderson, P. (Ed.), Rare Earth Element Geochemistry. Elsevier, Amsterdam, pp. 63–114.

Breiter, K., Fryda, J., Seltmann, R., Thomas, R., 1997. Mineralogical evidence for two magmatic stages in the evolution of an extremely fractionated P-rich rare-metal granite: the Podlesi stock, Krusne Hory, Czech Republic. Journal of Petrology 38, 1723–1739.

Burnham, C.W., 1997. Magmas and hydrothermal fluids. In: Barnes, H.L. (Ed.), Geochemistry of Hydrothermal Ore Deposits, 3rd edition. John Wiley & Sons, New York, pp. 63–123.

Candela, P.A., 1997. A review of shallow, ore-related granites: textures, volatiles, and ore metals. Journal of Petrology 38, 1619–1633.

Carten, R.B., Geraghty, E.P., Walker, B.M., 1988. Cyclic development of igneous features and their relationship to high-temperature hydrothermal features in the Henderson porphyry molybdenum deposit, Colorado. Economic Geology 83, 266–296.

Černý, P., 1991. Rare element granitic pegmatites: Part I. Anatomy and internal evolution of pegmatite deposits. Geoscience Canada 18, 49–67.

Clemens, J.D., 2003. S-type granitic magmas-petrogenetic issues, models and evidence. Earth-Science Reviews 61, 1–18.

Costi, H.T., Dall'Agnol, R., Moura, C.A.V., 2000. Geology and Pb–Pb geochronology of Paleoproterozoic volcanic and granitoid rocks of the Pitinge province, Amazonian craton, northern Brazil. International Geology Review 42, 832–849.

Edén, P., 1991. A specialized topaz-bearing rapakivi granite and associated mineralized greisen in the Ahvenisto complex, SE Finland. Bulletin of the Geological Society of Finland 63, 25–40.

Haapala, I., 1974. Some Petrological and geochemical characteristics of rapakivi granite varieties associated with greisen-type Sn, Be and W mineralization in the Eurajoki and Kymi areas, Southern Finland. In: Štemprok, M. (Ed.), Metallization Associated with Acid Magmatism. Geological Survey, Prague, pp. 159–169.

Haapala, I., 1977a. Petrography and geochemistry of the Eurajoki stock: a rapakivigranite complex with greisen-type mineralization in southwestern Finland. Bulletin Geological Survey of Finland 286, 128 pp.

Haapala, I., 1977b. The controls of tin and related mineralizations in the rapakivi–granite areas of south-eastern Fennoscandia. Geologiska Föreningens I Stockholm Förhandlingar 99, 130–142.

Haapala, I., 1988. Metallogeny of the Proterozoic rapakivi granites of Finland. In: Taylor, R.P., Strong, D.F. (Eds.), Recent Advances in the Geology of Granite-related Deposits. Canadian Institute of Mining and Metallurgy Special Volume, vol. 39, pp. 124–132.

Haapala, I., 1995. Metallogeny of the rapakivi granites. Mineralogy and Petrology 54, 149–160.

Haapala, I., 1997. Magmatic and postmagmatic processes in tin-mineralized granites: topaz-bearing leucogranites in the Eurajoki rapakivi granite stock, Finland. Journal of Petrology 38, 1645–1659.

Haapala, I., Ojanperä, P., 1972. Genthelvite-bearing greisens in southern Finland. Bulletin - Geological Survey of Finland, 259,

Haapala, I., Thomas, R., 1999. Melt inclusions in quartz and topaz of the topaz granite from Eurajoki, Finland. Journal of the Czech Geological Society 45, 149–154.

Haapala, I., Rämö, O.T., Frindt, S., 2005. Comparison of Proterozoic and Phanerozoic rift-related basaltic-granitic magmatism. Lithos 80, 1–32 (this issue).

Horbe, M.A., Horbe, A.C., Costi, H.T., Teixeira, J.T., 1991. Geochemical characteristics of cryolite–tin-bearing granites from Pitinga mine, northwestern Brazil—a review. Journal of Geochemical Exploration 40, 227–249.

Huppert, H.E., Sparks, R.S.J., Wilson, J.R., Hallworth, M.A., 1986. Cooling and crystallization at an inclined plate. Earth and Planetary Science Letters 79, 319–328.

Hytönen, K., 1999. Suomen mineraalit (minerals of Finland). Geologian tutkimuskeskus. Geological Survey of Finland (in Finnish), 400 pp.

Irber, W., 1999. The lanthanide tetrad effect and its correlation with K/Rb, Eu/Eu*, Sr/Eu, Y/Ho, and Zr/Hf of evolving peraluminous granite suites. Geochimica et Cosmochimica Acta 63, 439–508.

Jahn, B., Wu, F., Capdevila, R., Martineau, F., Zhao, Z., Wang, Y., 2001. Highly evolved juvenile granites with tetrad REE patterns: the Woduhle and Baerzhe granites from the Great Xing'an Mountains in NE China. Lithos 45, 147–175.

James, R.S., Hamilton, D.L., 1969. Phase relations in the system NaAlSi$_3$O$_8$–KAlSi$_3$O$_8$–CaAlSi$_2$O$_8$–SiO$_2$. Contributions to Mineralogy and Petrology 21, 111–141.

Kaartamo, K., 1996. Kymin stoking reunapegmatiittimuodostuman (stockscheider) rakenteesta ja mineralogiasta. MSc thesis, Department of Geology, University of Helsinki, Finland (in Finnish).

Kovalenko, V.I., Kuzmin, M.I., Antipin, V.S., Petrov, L.L., 1971. Topaz-bearing quartz keratophyre (ongonite), a new variety of subvolcanic igneous dike rocks. Doklady Akademii Nauk SSSR, Earth Science Section 199, 132–135.

Laitakari, A., 1928. Palingenese am Kontakt des postjotnischen Olivindiabases. Fennia 50, 1–25.

Lehmann, B., 1990. Metallogeny of Tin. Springer-Verlag, Berlin, Heidelberg, 211 pp.

Lenharo, S.L.R., Pollard, P.J., Born, H., 2003. Petrology and textural evolution of granites associated with tin and rare-metals mineralization at the Pitinga mine, Amazonas, Brazil. Lithos 66, 37–61.

Lukkari, S., 2002. Petrography and geochemistry of the topaz-bearing granite stocks in Artjärvi and Sääskjärvi, western margin of the Wiborg rapakivi granite batholith. Bulletin of the Geological Society of Finland 74, 115–132.

Mahood, G.A., Cornejo, P.C., 1992. Evidence for ascent of differentiated liquids in a silicic magma chamber found in a granitic pluton. Transactions of the Royal Society of Edinburgh. Earth Sciences 83, 63–69.

Manning, D.A.C., 1982. An experimental study of the effects of fluorine on the crystallization of granitic melts. In: Evans, A.M. (Ed.), Metallization Associated with Acid Magmatism, vol. 6. John Wiley & Sons, Chichester, New York, pp. 191–203.

McCarthy, T.S., Fripp, R.E.P., 1980. The crystallization history of a granitic magma, as revealed by trace element abundances. Journal of Geology 88, 211–224.

Miller, C.F., Meschter McDowell, S., Mapes, R.W., 2003. Hot and cold granites? Implications of zircon saturation temperatures and preservations of inheritance. Geology 31, 529–531.

Nilson, R.H., McBirney, A.R., Baker, B.H., 1985. Liquid fractionation: Part II. Fluid dynamics and quantitative implications for magmatic systems. In: Baker, B.H., McBirney, A.R. (Eds.), Processes in Magma Chambers. Journal of Volcanology and Geothermal Research, vol. 24, pp. 25–54.

Pearce, J.A., Harris, N.B.W., Tindle, A.G., 1984. Trace element discrimination diagrams for the tectonic interpretation of granitic rocks. Journal of Petrology 25, 956–983.

Phillips, W.J., 1973. Mechanical effects of retrograde boiling and its probable importance in the formation of some porphyry ore deposits. Transactions - Institution of Mining and Metallurgy 82, B, 90–98.

Raimbault, L., Cuney, M., Azencott, C., Duthou, J.L., Joron, J.L., 1995. Geochemical evidence for a multistage magmatic genesis of Ta–Sn–Li mineralization in the granite of Beavoir, French Massif Central. Economic Geology 90, 548–576.

Rämö, O.T., 1991. Petrogenesis of the Proterozoic rapakivi granites and related basic rocks of southeastern Fennoscandia: Nd and Pb isotopic and general geochemical constraints. Bulletin Geological Survey of Finland, 355, 161 pp.

Rieder, M., Haapala, I., Povondra, P., 1996. Mineralogy of dark mica from the Wiborg rapakivi batholith, southeastern Finland. European Journal of Mineralogy 8, 593–605.

Rosenbusch, H., 1896. Mikroskopische Physiographie der massigen Gesteine, 3rd Edition. E. Schweizerbart'sche Verlagshandlung (E. Koch), Stuttgart, 1360 pp.

Rundquist, D.V., 1980. Zoning of metallization associated with acid magmatism. In: Evans, A.M. (Ed.), Metallization Associated with Acid Magmatism, vol. 6. John Wiley & Sons, Chichester, New York, pp. 279–289.

Salonsaari, P., 1995. Hybridization in the bimodal Jaala-Iitti complex and its petrogenetic relation to rapakivi granites and associated mafic rocks of southern Finland. Bulletin of the Geological Society of Finland 67 (Part 1b), 104 pp.

Sawka, W.N., Chappel, B.W., Kistler, R.W., 1990. Granitoid compositional zoning by side-wall boundary layer differentiation: evidence from the Palisade Crest Intrusive suite, central Sierra Nevada, California. Journal of Petrology 31, 519–553.

Schmitz, C., Burt, D.M., 1990. The Black Pearl mine, Arizona: Wolframite veins and stockscheider pegmatite related to an albitic stock. Special Paper - Geological Society of America 246, 221–232.

Shannon, J.R., Walker, B.M., Carten, R.B., Geraghty, E.P., 1982. Unidirectional solidification textures and their significance in determining relative ages of intrusions at the Henderson Mine, Colorado. Geology 10, 293–297.

Shinohara, H., Hedenquist, J.W., 1997. Constrains on magma degassing beneth the far Southwest porphyry Cu–Au deposit, Philippines. Journal of Petrology 38, 1741–1752.

Štemprok, M., 1971. Petrochemical features of tin-bearing granites in the Krushne hory Mts., Czechoslovakia. Society of Mining Geologists of Japan Special Issue 2, 112–118.

Taylor, R.P., 1992. Petrological and geochemical characteristcs of the Pleasant Ridge zinnwaldite–topaz granite, southern New Brunswick, and comparison with other topaz-bearing felsic rocks. Canadian Mineralogist 30, 895–1921.

Vaasjoki, M., 1977. Rapakivi granites and other postorogenic rocksin Finland: their age and lead isotopic composition of certain associated galena mineralizations. Bulletin Geological Survey of Finland 294, 64 pp.

Viita, H., 1988. Kymin stokki, myöhäinen intruusiofaasi Viipurin rapakivimassiivissa. MSc thesis, Department of Geology, University of Helsinki, Finland (in Finnish).

Watson, E.B., Harrison, T.M., 1983. Zircon saturation revisted: temperature and composition effects in a variety of crustal magma types. Earth and Planetary Science Letters 64, 295–304.

Webster, J.D., Holloway, J.R., 1990. Partition of F and Cl between magmatic hydrothermal fluids and highly evolved magmas. In: Stein, H.J., Hannah, J.L. (Eds.), Ore-bearing Granitic Systems, Special Paper - Geological Society of America, vol. 246, pp. 21–33.

Whalen, J.B., Currie, K.C., Chappell, B.W., 1987. A-type granites: geochemical characteristics, discrimination and petrogenesis. Contributions to Mineralogy and Petrology 95, 407–419.

Whitney, J.A., 1988. The origin of granite: the role and source of water in the evolution of granitic magmas. Geological Society of America Bulletin 100, 1886–1897.

Worster, M.G., Huppert, H.E., Sparks, R.S.J., 1990. Convection and crystallization in magma cooled from above. Earth and Planetary Science Letters 101, 78–89.

Available online at www.sciencedirect.com

SCIENCE @ DIRECT°

Lithos 80 (2005) 363–386

www.elsevier.com/locate/lithos

Sn-polymetallic greisen-type deposits associated with late-stage rapakivi granites, Brazil: fluid inclusion and stable isotope characteristics

Jorge S. Bettencourt[a],*, Washington B. Leite Jr.[b], Claudio L. Goraieb[c], Irena Sparrenberger[a], Rosa M.S. Bello[a], Bruno L. Payolla[d]

[a]Instituto de Geociências, Universidade de São Paulo, São Paulo, SP, Brazil
[b]Instituto de Geociências e Ciências Exatas, Universidade Estadual Paulista, Rio Claro, São Paulo, SP, Brazil
[c]Instituto de Pesquisas Tecnológicas, São Paulo, SP, Brazil
[d]Eletronorte S/A, Brasília, Distrito Federal, DF, Brazil

Received 21 May 2003; accepted 9 September 2004
Available online 21 November 2004

Abstract

Tin-polymetallic greisen-type deposits in the Itu Rapakivi Province and Rondônia Tin Province, Brazil are associated with late-stage rapakivi fluorine-rich peraluminous alkali-feldspar granites. These granites contain topaz and/or muscovite or zinnwaldite and have geochemical characteristics comparable to the low-P sub-type topaz-bearing granites. Stockworks and veins are common in Oriente Novo (Rondônia Tin Province) and Correas (Itu Rapakivi Province) deposits, but in the Santa Bárbara deposit (Rondônia Tin Province) a preserved cupola with associated bed-like greisen is predominant. The contrasting mineralization styles reflect different depths of formation, spatial relationship to tin granites, and different wall rock/fluid proportions. The deposits contain a similar rare-metal suite that includes Sn ($\pm$W, $\pm$Ta, $\pm$Nb), and base-metal suite (Zn–Cu–Pb) is present only in Correas deposit. The early fluid inclusions of the Correas and Oriente Novo deposits are (1) low to moderate-salinity (0–19 wt.% NaCl eq.) CO_2-bearing aqueous fluids homogenizing at 245–450 °C, and (2) aqueous solutions with low CO_2, low to moderate salinity (0–14 wt.% NaCl eq.), which homogenize between 100 and 340 °C. In the Santa Bárbara deposit, the early inclusions are represented by (1) low-salinity (5–12 wt.% NaCl eq.) aqueous fluids with variable CO_2 contents, homogenizing at 340 to 390 °C, and (2) low-salinity (0–3 wt.% NaCl eq.) aqueous fluid inclusions, which homogenize at 320–380 °C. Cassiterite, wolframite, columbite–tantalite, scheelite, and sulfide assemblages accompany these fluids. The late fluid in the Oriente Novo and Correas deposit was a low-salinity (0–6 wt.% NaCl eq.) CO_2-free aqueous solution, which homogenizes at (100–260 °C) and characterizes the sulfide–fluorite–sericite association in the Correas deposit. The late fluid in the Santa Bárbara deposit has lower salinity (0–3 wt.% NaCl eq.) and characterizes the late-barren-quartz, muscovite and kaolinite veins. Oxygen isotope thermometry coupled with fluid inclusion data suggest hydrothermal activity at 240–450 °C, and 1.0–2.6 kbar fluid pressure at Correas and Oriente Novo. The hydrogen isotope composition of breccia-greisen, stockwork, and vein fluids

---

* Corresponding author. Tel.: +55 11 3091 4205; fax: +55 11 3091 4258.
  E-mail address: jsbetten@usp.br (J.S. Bettencourt).

($\delta^{18}O_{quartz}$ from 9.9‰ to 10.9‰, δD_{H_2O} from 4.13‰ to 6.95‰) is consistent with a fluid that was in equilibrium with granite at temperatures from 450 to 240 °C. In the Santa Bárbara deposit, the inferred temperatures for quartz-pods and bed-like greisens are much higher (570 and 500 °C, respectively), and that for the cassiterite-quartz-veins is 415 °C. The oxygen and hydrogen isotope composition of greisen and quartz-pods fluids ($\delta^{18}O_{qtz-H_2O}$=5.5–6.1‰) indicate that the fluid equilibrated with the albite granite, consistent with a magmatic origin. The values for mica ($\delta^{18}O_{mica-H_2O}$=3.3–9.8‰) suggest mixing with meteoric water. Late muscovite veins ($\delta^{18}O_{qtz-H_2O}$=−6.4‰) and late quartz ($\delta^{18}O_{mica-H_2O}$=−3.8‰) indicate involvement of a meteoric fluid. Overall, the stable isotope and fluid inclusion data imply three fluid types: (1) an early orthomagmatic fluid, which equilibrated with granite; (2) a mixed orthomagmatic-meteoric fluid; and (3) a late hydrothermal meteoric fluid. The first two were responsible for cassiterite, wolframite, and minor columbite–tantalite precipitation. Change in the redox conditions related to mixing of magmatic and meteoric fluids favored important sulfide mineralization in the Correas deposit.

Keywords: Rapakivi granite; Greisen deposits; Fluid inclusions; Stable isotopes; Brazil

1. Introduction

Proterozoic rapakivi granites in Brazil range in age from 1.88 to 0.59 Ga and are found in an anorogenic tectonic setting in the Amazonian craton (1.88–0.97 Ga) and the Tocantins Province (1.77–1.55 Ga), and in a post-orogenic tectonic setting in the Itu Rapakivi Province, which is a part of the Central Mantiqueira Province (800–540 Ma) (Almeida et al., 1981; Vlach et al., 1990; Wernick, 1992; Wernick et al., 1991; Ulbrich et al., 1991; Pimentel et al., 1991; Dall'Agnol et al., 1999; Bettencourt et al., 1995, 1999). Important Sn-polymetallic deposits are associated with some of them. The Rondônia and Pitinga provinces in the Amazonian craton produce tin with a total output of ~500,000 t Sn in concentrate. The current output comes mainly from the Pitinga Mining District (~230,000 t Sn by 2002), and the Bom Futuro Mine in the Rondônia Tin Province (170,000 t Sn by 2002).

Despite the progress achieved in the last few years, the magmatic–metallogenetic history of the rapakivi granites in Brazil is far from being fully understood and fundamental questions still require answers, before their origin, evolution, and relation with associated mineral deposits can be satisfactorily deciphered. The purpose of this work is to combine fluid inclusion and stable isotope data to estimate the temperature, pressure and source of fluids, and to present a comparative study of the Sn-polymetallic greisen-type deposits. These include the Correas deposit in the Itu Rapakivi Province, and the Santa Bárbara and Oriente Novo deposits in the Rondônia Tin Province.

2. Geologic setting

2.1. Itu Rapakivi Province

The Itu Rapakivi Province in the State of São Paulo, SE Brazil comprises a dozen post-collisional granite batholiths and stocks, which cross-cut the low to medium-grade metamorphic rocks of the Ribeira Fold Belt. The granites have U–Pb, Rb–Sr and K–Ar ages in the range of 0.6–0.5 Ga, and thus overlap and follow the late Brasiliano calc-alkaline magmatism (Janasi and Ulbrich, 1991; Campanha and Sadowski, 1999; Gimenez Filho et al., 2000; Prazeres Filho, 2000). Their composition varies from subalkaline to almost alkaline granitoids, and they are nearly similar to the Finnish rapakivi granites (Wernick et al., 1997). Some of them are locally associated with mineral deposits in the Itu Complex (W), Correas massif (Sn, W, Zn, Cu, Pb), and São Francisco massif (Sn, W, fluorite).

2.2. Rondônia Tin Province

The Rondônia Tin Province comprises mainly Paleo- to Mesoproterozoic medium- to high-grade metamorphic rocks, locally overlain by low-grade to underformed supracrustal sequences (Scandolara et al., 1999; Payolla et al., 2002). Six discrete episodes of magmatism have affected these rocks, and are represented by at least seven rapakivi suites with ages between 1.60 and 0.97 Ga (Bettencourt et al., 1999). Important Sn-polymetallic deposits are associated with at least the three youngest suites: the São

Lourenço-Caripunas Intrusive Suite (ca. 1.31 Ga), the Santa Clara Intrusive Suite (1.08–1.07 Ga), and the Younger Granites of Rondônia (1.00–0.97 Ga). The Santa Clara Intrusive Suite and the Younger Granites of Rondônia are found in the east-central part of the Rondônia Tin Province and show petrographic, geochemical and metallogenetic similarities (Leite et al., 2000, 2001). These suites are composed of several early- and late-stage intrusions. The late-stage intrusions are volumetrically minor, and comprise two rock groups: (1) a metaluminous to peralkaline group composed mainly of hornblende (±pyroxene, ±biotite) alkali feldspar syenite and microsyenite, trachyandesite, and trachyte, as well as biotite (±sodic amphibole) alkali feldspar granite and rhyolite; and (2) a peraluminous group formed by biotite-alkali feldspar granite, alaskite, Li-mica (±topaz) alkali-feldspar granite and Li-mica (±topaz) rhyolite porphyry (ongonite). The Sn-polymetallic greisen-type deposits are closely associated with these peraluminous rocks in both suites, such as in the Oriente Novo massif of the Santa Clara Intrusive Suite, and in the Santa Bárbara massif of the Younger Granites of Rondônia.

3. Methods

Microthermometry and Raman spectroscopy were carried out at the Instituto de Geociências, Universidade de São Paulo, Instituto de Geociências e Ciências Exatas, Universidade Estadual Paulista, and Instituto de Ciências Exatas, Universidade Federal de Minas Gerais. Microthermometry was performed using a CHAIXMECA MTM-85 stage and a Linkam THMS 600 heating–freezing stage fitted with a video system. Heating rates of 2 °C/min were used for Th determinations. The CHAIXMECA and Linkam THMS 600 stages were calibrated to the Merck Signotherm standard for low temperatures and Merck MSP standard for high temperatures; SYN FLINC synthetic inclusions (comercially available from FLUIDS, Denver, CO, USA) were also used. Precision estimates are ±0.2 °C for temperatures between −60 and 25 °C, and ±5 °C between 100 and 400 °C. The bulk composition of the fluid inclusions were calculated considering eutectic temperatures (Davis et al., 1990), ice, chlatrate and CO_2

melting temperatures and Raman spectrometry (Bodnar, 1993; Potter et al., 1978; Collins, 1979; Diamond, 1994; Heyen et al., 1982). The FLINCOR software (Brown, 1989) was used for isochore calculations (Correas and Oriente Novo deposits). This software is based on the Brown and Lamb (1989) and Bowers and Helgeson (1983) equations for the systems $H_2O+NaCl$ and H_2O+CO_2+NaCl, respectively. Raman spectrometry was performed on a micro-Raman type DILOR XY.

Oxygen analyses of granite whole-rock and mineral separates from granite, quartz-vein, stock work and greisen were carried out using the conventional fluorination technique (Clayton and Mayeda, 1963; Mcaulay et al., 2000) at Krueger Geochron Laboratories, Cambridge, OH, USA, and at the Scottish Universities Environmental Research Center (SUERC), East Kilbride, Scotland. Hydrogen isotope analyses were carried out by thermal degassing of minerals at 1400 °C (Fallick et al., 1993). Hydrogen was produced from water vapor using a chromium (Donnelly et al., 2001) or the conventional uranium furnace. Isotopic data are reported in the standard δ notation as per mil (‰) deviation from the Vienna SMOW standard for oxygen and hydrogen. The analytical precision is about ±0.2‰ for oxygen and ≤0.5‰ for hydrogen.

4. Correas deposit (Itu Rapakivi Province)

4.1. Geologic outline

The Correas massif is a post-tectonic, ellipsoid shaped, NE–SW trending stock (5 km^2), with ages of 603±7 Ma (Rb–Sr), and 619±11 Ma (U–Pb, monazite) (Goraieb, 2001). It comprises five units: biotite granite, biotite–muscovite granite, microgranite, topaz–muscovite–albite granite, and layered marginal pegmatite. The early-stage intrusions are metaluminous monzogranite to syenogranite. The late phases are evolved topaz–muscovite–albite syenogranites and alkali feldspar granites.

The tin-bearing topaz–muscovite–albite granite is porphyritic, containing muscovite, phengite and drop-like quartz phenocrysts. Granophyric and snow-ball textures are common. Laths of albite, non-perthitic microcline, quartz and muscovite are the primary minerals. Typical accessory minerals are topaz, cassiterite, fluorite, zircon, monazite, xenotime, iron oxides, sphalerite and pyrite (Goraieb, 2001). This tin granite facies is slightly peraluminous, has high SiO_2, Al_2O_3, Na_2O, F and Rb contents, and is poor in MgO and TiO_2. The REE pattern is characterized by slight enrichment of HREE and a marked negative Eu anomaly. The mineralogy, geochemical and isotope data point to a magmatic origin, and they are comparable to the low-P, topaz granite sub-type (cf. Taylor, 1992; Taylor and Fallick, 1997).

4.2. Mineralization

The Correas primary (Sn, W, Zn, Cu, Pb) deposit is dominantly a pipe-shaped stockwork and/or breccia-type deposit, hosted by schists, calc-silicate rocks, and granitic/granodioritic gneisses. Ore reserves are 5000 t Sn-in concentrate at 0.20% Sn and 1400 t W at a depth of 60 m. The tin deposit is represented by six ellipsoid or lens-shaped pipe-like ore bodies, which plunge steeply above the granite underlying cupola (Fig. 1). They are up to 50 m thick, at least 60 m deep, and trend WNW–ESE. The maximum width of the greisen zones is 150 m. These are genetically related to the topaz–muscovite–albite granite, and probably formed during hydraulic fracturing. This led to the following styles of mineralization: (a) cassiterite and wolframite-bearing pipe-like quartz-stockwork veins and veinlets

(ore bodies 1 and 4; Fig. 1); (b) pipe-shaped breccia (ore-bodies 2 and 3; Fig. 1) represented by tabular steeply dipping greisen columns developed within gneissic and calc-silicatic rocks. It includes cassiterite and minor wolframite mica–topaz–quartz–greisen, mica greisen, topaz–fluorite–mica greisen, milky quartz pods and veins. Breccia infill consists of sulfides (pyrite, chalcopyrite and sphalerite, followed by minor chalcocite, pyrrhotite as well as trace amounts of sellaite, stannite, and bismuthinite), and fluorite associated with mica. The greisens also include late fluorite and sericite, and are associated with cassiterite-rich topaz veins and veinlets (Fig. 2).

4.3. Fluid inclusions

Fluid inclusions were studied mainly in quartz crystals from the quartz stockwork and breccia greisen, and in topaz and quartz from mica–topaz–quartz greisen. All samples were taken from drill-cores of the ore bodies 1 and 4, and from outcrops of the marginal pegmatite. Microthermometric studies were performed on unmodified primary fluid inclusions (10–20 μm) with usually rounded, seldom irregular, negative crystal morphology. The primary fluid inclusions are randomly distributed in the inner part or along the growth-planes of the topaz and quartz. Three main types of fluid inclusions were distinguished based on morphology, composition, volume percentage of vapor phase, mode of homog-

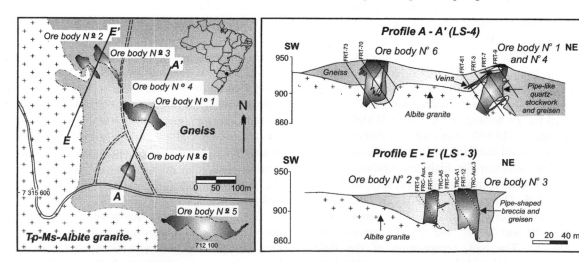

Fig. 1. Schematic geological map and styles of mineralization of the Correas (Sn, W, Zn, Cu, Pb) deposit as shown by the cross-sections through the LS-03 and LS-04 ore bodies. Abbreviations: Tp—topaz; Ms—muscovite. Location relative to Brazil is shown in left panel.

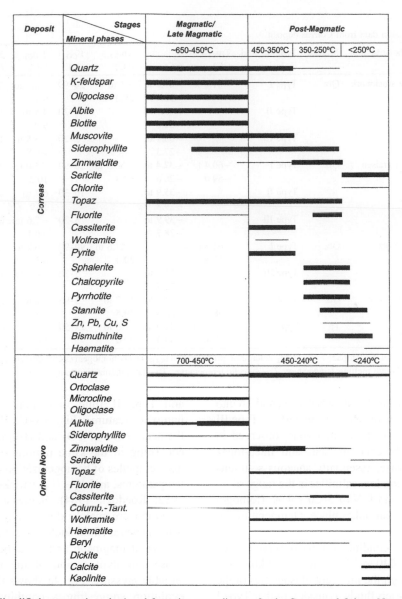

Fig. 2. Simplified paragenetic and mineral formation stage diagram for the Correas and Oriente Novo deposits.

enization (liquid/vapor/critical), and the presence of accidental minerals. Different fluid inclusions types are found together in the same samples. They have highly variable salinities, compositions, bulk densities of the carbonic phase, and show distinct homogenization characteristics. Microthermometric data are summarized in Table 1.

Type I fluid inclusions ($CO_2 \pm CH_4$–H_2O–$NaCl$–KCl–$FeCl_2 \pm CaCl_2$) were observed in quartz from quartz-stockwork, breccia-greisen, and the mica-

topaz-quartz greisen. The salinities are in the range of 0–19 wt.% NaCl eq., and the predominant volumetric proportion of the CO_2 phase varies from 20 to 90 vol.%. Homogenization temperature (Th) from primary inclusions into the H_2O or CO_2 phases in quartz has a range of 245–450 °C (Table 1). Type II (H_2O–$NaCl$–KCl–$FeCl_2 \pm CaCl_2$) inclusions are observed in quartz-stockwork and breccia greisen. They are two-phase, essentially aqueous inclusions with little CO_2, and the vapor bubble occupies 10–30 vol.%. The salinities are

Table 1
Summary of fluid inclusion data from Correas deposit

Stage	Morpho-structural types	Mineral	Type of fluid inclusion	Tm_{CO2} (°C)	T_e (°C)	Th_{CO_2} (°C)	V_{CO_2} (%)	Tm_{CL} (°C)	Salinity (wt.% NaCl eq.)	Th (°C)
Post-magmatic	Quartz stockwork	Qtz	Type I	−61.6 to −57.7	−47.6 to −32.4	3.3 to 29.3	20–99	−2.6 to 9.1	2 to 18	250 to 400
			Type II		−43.7 to −31.6		10–30	4.5 to 9.0	1 to 10	170 to 250
			Type III		−38.1 to −27.3		5–20	−3.5 to −0.8	1 to 6	100 to 150
	Breccia greisen	Qtz	Type I	−60.4 to −59.0	−37.4 to −28.6	−50.5 to 22.4	20–99	−5.0 to 10.2	2 to 19	260 to 450
			Type II		−35.9 to −29.8		10–30	6.7 to 10.9	0 to 4	130 to 230
			Type III		−39.9 to −28.7		5–20	−7.6 to −0.1	0 to 11	110 to 240
	Mica-tp-qtz greisen	Qtz	Type I	−64.6 to −57.8	–	−27.6 to 30.4	90–99	−6.2 to 6.3	–	–
			Type III		−34.7 to −29.7		5–20	−4.0 to −2.6	1 to 6	180 to 260
	Stockscheider	Qtz	Type I	−57.2 to −58.5	−33.3 to −38.4	15.7 to 29.5	5–95	3.1 to 7.5	5 to 12	245 to 420
		Tp	Type I	−57.0 to −58.1	−32.9 to −39.0	27.6 to 29.0	15–50	3.8 to 5.6	8 to 11	270 to 330

Abbreviations: Qtz—quartz; Tp—topaz; V_{CO_2}—CO₂ volumetric proportion; Th_{CO_2}—CO₂ homogenization temperature; Tm_{CO_2}—CO₂ melting temperature; Tm_{CL}—chlatrate melting temperature; T_e—eutectic temperature; Th—homogenization temperature.

lower, varying from 0 to 10 wt.% NaCl eq., and Th into the liquid phase lies between 130 and 250 °C. Type III (H₂O–NaCl±KCl) late fluid inclusions, observed in quartz-stockwork, breccia, greisens and mica-topaz greisen, are two-phase, essentially aqueous solutions without CO₂, showing low salinities in the range of 0–11 wt.% NaCl eq., and L/V of 5–20 vol.%. Homogenization of the inclusions takes place at temperatures of 100–260 °C.

The trapping P–T conditions of the observed fluids (Table 2) vary from 440 to 300 °C at 2.6–1.3 kbar, and were obtained through the intercept of isochores of the CO₂-rich and H₂O-rich fluids, considering that Type I

and Type II fluid inclusions exhibit some fluid unmixing features (Fig. 3A–C). This is indicated by the occurrence of CO₂ and H₂O-rich fluid inclusions coexisting in the same regions of the samples with distinct modes of homogenization at the same range of temperatures, and V_{CO_2}/V_{FI} variability (Ramboz et al., 1982; Roedder, 1984). Successive phases of fluid evolution are mainly depicted from the trapping of heterogeneous fluid inclusions as a result of partial mixing of magmatic and meteoric fluids in the two-phase subsolvus region of the fluid system under variable pressure. Fig. 4a shows three trends of fluid evolution suggesting a combination of processes

Table 2
Equilibrium temperatures and fluid pressures during hydrothermal activity in the Correas and Oriente Novo deposits

Deposit	Styles of mineralization	Rock	Sample	T_t (°C)	P_t (kbar)
Correas	Qtz-stockwork	Qtz vein	FRT-3a	340–440	2.0–2.6
	Breccia-greisen	Greisen	FRT-1	310–360	1.3–2.0
	Mica-tp-qtz greisen	Greisen	FRT-15	300–340	1.8–2.0
Oriente Novo	Stockwork	Greisen	W-23A	250–300	1.2–1.6
		Qtz vein	W-27A	270–370	1.4–2.4
	Endogreisen	Greisen	AW-28	290–360	1.6–2.1
	Sub-parallel vein	Qtz vein	AW-24G	240–290	1.0–1.5

Abbreviations: Qtz—quartz; tp—topaz; T_t—trapping temperature; P_t—trapping pressure.

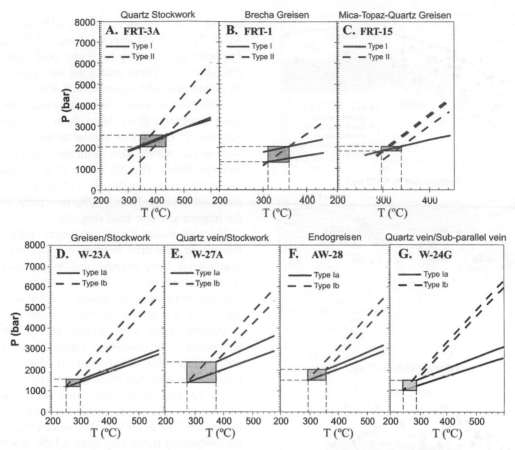

Fig. 3. *P–T* diagrams with representative isochores for the Type I and Type II fluid inclusions from the Correas deposit (A, B, C), and for Type Ia and Ib fluid inclusions from the Oriente Novo deposit (D, E, F, G). The studied rectangular boxes give the *P–T* conditions of mineralization at each deposit.

involving mixture/immiscibility of three types of fluids, marked by a progressive decrease in trapping temperatures under variable pressure conditions, from 400 to 100 °C. Trend I shows the immiscibility trend of an originally CO_2-rich aqueous carbonic fluid at $T\sim400$ °C and $X_{CO_2}=0.57$, which undergoes gas loss caused by hydraulic fracturing and associated pressure fluctuation. This is responsible for the reduction in temperature and important decline of the apparent salinity values related to solute concentrations in the solution, as well as the precipitation of CO_2 dissolved in the aqueous phase (cf. Hedenquist and Henley, 1985; Higgins, 1985a). Trend II represents the dilution of the original fluid by essentially lower T (100–125 °C) and lower salinity (3–6 wt.% NaCl eq.) aqueous solutions. The data points within the combined envelopes of the immiscibility and dilution trends may

indicate a combination of the two processes (Hedenquist and Henley, 1985). The dispersion Trend III reflects the isothermal fluid mixing processes (Fig. 4a).

4.4. Stable isotopes

Oxygen and hydrogen isotope analyses were made on quartz, K-feldspar and whole rock in granites, and quartz, topaz, muscovite, zinnwaldite (sense Rieder et al., 1998), cassiterite, and wolframite in quartz-stockwork, breccia greisen and topazite (Table 3). The $\delta^{18}O$ for quartz from the breccia greisen and stockwork are remarkably uniform, varying from 10.3‰ to 10.9‰, and indicate a magmatic origin for the fluids (Taylor, 1979; Zhang et al., 1982). They are higher than the values for the topaz–muscovite–albite granite, which range from 6.8‰ to 7.3‰ (Fig. 5). Cassiterite and

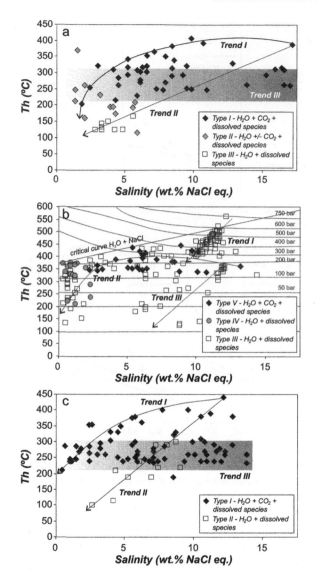

Fig. 4. Th (homogenization temperature) vs. salinity (wt.% NaCl eq.) plot for: (a) inclusions in topaz and quartz from Correas deposit; (b) quartz-pods, topaz–mica–quartz bed-like greisen, greisen–stockwork and cassiterite–quartz veins at Santa Bárbara deposit (with isobaric curves from Bodnar et al., 1985); and (c) greisen-pods, greisen veins and quartz veins at Oriente Novo deposit.

wolframite show a narrow range of $\delta^{18}O$, 1.9–3.0‰, much lower than those for coexisting quartz. Zinnwaldite from quartz-stockwork has $\delta^{18}O$ in the range of 4.7–5.2‰ and δD of −67‰ to −86‰, indicating a possible interaction with meteoric water. Muscovite

and topaz from topaz-veins have $\delta^{18}O$ in the range of 4.7–5.2‰, and δD varying from −53‰ to −50‰, respectively, denoting isotopic equilibrium during crystallization. These values are very similar to those of quartz and albite-granite, and are consistent with the formation of quartz, topaz and muscovite from the same magmatic fluid. Mica from mica-greisen has very low $\delta^{18}O$ of 4.9‰, similar to those of the quartz-stockwork, which also indicates interaction with meteoric fluids. The δD values of topaz and muscovite range from −44‰ to −65‰, while those of zinnwaldite range from −67‰ to −86‰, and fall in the magmatic water field (Fig. 5).

Selected mineral pairs of quartz, feldspar, wolframite and cassiterite from granite and hydrothermal quartz-veins were chosen, and the temperatures of crystallization were calculated using the isotopic fractionation curves (for quartz–feldspar after Fifarek, 1985; and for quartz–cassiterite–water and quartz–wolframite–water after Zhang et al., 1994). The temperatures for coexisting mineral pairs were also utilized for the calculation of the $\delta^{18}O_{H_2O}$ in equilibrium with the mineral associations. Temperatures for cassiterite-quartz equilibrium range from 441 to 459 °C, while those for wolframite-quartz range from 336 to 390 °C. The calculated $\delta^{18}O_{H_2O}$ of the tin-bearing fluids vary from 4.13‰ to 6.95‰, and are consistent with magmatic derived fluids (Taylor, 1979) and the magmatic-hydrothermal fluids of the Sn–W Xihuanshan-China deposit (Zhang et al., 1982).

5. Santa Bárbara deposit (Rondônia tin province)

5.1. Geologic outline

The Santa Bárbara massif is a semicircular stock with a diameter of 7 km and includes three subsolvus granite units (Fig. 6): (1) an early metaluminous porphyritic syenogranite; (2) a peraluminous porphyritic syenogranite; and (3) a central porphyritic albite-microcline granite (tin-granite). These have been dated by the conventional U–Pb monazite method at 993±5 and 989±13 Ma, and by the SHRIMP U–Pb zircon method at 978±13 Ma, respectively (Sparrenberger, 2003). The tin-granite encompasses two peraluminous granite facies, a pink medium-grained porphyritic albite-microcline granite and a pink to white fine-

Table 3
Summary of oxygen and hydrogen-isotope composition of granite types and mineral phases from the Correas, Santa Bárbara and Oriente Novo deposits

Deposit	Stage	Rock or Facies	Sample	δD SMOW (‰)				$\delta^{18}O$ SMOW (‰)						
				WR	Kfs	Mica	Kln	WR	Qtz	Tp	Kfs	Mica	Cas	Wolf Tp
Correas	Magmatic	Bt granite I	A-20	−68				7.8	10.3		7.0			
		Bt granite II	A-21	*−82/−78				9.7	9.6		9.3			
		Bt-ms granite	A-22	−91			−89	4.9	*−9.9/ −9.6		3.2	5.1		
		Ab granite I	A-23	−65				7.5	7.3					
		Ab granite II	A-24	−87				6.4	6.8					
	Post-Magmatic	Mica-tp-qtz greisen	A-11								10.5			
			A-12								9.9			
		Breccia-greisen	A-13								*10.9/ 10.6			
			A-14								10.3			
			A-15								10.5			1.9
			A-16								*10.6/ 10.3		2.8	2.8
		Qtz stockwork	A-17				−86				10.6	2.7		
			A-18				−67				10.5	*2.7/ 2.4/ 4.7		
			A-19								10.7	3.0		
		Topazitic veins	A-07				−65			−44		7.3	3.4	7.3
			A-08				−53			−50		8.4	2.7	7.5
		Mica-greisen (endogreisen)	A-10				*−53/ −51					4.9		
Santa Bárbara	Magmatic	Syenogranite	AM-52B			−98	−118	7.3	9.4		6.9	1.5		
		Syenogranite	AM-53			−74	−110		10.3		9.0	5.8		
		Ab-mc granite	AM-108				−92		10.4			6.7		
		Ab-mc granite	AM-145			−123	−93	8.7	9.3		7.8	4.8		
		White ab granite	AM-134				−125		11.3			7.9		
		Pegmatite	AM-213			−68	−106		10.0		4.5	7.3		
	Late-Magmatic	Quartz pod	AM-168				−124		9.8			7.9		7.5
		Salmon ab granite	AM-14a			−67	−127		9.8		*1.7/1.5	5.2		
		Bedded greisen	AM-35				−99		10.0			*1.4/ 2.0	2.9	
	Post-Magmatic	Greisen vein	AM-130				−124		11.8			4.7		
		Greisen vein	AM-174c				−133		11.7			5.9		
		Qtz-cas vein	AM-131						11.7				3.5	
		Greisen vein	AM-297				−102		9.7			*2.1/ 1.8		
		Barren qtz vein	AM-159a						*0.7/0.6					
		Mica greisen vein	AM-74c				−102					*−0.8/ −0.3		
		Ms vein	AM-307									*−6.2/ −6.1	3.3	
		Kln vein	AM-298	−69				15.7						

(continued on next page)

Table 3 (*continued*)

Deposit	Stage	Rock or Facies	Sample	δD SMOW (‰)				δ¹⁸O SMOW (‰)							
				WR	Kfs	Mica	Kln	WR Qtz	Tp	Kfs	Mica	Cas	Wolf Tp
Oriente Novo	Post-Magmatic	Li-mica alkali-feldspar granite	3524A			−92					6.4		
		Stockwork	W-23A			−101					6.3		
		Endogreisen	AW-8b			−91					6.6		
		Sub-parallel sheeted-veins	W-12b			−112					6.4		

Abbreviations: *—duplicate analysis. Bt—biotite; Cas—cassiterite; Kln—kaolinite; Kfs—K-feldspar; Ms—muscovite; Qtz—quartz; Tp—topaz; Wolf—wolframite; WR—whole rock.

grained equigranular to porphyritic albite-microcline granite; the latter occupies the apical part of the cupola. The main minerals are quartz, albite, microcline, and zinnwaldite; accessory minerals comprise fluorite, topaz, zircon, thorite, columbite, cerianite, and cassiterite. Pegmatoid pods (up to 2 m in diameter) and snowball, granophyric, and unidirectional solidification textures are common (Fig. 4). The tin granites are of magmatic origin, and they were affected by subsolidus reactions and hydrothermal alteration represented by greisenization, albitization, silicification,

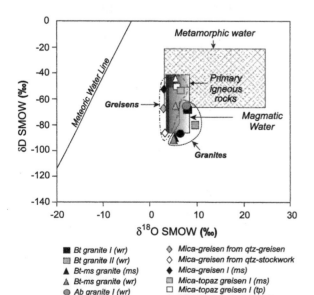

Fig. 5. δD vs. δ¹⁸O diagram showing the calculated δ¹⁸O and δD values for stockwork/veins and greisens from the Correas deposit. Abbreviations: Bt—biotite; Ab—albite; qtz—quartz; tp—topaz; ms—muscovite; wr—whole rock.

muscovitization, and argillization during late magmatic-hydrothermal and post-magmatic stages (Sparrenberger, 2003).

These granites are characterized by high SiO_2, K_2O, Na_2O, $Fe_2O_3/(Fe_2O_3+MgO)$, Sn, Rb, Ce, Nb, Ga, Y, F, Li, U and Th contents, and low concentrations of CaO, MgO, TiO_2, Al_2O_3, Ba, Sr and Zr. They also show high total REE contents, are slightly enriched in LREE, and exhibit a strong negative Eu anomaly. The white albite-microcline granite shows more evolved patterns characterized by higher Ga, Li, Ta, Y, and total REE, than the other tin-specialized granites (Sparrenberger, 2003).

5.2. Mineralization

The Santa Bárbara deposit covers an area of ~500 m by 150 m. The tin deposit encompasses two styles of mineralization (Fig. 6): (1) bed-like cassiterite-bearing topaz-zinnwaldite-quartz greisen bodies, up to 40 m thick, and salmon colored albitized granites (Taboquinha greisen); and (2) a vein-veinlet/stockwork, encompassing brittle fracture zones containing topaz-zinnwaldite-quartz greisen veins with cassiterite-wolframite, quartz–cassiterite veins, muscovite veins, and late kaolinite stockwork/veinlets (Sparrenberger, 2003). The mineral paragenesis is shown in Fig. 7.

5.3. Fluid inclusions

The fluid inclusions study was performed on topaz and quartz from the white albite-microcline granité, pegmatite, quartz pod, bedded greisen, salmon albitized granite, greisen stockwork, and quartz veins (Table 4). Microthermometric studies

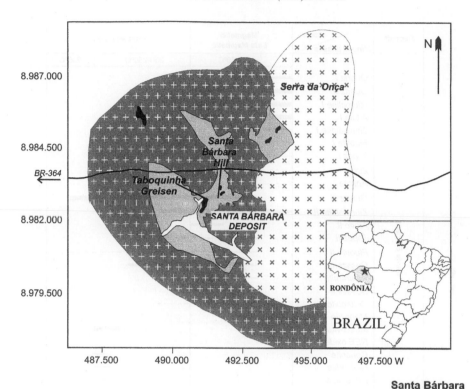

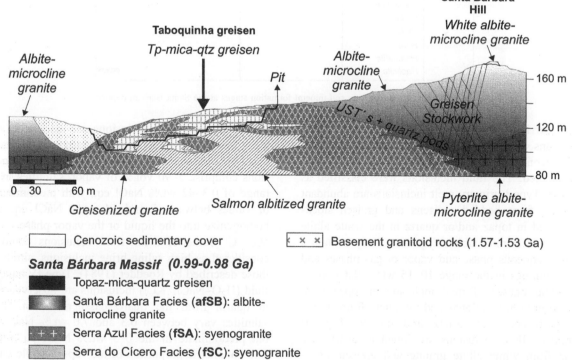

Fig. 6. Geological map of the Santa Bárbara massif showing the granitic facies, limits of the Santa Bárbara tin deposit (modified after Frank, 1990; Payolla et al., 2002), and styles of mineralization. Abbreviations: tp—topaz; qtz—quartz; USTs—unidirectional solidification textures. Inset in upper panel shows location relative to Brazil.

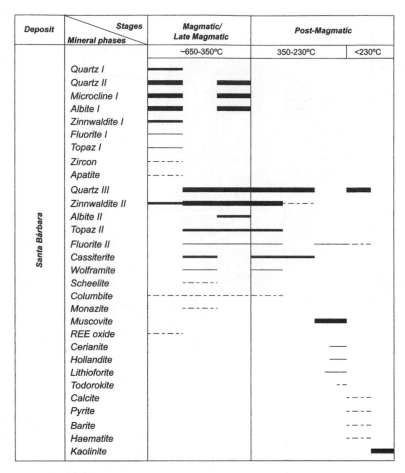

Fig. 7. Simplified paragenetic and mineral formation stages at the Santa Bárbara deposit.

were carried out on unmodified primary fluid inclusions with irregular or negative crystal morphology. Secondary inclusions located along trails were also studied. The fluid inclusions were classified as follows: Type I primary melt inclusions are abundant in topaz from bedded greisens and greisen stockwork, and in topaz and/or quartz in the white albite granite and salmon albitized granite. They contain a Na-rich aqueous phase and vapor or gas phases and have salinities in the range 10–15 wt.% NaCl eq. A few solid phases of melt inclusion in topaz from bedded greisen were analyzed via micro-Raman, and are represented by topaz and quartz. Type II composite fluid inclusions are found in quartz and topaz from white albite granite and greisen stockwork and constitute accidentally trapped solid minerals together with minor proportions of vapor/gas or liquid. Type III (H_2O–NaCl±KCl±CaCl$_2$±

FeCl$_2$) aqueous fluid inclusions are associated with topaz and quartz in all analyzed samples, and range from monophase to multiphase. The primary inclusions comprise 5–90 vol.% of vapor, salinities in the range of 0.3–42 wt.% NaCl eq., with predominance of values between 5 and 14 wt.% NaCl eq., and homogenize into the liquid or the vapor phase at 89–506 °C. Some of these fluid inclusions could be considered as high saline brine inclusions, similar to those described by Roedder (1984). Type IV aqueous fluid (H_2O–NaCl±KCl) inclusions are observed only in quartz from the quartz–cassiterite vein. Their salinities vary between 0.3 and 7.0 wt.% NaCl eq., but predominate in the 0–3 wt.% NaCl eq. range. The volumetric proportion of the vapor bubble varies from 10 to 90 vol.%, and homogenization into the liquid or vapor phase occurs between 153 and 375 °C, but predominantly at 320–380 °C; Type V

Table 4
Summary of fluid inclusion data from Santa Bárbara deposit

Stage	Granites and morpho-structural types	Mineral	Type of fluid inclusion	Tm_{CO_2} (°C)	T_e (°C)	Th_{CO_2} (°C)	Salinity (wt.% NaCl eq.)	Th (°C)	Fluid behaviour T_t (°C)
Magmatic	White Albite-Microcline Granite	Qtz+Tp	Type I, Type III, Type V ($H_2O\pm CO_2\pm CH_4$)	−58.2 to −56.8	Qz (p1) −50 to −22, Qz (p2) −54 to −44, Qz (ps) −42.3, Qz (s) −40 to −21, Tp (s) −42 to −25 (Na±Ca±Fe±K)	10.9	Qz (p1) 5 to 12, Qz (p2) 14 to 42, Qz (ps) 34, Qz (s) 0 to 3, Tp (s) 3 to 19	(I)>600, Qz (p1) 123–452 (III, V), Qz (p2) 90–216 (III), Qz(p3) 330 (III), Qz (s) 164–380 (III), Tp (s) 356–553 (<390) (III,V)	Immiscibility $T_t(s)$=350–370
	Pegmatite	Qtz	Type III, Type V ($H_2O\pm CO_2\pm CH_4$)	–	(s) −40 to −28 (Na±Ca±Fe±K)	–	(s) 0 to 6	(s1) 327–370, (s2) 114–302 (>230) (III), 123–304 (240–250) (V)	Immiscibility in secondary inclusions $T_t(s)$=350–370
Magmatic–hydrothermal	Quartz–Pods	Qtz+Tp	Type III, Type V ($H_2O\pm CO_2\pm CH_4$)	−58.6 to −57.8	Qz (p) −49 to −24, Qz (s) −50 to −21, Tp (p, ps) −51 to −23, Tp (s) −47 to −25 (Na±Ca±Fe±K)	15.9 to 30.3	Qz (p) 4 to 21, Qz (s) 6 to 17, Tp (p, ps) 7 to 15, Tp (s) 10 to 23	Qz (p) 97–417, (100–140) (III), Qz (p) 206–524 (V), Tf 400–500, (450–490) (III), Qz (s) 100–410 (III, V), Tp (s) 446–484, (450–470) (III), Tp (ps) 375–573, (400–490) (V)	Immiscibility ($T_t(p)$=500–510); Fluid mixing
	Bedded–Greisen	Qtz+Tp	Type I, Type III, Type V ($H_2O\pm CO_2\pm CH_4$)	−57.5 to −56.7	Qz (p) −41 to −28, Qz (ps) −54 to −28, Qz (s1) −53 to −23, Qz (s2) −36, Tp (p) −43 to −24 (Na±Ca±Fe±K)	–	Qz (p) 3 to 9, Qz (s1) 0 to 22, Qz (s2) 0 to 2, Tp (p) 2 to 14	(I)>600, Qz (p) 355–379 (V), Tp (p) 211–353 (III), Tp (p) 337–424, (340–370) (V), Qz (ps) 127–424, (330–370) (III), Qz (s) 102–332	Immiscibility ($T_t(p)$=350–370); Fluid mixing
	Salmon Albitized Granite	Qtz	Type I, Type III ($H_2O\pm CO_2$–traces)	–	(p, ps) −47 to −31 (Na±Ca±Fe±K)	–	(I) 9 to 12, (p, ps) 1 to 8	(I) >600, (p, ps) 150–404	Immiscibility ($T_t(p)$=350–370); Fluid mixing (final stages)
Post-magmatic	Greisen–Vein (Stockwork)	Qtz+Tp	Type III ($H_2O\pm CO_2$–traces), Type V ($H_2O\pm CO_2\pm CH_4$)	−57.8 to −56.6	Qz (p) −52 to −11, Qz (s) −48 to −28, Tp (p) −48 to −32 (Na±Ca±Fe±K)	16.7 to 19.8	Qz (p) 2 to 18, Qz (s) 0 to 6, Tp (p) 1 to 22	Qz (p) 78–435 (>320) (III), Tp (p) 272–446 (III, V), Qz (s) 124–415 (III)	Immiscibility ($T_t(p)$=350–370); Fluid mixing (final stages)
	Cassiterite–Quartz–Vein	Qtz	Type III, Type IV, H_2O	–	(p) −31 to −26 (Na±K), (s) −48 to −28 (Na±Ca±Fe±K)	–	(s) 0 to 7, (s) 2 to 15	(s) 153–375, (320–380) (IV), (s) 110–177 (III)	Boiling ($T_t(p)$=350–370); Evolution to (T_t) salinity fluids
	Late Quartz–Vein (barren)	Qtz	Type III, H_2O	–	(p) −42 to −34 (Na±Ca±Fe±K), (s) −58 to −47 (Na±Ca±K)	–	(p) 1 to 6, (s) 15 to 19	(p) 145–415, (s) 94–147	Boiling involving meteoric fluids; Evolution to (T_t) salinity fluids

Abbreviations: Qtz—quartz; Tp—topaz; p—primary inclusions; s—secondary inclusions; ps—pseudosecondary inclusions; Tm_{CO_2}—CO_2 melting temperature; T_e—eutectic temperature; Th_{CO_2}—CO_2 homogenization temperature; Th—homogenization temperature; T_t—trapping temperature.

(H_2O–CO_2±CH_4–$NaCl$±KCl±$CaCl_2$±$FeCl_2$) primary aqueous-carbonic fluid inclusions are found in topaz from the white albite granite, quartz pod, bedded greisen and greisen stockwork. They are biphase (H_2Ol–CO_2 g), rarely triphase (H_2Ol–CO_2l–CO_2 g) or multiphase (H_2Ol±CO_2l–CO_2 g-solid phases). The proportion of the CO_2 phase varies between 25 and 90 vol.%. The salinities vary in the range of 2–25 wt.% NaCl eq., but predominantly between 5 and 12 wt.% NaCl eq. Th into the CO_2 or H_2O phases occurs between 128 and 553 °C, mainly in the 340–390 °C range. They also have varying concentrations of CH_4, and include very high X_{H_2O}, and high X_{H_2O} fluid inclusion sub-types (cf. Diamond, 1994, 2001). Secondary inclusions, related to meteoric fluids, are aqueous and have low salinity (0–3 wt.% NaCl eq.) and Th (94–380 °C). Proportion of the vapor phase is between 10 and 55 vol.% and the composition is similar to Type III inclusions.

The CO_2- and H_2O-rich inclusions have the immiscibility characteristics cited by Ramboz et al. (1982). Their trapping temperatures were obtained from several diagrams of the type homogenization temperature vs. the ratio of gas or vapor volume to total volume (Th×$V_{v/g}$/V_{FI}), because the low molar fraction of CO_2 precluded measurement of the homogenization temperature of the phase and thus of the isochore calculations. The white albite-microcline granite (Table 4) shows primary melt-inclusions, primary aqueous and aqueous-carbonic fluid inclusions in re-equilibrated quartz, and secondary aqueous-carbonic inclusions in topaz, at trapping temperatures of 350–370 °C. The early evolution of the pervasive late magmatic hydrothermal fluid started at ~510 °C, and is represented by the quartz pods. Primary and pseudo-secondary aqueous inclusions observed in topaz from these pods are rich in calcite crystals and are thought to provide evidence for CO_2 degassing and of immiscibility near the critical point of the system. These processes would cause extensive calcite precipitation and diminution of CO_2 contents in the fluid. At a later stage, the system evolved into immiscible solutions trapped at lower temperatures (350–370 °C) represented by the primary fluid inclusions in bedded-greisens (magmatic-hydrothermal stage), affected by fluid mixture with meteoric hydrothermal fluids; the latter are represented by low

salinity secondary fluid inclusions, at even lower temperatures (Table 4). Post-magmatic fluids related to fissural alteration were trapped at similar temperatures with minor meteoric hydrothermal fluid influx, and are recorded within the greisen stockwork and quartz–cassiterite vein systems (Table 5). It is worth noting that the inclusions present in quartz–cassiterite veins are the only representatives of Type IV, and reflect particular conditions of cassiterite deposition. Almost pure meteoric hydrothermal fluids were trapped in late barren quartz-veins at temperatures as low as 145 °C. High salinity Ca-rich fluids, trapped at temperatures of 120–90 °C, were also observed in many samples (Table 4), and seem to reflect decrease of temperatures and correspondent increase of salinities and Ca content in the solutions.

The temperature vs. salinity diagram (Fig. 4b) shows a fluid evolution model for the quartz pods, bedded-greisen, greisen stockwork, and quartz–cassiterite veins. It comprises three trends, suggesting a combination of distinct progressive immiscibility, degassing and dilution processes (Hedenquist and Henley, 1985; Wilkinson, 2001). These are marked by decrease in trapping temperatures under variable pressure conditions: Trend I represents the immiscibility trend of an originally aqueous-carbonic fluid associated with CO_2 degassing near the critical point of the system, at T=500 °C and P~500 bar; Trend II shows the immiscibility trend of an originally CO_2 poor aqueous-carbonic fluid with very low X_{CO_2} (not measurable), accompanied by gas loss at T=350–370 °C and P~200 bar; and Trend III shows the effect of dilution of the original fluid by less saline, cooler water from 350 to 370 °C to 130 °C at ~10 bar.

5.4. Stable isotopes

Isotope analyses were made on K-feldspar, quartz and mica from granites and pegmatite, and quartz, mica, topaz, kaolinite, and muscovite, separated from the quartz pods, bedded-greisens, salmon albitized granites, greisen stockwork, quartz–cassiterite and muscovite veins, and kaolinite stockwork (Table 3). We determined equilibrium temperatures of quartz–feldspar, quartz–topaz and quartz–cassiterite pairs, utilizing the fractionation quartz–topaz and quartz–

Table 5
Summary of fluid-inclusion data of minerals from primary rare-metal deposit from Oriente Novo massif

Styles of mineralization	Rock/Sample	Mineral	Type of fluid Inclusion	Tm_{CO_2} (°C)	T_e (°C)	$Tm_{clathrate}$	Th_{CO_2} (°C)	d_{CO_2} (g/cm^3)	Salinity (wt.% NaCl eq.)	Th_{aq}/Th_{total} (°C)
Post-magmatic										
Stockwork	Greisen vein/W-23A	Qtz Tp	Ia	−57.9 to −56.8	−28.8 to −23.0	1.5–8.2	25.6–29.9	0.26–0.70	3.6–13.8	190–440
			Ib	−57.7 to −57.0	–	2.9–6.4	24.9–30.6	0.24–0.68	6.8–12.1	210–360
			II	–	−29.9 to −22.9	1.9–6.0	–	–	7.4–13.3	155–335
Stockwork	Qtz vein/W-27A	Qtz	Ia	−57.8 to −56.6	27.6 to −23.3	4.1–8.6	12.0–26.7	0.68–0.85	2.8–10.4	190–410
			Ib	−57.8 to −56.8	–	–	16.4–28.8	0.17–0.81	–	240–370
			II	–	−27.6 to −25.6	7.5–8.0	–	–	4.0–4.9	120–320
Endogreisen	Greisen pod/AW-28	Qtz Tp	Ia	−59.0 to −57.1	−27.8 to −23.2	5.3–9.8	14.5–25.8	0.21–0.83	0.4–8.6	210–420
			Ib	−59.1 to −57.3	–	8.4–9.1	10.8–25.9	0.14–0.86	1.8–3.2	245–380
			II	–	−27.0 to −25.3	5.4–8.9	–	–	2.4–8.4	100–305
Sub-parallel vein	Qtz vein/W-24G	Qtz	Ia	−57.3 to −56.6	−27.6 to −27.0	4.4–7.9	26.0–30.2	0.26–0.66	4.1–9.9	190–405
			Ib	−57.8 to −57.0	–	–	26.1–29.3	0.26–0.69	–	220–320
			II	–	−29.8 to −26.8	4.4–8.2	–	–	3.6–9.9	140–290

Abbreviations: Qtz—quartz; Tp—topaz; Ia—aqueous-carbonic fluid (H_2O–CO_2–CO_2–CH_4–NaCl); Ib—aqueous-carbonic fluid (CO_2–H_2O–CH_4–NaCl); II—CO_2-bearing water-rich fluid; Tm_{CO_2}—CO_2 melting temperature; T_e—eutectic temperature; $Tm_{clathrate}$—temperature of disassociation of a gas clathrate; Th_{CO_2}—CO_2 homogenization temperature; d_{CO_2}—CO_2 density; Th aq—vapour-liquid phase homogenization temperature; Th_{total}—total homogenization temperature.

feldspar curves of Zheng (1993a,b), and quartz–cassiterite curves of Zhang et al. (1994). The $\delta^{18}O$ and δD of the coexisting hydrothermal fluids in equilibrium with these minerals were calculated by using the quartz–water curve of Matsuhisa et al. (1979), the feldspar–water curves of O'Neil and Taylor (1967), Bottinga and Javoy (1973) and Zheng (1993a), the biotite–water curve of Zheng (1993b), topaz–water curve of Zheng (1993a), cassiterite–water curve of Zhang et al. (1994), kaolinite–water curves of Sheppard and Gilg (1996) and Gilg and Sheppard (1996), muscovite–water curve of Zheng (1993b), and the calculated curve of Jenkin (1998) for micas. All the oxygen and hydrogen data are shown in Table 3. Oxygen isotope geothermometry indicates that the minimal retrograde reaction temperature for a quartz–feldspar pair is in the order of 603 °C, for the syenogranites. Formation of the quartz pods and bed-like greisen bodies took place at ~570 and 500 °C, respectively, and that of the cassiterite-quartz veins at 415 °C. These temperatures are 45–130 °C higher than the fluid inclusion trapping temperatures.

The calculated $\delta^{18}O_{H_2O}$ of the tin-bearing fluids in equilibrium with host metasomatites ($\delta^{18}O_{H_2O}$=5.2–7.3‰; quartz–water pair) indicates that the fluids equilibrated with an evolving residual granitic magma or with a high temperature albite-granite, consistent with a magmatic origin (cf. Sheppard, 1986). The calculated $\delta^{18}O_{H_2O}$ deduced from the equilibrium mica-water indicates mixture with meteoric water, already at the pervasive magmatic-hydrothermal stage ($\delta^{18}O_{H_2O}$=1.1–9.8‰). The late muscovite veins ($\delta^{18}O_{H_2O}$=−6.4‰ at 380 °C) and late quartz ($\delta^{18}O_{H_2O}$=−3.8‰ at 380 °C) formed at the post-magmatic fissural hydrothermal alteration stage, implying a dominant meteoric water component (Fig. 8).

Measured $\delta^{18}O$ and δD and the data for mineralizing fluids in equilibrium with kaolinite at 100 °C are also shown in Table 3. The data plot near the magmatic water field, and the fluid is enriched in heavy hydrogen compared with the host granites (Fig. 8). Also some data points for the greisen seem to plot along a tie line leading to the composition of the kaolinite veins. This might suggest a mica-clay mixing line, resulting from a dominant magmatic fluid mixed with meteoric water, or deuterium enrich-

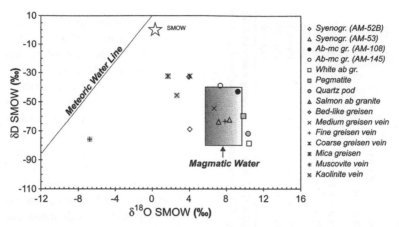

Fig. 8. δD vs. $\delta^{18}O$ diagram showing the fluid isotopic composition of the Santa Bárbara deposit, based on calculated δD_{H_2O} and $\delta^{18}O_{H_2O}$ values. Abbreviations: Ab—albite; mc—microcline; gr—granite. AM-52B, AM-53, AM-108 and AM-145 denote sample numbers.

ment of the waters caused by deuterium modifying processes.

6. Oriente Novo deposit (Rondônia Tin Province)

6.1. Geologic outline

The Oriente Novo massif occupies approximately 80 km², is subcircular in shape and intrudes Proterozoic (1.75–1.43 Ga) basement rocks and the Santa Clara batholith (Fig. 9). The massif consists of a central porphyritic biotite (±hornblende) monzogranites surrounded by a medium-grained porphyritic biotite syenogranite. Fine-grained porphyritic biotite syenogranite and muscovite–biotite microsyenogranite are found along the western border of the massif. Small late-stage bodies of siderophyllite–alkali–feldspar granite, alaskite, biotite (±alkali amphibole) microgranite, zinnwaldite–alkali–feldspar granite, and rhyolite porphyry are found in the west-central part. Alkali feldspar microsyenite, trachyte, trachyandesite, microgranite and, more rarely, aplite and pegmatite occur as dykes within and/or close to the massif. The primary rare-metal (Sn, Ta, Nb, W) deposits are associated with siderophyllite-alkali-feldspar granite and alaskite, as well as with zinnwaldite-alkali feldspar granite and rhyolite porphyry (Leite, 2002). The siderophyllite-alkali feldspar granite and alaskite are composed of orthoclase perthite and microcline, round and irregular quartz, albite-oligoclase, and siderophyllite. Zircon, monazite and ilmenite are the primary accessory minerals. The zinnwaldite-alkali feldspar granite and rhyolite porphyry show porphyritic texture, and exhibit poikilitic microcline and quartz megacrystals (<0.5 cm) dispersed in an albite-dominated fine- to very fine-grained matrix. The megacrystals include albite laths sometimes with a snowball texture. Zinnwaldite is present and includes zircon, fluorite, rutile, cassiterite, and columbite–tantalite crystals. The siderophyllite–alkali feldspar granite and alaskite are high in SiO₂, F, Rb, U, Y, and low in TiO₂, Al₂O₃, Fe₂O₃ total, MgO, CaO, P₂O₅, Ba, Sr, and Eu, whereas the zinnwaldite–alkali feldspar granite and rhyolite porphyry are low in SiO₂, TiO₂, Fe₂O₃ total, CaO, K₂O, and high in Al₂O₃, Na₂O, Li, Rb, Ga, Nb, Sn, and Ta (Leite, 2002). These geochemical characteristics are similar to those of rare-metal peraluminous granites (see, e.g., Pollard, 1989a).

6.2. Mineralization

The primary deposit occupies an area of ~1.5 km² in the west-central part of the Oriente Novo massif. Major styles of mineralization can be divided into three groups (Fig. 9): (1) stockwork of veins/veinlets of greisen with cassiterite and of quartz with cassiterite and wolframite hosted by older medium-grained porphyritic biotite syenogranite; (2) disseminated cassiterite and columbite–tantalite in the zinnwaldite alkali feldspar granite with cassiterite-bearing greisen pods; and (3) sub-parallel veins/veinlets of greisen with cassiterite and of quartz with cassiterite and wolframite having a general strike of N30–50° E

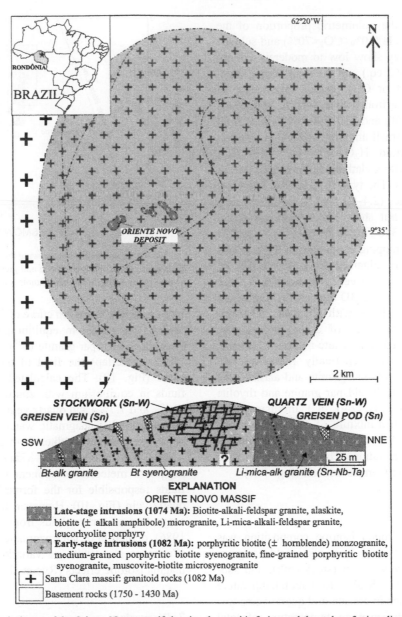

Fig. 9. Schematic geological map of the Oriente Novo massif showing the granitic facies, and the styles of mineralization of the Oriente Novo deposit (modified after Leite, 2002). Abbreviations: Bt—biotite; alk—alkali. Inset shows location relative to Brazil.

and a dip of 50–80° SE. The stockwork mineralization is interpreted to be genetically related to the side-rophyllite-alkali feldspar granite and alaskite, whereas the disseminated, endogreisen and sub-parallel vein/veinlet mineralization systems are related to the zinnwaldite-alkali feldspar granite and rhyolite porphyry (Leite, 2002). The mineral paragenesis is illustrated in Fig. 2.

6.3. Fluid inclusions

Fluid inclusions have been studied in samples of greisen pods and veins, and quartz veins. Two types of primary fluid inclusion are found in topaz and quartz at room temperature (Table 5). Type I aqueous carbonic fluid inclusions ($H_2O \pm CO_2$–CH_4–NaCl) are the most abundant. Two subtypes are distinguished based on the volumetric proportion of the CO_2 phase: subtype Ia ($10\%<CO_2<70\%$) and subtype Ib ($CO_2>80\%$). They show low to moderate salinity (0.4–13.8 wt.% NaCl eq.) and the total homogenization, into the H_2O or CO_2 phases, takes place at temperatures varying from 190 to 440 °C, with peaks at 240–290, 280–330, and 310–360 °C. Type II fluid inclusions contain small amounts of CO_2 in addition to the water-rich fluid (H_2O–CO_2–NaCl). The CO_2 was detected by melting clathrate temperatures during heating experiments. The type II inclusions have low to moderate salinities (2.4–14.1 wt.% NaCl eq.) very similar to type I, but show lower homogenization temperatures (100–340 °C), with peaks at 200–250 and 250–300 °C. Trapping conditions of Type Ia and Ib immiscible fluids, obtained from intersection of isochores, took place between 240 and 370 °C and 1.0–2.4 kbar (Table 2; Fig. 3D through G). The type I and II inclusions (those with Th>240 °C) have resulted from effervescence of a similar aqueous-carbonic fluid derived from a late-magmatic parental fluid phase. This was genetically related to the siderophyllite-alkali feldspar granite and alaskite and to the zinnwaldite-alkali feldspar granite and rhyolite porphyry. Type II (Th<240 °C) inclusions are probably products of a mixture of a magmatic CO_2-bearing water-rich fluid and some meteoric water in all styles of mineralization. The fluid evolution is similar to that of the Correas deposit (Fig. 4a,c).

6.4. Stable isotopes

Oxygen and hydrogen isotope composition was determined for zinnwaldite from alkali-feldspar granite, greisen pod, greisen vein (stockwork), and sub-parallel quartz vein (Table 3). A formation temperature of 300 °C for Li-mica in greisen and quartz veins was calculated using the average temperature of total homogenization of fluid inclusions in cogenetic topaz and/or quartz. For the zinnwaldite from the alkali feldspar granite, we applied the estimated temperature of 650 °C of Manning (1982), which is the minimum experimental crystallization temperature for F-rich granites at pressure of 1 kbar. The $\delta^{18}O_{H_2O}$ in equilibrium with the zinnwaldites was calculated using the biotite–water fractionation curve of Zheng (1993b). The δD_{H_2O} in equilibrium with zinnwaldite from alkali-feldspar granite was calculated using the

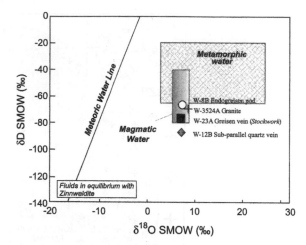

Fig. 10. δD vs. $\delta^{18}O$ diagram showing the calculated $\delta^{18}O$ and δD for Li-mica alkali feldspar granite and greisens from Oriente Novo deposit. Abbreviations indicate sample numbers.

mica-water fractionation curve of Suzuoki and Epstein (1976). The water in equilibrium with zinnwaldite from alkali-feldspar granite ($\delta^{18}O_{H_2O}$=8.9‰; δD_{H_2O}= −6.9‰) overlaps the field of the typical magmatic water (Fig. 10). The calculated $\delta^{18}O_{H_2O}$ values of fluids in equilibrium with zinnwaldite from greisen (7.4‰ and 7.7‰) and quartz vein (7.5‰) also overlap with the field of magmatic water. The calculated δD values for the zinnwaldites from greisen (−66‰ and −76‰) and quartz vein (−87‰) are low enough to suggest a meteoric component in the hydrothermal fluids responsible for the formation of greisen and quartz vein (Fig. 10). However, this may also reflect outgassing of earlier fluids from the magma (Taylor, 1986), depletion of late-magmatic fluids in deuterium (Carten et al., 1988), or variations linked to fluid-magma isotope effects (Rye et al., 1990).

7. Discussion

The geological, petrographic, fluid inclusion, and stable-isotope studies point to a number of common and contrasting features of the studied Sn-polymetallic greisen-type deposits. The tin-granites of the Itu Rapakivi Province are related to the post-collisional stage of the Brasiliano orogeny (625±5 Ma) (Wernick et al., '1997; Campos Neto, 2000), whereas those of the Rondônia Tin Province are considered anorogenic like the Younger Granites of Nigeria (Kloosterman,

1970; Priem et al., 1989). More recently, however, the tin-granites and related rocks of the Rondônia Tin Province were interpreted as inboard rapakivi magmatism temporally related to the Rondonian-San Ignacio (1.50–1.30 Ga) and Sunsás (1.25–1.00 Ga) orogenies (Bettencourt et al., 1999). In both provinces, the tin-granites are fluorine-rich peraluminous alkali feldspar granites and contain topaz and/or muscovite or zinnwaldite; the porphyritic facies shows "snowball" texture, indicative of magmatic origin (Pollard, 1989b; Sun and Yu, 1992). Their geochemical characteristics are comparable to the low P_2O_5 (<0.10 wt.%) sub-type of topaz-bearing granites (Taylor, 1992). A meta-igneous source for these granites is considered possible.

The Santa Bárbara and Oriente Novo deposits occur within granites, whereas the Correas deposit is hosted in gneisses and calc-silicate rocks. An important point is the common occurrence of stockworks and veins at Correas and Oriente Novo deposits. In contrast, the Santa Bárbara deposit is dominantly a preserved mineralized cupola with associated bed-like greisen bodies. These deposits contain a similar rare-metal suite that includes Sn ($\pm$W, $\pm$Ta, $\pm$Nb), but Zn–Cu–Pb mineralization is present only in Correas deposit.

High temperature fluid inclusions were identified only at the Santa Bárbara deposit. They are represented by primary silicate-melt inclusions with a Na-rich aqueous-phase, with or without CO_2, and salinity from 0.5 to 15 wt.% NaCl eq. They indicate a minimum crystallization temperature for the albite granite at Tt$\geq$600 °C.

The early primary fluids related to mineralization in the Correas and Oriente Novo deposits have common features: they are low-salinity (0–19 wt.% NaCl eq.) CO_2-bearing aqueous fluids, which homogenize between 245 and 450 °C. In the Santa Bárbara deposit, this early fluid is represented by low-salinity (5–12 wt.% NaCl eq.), low CO_2-bearing aqueous fluids, which homogenize at a more restricted temperature range (340–390 °C). In the Santa Bárbara, Oriente Novo and Correas deposits, this type of fluid may account for the deposition of early-disseminated cassiterite and columbite–tantalite, either directly from a melt or from a coexisting vapor, as well as from a late-magmatic fluid (see Haapala, 1997; Linnen, 1998). The compositional range and homogenization

behavior in the Santa Bárbara and Oriente Novo deposits is consistent with immiscibility around 350–370 °C and 350–450 °C, respectively. This process is responsible for the successive precipitation of cassiterite, wolframite and minor columbite–tantalite, down to the temperature of ~250 °C. These oxides continued to grow up to the post-magmatic stage in the presence of a magmatic-derived fluid mixed with external fluids. The occurrence of scheelite in the Santa Bárbara deposit is better explained by remobilization of tungsten and its subsequent deposition as scheelite in the late-magmatic stage.

The second type of fluid inclusions, also related to deposition of cassiterite, wolframite and minor columbite–tantalite (inclusions in cassiterite) in Oriente Novo deposit (Th>240 °C) and sulfide in Correas deposit, is essentially represented by aqueous solutions with minor CO_2 and low to moderate-salinities (0–14.1 wt.% NaCl eq.), which homogenize between 100 and 340 °C. In the Santa Bárbara deposit these fluids are represented by aqueous fluid inclusions in quartz–cassiterite veins, which have low salinities (0.3–3.0 wt.% NaCl eq.) and homogenize at 320–380 °C.

The trapping P–T conditions (P_t, T_t) of the observed early fluids in the Correas and Oriente Novo deposits, obtained from fluid-inclusions, suggest hydrothermal activity at 240–440 °C and 1.0–2.6 kbar, close to lithostatic. The observed progressive decrease in T_t and P_t indicate either erosion of overlying rock or an open hydrostatic system and hydraulic fracturing of the host-rocks. The contrasting fact is that in the Santa Bárbara deposit, the early magmatic and late post-magmatic fluids indicate similar T_t from 350 to 400 °C and a P_t from 300 to 200 bar, near hydrostatic. The depth of emplacement at Oriente Novo and Correas deposit is roughly the same (maximum=9.8 km) and is compatible with a transitional hypabyssal to plutonic environment. The Santa Bárbara deposit is more shallow (maximum=1.8 km), characteristic of subvolcanic environment. These fluids affected the magmatic-stage rocks and modified the primary fluid inclusions of quartz from granites, attesting some hydro-fracturing and mineral equilibrium at the post-magmatic stage.

The late post-magmatic fluid observed in Correas deposit was a CO_2-free aqueous solution, with low salinity (0–6 wt.% NaCl eq.), and a homogenization temperature between 100 and 260 °C. It characterizes the sulfide–fluorite–sericite Correas deposit association. In the Santa Bárbara deposit, later-stage post-magmatic aqueous fluids show a much larger range of homogenization temperatures (94–380 °C) than in the other deposits, lower salinity (0–3 wt.% NaCl eq.), and include late-barren-quartz, muscovite, and kaolinite veins. The low temperature (120–90 °C) late fluid type at the Santa Bárbara deposit is also high-saline Ca-rich, and was associated with meteoric hydrothermal water that percolated along fissures (open system) at shallow depths. A cold low-salinity fluid and neutralization of acid condensates from the previous acid fluids are responsible for the kaolinite veins.

In all cases the post-magmatic late fluids resulted from decreasing temperature as mixing and dilution proceeded below 290 °C. The switch-over of the process would lead to the predominance of an essentially aqueous solution (H_2O–NaCl$\pm$KCl) system. These fluids may represent a mixture of magmatic derived fluids and hydrothermal meteoric water (dominant) components, or different overprints on granites of diverse chemical composition.

The oxygen and hydrogen stable isotope data indicate that the fluids contained a magmatic component mixed with meteoric water. Hydrofracturing, immiscibility, greisenization and loss of CO_2 ($\pm CH_4$) with falling temperature, are the principal processes responsible for Sn–W deposition. In the Santa Bárbara deposit, oxygen isotope geothermometry indicates temperatures of 500 °C for the bed-like greisen bodies, and 400 °C for the cassiterite-quartz veins. The calculated isotopic composition of water in equilibrium with host metasomatites ($\delta^{18}O_{H_2O}$=1.1–9.8‰) is consistent with a magmatic origin but with a variable hydrothermal-meteoric component for the fluids, except for muscovite ($\delta^{18}O_{H_2O}$=−6.4‰) and late quartz ($\delta^{18}O_{H_2O}$=−3.8‰) veins, which show a dominant meteoric water component. A magmatic fluid, partly mixed with meteoric fluids ($\delta^{18}O_{qtz}$=9.9–10.9‰; $\delta^{18}O_{H_2O}$=4.13–6.95 ‰; $\delta^{18}O_{mica}$=4.7–5.2‰) was involved in the genesis of the Correas deposit. The small range of variation in the $\delta^{18}O_{H_2O}$ values indicates re-equilibration of the fluid phase with granite at subsolidus temperatures, and CO_2–H_2O fractionation, mostly during the vein formation, in the same manner as proposed by Higgins (1985b). The isotopic temperatures calculated for the mineral pairs were in the range of 441–459 °C (quartz–cassiterite)

and 336–390 °C (quartz-wolframite), in agreement with temperatures of 300–440 °C, at fluid pressures of 1.3–2.6 kbar, respectively, obtained via fluid-inclusion microthermometry (Table 2 and Fig. 3A–C). A linear trend in δD values extending to lighter values for the same homogeneous range of $\delta^{18}O$ values is observed in the Oriente Novo and Correas deposit. Similar $\delta^{18}O/\delta D$ trends were interpreted as a result of CO_2–H_2O and CH_4–H_2O fractionation (Bottinga, 1969; Higgins, 1985b), outgassing of earlier fluids (Taylor, 1986), changing salinity in the magmatic fluids (Carten et al., 1988), minor admixture of meteoric water (Heinrich, 1990), and fluid-magma isotope effects (Rye et al., 1990), among other explanations. A contrasting $\delta^{18}O/\delta D$ behavior is observed in the Santa Bárbara deposit: the δD values for micas from greisens indicate a narrow range of variation for a wide range of $\delta^{18}O$ values, as well as deuterium enrichment in the fluids. This may reflect re-equilibration of aqueous fluid with granite at low water–rock ratios and progressively lower temperatures, as proposed by Smith et al. (1996) for similar vein fluids from the Cligga Head Sn–W deposit (southwestern England). Other explanations for similar $\delta^{18}O/\delta D$ trends at the Zaaiplaats Tin Mine (South Africa) and at the Hemerdon Ball Sn–W deposit (SW England) are provided by Pollard et al. (1991) and Shepherd et al. (1985), respectively.

8. Conclusions

The contrasting mineralization styles and fluid characteristics of the Santa Bárbara deposit in relation to the Correas and Oriente Novo deposits reflect different depths of formation, spatial relationships to tin granites and different wall rock/fluid proportions.

The fluid inclusion studies favor late-magmatic to early post-magmatic models of mineralization. The main processes involved in Sn–W mineralization are fluid-melt immiscibility, immiscibility of an original orthomagmatic CO_2–H_2O fluid, mixture, dilution, CO_2 degassing, accompanied by hydro-fracturing, and different styles of hydrothermal alteration, mainly feldspathization, greisenization, silicification, and argilization.

The fluids associated with the mineralization are aqueous with variable concentrations of CO_2 as the dominant volatile component. The P–T trapping conditions allow to consider two different mineralization emplacement levels: transitional from hypabyssal to plutonic, corresponding to a maximum depth of 9.8 km (Correas and Oriente Novo deposits) and at more shallow subvolcanic level, i.e., at a depth of 1.8 km (Santa Bárbara deposit).

Stable isotope and fluid inclusion data reveal three fluid types: (1) an orthomagmatic fluid, which has equilibrated with granite; (2) an orthomagmatic derived fluid mixed with hydrothermal meteoric water; and (3) a dominant late hydrothermal meteoric fluid. The first two types are responsible for cassiterite, wolframite, and columbite–tantalite precipitation. Changes in the redox conditions related to mixing of magmatic and meteoric fluids favored sulfide mineralization in Correas deposit at the post-magmatic stage.

Acknowledgements

This research was funded by FAPESP-Fundação de Amparo à Pesquisa do Estado de São Paulo (grant awarded to J.S. Bettencourt, Proc. no. 2000/08033-5, and scholarship to I. Sparrenberger, Proc. no. 1998/04469-1), PADCT-FINEP (research grant to J.S. Bettencourt, convênio PADCT/FINEP no. 64.99.0276.00), and CNPq (research grant to R.M.S. Bello, Proc. no. 303872/1985-3). We would like to express our appreciation to M. Poutiainen (Helsinki, Finland) and V.T. McLemore (Socorro, USA) for constructive and thorough reviews that improved the manuscript and R.L. Linnen (Waterloo, Canada) for his comments on the paper. We particularly thank Tapani Rämö for his fruitful suggestions, support, and invitation to contribute to this volume. The authors also thank L.V.S. Monteiro and T. Benevides for the drafts. The paper forms parts of Ph.D. studies carried out by W.B. Leite Jr., C.L. Goraieb, and I. Sparrenberger under the supervision of J.S. Bettencourt.

References

Almeida, F.F.M., Hasui, Y., Brito Neves, B.B., Fuck, R.A., 1981. Brazilian structural provinces: an introduction. Earth-Science Reviews 17, 1–29.

Bettencourt, J.S., Leite Jr., W.B., Payolla, B.L., Scandolara, J.E., Muzzolon, R., Vian, J.A.J., 1995. The rapakivi granites of the Rondônia tin province and associated mineralization. In: Bettencourt, J.S., Dall'Agnol, R. (Eds.), Excursion Guide: The Rapakivi Granites of Rondônia Tin Province and Associated Mineralization. Symposium Rapakivi Granites and Related Rocks, Belém, Brazil. Federal University of Pará, Center for Geosciences, pp. 5–16.

Bettencourt, J.S., Tosdal, R.M., Leite Jr., W.B., Payolla, B.L., 1999. Mesoproterozoic rapakivi granites of Rondônia tin province, southwestern border of the Amazonian craton, Brazil: I. Reconnaissance U–Pb geochronology and regional implications. Precambrian Research 95, 41–67.

Bodnar, R.J., 1993. Revised equation and table for determining the freezing point depression of H_2O NaCl solutions. Geochimica et Cosmochimica Acta 57, 683–684.

Bodnar, R.J., Burnham, C.W., Sterner, S.M., 1985. Synthetic fluid inclusions in natural quartz: III. Determinations of phase equilibrium properties in the system $H_2O+NaCl$ to 1000 °C and 1500 bars. Geochimica et Cosmochimica Acta 49, 1861–1873.

Bottinga, Y., 1969. Calculated of fractionation factors for carbon and oxygen isotope exchange in the system calcite–CO_2–C–CH_4–H–H_2O vapor. Geochimica et Cosmochimica Acta 33, 49–65.

Bottinga, Y., Javoy, M., 1973. Comments on oxygen isotope geothermometry. Earth and Planetary Science Letters 20, 250–265.

Bowers, T.S., Helgeson, H.C., 1983. Calculations of the thermodynamic and geochemical consequences of nonideal mixing in the system H_2O–CO_2–NaCl on phase relations in geologic systems: metamorphic equilibria at high pressures and temperatures. American Mineralogist 68, 1059–1075.

Brown, P.E., 1989. FLINCOR: a microcomputer program for the reduction and investigation of fluid inclusion data. American Mineralogist 74, 1309–1393.

Brown, P.E., Lamb, W.M., 1989. P–V–T properties of fluid in the system $H_2O\pm CO_2\pm NaCl$: new graphical presentation and implications for fluid inclusion studies. Geochimica et Cosmochimica Acta 53, 1209–1221.

da Campanha, G.A.C., Sadowski, G.R., 1999. Tectonics of the southern portion of the ribeira belt (apiaí domain). Precambrian Research 98, 31–51.

Campos Neto, M.C., 2000. Orogenic systems from southwestern Gondwana: an approach to Brasiliano-Pan African cycle and orogenic collage in southearstern Brazil. In: Cordani, U.G., Milani, E.J., Thomaz Filho, A., Campos, D.A. (Eds.), Tectonic Evolution of South America 31st International Geological Congress, 2000, Rio de Janeiro, Brazil, pp. 335–365.

Carten, R.B., Rye, R.O., Landis, G.P., 1988. Effects of igneous and hydrothermal processes on the composition of ore-forming fluids; stable isotope and fluid inclusion evidence, Henderson molibdenum deposit, Colorado. Abstracts with Programs-Geological Society of America 19, A94.

Clayton, R.N., Mayeda, T.K., 1963. The use of bromine pentafluoride in the extraction of oxygen from silicates for isotopic analysis. Geochimica et Cosmochimica Acta 27, 43–52.

Collins, P.C., 1979. Gas hydrates in CO_2-bearing fluid inclusions and the use of freezing data for estimations of salinity. Economic Geology 74, 1435–1444.

Dall'Agnol, R., Costi, H.T., Leite, A.A.S., Magalhães, M.S., Teixeira, N.P., 1999. Rapakivi granites from Brazil and adjacent areas. Precambrian Research 95, 9–39.

Davis, D.W., Lowenstein, T.K., Spencer, R.J., 1990. Melting behavior of fluid inclusions in laboratory-grown halite crystals in the systems $NaCl–H_2O$, $NaCl–KCl–H_2O$, $NaCl–MgCl_2–H_2O$ and $NaCl$, $CaCl_2–H_2O$. Geochimica et Cosmochimica Acta 54, 591–601.

Diamond, L.W., 1994. Introduction to phase relations of $CO_2–H_2O$ fluid inclusions. In: De Vivo, B., Frezzotti, M.L. (Eds.), Fluid Inclusions in Minerals: Methods and Apliccations. Short Course of the Working Group (IMA) 'Inclusions in minerals'. Virginia Polytechnic Institute, pp. 131–158.

Diamond, L.W., 2001. Review of the systematics of $CO_2–H_2O$ fluid inclusions. Lithos 55, 69–99.

Donnelly, T., Waldron, S., Tait, A., Dougans, J., Bearhop, S., 2001. Hydrogen isotope analysis of natural abundance and deuterium-enriched waters by reduction over chromium on-line to a dynamic dual inlet isotope-ratio mass spectrometer. Rapid Communications in Mass Spectrometry 15, 1297–1303.

Fallick, A.E., Macaulay, C.L., Haszeldine, R.S., 1993. Implications of linearly correlated oxygen and hydrogen isotopic compositions for kaolinite and illite in the Magnus sandstone, North Sea. Clays and Clay Minerals 41, 184–190.

Fifarek, R.H., 1985. Alteration geochemistry, fluid inclusion, and stable isotope study of the Red Ledge volcanogenic massif sulfide deposits, Idaho. Ph.D. thesis, Oregon State University, U.S.A.

Frank, R.E., 1990. Geologia, petrologia e mineralizações estaníferas do Complexo Granítico de Santa Bárbara, Rondônia, Brasil. M.Sc. thesis, Instituto de Geociências, Universidade Federal do Rio de Janeiro, Rio de Janeiro, Brazil (in Portuguese).

Gilg, H.A., Sheppard, S.M.F., 1996. Hydrogen isotope fractionation between kaolinite and water revisited. Geochimica et Cosmochimica Acta 60, 529–533.

Gimenez Filho, A., de Janasi, V.A., da Campanha, G.A.C., Teixeira, W., Trevizoli Jr., L.M., 2000. U–Pb dating and Rb–Sr isotope geochemistry of the eastern portion of the Três Côrregos batholith, Ribeira fold belt, São Paulo. Brazilian Contributions to 31st International Geological Congress, Rio de Janeiro, Brazil, Revista Brasileira de Geociências 30, pp. 45–50.

Goraieb, C.L., 2001. Contribuição à gênese do depósito primário polimetálico (Sn, W, Zn, Cu, Pb) Correas, Ribeirão Branco (SP). Ph.D. thesis, Instituto de Geociências, Universidade de São Paulo, São Paulo, Brazil (in Portuguese).

Haapala, I., 1997. Magmatic and postmagmatic processes in tin-mineralized granites: topaz-bearing leucogranite in the eurajoki rapakivi granite stock, Finland. Journal of Petrology 38 (12), 1645–1659.

Hedenquist, J.W., Henley, R.W., 1985. The importance of CO_2 on freezing point measurements of fluid inclusions: evidence from active geothermal systems and implications for epithermal ore deposition. Economic Geology 80, 1379–1406.

Heinrich, C.H., 1990. The chemistry of hydrothermal tin (-tungsten) ore deposition. Economic Geology 85, 457–481.

Heyen, N.C., Ramboz, C., Dubessy, J., 1982. Simulations des équilibres de phases dans le système $CO_2–CH_4$ en dessous de 50 °C et de 100 bar: application aux inclusions fluids. Comptes Rendus de la Académie des Sciences, Paris 294, 203–206.

Higgins, N.C., 1985a. Wolframite deposition in a hydrothermal vein system: the grey river tungsten prospect, Newfoundland, Canada. Economic Geology 80, 1297–1327.

Higgins, N.C., 1985b. Moderately depleted oxygen isotope composition of waters associated with tin-and tungsten-bearing quartz veins: an evaluation of isotopic models. In: Herbert, H.K., Ho, S.E. (Eds.), Conference on Stable Isotopes and Fluid Processes in Mineralization, Queensland, 1985. Geology Department and University Extension of the University of Western Australia Publication, vol. 23, pp. 204–214.

de Janasi, V.A., Ulbrich, H.H.G.J., 1991. Late proterozoic granitoid magmatism in the state of São Paulo, Southeastern Brazil. Precambrian Research 51, 351–374.

Jenkin, G.R.T., 1998. Stable isotope studies in the Caledonides of S.W. Connemara, Ireland. Glasgow, Ph.D. thesis, University of Glasgow, United Kingdom.

Kloosterman, J.B., 1970. A two fold analogy between the Nigerian and Amazonian tin provinces. Technical Conference on Tin 2, Bangkok, vol. 1, pp. 193–221.

Leite, W.B. Jr., 2002. A Suíte Intrusiva Santa Clara (RO) e a mineralização primária polimetálica (Sn, W, Nb, Ta, Zn, Cu e Pb) associada. Ph.D. Thesis, Instituto de Geociências, Universidade de São Paulo, São Paulo, Brazil (in Portuguese).

Leite Jr., W.B., Payolla, B.L., Bettencourt, J.S., 2000. Petrogenese of two Grenvillian tin-bearing rapakivi suites in the Rondônia tin province, SW Amazonian craton, Brazil. The 31st International Geological Congress, Rio de Janeiro, Brazil, Abstracts volume, Rio de Janeiro. Geological Survey, Brazil. [CD-ROM].

Leite Jr., W.B., Bettencourt, J.S., Payolla, B.L., 2001. 1.08–1.07 Ga A-type granite magmatism in the Rondônia tin province, SW Amazonian craton, Brazil: petrologic and geochemical constraints. Abstracts, vol. 26. Geological Association of Canadá/Mineralogical Association of Canadá Joint Annual Meeting, St John's, Newfoundland, Canadá, p. 85.

Linnen, R.L., 1998. Depth of emplacement, fluid provenance and metallogeny in granitic terranes: a comparison of western Thailand with other tin belts. Mineralium Deposita 33, 461–476.

Manning, D.A.C., 1982. An experimental study of the effects of fluorite on the crystallization of granitic melts. In: Evans, A.M. (Ed.), Metallization Associated with Acid Magmatism. John Wiley, Chichester, pp. 191–203.

Matsuhisa, Y., Goldsmith, J.R., Clayton, R.N., 1979. Oxygen isotopic fractionation in the system quartz-albite-anorthite-water. Geochimica et Cosmochimica Acta 43, 1131–1140.

Mcaulay, C.L., Fallick, A.E., Haszeldine, R.S., Graham, C.M., 2000. Methods of laser-based stable isotope measurement applied do diagenetic cements and hydrocarbon reservoir quality. Clay Minerals 35, 313–322.

O'Neil, J.R., Taylor, H.P., 1967. The oxygen isotope and cation exchange chemistry of feldspars. American Mineralogist 52, 1414–1437.

Payolla, B.L., Bettencourt, J.S., Kozuch, M., Leite Jr., W.B., Fetter, A.H., Van Schmus, W.R., 2002. Geological evolution of the basemente rocks in the east-central part of the Rondônia tin province, SW Amazonian craton, Brazil: U–Pb and Sm–Nd isotopic constraints. Precambrian Research 119, 141–169.

Pimentel, M.M., Heaman, L., Fuck, R.A., Marini, O.J., 1991. U–Pb zircon geochronology of precambrian tin-bearing continental type acid magmatism in central Brazil. Precambrian Research 52, 321–335.

Pollard, P.J., 1989a. Geochemistry of granites associated with tantalum and niobium mineralization. In: Möller, P., Černý, P., Saupé, F. (Eds.), Lanthanides, Tantalum and Niobium. Springer-Verlag, Berlin, pp. 145–158.

Pollard, P., 1989b. Geologic characteristics and genetic problems associated with the development of granite-hosted deposits of tantalum and niobium. In: Möller, P., Černý, P., Saupé, F. (Eds.), Lanthanides, Tantalum and Niobium. Springer-Verlag, Berlin, pp. 240–256.

Pollard, P.J., Andrew, A.S., Taylor, R.G., 1991. Fluid inclusion and stable isotope evidence for interaction between granites and magmatic hydrothermal fluids during formation of disseminated and pipe-style mineralization at the zaaiplaats tin mine. Economic Geology 86, 121–141.

Potter III, R.W., Clyne, M.A., Brown, D.L., 1978. Freezing point depression of aqueous sodium chloride solutions. Economic Geology 73, 284–285.

dos Prazeres Filho, H.J., 2000. Litogeoquímica, Geocronologia (U–Pb) e Geologia Isotópica dos Complexos Graníticos Cunhaporanga e Três Córregos, Estado do Paraná. M.Sc. thesis, Instituto de Geociências, Universidade de São Paulo, São Paulo, Brazil (in Portuguese).

Priem, H.N.A., Bon, E.H., Verdurmen, E.A.T., Bettencourt, J.S., 1989. Rb–Sr chronology of precambrian crustal evolution in Rondônia (western margin Brasilian craton). Journal of South American Earth Sciences 2, 163–170.

Ramboz, C., Pichavant, M., Weisbrod, A., 1982. Fluid immiscibility in natural processes: use and misuse of fluid inclusion data; II: interpretation of fluid inclusion data in terms of immiscibility. Chemical Geology 37, 29–48.

Rieder, M., Cavazzini, G., D'Yakonov, Y., Frank-Kamenetskii, V.A., Gottardi, G., Guggenheim, S., Koval, P.V., Müller, G., Neiva, A.M.R., Radoslovich, E.W., Robert, J.-L., Sassi, F.P., Takeda, H., Weiss, Z., Wones, D.R., 1998. Nomenclature of the micas. The Canadian Mineralogist 36, 905–912.

Roedder, E., 1984. Fluid Inclusions. Reviews in Mineralogy vol. 12. Mineralogical Society of America.

Rye, R.O., Lufkin, J.L., Wasserman, M.D., 1990. Genesis of the rhyolite-hosted tin ocurrences in the black range, New Mexico, as indicated by stable isotope studies. Geological Society of America Special Paper 246, 233–250.

Scandolara, J.E., Rizzotto, G.J., Bahia, R.B.C., Quadros, M.L.E.S, Amorim, J.L., Dall'Igna, L.G., 1999. Geologia e Recursos Minerais do Estado de Rondônia: texto explicativo e mapa geológico na escala 1:1,000,000. Programa Levantamentos Geológicos do Brasil. CPRM-Serviço Geológico do Brasil, Brasília, Brazil. In Portuguese.

Shepherd, T.J., Miller, M.F., Scrivener, R.C., Darbyshire, D.F.P., 1985. Hydrothermal fluid evolution in relation to mineralization in southwest England with special reference to the dartmoor-bodmin area. High heat production (HHP) granites, hydro-thermal circulation and ore genesis. The Institution of Mining and Metallurgy, London, pp. 345–364.

Sheppard, S.M.F., 1986. Characterization and isotopic variations in natural waters. In: Valley, J.W., Taylor, H.P., O'Neil, J.R. (Eds.), Stable Isotopes in High Temperature Geological Processes, Reviews in Mineralogy, vol. 16. Mineralogical Society of America, pp. 165–183.

Sheppard, S.M.F., Gilg, H.A., 1996. Stable isotope geochemistry of clay minerals. Clay Minerals 31, 1–24.

Smith, M., Banks, D.A., Yardley, B.W.D., Doyce, A., 1996. Fluid inclusion and stable isotope constraints on the genesis of the cligga head Sn–W deposit, S.W. England. European Journal of Mineralogy 8, 961–974.

Sparrenberger, I., 2003. Evolução da mineralização primária estanífera associada ao maciço granítico Santa Bárbara, Rondônia. Ph.D. thesis, Instituto de Geociências, Universidade de São Paulo, São Paulo, Brazil (in Portuguese).

Sun, S., Yu, J., 1992. Euhedral crystals of quartz in Ta-granite: eocrystal, not porphyroblastic. 29th International Geological Congress, Kyoto, Japan, 1992. Abstracts, vol. 2. IUGS, Kyoto, p. 539.

Suzuoki, T., Epstein, S., 1976. Hydrogen isotope fractionation between OH-bearing minerals and water. Geochimica et Cosmochimica Acta 40, 1229–1240.

Taylor Jr., H.P., 1979. Oxygen and hydrogen isotope relationships in hydrothermal mineral deposits. In: Barnes, H.L. (Ed.), Geochemistry of Hydrothermal Ore Deposits, 2nd ed. John Wiley and Sons, New York, pp. 236–277.

Taylor, B.E., 1986. Magmatic volatiles; isotopic variation of C, H, and S. In: Valley, J.W., Taylor Jr., H.P., O'Neil, J.R. (Eds.), Stable Isotopes in High Temperature Geological Processes, Reviews in Mineralogy, vol. 16. Mineralogical Society of America, pp. 185–220.

Taylor, R.P., 1992. Petrological and geochemical characteristics of the pleasant ridge zinnwaldite-topaz granite, southern New Brunswick, and comparisons with other topaz-bearing felsic rocks. Canadian Mineralogist 30, 895–921.

Taylor, R.P., Fallick, A.E., 1997. The evolution of fluorine-rich felsic magmas: source dichotomy, magmatic convergence and the origins of topaz granite. Terra Nova 9 (3), 105–108.

Ulbrich, H.H.G.J., Janasi, V., Vlach, S.R.F., 1991. Contrasted granitoid occurrences with rapakivi affinities in basement areas of the states of São Paulo and Minas Gerais, southeastern Brazil. In: Haapala, I., Rämö, O.T. (Eds.), Abstract Volume, Symposium on Rapakivi Granites and Related Rocks, IGCP Project 315, Guide-Geological Survey of Finland vol. 34, p. 54.

Vlach, S.R.F., de Janasi, V.A., Vasconcellos, A.C.B.C., 1990. The Itu belt: associated calc-alkaline and aluminous A - type late Brasiliano granitoids in the states of São Paulo and Paraná, southern Brazil. Congresso Brasileiro de Geologia 36,1990, Natal, Anais Natal SBG, vol. 4, pp. 1700–1711.

Wernick, E., 1992. Rapakivi granites related to post-collisional relaxing stage: the Itu province (late precambrian) of SE Brazil.

29th International Geological Congress, Kyoto, Japan, 1992. Abstracts, vol. 2. IUGS, Kyoto, p. 563.

Wernick, E., Godoy, A.M., Galembeck, T.M.B., 1991. The São Francisco, Sorocaba and Itu rapakivi complexes (late precambrian, state of São Paulo, Brazil): geological, petrographic and chemical aspects. In: Haapala, I., Rämö, O.T. (Eds.), Abstract Volume, Symposium on Rapakivi Granites and Related Rocks, IGCP Project 315, Guide-Geological Survey of Finland, vol. 34, pp. 61–62.

Wernick, E., Galembeck, T.M.B., Godoy, A.M., Hörmann, P.K., 1997. Geochemical variability of the rapakivi Itu rovince, state of São Paulo, SE Brazil. Anais da Academia Brasileira de Ciências 69, 395–413.

Wilkinson, J.J., 2001. Fluid inclusions in hydrothermal ore deposits. Lithos 55, 229–272.

Zhang, L.G., Zhuang, L., Quian, Y., Guo., Y., Qu, P., 1982. Stable isotope geochemistry of granites and tungsten-tin deposits in Xihuashan - Piaotang area, Jiangxi Province. Tungsten Geol. Symposium, Jiangxi, China. ESCAP RMRDC, Bandung, Indonesia. Geological Publishing House, Beijing, China, pp. 553–566.

Zhang, L.G., Liu, J.X., Chen, Z.S., Zhou, H.B., 1994. Experimental investigations of oxygen isotope fractionation in cassiterite and wolframite. Economic Geology 89, 150–157.

Zheng, Y.F., 1993a. Calculation of oxygen isotope fractionation in anhydrous silicate minerals. Geochimica et Cosmochimica Acta 57, 1079–1091.

Zheng, Y.F., 1993b. Calculation of oxygen isotope fractionation in hydroxyl-bearing silicates. Earth and Planetary Science Letters 120, 247–263.

Available online at www.sciencedirect.com

SCIENCE DIRECT®

Lithos 80 (2005) 387–400

ELSEVIER

LITHOS

www.elsevier.com/locate/lithos

Solubility of cassiterite in evolved granitic melts: effect of T, fO_2, and additional volatiles

P. Bhalla[a],*, F. Holtz[a], R.L. Linnen[b], H. Behrens[a]

[a]*Institut für Mineralogie, Universität Hannover, Welfengarten 1, 30167 Hannover, Germany*
[b]*Department of Earth Sciences, University of Waterloo, Waterloo, Ontario, Canada N2L 3G1*

Received 29 April 2003; accepted 9 September 2004
Available online 11 November 2004

Abstract

The aim of this experimental study was to determine the solubility of cassiterite in natural topaz- and cassiterite-bearing granite melts at temperatures close to the solidus. Profiles of Sn concentrations at glass–crystal (SnO_2) interface were determined following the method of (Harrison, T.M., Watson, E.B., 1983. Kinetics of zircon dissolution and zirconium diffusion in granitic melts of variable water content. Contributions to Mineralogy and Petrology 84, 66–72). The cassiterite concentration calculated at the SnO_2–glass interface is the SnO_2 solubility. Experiments were performed at 700–850 °C and 2 kbar using a natural F-bearing peraluminous granitic melt with 2.8 wt.% normative corundum. Slightly H_2O-undersaturated to H_2O-saturated melt compositions were chosen in order to minimize the loss of Sn to the noble element capsule walls. At the nickel–nickel oxide assemblage (Ni–NiO) oxygen fugacity buffer, the solubility of cassiterite in melts containing 1.12 wt.% F increases from 0.32 to 1.20 wt.% SnO_2 with an increasing temperature from 700 to 850 °C. At the Ni–NiO buffer and a given corundum content, SnO_2 solubility increases by 10% to 20% relative to an increase of F from 0 to 1.12 wt.%. SnO_2 solubility increases by ~20% relative to increasing Cl content from 0 to 0.37 wt.% in synthetic granitic melts at 850 °C. We show that Cl is at least as important as F in controlling SnO_2 solubility in evolved peraluminous melts at oxygen fugacities close to the Ni–NiO buffer. In addition to the strong effects of temperature and fO_2 on SnO_2 solubility, an additional controlling parameter is the amount of excess Al (corundum content). At Ni–NiO and 850 °C, SnO_2 solubility increases from 0.47 to 1.10 wt.% SnO_2 as the normative corundum content increases from 0.1 to 2.8 wt.%. At oxidizing conditions (Ni–NiO +2 to +3), Sn is mainly incorporated as Sn^{4+} and the effect of excess Al seems to be significantly weaker than at reducing conditions.

Keywords: Granite; Cassiterite; Concentration; SnO_2 solubility; Cl; F

* Corresponding author. Tel.: +49 511 762 5662; fax: +49 511 762 3045.

E-mail address: P.Bhalla@mbox.mineralogie.uni-hannover.de (P. Bhalla).

0024-4937/$ - see front matter © 2004 Elsevier B.V. All rights reserved.
doi:10.1016/j.lithos.2004.06.014

1. Introduction

Tin deposits display a wide range of structural and mineralogical types (Taylor, 1979a) and are generally spatially related to evolved granite intrusions. Sn and associated rare-element mineralisation (W, Be, Nb, Ta) are frequently associated with chemically specialized granites, with high contents of SiO_2, alkalis, Sn, W, Mo, Be, B, Li, and F, and low contents of CaO, MgO, Ba, and Sr (Tischendorf, 1977). The deposits themselves characteristically contain fluorite, topaz, tourmaline and Li-micas (Pollard et al., 1987). In a number of natural processes, F and B contribute to the magmatic and post-magmatic evolution of the tin systems, including fractionation/crystallization, fluid phase evolution, wall-rock alterations, metal transportation, and deposition (Pollard et al., 1987).

A fundamental concept in understanding the relationship between Sn mineralisation and magmatism is that variations in the basicity of a melt (i.e., the activity of free oxygens) play an important role (Taylor, 1988). Peralkaline magmas display high melt basicities, and, consequently, will generate fluids during resurgent boiling that are markedly less acid than fluids associated with peraluminous melt systems (Taylor, 1988). Hence, peralkaline magmas have the capacity to store a far greater concentration of Sn than peraluminous magmas and Sn would be less prone to remobilization in peralkaline magmas by a fluid phase. Sn will be most effectively partitioned from reduced, high-temperature, potassium-rich, aluminous magmas (Taylor, 1988). These characteristics describe a major portion of the Sn-mineralized granitoids (see also Štemprok, 1971; Štemprok and Skvor, 1974). Bivalent Sn is a relatively large cation and is considered to behave as an incompatible element during the evolution of granitoid suites (Taylor, 1979a,b; Eugster, 1985). Crystal fractionation, crystallochemical dispersion, magma mixing, and assimilation are the main processes that account for the generally systematic trends in Sn distribution in natural granitoid systems (Barsukov and Durasova, 1966; Kovalenko et al., 1968; Tauson, 1968, 1974; Groves, 1972, 1974; Tauson, 1974; Groves and McCarthy, 1978; Taylor, 1979a; Lehmann, 1982; Taylor and Wall, 1992).

Taylor (1988) concluded that, across a broad range of physical and chemical conditions, resurgent boiling is a viable mechanism for mobilising substantial amounts of Sn that can be readily accommodated in many granitoid magmas. Taylor (1988) suggested that large quantities of Sn can be efficiently extracted from the magmas by relatively dilute, supercritical Cl

solutions, and transported as a complex series of Cl-bearing stannous species. However, if the chlorinity of the fluid coexisting with melt is low, less Sn will be partitioned into the fluid. This may be an important parameter influencing the formation of magmatic cassiterite (Linnen, 1998). Without a good understanding of the solubility systematics of Sn in natural granitic magmas, and a knowledge of the solubility of Sn in Cl-bearing melts (+Cl-bearing fluid), it is difficult to predict the influence of these or other mechanisms that control the distribution of Sn in granitic rocks. In this respect, the solubility behavior of Sn in natural granitic systems containing halogens (Cl, F) has received little systematic attention.

Studies on ongonites (subvolcanic analogues of Li–F granites) show that Sn accumulates in the residual magma and may reach concentrations high enough to promote SnO_2 precipitation. Other examples of this level of accumulation include the Beauvoir Granite (Cuney et al., 1992), which is an example of a Li–F granite containing disseminated SnO_2, and the Macusani glasses (Pichavant et al., 1987). The presence of B and F in granite magmas results in the reorganisation of melt structure (Manning, 1981) by promoting depolymerisation and the creation of new complex species in the melt. In this process, the number of structurally favorable sites for incorporation of lithophile elements such as Sn increases, which may result in enhanced solubility of these elements (Manning, 1981; Pollard et al., 1987).

Ryabchikov et al. (1978a,b) were the first to study the solubility of SnO_2 in a synthetic granitic system under geologically relevant conditions. They determined that at 750 °C, 1.5 kbar, and fO_2 buffered by the nickel–nickel oxide assemblage (Ni–NiO), a haplogranite melt saturated with SnO_2, fluorite, and topaz, and in equilibrium with a 4.0 M chloride solution contained 1100 (±500) ppm SnO_2. Štemprok and Voldán (1978) determined the concentration of Sn in SnO_2-saturated natural granite and sodium disilicate glass at 1300–1600 °C, at oxygen fugacities controlled by atmospheric pressure. They found that the Sn concentration in equilibrium with cassiterite ranged from 5 to 7 wt.% SnO_2 at 1300 °C, up to 20 wt.% at 1600 °C, and was directly proportional to the Na content of the peralkaline glasses. Equal concentrations of stannic and stannous ions in basaltic liquids (hawaiite) at 1200 °C and at oxygen fugacity conditions close to the fayalite–magnetite–quartz buffer (FMQ) were found by Durasova et al. (1984). They also observed an increase in SnO_2 solubility in melts under reducing conditions. To examine the effect of melt composition and fO_2 on SnO_2 solubility, Linnen et al. (1995, 1996) investigated the solubility of cassiterite in peralkaline and peraluminous melts. In addition, Linnen et al. (1996) confirmed a significant increase in the solubility of SnO_2 with decreasing fO_2.

The effect of fO_2 and melt composition on the solubility of cassiterite in granitic melts were modeled by Taylor and Wall (1992) and Linnen et al. (1995, 1996). On the other hand, the effect of temperature and F and Cl content of the melt to cassiterite solubility has received less attention Štemprok (1990a,b). In partitioning experiments between haplogranitic melts and aqueous fluids, Keppler and Wyllie (1991) observed a strong increase of the fluid/melt partition coefficient of Sn with increasing Cl concentration in the fluid, while F content had no effect. This can be explained by the formation of Cl complexes of Sn and by F complexes being less stable. Dissolution experiments with various minerals in haplogranitic melts show that dissolved F can strongly enhance the solubility of the high field strength elements (HFSEs) in the melt (Keppler, 1993). Keppler (1993) suggested that this behaviour is due to the formation of HFSE complexes with nonbridging oxygen atoms which are expelled from coordination with Al by reaction with F (Keppler found that the solubility of TiO_2 increased from 0.26 ± 0.01 to 0.33 ± 0.01 wt.% TiO_2 when 2 wt.% of F was added into a water-saturated haplogranite melt). In the present study, a series of experiments has been performed to investigate the effects of temperature, F, and, to some extent, Cl on the solubility of SnO_2 in natural granitic melts.

2. Experimental and analytical methods

2.1. Starting materials

Five starting glass compositions were used. One was a remelted natural topaz- and cassiterite-bearing granite from Kymi (Finland) containing 1.12 wt.% fluorine and approximately 200 ppm SnO_2 (Glass 1,

Table 1), provided by Ilmari Haapala. A second composition (Glass 2) was prepared by adding Na_3AlF_6 and AlF_3 to Glass 1 so as to obtain a F-content of 2.12 wt.%. In this glass, F was introduced so that F replaced a stoichiometrically equivalent amount of O (2F=O) while the molar Al/(Na+K) ratio was kept constant. The starting glass 3 was prepared with a mixture of oxides of Si, Al, Ca, Na, K, and Fe, and is nearly same as glass 1, but F-free. The starting material was placed in an open Pt crucible and melted at 1600 °C/1 atm for 2–3 h in a furnace. The obtained glass was crushed and melted again at 1600 °C/1 atm for 1–2 h to obtain a homogeneous glass. These dry starting glasses were hydrated at high pressure and temperature to obtain water-bearing glasses, which were subsequently used as starting materials. The two additional Cl-bearing glasses (glasses 4 and 5) were also synthesized at high pressure and temperature by adding diluted $HCl–H_2O$ fluid (slightly water-under-saturated) to glass 3. The glasses were synthesized in cold seal pressure vessels at 850 °C and 2 kbar for 6–7 days. The fO_2 conditions were buffered by a Ni—NiO

Table 1
Microprobe analysis of starting material (natural granitic melt and synthetic granitic melt)

Starting materials (wt.%)	Glass 1[a]	Glass 2[a]	Glass 3	Glass 4	Glass 5
SiO_2	73.87	72.57	75.79	75.06	77.10
Al_2O_3	15.46	15.88	14.66	15.06	13.95
$\sum FeO$	0.99	0.95	0.99	1.06	0.38
CaO	0.80	0.79	0.73	0.80	0.69
Na_2O	4.03	4.33	3.65	3.64	3.56
K_2O	4.30	4.24	4.18	4.18	3.95
SnO_2	0.02	0.02	0.00	0.00	0.00
F	1.12	2.12	0.00	0.00	0.00
F=O	0.47	0.88	0.00	0.00	0.00
Cl	0.00	0.00	0.00	0.20	0.37
Total	100.17	100.06	100.00	100.00	100.00
Quartz	35.2	32.3	37.9	37.2	41.0
Orthoclase	27.7	27.5	25.9	26.1	24.2
Albite	37.1	40.2	36.2	36.7	34.8
Corundum	2.8	2.8	2.8	3.1	2.6
A/CNK	1.21	1.21	1.24	1.26	1.23

Glass compositions 3, 4, and 5 are normalized to 100%. The concentration of TiO_2, MgO, and MnO are lower than 0.02, 0.01, and 0.03 wt.%, respectively, in the natural glasses (glasses 1 and 2). P_2O_5 is below detection limit. Glasses 3, 4, and 5 are synthetic and free of these oxides.

[a] Data from Lukkari and Holtz (2000).

assemblage. For these experiments, deionised H$_2$O or dilute HCl (33.5 wt.%) was loaded in gold capsules (inner diameter 5.0 mm, outer diameter 5.4 mm, length 45 mm), together with the dry starting glasses (glasses 1 and 3). The samples were sealed in gold capsules by arc welding. Water contents in the hydrous starting materials were determined by Karl Fischer Titration (KFT; cf. Behrens et al., 1996). For each synthesis, two KFT analyses were done. The variation of the duplicate analyses was less than 0.03 wt.%. The water content of the hydrous glasses varied between 5.1 and 6.2 wt.% and the glasses were either slightly water-undersaturated or water-saturated.

2.2. Experimental method

The experimental technique used was similar to that of Linnen et al. (1995), which was modified from Harrison and Watson (1983). The experiments produced a concentration gradient of SnO$_2$ in a melt adjacent to a SnO$_2$ crystal. Care was taken that the crystal was surrounded by the melt to avoid any contact between cassiterite and the noble metal capsule. This minimizes the formation of Au–Sn alloy, which may result in experimental problems. Hydrous starting glasses were crushed and inserted in a ~15-mm-long gold capsule that was closed at the base by arc welding. A natural tabular SnO$_2$ crystal (~1×1×2 mm) was placed in the center of the capsule. The top end of the capsule was crimped and welded shut. The experiments were performed in a horizontal cold seal pressure vessel (CSPVs) at 2 kbar and in the temperature range of 700–850 °C for 5–22 h, except for one experiment, which was conducted for 145 h.

The CSPVs used for the experiments were made of a Ni-rich alloy. The temperature was controlled by type K-thermocouple placed outside of the vessel close to the hot-spot zone. The vessels were calibrated periodically and the temperature was known within ±10 °C. The experiments were quenched by removing the vessel from the furnace and blowing compressed air around the vessel—the temperature inside the vessel dropped to ~200 °C after 5 min. Pressure was measured with a strain-gauge manometer (accuracy ±0.03 kbar), using two pressure media that allowed different fO_2 ranges. In one case, water was used as pressure medium and the buffer was close to

that of the Ni–NiO assemblage. In these experiments, the Ni–NiO powder was placed around the sample capsules. Because of the Ni-rich composition of the vessel, this buffer had a long lifetime and the buffer assemblage comprised Ni–NiO after the experiment (the intrinsic fO_2 of the vessel was close to that of the Ni–NiO buffer). In other experiments, vessels were pressurized with argon and the oxygen fugacity was significantly higher. At 800–840 °C and 2 kbar, the oxygen fugacity measured by sensor capsules (Taylor et al., 1992) was Ni–NiO +2.3 log units (Berndt et al., 2001). Hereafter, the runs buffered by the Ni–NiO assemblage will be referred to as "NNO experiments" or runs under reducing conditions, and the experiments with argon as the pressure medium will be referred to as "NNO+2.3 experiments" or runs under oxidizing conditions.

After quenching, the capsules were weighed and cut perpendicular to the base of the SnO$_2$ crystal. The experimental products were always glass+cassiterite, except in run GGC 11 in which quench crystals (feldspar) were formed near the cassiterite due to slow cooling rate. Because the exact orientation of the SnO$_2$ crystal in the capsule was not known, the studied sections were not always strictly perpendicular to the SnO$_2$ crystal. This may have influenced the calculation of diffusivity of SnO$_2$, but is not expected to have influenced the solubility of SnO$_2$, which corresponds to the concentration of SnO$_2$ at the crystal–melt interface. In three experiments (GGC 3, GGC 4, and GGC 5), two sections were prepared, one longitudinally and one transversely relative to the axis of the capsule. This procedure allowed us to investigate concentration profiles of Sn along three orthogonal directions.

2.3. Analytical techniques

2.3.1. Electron microprobe

The compositions of the glasses were determined with a Cameca SX-100 electron microprobe with analytical conditions of 15 kV and 6 nA. Counting time for Na and K was 2 s for all starting glasses and run products and 5 seconds for other elements. The beam was defocused at least 10 μm for starting glasses and run products. To measure Cl in the two starting glasses, profiles of more than 40 analyses were performed. Peak count time for each analysis of Cl

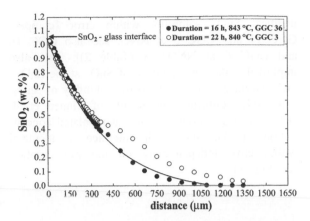

Fig. 1. Typical concentration profiles of SnO₂ for experiments with run durations of 16 (●) and 22 h (○). The solubility of SnO₂ is defined by the intercept between the concentration profiles and *y*-axis. The two profiles obtained at ~840 °C show that the concentration of SnO₂ at the interface is not time dependent. A typical concentration profile of SnO₂ fitted from the error function (—; see text) is also shown (experiment GGC 36).

in the melt was 30 s for the starting glasses. The Cl content was homogeneous in the run product glasses (samples containing SnO₂) this was confirmed by traverses across the final run products. The fluorine content of the starting glasses were also analyzed by electron microprobe (see Lukkari and Holtz, 2000).

All the analyses of Sn in the glasses were conducted with an electron beam (2 μm in diameter) and the peak counting time for determining SnO₂ was 120 s. Concentration profiles were determined at a 90° angle to the SnO₂–glass interface. To avoid fluorescence from the SnO₂ crystal, the first analysis was done at a distance of 10 to 15 μm from the SnO₂–crystal boundary. One to four profiles with 35 to 50 analytical points each were determined per sample. In most of the samples, the length of the profiles was from 900 to 1400 μm. A typical concentration profile obtained from the experiment at 843 °C, 16 h, and NNO is shown in Fig. 1. This profile fits well with the solution of Fick's second law

$$C = C_0 \left(1 - \mathrm{erf}\ x/(4Dt)^{1/2} \right) \qquad (1)$$

assuming infinite one dimensional diffusion, a concentration independent diffusivity D of Sn and a constant concentration C_0 of SnO₂ at the surface of the crystal. x is the distance to the surface and t is the

run duration. The solubility of SnO₂ in the melt can be calculated by extrapolating the error function to the SnO₂–glass interface. Not all concentration profiles could be fitted accurately by an error function, however. In these cases, the run duration may have been too long and concentration of SnO₂ along the profile never reached a value close to zero, or the profile may have been longer than the half-size of the crystal so that the source cannot be considered as an infinite plane. Furthermore, the SnO₂ crystal may have descended in the melt during the experiment and the experimental section may not have been cut perpendicular to the SnO₂ surface.

In any case, even if the concentration profiles had different shapes, the maximum SnO₂ concentration in the glass at the interface, corresponding to the equilibrium solubility value, should still be discernable. This is illustrated in Fig. 1, in which two profiles obtained at the same temperature in two different experiments have been plotted. The profile of sample GGC 36 can be fitted successfully by an error function. The profile of sample GGC 3 does not have the typical shape of a concentration profile (in fine planar diffusion), especially at 300–750 μm from the interface (Fig. 1). However, the concentration at the interface is identical within error [1.05±0.05 wt.% SnO₂ (GGC 36) and 1.03±0.05 wt.% SnO₂ (GGC 3)]. For concentration profiles that could not be fitted with an error function, the concentration at the interface was extrapolated by fitting data points close to the interface empirically. Such an empirical fit is shown in Fig. 2.

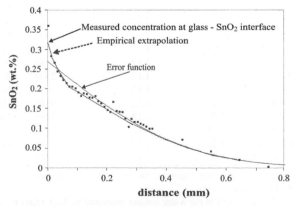

Fig. 2. Concentration profile of SnO₂ in experiment GGC 14. The concentration at the interface was extrapolated empirically from the data points close to SnO₂–glass interface.

3. Results

3.1. Attainment of saturation values and diffusion problems

3.1.1. Run duration

To know the effect of time on the SnO_2 concentration profile, experiments were performed at different run durations. Experiments conducted for 145 and 22 h at identical P, T, and fO_2 also yielded identical SnO_2 solubility within error [compare GGC 11 and GGC 5 at NNO; see also GGC 16 and GGC 6 at NNO+2.3 (Table 2)]. This also implies that the concentration of SnO_2 at the SnO_2–melt interface corresponds to the solubility of SnO_2. Our study confirmed the results of Linnen et al. (1995) who suggested that the extrapolated concentration at the SnO_2–melt interface is identical within error independently on the run duration (cf. Fig. 1).

Table 2
Experimental run conditions and run products (F-bearing melts)

Exp. Ref. No.	Starting glass		P (bar)	T (°C)	t (h)	Buffer	Solubility of SnO_2 (in log ppm)		
	Initial H_2O (wt.%)[a]	Fluorine content (wt.%)[b]					Average	Max	Min
GGC 3	6.20	1.12	2000	840	22	NNO	4.01	4.03	4.00
GGC 4	6.20	1.12	2000	840	22	NNO	4.02	4.03	4.00
GGC 5	6.20	1.12	2000	845	22	NNO	3.95	3.98	3.93
GGC 11	6.88	1.12	2010	840	145	NNO	3.93	3.93	3.93
GGC 18	6.26	1.12	2000	840	18	NNO	4.05	–	–
GGC 19	6.26	1.12	2000	840	18	NNO	4.08	–	–
GGC 30	5.28	1.12	2000	840	16	NNO	4.08	–	–
GGC 32	5.17	1.12	2000	843	16	NNO	4.07	–	–
GGC 36	5.14	1.12	2000	843	16	NNO	4.03	–	–
GGC 31	5.28	1.12	2000	788	5	NNO	3.74	–	–
GGC 34	5.28	1.12	2080	795	5	NNO	3.76	–	–
GGC 35	5.17	1.12	2000	790	5	NNO	3.77	–	–
GGC 12	6.26	1.12	2005	755	22	NNO	3.67	–	–
GGC 21	5.14	1.12	2015	750	20	NNO	3.79	–	–
GGC 22	5.14	1.12	2015	755	20	NNO	3.75	3.78	3.75
GGC 23	5.14	1.12	1990	750	18	NNO	3.86	–	–
GGC 24	5.14	1.12	1990	745	18	NNO	3.78	–	–
GGC 27	6.26	1.12	2000	760	20	NNO	3.74	–	–
GGC 28	5.14	1.12	2000	755	20	NNO	3.70	–	–
GGC 10	6.20	1.12	2000	708	24	NNO	3.51	3.51	3.51
GGC 13	6.26	1.12	2000	708	16	NNO	3.57	–	–
GGC 14	6.88	1.12	2000	710	16	NNO	3.52	–	–
GGC 16	6.88	1.12	2000	704	16	NNO	3.45	–	–
GGC 25	5.14	1.12	2000	705	48	NNO	3.54	–	–
GGC 29	5.17	1.12	2000	705	16	NNO	3.60	–	–
GGC 39	5.28	1.12	2030	836	16	NNO+2.3	3.13	3.18	3.09
GGC 40	5.28	1.12	2030	750	16	NNO+2.3	2.78	2.78	2.78
GGC 43	5.28	1.12	2030	746	16	NNO+2.3	2.80	2.82	2.78
GGC 44	5.28	1.12	2010	836	6	NNO+2.3	3.19	–	–
GGC 20[c]	5.20	2.12	2015	750	20	NNO	3.65	3.68	3.64

All run products were glasses. Sample GGC 11 was partially crystallized after the experiment (not included in average solubility data at 850 ± 10 °C).

[a] Initial H_2O is the water content measured by KFT (Karl Fischer Titration).

[b] F content is the F measured by electron microprobe.

[c] No prehydrated starting glass was used; water content was determined by KFT. The minimum and maximum values are obtained from different concentration profiles.

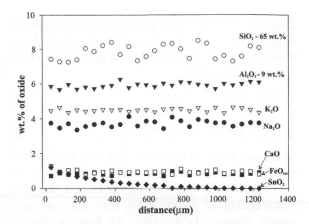

Fig. 3. Representative electron microprobe analyses of major elements as a function of distance from SnO_2 crystal in experiment GGC 19. All analyses are normalized to 100%. SnO_2 concentration at the interface is 1.2 wt.%. To report the SiO_2 and Al_2O_3 contents on the same scale, 65 and 9 wt.% have been subtracted from the analyzed SiO_2 (O) and Al_2O_3 (▼) contents, respectively.

3.1.2. Interdiffusion of major elements

Dissolution of SnO_2 in the melt is a complex multicomponent diffusion process and the concentration of other melt components may change relative to each other along the SnO_2 diffusion profiles. Such compositional variations in the melt were observed by Linnen et al. (1995, 1996) especially at reducing conditions when the solubility of SnO_2 is high. In this case, alkalis were depleted near the surface of the crystal. Solubility of SnO_2 varies with the alkali/Al ratio of the melt (Taylor and Wall, 1992; Linnen et al., 1995, 1996; Ellison et al., 1998) and, hence, the measured concentration of SnO_2 at the surface is not representative of the initial melt but for a more peraluminous composition. In our study, the SnO_2 concentration was at most 1.20 wt.%. Major-element concentrations along the profiles were analysed on selected samples with the highest SnO_2 concentrations. Typical results for run product GGC 19 are shown in Fig. 3. There is no significant variation in alkalis and Al; thus, coupled diffusion would not be considered a problem in this study.

3.1.3. Reproducibility of experiments

The diffusion profile method used in this study is well suited to investigate SnO_2 solubility and to minimize problems related to interaction of Sn and noble metal capsules. However, several factors may influence the concentration of SnO_2 at the SnO_2–melt interface. Concentration profiles cannot always be fitted by an error function, making it difficult to extrapolate SnO_2 concentration at the crystal–melt interface. Beside analytical errors, the scatter of the solubility data may be due to variations in water concentration and temperature. Several experiments were conducted at identical conditions (P, T, fO_2, and run duration) to establish the reproducibility of our results. The calculated SnO_2 solubilities for the nine experiments performed at NNO and at 840 °C vary by up to $\pm15\%$ relative, and are independent of the run duration (Fig. 4). However, for seven of these experiments, the relative variation is less than $\pm10\%$. Long run durations and resulting interaction with gold capsule may explain the low solubility values of the experiments GGC 11 and GGC 5. Furthermore, crystals were observed in the product of run GGC 11, probably as a result of slow quench (diminished air flux around the vessel), and this may have affected the solubility determination. We consider that a variation of $\pm10\%$ relative corresponds to the precision of our solubility data at 840 ± 10 °C. At 700 and 750 °C, the relative variation in SnO_2 solubility is larger due to lower absolute concentration of SnO_2 ($\pm20\%$ at 750 °C, $\pm18\%$ at 700 °C). The error bars shown in Fig. 5 correspond to this variation.

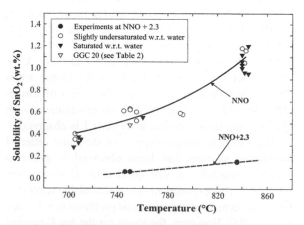

Fig. 4. Solubility of SnO_2 as a function of temperature in a natural F-bearing composition in reduced (NNO) and oxidized (NNO+2.3) conditions. Experiment GGC 20 was performed using glass 2, all other with glass 1.

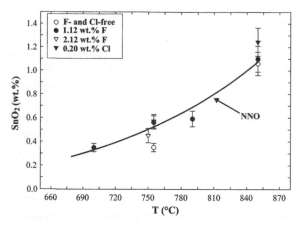

Fig. 5. Effect of F and Cl content on SnO_2 solubility as a function of temperature in reduced conditions (NNO). Experiments performed with glasses 1, 2, 3, 4, and 5. Average solubility values are shown (see Table 2).

3.2. Effect of temperature and fO_2 on SnO_2 solubility in natural granitic melts

The experimental results on the solubility of SnO_2 in the melt with composition of glass 1 are shown in Table 2 and Fig. 4. At NNO, the solubility of SnO_2 increases from 0.28 wt.% SnO_2 at 700 ± 10 °C to 1.20 wt.% SnO_2 at 850 ± 10 °C (corresponding to GGC 16 and GGC 19, respectively). At NNO+2.3, the solubility of SnO_2 is significantly lower (see also Linnen et al., 1995) and varies from 0.06 wt.% SnO_2 at 750 ± 10 °C to 0.14 wt.% SnO_2 at 840 ± 10 °C. However, the effect of temperature is of the same order of magnitude as at NNO. When plotted in a diagram log C_{SnO_2} versus $1/T \times 10^3$ (Fig. 6), the slope of the linear regression between 700 and 840 °C at NNO is -3.51. At NNO+2.3, the slope between 736 and 836 °C is -4.25 (not shown in Fig. 6). Thus, the slopes are similar at both oxygen fugacities. The effect of temperature can also be estimated from experiments with glass 3—the same trend as with glass 1 is observed.

The effect of temperature is of the same order of magnitude as what has been observed in previous studies, especially in dry conditions. High-pressure (hydrothermal) experiments are not available (except at 850 °C), because of experimental problems (see Linnen et al., 1995). However, the slope for the dry F-bearing peralkaline composition at 1100–1600 °C (Štemprok, 1990b) in Fig. 6 is identical within error to the slope of glass 1 with peraluminous composition. Thus, the

temperature effect seems to be of the same order of magnitude for all geologically relevant melt compositions. The dissolution of SnO_2 in the melt can be treated as a two-step reaction:

$$\text{Step}(1): SnO_2^{cas} = SnO_2^{melt} \tag{2}$$

$$\text{Step (2)}: SnO_2^{melt} = SnO^{melt} + \frac{1}{2}O_2 \tag{3}$$

where *cas* refers to cassiterite and *melt* to dissolved species in the melt. At oxidizing conditions, the concentration of SnO is negligible and the total concentration of dissolved Sn does not vary with oxygen fugacity. In this case, the slope in plots of log C_{SnO2} versus $1/T$ corresponds to $\Delta H°/2.303R$, where $\Delta H°$ is the standard enthalpy of reaction (2) and R is the gas constant. At moderately reducing conditions, a slope of -0.5 is observed in plots of log C_{SnO2} versus log fO_2. This can be explained by the increasing abundance of divalent Sn with decreasing oxygen fugacity (see Linnen et al., 1995). In our experiments, the fO_2 was buffered for a given solid buffer, allowing fO_2 to change with temperature. Therefore, the slope in plots of log C_{SnO2} versus $1/T$ does not strictly represent the dissolution enthalpy.

3.3. Effect of F on SnO_2 solubility

The comparison of data from glass 1 with the subaluminous composition of Linnen et al. (1995)

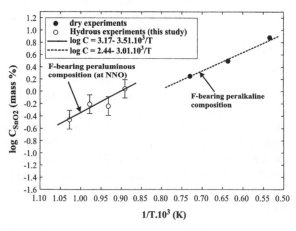

Fig. 6. The logarithm of wt.% cassiterite solubility (C_{SnO2}) versus $1/T$ (K) for hydrous (O) and dry granitic melts (●). Data for F-bearing peraluminous composition at 2 kbar are from this study. F-bearing peralkaline composition at 1 atm and oxidizing conditions are from Štemprok (1982), Štemprok and Voldan (1978) and Nekrasov (1984).

may indicate that F has a significant effect on SnO_2 solubility, as has been often assumed in natural granitic systems (e.g., Pichavant et al., 1987). The solubility of SnO_2 at NNO increases from 0.30 wt.% SnO_2 in an F-free granitic system (at 850 °C) to 1.10 wt.% (average value) SnO_2 in an F-bearing system (glass 1 at 840±10 °C). However, increasing F content from 1.12 to 2.12 wt.% (glasses 1 and 2) does not lead to a further increase in SnO_2 solubility. The measured SnO_2 concentration was even slightly lower in the more F-rich melt [0.45 wt.% SnO_2 at 750 °C in glass 2, 0.55 wt.% SnO_2 (average value) in glass 1]. The results obtained with the F-free glass 3, which has the same corundum content as glasses 1 and 2, confirm that the effect of F is low or even negligible at an fO_2 of NNO. Data at 840 °C for glasses 1 and 3 are almost identical [1.10 wt.% SnO_2 (average value) and 1.06 wt.% SnO_2, respectively]. There is a small difference at 750 °C for glasses 1 and 3 (0.35 and 0.56 wt.% SnO_2), but the variation is still with in the ±20% error. Based on the effect of temperature on F-free and F-bearing glasses (Fig. 5), we conclude that F does not significantly effect SnO_2 solubility.

We also conclude that, at reducing conditions, F has a minor influence on Sn solubility and that the difference between the data of Linnen et al. (1995) for subaluminous haplogranite and the data obtained in our study is related to another compositional factor. The difference in Q–Ab–Or proportions does not seem to affect the SnO_2 solubility significantly. Glass 3 does not have exactly the same composition as glasses 1 and 2, and Linnen et al. (1996) observed very similar SnO_2 solubilities in haplo-granite, quartz–albite, and quartz–feldspar melts at same P, T, and fO_2. The main difference is the normative corundum content as our melts are all peraluminous. Under oxidizing conditions, there may be a larger effect of F on Sn solubility. In run GGC 39 at 836 °C, 0.14 wt.% of SnO_2 dissolved in the F-bearing melt, whereas according to Linnen et al. (1996), less than 0.10 wt.% SnO_2 can be dissolved in subaluminous to peraluminous melts containing no F. This result is consistent with the observation of Keppler (1993) that dissolved F promotes the solubility of HFSE because, under oxidizing conditions, Sn is present dominantly in the +4 valence state.

3.4. Effect of corundum content on SnO_2 solubility

Although our study does not systematically evaluate the effect of normative corundum (C), it is evident by comparing the data of Linnen et al. (1996) with this study that SnO_2 solubility increases with increasing corundum content. SnO_2 solubility at NNO increases progressively at 850 °C from 0.47 wt.% SnO_2 at C=0.2 wt.% to 0.57 wt.% SnO_2 at C=1.14 wt.% and to 0.85 wt.% SnO_2 at C=2.3 wt.% (Fig. 7).

In this study, we found an average SnO_2 solubility of 1.05 wt.% SnO_2 at NNO for glasses 1, 2, and 3. This value is higher than the maximum value found by Linnen et al. (1996). At 850 °C, our results can be fitted with those of Linnen et al. (1996) and they show that there is a slight nonlinear increase in SnO_2 solubility with normative corundum of the melt (Fig. 8). However, consider-ing a precision of ±10% relative, the points at NNO in Fig. 8 could also be fitted by a linear trend. This effect of normative corundum may also be observed at more oxidizing (NNO+2.3) con-ditions. Because of the low concentrations and relatively high uncertainty, it is difficult to distin-guish between a linear and nonlinear increase in SnO_2 solubility as a function of corundum content. In any case, the effect of normative corundum on SnO_2 solubility is significantly lower at Ni–NiO+2.3 than at NNO.

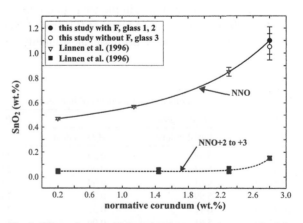

Fig. 7. Effect of excess Al (expressed as normative corundum) on SnO_2 solubility in silicate melt at 850 °C for two oxygen fugacity ranges (~Ni–NiO, and Ni–NiO +2 to +3). The data obtained in this study for F-bearing glasses are average values.

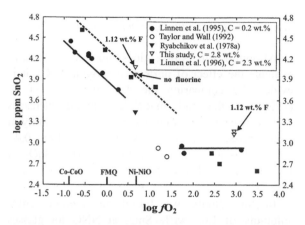

Fig. 8. Log solubility of Sn (ppm SnO$_2$) versus log fO$_2$ in granitic melts at 2 kbar. All data are obtained at 850 °C, except for the study of Taylor and Wall (1992) which occurred at 800 °C. Solid lines connecting the data of Linnen et al. (1995) are for samples with 0.2 wt.% normative corundum (C). The dashed lines connect the data of Linnen et al. (1996) obtained from samples with 2.3 wt.% C.

3.5. Effect of Cl on SnO$_2$ solubility

The effect of Cl on the SnO$_2$ solubility in granitic melts was also determined (Fig. 5). We found a solubility of 0.59 wt.% SnO$_2$ at 760 °C and 1.24 wt.% SnO$_2$ at 860 °C in a melt containing 0.20 wt.% Cl at NNO (glass 4). In a Cl-free melt at NNO (glass 3), these values were 0.35 wt.% SnO$_2$ at 750±10 °C and 1.06 wt.% SnO$_2$ at 860 °C. Thus, our results indicate that SnO$_2$ solubility increases slightly with Cl content of the melt. This contrasts with the data of Taylor and Wall (1992) who performed experiments in both Cl-free and Cl-bearing systems, in the temperature range of 750–800 °C and in a wide range of fO$_2$ from NNO to Co–CoO, and found a decrease in solubility of

SnO$_2$ when Cl was added to the system. At 750 °C, 2 kbar and ~NNO, Taylor and Wall (1992) found 0.085 wt.% Sn in a Cl-free melt and 0.068 wt.% Sn with 0.11 wt.% Cl in the melt. This corresponds to a 20% relative decrease with the addition of 0.11 wt.% Cl, whereas we observed an increase of about 40% relative with addition of 0.20 wt.% at 750–760 °C. Our results also show that further addition of Cl leads to enhanced SnO$_2$ solubility (0.61 wt.% SnO$_2$ at 760 °C in a melt containing 0.37 wt.% Cl; Table 3).

These discrepancies may be related to experimental problems encountered by Taylor and Wall (1992). Linnen et al. (1995) demonstrated that the experimental methods applied by Taylor and Wall (1992) were not appropriate to determine the SnO$_2$ solubility at reducing conditions. In the presence of a fluid phase, transport of Sn to the capsule wall occurs at a faster rate than the diffusivity of Sn in silicate melts. Therefore, equilibrium between SnO$_2$, silicate melt, and Sn–Au alloy was not attained in the study of Taylor and Wall (1992).

4. Discussion

Linnen et al. (1995, 1996) investigated the effects of fO$_2$, peralkanity, and excess Al on Sn solubility in granitic melts at 850 °C and found that cassiterite solubility ranges from 2.8 wt.% SnO$_2$ at FMQ −0.84 to approximately 0.08 wt.% SnO$_2$ at FMQ +3.12 in a synthetic granitic melt composition with 0.4 wt.% C at 850 °C and 2 kbar. At fO$_2$ higher than FMQ +1.5, fO$_2$ was not observed to affect SnO$_2$ solubility, implying that SnO$_2$ dissolves into the melt largely

Table 3
Experimental run conditions and run products (F-free melts)

Exp. Ref. No.	Starting glass		P	T	t	Buffer used	Solubility of SnO$_2$ (in log ppm)
	Initial chlorine	H$_2$O content					
GGC 51	5.59	0.20	2000	860	18	NNO	4.093
GGC 57	5.73	0.00	2000	855	18	NNO	4.020
GGC 59	5.61	0.00	2000	852	18	NNO	4.030
GGC 53	5.59	0.20	2030	760	18	NNO	3.770
GGC 56	6.12	0.37	2030	760	18	NNO	3.785
GGC 60	5.61	0.00	2030	754	18	NNO	3.550
GGC 58	5.73	0.00	2030	752	18	NNO	3.530

as Sn^{4+} (see Fig. 8). At more reduced conditions, log SnO_2 solubility decreases with increasing log fO_2 at a slope of -0.5, implying that Sn dissolves in the melt as Sn^{2+} (cf. Paparoni, 2000).

This study extends the data set of Linnen et al. (1995, 1996) to lower temperatures and also confirms the data of Štemprok (1990b). We show that the temperature effect can be modeled by assuming a linear dependence of log C_{SnO2} versus $1/T$. In subaluminous to peraluminous compositions, F has only a minor effect on SnO_2 solubility at an fO_2 of NNO, but the normative corundum of the melt is a compositional parameter, which needs to be taken into account to model cassiterite solubility in evolved granitic melts. A possible explanation for the relatively low effect of F on the SnO_2 solubility at NNO is that Sn is dissolved mainly as Sn^{2+}. Thus, the effect of F on solubility of HFSE shown by Keppler (1993) is not expected if Sn is not dissolved as Sn^{4+} (Sn^{4+} is a HFSE whereas Sn^{2+} is not). For a peraluminous melt composition, we also found a significant effect of Cl. Our dataset, combined with that of Linnen et al. (1995, 1996), allows the prediction of SnO_2 solubility in natural granitic melts, peraluminous melts in particular, as a function of fO_2, temperature, and melt composition.

By extrapolating our data close to the solidus temperatures of natural topaz and cassiterite-bearing granite (<680 °C), SnO_2 concentrations required to produce magmatic cassiterite are approximately 0.30 wt.% (at NNO). Even if solidus temperatures as low as 600 °C were assumed (cf. Pichavant et al., 1987), our results show that a relatively high concentration of SnO_2 (~0.20 wt.%) can be dissolved in such melts. In our study, the SnO_2 concentrations obtained at 700 °C are of the same order of magnitude than those observed in natural systems (1400 ppm in the Beauvoir granite; Cuney et al., 1992). As emphasized by Linnen et al. (1995), changing fO_2 may result in the crystallization of a significant amount of SnO_2 in the granitic melts, even at low temperatures.

One of the most surprising implications of this study is that corundum (excess Al) content may be a compositional parameter controlling the SnO_2 solubility. To test this hypothesis, we compared glass inclusion compositions in quartz from a Sn-bearing, F- and P-rich evolved peraluminous pegmatite from Ehrenfriedersdorf, Central Erzgebirge in southeastern

Germany (Webster et al., 1997). All melt inclusions analyses by Webster et al. (1997), except one with a high CaO (3.02 wt.%), were used for this comparison. The normative corundum of all melt inclusions analyzed was calculated for the compositions given by Webster et al. (1997). Fig. 9 shows the amount of F and normative corundum as a function of the Sn concentration in the glass inclusions. There is no evident correlation between F and Sn, but there is a positive correlation between normative corundum and Sn. This observation seems to confirm that the normative corundum (excess alumina) influences SnO_2 solubility in natural magmatic systems.

Systematic variations between Sn and other elements cannot be observed from the data set of Webster et al. (1997). As emphasized by Webster et al., the P content of the inclusions is exceptionally high, yet there is no obvious correlation between P and Sn in the melt. Except for F, the volatile concentrations are low (average 1.2 wt.% H_2O and 0.1 wt.% Cl) in the melt inclusions, compared to melt inclusions from high-silica, high-Sn topaz rhyolites (Webster and Duffield, 1991, 1994; Webster et al., 1991; Lowenstern, 1995). No systematic correlation is observed between Sn and Cl or H_2O. The low H_2O and Cl contents suggest that the residual liquids entrapped in quartz were already highly degassed, suggesting in

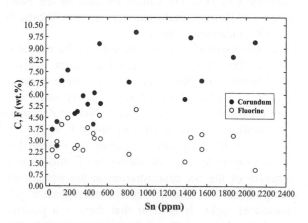

Fig. 9. Effect of fluorine (F, ○) and normative corundum (C, ●) on the solubility of Sn in natural melt inclusions in quartz from an evolved peraluminous pegmatite (data from Webster et al., 1997). All the data listed in Webster et al. (1997) have been plotted except one inclusion containing >3 wt.% CaO (other inclusions, in total 20, have <0.36 wt.% CaO). Note that there is no obvious correlation between F and Sn content, whereas normative corundum and Sn are positively correlated.

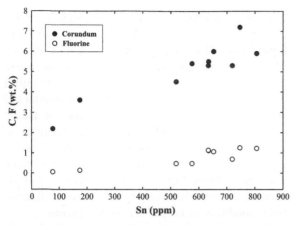

Fig. 10. Effect of fluorine (F, O) and normative corundum (C, ●) on the Sn content in the Argemela microgranite from Portugal (data from Charoy and Noronha, 1996). Note that there is a positive correlation between Sn concentration and normative corundum of the melt.

turn that Sn is not strongly partitioned to the fluid phase (cf. Taylor, 1988). Another explanation for variable content of SnO_2 in the melt inclusions would be that melt inclusions with high SnO_2 were trapped from melts at low fO_2. However, in the dataset of Webster et al. (1997), there is no compositional parameter that may reflect change in fO_2. One such parameter would be the Fe/(Fe+Mg) ratio, which is dependent of fO_2, but cannot be used as the MgO content of the melt inclusions reported by Webster et al. (1997) is close to zero.

Another dataset, which may confirm the role of C in controlling Sn solubility in silicate melts, is given by Charoy and Noronha (1996). These authors investigated a small volatile-rich, highly sodic and strongly peraluminous microgranite body in Argemela (Portugal) with extreme enrichments in F, P, Rb, Cs, Li, Sn, and Be. Magmatic cassiterite was observed in the microgranite. Fig. 10 shows the F and C as a function of the Sn concentration in the whole-rock compositions of the Argemela microgranite and associated rocks. It is clear that there is a positive correlation between C and Sn concentrations. In this example, there is also a positive, although less pronounced, correlation between F and Sn, as well as between F and other incompatible elements such as Li, Rb, and Cs (Charoy and Noronha, 1996). Charoy and Noronha emphasized that the Argemela microgranite may represent mixing of two magmas (one of

them enriched in F and incompatible elements) and, thus, the positive correlation of F and Sn and C and Sn may be explained by the mixing hypothesis. In this case, however, high Sn and F concentrations are observed in the peraluminous end-member. This is consistent with our results that imply a substantial role for Al in the solubility of Sn in evolved granitic melts, while the effect of F is relatively small.

Acknowledgements

The authors appreciate the help of J. Koepke while conducting the electron microprobe analyses at the Institut für Mineralogie, Universität Hannover, Germany. Technical assistance was provided by Otto Dietrich. Helpful comments to the manuscript made by N. Chaudhri, A. Wittenberg, J. Webster, F. Farges, and O.T. Rämö are greatly appreciated. We are also greatly thankful for the Georg-Christoph-Lichtenberg fellowship from the Ministry of Science and Culture, Lower Saxony, Germany.

References

Barsukov, V.L., Durasova, N.A., 1966. Metal content and metallogenic specialization of intrusive rocks in the regions of sulphide-cassiterite deposits (Miao-Chang and Sikhote Alin.). Geochemistry International 3, 97–107.

Behrens, H., Romano, C., Nowak, M., Holtz, F., Dingwell, D.B., 1996. Near-infrared spectroscopic determination of water species in glasses of the system $MAlSi_3O_8$ (M=Li, Na, K): an interlaboratory study. Chemical Geology 128, 41–64.

Berndt, J., Holtz, F., Koepke, J., 2001. Experimental constraints on storage conditions in the chemically zoned phonolitic magma chamber of the Laacher See volcano. Contributions to Mineralogy and Petrology 140, 469–486.

Charoy, B., Noronha, F., 1996. Multistage growth of a rare-element, volatile-rich microgranite at Argemela (Portugal). Journal of Petrology 37, 73–94.

Cuney, M., Marignac, C., Weisbrod, A., 1992. The Beauvoir topaz–lepidolite albite granite (Massif Central, France): the disseminated magmatic Sn–Li–Ta–Nb–Be mineralization. Economic Geology 87, 1766–1794.

Durasova, N.A., Barsukov, V.L., Ryabchikov, I.O., Khramov, D.A., Kravtsova, R.P., 1984. The valency states of tin in basalts at various oxygen fugacities. Geochemistry International 21, 7–8.

Ellison, A.J., Hess, P.C., Naski, G.C., 1998. Cassiterite solubility in high-silica $K_2O–Al_2O_3–SiO_2$ liquids. Journal of the American Ceramic Society 81, 3215–3220.

Eugster, H.P., 1985. Granites and hydrothermal ore deposits: a geochemical framework. Mineralogical Magazine 49, 7–23.

Groves, D.I., 1972. The geochemical evolution of tin-bearing granites in the Blue Tier batholith, Tasmania. Economic Geology 67, 445–457.

Groves, D.I., 1974. Geochemical variation within tin-bearing granites, Blue Tier batholith, NE Tasmania. In: Štemprok, M. (Ed.), Metallization Associated with Acid Magmatism. Wiley, New York, pp. 154–158.

Groves, D.I., McCarthy, T.S., 1978. Fractional crystallization and origin of tin deposits in granitoids. Mineralium Deposita 13, 11–26.

Harrison, T.M., Watson, E.B., 1983. Kinetics of zircon dissolution and zirconium diffusion in granitic melts of variable water content. Contributions to Mineralogy and Petrology 84, 66–72.

Keppler, H., 1993. Influence of fluorine on the enrichment of high field strength trace elements in granitic rocks. Contributions to Mineralogy and Petrology 114, 479–488.

Keppler, H., Wyllie, P.J., 1991. Partitioning of Cu, Sn, Mo, W, U and thorium between melts and aqueous fluid in the systems haplogranite-H_2O–HCl and haplogranite-H_2O–HF. Contributions to Mineralogy and Petrology 109, 139–150.

Kovalenko, V.I., Legeydo, V.A., Petrov, L.L., Popolitov, E.I., 1968. Tin and beryllium in alkali granitoids. Geochemistry International 5, 883–892.

Lehmann, B., 1982. Metallogeny of tin: magmatic differentiation versus geochemical heritage. Economic Geology 77, 50–59.

Linnen, R.L., 1998. Depth of emplacement, fluid provenance and metallogeny in granitic terrains: a comparison of western Thailand with other Sn–W belts. Mineralium Deposita 33, 461–476.

Linnen, R.L., Pichavant, M., Holtz, F., Burgess, S., 1995. The effect of fO_2 on the solubility, diffusion, and speciation of tin in haplogranitic melt at 850 °C and 2 kbar. Geochimica et Cosmochimica Acta 59, 1579–1588.

Linnen, R.L., Pichavant, M., Holtz, F., 1996. The combined affect of fO_2 and melt composition on SnO_2 solubility and tin diffusivity in haplogranitic melts. Geochimica et Cosmochimica Acta 60, 4965–4976.

Lowenstern, J.B., 1995. Applications of silicate melt inclusions to the study of magmatic volatiles. In: Thompson, J.F.H. (Ed.), Magmas, Fluids, and Ore Deposits vol .23. Mineralogical Association of Canada, pp. 71–99.

Lukkari, S., Holtz, F., 2000. Phase relations on F-enriched leucogranitic melts at 200 MPa: an experimental investigation. IGCP Project 373 Field Conference in southern Finland, July 3–7, 2000, Excursion Guide and Abstracts, p. 43.

Manning, D.A.C., 1981. The effect of fluorine on liquidus phase relations in the system Qz–Ab–Or with excess water at 1 kb. Contributions to Mineralogy and Petrology 76, 206–215.

Nekrasov, I.Ya., 1984. Tin in magmatic and postmagmatic processes. Doklady Akademii Nauk SSSR, Izdanja Nauka, Moscow. 238 pp. (in Russian).

Paparoni, G., 2000. Tin in silicate melts. PhD dissertation. Columbia University, New York, U.S.A.

Pichavant, M., Herrera, J.V., Boulmier, S., Briqueu, L., Joron, J., Juteau, J., Marin, L., Michard, A., Sheppard, S.M.F., Treuil, M., Vernet, M., 1987. The Macusani glasses, SE Peru: evidence of chemical fractionation in peraluminous magmas. In: Mysen,

B.O. (Ed.), Magmatic Processes: Physicochemical Principles, Special Publication vol. 1. Geochemical Society, University Park, Pennsylvania, pp. 359–373.

Pollard, P.J., Pichavant, M., Charoy, B., 1987. Contrasting evolution of fluorine- and boron-rich tin systems. Mineralium Deposita 22, 315–321.

Ryabchikov, I.D., Durasova, N.A., Barsukov, V.I., Laputina, I.P., Efimov, A.S., 1978a. Role of volatiles for the mobilization of tin from granitic magmas. In: Štemprok, M., Burnol, L., Tischendorf, B. (Eds.), Metallization Associated with Acid Magmatism, vol. 3. John Wiley and Sons, New York, pp. 94–109.

Ryabchikov, I.D., Durasova, N.A., Barsukov, V.I., Efimov, A.S., 1978b. Oxidation reduction potential as a factor of an ore-bearing capacity of acid magmas. Geokhimiya 8, 832–834. (in Russian).

Štemprok, M., 1971. Petrochemical features of tin-bearing granites in Krušné Hory Mountains, Czechoslovakia. Society of Mining Geologists of Japan Special Issue 2 (Proc IMA-IAGOD Meeting), pp. 112–118.

Štemprok, M., 1982. Tin–fluorine relationship in ore bearing assemblage. In: Evans, A.M. (Ed.), Metallization Associated with Acid Magmatism. John Wiley and Sons, New York, pp. 321–338.

Štemprok, M., 1990a. Intrusion sequences with ore-bearing granitoid plutons. Geological Journal 25, 413–417.

Štemprok, M., 1990b. Solubility of tin, tungsten, and molybdenum oxides in felsic magmas. Mineralium Deposita 25, 205–212.

Štemprok, M., Skvor, P., 1974. Composition of tin-bearing granites from the Krusne hory metallogenic province of Czechoslovakia. Sbornik Geologickych Vied-Rad 6, 7–83.

Štemprok, M., Voldán, J., 1978. Solubility of tin oxide in dry sodium rich granite melts: Mineralogical criteria for the relationship between magmatism and ore mineralization. Proceedings of the XI General Meeting of the International Mineralogical Association, Novosibirsk 4–10, September 1970. Izd. Nauka, Leningrad, pp. 125–133.

Tauson, L.V., 1968. Distribution regularities of trace elements in grantoid intrusions of the batholith and hypabyssal types. In: Ahrens, L.H. (Ed.), Origin and Distribution of Elements. Pergamon, Oxford, pp. 629–639.

Tauson, L.V., 1974. The geochemical types of granitoids. In: Štemprok, M. (Ed.), Metallization Associated with Acid Magmatism. John Wiley and Sons, New York, pp. 221–227.

Taylor, R.G., 1979a. Geology of Tin Deposits. Elsevier, Amsterdam.

Taylor, R.G., 1979b. Some observations upon the primary tin deposits of Australia. Bulletin of the Geological Society Malaysia 11, 181–207.

Taylor, J.R.P., 1988. Experimental studies on tin in magmatic-hydrothermal systems. PhD thesis, Monash University, Australia.

Taylor, J.R., Wall, V.J., 1992. The behavior of tin in granitoid magmas. Economic Geology 87, 403–420.

Taylor, J.R., Wall, V.J., Pownceby, M.I., 1992. The calibration and application of accurate redox sensors. American Mineralogist 77, 284–295.

Tischendorf, G., 1977. Geochemical and petrographic characteristic of silicate magmatic rocks associated with rare-element mineralisation. In: Štemprok, M., Burnol, L., Tischendorf, G. (Eds.),

Metallization Associated with Acidic Magmatism. John Wiley and Sons, New York, pp. 41–96.

Webster, J.D., Duffield, W.A., 1991. Volatiles and lithophile elements in Taylor Creek Rhyolite: constraints from glass inclusion analysis. American Mineralogist 76, 1628–1645.

Webster, J.D., Duffield, W.A., 1994. Extreme halogen abundances in tin-rich magmas of the Taylor Creek Rhyolite, New Mexico. Economic Geology 89, 840–850.

Webster, J.D., Burt, D.M., Duffield, W.A., Nash, W.P., Gavigan, T., Augillon, R.A., 1991. Constraints from glass inclusions in tin/topaz rhyolites on magma evolution, volatile element degassing, and mineralization. Abstracts with Programs - Geological Society of America 23, A46.

Webster, J.D., Thomas, R., Rhede, D., Forster, H.-J., Seltmann, R., 1997. Melt inclusions in quartz from an evolved peraluminous pegmatite: geochemical evidence for strong tin enrichment in fluorine-rich and phosphorous-rich residual liquids. Geochimica et Cosmochimica Acta 61, 2589–2604.

Available online at www.sciencedirect.com

SCIENCE @ DIRECT®

Lithos 80 (2005) 401–402

ELSEVIER

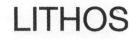
LITHOS

www.elsevier.com/locate/lithos

Author index to volume 80

oi:10.1016/S0024-4937(05)00016-2